Volume III

COMPOUNDS OF ARSENIC, ANTIMONY, AND BISMUTH

FIRST SUPPLEMENT

ORGANOMETALLIC COMPOUNDS

METHODS OF SYNTHESIS
PHYSICAL CONSTANTS AND CHEMICAL REACTIONS

Volume III

COMPOUNDS OF ARSENIC, ANTIMONY, AND BISMUTH

Second Edition

FIRST SUPPLEMENT

Covering the Literature from 1965 to 1968

Edited by
MICHAEL DUB
Central Research Department, Monsanto Company

SPRINGER SCIENCE+BUSINESS MEDIA, LLC
1972

© 1972 by Springer Science+Business Media New York
Originally published by Springer-Verlag New York Inc. in 1972

ISBN 978-3-642-50289-7   ISBN 978-3-642-50287-3 (eBook)
DOI 10.1007/978-3-642-50287-3

PREFACE TO THE FIRST SUPPLEMENT VOLUMES

The output of chemical literature has increased to such an extent that the four-year supplements to Volumes I and II cover more references than the respective main volumes contained for the period from 1937 through 1964. The supplement to the third volume includes almost three fourths as many references as the main volume.

In view of this enormous increase in the publications, the preparation of the supplements was only possible through the cooperation of Dr. Richard W. Weiss, Fachhochschule Lippe, Lemgo, Germany, who further carried out the work on Volume II, and of Drs. Klaus Bauer and Gilbrecht Haller, University of Vienna, Austria, who took over the coverage of Volume I.

This supplement series covers the literature and patents abstracted by Chemical Abstracts, Volumes 62 through 69, and includes some references from Volumes 60 and 61 that were omitted in the main volumes of the series. While the collection of references was based on Chemical Abstracts, for the most part original articles and patent specifications were used in compiling the supplements.

The classification systems of the main volumes have been retained and expanded by inclusion of some new types of compounds unknown through 1964. Each supplement volume was prepared independently; therefore some slight, inadvertent variations in their forms probably occurred.

Numerous patents and several articles from Soviet journals were obtained from the USSR National Public Library for Science and Technology in exchange for our publications. The cooperation of Messrs. V. Orlov and N. Tyshkevich, directors of the Moscow library, in the exchange is very much appreciated.

Support provided by the Monsanto Company Information Center is also acknowledged with thanks.

Michael Dub

St. Louis, Missouri
March 1972

INTRODUCTION

This supplement is arranged in the same manner as the main volume; the classification numbers are identical. In addition to methods of synthesis, physical properties, and chemical reactions, biological properties and uses are also included, with the exception of pharmacological activities and medicinal uses.

As before, in numerous sections of the supplement, compounds for which much literature data was reported are described individually. Others for which only a few data are available are compiled in tables. The individual entries of each section, as well as those in the tables, are arranged in the order of increasing numbers of carbon and hydrogen atoms. For compounds which also appeared in the main volume, references to the latter are given by the parenthetic entries of M hyphenated with the page numbers of the main volume.

Attempts were made to adhere to the nomenclature used in the subject indexes to Chemical Abstracts.

Linear simplified formulae are used for coordination complexes. The coordinated and covalent ligands, together with the complex-forming atoms or moieties, are enclosed in brackets, and/or braces. In some cases, a distinction of nonionic from ionic ligands is uncertain due to insufficient experimental evidence.

In the interest of brevity, certain numerical data are omitted, but references to the original literature are given. Yield data are rounded to two significant figures.

A few entries missed in the main body of this volume appear on pages 547-49.

In the Bibliography section, references with not more than three authors contain the names of all the authors, but those with more than three authors give only the name of the first author followed by "et al."

My thanks are due to Dr. Oliver St. C. Headley, University of the West Indies, Trinidad, for a loan of his Ph. D. thesis.

I am very grateful to Mrs. Mary Alice Doiron for the perseverance in her preparation of the typescript.

Michael Dub

A B B R E V I A T I O N S

The following abbreviations are used in addition to, or at variance with, those in Chemical Abstracts:

BiPy	bipyridyl
(1- or 2-)$C_{10}H_7$-	naphthyl radical
-$C_{10}H_6$-	naphthylene diradical
corr.	corresponding
dec.	decomposition, decomposes, decomposed, decomposing
DMF	dimethylformamide
Fc	ferrocenyl, $C_5H_5FeC_5H_4$-
Fcd	1,1'-ferrocenylene biradical, -$C_5H_4FeC_5H_4$-
GLC	gas-liquid chromatography
i-Bu	isobutyl
i-Pr	isopropyl
IR	infrared
liq.	liquid (adjective)
m.	melting point, melts at, melted at
M	a metal atom in generic formulae
matl.	material(s)
obtd.	obtained
Pip	piperidine
prod.	product(s)
psi	pound per square inch pressure
Py	pyridine
ref.	reference(s)
r. t.	room temperature
rxn.	reaction(s)
s-Bu	secondary butyl
subl.	sublimes at, sublimed at, sublimation
substd.	substituted
t-Bu	tertiary butyl
THF	tetrahydrofuran
TLC	thin-layer chromatography
UV	ultraviolet

Parenthetic entries of M hyphenated with a number, e.g. (M-223), indicate the page number where the same compound can be found in the main volume.

CONTENTS

A R S E N I C

1.   COMPOUNDS OF TRIVALENT ARSENIC

1.1   TERTIARY ARSINES

1.1.1 SYMMETRIC DERIVATIVES

The methods of preparation of symmetric tertiary arsines reported in the period 1937-1964 were summarized in the main volume (M-1). Since then a few more methods were described.

Trimethylarsine was prepared from a mixture of arsenic trichloride with methyl chloride by a dehalogenation with finely divided aluminum produced by an electrolytic reduction. A reaction of arsenic trichloride with trialkyl-alkynyltin gave trialkynylarsines.

Arsine, AsH_3, is a useful starting material for the synthesis of tertiary arsines through a condensation with formaldehyde and through a reaction with activated olefins. The condensation with formaldehyde in the presence of cadmium chloride gave arsylidynetrimethanol. In a reaction of arsine and mercury oxide with allyl chloride or methallyl chloride in the presence of organic or inorganic acids, $2,2',2''$-arsylidynetris(3-chloroalkylmercury salts) were obtained.

A cleavage of tetraphenyldiarsine with phenyllithium gave triphenylarsine and diphenylarsinolithium. A conversion of diphenylarsinopotassium with benzenesulfonic acid salts in high boiling ethers gave triphenylarsine.

Tris(trifluoromethyl)arsine, AsC_3F_9, $(CF_3)_3As$,   (M-2)
<u>Synth.</u>: From $(CF_3)_2AsI$ or CF_3AsCl_2 by shaking with CF_3I at $20°$ in the presence of Hg, 62% yield along with $[(CF_3)_2As]_2$ (414).
<u>Prop.</u>: IR and Raman spectra (409); mass spectrum (485).
<u>Rxn. with</u>: $(CF_3)_2NCl \rightarrow (CF_3)_2NAs(CF_3)_2$ and $[(CF_3)_2N]_2AsCF_3$ (71, 72).
$AsCl_3$ at $210° \rightarrow CF_3AsCl_2$ and $(CF_3)_2AsCl$ (413).
AlI_3 at $115° \rightarrow CF_3I$ (413).
AlI_3 at $150° \rightarrow CF_4$, C_2F_6, AsF_3, and SiF_4 (413).
BI_3 at $130° \rightarrow CF_3I$, $(CF_3)_2AsI$, and BF_3 (413).
HgO at $95° \rightarrow [(CF_3)_2As]_2O$ and CF_3NO (413).
NO at $70° \rightarrow CF_3NO$, CF_3NO_2, and some unidentified compds. (413).

Tricyanoarsine, AsC_3N_3, $As(CN)_3$,   (M-3)
<u>Synth.</u>: (M-3)
<u>Prop.</u>: Potential field and force consts. data (1119).

Trimethylarsine, AsC_3H_9, Me_3As,   (M-3)
<u>Synth.</u>: From $AsCl_3$ and MeMgI in Bu_2O, 70% yield (349, 1167); when CD_3MgI

was used instead of MeMgI, $(CD_3)_3As$ was obtd. (1167).

From $AsCl_3$ by adding slowly to MeCl in the presence of finely divided Al produced by electrolysis of a $NaCl + AlCl_3$ (1:1) mixt. at 220° (1465).

By-prod. formed in the rxn. of $Me_2AsSiMe_3$ with MeI (1309).

Prop.: b. 52° (349), b. 55.4° calcd. by the Kinney equation (66). Dipole moment (349). Force const. data (349a). Interatomic distances detd. (1026). IR spectrum: As-C and C-H vibration data (209); the effect of adjacent groups on the rocking and symmetric deformation frequencies of the Me group and of the $(CD_3)_3As$ analog (1491). Raman spectrum - revision of the former assignments (212). NMR spectrum (594) and its correlation with those of Me_3N, Me_3P, Me_3Sb, and Me_3Bi (880). Mass spectrum (881).

Rxn. with: HgO (1:1) → Me_3AsO (1057).

H_2O_2 → Me_3AsO (1057).

Halogens (X = Cl, Br, or I) → Me_3AsX_2 (1167).

Use: A complex with $PhCH_2AlCl_2$ and $Zr(OBu)_4$ catalyzes the polymerization of C_3H_6 (383).

Coordination Deriv.: $Me_3As \cdot Al(BH_4)_3$, crystals m. 57°, prepd. from Me_3As and $Al(BH_4)_3$ in $n\text{-}C_7H_{16}$ at -78°, highly sensitive to, and reactive with, O and H_2O; IR and NMR spectra (186); interpretation of IR and NMR spectra (442).

$Me_3As \cdot BH_3$ (M-4), prepd. from $Me_3As + Ca(BH_4)_2 + (p\text{-}t\text{-}BuC_6H_4O)_3B$ in THF at 140° (97). On heating with B_2H_6, B_5H_9, $B_{10}H_{14}$, or $B_{18}H_{22}$ at 175° → $[B_{12}H_{11}AsMe_3]^-$ 40%, $B_{12}H_{10}AsMe_3$ 5%, and $[H_2B(AsMe_3)_2]^+$ 4% (1065).

$(Me_3As)_2B_{12}H_{10}$ (M-4), prepd. from $Me_3As \cdot BH_3$ by heating with B_2H_6 or with higher boranes at 175° (1065).

$Me_3S[B_{12}H_{11}AsMe_3]$ (M-4), feathery colorless crystals, obtd. as a by-prod. from the rxn. of Me_3As with B_2H_6 at 175°, followed by extn. of $[B_{12}H_{10}(AsMe_3)_2]$ with Diglyme, dec. of the solid residue with H_2O, isolation of $[BH_2(AsMe_3)_2][B_{12}H_{11}AsMe_3]$, and neutralization of the mother liquor with $[Me_3S]OH$ (1067).

$Me_4N[B_{12}H_{11}AsMe_3]$ (M-4), colorless crystals, prepd. as described above by neutralizing the mother liquor contg. $[B_{12}H_{11}AsMe_3]^-$ with aq. $[Me_4N]OH$ (1067).

$X[B_{12}H_{11}AsMe_3]$, where $X = H_4N^+$, Na^+, Cs^+, $[Cu(NH_3)_4]^+$, Mg^{++}, $[MeNH_3]^+$, $[MePh_3P]^+$, $[PyH]^+$, and $[Et_3(C_{12}H_{25})As]^+$, were prepd. as described in the preceding paragraph by replacing $[Me_4N]OH$ with the appropriate base (1067).

$H_3O[B_{12}H_{11}AsMe_3]$ was obtd. as a by-prod. from the rxn. of Me_3As with B_2H_6 at 175° in a sealed tube, after isolation of $[B_{12}H_{10}(AsMe_3)_2]$, by dec. of the residue with H_2O. The complex is useful in etching metals and for the absorption of NH_3 and of amines from air (1067).

$H_3O[B_{12}H_{10}ClAsMe_3]$, a cryst. solid, prepd. from $H_3O[B_{12}H_{11}AsMe_3]$ by treating with Cl (1067).

$[B_{12}H_{10}(AsMe_3)_2]$ (M-4), colorless crystals, prepd. from Me_3As and B_2H_6 by heating in a sealed tube at 175°; soly. data; a rxn. with halogens leads to a replacement of the hydride atoms with halogens (1067).

$[BH_2(AsMe_3)_2][B_{12}H_{11}AsMe_3]$ (M-4), feathery colorless crystals, m. 171-173°, prepd. from Me_3As and B_2H_6 by heating in a sealed tube at 175°, extracting the rxn. mixt. with Diglyme, treating the residue with hot H_2O and, after cooling, septg. the solid from the mother liquor (1067).

$[Cu(AsMe_3)Cl]_4$, a discussion of its electronic structure, using a simple topological equivalent orbital approach (846).

$[Mn(AsMe_3)_2(CO)_3Br]$ (M-4), prepd. from $[Mn(CO)_4Br]_2$ + Me_3As, sol. in hydrocarbons, useful as an antiknock agent (1612).

$[Ni(AsMe_3)(CO)_3]$, IR and Raman spectra (183, 213).

$[Ni(AsMe_3)_2(CO)_2]$, IR and Raman spectra (183, 213).

$[Ni(AsMe_3)_3(CO)]$, IR and Raman spectra (183, 213).

cis-$[Pd(AsMe_3)_2Cl_2]$ (M-5), yellow, m. 234-235°, prepd. from Na_2PdCl_4 + Me_3As in EtOH; IR spectra in the solid state and in soln. were recorded (629).

trans-$[Pd(AsMe_3)_2Br_2]$ (M-5), orange, m. 230-231°, prepd. from Na_2PdCl_4 + Me_3As in EtOH in the presence of an excess of NaBr; IR spectra in the solid state and in soln. were recorded (629).

trans-$[Pd(AsMe_3)_2I_2]$, red, m. 178-179°, prepd. from Na_2PdCl_4 + Me_3As in EtOH in the presence of an excess of NaI; IR spectra in the solid state and in soln. were recorded (629).

$[Pd(AsMe_3)Cl_2]_2$ (M-5), IR spectrum (629, 631), a Cl-bridged structure with a trans configuration postulated (631).

$[Pd(AsMe_3)Br_2]_2$ (M-5), brown, m. 270-277° (dec.), IR spectrum (629).

$[Pd(AsMe_3)I_2]_2$, black, m. 240-244° (dec.), IR spectrum (629).

$[Pr_4N][Pd(AsMe_3)Cl_3]$, prepd. from $[Pd(AsMe_3)Cl_2]_2$ + $[Pr_4N]Cl$ in CH_2Cl_2; IR spectrum (630).

cis-$[Pt(AsMe_3)_2Cl_2]$, white, m. 272-274°, prepd. from Na_2PtCl_4 + Me_3As in EtOH; IR spectra in the solid state and in soln. were recorded (629); NMR spectrum (594).

trans-$[Pt(AsMe_3)_2Cl_2]$, yellow, m. 270-275° (dec.), prepd. from Na_2PtCl_4 + Me_3As in EtOH; IR spectra in the solid state and in soln. were obtd. (629).

cis-$[Pt(AsMe_3)_2Br_2]$, yellow, m. 238-241°, prepd. from Na_2PtCl_4 + Me_3As in EtOH in the presence of an excess of NaBr; IR spectrum (629).

trans-$[Pt(AsMe_3)_2I_2]$, orange, m. 198-200°, prepd. from Na_2PtCl_4 + Me_3As in EtOH in the presence of an excess of NaI; IR spectra in the solid state and in soln. were recorded (629).

$[Pt(AsMe_3)Cl_2]_2$, orange red crystals, dec. 220-225°, prepd. from finely ground $[Pt(AsMe_3)_2Cl_2]$ + $PtCl_2$ (1:1) in a xylene-naphthalene mixt. by heating at 155°, 79% yield (632); IR spectrum (629, 631), a Cl-bridged trans configuration postulated (631); crystal data by x-ray analysis - discussion of the molec. structure (1587).

[Pt(AsMe$_3$)Br$_2$]$_2$, orange, m. 230-239° (dec.), prepd. from [Pt(AsMe$_3$)$_2$Br$_2$] + PtBr$_2$ by heating in a xylene-naphthalene mixt. at 155°; IR spectrum (629).

[Pt(AsMe$_3$)I$_2$]$_2$, red, m. 275-279° (dec.), prepd. from [Pt(AsMe$_3$)$_2$I$_2$] + PtI$_2$ as described above; IR spectrum (629).

[Pr$_4$N][Pt(AsMe$_3$)Cl$_3$], yellow plates, m. 147-148°, prepd. from [Pt(AsMe$_3$)Cl$_2$]$_2$ + [Pr$_4$N]Cl in CH$_2$Cl$_2$ by pptn. with Et$_2$O; IR spectrum (630).

cis-[Me$_2$Pt(AsMe$_3$)$_2$], NMR spectrum (594).

Arsylidynetrimethanol, Tris(hydroxymethyl)arsine, AsC$_3$H$_9$O$_3$, (HOCH$_2$)$_3$As
Synth.: From AsH$_3$ by passing into aq. HCHO soln. in the presence of CdCl$_2$ at 20-45°/150-200 mm, 56% yield; with 35% HCl as catalyst the yield was 28% (1265).
Prop.: b$_{0.07}$ 74-76°, d$_4^{20}$ 1.7213, n$_D^{20}$ 1.5942, MR$_D$ 33.28 (1265).
Rxn. with: MeI at 60° → [Me(HOCH$_2$)$_3$As]I (1265).

Triethynylarsine, AsC$_6$H$_3$, (HC≡C)$_3$As
Synth.: From AsCl$_3$ in Et$_2$O by adding dropwise HC≡CMgBr in THF at -30°, then warming to r.t., 62% yield (1564).
Prop.: Colorless crystals, m. 49-50°, subl$_{13}$ 40°, IR and NMR spectra (1564). IR and Raman spectra, frequency assignments and mol. symmetry (950, 1064).

Trivinylarsine, AsC$_6$H$_9$, (CH$_2$=CH)$_3$As, (M-6)
Synth.: see (M-6)
Prop.: b$_{40}$ 49-50°. UV spectra in isooctane, MeOH, and in the gas phase (1594). IR spectrum (209, 210, see also 345).

Triethylarsine, AsC$_6$H$_{15}$, Et$_3$As, (M-7)
Synth.: From AsCl$_3$ by the Grignard rxn. with EtMgX (508).
Prop.: b$_{24}$ 48°, d$_4^{20}$ 1.0695, n$_D^{20}$ 1.4735, [ρ]$_M$ 1031.4, χ$_M$ -101.3, R$_M$ 42.55 (508), b. 138.6° calcd. by the Kinney equation (66). Spectra: IR and Raman (209, 210), NMR (343), UV in isooctane, MeOH, and in the gas state (1594).
Rxn. with: HgO (1:1) in Me$_2$CO → Et$_3$AsO (1057).
Aq. H$_2$O$_2$ → Et$_3$AsO (1057).
H$_2$O$_2$ excess → Et$_2$AsO$_2$H (1057).
Atm. O and H$_2$O → Et$_2$AsO$_2$H (1057).
MeI in MeOH at 25°, the rxn. kinetics and rel. nucleophilic reactivity parameter were detd. by measuring the change in elec. cond. (1203).
trans-[Pt(Py)$_2$Cl$_2$] in MeOH at 25°, the rxn. kinetics and rel. nucleophilic reactivity parameter were detd. by UV spectroscopy (1203).
Uses: A stabilizer for the polymerization catalysts such as halides or oxy-halides of Group IV, V, or VI transition metals deposited on SiO$_2$ or Al$_2$O$_3$ (41). A complex formed from the arsine with [Co(CO)$_4$]$_2$ is an oxo catalyst favoring the formation of alcohols over aldehydes (291, 1424). A complex

with $PhCH_2AsCl_2$ and $(BuO)_4Ti$ catalyzes the polymerization of C_3H_6 (383). Complexes with $M[AlR_4]$ (M = alkali metal) and Ti, V, Zr, Cr, or Mo halides catalyze the polymerization of α-olefins (385). Complexes with Ni^0 and non-aromatic hydroxy compds. catalyze the oligomerization of linear 1,3-dienes (1382a).

Coordination Deriv.: $[Et_3As \cdot BH_3]$ (M-7), prepd. from $Et_3As + B_2H_6$ (2:1) at r.t.; a rxn. of the complex with $B_5H_9 > 75°$ gave $[(Et_3As)_2BH_2]B_{12}H_{12}$, $[(Et_3As)_2BH_2][B_{12}H_{11}AsEt_3]$, and $B_{12}H_{10}(AsEt_3)_2$; the latter is useful as a scavenger of halogens (1067).

$[(Et_3As)_2B_{12}H_{10}]$, prepd. from $Et_3As + B_2H_6 > 75°$; useful as a scavenger of halogens (1067).

$[Co(AsEt_3)(CO)_3]$, red-brown crystals, m. 195° (dec.), prepd. from the arsine and $[Co(CO)_4]_2$ (2:1) in C_6H_6 on a steam bath; IR spectrum (1021). The effect of various solvents on the IR spectrum was detn. The structure of the complex in soln. is interpreted on the basis of an equilibrium involving a non-bridged and two bridged isomers (1022).

$[Co(AsEt_3)_n(CO)_{4-n}]$, a catalyst for the oxonation of olefins to aldehydes and alcohols (1424).

$Hg[Co(AsEt_3)(CO)_3]_2$, crystals, m. 146-147°, prepd. from $Hg[Co(CO)_4]_2$ and the arsine in C_6H_6; IR spectrum (1020).

trans-$[Ir(AsEt_3)_3Cl_3]$ (M-8), far-IR spectrum; the trans configuration confirmed (799).

$[Ir(AsEt_3)(PMe_2Ph)_2Cl_3]$, yellow prisms, m. 178-183°, prepd. from mer-$[Ir(AsEt_3)_3Cl_2]$ and Me_2PhP in C_6H_6 under reflux and pptn., after cooling, with light petr. IR spectrum; a structure with the arsine in an equatorial position trans to Cl and both phosphine groups occupying the other equatorial positions postulated (1250).

$[IrH(AsEt_3)_3Cl_2]$ (M-8), far-IR spectrum (790).

$[IrH(AsEt_3)_2(PMe_2Ph)Cl_2]$, yellow prisms, m. 99-101°, prepd. from $[IrH(PMe_2Ph)_3-Cl_2] + Et_3As$ by equilibration, pouring the rxn. mixt. into MeOH, removing the solvent, treating the residue with a MeOH-light petr. mixt., and crystg. from MeOH. IR and NMR spectra; an octahedral structure with the Cl atoms in the opical positions and one arsine group in a trans position to the H atom postulated (1250).

$[(Et_3As)_2(HgCl_2)_3]$ (M-9), colorless needles, m. 114°, prepd. from Et_3As and $HgCl_2$ in H_2O; IR spectrum (369).

$[Hg(AsEt_3)Cl_2]_2$ (M-8), m. 163°, IR spectrum (369).

$[Mo(AsEt_3)(CO)_5]$ (M-9), IR spectrum (435).

$[Pd(AsEt_3)Cl_2]_2$ (M-9), IR spectra in Nujol and in cis-$ClCH=CHCl$; a Cl-bridged, trans configuration postulated (631).

[Pr$_4$N][Pd(AsEt$_3$)Cl$_3$], prepd. from [Pd(AsEt$_3$)Cl$_2$]$_2$ + [Pr$_4$N]Cl in CH$_2$Cl$_2$; IR spectrum (630).

trans-[Pt(AsEt$_3$)$_2$Cl$_2$] (M-9), kinetics of the anion exchange with Br$^-$, I$^-$, N$_3$$^-$, NO$_2$$^-$, NH$_3$, SCN$^-$, and SeCN$^-$ were measured by UV spectroscopy (148, 308).

trans-[Pt(AsEt$_3$)$_2$Br$_2$] (M-9), crystals, m. 120°, prepd. from trans-[Pt(AsEt$_3$)$_2$-Cl$_2$] + [Et$_4$N]Br in MeOH at 30° (148), IR spectrum (802).

trans-[Pt(AsEt$_3$)$_2$I$_2$] (M-10), crystals, m. 93°, prepd. from trans-[Pt(AsEt$_3$)$_2$-Cl$_2$] + [Bu$_4$N]I in MeOH at 30° (148), IR spectrum (803).

trans-[Pt(AsEt$_3$)$_2$(NCS)$_2$], m. 125°, prepd. from trans-[Pt(AsEt$_3$)$_2$Cl$_2$] + NaSCN in MeOH at 30°, followed by H$_2$O; the rxn. kinetics and activation energy were detd. (308).

trans-[Pt(AsEt$_3$)$_2$(N$_3$)$_2$], crystals, m. 92°, prepd. from trans-[Pt(AsEt$_3$)$_2$Cl$_2$] + NaN$_3$ in MeOH at 30° (148).

trans-[Pt(AsEt$_3$)$_2$(NO$_2$)$_2$] (M-10), crystals, m. 199°, prepd. from trans-[Pt-(AsEt$_3$)$_2$Cl$_2$] + NaNO$_2$ in MeOH at 30° (148).

trans-[HPt(AsEt$_3$)$_2$X] (M-10), where X = Cl, Br, or I; NMR spectra (1247).

trans-[HPt(AsEt$_3$)$_2$SCN] (M-10), the NMR spectrum of the complex previously described in (M-10) indicated the presence of the thiocyanato and isothio-cyanato complexes: [HPt(AsEt$_3$)$_2$SCN] and [HPt(AsEt$_3$)$_2$NCS] in a ratio of 1:2 (1247).

[Pr$_4$N][Pt(AsEt$_3$)Cl$_3$], prepd. from [Pt(AsEt$_3$)Cl$_2$]$_2$ + [Pr$_4$N]Cl in CH$_2$Cl$_2$; IR spectrum (630).

[Pt(AsEt$_3$)Cl$_2$]$_2$ (M-10), orange-red crystals, dec. 208-209°, prepd. from a finely ground mixt. of [Pt(AsEt$_3$)$_2$Cl$_2$] and PtCl$_2$ in xylene at 140° (86% yield) (632). IR spectra in Nujol and in cis-ClCH=CHCl (631) and far-IR spectrum (1500) indicated a Cl-bridged complex with a trans configuration.

trans-{Pt(AsEt$_3$)$_2$[P(O)(OPh)$_2$]$_2$}, m. 143-145°, prepd. from cis-[Pt(AsEt$_3$)$_2$Cl$_2$] by treating with excess AgOP(OPh)$_2$ in C$_6$H$_6$ under reflux (1227).

cis-[Pt(AsEt$_3$)$_2$Cl$_4$] (M-10), far-IR spectrum (29).

trans-[Pt(AsEt$_3$)$_2$Cl$_4$] (M-10), far-IR spectrum (29). A rxn. with NaI in MeOH → trans-[Pt(AsEt$_3$)$_2$I$_2$] and I$_3$$^-$; the rxn. was followed by UV spectro-scopy and the presence of I$_3$$^-$ was established polarographically. Attempts to isolate the trans-[Pt(AsEt$_3$)$_2$I$_2$] by adding H$_2$O gave only trans-[Pt(AsEt$_3$)$_2$I$_4$]. A rxn. of the tetrachloro complex with Na$_2$S$_2$O$_3$ in MeOH contg. 4.3% H$_2$O gave trans-[Pt(AsEt$_3$)$_2$Cl$_2$] which was isolated by adding more H$_2$O. The redn. was followed by UV spectroscopy (1205). A redn. of the tetrachloro complex with KSeCN in MeOH at 0° gave nonreproducible results; a spectrophotometric study of the redn. indicated that the rxn. proceeded through more than one step (1208).

cis-[Pt(AsEt$_3$)$_2$Br$_4$], far-IR spectrum (29).

trans-[Pt(AsEt$_3$)$_2$Br$_4$], (M-10), far-IR spectrum (29). A redn. of the complex with Na$_2$S$_2$O$_3$ in MeOH gave trans-[Pt(AsEt$_3$)$_2$Br$_2$] which was isolated by addn. of H$_2$O and immediate extn. with Et$_2$O. When the redn. prod. was allowed to stand for a while after addn. of H$_2$O, the Br ligands were replaced with S$_2$O$_3$$^{--}$ giving trans-[Pt(AsEt$_3$)$_2$(S$_2$O$_3$)$_2$]$^{--}$ (1205). A redn. of the tetrabromo complex with KSeCN in MeOH at 40° gave nonreproducible results. The rate of the rxn. was followed spectrophotometrically. The redn. occurred stepwise (1208).

cis-[Pt(AsEt$_3$)$_2$I$_4$], far-IR spectrum (29).

trans-[Pt(AsEt$_3$)$_2$I$_4$] (M-10), far-IR spectrum (29).

trans-[PhPt(AsEt$_3$)$_2$SnCl$_3$], pale yellow crystals, m. 147-149°, prepd. from trans-[PhPt(AsEt$_3$)$_2$Cl] in EtOH + SnCl$_2$·2H$_2$O in MeOH at r.t.; IR spectrum was recorded. The rates of the rxn. with CS(NH$_2$)$_2$ and KSCN in MeOH and MeCN at different temp. were compared. The rxns. lead to the displacement of the SnCl$_3$ ligand with SC(NH$_2$)$_2$ and SCN, resp. UV spectra of all three complexes were obtd. (548).

$$\left[\begin{array}{c} \text{PhO} \quad \text{OPh} \\ \diagdown \text{P} \diagup \quad \text{HO} \\ \quad \diagdown \quad \diagdown \text{P(OPh)}_2 \\ \text{Et}_3\text{As} \quad \text{O} \quad \nearrow \\ \quad \searrow \text{Pt} \quad \text{Pt} \nwarrow \\ \text{(PhO)}_2\text{P} \diagup \quad \diagdown \text{O} \diagup \quad \nwarrow \text{AsEt}_3 \\ \quad \text{OH} \quad \diagdown \text{P} \diagup \\ \quad \quad \text{PhO} \quad \text{OPh} \end{array} \right]$$ Cl$_2$, m. 118-120°, prepd. from trans-[Pt-(AsEt$_3$)Cl$_2$]$_2$ + (PhO)$_2$PHO (1:2) in Me$_2$CO-C$_6$H$_6$ mixt. under reflux (1227).

trans-[Rh(AsEt$_3$)$_2$(CO)Cl], yellow crystals, m. 56-60°, prepd. from RhCl$_3$·3H$_2$O in EtOH under reflux by bubbling CO and then adding the arsine; IR spectrum (330). The complex catalyzes the formation of urethans from alcohols or phenols, CO, and nitro compds. (775).

[Rh(AsEt$_3$)$_2$(CO)Cl$_3$], yellow crystals, prepd. from [Rh(AsEt$_3$)$_2$(CO)Cl] by treating with CCl$_4$; far-IR spectrum, an octahedral structure postulated (160).

[Rh(AsEt$_3$)$_2$(CO)ClBr$_2$], orange crystals, prepd. from [Rh(AsEt$_3$)$_2$(CO)Cl] by treating with Br in CHCl$_3$; far-IR spectrum, an octahedral structure postulated (160).

trans-[Rh(AsEt$_3$)$_3$Br$_3$], dark red needles, prepd. from trans-[Rh(AsEt$_3$)$_3$Cl$_3$] by treating with LiBr in Me$_2$CO under reflux; far-IR spectrum, an octahedral structure postulated (160).

[Ru(AsEt$_3$)$_2$(NO)Cl$_3$] (M-11), brown prisms, m. 145-146.5°, prepd. from RuCl$_3$-(NO)·5H$_2$O by heating with Et$_3$As in MeOCH$_2$CH$_2$OH under reflux; IR spectrum and dipole moment indicated a trans arrangement of the arsine ligands (331).

[Ru(AsEt$_3$)$_2$(NO)Br$_3$], brown prisms, m. 142-143.5°, prepd. from [Ru(AsEt$_3$)$_2$(NO)-Cl$_3$] by boiling with LiBr in MeOCH$_2$CH$_2$OH; IR spectrum (331).

[Ru(AsEt$_3$)$_2$(NO)I$_3$] (M-11), red-brown needles, m. 131.5-132.5°, prepd. from the trichloro complex and NaI in MeOCH$_2$CH$_2$OH under reflux; IR spectrum (331).

Tripropynylarsine, AsC_9H_9, $(MeC\equiv C)_3As$
Synth.: From AsX_3 + MeC≡CMgX in THF at r.t., 92% yield (149).
Prop.: m. 131°, IR and NMR spectra (149).

Triallylarsine, AsC_9H_{15}, $(CH_2=CHCH_2)_3As$, (M-11)
Prop.: IR spectrum (209, 210, see also 345).

Tri(cis-propenyl)arsine, AsC_9H_{15}, $(MeCH=CH)_3As$, (M-12)
Prop.: b_{18} 87-88°, n_D^{20} 1.5257, IR spectrum (346).

Tri(trans-propenyl)arsine, AsC_9H_{15}, $(MeCH=CH)_3As$, (M-12)
Prop.: b_6 72°, n_D^{20} 1.5209, IR spectrum (346).

Tripropylarsine, AsC_9H_{21}, Pr_3As, (M-12)
Prop.: b_{18} 88°, d_4^{20} 1.0102, n_D^{20} 1.429, $[\rho]_M$ 1253.0, X_M -135.5, R_M 56.69 (508).
Rxn. with: HgO (1:1) in $Me_2CO \rightarrow Pr_3AsO$ (1057).
Aq. H_2O_2 in $Et_2O \rightarrow Pr_3AsO$ (1057).
Aq. H_2O_2, excess $\rightarrow Pr_2AsO_2H$ (1057).
Moist air $\rightarrow Pr_2AsO_2H$ (1057).
Coordination Deriv.: $[Os(AsPr_3)_2Cl_4]$, NMR spectrum and magnetic data (329).
trans-$[Pd(AsPr_3)(Py)Cl_2]$, yellow crystals, m. 72-73.5°, prepd. from $[Pd(AsPr_3)Cl_2]_2$ by treating with Py in CH_2Cl_2; NMR spectrum (332).

cis-$[Pt(AsPr_3)_2Cl_2]$ (M-13), NMR spectrum - ^{195}Pt chem. shift in $CHCl_3$ (1226).

trans-$[Pt(AsPr_3)_2Cl_2]$ (M-13), NMR spectrum - ^{195}Pt chem. shift in $CHCl_3$ (1226). The complex catalyzes the polymerization of 1,1,3,3-tetramethyl-1,3-disilacyclobutane at 120° (114, 983).

trans-$[Pt(AsPr_3)_2Br_2]$, NMR spectrum - ^{195}Pt chem. shift in $CHCl_3$ and electronic spectrum in C_6H_{14} (1226).

trans-$[Pt(AsPr_3)_2I_2]$, NMR spectrum - ^{195}Pt chem. shift in $CHCl_3$ and electronic spectrum (1226).

trans-$[Pt(AsPr_3)_2(CN)_2]$, NMR spectrum - ^{195}Pt chem. shift in $CHCl_3$ (1226).

trans-$[Pt(AsPr_3)(Py)Cl_2]$, yellow crystals, m. 82-83.5°, prepd. from $[Pt(AsPr_3)Cl_2]_2$ by treating with Py in CH_2Cl_2; NMR spectrum (332).

2,2′,2″-Arsylidynetris(3-chlorobutylmercury) Phosphate, $AsHg_3C_{12}H_{21}Cl_3O_4P$,
    $As(\overset{|}{C}HCH_2HgO)_3PO$
     MeCHCl
Synth.: From AsH_3 by a condensation with methallyl chloride, HgO, and H_3PO_4 (1:3:3:1) in AcOH at a low temp. (1447).
Biol. Prop.: A fungicide and bactericide (1447).

Tributylarsine, $AsC_{12}H_{27}$, Bu_3As, (M-15)
Prop.: b_{24} 134°, d_4^{20} 0.9798, n_D^{20} 1.4739, $[\rho]_M$ 1475.0, X_M -170.3, R_M 70.61

(508). IR spectrum interpretation (1223).

Rxn. with: HgO (1:1) in $Me_2CO \rightarrow Bu_3AsO$ (1057).

Aq. H_2O_2 in $Et_2O \rightarrow Bu_3AsO$ (1057).

Aq. H_2O_2, excess $\rightarrow Bu_2AsO_2H$ (1057).

Moist air $\rightarrow Bu_2AsO_2H$ (1057).

$(p\text{-}MeOC_6H_4)_2Si(N_3)_2$ (2:1) $\rightarrow (p\text{-}MeOC_6H_4)_2Si(N=AsBu_3)_2$ (1584, 1586).

Uses: A catalyst for the prodn. of polyurethan foam from polyesters and diisocyanates (757). A catalyst for the polymerization of $\overline{CH_2CH_2S}$ (801). Mixtures with $M[AlR_4]$ (where M = alkali metal) and halides of Ti, V, Zr, Cr, or Mo catalyze the polymerization of α-olefins (385). Mixtures with Na, Li, K, Mg, or Zn and halides of Ti or V catalyze the polymerization of α-olefins (387). Mixtures with organoaluminum halides and halides of Ti, V, Zr, Cr, or Mo catalyze the polymerization of α-olefins (388). Complexes with Ni^0 compds. and nonaromatic hydroxy compds. catalyze the oligomerization of linear 1,3-dienes (1382a). Complexes with Co, Rh, or Ir carbonyls catalyze the oxonation of olefins to aldehydes and alcohols (1424, 1425).

Coordination Deriv.: $[Co(AsBu_3)(CO)_2(NO)]$, the formation of the complex from $[Co(CO)_3(NO)]$ and Bu_3As in MePh was detected by observing the changes in the IR spectrum of the rxn. mixt. and the rate of CO evolution (1515).

$[Pd(AsBu_3)_2(SCN)_2]$ (M-15), IR spectrum (1314).

trans-$[Pt(AsBu_3)_2Cl_2]$ (M-16), NMR spectrum - ^{195}Pt chem. shift in $CHCl_3$ (1226).

Triisobutylarsine, $AsC_{12}H_{27}$,   i-Bu_3As,   (M-16)
Uses: A stabilizer for the polymerization catalysts formed from halides or oxyhalides of the transition metals of Group IV, V, or VI deposited on SiO_2 or Al_2O_3 (41).

2,2′,2″-Arsylidynetris(3-chloropropylmercury Acetate), $AsHg_3C_{15}H_{24}Cl_3O_6$,
    $As[CH(CH_2Cl)CH_2HgOAc]_3$
Synth.: From AsH_3 by a condensation with allyl chloride and HgO in AcOH (1447).
Biol. Prop.: A fungicide and bactericide (1447).

Tripentylarsine, $AsC_{15}H_{33}$, $(C_5H_{11})_3As$,   (M-30)
Prop.: $b_{0.4}$ 103°, d_4^{20} 0.9588, n_D^{20} 1.4742, $[\rho]_M$ 1690.8, χ_M -204.4, R_M 84.54 (508).
Rxn. with: H_2O_2, excess $\rightarrow (C_5H_{11})_2AsO_2H$ (1057).
Moist air $\rightarrow (C_5H_{11})_2AsO_2H$ (1057).

Tris(pentafluorophenyl)arsine, $AsC_{18}F_{15}$, $(C_6F_5)_3As$
Synth.: From $AsCl_3 + C_6F_5MgBr$ in Et_2O at -10 to -20°, 39% yield (566). From $AsCl_3$ in Et_2O by adding dropwise to C_6F_5Li in the same solvent at -78° and warming to r.t., 75% yield (842).
Prop.: Colorless platelets, m. 106° (566), m. 104-106° (842), IR spectrum (566).

Rxn. with: Cl_2 or ClI → $(C_6F_5)_3AsCl_2$ (648).

Tris(difluorophenyl)arsine, $AsC_{18}H_9F_6$, $(C_6H_3F_2)_3As$
Synth.: From $AsCl_3$ in Et_2O by adding dropwise to $C_6H_3F_2Li$ in Et_2O at -78°
and warming the mixt. to r.t., 75% yield (842).
Prop.: m. 105-106° (842).

Tris(4-chlorophenyl)arsine, $AsC_{18}H_{12}Cl_3$, $(4-ClC_6H_4)_3As$, (M-17)
Synth.: From $AsCl_3$ + $p-ClC_6H_4Li$ (1:3) in Et_2O at -70°, followed by hydroly-
sis, 63% yield (723).
Prop.: Crystals, m. 104-107°, NMR spectrum (723).
Rxn. with: H_2O_2 in Me_2CO → $(4-ClC_6H_4)_3AsO$ (1392).
Chloramine T in THF → $(4-ClC_6H_4)_3As=NO_2SC_6H_4Me-4$ (723).
Coordination Deriv.: $\{Rh[As(C_6H_4Cl-4)_3]_2(CO)(NCS)\}$, a yellow solid, m. 158-
159° (dec.), prepd. from the arsine + $[Bu_4N][Rh(CO)_2(NCS)_2]$ in MeOH at -78°;
IR spectrum (802).

Triphenylarsine, $AsC_{18}H_{15}$, Ph_3As, (M-17)
Synth.: From $AsCl_3$ + PhMgBr, 90% yield (90).
From Ph_2AsK + $PhSO_3Na$ in $(EtOCH_2CH_2)_2O$ by heating up to 180°, 29% yield
(1347).
From $(Ph_2As-)_2$ + PhLi in Et_2O followed by Et_2O-dioxan mixt., 78% yield, along
with $Ph_2AsLi \cdot$ Dioxan complex (786).
$Ph_3{}^{77}As$ is formed by a recombination of the fragments from the β-decay of
$Ph_4{}^{77}Ge$ under various conditions (1277).
Prop.: m. 60.5-61° (90), m. 60° (1658). Microanalysis by gas-liquid chromatography after hydrogenolysis over Pd/Al_2O_3 at 200° to C_6H_{12} (1508). Half-
wave potential (467, 749). Force consts. (832). Soly. in styrene 30% at 20°
and 50% at 50° (123). Sepn. from mixts. by paper chromatography and paper
electrophoresis (1422). TLC and radiometric identification (169, 170). Di-
electric const. and loss data (699). Thermionic response of K, Rb, and Cs
salts to Ph_3As in gas chromatography (795). Inhibition of Fe corrosion in
HCl and H_2SO_4 soln. (981, 982). Spectra: electronic (423); UV in aq. MeOH
and absorption shift upon adding CF_3CO_2H, CF_3CO_2Ag, and $AgNO_3$, resp. (1175),
IR (418, 649, 803, 1001, 1167) qual. interpretation (1223). Far-IR (244).
Low-frequency IR (158). Raman displacements (649). K-edge x-ray absorption
(962).
Rxn. with: Na (1:2) in liq. NH_3 → Ph_2AsNa (525).
Na in liq. NH_3 followed by H_2O → Ph_2AsH (1017).
Na in liq. NH_3 followed by NH_4Br and R_nMCl_{4-n} (M = Ge, Sn, Pb) → $(Ph_2As)_{4-n}-$
MR_n (1371, 1374).
Li, followed by t-BuCl → Ph_2AsLi (43, 89).
KNH_2 in NH_3 → $K[As(NH)_2]$ + C_6H_6 (1362, 1364).
$p-MeC_6H_4Li$ in Et_2O at r.t. and pouring into a suspension of CO_2 in Et_2O → a
mixt. of BzOH and $p-MeC_6H_4CO_2H$ - the rxn. probably involves an equilibration:
Ph_3As + $MeC_6H_4Li \rightleftharpoons [Ph_3AsC_6H_4Me]Li \rightleftharpoons Ph_2AsC_6H_4Me$ + PhLi (1620).
30% aq. H_2O_2 in Me_2CO under reflux → Ph_3AsO (1347).

H_2O_2 in aq. soln. at r.t. or in MeOH → $Ph_3AsO \cdot H_2O_2$ (759, 1305).

I in $CHCl_3$ → $Ph_3As^+I_2^-$ (a charge transfer complex) → $[Ph_3AsI]^+I^-$ (179, see also 869).

I in CCl_4 or CH_2Cl_2 → $Ph_3As^+I_n^-$ - a charge transfer complex (99, 179, 869).

Cl in C_6H_6 at 20° → Ph_3AsCl_2 and Ph_2AsCl_3 (1421).

Cl in Et_2O at -78° → Ph_3AsCl_2 (1421).

Br in MeCN or CCl_4 → Ph_3AsBr_2 (177, 1167, 1521).

Br in C_6H_6 followed by RSO_2NH_2 in Et_3N → $Ph_3AsO \cdot H_2NSO_2R$ (342).

Br in C_6H_6 followed by RNH_2 in Et_3N → $Ph_3AsO \cdot H_2NR$ (342).

Br in CCl_4 followed by RSO_2NH_2 → $Ph_3As=NO_2SR$ (341).

I soln. applied to Ph_3As soln. on filter paper → $Ph_3AsI_2 \cdot xI_2$ - a yellow spot, an analytical test (555).

HNO_3 → $Ph_3AsO \cdot HNO_3$ (1521).

MeI → $[MePh_3As]I$ (90).

MeI in MeOH at 25°, the rate of rxn. and rel. nucleophilic reactivity parameter were detd. from the change in elec. cond. (1203).

Aryl halides (RX) under γ-radiation → $[RPh_3As]X$ (498).

H_2NCl in C_6H_6 → a mixt. of $[Ph_3AsNH_2]Cl$ and $[Ph_2As(NH_3)_2]Cl$ (1421).

S in C_6H_6 → Ph_3AsS (603, 1347).

RSSR → $Ph_3AsS + R_2S$ (1115).

$BrCH_2CO_2Me$ → $[Ph_3AsCH_2CO_2Me]Br$ (766).

H over Pd/Al_2O_3 at 200° → hydrogenolysis producing C_6H_{12}, detn. by gas-liquid chromatography - microanal. procedure (1508).

$p\text{-}MeC_6H_4SO_2NSO$ in C_6H_6 under reflux → $p\text{-}MeC_6H_4SO_2N=AsPh_3$ (1385).

Hexamethylbenzene → a solid 1:1 complex (1393).

Naphthalene, no complex formation (1393).

$Ph_2P(O)N_3$ → $Ph_2P(O)N=AsPh_3$ (1583).

$P(O)Cl_3$ under reflux → $[Ph_3AsP(O)Cl_2]Cl$ (968).

Perfluoropiperidine → Ph_3AsF_2 (118).

Diazotetraphenylcyclopentadiene → $Ph_3As=\overline{CCPh=CPhCPh=CPh}$ (974).

trans-$[Pt(Py)_2Cl_2]$ in MeOH at 25°, the rate of rxn. and rel. nucleophilic reactivity parameter were detd. by UV spectroscopy (1203).

$[(C_6Me_6)Cr(CO)_2(PhC \equiv CPh)]$ in C_6H_6 at 40°, rxn. kinetics (75).

$$\left\{ M \begin{Bmatrix} S-CCF_3 \\ S-CCF_3 \end{Bmatrix}_2 \right\}_2 \rightarrow Ph_3AsM \begin{Bmatrix} S-CCF_3 \\ S-CCF_3 \end{Bmatrix}_2 \text{, where } M = Co \text{ or } Fe \text{ (110)}.$$

1,4-$[PhGe(N_3)_2]_2C_6H_4$ → 1,4-$[(Ph_3As=N)_2GePh]_2C_6H_4$ (1584-1586).

$UO_2(NO_3)_2 \cdot 6H_2O$ → $[UO_2(Ph_3AsO)_2(NO_3)_2]$ (1645).

Polarographic oxidn. on a Hg anode in MeCN → $[Hg(AsPh_3)_2]^{++}$ (749).

Neutron irradiation (350).

^{60}Co γ-radiation → a radiosensitized dec. to Ph_2 and phenylcyclohexadienes (1219, 1220).

UV irradiation (2537 Å) in C_6H_6 soln. → Ph_2, rxn. mechanism (1220).

Uses: A sensitizer for Ag halides in films (53). A background fog inhibitor in photothermographic sheets (743, 744). An activator for photographic direct-positive emulsions (745). A photosensitizer for lithographic plates (746). A stabilizer for photochromic systems such as mixts. of 2,3-diphenylindenone

2,3-epoxide in poly(methyl methacrylate) against mol. degradation (1544).
A synergistic enhancer for the stabilizing effect of secondary aromatic
amines on polyesters contg. hydroperoxides (550). An initiator for the UV
light catalyzed polymerization of olefins (1024). An initiator for anionic
polymerization of HCHO (903). A moderate flammability inhibitor for epoxy
resins (1025). An inhibitor of the electrode rxn. in the Fe/N HCl and Fe/N
HCl-MeOH systems (1033). At the concn. of 2-5% in polyimide compns. makes
the latter resistant to high temp. corona discharge (1046). A stabilizer
for the polymerization catalysts consisting of halides or oxyhalides of the
transition metals of Group IV, V, or VI deposited on SiO_2 or Al_2O_3 (41).
A catalyst for the cyanoethylation of silanes contg. at least one H atom and
at least one hydrolyzable group bonded to Si (804). A polystyrene contg.
1.5% Ph_3As is useful as a scintillator for x-ray and γ-ray dosimeters (184).
A polystyrene contg. up to 50% Ph_3As, prepd. by a thermal polymerization of
styrene soln. contg. Ph_3As, a colorless transparent solid, upon addn. of
3% p-terphenyl and 0.08% 1,3,5-triphenyl-2-pyrazoline, is useful as a scintillator for the detection of γ-radiation (123). Mixtures with ethylene
oxide catalyze the formation of formates, HCO_2R, from alcohols and CO (224,
1431). Mixtures with $M[AlR_4]$, where M is an alkali metal, and Ti, V, Zr,
Cr, or Mo halides catalyze the polymerization of olefins (385). Mixtures
with organoaluminum halides and Ti, V, Zr, Cr, or Mo compds. catalyze the
polymerization of α-olefins (388). Mixtures with Na, Li, K, Zn, or Mg and
Ti or V halides catalyze the polymerization of α-olefins (387). Complex
compds. formed from Ph_3As, Ni acetylacetonate, and $Et_3Al_2Cl_3$ catalyze the
dimerization of C_2H_4 and C_3H_6 (1367). A complex formed with $EtAlCl_2$ and
$(BuO)_4Ti$ catalyzes the polymerization of C_3H_6 (383). A rxn. prod. formed
from Ph_3As, $EtAlCl_2$, and $PdCl_2$ (1:25:5.5) catalyzes the dimerization and
oligomerization of C_3H_6 (770). A complex with $CuCl_2$ (1:1) is useful as a
semiconductor for thermistors based on organophosphorus compds. (1433). A
complex with π-allyl-Ni deriv. catalyzes the oligomerization of olefins
(1462). Mixts. with Ni^0 and nonaromatic hydroxy compds. catalyze the oligomerization of linear 1,3-dienes (1382a). A promoter for the catalysts consisting of π-allyl derivs. of the transition metals of Groups IV-VII or of
π-allylmetal halides and Lewis bases such as organoaluminum halides; the complexes promote the oligomerization and polymerization of olefins (1610).
Coordination Deriv.: In the following entries (Ars) stands for Ph_3As.

$[Cr(Ars)(CO)_5]$ (M-19), the rate of rxn. of $Cr(CO)_6$ with Ars at the ratios
1:1 to 1:10 in n-decane-cyclohexane at $122.5 \pm 0.3°$ was studied by IR spectroscopy. Even for the 1:10 ratio the extent of a disubstitution was extremely limited; a S_N1 dissocn. mechanism postulated (1599).

$[π-C_5H_5Cr(Ars)(NO)]Cl$, paramagnetic crystals, m. 129-130°, prepd. from
$π-C_5H_5Cr(NO)_2Cl$ + Ars in C_6H_6 by heating under reflux for 16 hrs., purifying
by chromatography and recrystd., 7% yield. IR spectrum and magnetic data
(567).

$[Co_2(Ars)(CO)_7]$ (M-20), was detected by IR spectroscopy in an equil. mixt.

of $[Co(CO)_4]_2 + [Co_2(Ars)_2(CO)_6] \rightleftharpoons [Co_2(Ars)(CO)_7]$ (1467).

$[Co_4(Ars)(CO)_{11}]$, crystals, prepd. from $Co_4(CO)_{12}$ + Ars in petr. ether, 16% yield. IR spectrum indicated a substitution of a terminal CO with the arsine ligand (313).

$[Co(Ars)(NO)_2Cl]$ (M-19), IR spectrum - the solvent effect on the NO vibration and the stability of the complex in various solvents were detn. (136).

$[Co(Ars)(NO)_2NCS]$, black crystals, m. 121-123° (dec.), prepd. from $[Co(NO)_2$-$SCN]_2$ + Ars in THF at elevated temp. In solutions the complex dissociates into $[Co(NO)_2SCN]_2$ + Ars (729).

$[Co(Ars)(NO)_2SEt]$, black-brown crystals, m. 55-56° (dec.), prepd. from $[Co(NO)_2SEt]_2$ + Ars in n-C_5H_{12}. In solutions the complex dissociates into $[Co(NO)_2SEt]_2$ + Ars (729).

$[Co(Ars)(NO)_2SPh]$, black-brown crystals, m. 62-65° (dec.), prepd. from $[Co(NO)_2SPh]_2$ + Ars in C_6H_6 by heating. In soln. the complex dissociates into $[Co(NO)_2SPh]_2$ and Ars (729).

$[Co(Ars)(CO)_2NO]$ (M-19), a red solid, m. 110° (dec.), prepd. from $[Co(CO)_3$-$NO]$ by adding Ars in MePh at 50° (1516) or in C_6H_6 (297). The kinetics of the complex formation was followed by observing changes in the IR spectrum of the rxn. mixt. and by measuring the rate of the CO evolution (297, 1515). The complex is readily sol. in alcohols and ethers and moderately sol. in paraffins and aromatic solvents (1515). IR and UV spectra (297); IR spectrum - NO valence vibration (135), CO and NO valence vibrations and bond angles (137). A rxn. of the complex with Bu_3P in MePh at 40.4° leads to the displacement of the Ars ligand by the phosphine; kinetics of the rxn. was detn. (1516).

$[Co(Ars)_2(CO)NO]$ (M-20), IR spectrum - NO valence vibration (135).

$(Ars)CO\left\{\begin{matrix}S-CCF_3 \\ \| \\ S-CCF_3\end{matrix}\right\}_2$, brown, paramagnetic crystals, m. 186.5-188°, prepd. from $[CoS_4C_4(CF_4)_4]_2$ + Ars in dry C_5H_{12}. Polarographic data, electronic and ESR spectra; a square-pyramidal structure for the 5-coordinate Co was suggested (110).

$\{Co(Ars)[ON=C(Me)C(Me)=NOBPh_2]_2\}$, red-brown crystals, m. 192-195°, prepd. from $\{Co(Ars)[ON=C(Me)C(Me)=NOH]_2Cl\}$ in THF by treating the soln. with K (1:2) under reflux, followed by Ph_2BCl and completing the rxn. under reflux. A structure with 4 N and As atoms chelated with the 5-coordinate Co was postulated (1350).

$\{Ph_2BCo(Ars)[ON=C(Me)C(Me)=NOBPh_2]_2\}$, brown crystals, prepd. from $\{Co(Ars)$-$[ON=C(Me)C(Me)=NOBPh_2]_2\}$ by treating with K ($\sim$ 1:1) in THF under reflux, isolating the K salt and treating it with Ph_2BCl in THF. An octahedral structure with 4 N atoms in the equatorial and the As atom in one and the third B atom in the other opical position postulated (1350).

(In the following entries (Ars) stands for Ph_3As.)

{$(Ars)_2Co[ON=C(Me)C(Me)=NOH]_2$}, black, paramagnetic crystals, prepd. from $Co(OAc)_2 \cdot 4H_2O$ and dimethylglyoxime in MeOH in the presence of the arsine. IR spectrum (1370).

$$\begin{array}{l} \overset{\displaystyle CF_2}{\underset{\displaystyle CF}{\|}} \longrightarrow Co(Ars)(CO)_3 \\ \overset{\displaystyle CF}{\underset{\displaystyle CF_2}{\|}} \longrightarrow Co(Ars)(CO)_3 \end{array}$$

a yellow, paramagnetic substance, m. 110° (dec.), prepd. from $[-CF=CF_2Co(CO)_3]_2$ and the arsine (1:2) in petroleum under reflux. IR and ESR spectra. A π-bonding of the Co atoms to each olefinic bond postulated (772).

$CF_3-CCo(Ars)(CO)_3$
$CF_3-CCo(Ars)(CO)_3$, burgundy crystals, m. 130-135°, prepd. from $[(OC)_3CoC-(CF_3)=C(CF_3)Co(CO)_3]_2$ by heating with excess arsine in petroleum under reflux (60-80°), 70% yield. IR spectrum (1300).

$Hg[Co(Ars)(CO)_3]_2$, crystals, m. 198-199° (dec.), prepd. from $Hg[Co(CO)_4]_2$ + Ars in C_6H_6 at 50°. IR spectrum - vibration assignment (1020).

$Cl_2Sn[Co(Ars)(CO)_3]_2$, orange-red ppt., m. 180°, prepd. from $Cl_2Sn[Co(CO)_4]_2$ and the arsine in C_6H_6 under reflux. IR spectrum (203).

$[Cu(Ars)_2(B_3H_8)]$, white crystals, nonelectrolyte in $CHCl_3$, prepd. from the crude prod. formed by a rxn. of Cu_2Cl_2 with the arsine in $CHCl_3$ with CsB_3H_8 in Me_2CO at r.t. and pptn. with H_2O. The hydroborate is sol. in $CHCl_3$, Me_2CO, DMF, and MeCN, insol. in EtOH and H_2O. A soln. in a 1:1 $CHCl_3$-Me_2CO mixt. dec. at 24° giving a black metallic solid. IR spectrum of the complex was recorded (972).

$[Cu(Ars)Cl]$, crystals, prepd. from Cu_2Cl_2 and the arsine in $CHCl_3$, recrystd. from $CHCl_3$-EtOH. The compn. of the crude prod. before recrystn. corresponded to the $Cu(Ars)_2Cl$ formula (972).

$[Cu(Ars)_2CN]$, m. 179-181°, prepd. from CuCN and Ars (0.2:10) by heating in a sealed tube at 150°. IR spectrum - a 4-coordination of Cu by bridging through the N atom with the adjacent Cu atom was postulated (382).

$[Au(Ars)Cl_3]$, pale yellow crystals, prepd. from $HAuCl_3$ and the arsine and from $AuCl_3$ and Ph_3AuS (851).

$[(Ars)AuMn(CO)_5]$, m. 64° (dec.), diamagnetic, nonelectrolyte in $PhNO_2$, prepd. from $Au(Ars)Cl$ and $Na[Mn(CO)_5]$ in THF, stable in air and to moisture (832). IR spectrum (226, 832).

$[(Ars)$-Hexamethylbenzene], a 1:1 complex, m. 144°, prepd. by the phase-diagram technique (1393).

$[In(Ars)_4](ClO_4)_3$, white powder, prepd. from $In(ClO_4)_3$ and the arsine in a sealed tube at 120°. IR spectrum. The complex salt is insol. in H_2O, C_6H_6, Me_2CO, and dioxan (304).

$[(Ars)^+I_x^-]$, a charge-transfer complex, prepd. from the arsine and iodine in CCl_4 and CH_2Cl_2. UV, visible, IR, and ESR spectra and thermodynamic data were detd. The complex was previously formulated (M-345) as the $[Ph_3AsI]^+I_3^-$ salt (99).

$[Ir(Ars)(COD)Cl]$ (where COD = 1,5-cyclooctadiene), a dark orange substance (1572), m. 130-140° (1573), monomeric and nonconducting in $PhNO_2$ and $ClCH_2CH_2Cl$ (1572, 1573), prepd. from $[Ir(COD)Cl]_2$ and the arsine (1:2) (1573). IR spectrum indicated the presence of both olefinic bonds coordinated to the metal. NMR spectrum in $CDCl_3$ showed nonequivalent olefinic protons, due to asymmetry of the complex. At rising temp. the proton signals gradually widen and finally coalesce to one signal (1572, 1573). The rate of coalescence was detd. (1573). It arises from the interaction of the complex molecules exchanging their Ars ligands (1572). When free Ars was present besides the complex in $PhNO_2$ soln., elec. cond. measurements indicated the presence of the equilibrium mixt.: $[Ir(COD)(Ars)Cl] + Ars \rightleftarrows [Ir(COD)(Ars)_2]^+Cl^-$. The concn. of the ionic species increases as the temp. decreases or more Ars is added. The mechanism of the ligand exchange is considered (1572, 1573). The ligand exchange at -70 to -30° may involve a 5-coordinate Ir species in equilibrium: $[Ir(Ars)_2(COD)Cl] \rightleftarrows [Ir(Ars)_2(COD)]^+Cl^-$. The intermediates were detected by NMR spectroscopy and by elec. cond. measurement. The rate of the rxn. was calcd. At -80° the soln. of the $[Ir(Ars)(COD)Cl] + Ars$ mixt. is deep red, but at 0° it turns yellow-orange. At sufficiently high temp. $[Ir(Ars)(COD)Cl]$ dissociates and forms an equil. mixt.: $2[Ir(Ars)(COD)Cl] \rightleftarrows [Ir(COD)Cl]_2 + 2Ars$. The dissocd. Ars reacts with the dimer and the monomer complex, resp., as follows: $[Ir(COD)Cl]_2 + Ars \rightleftarrows [(COD)IrCl(Ars)IrCl(COD)]$; $[(COD)IrCl(Ars)IrCl(COD)] + Ars \rightleftarrows 2[Ir(Ars)(COD)Cl]$; $[Ir(Ars)(COD)Cl] + Ar\overset{*}{s} \rightleftarrows [Ir(A\overset{*}{r}s)(COD)Cl] + Ars$; $[Ir(Ars)(COD)Cl] + Cl^- \rightleftarrows [Ir(Ars)(COD)Cl_2]^- \rightleftarrows [Ir(COD)Cl_2]^- + Ars$ (1561).

$[Ir(Ars)_2Br_3]$, yellow-maroon crystals, m. 197° (dec.), prepd.: (a) from $[Ir(Ars)_3H_3] + Br$ (stoichiometric amt.) in C_6H_6 soln.; (b) from $[Ir(Ars)_3H_3]$ in C_6H_6 soln. by bubbling HBr through it; (c) from $IrBr_3 + Ars$ excess in EtOH under reflux; and (d) from $K_2IrBr_6 + Ars$ (1:2) in aq. EtOH at 50-60°. Soly. data (293).

$[Ir(Ars)_2(COD)Cl]$, m. 97-99°, a nonelectrolyte in $ClCH_2CH_2Cl$, prepd. from $[Ir(Ars)(COD)Cl] + Ars$ in EtOH (1573).

$[Ir(Ars)_2(COD)SnCl_3]$, yellow crystals, m. 171-174°, prepd. from $[Ir(COD)_2-SnCl_3]$ suspension in CH_2Cl_2 by treating with Ars. Molar cond. in $MeNO_2$ was detd. A 5-coordinate Ir complex (1637).

$[Ir(Ars)_2Cl_3]$, yellow crystals, m. 180° (dec.), prepd. from $[Ir(Ars)_3H_3]$ in THF by bubbling HCl gas. A nonelectrolyte in soln. Soly. data (293).

$[Ir(Ars)_2I_3]$, orange-maroon ppt., m. 170° (dec.), prepd.: (a) from $[IrH_3-(Ars)_3]$ by heating with iodine soln. in C_6H_6 and (b) from the trihydrido complex by bubbling HI in THF suspension; soly. data (293).

(In the following entries (Ars) stands for Ph$_3$As.)

[Ir(Ars)$_2$(CO)Cl], canary-yellow crystals, dec. 251°, prepd. from [Ir(CO)$_2$-(p-H$_2$NC$_6$H$_4$Me)Cl] + Ars (1:2) in C$_6$H$_6$; IR spectrum (730).

[Ir(Ars)$_2$(CO)Br], yellow, diamagnetic crystals, m. 236°, sol. in C$_6$H$_6$, Me$_2$CO, and THF; a nonelectrolyte in Me$_2$CO; prepd. from [Ir(CO)$_3$Br] + Ars in MePh or in C$_6$H$_6$ under reflux; IR spectrum, a trans configuration in a square planar structure postulated (293, 294). Dipole moment (293).

[Ir(Ars)$_2$(CO)(O$_2$)Br], pink, diamagnetic crystals, m. 156°, sol. in C$_6$H$_6$, CHCl$_3$, CH$_2$Cl$_2$, THF, and Me$_2$CO; a nonelectrolyte in Me$_2$CO; prepd. from [Ir(Ars)$_2$(CO)Br] by refluxing in C$_6$H$_6$ in the presence of oxygen. IR spectrum (294).

[Ir(Ars)$_2$(CO)Br$_3$], yellow-green crystals, m. 304° (dec.), slightly sol. in common solvents, prepd. from [Ir(Ars)$_2$(CO)Br] by treating with Br in C$_6$H$_6$. IR spectrum, an octahedral structure with the arsine ligands in trans positions postulated (294).

[Ir(Ars)$_2$(CO)I$_3$] (M-20), yellow-orange crystals, m. 292°, slightly sol. in common org. solvents, prepd. from [Ir(Ars)$_2$(CO)H$_3$] by treating with iodine in C$_6$H$_6$. IR spectrum (294).

[Ir(Ars)$_2$(CO)Cl(SO$_4$)], lime-green ppt., m. 300°, prepd. from [Ir(Ars)$_2$(CO)Cl-(SO$_2$)] by the rxn. with oxygen (956).

[Ir(Ars)$_2$(CO)I(NO$_3$)$_2$], buff ppt., m. 200°, prepd. from [Ir(Ars)$_2$(CO)I(NO)$_2$] by the rxn. with oxygen (956).

{Ir(Ars)$_2$(CO)[(NC)$_2$C=C(CN)$_2$]Cl}, pale yellow crystals, m. 295-300° (dec.), prepd. from [Ir(Ars)$_2$(CO)Cl] + (NC)$_2$C=C(CN)$_2$ in C$_6$H$_6$. IR spectrum, a trigonal bipyramidal structure postulated (105).

[Ir(Ars)$_2$H$_3$], white crystals, m. 143° (dec.), moderately stable in air (293, 296), sol. in chlorinated solvents (296), insol. in common solvent (293), prepd. from [Ir(Ars)$_2$Br$_3$] in THF by treating with ethanolic NaBH$_4$ (293) and from [Ir(Ars)$_3$HBr$_2$] in THF by reducing with ethanolic NaBH$_4$ in the cold and sepg. the remaining starting matl. by extraction with C$_6$H$_6$ (296). IR spectrum - a trigonal bipyramidal structure suggested (293). A rxn. of the complex with Ars in C$_6$H$_6$ at 35-40° → [Ir(Ars)$_3$H$_3$] (296).

[Ir(Ars)$_2$(CO)H$_3$], ivory-white crystals, m. 150° (dec.) (293), white crystals, m. 156° (dec.), prepd. from [Ir(Ars)$_2$(CO)Br] in C$_6$H$_6$ by reducing with ethanolic NaBH$_4$ (294). The trihydrido complex was obtd. as a by-prod. from the mother liquor remaining from the rxn. of [Ir(Ars)$_2$H$_3$] with CO in C$_6$H$_6$ at 50-60° in the prepn. of [Ir(Ars)$_2$(CO)H] by concn. and pptn. with C$_5$H$_{12}$ (293). Soly. data and IR spectrum (293, 294). An octahedral structure with the Ars ligands in cis positions postulated (293).

[Ir(Ars)$_2$(Py)H$_3$], white crystals, m. 138° (dec.), sol. in Py, C$_6$H$_6$, THF, and chlorinated solvents, prepd. from [Ir(Ars)$_2$H$_3$] by dissolving in Py and allowing to stand for 20 days (296).

[Ir(Ars)$_2$(CO)H], light yellow crystals, m. 162° (dec.), prepd. from [Ir(Ars)$_2$-H$_3$] in C$_6$H$_6$ by bubbling CO at 50-60°. Soly. data; dipole moment in C$_6$H$_6$ 4.06 D; IR spectrum - a square planar structure suggested (293).

[Ir(Ars)$_3$(CO)H], yellow crystals, m. 175°, prepd. from [Ir(Ars)$_2$(CO)H$_3$] in C$_6$H$_6$ + Ars at 80-90° and from [Ir(Ars)$_2$(CO)H$_2$]ClO$_4$ by treating with ethanolic KOH; IR spectrum (294).

[Ir(Ars)$_2$(CO)H$_2$]ClO$_4$, white diamagnetic crystals, m. 250° (dec.), a uni-univalent electrolyte in Me$_2$CO, prepd. from [Ir(Ars)$_2$(CO)H$_3$] by treating with HClO$_4$ in EtOH. IR spectrum (294).

[Ir(Ars)$_3$H$_2$Br], orange-yellow crystals, m. 234° (dec.), prepd. from IrBr$_3$ + Ars in EtOH under reflux and from [Ir(Ars)$_3$H$_3$] by bubbling HBr gas. IR spectrum (296).

[Ir(Ars)$_3$H$_2$I], light yellow cryst. ppt., m. 150° (dec.), a nonelectrolyte in Me$_2$CO, prepd. from [Ir(Ars)$_3$H$_3$] + stoichiometric amt. iodine in C$_6$H$_6$ at r.t. Soly. data, IR spectrum (293).

[Ir(Ars)$_3$H$_2$]ClO$_4$, white crystals, m. 120° (dec.), sol. in org. solvents, prepd. from [Ir(Ars)$_3$H$_3$] in THF + 70% HClO$_4$ (296).

[Ir(Ars)$_3$HBr$_2$], yellow cryst. ppt., m. 240° (dec.) (293), yellow platelets, m. 244°, prepd. from IrBr$_3$ in alc. soln. by treating with Ars on a steam bath and from [Ir(Ars)$_3$H$_3$] in C$_6$H$_6$ soln. by bubbling HBr gas (296) or adding a stoichiometric amt. Br (293). A nonelectrolyte in Me$_2$CO (293). Soly. data and IR spectrum (293, 296). Useful as a homogeneous catalyst for the hydrogenation of α-olefins (1628).

[Ir(Ars)$_3$HCl$_2$], yellow crystals, m. 195° (dec.), prepd. from IrCl$_3$ + Ars in EtOH soln. under reflux. Soly. data and IR spectrum (296). Useful as a homogeneous catalyst for the hydrogenation of unsatd. nonaromatic org. compds. (582).

[Ir(Ars)$_3$HI$_2$], orange cryst. ppt., m. 155° (dec.), a nonelectrolyte in Me$_2$CO, prepd. from [Ir(Ars)$_3$H$_3$] and a stoichiometric amt. of iodine in C$_6$H$_6$ soln. Soly. data and IR spectrum (293). Useful as a homogeneous catalyst for the hydrogenation of α-olefins (1628).

[Ir(Ars)$_3$H$_3$], the complex was obtd. in two isomeric forms. The cis isomer, white crystals, m. 210° (dec.), prepd. from [Ir(Ars)$_3$HBr$_2$] in THF by reducing with NaBH$_4$ in the cold. Dipole moment in C$_6$H$_6$ at 30° 5.2 D (296). The trans isomer, white crystals, m. 223° (dec.), prepd. from [Ir(Ars)$_3$HBr$_2$] in warm Py by reducing with ethanolic NaBH$_4$. Soly. data (296). A rxn. of the complex with HBr gas in C$_6$H$_6$ gave a mixt. of [Ir(Ars)$_3$H$_2$Br] and [Ir(Ars)$_3$HBr$_2$]. Analogous rxn. with HCl gas in C$_6$H$_6$ gave [Ir(Ars)$_3$HCl$_2$] (296). A rxn. of the trihydrido complex with halogens leads to a step-wise substitution of the hydride atoms with halogens (293).

[Ir(Ars)$_4$H$_2$]ClO$_4$, ivory white crystals, m. 203° (dec.), prepd. from [Ir(Ars)$_3$-H$_2$]ClO$_4$ in C$_6$H$_6$ soln. by treating with Ars (296).

(In the following entries (Ars) stands for Ph_3As.)

[Ir(Ars)$_2$H$_2$(CO)Br], light yellow crystals, m. 270°, soly. data, prepd. from
[Ir(Ars)$_2$(CO)Br] by hydrogenation in C_6H_6 at 5-6° and adding EtOH. IR spec-
trum, an octahedral structure with the Ars ligands in trans positions, one
H ligand in the apical and the other in equatorial position suggested (294).

[Ir(Ars)$_2$H(CO)Br$_2$], light yellow cryst. ppt., m. 309°, readily sol. in THF,
CHCl$_3$, slightly sol. in C_6H_6 and Me_2CO, insol. in EtOH and C_5H_{12}, prepd.
from [Ir(Ars)$_2$(CO)H$_3$] by treating with HBr in C_6H_6 and from [Ir(Ars)$_2$(CO)Br]
in C_6H_6 by adding HBr followed by EtOH. IR spectrum, a structure with H
trans to Co suggested (294).

[Ir(Ars)$_2$H(CO)Cl$_2$], ivory white ppt., m. 295°, soly. data, prepd. from
[Ir(Ars)$_2$H$_3$(CO)] in C_6H_6 + HCl at 10°, IR spectrum (294).

[Ir(Ars)$_2$D(CO)Cl$_2$], prepd. from [Ir(Ars)$_3$(CO)Cl] in C_6H_6 + DCl in D_2O (731).

[π-CH$_2$=C(Me)CH$_2$Ir(Ars)(CO)Br$_2$], pale yellow prisms, m. 175-179° (dec.),
prepd. from the corr. dichloro complex by treating with LiBr in CH_2Cl_2-Me_2CO
mixt. IR and NMR spectra (1390).

[π-CH$_2$=C(Me)CH$_2$Ir(Ars)(CO)Cl$_2$], pale yellow prisms, m. 168-170° (dec.), prepd.
from [π-CH$_2$=C(Me)CH$_2$Ir(CO)(cyclooctene)Cl$_2$] in CH_2Cl_2 by warming with Ars,
adding MeOH and evapg. CH_2Cl_2. IR and NMR spectra, a symmetrically π-bonded
methallyl group with the C atom trans to the Ars ligand more weakly bonded
than the other terminal C atom of the methallyl group (1390).

[Fe(Ars)(CO)$_4$] (M-21), yellow, monoclinic needles, m. 178° (dec.), prepd.
from Fe$_3$(CO)$_{12}$ + Ars in THF under reflux or in dioxan at 80-90°. IR spectrum (367). Low-frequency IR spectrum; a trigonal-bipyramidal configuration
with the Ars ligand in one of the apical positions postulated (1419). A rxn.
of the complex with Ph$_2$CCl$_2$ in C_6H_6 under reflux → Ph$_2$C=CPh$_2$ + FeCl$_2$ (1417).

[Fe(Ars)$_2$(CO)$_3$] (M-21), yellow crystals, m. 193-195°, prepd. from Fe$_3$(CO)$_{12}$
+ Ars in THF under reflux or in dioxan at 80-90°. IR spectrum (367). Low-
frequency IR spectrum, a trigonal-bipyramidal configuration with the Ars
ligands in the apical positions postulated (1419). Rxn. with HgX$_2$ (X = Cl or
Br) in EtCOMe-C_6H_6 → [Fe(Ars)$_2$(CO)$_3$·HgX$_2$] (31).

[Fe(Ars)$_2$(CO)$_3$·HgCl$_2$], yellow crystals, prepd. from [Fe(Ars)$_2$(CO)$_3$] + HgCl$_2$
in EtCOMe-C_6H_6. IR spectrum (31).

[Fe(Ars)$_2$(CO)$_3$·HgBr$_2$], yellow crystals, dec. 96-98°, prepd. from [Fe(Ars)$_2$-
(CO)$_3$] + HgBr$_2$ in EtCOMe-C_6H_6. IR spectrum (31).

[Fe(Ars)(CO)(NO)$_2$] (M-21), red-brown cryst. solid (1085), prepd. from
[Fe(CO)$_2$(NO)$_2$] by the rxn. with Ars in various solvents. The rate of the CO
displacement with Ars and the activation energy were detd. (1083a, 1084).
The changes in the complex compn. were followed by IR spectroscopy and the
rxn. mechanisms were considered (1083a, 1085). The displacement of CO with

Ars is catalyzed by [Bu$_4$N]Cl, amines, and NaOMe in MePh (1085). In THF or MeOH the displacement occurs without the catalysts (1083a). IR spectra: NO valence vibration (135), NO and CO valence vibration data in CH$_2$Cl$_2$ (1084).

[Fe(Ars)Cl$_3$], dark yellow crystals, m. 130° (dec.), prepd. from Fe$_3$(CO)$_{12}$ + Ars in CHCl$_3$ under reflux, concg. the liquid phase, pptg. the prod. with n-C$_6$H$_{14}$, and recrystg. from EtOH. IR spectrum (1417).

$$\left[(Ars)Fe\left(\begin{array}{c} S-CCN \\ \| \\ S-CCN \end{array}\right)_2\right]$$, polarography (110).

$$\left[(Ars)Fe\left(\begin{array}{c} S-CCF_3 \\ \| \\ S-CCF_3 \end{array}\right)_2\right]$$, black diamagnetic prisms, m. 158-159.5°, prepd. from [FeS$_4$C$_4$(CF$_3$)$_4$]$_2$ + Ars in pentane and from Et$_4$N[FeS$_4$C$_4$-(CF$_3$)$_4$]$_2$ + Ars in CH$_2$Cl$_2$. Polarographic data. Electronic and ESR spectra; a square-pyramidal structure of the 5-coordinate Fe complex postulated (110).

[Fe(Ars)$_2$(NO)$_2$] (M-21), dark brown crystals, prepd. from Na[Fe(CO)$_3$(NO)] + Ars in Et$_2$O-CF$_3$CO$_2$H mixt. at -65° (733). The complex was also formed from [Fe(CO)$_2$(NO)$_2$] + Ars in the presence of NaOMe in a rxn. lasting one day (1085). IR spectrum (135, 1084, 1085).

Na$_3$[Fe(CN)$_5$Ars], light yellow, fine crystals, prepd. from Na$_3$[Fe(CN)$_5$NH$_3$]·2H$_2$O + Ars in MeOH. Elec. cond. in H$_2$O was detd. IR spectrum (1124). Mössbauer spectrum (573, 1124).

Na$_2$[Fe(CN)$_5$Ars], dark purple, fine, paramagnetic crystals, prepd. from Na$_3$[Fe(CN)$_5$Ars] dissolved in MeOH by treating with a 0.1 N Br soln. Elec. cond. in H$_2$O and magnetic prop. were detd. IR spectrum (1124). Mössbauer spectrum (573, 1124).

[Mn(Ars)(CO)$_4$]$_2$, orange red crystals, prepd. from [Mn(CO)$_5$]$_2$ + Ars in xylene at 140° for 80 hrs. (1540), dark red, slightly paramagnetic substance, m. 155°, nonconducting in PhNO$_2$, prepd. from [Mn(CO)$_5$]$_2$ + Ars in cyclohexane under UV light (1181). The complex was also prepd. from H$_3$Mn$_3$(CO)$_{12}$ + Ars in cyclohexane under reflux (808). IR spectrum (1181, 1194), the solvent effect on the C-O stretching was detd. (1194). A dimeric structure with a Mn-Mn bond was postulated (1181). See also (M-22).

[Mn(Ars)(CO)$_4$Cl] (M-22), an octahedral Mn complex, low-frequency IR spectrum (158), IR spectrum (832).

[Mn(Ars)(CO)$_4$Br] (M-22), an octahedral Mn complex, low-frequency IR spectrum (158).

[Mn(Ars)(CO)$_4$NO$_3$], dark yellow, diamagnetic powder, a nonelectrolyte, very sol. in polar org. solvents, prepd. from Mn(CO)$_5$NO$_3$ + Ars (1:1) in CHCl$_3$ at 20°. IR spectrum in soln. and in the solid state (38).

[Mn(Ars)(CO)$_4$SCN], yellow solid, prepd. from Mn(CO)$_5$SCN + Ars in CHCl$_3$ at 25° and from [Mn(Ars)(CO)$_4$Cl] by metathesis with excess KSCN. IR spectrum (553).

cis-[Mn(Ars)$_2$(CO)$_3$SCN], yellow solid, prepd. from Mn(CO)$_5$SCN + Ars in CHCl$_3$ at 25°. IR spectrum (553).

(In the following entries (Ars) stands for Ph_3As.)

trans-[Mn(Ars)$_2$(CO)$_3$NCS], yellow solid, prepd. from Mn(CO)$_5$SCN + Ars in CHCl$_3$ at 80°; the rxn. mixt. also contd. cis-[Mn(Ars)$_2$(CO)$_3$SCN] and [Mn(Ars)(CO)$_4$-SCN]. IR spectrum (553).

[Mn(Ars)(NO)$_3$] (M-22), prepd. from [Mn(CO)(NO)$_3$] + Ars in xylene or in C_6H_{12} (135, 1588). The rxn. rate was followed by means of the changes in the IR spectrum of the reacting mixt. (1588). IR spectrum (135).

[Mn(Ars)(CO)$_3$(NO)], prepd. from [Mn(CO)$_4$NO] + Ars in xylene at 120°. The rate of the rxn. was followed by means of the changes in the IR spectrum of the reacting mixt.; a 2nd order kinetics was deduced (1588).

[Mn(Ars)$_2$(CO)$_3$Br] (M-23), orange crystals, prepd. from Mn(CO)$_5$Br + Ars under N at 120°. IR spectrum. The complex is useful as an antiknock additive (1612).

[Mn(Ars)$_2$(CO)$_3$Cl] (M-23), yellow crystals, prepd. from [Mn(CO)$_4$Cl]$_2$ + Ars at 100°. IR spectrum. The complex is useful as an antiknock additive (1612).

[Mn(Ars)$_2$(CO)$_3$I] (M-23), dark orange crystals, prepd. from [Mn(CO)$_4$I]$_2$ + Ars at 100°. IR spectrum. The complex is useful as an antiknock additive (1612).

[Mn(Ars)(CF$_3$COCHCOCF$_3$)(CO)$_3$], orange solid, m. 64°, prepd. from [Mn(CO)$_4$-(CF$_3$COCHCOCF$_3$)] in CHCl$_3$ + Ars at 30° and from [Mn(Ars)(CO)$_4$Cl] in CHCl$_3$ + Tl(CF$_3$COCHCOCF$_3$). IR and NMR spectra. An octahedral structure with all 3 CO ligands cis to each other and with the CF$_3$COCHCOCF$_3$ ligand coordinated through both O atoms postulated (692).

cis-[MeMn(Ars)(CO)$_4$], orange crystals, m. 102.5-103.5°, prepd. from [MeMn(CO)$_5$] + Ars in C_6H_6 under reflux. IR and NMR spectra in various solvents (893).

cis-[AcMn(Ars)(CO)$_4$], the complex was prepd. along with the trans isomer and cis-[MeMn(Ars)(CO)$_4$] from [MeMn(CO)$_5$] + Ars in C_6H_6 at r.t. IR and NMR spectra (893).

trans-[AcMn(Ars)(CO)$_4$], m. 79.5-80°, prepd. from [MeMn(CO)$_5$] + Ars in C_6H_6 at r.t. (119, 892). IR spectrum (119), NMR spectra in various solvents (893).

[π-C$_5$H$_5$Mn(Ars)(CO)$_2$], (M-23), pale yellow ppt., stable in air, prepd. from [π-C$_5$H$_5$Mn(CO)$_3$] + Ars in THF in UV light at 75°. IR and NMR spectra and magnetic moment were detd. (120).

[π-C$_5$H$_5$Mn(Ars)$_2$(CO)], red-brown crystals, prepd. from [π-C$_5$H$_5$Mn(CO)$_3$] + Ars in EtOH in UV light. The complex dec. on exposure to O. IR and NMR spectra and dipole moment were detd. (120).

[PhMn(Ars)(CO)$_4$], m. 129-130°, prepd. from [PhMn(CO)$_5$] + Ars in CH$_2$Cl$_2$ at r.t. IR spectrum (119).

[Ph$_3$GeMn(Ars)(CO)$_4$], light yellow, m. 229.5-231°, prepd. from [Ph$_3$GeMn(CO)$_5$] + Ars at 170° under Ar. IR spectrum (1138).

[(Ph₃P)AuMn(Ars)(CO)₄], diamagnetic solid, m. 135° (dec.), a nonelectrolyte in PhNO₂, prepd. from [(Ph₃P)AuMn(CO)₅] + excess Ars in a sealed tube at 120° and from [(Ph₃P)AuCl] + Na[Mn(Ars)(CO)₄] in THF. IR spectrum (832).

[Hg(Ars)Cl₂]₂ (M-23), m. 258°, IR spectrum (369). Spectroscopic study of the complex formation (187). A complex corr. to the empirical formula Hg(Ars)Cl₂ was obtd. in the form of a white ppt. from equimolar amts. of HgCl₂ and Ars in 74% MeOH. UV spectrum; a Cl-bridged dimeric structure was suggested (1175). Another complex corr. to the empirical formula Hg(Ars)₁.₅-Cl₂ was obtd. in a solid form by adding solid HgCl₂ to Ars (1:3) in methanolic soln. UV spectrum; a Cl-bridged dimeric structure for the complex was suggested (1175).

[Hg(Ars)₂](ClO₄)₂, m. 222° (explosive dec.), prepd. from Ars during an anodic oxidn. at a Hg electrode in MeCN contg. LiClO₄ electrolyte (749).

[Mo(Ars)₂Cl₄] (M-23), brick-red, paramagnetic ppt., prepd. from [(PrCN)₂Mo]-Cl₄ + Ars in CHCl₃. Visible reflectance spectrum and magnetic susceptibility data; a 6-coordination of Mo postulated (58).

[Mo(Ars)(CO)₅] (M-23), prepd. from Mo(CO)₆ + Ars in decalin at 112°. The rate of the CO displacement by Ars was followed by IR spectroscopy and the rxn. mechanism was considered (643). Likewise, the rxn. kinetics in n-decane-cyclohexane at 97.8 ± 0.2° and in dioxan-THF at 82.3 ± 0.2° were followed by IR spectroscopy. Even at a Mo(CO)₆:Ars ratio of 1:10 extremely limited disubstitution occurred; a S$_N$1 dissocn. mechanism was postulated (1599). The inductive effect of the ligand-metal σ-bond and ligand-metal π-bond on the CO stretching force const. is considered (644). Spectra: IR (435, 439), electronic (439). Low-frequency IR spectrum; a tetragonal bipyramidal configuration with the Ars ligand in one of the apical positions postulated (1419). The rate of the CO substitution with Ars in [Mo(Ars)(CO)₅] was found to be too slow for a meaningful detn. (1653).

cis-[Mo(Ars)₂(CO)₄] (M-23), prepd. from [Mo(CO)₄(Cyclooctadiene)] + Ars in n-heptane. IR spectrum (646, 1651). The rate of the diene displacement with Ars in C₆H₆ was followed by recording the changes of the UV and visible spectra. The rate const. and activation parameters were calcd. (1651). Rxn. of the complex with α,α′-bipyridyl: the rate of the Ars groups displacement with bipyridyl was measured (646).

[Mo(Ars)₂(PrCN)Br₃], orange yellow solid, prepd. from [Mo(PrCN)₃Br₃] + Ars in C₆H₆. Electronic spectrum - a 6-coordinate Mo complex (58).

[Mo(Ars)₂(CO)₃]Br₂, yellow orange, diamagnetic ppt., prepd. from Mo(CO)₄Br₂ + Ars in Me₂CO. IR spectrum (376).

[Mo(Ars)₂(CO)₃Cl₂], dark yellow, diamagnetic crystals, prepd. from [Mo(CO)₄-Cl₂] + Ars in CH₂Cl₂. Soly. data. IR spectrum (375).

(In the following entries (Ars) stands for Ph_3As.)

$[Mo(Ars)_2(NO)_2Br_2]$, green crystals, m. 236-238° (807); a dark green solid
(77), prepd. from $Mo(NO)_2Br_2$ + Ars in C_6H_6 under reflux, from $[Mo(Ars)_2(CO)_4]$
+ NOBr in CH_2Cl_2 (807), and from $[Mo(Ars)_2(CO)_3Br_2]$ + NO in C_6H_6 or CH_2Cl_2;
chromatographic sepn. gave 3 cis isomers distinguishable by IR spectroscopy
(77). IR spectrum (807). A tetragonal bipyramidal structure suggested (77,
807).

$[Mo(Ars)_2(NO)_2Cl_2]$ (M-24), green crystals, prepd. from $[Mo(Ars)_2(CO)_3Cl_2]$
+ NO in CH_2Cl_2. Three cis isomers detected by IR spectroscopy (77).

$[Mo(Ars)_2(NO)_2I_2]$, purple crystals, m. 242-243°, prepd. from $[Mo(NO)_2I_2]$ +
Ars in C_6H_6 under reflux and from $[Mo(Ars)_2(CO)_4]$ + NOI in CH_2Cl_2. IR spectrum - an octahedral structure suggested (807).

$[Mo(Ars)(CO)_3(o-Phenanthroline)]$, purple black crystals, prepd. from $[Mo(CO)_4-$
$(o-Phenanthroline)]$ + Ars in xylene (758). IR spectrum (436, 758).

$[\pi-C_5H_5Mo(Ars)(CO)_2Br]$, crystals, m. 147°, prepd. from $[\pi-C_5H_5Mo(CO)_3Br]$ +
Ars in C_6H_6 under reflux and purified by chromatography. IR spectrum (1522).

$[\pi-C_5H_5Mo(Ars)(CO)_2Cl]$, orange crystals, dec. 172°, a nonelectrolyte (670).
m. 154° (1522), prepd. from $[\pi-C_5H_5Mo(CO)_3Cl]$ + Ars in C_6H_6 under reflux
(1522) or in C_6H_6-cyclohexane in UV light (670), purified by chromatography
(1522). Soly. data (670). Spectra: IR (670, 1522), visible and NMR (670).

$[\pi-C_5H_5Mo(Ars)(CO)_2I]$, red solid, dec. 158° (670), crystals, m. 151° (1522),
a nonelectrolyte, prepd. from $[\pi-C_5H_5Mo(CO)_3I]$ + Ars in C_6H_6-cyclohexane by
UV irradiation (670) or in C_6H_6 under reflux and purified by chromatography
(1522). Soly. data (670). Spectra: IR (670, 1522), visible and NMR (670).

$[\pi-C_5H_5Mo(Ars)_2(CO)Cl]$, red crystals, dec. 141°, a nonelectrolyte, prepd.
from $[\pi-C_5H_5Mo(CO)_3Cl]$ + excess Ars in C_6H_6-cyclohexane by UV irradiation.
Soly. data; IR, visible, and NMR spectra (670).

$[\pi-C_5H_5Mo(Ars)(CO)_3]PF_6$, yellow crystals, m. 223°, prepd. from $[\pi-C_5H_5Mo(CO)_3-$
$Cl]$ + Ars + $AlCl_3$ (3.5:2.68:5.62) in C_6H_6 by stirring at r.t., treating the
mixt. with H_2O, dissolving the solid prod. in Me_2CO and pptg. with $[NH_4]PF_6$.
The aq. mother liquor contd. $[\pi-C_5H_5Mo(CO)_3Cl]$ and $[\pi-C_5H_5Mo(Ars)(CO)_2Cl]$.
IR spectrum (1522).

$[\pi-C_5H_5Mo(Ars)_2(CO)_2]PF_6$, yellow crystals, m. 274°, prepd. from $[\pi-C_5H_5Mo-$
$(Ars)(CO)_2Cl]$ + Ars + $AlCl_3$ 1.4:4.3:4.3 in C_6H_6 by stirring at r.t., treat-
ing the rxn. mixt. with H_2O, dissolving the ppt. in Me_2CO, and pptg. with
$[NH_4]PF_6$. IR spectrum (1522).

$[Mo(Ars)(Ph_2PCH_2CH_2PPh_2)(CO)_3]$, prepd. from $[Mo(Ph_2PCH_2CH_2PPh_2)(CO)_4]$ + excess
Ars in xylene under reflux for 41 hrs. IR spectrum, an octahedral Mo complex
(486).

$[Mo(Ars)_2(Ph_2PCH_2CH_2PPh_2)(CO)_2]$, yellow ppt., prepd. from $[Mo(Ph_2PCH_2CH_2PPh_2)-$
$(CO)_4]$ + excess Ph_3As in xylene under reflux for 72 hrs. IR spectrum, a cis-

configuration in an octahedral structure postulated (486).

[Ni(Ars)$_2$(CO)$_2$] (M-24), useful as a catalyst for the oligomerization of buta-diene to 1,3,6-octatriene, 2,6-octadiene, vinylcyclohexene, cyclooctadiene, and cyclododecatriene (556).

[Ni(Ars)$_4$] (M-24), yellow, diamagnetic crystals (147), brown-yellow crystals (1611); a nonelectrolyte, prepd. from K$_4$[Ni(CN)$_4$] + Ars (1:6) at 60° or 120-130° (147) and from Ni acetylacetonate in C$_6$H$_6$ by a redn. with Et$_2$AlOEt in the presence of Ars. Useful as a catalyst for the oligomerization of 1,3-dienes (1611) and for the rxn. of nonterminal olefins with low mol.-wt. alkyl-Al compds. at 100-200° to form α-olefins by displacement:

$$3C_nH_{2n} + (RCH_2CH_2)_3As \xrightarrow[100-120°]{Cat.} (C_nH_{2n+1})_3Al + 3RCH=CH_2 \quad (1654).$$

[π-C$_5$H$_5$Ni(Ars)Br], dark violet crystals, m. 109-110° (dec.), prepd. from NiBr$_2$ + (π-C$_5$H$_5$)$_2$Ni + Ars (1:1:2) in THF at 50° (66% yield) and from Ni(CO)$_4$ + Ars (1:2) in THF at r.t. until the gas evolution subsides, heating the soln. under reflux and, after cooling, adding dropwise (π-C$_5$H$_5$)$_2$Ni, followed by Br in THF, stirring at r.t. and then heating under reflux. A rxn. of the complex with RMgX → [π-C$_5$H$_5$Ni(Ars)R] (1629).

[π-C$_5$H$_5$Ni(Ars)I], dark red-violet crystals, m. 118-119° (dec.), prepd. from Ni(CO)$_4$ by mixing with Ars in THF at r.t. and, after completed gas evolution, heating under reflux, then cooling with ice, adding dropwise (π-C$_5$H$_5$)$_2$Ni followed by I in THF at r.t. and then heating under reflux (53% yield). A rxn. with RMgX → [π-C$_5$H$_5$Ni(Ars)R] (1629).

[π-C$_5$H$_5$Ni(Ars)Me], dark green, diamagnetic crystals, m. 101-104° (dec.), prepd. from [π-C$_5$H$_5$Ni(Ars)X] in C$_6$H$_6$ + MeMgI in Et$_2$O (50% yield) (1629).

[π-C$_5$H$_5$Ni(Ars)Ph], dark green, diamagnetic crystals, m. 126-127° (dec.), prepd. from [π-C$_5$H$_5$Ni(Ars)I] in C$_6$H$_6$ + PhMgX in Et$_2$O (36% yield) (1629).

[Nb(Ars)Cl$_5$], colored ppt., prepd. from NbCl$_5$ + Ars in CCl$_4$, n-C$_6$H$_{14}$, or cyclohexane (466).

[Nb(Ars)Cl$_5$], colored ppt., prepd. from NbCl$_5$ + Ars in C$_6$H$_6$ (466).

[π-C$_5$H$_5$Nb(Ars)(CO)$_3$], red cryst. substance, dec. 180° without melting when heated in a sealed tube, prepd. from [π-C$_5$H$_5$Nb(CO)$_4$] + Ars in n-C$_5$H$_{12}$ under UV light (68% yield). Soly. data; IR and NMR spectra (869a).

[Os(Ars)$_2$(CO)$_2$Br$_2$], white crystals, prepd. from [Os(CO)$_3$Br$_2$] + Ars in C$_6$H$_6$ under reflux. IR spectrum (674), solvent effect on IR spectrum (673).

[Os(Ars)$_2$(CO)$_2$Cl$_2$], white crystals, prepd. from [Os(CO)$_3$Cl$_2$] + Ars as described for the dibromo deriv. IR spectrum (674), solvent effect on IR spectrum (673).

[Os(Ars)$_2$(CO)$_2$I$_2$], white crystals, prepd. from [Os(CO)$_3$I$_2$] as described for the dibromo deriv. IR spectrum (674), solvent effect on IR spectrum (673).

(In the following entries (Ars) stands for Ph$_3$As.)

[OsH(Ars)$_3$Cl$_2$], useful as a homogeneous catalyst for the hydrogenation of unsatd. nonaromatic org. compds. (582).

cis-[Pd(Ars)$_2$Cl$_2$] (M-24), yellow crystals, prepd. from [Pd$_2$Cl$_4$(trans-o-H$_2$N-C$_6$H$_4$SCH$_2$CH=CHCH$_2$SC$_6$H$_4$NH$_2$-o)] + Ars in EtOH by heating under reflux. IR spectrum (628).

[Pd(Ars)$_2$Cl$_2$] with undetermined ligand configuration, orange crystals, m. 278-280°, prepd. from PdCl$_2$ in H$_2$O + Ars in EtOH and repptd. from CHCl$_3$ soln. with Et$_2$O (793, see also 1151). Readily sol. in CHCl$_3$, sparingly sol. in C$_6$H$_6$ and MeOH, insol. in H$_2$O, Me$_2$CO, CCl$_4$, and n-C$_7$H$_{16}$. The complex catalyzes the hydrogenation of soybean oil methyl ester to a monoolefinic ester and its isomerization to trans,trans and cis,trans conjugated diolefinic esters. Addn. of SnCl$_2$·2H$_2$O to the Pd complex inhibits somewhat the redn. and formation of the trans isomers (793).

[Pd(Ars)$_2$(CN)$_2$], white crystals, m. 209° (dec.), prepd. from [Pd(Ars)$_2$Cl$_2$] + KCN in C$_6$H$_6$-MeOH soln. IR spectrum. The complex catalyzes the hydrogenation of soybean oil methyl ester to the monoolefinic ester and its isomerization to trans,trans and cis,trans conjugated diolefinic ester (793).

[Pd(Ars)$_2$(NCO)$_2$], crystals, a nonelectrolyte in MeNO$_2$, prepd. from [Pd(Ars)$_2$-Cl$_2$] by metathesis with AgNCO in Me$_2$CO (1156, 1157). Spectra: IR (1156, 1157) and electronic absorption (1157). A coordination of the NCO group through the N atom postulated (1156, 1157).

[Pd(Ars)$_2$(SCN)$_2$] (M-25), IR spectra (637, 841, 1314).

[Pd(Ars)$_2$(NCS)$_2$] (M-25), IR spectra (637, 841, 1314).

[Pd(Ars)$_2$(NO$_2$)$_2$], prepd. from K$_2$[Pd(NO$_2$)$_4$] in H$_2$O + Ars (1:2) in EtOH at 70°. IR spectrum indicated a N-bonding of the NO$_2$ groups (268).

[Pd(Ars)$_2$(O$_2$CCF$_3$)$_2$], bright yellow crystals, m. 192-193°, prepd. from Pd(O$_2$CCF$_3$)$_2$ + excess Ars in C$_6$H$_6$ (1452).

[Pd(Ars)$_2$(O$_2$CC$_2$F$_5$)$_2$], bright yellow crystals, m. 204-205°, prepd. from Pd(O$_2$CC$_2$F$_5$)$_2$ + excess Ars in C$_6$H$_6$. IR spectrum (1452).

[Pd(Ars)$_2$(OBz)$_2$], bright yellow crystals, m. 198-199°, prepd. from Pd(OBz)$_2$ + excess Ars in C$_6$H$_6$. IR spectrum (1452).

[π-CH$_2$=CClCH$_2$Pd(Ars)Cl], pale yellow prisms, dec. 152-160°, prepd. from [π-CH$_2$=CClCH$_2$PdCl]$_2$ + Ars in warm C$_6$H$_6$. NMR spectrum (1248).

[π-CH$_2$=CHCH$_2$Pd(Ars)Cl], pale yellow prisms, dec. 160-170°, prepd. from [π-CH$_2$=CHCH$_2$PdCl]$_2$ + Ars in warm C$_6$H$_6$. IR and NMR spectra (1248). UV spectrum in CHCl$_3$ (717). The changes in the NMR spectrum in CDCl$_3$ affected by the temp. changes between -50 and +70° were interpreted as an evidence for the position exchange of the Ars ligand with Cl at rising temp. and equilibration of the π ⇌ σ bonding of the allyl ligand at still higher temp. (1261).

24

[π-CH$_2$=C(Me)CH$_2$Pd(Ars)Cl], pale yellow prisms, m. 205-208° (dec.), prepd. from [π-CH$_2$=C(Me)CH$_2$PdCl]$_2$ + Ars in warm C$_6$H$_6$. NMR and IR spectra; an asymmetric bonding of the allyl ligand postulated (1248). UV spectrum in CHCl$_3$ (717). The formation of the complex in the rxn. of [π-CH$_2$=C(Me)CH$_2$PdCl]$_2$ with Ars (1:2) in CDCl$_3$ at -73° was followed by NMR spectroscopy - the proton chem. shifts and coupling consts. were detd. At higher temp. with excess Ars the spectrum changes were interpreted as an indication of the π-allyl group rotation in its own plane. In the systems contg. only a part of the dimer converted to the monomer with the Ars ligand, NMR spectra indicated that both monomer-monomer and monomer-dimer rxns. proceeded. At the lowest temp. the spectra of both the dimer and monomer were distinctly recognizable, but with rising temp. these spectra broadened and eventually coalesced. At still higher temp. all four allylic protons formed a single line (1567, 1569, also 1568). At high temp. and/or higher Ars/Pd ratios the formation of σ-allyl deriv. was detected. The rxn. kinetics and mechanisms are discussed (1568). In the rxn. of the dimer with Ars present in an excess over the 1:2 ratio in CDCl$_3$ at -70 to -40°, the NMR spectrum indicated a proton exchange between 2 equally populated sites of different shifts. The rxn. was 1st order with respect to Ars. The possibility of the ionic species [π-CH$_2$=C(Me)CH$_2$Pd-(Ars)$_2$]$^+$Cl$^-$ was taken into consideration. With a small excess of Ars at + 20 to + 80° the reaction may involve a σ-allyl intermediate. A 1st order kinetics in both Ars and [π-CH$_2$=C(Me)CH$_2$Pd(Ars)Cl] was postulated. With a large Ars excess between -80° and +80° a 3rd rxn. occurred: [π-CH$_2$=C(Me)CH$_2$Pd(Ars)-Cl] + Ars $\rightleftharpoons$ [π-CH$_2$=C(Me)CH$_2$Pd(Ars)$_2$]$^+$Cl$^-$. Elec. cond. measurements at -70° showed a very low concn. of the ionic species; at r.t. its concn. was negligible. With the Ars/Pd ratio < 1 in CDCl$_3$ below -50° the monomer and dimer were detected spectroscopically. The signals of the monomer as well as dimer broadened with rising temp. and finally coalesced. The exchange rates of the monomer and dimer protons at varying amts. of both were detn. for the -30° to 0° range. Addn. of a large excess of halide ions to the mixts. with a Ars/Pd ratio > 1 and < 1 had very little effect on the exchange (1567).

[π-CH$_2$=C(Me)CH$_2$Pd(Ars)Br], pale yellow prisms, m. 198-205°, prepd. from [π-CH$_2$=C(Me)CH$_2$PdBr]$_2$ + Ars in warm C$_6$H$_6$. IR and NMR spectra (1248).

[π-CH$_2$=C(Me)CH$_2$Pd(Ars)I], yellow prisms, m. 195-199° (dec.), prepd. from [π-CH$_2$=C(Me)CH$_2$PdI]$_2$ + Ars in warm C$_6$H$_6$. NMR spectrum in CHCl$_3$ indicated a considerable dissocn. of the Ars ligand (1248).

[π-MeCH=CHCH$_2$Pd(Ars)Cl], UV spectrum in CHCl$_3$ (717).

[π-CH$_2$=CHCMe$_2$Pd(Ars)Cl], yellow prisms, m. 177-183° (dec.), prepd. from [π-CH$_2$=CHCMe$_2$PdCl]$_2$ + Ars in warm C$_6$H$_6$. IR and NMR spectra; asymmetric bonding of the allyl ligand postulated (1248).

[π-Me$_2$C=C(Me)CH$_2$Pd(Ars)Cl], the formation of the complex from [π-Me$_2$C=C(Me)-CH$_2$PdCl]$_2$ + Ars in CDCl$_3$ at low temp. was detected by NMR spectroscopy. The Ars ligand always occupied a trans position with respect to dimethyl-substd. C atom. The proton chem. shifts and coupling const. were detd. With rising

(In the following entries (Ars) stands for Ph$_3$As.)

temp. and/or increasing Ars concn., at Ars/Pd > 1, the proton signals of the CH$_2$ group broadened and coalesced, while no perceptible broadening of the 1,1-dimethyl proton signals could be detected even at + 80°. The same NMR signal changes were observed at the Ars/Pd ratios < 1. The suggestion was made that the protons of the CH$_2$ group which is cis to the Ars ligand interchange their position probably via a σ-allyl intermediate, while no interchanges of the 1,1-dimethyl protons take place (1570, see also 1248).

[π-syn-MeCH=C(Me)CH(syn-Me)Pd(Ars)Cl], the formation of the complex from [π-syn-MeCH=C(Me)CH(syn-Me)PdCl]$_2$ + Ars in CDCl$_3$ at - 80° was detected by NMR spectroscopy. When the temp. increased to + 10° the signals of both protons at C-1 and C-3 and of those of the Me group at C-1 and C-3 coalesced at their weighed mean if Ars/Pd ⩾ 1. At Ars/Pd < 1 the allylic ligand in the monomeric complex isomerized via a σ-allyl intermediate at - 30°. The Me groups and H cis to the Ars ligand may isomerize to give the 1-anti-Me deriv. (1570).

[π-PhCH=CHCH$_2$Pd(Ars)Cl], UV spectrum in CHCl$_3$ (717).

[Pd(Ars)(OAc)$_2$]$_2$, orange-red crystals, prepd. from [Pd(OAc)$_2$]$_3$ in C$_6$H$_6$ by adding Ars (1:6). IR spectrum - a dimeric structure contg. both bridged and unidentate carboxylate groups postulated (1454).

[Pd(Ars)(SeCN)$_2$]$_2$, a Se-bridged compd., prepd. from [Pd(Ars)$_2$Cl$_2$] by treating with AgNO$_3$ (1:2) in DMF and converting the resulting [Pd(Ars)$_2$(DMF)$_2$](NO$_3$)$_2$ with KSeCN. IR spectrum (267).

[Pd$_2$(trans-o-H$_2$NC$_6$H$_4$SCH$_2$CH=CHCH$_2$SC$_6$H$_4$NH$_2$-o)(Ars)Cl$_4$], pale yellow complex, m. 140-146° (dec.), a nonelectrolyte, prepd. from [Pd$_2$(trans-o-H$_2$NC$_6$H$_4$SCH$_2$-CH=CHCH$_2$SC$_6$H$_4$NH$_2$-o)Cl$_4$] + Ars in EtOH at 20°. IR spectrum (628).

[Pt(Ars)$_4$], rxn. with HCl in C$_6$H$_6$ or EtOH → [HPt(Ars)$_2$Cl] (298, 299).

[HPt(Ars)$_2$Cl] (M-25), white needles, m. 187-190° (109), m. 296-297° (298, 299), prepd. from [Pt(Ars)$_4$] in C$_6$H$_6$ + MeOH-HCl on a steam bath (109) and from [Pt(Ars)$_4$] + HCl in C$_6$H$_6$ or EtOH. A nonelectrolyte in PhNO$_2$ (298, 299). IR spectrum (109, 298, 299).

cis-[Pt(Ars)$_2$Cl$_2$] (M-25), prepd. from [Pt$_2$(trans-o-H$_2$NC$_6$H$_4$SCH$_2$CH=CHCH$_2$SC$_6$H$_4$-NH$_2$-o)Cl$_4$] + Ars in EtOH by heating under reflux. IR spectrum (628). The complex was purified by dissolving in CH$_2$Cl$_2$ and pptg. with n-C$_6$H$_{14}$ (1499). The complex is an effective catalyst for the hydrogenation and hydroformylation of olefins (1613) and for the hydrogenation and isomerization of Me linoleate to a monoolefinic ester (109, 586). Addn. of SnCl$_2$·2H$_2$O enhanced the hydrogenating and isomerizing activity (109). A mixt. of the complex with SnCl$_2$·2H$_2$O isomerized 1,5-hexadiene to a mixt. of 1,4-, 1,3-, and 2,4-hexadiene and catalyzed the redn. of CH$_2$=CHCN to EtCN (37). A mixt. of the complex with a 10 molar excess SnCl$_2$·2H$_2$O catalyzed the hydrogenation of 1,3- and 1,5-cyclooctadiene and of soybean oil methyl esters to the corr. monoolefins (1499).

[Pt(Ars)$_2$(CN)$_2$], crystals, m. 293-294°, prepd. from [Pt(Ars)$_2$Cl$_2$] + KCN (1:2) in C$_6$H$_6$-MeOH. Sol. in C$_6$H$_6$-MeOH; IR spectrum (109).

[Pt(Ars)$_2$(NCO)$_2$], crystals, a nonelectrolyte in MeNO$_2$, prepd. from [Pt(Ars)$_2$-Cl$_2$] + AgNCO in Me$_2$CO. IR and electronic spectra; a coordination of the NCO groups through the N atom postulated (1157).

trans-[Pt(Ars)$_2$(NCS)$_2$], (M-25), prepd. from trans-[Pt(Ars)$_2$Cl$_2$] by metathesis with SCN$^-$. IR spectrum, a coordination of the NCS ligands through the N atom postulated (1204).

[Pt(Ars)$_2$(NO$_2$)$_2$], prepd. from K$_2$[Pt(NO$_2$)$_4$] in H$_2$O + Ars in EtOH at 70°. IR spectrum indicated the Pt-N bonds of the NO$_2$ groups (268).

[Pt(Ars)$_2$Cl(SnCl$_3$)], orange crystals, m. 150-151°, prepd. from Na$_2$PtCl$_4$·4H$_2$O and SnCl$_2$ in EtOH under N by adding Ars in EtOH and from Na$_2$PtCl$_4$·4H$_2$O in 3 N HCl by adding SnCl$_2$ followed by Ars in EtOH. The complex is sol. in Me$_2$CO and CHCl$_3$. Molar cond. measurements in MeNO$_2$ indicated a slight dis-socn. Rxn. with dild. HCl in Me$_2$CO → [Pt(Ars)$_2$Cl$_2$] (1637).

[Pt(Ars)$_2$(Py)$_2$](O$_3$SC$_6$H$_4$Me-o)$_2$, prepd. from trans-[Pt(Py)$_2$Cl$_2$] + Ars in MeOH, followed by o-MeC$_6$H$_4$SO$_3$Ag. The nucleophilic reactivity const. and the rxn. kinetics were detd. by means of the changes in the optical density in the UV region (1203).

[(CF$_3$CF=CFCF$_3$)Pt(Ars)$_2$], crystals, m. 198-200°, prepd. from [Pt(Ars)$_4$] suspended in CH$_2$Cl$_2$ + excess CF$_3$CF=CFCF$_3$ in a sealed tube. IR and NMR spectra (1300).

[Ac$_2$CHPt(Ars)(COD)Br]*, crystals, m. 196° (dec.), a nonelectrolyte in PhNO$_2$, prepd. from [Ac$_2$CHPt(COD)Br]$_2$ in Et$_2$O + Ars. IR and far-IR spectra (809).

[Ac$_2$CHPt(Ars)(COD)Cl]*, crystals, m. 204-205°, a nonelectrolyte in PhNO$_2$, prepd. from [Ac$_2$CHPt(COD)Cl]$_2$ in Et$_2$O + Ars. IR and far-IR spectra (809).

[Ac$_2$CHPt(Ars)(COD)I]*, crystals, m. 174-175°, a nonelectrolyte in PhNO$_2$, prepd. from [Ac$_2$CHPt(COD)I]$_2$ in Et$_2$O + Ars. IR, far-IR, and mass spectra (809).

[Pt(Ars)Cl$_2$]$_2$, a Cl-bridged complex; far-IR spectrum (1500).

[Pt$_2$(Ars)(trans-o-H$_2$NC$_6$H$_4$SCH$_2$CH=CHCH$_2$SC$_6$H$_4$NH$_2$-o)Cl$_4$], pale yellow complex, m. > 250°, prepd. from [Pt$_2$(trans-o-H$_2$NC$_6$H$_4$SCH$_2$CH=CHCH$_2$SC$_6$H$_4$NH$_2$-o)Cl$_4$] + Ars in EtOH at 20°, IR spectrum (628).

[Re(Ars)Cl$_3$]$_n$ (M-25), yellow-green solid, contaminated with [Re(O)(Ars)$_2$Cl$_3$], prepd. from ReCl$_4$ + Ars in Me$_2$CO at r.t. (403).

[Re(Ars)$_2$Cl$_4$], scarlet crystals, prepd. from [Re(MeCN)$_2$Cl$_4$] + Ars in sulfolan by heating to 100° and from [Re(MeCN)$_2$Cl$_4$] by fusion with Ars, 65 and 10% yield, resp. (1302, see also 1301).

* COD = cyclooctadiene.

(In the following entries (Ars) stands for Ph_3As.)

[Re(Ars)$_2$(O)Br$_3$], (M-25), yellow microcrystals, prepd. from [Bu$_4$N][Re(O)Br$_4$] + Ars in MeCN (399).

cis-[Re(Ars)(CO)$_4$Cl], m. 168°, prepd. from [Re(CO)$_4$Cl]$_2$ + Ars in CCl$_4$ at r.t. IR spectrum. A rxn. with Ph$_3$P → displacement of one CO group (1652).

cis-[Re(Ars)(PPh$_3$)(CO)$_3$Cl], prepd. from cis-[Re(Ars)(CO)$_4$Cl] + Ph$_3$P in MePh at 70°. The rate of rxn. was detn. by means of the changes of the IR spectrum of the rxn. mixt. (1652).

[ReH$_5$(Ars)$_3$], yellow ppt., formed from [Et$_4$N]$_2$[ReH$_9$] + Ars in i-PrOH under reflux; the liq. phase cont. [Et$_4$N][ReH$_8$(Ars)] (619).

Et$_4$N[ReH$_8$(Ars)], white cryst. compd., a 1:1 electrolyte in MeCN, prepd. as described above along with [ReH$_5$(Ars)$_3$] and sepd. from the liq. phase by evapn. to dryness. IR and NMR spectra; a 9-coordinate Re deriv. (619).

[Ph$_3$GeRe(Ars)(CO)$_4$], white solid, m. 225-227°, prepd. from [Ph$_3$GeRe(CO)$_5$] by heating with Ars at 210-215°. IR spectrum (1143).

[Ph$_3$SnRe(Ars)(CO)$_4$], white crystals, m. 213-215°, prepd. from [Ph$_3$SnRe(CO)$_5$] by heating with Ars at 220°. IR spectrum (1143).

[Rh(Ars)$_3$Cl], brown crystals, m. 150-155° (dec.), very sensitive to air (1006), m. 88-92° (dec.) (774), prepd. from [Rh(CH$_2$=CH$_2$)$_2$Cl]$_2$ + Ars in MeOH under reflux (1006, 1559) and from RhCl$_3$ + Ars in MeOCH$_2$CH$_2$OH under reflux (774). IR spectrum (1006). Useful as a catalyst for the hydrogenation of olefins and acetylenes (370, 774).

[Rh(Ars)$_3$Cl$_3$], m. 219-220°, useful as a catalyst for the oxidation of EtPh to BzMe (195).

[Rh(Ars)$_2$(CO)Cl] (M-25), orange solid, m. 250-253° (dec.) (1252), yellow microcrystals, m. 242-244° (538), prepd. from [Rh(CO)$_2$Cl]$_2$ + Ars (1:4) in C$_6$H$_6$ (160, 1038), from [Rh(Ars)$_3$Cl] in CHCl$_3$ by treating with CO at 25° or with aldehydes in refluxing C$_6$H$_6$, from [Rh(Ars)$_2$(C$_2$H$_4$)Cl] by treating with CO (1006), and from RhCl$_3$·3H$_2$O in EtOH + excess Ars in EtOH, followed by 37% HCHO soln. (538). IR and NMR spectra (1252). Far-IR spectrum, a square planar, trans configuration postulated (160).

[Rh(Ars)$_2$(CO)NCS], yellow substance, m. 188-190° (dec.), prepd. from [Rh(Ars)$_2$(CO)Cl] + KSCN in Me$_2$CO under reflux. IR spectrum (802).

[Rh(Ars)(CO)(S$_2$PF$_2$)]$_n$, brownish red substance stable in air, prepd. from [Rh(CO)$_2$(S$_2$PF$_2$)]$_2$ by treating with Ars at -196°, adding C$_6$H$_6$ and warming to r.t. IR spectrum; mol. wt. detn. indicated a polymeric structure (693).

[Rh(Ars)$_2$(CO)(BBr$_3$)Cl], orange-brown, m. 261-265° (dec.), prepd. from [Rh(Ars)$_2$(CO)Cl] + BBr$_3$ in C$_6$H$_6$. IR spectrum, a formulation of a 5-coordinate Rh proposed (1252).

[RhH(Ars)$_2$Cl$_2$], a solid, prepd. from [Rh(Ars)$_3$Cl] in CHCl$_3$ or CH$_2$Cl$_2$ by treating with HCl. The complex is unstable in the cryst. state, loses HCl and the solvent of crystn. NMR spectrum in CH_2Cl_2. Anal. data indicated that the complex may be a binuclear species contg. CH_2Cl_2: $[RhH(Ars)_2Cl_2]_2 \cdot CH_2Cl_2$ (1006).

$[RhH_2(Ars)_2Cl]$, yellow ppt., formed a yellow-orange soln. in C_6H_6 and CH_2Cl_2 and greenish yellow in $CHCl_3$; stable in soln. and in the solid state, prepd. from $[Rh(Ars)_3Cl]$ by dissolving in CH_2Cl_2 satd. with H and concg. in a H atm. IR and NMR spectra; a cis coordination of both H atoms postulated. An octahedral configuration with the sixth position filled with a molecule of CH_2Cl_2 was suggested. The complex is unable to transfer the H ligand to olefins. Addn. of Ph_3P to the complex in $CHCl_3$ soln. gives mixed species (1006).

$[RhD(Ars)_2(CO)Cl_2]$, prepd. from $[Rh(Ars)_3(CO)Cl]$ in C_6H_6 + DCl in D_2O (731).

$[RhH(Ars)_3(PPh_3)]$, yellow crystals, m. 168° (dec.), prepd. from $[PhRh(1,5-cyclooctadiene)(PPh_3)]$ by treating with Ars (1:5) in MePh followed by H at r.t. IR spectrum (1492).

$[Rh(Ars)_2(PPh_3)O_2]$, m. 137-139°, prepd. from $[PhRh(1,5-cyclooctadiene)(PPh_3)]$ by treating with Ars (1:5) in MePh followed by O. IR spectrum (1492).

$[Rh(Ars)_2(O_2)Cl]$, brown crystals, prepd. from $[Rh(Ars)_3Cl]$ in CH_2Cl_2 by treating with oxygen and from Ars soln. in hot MeOH by adding $[Rh(CH_2=CH_2)_2Cl]_2$ in the presence of air. IR spectrum (1006).

$[Rh(Ars)_2(CH_2=CH_2)Cl]$, yellow microcrystals, m. 163° (dec.), prepd. from $[Rh(Ars)_3Cl]$ by adding CH_2Cl_2 satd. with $CH_2=CH_2$. IR and NMR spectra. The C_2H_4 group is readily displaced by CO (1006).

$[Rh(Ars)_2(CF_2=CF_2)Cl]$, yellow-orange ppt., m. 86-88°, prepd. from $[Rh(Ars)_3Cl]$ in CH_2Cl_2 + excess $CF_2=CF_2$ in a sealed tube at 25°. IR spectrum (1006).

$[Rh(Ars)_2(CF_3C{\equiv}CCF_3)_2Cl]$, bright yellow crystals, m. 195-197°, prepd. from $[Rh(Ars)_3Cl]$ in C_6H_6 + $CF_3C{\equiv}CCF_3$ in a sealed tube at 75-80°. IR and NMR spectra; a 5-coordinate Rh(III) structure postulated in which the 2 $CF_3C{\equiv}CCF_3$ groups are σ-bonded to the Rh atom and to each other, forming a rhodiacyclo-pentadiene ring (1007).

$[Rh(Ars)_2(MeO_2CC{\equiv}CCO_2Me)_2Cl]$, orange crystals, m. 238-240°, prepd. from $[Rh(Ars)_2(CO)Cl]$ suspension in MePh in the presence of furoyl azide by a rxn. with $MeO_2CC{\equiv}CCO_2Me$ at -5°, followed by warming to 35-40°. IR and NMR spectra. The complex catalyzes the trimerization of $MeO_2CC{\equiv}CCO_2Me$ to hexakis(methoxycarbonyl)benzene (373).

$[Rh(Ars)_2(PhC{\equiv}CPh)Cl]$, orange crystals, m. 218-220° (dec.), prepd. from $[Rh(Ars)_3Cl]$ + $PhC{\equiv}CPh$ in CH_2Cl_2. IR spectrum (1006).

(In the following entries (Ars) stands for Ph$_3$As.)

[Rh(Ars)(COD)Cl]*, dark orange crystals, m. 133-136° (1573), orange yellow crystals, m. 137-141° (dec.) (450), a monomeric structure postulated; a non-electrolyte in PhNO$_2$ and ClCH$_2$CH$_2$Cl (1572, 1573); prepd. from [Rh(COD)Cl]$_2$ in CH$_2$Cl$_2$ + Ars (1:2) in light petr. (450, 1573). IR spectrum (450, 1572, 1573) indicated both olefinic bonds coordinated with the Rh atom. NMR spectrum in CDCl$_3$ was temp. dependent and showed nonequivalent "olefinic" protons, due to asymmetry of the complex. At rising temp. the proton signals broadened and finally coalesced to one signal; the rate of the coalescence was detd. The latter phenomenon may be caused by the interaction of the monomeric, dimeric, and two monomeric complex molecules exchanging the Ars ligands. The rate of the exchange increased with increasing Ars ligand concn. Addn. of Cl$^-$ accelerated the coalescence, while addn. of COD had no effect. The exchange rxn. may involve a 5-coordinate Rh intermediate [Rh(Ars)$_2$(COD)Cl] (1572, 1573).

[Rh(Ars)(COD)Cl]·2HgCl$_2$, red crystals, m. 106°, prepd. from [Rh(Ars)(COD)Cl] + HgCl$_2$ in AcEt-C$_6$H$_6$ mixt. (450).

[Rh(Ars)(Nor)Cl]**, a dark orange substance, monomeric, nonelectrolyte in PhNO$_2$ and ClCH$_2$CH$_2$Cl, prepd. from [Rh(Nor)Cl]$_2$ + Ars. IR spectrum indicated both olefinic bonds coordinated to the metal (1572). Far-IR spectrum; a square planar structure postulated (160). NMR spectrum in CDCl$_3$ showed nonequivalent "olefinic" protons. At rising temp. the proton signals broadened and finally coalesced to one signal. When free Ars was present, a rapid Ars ligand exchange occurred. The rate of the exchange increased with rising temperature and increasing Ars concn. The rxn. kinetics were detd. for the -70 to -10° temp. range. The exchange may proceed via a 5-coordinate short-lived [Rh(Ars)$_2$(Nor)Cl] complex (1572, 1574).

[Rh(Ars)$_2$(Nor)SnCl$_3$], orange crystals, dec. 177-179°, a nonelectrolyte in MeNO$_2$, prepd. from [Rh(Nor)$_2$SnCl$_3$] + Ars in CH$_2$Cl$_2$. A 5-coordinate Rh complex postulated (1637).

[Rh(Ars)(CO)(8-quinolinolate)], yellow solid, m. 168°, prepd. from dicarbonyl-8-quinolinolatorhodium + Ars in C$_6$H$_6$. IR spectrum; a 4-coordinate Rh complex with the 8-quinolinolate ligand covalently bonded through the O atom and chelated through the N atom (1542).

[π-CH$_2$=CHCH$_2$Rh(Ars)$_2$Cl$_2$], yellow, paramagnetic substance, m. 188-189°, a non-electrolyte in ClCH$_2$CH$_2$Cl and MeNO$_2$, prepd. from [Rh(Ars)$_3$Cl] by slowly adding CH$_2$=CHCH$_2$Cl. IR spectrum indicated a π-bonding of the allyl group (1559).

[π-CH$_2$=C(Me)CH$_2$Rh(Ars)$_2$Cl$_2$], orange ppt., m. 175-180°, a nonelectrolyte in ClCH$_2$CH$_2$Cl and MeNO$_2$, prepd. from [Rh(Ars)$_3$Cl] by slowly adding to CH$_2$=C(Me)-CH$_2$Cl (1559) or by heating the reactants under reflux (940). IR and NMR

---

* COD = 1,5-cyclooctadiene.
** Nor = bicyclo[2.2.1]hepta-2,5-diene.

spectra and dipole moment were detd. A square pyramidal configuration with
the π-allylic ligand in a trans position to one Ars ligand suggested (1559).
The NMR signals of the protons at C-1 and C-3 in CDCl₃ broadened and finally
coalesced at rising temp. due to their position interchange, which might involve a σ-allyl intermediate. No changes in the spectrum that could be attributed to the phenyl protons were observed. Addn. of Ph₃P to the complex displaced the Ars ligand (1571). A rxn. with CH₂=CH₂ in CHCl₃ proceeded first
through the absorption of the olefin, followed by evolution of a gas contg.
1-C₄H₈, cis- and trans-2-C₄H₈, i-C₄H₈, and CH₂=CH₂. The i-C₄H₈ may originate
from the methallyl ligand. The complex catalyzed the hydrogenation of olefins.
The rxn. with CH₂=CH₂ proceeds through the displacement of one of the Ars
groups with CH₂=CH₂, followed by addn. of another CH₂=CH₂ mol. and a rearrangement of the π-methallyl to σ-methallyl. The formation of the intermediates
was postulated on the basis of the NMR spectra and elec. cond. data. Further
rxn. steps are uncertain. CO displaces the π-methallyl ligand and one of the
Cl ligands from [π-CH₂=C(Me)CH₂Rh(Ars)₂Cl₂] to form trans-[Rh(Ars)₂(CO)Cl]
and methallyl chloride. The displacement with CO was followed by NMR and
elec. cond. measurements (1560).

[(π-CH₂=CHCH₂)₂Rh(Ars)Cl], yellow prisms, m. 125-128°, prepd. from [(π-CH₂=CH-
CH₂)₂RhCl]₂ + Ars in CH₂Cl₂. NMR spectrum (1249). The rxn. of the dimeric
complex with Ars (1:2) in CDCl₃ above 35° was followed by NMR spectroscopy;
unequal bonding of the C-1 and C-3 atoms to Rh was detected (1260).

[$\overline{CF=CFCF=CF}$Rh(Ars)Cl], yellow needles, m. 182-183°, prepd. from [Rh(Ars)₃Cl]
+ excess CF₂=CFCF=CF₂ in C₆H₆ at 60° in a sealed tube. IR and NMR spectra
(1299).

[$\overline{OCCH₂CH₂CH₂}$Rh(Ars)₂(CO)Cl], colorless needles, m. 247° (dec.), prepd. from
[$\overline{OCCH₂CH₂CH₂}$Rh(CO)Cl]₂ in CH₂Cl₂ + Ars at r.t., followed by boiling with
light petr. On standing in air the complex becomes yellow. The complex is
insol. in H₂O, sol. in CH₂Cl₂, CHCl₃, and C₆H₆. IR and NMR spectra. An octahedral configuration with the Ars ligands probably trans to each other (1299).

[Ru(Ars)₂(CO)₃], prepd. from RuCl₃·xH₂O by a treatment with CO in MeOH under
60 psi pressure, followed by Ars at 65° and redn. with Zn dust in hot DMF.
IR spectrum, a trigonal bipyramidal structure postulated (374).

cis-[Ru(Ars)₂(CO)₂Br₂], white crystals, m. 268°, prepd. from RuCl₃·3H₂O in
EtOH contg. excess LiBr under reflux by passing CO, adding 4-fold excess Ars
in Et₂O, dilg. with EtOH, boiling gently under N, filtering off the cryst.
prod. from the hot soln. and recrystg. from a CH₂Cl₂-MeOH mixt. IR spectrum
(1453).

cis-[Ru(Ars)₂(CO)₂Cl₂], white crystals, m. 295°, prepd. from RuCl₃·3H₂O soln.
in EtOH by bubbling CO, adding 4-fold excess Ars in Et₂O, dilg. with EtOH,
boiling gently under N, filtering off the pale yellow crystals from the hot
soln. and recrystg. from a CH₂Cl₂-MeOH mixt. The hot ethanolic mother liquor
contained the monocarbonyl complex, described below, which crystd. upon cooling (1453). The title complex was also prepd. from [Ru(CO)₃(THF)Cl₂] + Ars in
THF (249). IR spectrum (249, 1453).

(In the following entries (Ars) stands for Ph₃As.)

[Ru(Ars)₃(CO)Cl₂], fawn-colored powder, m. 202°, prepd. from RuCl₃·3H₂O in boiling EtOH by bubbling CO and, after cooling, adding 6-fold excess of Ars in Et₂O. The complex was also obtd. as by-prod. in the prepn. of the preceding dicarbonyl complex. The title compd. is readily sol. in CH₂Cl₂, MeNO₂, and warm Me₂CO. IR spectrum (1453).

[Ru(Ars)₂(MeOH)Br₃], dark brown, paramagnetic crystals, m. 166-167°, prepd. from RuCl₃·3H₂O in MeOH by shaking with excess LiBr and treating with excess Ars (1453).

[Ru(Ars)₂(MeOH)Cl₃], green crystals, m. 149°, prepd. from RuCl₃·3H₂O + 6-fold excess Ars in MeOH under reflux (1453). Useful as a catalyst for the redn. of olefins (1613).

[Ru(Ars)₂(Me₂CO)Cl₃], yellow-brown crystals, m. 195°, prepd. from [Ru(Ars)₂-(MeOH)Cl₃] by shaking with excess Me₂CO (1453).

[Ru(Ars)₂(NO)Cl₃] (M-26), yellow-orange, paramagnetic solid, dec. > 320°, prepd. from Ru(NO)Cl₃·3H₂O + Ars in EtOH at 60°. Soly. data; IR spectrum (543).

[Ag(Ars)]O₂CCF₃, the formation of the complex from Ars and AgO₂CCF₃ in aq. MeOH was detected by UV spectroscopy (1175).

[Tc(Ars)₂(CO)₃Cl], colorless crystals, a nonelectrolyte, prepd. from Tc(CO)₅Cl + Ars (1:2) in EtOH at 60-70° (737).

[Tc(Ars)₂Cl₄], brown, paramagnetic substance, prepd. from TcCl₄ + Ars in EtOH under reflux and pptd. with Et₂O. An octahedral stereochemistry postulated (562).

[(C₆F₅)₂Tl(Ars)Br], white microcrystals, dec. ~ 130°, prepd. from (C₆F₅)₂TlBr + Ars in boiling Et₂O, evapg. to dryness, and crystg. from i-Pr₂O-C₆H₁₄. A nonelectrolyte in Me₂CO (458). IR spectrum (454, 458). A tetrahedral structure suggested (458).

[(C₆F₅)₂Tl(Ars)Cl], white microcrystals, m. 149.5°, a nonelectrolyte in Me₂CO, prepd. from (C₆F₅)₂TlCl + Ars in boiling Et₂O and adding boiling C₆H₁₄ (458). IR spectrum (454, 458). A tetrahedral structure suggested (458).

[Sn(Ars)₂Br₄], orange crystals, prepd. from Ars in CH₂Cl₂ by mixing with SnBr₄. Far-IR spectrum (1280).

[Sn(Ars)₂Cl₄], white powdery substance, prepd. from Ars in n-C₆H₁₄ by adding SnCl₄. Far-IR spectrum (1280).

[Ti(Ars)Br₄] (M-26), a monomeric adduct, very sol. in C₆H₆, prepd. by mixing TiBr₄ with Ars in C₆H₄. IR spectrum (585).

[Ti(Ars)Cl₄] (M-26), red-black crystals, m. 126-128°, sensitive to moisture (1603), very sol. in C₆H₆ (585), prepd. by mixing Ars with TiCl₄ (1:1) in C₆H₆. IR spectrum (585, 1603). The compn. corr. either to [Ti(Ars)Cl₄] or to

[Ti(Ars)$_2$Cl$_4$] (585). The suggestion was made that the substance may be a 5-coordinate monomer or a more highly coordinated polymer (1603).

[Ti(Ars)$_2$Cl$_4$], deep red complex, prepd. from TiCl$_4$ + Ars in petr. ether or C$_6$H$_6$, or in vapor phase (65).

[V(Ars)(CO)$_5$], rxn. with Na-Hg in EtOH → [V(Ars)(CO)$_5$]$^-$ (1601).

[V(Ars)$_2$(CO)$_4$], paramagnetic ppt., prepd. from V(CO)$_6$ + Ars in n-C$_6$H$_{14}$ at r.t. Useful for a vapor-phase plating of metals with V at ~ 500-550° under a reduced pressure (1600). Rxn. with NO in petr. ether → [V(Ars)(NO)(CO)$_4$] (1601).

[V(Ars)$_2$Cl$_4$], prepd. from VCl$_4$ + Ars in CCl$_4$ or in C$_6$H$_6$. The complex was not identified (225).

[W(Ars)(CO)$_5$], the complex was formed from W(CO)$_6$ + Ars (1:1 to 1:10) in a n-decane-cyclohexane mixt. at 149.0 $\pm$ 0.3°. The rate of the complex formation was followed by IR spectroscopy. Even for the 1:10 ratio the extent of a disubstitution was extremely limited. A S$_N$1 dissocn. mechanism postulated (1599).

[W(Ars)$_2$(CO)$_3$Br$_2$], yellow, diamagnetic crystals, prepd. from W(CO)$_6$ by treating with Br and mixing with Ars in CH$_2$Cl$_2$. IR spectrum; a pentagonal bipyramidal structure with the Br atoms in the apical positions suggested (76). Rxn. with NO → [W(Ars)$_2$(NO)$_2$Br$_2$] (77).

[W(Ars)$_2$(CO)$_3$Cl$_2$], yellow, diamagnetic crystals, stable in air, prepd. from W(CO)$_6$ by treating with liquid Cl, exposing to air, extracting with Me$_2$CO and adding Ars to the soln. IR spectrum; a pentagonal bipyramidal structure with the Cl atoms in the apical positions suggested (76).

[W(Ars)$_2$(NO)$_2$Br$_2$], green crystals, m. > 300°, prepd. from W(NO)$_2$Br$_2$ + Ars in C$_6$H$_6$ under reflux, from [W(Ars)$_2$(CO)$_4$] + NOBr in CH$_2$Cl$_2$ (807), and from [W(Ars)$_2$(CO)$_3$Br$_2$] + NO in CH$_2$Cl$_2$ (77). IR spectrum (77, 807) - two cis isomers were identified by IR spectrum (77). An octahedral structure postulated (77, 807).

[W(Ars)$_2$(NO)$_2$Cl$_2$] (M-26), green crystals, prepd. from [W(Ars)$_2$(CO)$_3$Cl$_2$] + NO in CH$_2$Cl$_2$. IR spectrum - one cis isomer was detected. An octahedral structure postulated (77).

[π-C$_5$H$_5$W(Ars)(CO)$_2$Cl], crystals, m. 165°, prepd. from [π-C$_5$H$_5$W(CO)$_3$Cl] + excess Ars in C$_6$H$_6$ under reflux. IR spectrum (1522).

[π-C$_5$H$_5$W(Ars)(CO)$_2$COC$_6$F$_5$], yellow crystals, m. 125°, prepd. from [π-C$_5$H$_5$W-(Ars)(CO)$_3$]PF$_6$ in THF by treating with C$_6$F$_5$Li (slight excess) and purifying by chromatography on Al$_2$O$_3$. IR and NMR spectra (1523).

[π-C$_5$H$_5$W(Ars)(CO)$_3$]PF$_6$, yellow crystals, m. 237°, prepd. from [π-C$_5$H$_5$W(CO)$_3$Cl] + Ars + AlCl$_3$ (1:1:1) in C$_6$H$_6$ at r.t., treating the prod. with H$_2$O, dissolving the solid deposit in Me$_2$CO and pptg. with aq. [NH$_4$]PF$_6$. IR spectrum (1522).

(In the following entry (Ars) stands for Ph$_3$As.)

[π-C$_5$H$_5$W(Ars)$_2$(CO)$_2$]PF$_6$, yellow crystals, m. 270°, prepd. from [π-C$_5$H$_5$W(Ars)-(CO)$_2$Cl] + Ars + AlCl$_3$ (1:3:3) in C$_6$H$_6$ at r.t., treating the mixt. with H$_2$O, dissolving the solid prod. in Me$_2$CO, and pptg. with aq. [NH$_4$]PF$_6$. IR spectrum (1522).

Tricyclohexylarsine, AsC$_{18}$H$_{33}$, (C$_6$H$_{11}$)$_3$As
Synth.: From AsCl$_3$ + C$_6$H$_{11}$MgBr (90).
Coordination Deriv.: {Mn[As(C$_6$H$_{11}$)$_3$]$_2$(CO)$_3$Cl}, prepd. from Mn(CO)$_5$Cl by heating with the arsine. Sol. in hydrocarbons, useful as an antiknock additive (1612).

Trihexylarsine, AsC$_{18}$H$_{39}$, (C$_6$H$_{13}$)$_3$As, (M-27)
Prop.: b$_{0.6}$ 131-132°, d$_4^{20}$ 0.9424, n$_D^{20}$ 1.4739, [ρ]$_M$ 1909.0, χ_M -237.8, R$_M$ 98.51 (508).
Coordination Deriv.: {Mn[As(C$_6$H$_{13}$)$_3$]$_2$(CO)$_3$F}, prepd. from [Mn(CO)$_4$F]$_2$ by heating with the arsine in hydrocarbon soln. Useful as an antiknock agent (1612).

Tris[m-(trifluoromethyl)phenyl]arsine, AsC$_{21}$H$_{12}$F$_9$, (3-CF$_3$C$_6$H$_4$)$_3$As, (M-31)
Prop.: Electronic spectrum - the effect of the CF$_3$ group and of solvents (423).
Coordination Deriv.: {Rh[Ars(C$_6$H$_4$CF$_3$)$_3$]$_2$(CO)Cl}, yellow solid, m. 112-115° (dec.); IR spectrum (802).

Tris[p-(trifluoromethyl)phenyl]arsine, AsC$_{21}$H$_{12}$F$_9$, (4-CF$_3$C$_6$H$_4$)$_3$As, (M-31)
Prop.: Electronic spectrum - the effect of the CF$_3$ group and of solvents (423).

2,2',2"-Arsylidynetribenzoic Acid, AsC$_{21}$H$_{15}$O$_6$, (2-HO$_2$CC$_6$H$_4$)$_3$As
Prop.: IR spectrum, qual. interpretation (1223).

Tribenzylarsine, AsC$_{21}$H$_{21}$, (PhCH$_2$)$_3$As, (M-27)
Uses: Mixtures with Na, Li, K, Mg, or Zn and Ti or V halides catalyze the polymerization of α-olefins (387).
Coordination Deriv.: {Mn[As(CH$_2$Ph)$_3$]$_2$(CO)$_3$Br}, prepd. by heating [Mn(CO)$_4$-Br]$_2$ with the arsine; sol. in hydrocarbons, useful as an antiknock additive (1612).

Tri-o-tolylarsine, AsC$_{21}$H$_{21}$, (2-MeC$_6$H$_4$)$_3$As, (M-27)
Prop.: m. 112°, electronic spectrum - the effect of the Me group and of solvents (423). IR spectrum - qual. interpretation (1223).

Tri-m-tolylarsine, AsC$_{21}$H$_{21}$, (3-MeC$_6$H$_4$)$_3$As, (M-27)
Prop.: m. 96°, electronic spectrum - the effect of the Me group and of solvents (423).

Tri-p-tolylarsine, $AsC_{21}H_{21}$, $(4-MeC_6H_4)_3As$, (M-27)

Synth.: From $AsCl_3$ + $4-MeC_6H_4Li$ in Et_2O at $-70°$ and heating to boiling, 86% yield (723).

Prop.: m. 151° (423), m. 149-150° (723). Electronic spectrum - the effect of the Me group (423). NMR spectrum (723).

Rxn. with: Br in CCl_4, followed by RSO_2NH_2 → $(4-MeC_6H_4)_3As=NO_2SR$ (341). Br in C_6H_6, followed by RSO_2NH_2 in Et_3N → $(4-MeC_6H_4)_3AsO\cdot H_2NO_2SR$ (342). Chloroamine T (1:1) in THF → $(4-MeC_6H_4)_3As=NO_2SC_6H_4Me-4$ (723).

Uses: A stabilizer for photochromic systems such as the mixt. of 2,3-di-phenylindenone 2,3-epoxide and poly(methyl methacrylate) against mol. degra-dation (1544).

Tri(p-methoxyphenyl)arsine, Tri-p-anisylarsine, $AsC_{21}H_{21}O_3$, $(4-MeOC_6H_4)_3As$, (M-28)

Rxn. with: Br in CCl_4, followed by RSO_2NH_2 → $(4-MeOC_6H_4)_3As=NO_2SR$ (341). Br in C_6H_6, followed by RSO_2NH_2 in Et_3N → $(4-MeOC_6H_4)_3AsO\cdot H_2NO_2SR$ (342).

Coordination Deriv.: $\{Mn[As(C_6H_4OMe-4)_3](NO)_3\}$, prepd. from $[Mn(NO)_3CO]$ + the arsine in C_6H_{12}. IR spectrum - NO valence vibration (135).

Tris(o-methylthiophenyl)arsine, $AsC_{21}H_{21}S_3$, $(2-MeSC_6H_4)_3As$

Synth.: From $2-BrC_6H_4SMe$ in EtOH + BuLi in hexane, followed by a slow addn. of $AsCl_3$ and hydrolysis with 0.2 N HCl, 87% yield (512).

Prop.: Colorless crystals, dec. 164-166°, a nonelectrolyte in $MeNO_2$ (512).

Rxn. with: Ni^{++} salts, no rxn. (512, 1052).

Coordination Deriv.: $\{Pd[As(C_6H_4SMe-2)_3]Cl_2\}$, orange crystals, a nonelectro-lyte in $MeNO_2$, prepd. by adding the arsine in EtOH soln. to Na_2PdCl_4 in EtOH, removing the first ppt. of $\{Pd_2[As(C_6H_4SMe-2)_3]Cl_4\}$, and recovering the title complex ppt. after a longer standing. Electronic spectrum - a 4-coordinate Pd complex (515).

$\{Pd[As(C_6H_4SMe-2)_3]_2\}(ClO_4)_2$, orange-brown crystals, a 1:1 electrolyte in $MeNO_2$, prepd. from $Na_2PdCl_4\cdot 4H_2O$ in EtOH + the arsine in EtOH under reflux in the presence of $LiClO_4$. Electronic absorption spectrum (515).

$\{Pd_2[As(C_6H_4SMe-2)_3]Cl_4\}$, yellow brown crystals, insol. in $MeNO_2$, prepd. from the arsine and Na_2PdCl_4 in EtOH. The arsine functions as a tetradentate coordinated to two different Pd atoms (515).

Tris(p-nitrophenylethynyl)arsine, $AsC_{24}H_{12}N_3O_6$, $(4-O_2NC_6H_4C\equiv C)_3As$

Prop.: Crystal structure by x-ray diffraction (1081).

Tris(phenylethynyl)arsine, $AsC_{24}H_{15}$, $(PhC\equiv C)_3As$, (M-28)

Synth.: From $R_3SnC\equiv CPh$ (R = Me, Et, or Pr) + $AsCl_3$ in C_6H_6, 91-96% yield (694).

Tris(2-phenylethyl)arsine, $AsC_{24}H_{27}$, $(PhCH_2CH_2)_3As$

Use: Complexes with $M[AlR_4]$ (where M = alkali metal) and Ti, V, Zr, Cr, or Mo halides catalyze the polymerization of α-olefins (385).

Tris(2,3-dimethylphenyl)arsine, $AsC_{24}H_{27}$, $(2,3-Me_2C_6H_3)_3As$
Prop.: m. 162°, electronic spectrum - the effect of the Me groups and of solvents (423).

Trixylylarsine, $AsC_{24}H_{27}$
Use: A stabilizer for photochromic systems such as mixts. of 2,3-diphenylindenone 2,3-epoxide and poly(methyl methacrylate) against mol. degradation (1544).

Tris[p-(dimethylamino)phenyl]arsine, $AsC_{24}H_{30}N_3$, $(4-Me_2NC_6H_4)_3As$
Synth.: From $AsCl_3$ in Et_2O by adding to a soln. of $4-Me_2NC_6H_4Li$ in Et_2O at -20°, warming to r.t., and heating under reflux for 2 hrs., 21% yield (1519).
Prop.: m. 240-242° (1519).
Rxn. with: $Pb(OAc)_4$ or with N/20 $AgClO_4$ soln. in C_6H_6 → a green colored radical cation (1519).
Coordination Deriv.: $\{Ir[As(C_6H_4NMe_2-4)_3](COD)Cl\}$*, m. > 180°, a monomer, nonconducting in $ClCH_2CH_2Cl$, prepd. from $[Ir(COD)Cl]_2$ and the arsine (1:2). IR spectrum indicated the presence of both olefinic bonds coordinated with the metal atom. NMR spectrum showed a nonequivalence of the "olefinic" protons in the asymmetric complex. With rising temp. the proton signals broadened and finally coalesced. When the free arsine was present besides the complex in $PhNO_2$ soln., conductance measurements indicated an equilibrium mixt.: $\{Ir[As(C_6H_4NMe_2-4)_3]Cl\} + (4-Me_2NC_6H_4)_3As \rightleftharpoons \{Ir[As(C_6H_4NMe_2-4)_3]_2\}^+Cl^-$. The concn. of the ionic species increased when the temp. was lowered or the arsine was added to the mixt. Addn. of Cl^- ions accelerated the coalescence, but COD had no effect. A mechanism of the ligand exchange is discussed (1373).

$\{Ir[As(C_6H_4NMe_2-4)_3]_2(COD)Cl\}$, m. 220°, a nonelectrolyte in $ClCH_2CH_2Cl$, prepd. from $\{Ir[As(C_6H_4NMe_2-4)_3](COD)Cl\}$ + the arsine in EtOH (1575).

$\{Rh[As(C_6H_4NMe_2-4)_3]_3Cl\}$, prepd. from $[Rh(CH_2=CH_2)_2Cl]_2$ and the arsine in EtOH (1559).

$\{\pi-CH_2=C(Me)CH_2Rh[As(C_6H_4NMe_2-4)_3]_2Cl_2\}$, orange, paramagnetic ppt., dec. 250°, a nonelectrolyte in $ClCH_2CH_2Cl$ and Me_2NO_2, prepd. from $\{Rh[As(C_6H_4NMe_2-4)_3]_3-Cl\}$ by slowly adding to $CH_2=C(Me)CH_2Cl$. IR and NMR spectra indicated a π-bonding of the allyl ligand in a trans position to the arsine ligands (1559). NMR spectrum in $CDCl_3$ showed the signals of the C-1 and C-3 protons which broadened and coalesced at rising temp., due to an interchange of their positions which might involve a σ-allyl intermediate (1571).

Tris(2-phenyl-1,2-dicarbadodecaboran(12)-1-yl)arsine, $AsB_{30}C_{24}H_{45}$,
Synth.: From $AsCl_3$ in Et_2O + 2-phenyl-1,2-dicarbado-decaboran(12)-1-yllithium in C_6H_6 under reflux, 55% yield (1642).
Prop.: m. 175-176° (1642). IR spectrum (222).

$$(PhC-C)_3As$$
$$\underset{B_{10}H_{10}}{\diagdown\diagup}$$

* COD = 1,5-cyclooctadiene.

Trioctylarsine, $AsC_{24}H_{51}$, $(n-C_8H_{17})_3As$, (M-29)
Rxn. with: HgO (1:1) in $Me_2CO \rightarrow (n-C_8H_{17})_3AsO$ (1057).
Aq. H_2O_2 in $Et_2O \rightarrow (n-C_8H_{17})_3AsO$ (1057).

Tris(2-ethylhexyl)arsine, $AsC_{24}H_{51}$, $(BuEtCHCH_2)_3As$, (M-31)
Use: Complexes with Co carbonyls catalyze the hydroformylation of olefins to
aldehydes and alcohols (1424).

Trimesitylarsine, $AsC_{27}H_{33}$, $(2,4,6-Me_3C_6H_2)_3As$, (M-29)
Prop.: m. 177°, electronic spectrum - the effect of the Me groups and of
solvents (423).
Coordination Deriv.: $\{Mn[Ars(C_6H_2Me_3)_3]_2(CO)_3I\}$, prepd. from $Mn(CO)_5I$ by
heating with the arsine. Sol. in hydrocarbons, useful as an antiknock additive (1612).

Trinaphthylarsine, $AsC_{30}H_{21}$, $(C_{10}H_7)_3As$, (M-32)
Uses: A stabilizer for photochromic systems such as the mixts. of 2,3-di-
phenylindenone 2,3-epoxide in poly(methyl methacrylate) against mol. degradation (1544). Complexes with $M[AlR_4]$ (M = alkali metal) and Ti, V, Zr, Cr,
or Mo halides catalyze the polymerization of α-olefins (385).

Tris[4-(dimethylamino)-1-naphthyl]arsine, $AsC_{36}H_{36}N_3$, $(4-Me_2NC_{10}H_3)_3As$,
(M-32)
Synth.: From $AsCl_3$ in Et_2O by adding dropwise to $4-Me_2NC_{10}H_6MgBr$ in Et_2O
under reflux, 35% yield (1519).
Prop.: m. 339-342° (1519).
Rxn. with: $AgClO_4$ in $C_6H_6 \rightarrow$ a gray-colored soln. (1519).

1.1.2 UNSYMMETRIC R₂R′As DERIVATIVES

The methods of synthesis of the unsymmetric arsines reported through 1964
were summarized in the main volume (M-33). Since then several new methods
were described. Secondary arsines and, particularly, arsino- and arsylene-
alkali metal derivatives have been extensively used as the starting materials
in the preparation of the arsines.

Secondary arsines have been converted to the unsymmetric arsines by the
condensations with acetylene, alkyl halides, alkenyl halides, acyl halides,
and ketones, respectively, according to the following formulae:

$$R_2AsH + HC\equiv CH \rightarrow R_2AsCH=CH_2$$

$$R_2AsH + R'X \rightarrow R_2AsR'$$

$$R_2AsH + R'COX \rightarrow R_2AsCOR'$$

$$R_2AsH + R'_2CO \rightarrow R_2AsC(OH)R'_2$$

A conversion of secondary arsines with butyllithium leads to secondary
arsinolithiums which cleave ethers to form unsymmetric tertiary arsines.

$$R_2AsH + BuLi \rightarrow [R_2AsLi] \rightarrow +PhOR' \rightarrow R_2AsR'$$

The condensation of primary and secondary arsines with isocyanates affords arsylenebisformamides and arsinoformamides, respectively. The reaction is catalyzed by dialkyltin diacetates.

$$R_{3-n}AsH_n + R'NCO \rightarrow R_{3-n}As(CONHR')_n \qquad n = 1 \text{ or } 2$$

The same results are obtained by the condensation of organoarsinolithium and organoarsylenedilithium, respectively, with isocyanates followed by hydrolysis of the lithium derivatives. Isothiocyanates yield arsinothio-

$$R_{3-n}AsLi_n + R'NCO \rightarrow R_{3-n}As(CONLiR')_n \xrightarrow{H_2O} R_{3-n}As(CONHR')_n \qquad n = 1 \text{ or } 2$$

formamides and arsylenebis(thioformamides), respectively. When the condensation of the organoarsinolithiums with isothiocyanates is followed by alkyl halides, the corresponding thioformamidic acid esters are formed.

$$R_{3-n}AsLi_n + R'NCS \rightarrow [R_{3-n}As(CSNLiR')_n] \xrightarrow{R''Cl} R_{3-n}As[C(SR'')=NR']_n$$

$$n = 1 \text{ or } 2$$

Carbodiimides condense with secondary arsinolithiums and primary arsylene-dilithiums to form the corresponding arsino- and arsylenebis-formamidine lithium salts, respectively.

$$R_{3-n}AsLi_n + R'N=C=NR' \rightarrow R_{3-n}As[C(=NR')N(Li)R']_n \qquad n = 1 \text{ or } 2$$

A reaction of a secondary arsinolithium with acetylene in the presence of primary or secondary amines gives the corresponding ethynylarsine, $R_2AsC\equiv CH$.

A condensation of secondary arsino-alkali metals with epoxides and cyclic ethers, respectively, followed by hydrolysis yields ω-hydroxyalkylarsines.

$$R_2AsM + \overline{CH_2(CH_2)_nO} \xrightarrow{H_2O} R_2As(CH_2)_{n+1}OH$$

Diarylarsinoalkali metals on heating with alkali metal arenesulfonates yield unsymmetric triarylarsines.

Primary arsines react with formaldehyde to form the corresponding arsy-lenedimethanols.

A condensation of dialkylhaloarsines with ketene in ether affords acetyl-dialkylarsines. A conversion of chlorodiphenylarsine with dichloropropargyl-aluminum yields diphenylpropargylarsine along with diphenylpropadienyl- and diphenylpropynyl-arsine. The reaction of secondary haloarsines with acetylene is catalyzed by ultraviolet light and yields 2-halovinylarsines, $R_2AsCH=CHCl$.

A metathetic reaction of primary haloarsines with triethyl(trifluoromethyl)-tin produces bis(trifluoromethyl)arsines.

$$RAsX_2 + Et_3SnCF_3 \rightarrow RAs(CF_3)_2$$

Secondary arsinoamines insert ketene between the arsenic and nitrogen atoms to form secondary arsinoacetamides. Acetylene derivatives displace the amino group and form the corresponding alkynylarsines.

$$R_2AsNR'_2 + CH_2=C=O \rightarrow R_2AsCH_2CONR'_2$$

$$R_2AsNR'_2 + HC\equiv CR'' \rightarrow R_2AsC\equiv CR''$$

Dialkylsilylarsines, $R_2AsSiR'_3$, react with carbon disulfide to form silyl dialkylarsinodithioformates.

A metathetic reaction of thioarsenoso compounds with dialkylmercury under reflux leads to the exchange of the sulfur for two alkyl groups.

$$RAsS + R'_2Hg \rightarrow RAsR'_2$$

Oxybis(secondary arsines), $(R_2As)_2O$, dialkyl(alkylthio)arsines, R_2AsSR', and tetraorganodiarsines, R_2AsAsR_2, react with acetylene derivatives to form the corresponding alkynylarsines, $R_2AsC\equiv CR$.

Tetraorganoarsines are cleaved by acyl chlorides and carboxylic acid anhydrides, respectively, to form acylarsines, R_2AsCOR'.

Arsinic acids on heating with acetylene derivatives, while azeotropically removing water from the reaction mixture, yield the corresponding alkynylarsines, $R_2AsC\equiv CR'$.

Secondary arsinomagnesium halides react with α-olefins to form 1-alkenylarsines, $R_2AsCH=CHR$.

Depending on the volume of information on the arsines of this class, some of them were described individually hereafter, while others are listed in Table 1. Both the individual entries and the compounds in the table are arranged in the order of the increasing numbers of carbon and hydrogen atoms.

Dimethyl(trifluoromethyl)arsine, $\quad AsC_3H_6F_3$, $\quad Me_2AsCF_3$, $\quad$ (M-35)
Synth.: By-prod. from the rxn. of $Me_2AsAsMe_2$ with CF_3COCl or $(CF_3CO)_2O$ at 20° (427).
Prop.: b. 56-60° (427).

Dimethyl(trifluoroacetyl)arsine, $\quad AsC_4H_6F_3O$, $\quad Me_2AsCOCF_3$
Synth.: From $Me_2AsAsMe_2 + CF_3COCl$ at 20°, 96% yield, along with some Me_2AsCF_3 (427).
From $Me_2AsH + CF_3COCl$ at 20°, a mixt. of the title compd. with Me_2AsCl was obtd. (427).
Prop.: Pale yellow liquid, dec. > 150°, IR and NMR spectra (427).

(1,1,2,2-Tetrafluoroethylarsylene)dimethanol, $\quad AsC_4H_7F_4O_2$, $\quad HCF_2CF_2As(CH_2OH)_2$
Synth.: From $HCF_2CF_2AsH_2$ by treating with aq. HCHO contg. 35% HCl, 56% yield (1266).
Prop.: Yellow oil dec. during distn., d_4^{20} 1.7520, n_D^{20} 1.4570 (1266).

Methylbis(1,1,2,2-tetrafluoroethyl)arsine, $\quad AsC_5H_5F_8$, $\quad (HCF_2CF_2)_2AsMe$
Synth.: By-prod. formed in the rxn. of $MeAsH_2$ with $CF_2=CF_2$ at 75° in an auto-

clave, 11.4% yield besides $Me(HCF_2CF_2)AsH$ (1266).
Prop.: b_{45} 48-50°, d_4^{20} 1.7392, n_D^{20} 1.3790, MR_D 38.80 (1266).

(2-Chlorotetrafluoropropenyl)dimethylarsine, $AsC_5H_6ClF_4$, $Me_2AsCF=CClCF_3$
Synth.: From $CF_2=CClCF_2AsMe_2$ by isomerization in contact with CsF in a
sealed tube at 30°, 91% yield (627).
Prop.: b. 100° (extrapolated), IR and NMR spectra (627).

(2-Chlorotetrafluoroallyl)dimethylarsine, $AsC_5H_6ClF_4$, $Me_2AsCF_2CCl=CF_2$
Synth.: From $Me_2AsH + CF_2ClCCl=CF_2$ in a sealed tube at r.t., 57% yield (627).
Prop.: b. 105° (extrapolated), IR and NMR spectra. In contact with CsF in a
sealed tube at 30° → $Me_2AsCF=CClCF_3$ (627).

Dimethyl(3,3,3-trifluoropropynyl)arsine, $AsC_5H_6F_3$, $Me_2AsC\equiv CCF_3$
Synth.: From $Me_2AsAsMe_2 + CF_3C\equiv CH$ at 20°, ~ 80% yield along with trans-
$CF_3CH=CHAsMe_2$ (425).
From $Me_2AsNMe_2 + CF_3C\equiv CH$ at 20°, ~ 100% yield (425).
From $(Me_2As)_2O + CF_3C\equiv CH$ at 20°, 44% yield (425).
From $Me_2AsI + CF_3C\equiv CMgI$ in Et_2O at 20° (425).
From $Me_2AsSEt + CF_3C\equiv CH$ at 125°, trace amt. (425).
By-prod. from the rxn. of $Me_2AsCl + CF_3C\equiv CH$ by heating at 155°, a mixt. of
the title compd. with trans-$CF_3CH=CHAsMe_2$ and $CF_3C(=CHCl)AsMe_2$ was obtd.
(425).
Prop.: b. 98°, NMR spectrum (425).
Rxn. with: Me_3SnCF_3 at 140° → Me_3SnF and $Me_2As\overline{C=C(CF_3)C}F_2$ (425).
Aq. 10% NaOH → $CF_3C\equiv CH$ (425)

[2-Chloro-1-(trifluoromethyl)vinyl]dimethylarsine, $AsC_5H_7ClF_3$,
    $Me_2AsC(CF_3)=CHCl$
Synth.: From $Me_2AsCl + CF_3C\equiv CH$ by heating at 155° or in UV light at 110°
(425).
Prop.: A liquid b. 142-144°, consisting of 2 isomers in a ratio of 1:9. The
isomers were sepd. by vapor phase chromatography and were identified by IR
and NMR spectra. Both isomers are stable up to 140° in the absence of air.
The major prod. isomerizes above 155° (425).

1,1,2,3,3,3-Hexafluoropropyldimethylarsine, $AsC_5H_7F_6$, $Me_2AsCF_2CFHCF_3$,
    (M-35)
Synth.: From $Me_2AsH + CF_3CF=CF_2$ at 100°, 90% yield (421).
From $Me_2AsMgBr + CF_3CF=CF_2$ in Et_2O at 20°, a mixt. of the title compd. with
$Me_2AsCF=CFCF_3$ was obtd. (421).
Prop.: b_{739} 110-111°, stable at 180°, IR spectrum (421).
Rxn. with: Br in CCl_4 → CF_3CHFCF_2Br (421).

Dimethyl(trifluoropropenyl)arsine Isomers, $AsC_5H_8F_3$, cis-$Me_2AsCH=CHCF_3$,
    trans-$Me_2AsCH=CHCF_3$, and $Me_2AsC(CF_3)=CH_2$
Synth.: From Me_2AsH and $CF_3C\equiv CH$ at 20° (416).

By-prod. from the rxn. of Me_2AsCl with $CF_3C{\equiv}CH$ at 155° (425).
By-prod. from the rxn. of $Me_2AsAsMe_2$ with $CF_3C{\equiv}CH$ at 20° (425).
Prop.: Crude prod. b. 109-110° containing the cis, trans, and iso isomers
in a ratio of 3:3:2. Vapor phase chromatography gave a fraction contg. the
trans and iso isomers and another fraction being the cis isomer. IR and NMR
spectra were obtd. The cis isomer on heating to 175° and under UV light
isomerized to the trans form (416).

Diethylmethylarsine, AsC_5H_{13}, Et_2AsMe,   (M-36)
Prop.: b. 113.2° calcd. by the Kinney equation (66).
Rxn. with: HgO in $Me_2CO \rightarrow Et_2MeAsO$ (1057).
Aq. H_2O_2 in $Et_2O \rightarrow Et_2MeAsO$ (1057).
Coordination Deriv.: $[B_{12}H_8BrCl(AsEt_2Me)_2]$, useful as a halogen scavenger
(1067).

$K[B_{12}H_9I_2(AsEt_2Me)]$, cryst. solid, prepd. from $K[B_{12}H_{11}(AsEt_2Me)]$ by a treatment with iodine (1067).

(2-Chloro-3,3,4,4-tetrafluoro-1-cyclobutenyl)dimethylarsine, $AsC_6H_6ClF_4$,
     $Me_2As\overline{C=CClCF_2CF_2}$,   (M-37)
Synth.: From $Me_2AsAsMe_2$ and $\overline{CF_2CF_2CCl=CCl}$ at 20°, 85% yield, along with
Me_2AsCl (421).
From Me_2AsH and $\overline{CF_2CF_2CCl=CCl}$ at 100°, 97% yield (421).
Prop.: b. 154°, IR and NMR spectra (421).
Rxn. with: Me_2AsH at 140° $\rightarrow Me_2As\overline{C=C(AsMe_2)CF_2CF_2}$ (421).

[2-Chloro-3,3,3-trifluoro-1-(trifluoromethyl)propenyl]dimethylarsine,
     $AsC_6H_6ClF_6$, $Me_2AsC(CF_3)=CClCF_3$,   (M-64)
Synth.: From Me_2AsCl and $CF_3C{\equiv}CCF_3$ in UV light or by heating at 140° (415).
Prop.: b_{50} 79-80°, b_{83} 131-137°, IR and NMR spectra. Dec. at 220° (415).
Rxn. with: Br (1:1 mole) in $CCl_4 \rightarrow$ a cream cryst. solid which upon heating
at 100° $\rightarrow$ MeBr and $CF_3CCl=C(CF_3)AsBrMe$ (415).
Br (1:2 moles) in $CCl_4 \rightarrow$ MeBr and $CF_3CCl=C(CF_3)AsBr_2$ (415).

Dimethyl(pentafluorocyclobutenyl)arsine, $AsC_6H_6F_5$, $Me_2As\overline{C=CFCF_2CF_2}$,
     (M-37)
Synth.: From Me_2AsH and $\overline{CF_2CF_2CF=CF}$ at 20°, 94% yield (421).
From $Me_2AsMgBr$ and $\overline{CF_2CF_2CF=CF}$ at 20°, 50% yield (421).
Prop.: b. 127-129° (421).
Rxn. with: $R_2NH \rightarrow Me_2As\overline{C=C(NR_2)CF_2CF_2}$ (420).

Dimethyl(3,3,4,4-tetrafluoro-1-butenyl)arsine, $AsC_6H_7F_4$, $Me_2As\overline{C=CHCF_2CF_2}$
Synth.: From Me_2AsH and $\overline{CF_2CF_2CH=CCl}$ by heating at 150°, a mixt. of the
title compd. with Me_2AsCl was obtd. (420).
Prop.: b. 153°, IR and NMR spectra (420).

Dimethyl[3,3,3-trifluoro-1-(trifluoromethyl)propenyl]arsine, $AsC_6H_7F_6$,
     $Me_2AsC(CF_3)=CHCF_3$

Synth.: From Me_2AsH and $CF_3C\equiv CCF_3$ by condensing the reagents at a low temp. and warming up to 20°, a mixt. of the cis and trans isomers (5:95) was obtd. (416).
Prop.: b. 109-110°, IR and NMR spectra. The cis isomers on standing for 36 hrs. underwent a spontaneous rearrangement to the trans form (416).
Rxn. with: 10% NaOH → a rearrangement of the trans to cis isomer (416).
Me_3SnCF_3 at 140° → $Me_2As\overline{C(CF_3)CF_2}CHCF_3$ (425).

Diethylethynylarsine, AsC_6H_{11}, $Et_2AsC\equiv CH$
Synth.: From Et_2AsI in Et_2O and $CH\equiv CMgBr$ (1:1) in THF mixed at the ice temp. and completed on a steam bath, 33% yield (913).
Prop.: b. 133-133.5°, n_D^{20} 1.4940, d_4^{20} 1.1430, MR_D 40.20, AR_{As} 11.19, IR spectrum (913).

Diethylvinylarsine, AsC_6H_{13}, $Et_2AsCH=CH_2$
Synth.: From Et_2As-halides + $CH_2=CHMgBr$ or $CH_2=CHLi$ in THF (1594).
Prop.: b_{39} 58°, UV spectra in i-octane, MeOH, and in the vapor phase (1594).
Rxn. with: MnO_2 in C_6H_{14} or dioxan at 50° → $Et_2(CH_2=CH)AsO$ (1594).

Isobutyldimethylarsine, AsC_6H_{15}, Me_2AsBu-i
Coordination Deriv.: [i-$BuMe_2AsBH_3$], prepd. by mixing Me_2AsBu-i with B_2H_6 (2:1) at r.t. On heating with excess B_5H_9 the following compds. were formed: [(i-$BuMe_2As)_2BH_2]B_{12}H_{12}$], [(i-$BuMe_2As)_2BH_2$][i-$Bu(Me_2As)B_{12}H_{11}$], and [(i-$BuMe_2As)_2B_{12}H_{10}$] (1067).

Methylbis(3,3,3-trifluoropropynyl)arsine, $AsC_7H_3F_6$, $(CF_3C\equiv C)_2AsMe$
Synth.: From $MeAsCl_2$ + $CF_3C\equiv CMgI$ in Et_2O at 20° (425).
Prop.: b. 122-124°, dec. 140°, discolors in UV light (425).
Rxn. with: 10% aq. NaOH → $CF_3C\equiv CH$ (425).

(2-Chloro-3,3,4,4,5,5-hexafluoro-1-cyclopentenyl)dimethylarsine, $AsC_7H_6ClF_7$,
    $Me_2As\overline{C=CClCF_2CF_2CF_2}$,   (M-38)
Synth.: From $Me_2AsAsMe_2$ + $\overline{CF_2CF_2CF_2CCl=CCl}$ at 20°, 64% yield, along with Me_2AsCl (421).
From Me_2AsH + $\overline{CF_2CF_2CF_2CCl=CCl}$ at 100°, 55% yield (421).
Prop.: b_{103} 108°, IR and NMR spectra (421).

Dimethyl-4-pentenylarsine, AsC_7H_{15}, $Me_2As(CH_2)_3CH=CH_2$,   (M-38)
Synth.: From Me_2AsI + $CH_2=CH(CH_2)_3MgBr$ in Et_2O-C_6H_6 (164).
Prop.: Colorless liquid, sensitive to air, b_{747} 148° (164).
Coordination Deriv.: In the following complex compds. (Ars) stands for $Me_2As(CH_2)_3CH=CH_2$.

[Hg(Ars)Br_2]$_2$, white crystals, m. 95°, prepd. from $HgBr_2$ and Ars in EtOH (164).

[Hg(Ars)I_2]$_2$, pale green crystals, m. 93°, prepd. from HgI_2 and the arsine in EtOH (164).

[Pt(Ars)Cl$_2$], colorless, diamagnetic crystals, m. 196° (dec.), a nonelectrolyte in PhNO$_2$, prepd. from Na$_2$[PtCl$_4$] and Ars in aq. EtOH and from PtCl$_2$ and Ars in aq. EtOH and from PtCl$_2$ and Ars in CHCl$_3$. IR spectrum - a square planar structure with the As atom and the olefinic bond coordinated with Pt (164).

[Pt(Ars)Br$_2$], pale yellow, diamagnetic crystals, m. 197° (dec.), a nonelectrolyte in PhNO$_2$, prepd. from [Pt(Ars)Cl$_2$] + LiBr and from PtBr$_2$ and Ars in CHCl$_3$. IR spectrum - a square planar structure with the As atom and the olefinic bond coordinated with Pt (164).

[Pt(Ars)I$_2$], yellow, diamagnetic crystals, m. 137° (dec.), a nonelectrolyte in PhNO$_2$, prepd. from [Pt(Ars)Cl$_2$] and NaI in alcohol. IR spectrum - a square planar structure with the As atom and the olefinic bond coordinated with Pt (164).

[Pt(Ars)$_2$(NO$_2$)$_2$], white, diamagnetic crystals, m. 94°, a nonelectrolyte in PhNO$_2$, prepd. from [Pt(Ars)Cl$_2$] + NaNO$_2$ in EtOH, followed by evapn. IR spectrum - a square planar structure with noncoordinated olefinic bonds postulated (164).

[Pt(Ars)(p-H$_2$NC$_6$H$_4$Me)Cl$_2$], yellow needles, m. 98°, prepd. from [Pt(Ars)Cl$_2$] + p-MeC$_6$H$_4$NH$_2$ in CHCl$_3$ under reflux. IR spectrum, a square planar structure postulated (164).

[Pt(Ars)$_2$Cl$_2$], pale yellow crystals, m. 55°, prepd. from [Pt(Ars)Cl$_2$] + Ars in C$_6$H$_6$ under reflux. IR spectrum, a square planar structure postulated (164).

[Pt(Ars)(Ph$_2$AsMe)Cl$_2$], colorless crystals, m. 162°, prepd. from [Pt(Ars)Cl$_2$] + Ph$_2$AsMe in CHCl$_3$. IR spectrum (164).

(2-Bromophenyl)dimethylarsine, AsC$_8$H$_{10}$Br, 2-BrC$_6$H$_4$AsMe$_2$, (M-39)
Synth.: From o-LiC$_6$H$_4$AsMe$_2$ + o-C$_6$H$_4$Br$_2$ in THF under reflux (1016).
Prop.: b$_{11}$ 124-125° (1016), b$_{0.8}$ 85-86°, sepn. from the m- and p-isomers by gas liquid chromatography (1303).
Rxn. with: n-BuLi in Et$_2$O → 2-LiC$_6$H$_4$AsMe$_2$ (1016).
n-BuLi in Et$_2$O, followed by MeAsCl$_2$ → [2-(Me$_2$As)C$_6$H$_4$]$_2$AsMe (428).

Dimethylphenylarsine, AsC$_8$H$_{11}$, Me$_2$AsPh, (M-39)
Synth.: From Me$_2$AsNa or Me$_2$AsK + PhBr in THF, 85 and 16% yield, resp. (1229).
Prop.: b$_{18}$ 88-90° (1229), b. 199.6° calcd. by the Kinney equation (66). IR spectrum (418, 649), Raman spectrum (649), NMR spectrum (882).
Rxn. with: R$_2$NCl → [Me$_2$PhAsNR$_2$]Cl (1420).
Ph$_2$P(O)N$_3$ in Py under reflux → Ph$_2$P(O)N=AsMe$_2$Ph (1583).
Coordination Deriv.: (In the following entries (Ars) stands for Me$_2$AsPh.)

[Ir(Ars)$_3$Br$_3$], orange prisms, m. 206-208°, prepd. from Na$_3$IrBr$_6$ in H$_2$O-AcEt. Far-IR spectrum, a trans-configuration postulated (799).

[Ir(Ars)$_3$Cl$_3$], far-IR spectrum, a trans configuration postulated (799).

(In the following entries (Ars) stands for Me₂AsPh.)

[Ir(Ars)(PEt₃)₂Cl₃], yellow prisms, m. 104-111°, prepd. from mer-[Ir(PEt₃)₂-Cl₃] in C₆H₆ + Ars under reflux, followed, after cooling, by pptn. with light petr. IR spectrum. An octahedral structure with the arsine ligand trans to a Cl atom and both phosphine groups occupying the other two equatorial positions postulated (1250).

[IrH(Ars)₂(CO)Cl₂], white prisms, m. 132-136°, prepd. from NaIrCl₄ by treating with CO in MeOCH₂CH₂OH under reflux, followed by Ars at 60° and aq. HCl, 68% yield. IR and NMR spectra. An octahedral structure postulated (460).

[Ir(Ars)₂(CO)Cl], yellow needles, m. 111-115° (dec.), prepd. from [IrH(Ars)₂-(CO)Cl₂] + NaOMe in MeOH. IR spectrum; a trans configuration postulated (460).

[Os(Ars)₂Cl₄], NMR spectrum and magnetic data (329).

cis-[Pd(Ars)₂Cl₂], yellow crystals, prepd. from [Pd₂(trans-o-H₂NC₆H₄SCH₂CH=CH-CH₂SC₆H₄NH₂-o)Cl₄] + Ars in EtOH by heating under reflux. IR spectrum (628).

[π-CH₂=C(Me)CH₂Pd(Ars)₂]BPh₄, white needles, m. 92-96°, prepd. by adding Ars to a suspension of [π-CH₂=C(Me)CH₂PdCl]₂ in H₂O, followed by NaBPh₄. NMR spectrum in CHCl₃ at 34°, a symmetrical bonding of the π-methallyl ligand postulated (1251).

[Pd₂(Ars)(trans-o-H₂NC₆H₄SCH₂CH=CHCH₂SC₆H₄NH₂-o)Cl₄], yellow solid, m. 128-137° (dec.), a nonelectrolyte, prepd. from [Pd₂(trans-o-H₂NC₆H₄SCH₂CH=CHCH₂S-C₆H₄NH₂-o)Cl₄] + Ars in EtOH at 20°. IR spectrum (628).

cis-[Pt(Ars)₂Cl₂], colorless crystals, m. 160° (164), prepd. from [Pt₂(trans-o-H₂NC₆H₄SCH₂CH=CHCH₂SC₆H₄NH₂-o)Cl₄] + Ars in EtOH by heating under reflux (628) and from [Pt(Me₂AsCH₂CH₂CH=CH₂)Cl₂] + Ars in CHCl₃ (164). IR spectrum (164, 628).

[Pt(Ars)₂Br₄], UV spectrum (157).

[Pt₂(Ars)(trans-o-H₂NC₆H₄SCH₂CH=CHCH₂SC₆H₄NH₂-o)Cl₄], yellow substance, m. 220-231° (dec.), a nonelectrolyte, prepd. from [Pt₂(trans-o-H₂NC₆H₄SCH₂CH=CH-CH₂SC₆H₄NH₂-o)Cl₄] + Ars in EtOH at 20°. IR spectrum (628).

[π-CH₂=C(Me)CH₂Pt(Ars)₂]BPh₄, white prisms, m. 94-95°, prepd. by adding cis-[Pt(Ars)₂Cl₂] to CH₂=C(Me)CH₂MgCl in Et₂O, refluxing the mixt., cooling to -40°, hydrolyzing with H₂O, and treating with Na[BPh₄]. NMR spectrum (1251).

cis-[Me₄Pt(Ars)₂], prepd. from cis-[Pt(Ars)₂Cl₂] by treating with MeLi to form cis-[Me₂Pt(Ars)₂], converting the latter with MeBr to [Me₃Pt(Ars)₂Br], and treating the bromo compd. with MeLi. The dipole moment of 5.4 D and NMR spectrum indicate a cis configuration (1304).

[Rh(Ars)₃Br₃], red needles, m. 192-200° (dec.), prepd. from mer- or fac-[Rh-(Ars)₃Cl₃] + LiBr in EtOH and Me₂CO, resp., under reflux. Far-IR and NMR spectra, a mer-configuration postulated (241).

mer-[Rh(Ars)₃Cl₃], far-IR and NMR spectra (241).

mer-[Rh(Ars)$_3$I$_3$], red-brown prisms, m. 192-201° (dec.), prepd. from mer- or
fac-[Rh(Ars)$_3$Cl$_3$] + NaI by a prolonged heating in EtOH and Me$_2$CO, resp. Far-
IR and NMR spectra (241).

[Rh(Ars)$_3$BrCl$_2$], orange prisms, m. 191-195°, prepd. from mer-[Rh(Ars)$_3$Cl$_3$] +
LiBr in Me$_2$CO under reflux. Far-IR and NMR spectra (241).

[Rh(Ars)$_3$ICl$_2$], blood-red prisms, m. 167-173° (dec.), prepd. from mer-[Rh-
(Ars)$_3$Cl$_3$] + NaI in EtOH by a short heating. Far-IR and NMR spectra (241).

[(π-CH$_2$=CHCH$_2$)$_2$Rh(Ars)Cl], pale yellow needles, m. 100-107°, prepd. from
[(π-CH$_2$=CHCH$_2$)$_2$RhCl]$_2$ + Ars in CH$_2$Cl$_2$. NMR spectrum in CDCl$_3$ indicated that
one of the allyl groups is unsymmetrically bonded, the other allyl group may
be trans to the Ars ligand and C and is rotating and thus causing the 2 syn-
and 2-anti-protons to interchange (1249).

[π-CH$_2$=CHCH$_2$Rh(Ars)Cl$_2$]$_2$, orange-yellow microprisms, m. 153-160°, prepd. from
[(π-CH$_2$=CHCH$_2$)$_2$Rh(Ars)Cl] soln. in allyl chloride and MeOH (2:5) under reflux.
A dimeric structure with 2 Cl bridges postulated (1249).

[RuH(Ars)$_3$(CO)Cl], cream-colored prisms, m. 91-92°, prepd. from [Ru(Ars)$_3$(CO)-
Cl$_2$] in EtOH by adding aq. KOH and heating under reflux. IR spectrum; an
octahedral configuration with the Cl and CO ligands in the apical positions
suggested. On heating the complex with concd. HCl in MeOH → [Ru(Ars)$_3$(CO)Cl$_2$]
(995).

[Ru(Ars)$_3$(CO)Cl$_2$], yellow prisms, m. 181-189°, prepd. from RuCl$_3$·3H$_2$O in EtOH
+ Ars under reflux and then passing CO through the boiling soln. IR spectrum;
an octahedral configuration with one Cl and CO in the apical positions postulated (995).

[Ru(Ars)$_2$(NO)Cl$_3$], brown prisms, m. 145-146.5°, prepd. from [RuCl$_3$(NO)·5H$_2$O]
by boiling with Ars in MeOCH$_2$CH$_2$OH. IR spectrum (331).

Dimethyl-o-tolylarsine,   AsC$_9$H$_{13}$,   Me$_2$AsC$_6$H$_4$Me-2,   (M-43)
Prop.: b$_{0.6}$ 54-56° (1163).
Coordination Deriv.: [Ir(Me$_2$AsC$_6$H$_4$Me-2)$_3$Cl$_3$], yellow crystals, a nonconductor
in MeNO$_2$, monomeric in ClCH$_2$CH$_2$Cl, prepd. from K$_3$[IrCl$_6$] + the arsine in dild.
HCl. UV spectrum; crystallographic data (201).

(m-Methoxyphenyl)dimethylarsine,   AsC$_9$H$_{13}$O,   3-MeOC$_6$H$_4$AsMe$_2$,   (M-66)
Synth.: From Me$_2$AsI by treating with Na in THF, followed by 3-MeOC$_6$H$_4$Br at
r.t. (1414).
Prop.: b$_1$ 75-77°, UV spectrum (1414).
Coordination Deriv.: [Hg(3-MeOC$_6$H$_4$AsMe$_2$)]Cl$_2$, m. 172°, prepd. from HgCl$_2$ +
the arsine in abs. EtOH (1414).

(o-Methoxyphenyl)dimethylarsine,   AsC$_9$H$_{13}$O,   2-MeOC$_6$H$_4$AsMe$_2$
Synth.: From Me$_2$AsNa in THF + 2-MeOC$_6$H$_4$Br at r.t. (1415).
From 2-MeOC$_6$H$_4$Br in Et$_2$O by treating with Li, followed by Me$_2$AsI (1414, 1415).

Prop.: Colorless liquid, b_1 70-72°, with a nauseating odor, unstable in air. UV spectrum (1414, 1415).
Coordination Deriv.: (In the following entries (Ars) stands for 2-MeOC$_6$H$_4$-AsMe$_2$.)

[Cd(Ars)]Br$_2$, prepd. from CdBr$_2$·2.5H$_2$O + Ars in abs. EtOH (1414).

[Cd(Ars)]Cl$_2$, prepd. from CdCl$_2$·2.5H$_2$O + Ars in abs. EtOH. At 275-280° the complex was transformed to a violet substance (1414).

[Cd(Ars)]I$_2$, m. 154-155°, prepd. in the same manner as the preceding complex (1414).

[Cd(Ars)$_2$]I$_2$, m. 100-102°, prepd. from CdI$_2$ and the arsine in H$_2$O (1414).

[Hg(Ars)Br$_2$], m. 174°, prepd. from HgBr$_2$ + Ars (1:1) in Et$_2$O (1415).

[Hg(Ars)Cl$_2$], white microcrystals, m. 178-179°, prepd. from HgCl$_2$ + Ars (1:1) in EtOH (1415).

[Hg(Ars)I$_2$], m. 157°, prepd. from HgI$_2$ + Ars in EtOH (1415).

[Rh(Ars)$_2$Cl$_3$], dark red, diamagnetic crystals, m. 220° (dec.), a nonelectrolyte in MeNO$_2$, prepd. from aq. RhCl$_3$ soln. and alcoholic Ars soln. under reflux. Sol. in org. solvents, insol. in H$_2$O. Electronic absorption spectrum (1193). X-ray analysis; a distorted octahedral structure postulated in which both As atoms are coordinated in the cis positions, while one O atom and three Cl occupy the other positions (647, 1193).

[Rh(Ars)$_2$Cl$_2$]NO$_3$, orange yellow crystals, m. 158°, prepd. from [Rh(Ars)$_2$Cl$_3$] by treating with AgNO$_3$ in MeOH. Mol. cond. data in Me$_2$CO, MeNO$_2$, and MeOH indicated an ionic structure. The nitrate did not react with Na[BPh$_4$] (1193).

(p-Methoxyphenyl)dimethylarsine,   AsC$_9$H$_{13}$O,   4-MeOC$_6$H$_4$AsMe$_2$
Synth.: From Me$_2$AsI + Na in THF by treating with 4-MeOC$_6$H$_4$Br at r.t., 65-70% yield (1414).
From 4-MeOC$_6$H$_4$Br in Et$_2$O by treating with Li, followed by Me$_2$AsI at r.t. and completing under reflux, 70% yield (1414).
Prop.: b_1 73-75°, UV spectrum (1414).
Coordination Deriv.: [Cd(4-MeOC$_6$H$_4$AsMe$_2$)]I$_2$, m. 147°, prepd. from CdI$_2$ and the arsine in abs. EtOH (1414).

[Hg(4-MeOC$_6$H$_4$AsMe$_2$)]Cl$_2$, m. 189°, prepd. from HgCl$_2$ and the arsine (1:1) in abs. EtOH (1414).

Dimethyl(2-methylthiophenyl)arsine,   AsC$_9$H$_{13}$S,   2-MeSC$_6$H$_4$AsMe$_2$,   (M-44)
Coordination Deriv.: (In the following entries (Ars) stands for 2-MeSC$_6$H$_4$-AsMe$_2$.)

[Ni(Ars)$_2$](ClO$_4$)$_2$, brown crystals, prepd. from Ni(ClO$_4$)$_2$·6H$_2$O in hot EtOH + Ars in hot EtOH (970).

[Pd(Ars)Cl$_2$] on heating in THF under reflux undergoes a demethylation and formation of the S-Pd bonds giving [(PdSC$_6$H$_4$AsMe$_2$)Cl]$_2$, orange flakes, a nonelectrolyte in MeNO$_2$ (971, see also 969).

[Pd(Ars)Br$_2$], attempted demethylation by heating in Me$_2$CO (971).

[Pd(Ars)I$_2$], red crystals, prepd. from [Pd(SC$_6$H$_4$AsMe$_2$)$_2$] in CHCl$_3$ by a treatment with MeI (971).

[Pt(Ars)I$_2$], yellow crystals, prepd. from [Pt(SC$_6$H$_4$AsMe$_2$)$_2$] in CHCl$_3$ by a treatment with MeI (971).

Phenylarsylenediacetic Acid,   AsC$_{10}$H$_{11}$O$_4$,   PhAs(CH$_2$CO$_2$H)$_2$,   (M-47)
Prop.: Dissocn. const. in 20% aq. dioxan at 20° was detd. (1222). IR spectrum, qual. interpretation (1223). The acid formed very weak complexes with transition metals; a strong 1:1 complex and a protonated 1:1 complex with Ag$^+$ were obtd. (1222).

Dimethyl(m-vinylphenyl)arsine,   AsC$_{10}$H$_{11}$,   3-CH$_2$=CHC$_6$H$_4$AsMe$_2$
Synth.: From 3-CH$_2$=CHC$_6$H$_4$MgBr in Et$_2$O + Me$_2$AsI in C$_6$H$_6$ (157).
Prop.: White oil, b$_{0.3}$ 64°, IR spectrum (157).
Rxn. with: Br in CHCl$_3$ → an intractable oil (157).
Coordination Deriv.: [Pt(3-CH$_2$=CHC$_6$H$_4$AsMe$_2$)$_2$Br$_2$], yellow crystals, m. 164°, prepd. from K$_2$[PtBr$_4$] + the arsine (1:2) in H$_2$O, followed by a recrystn. from EtCOMe. IR spectrum, a trans configuration postulated (157).

[Pt(3-CH$_2$=CHC$_6$H$_4$AsMe$_2$)$_2$Br$_4$], orange crystals, m. 188°, prepd. from the preceding dibromo complex by a rxn. with Br in CCl$_4$. IR spectrum (157).

[Rh(3-CH$_2$=CHC$_6$H$_4$AsMe$_2$)$_3$Br$_3$], orange crystals, m. 100-110° (dec.), prepd. from RhBr$_3$·3H$_2$O + LiBr + the arsine (1:20:4) in EtOH, evapn., and crystn. from C$_6$H$_6$-petroleum mixt. IR spectrum. Rxn. with Br leads to the bromination of the olefinic bond (157).

Dimethyl(o-vinylphenyl)arsine,   AsC$_{10}$H$_{13}$,   2-CH$_2$=CHC$_6$H$_4$AsMe$_2$
Synth.: From 2-CH$_2$=CHC$_6$H$_4$MgBr in THF + Me$_2$AsI in C$_6$H$_6$, ~60% yield (157).
Prop.: White oil, b$_{0.5}$ 70°, IR spectrum (157). NMR spectrum (161).
Rxn. with: Br in CHCl$_3$ → an intractable oil (157).
Coordination Deriv.: (In the following entries (o-SA) stands for 2-CH$_2$=CHC$_6$H$_4$AsMe$_2$.)

[Pt(o-SA)$_2$Br$_2$], the complex occurs in cis- and trans-configurations. The cis-isomer, yellow crystals, m. 190-192°, while the trans-isomer, m. 177-181°. Both isomers were prepd. from K$_2$[PtBr$_4$] + o-SA (1:2) in H$_2$O as a pale yellow solid. Extn. of the prod. with EtOH-Et$_2$O mixt. gave a soln. contg. the trans-isomer and a solid residue - the cis-isomer. The trans-isomer was recovered by evapn. of the solvents. IR and UV spectra of both isomers were obtd. (157).

[Pt(o-SA)Br$_2$], white crystals, m. 235-237°, prepd. from PtBr$_2$ + o-SA (2:1) in CHCl$_3$ and recrystd. from CH$_2$Cl$_2$. IR spectrum (157). NMR spectrum (161). A 4-coordinate Pt complex with o-SA coordinated through the As atom and the vinyl group (157).

(In the following entries (o-SA) stands for 2-CH$_2$=CHC$_6$H$_4$AsMe$_2$.)

[Pt(o-SA)(o-SABr)Br$_3$], yellow crystals, m. 132°, prepd. from [Pt(o-SA)$_2$Br$_2$] by treating with Br$_2$ in CCl$_4$. IR spectrum - a 6-coordinate Pt complex is considered in which the o-SA ligand is chelated through the As atom only, while o-SABr contains the Br atom bonded to one of the "olefinic" C atoms, the other C atom forms a σ-bond with Pt and the As atom is chelated (157).

[Pt(o-SABr)(o-SABr$_2$)Br$_3$], yellow crystals, m. 146-149°, prepd. from [Pt(o-SA)$_2$-Br$_2$] + 2Br$_2$ in C$_6$H$_6$-petr. and from [Pt(o-SA)(o-SABr)Br$_3$] + Br$_2$ in C$_6$H$_6$-petr. IR and UV spectra. A structure contg. Pt(IV) chelated to the As atoms of both arsine ligands and σ-bonded to the β-C atom of the (o-SABr) ligand (157).

[PtCH$_2$CH(OMe)C$_6$H$_4$AsMe$_2$(o-SA)Br$_3$], yellow, m. 176-178°, prepd. from [PtCH$_2$CHBr-C$_6$H$_4$AsMe$_2$)(o-SA)Br$_3$ + MeOH. IR spectrum (157).

[PtCH$_2$CH(OEt)C$_6$H$_4$AsMe$_2$-o(o-SA)Br$_3$]*, yellow, m. 175-176°, prepd. from [Pt-(o-SA)(o-SABr)Br$_3$] + EtOH. UV spectrum (157).

[PtCH$_2$CH(OPr)C$_6$H$_4$AsMe-o(o-SA)Br$_3$]*, yellow, m. 215°, prepd. from [Pt(o-SA)-(o-SABr)Br$_3$] + PrOH (157).

[PtCH$_2$CH(OBu)C$_6$H$_4$AsMe$_2$-o(o-SA)Br$_3$]*, yellow, m. 169-171°, prepd. from [Pt-(o-SA)(o-SABr)Br$_3$] + BuOH (157).

[Pt(o-SA)Br$_4$], deep orange, m. 160° (dec.), prepd. from [Pt(o-SA)Br$_2$] + Br in C$_6$H$_6$. IR spectrum had no bands characteristic of the C=C stretching, but it showed some bands due to a C-Br or Pt-C vibration. A dimeric, Br-bridged, octahedral structure with a Pt-C bond suggested. The position of the Br atom in the alkyl group has not been ascertained, but more probably the Br may be in the α-position. A rxn. with methanol produced HBr and the [Pt(MeOCH=CH-C$_6$H$_4$AsMe$_2$)Br$_2$] complex (161).

[Rh(o-SA)$_3$Br$_3$], orange crystals, m. 135-140° (dec.), prepd. from RhCl$_3$·3H$_2$O + LiBr + o-SA (1:20:4) in EtOH by heating, evapn., and recrystn. from C$_6$H$_6$-petr. IR spectrum. Rxn. with Br → bromination of the olefinic bond (157).

Dimethyl(p-vinylphenyl)arsine,   AsC$_{10}$H$_{13}$,   4-CH$_2$=CHC$_6$H$_4$AsMe$_2$
Synth.: From 4-CH$_2$=CHC$_6$H$_4$MgBr in Et$_2$O + Me$_2$AsI in C$_6$H$_6$, ~ 60% yield (157).
Prop.: White oil, b$_1$ 88-90°, IR spectrum (157).
Rxn. with: Br in CHCl$_3$ → an intractable oil (157).
Coordination Deriv.: (In the following entries (p-SA) stands for 4-CH$_2$=CH-C$_6$H$_4$AsMe$_2$.)

[Pt(p-SA)$_2$Br$_2$], yellow crystals, m. 153-156°, prepd. from K$_2$[PtBr$_4$] + p-SA (1:2) in H$_2$O and recrystd. from a mixt. of C$_6$H$_6$ with petr. ether. IR spectrum - a trans configuration postulated. No other isomers could be obtd. from the rxn. mixt. (157).

* The position of the alkoxy group has not been ascertained.

[Pt(p-SA)$_2$Br$_4$], orange crystals, m. 168-171°, prepd. from [Pt(p-SA)$_2$Br$_2$] + Br$_2$ in C$_6$H$_6$-petr. mixt. IR spectrum (157).

[Rh(p-SA)$_3$Br$_3$], orange crystals, m. 172° (dec.), prepd. from RhBr$_3$·3H$_2$O + LiBr + p-SA (1:20:4) by heating in EtOH, followed by evapn. and crystn. from C$_6$H$_6$-petr. mixt. IR spectrum. A rxn. with Br → bromination of the olefinic bond (157).

[m-(1,2-Dibromoethyl)phenyl]dimethylarsine,   AsC$_{10}$H$_{13}$Br$_2$,
    3-(CH$_2$BrCHBr)C$_6$H$_4$AsMe$_2$
The arsine was prepared and isolated only in complexes with platinum and rhodium (157).
Coordination Deriv.: [Pt(Me$_2$AsC$_6$H$_4$CHBrCH$_2$Br-m)$_2$Br$_4$], orange crystals, m. 180° (dec.), prepd. from [Pt(Me$_2$AsC$_6$H$_4$CH=CH$_2$-m)$_2$Br$_2$] + 3Br$_2$ in C$_6$H$_6$. IR spectrum (157).

[Rh(Me$_2$AsC$_6$H$_4$CHBrCH$_2$Br-m)$_3$Br$_3$], orange crystals, m. 105-110° (dec.), prepd. from [Rh(m-CH$_2$=CHC$_6$H$_4$AsMe$_2$)$_3$Br$_3$] + excess Br in CCl$_4$. IR spectrum (157).

[o-(1,2-Dibromoethyl)phenyl]dimethylarsine,   AsC$_{10}$H$_{13}$Br$_2$,
    2-(CH$_2$BrCHBr)C$_6$H$_4$AsMe$_2$
The arsine was prepared and isolated only in complexes with platinum and rhodium.
Coordination Deriv.: (In the following entries (Br$_2$EA) stands for 2-(CH$_2$Br-CHBr)C$_6$H$_4$AsMe$_2$.)

[Pt(Br$_2$EA)$_2$Br$_2$], yellow, m. 167-169°, prepd. from [Pt(Br$_2$EA)$_2$Br$_4$] by a redn. with SnCl$_2$ in Me$_2$CO. IR and UV spectra (157).

[Pt(Br$_2$EA)$_2$Br$_4$], orange crystals, m. 181-183°, prepd. from [Pt(o-CH$_2$=CHC$_6$H$_4$-AsMe$_2$)$_2$Br$_2$] + 3Br$_2$ in CH$_2$Cl$_2$-petr. ether. IR and UV spectra (157).

[o-Me$_2$AsC$_6$H$_4$CHBrCH$_2$Pt(Br$_2$EA)Br$_3$]*, yellow crystals, m. 146-149°, prepd. from [Pt(o-CH$_2$=CHC$_6$H$_4$AsMe$_2$)$_2$Br$_2$] + 2Br$_2$ in C$_6$H$_6$-petr. mixt. and from [PtCH$_2$CHBr-C$_6$H$_4$AsMe$_2$(o-CH$_2$=CHC$_6$H$_4$AsMe$_2$)Br$_3$] + Br$_2$ in C$_6$H$_6$-petr. mixt. IR and UV spectra. A structure with both As atoms chelated to the Pt atom which forms a σ-bond with the arsinophenylbromoethyl group was suggested.

[o-Me$_2$AsC$_6$H$_4$CH(OMe)CH$_2$Pt(Br$_2$EA)Br$_3$]**, yellow, m. 147-150°, prepd. from [o-Me$_2$AsC$_6$H$_4$CHBrCH$_2$Pt(Br$_2$EA)Br$_3$] + MeOH (157).

[Rh(Br$_2$EA)$_3$Br$_3$], orange crystals, m. 103-105° (dec.), prepd. from Rh(o-CH$_2$=CH-C$_6$H$_4$AsMe$_2$)$_3$Br$_3$] + 3Br$_2$ in CCl$_4$. IR spectrum (157).

---

* The position of the Br atom at the α-C and the σ-bond of the β-C with As have not been ascertained.
** Similarly as in the bromoethyl deriv., the position of the MeO group at the α-C and the σ-bond of the β-C with As have not been ascertained.

[p-(1,2-Dibromoethyl)phenyl]dimethylarsine, $AsC_{10}H_{13}Br_2$,
     $4-(CH_2BrCHBr)C_6H_4AsMe_2$

The arsine was prepared and isolated only in complexes with platinum and rhodium.

<u>Coordination Deriv.</u>: $[Pt(4-Me_2AsC_6H_4CHBrCH_2Br)(p-Me_2AsC_6H_4CH=CH_2)Br_4]$, orange, m. 194-195°, prepd. from $[Pt(p-Me_2AsC_6H_4CH=CH_2)_2Br_2]$ + $2Br_2$ in Me_2CO. IR spectrum (157).

$[Pt(p-Me_2AsC_6H_4CHBrCH_2Br)_2Br_4]$, orange, m. 209-211°, prepd. from $[Pt(p-Me_2As-C_6H_4CH=CH_2)_2Br_2]$ + $3Br_2$ in C_6H_6. IR spectrum (157).

$[Rh(p-Me_2AsC_6H_4CHBrCH_2Br)_3Br_3]$, orange crystals, m. 110-125° (dec.), prepd. from $[Rh(p-Me_2AsC_6H_4CH=CH_2)_3Br_3]$ + excess Br in CCl_4. IR spectrum (157).

Diethylphenylarsine, $AsC_{10}H_{15}$, Et_2AsPh,   (M-47)
<u>Synth.</u>: From Et_2AsCl + $PhMgBr$, 88% yield (340).
From $PhAsLi_2$ in THF by treating with $EtBr$, 14% yield (1514).
<u>Prop.</u>: b_{25} 118°, d_4^{20} 1.1701, n_D^{20} 1.5563 (340), b.238.7°, calcd. by the Kinney equation (66).
<u>Rxn. with</u>: H_2O_2 in Me_2CO → Et_2PhAsO (340).
S in C_6H_6 → Et_2PhAsS (603).
<u>Use</u>: Complexes with Co or Pt carbonyls catalyze the hydroformulation of olefins to aldehydes and alcohols (1424, 1425).
<u>Coordination Deriv.</u>: trans-$[Ir(AsEt_2Ph)_2(CO)Br]$, yellow prisms, m. 105-114°, prepd. by bubbling CO through a boiling mixt. of $Na_3[IrBr_6]$ in $MeOCH_2CH_2OH$ in the presence of $RhBr_3$, adding the arsine and boiling again. IR spectrum and dipole moment (328).

$[Ir(AsEt_2Ph)(PMe_2Ph)_2Cl_3]$, yellow prisms, m. 173-183°, prepd. from mer-$[Ir-(AsEt_2Ph)_3Cl_3]$ + Me_2PhP in C_6H_6 under reflux, followed by cooling and pptn. with light petr. IR spectrum. An octahedral structure with the arsine ligand in an equatorial position trans to the Cl atom and both phosphine ligand in the other equatorial positions postulated (1250).

$[IrH_3(AsEt_2Ph)_3]$, prepd. from $[HIr(AsEt_2Ph)_3Cl_2]$ by a redn. with $LiBH_4$ in THF, followed by hydrolysis with MeOH. The complex catalyzes the hydrogenation of olefins and ketones and promotes hydroformylation (777).

trans-$[AcIr(AsEt_2Ph)_2(CO)Br_2]$, pale yellow, m. 163-166°. IR spectrum (328).

trans-$[Rh(AsEt_2Ph)_2(CO)Cl]$, yellow needles, m. 94-96°, prepd. from $RhCl_3 \cdot 3H_2O$ in boiling EtOH by bubbling CO and then treating with the arsine. IR spectrum (330).

trans-$[Rh(AsEt_2Ph)_2(CO)Cl_3]$, orange yellow needles, m. 145-153°, prepd. by heating the arsine in $MeOCH_2CH_2OH$ with $RhCl_3 \cdot 3H_2O$ and bubbling CO through the soln. under reflux. IR spectrum and dipole moment (330).

$[Rh(AsEt_2Ph)_3Cl_3]$, the complex in the presence of Ph_3P catalyzes the hydrogenation of the unsatd. butadiene/piperylene/SO_2 terpolymers (1395).

[2-(Dimethylarsino)-3,3,4,4-tetrafluorocyclo-1-butenyl]diethylamine,
AsC$_{10}$H$_{16}$F$_4$N, $\overline{CF_2CF_2C(NEt_2)=C}$AsMe$_2$
Synth.: From $\overline{CF_2CF_2CF}$=CAsMe$_2$ + Et$_2$NH at 20°, 59% yield (420).
Prop.: Colorless liquid, b$_{0.01}$ 68°, IR and NMR spectra (420).

4-(Dimethylarsino)-N,N-dimethylaniline, AsC$_{10}$H$_{16}$N, 2-(Me$_2$N)C$_6$H$_4$AsMe$_2$,
(M-49)
Coordination Deriv.: [Ir(o-Me$_2$AsC$_6$H$_4$NMe$_2$)Cl$_3$], orthorhombic crystals, m.
200-204° (dec.), prepd. from K[IrCl$_4$] + the arsine in aq. EtOH under reflux.
Crystallographic data (1192).

[Rh(o-Me$_2$AsC$_6$H$_4$NMe$_2$)$_2$Cl$_3$], crystallographic data; a 6-coordinate Rh complex
with one arsine ligand chelated through As and N and the other through the
As atom only (202).

8-(Dimethylarsino)quinoline, AsC$_{11}$H$_{12}$N,
Synth.: From Me$_2$AsNa + 8-chloroquinoline in THF (121a).
Prop.: Colorless liquid, b$_{0.8}$ 118°, n$_D^{20}$ 1.6585 (121a).
Rxn. with: HClO$_4$ → pale yellow, H$_2$O-sol. cryst. per-
chlorate (121a).

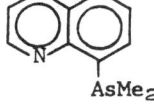

Coordination Deriv.: (In the following entries (AQ) stands for 8-(dimethylarsino)quinoline.)

[Cd(AQ)$_2$](ClO$_4$)$_2$, colorless, diamagnetic crystals, a 1:2 electrolyte in
MeNO$_2$, prepd. from Cd(ClO$_4$)$_2$ + AQ in EtOH (121a).

[Co(AQ)$_2$](ClO$_4$)$_2$·2H$_2$O, orange, paramagnetic crystals, a 1:2 electrolyte in
MeNO$_2$, prepd. from Co(ClO$_4$)$_2$ + AQ in aq. EtOH (121a).

[Cu(AQ)$_2$I], orange, diamagnetic crystals, slightly conductive in MeNO$_2$, prepd.
from Cu$_2$I$_2$ + AQ in EtOH (121a).

[Cu(AQ)$_2$]ClO$_4$, yellow, diamagnetic crystals, a 1:1 electrolyte in MeNO$_2$,
prepd. from Cu(ClO$_4$)$_2$ + AQ in EtOH (121a).

[Au(AQ)I], colorless, diamagnetic crystals, a nonelectrolyte in MeNO$_2$, prepd.
from [Au(AQ)$_2$]ClO$_4$ + I⁻ in aq. alc. (121a).

[Au(AQ)$_2$]ClO$_4$, colorless, diamagnetic crystals, a 1:1 electrolyte in MeNO$_2$,
prepd. from Au(ClO$_4$)$_3$ + AQ in EtOH (121a).

[Fe(AQ)$_3$](ClO$_4$)$_2$·H$_2$O, red, diamagnetic crystals, a 1:2 electrolyte in MeNO$_2$,
prepd. from [Fe(AQ)$_3$][FeCl$_4$]$_2$ + [Fe(H$_2$O)$_6$](ClO$_4$)$_2$ in aq. EtOH and from
[Fe(H$_2$O)$_6$](ClO$_4$)$_2$ + AQ (121a).

[Fe(AQ)$_3$](ClO$_4$)$_3$, dark green, paramagnetic crystals, a 1:3 electrolyte in
MeNO$_2$, prepd. from [Fe(AQ)$_3$](ClO$_4$)$_2$ by dissolving in 5 N HNO$_3$ and pptg. with
HClO$_4$ (121a).

[Fe(AQ)$_3$][FeCl$_4$]$_2$, red, diamagnetic crystals, a 1:2 electrolyte in MeNO$_2$,
prepd. from FeCl$_2$ + AQ in anhyd. EtOH (121a).

(In the following entries (AQ) stands for 8-(dimethylarsino)quinoline.)

[Ni(AQ)$_2$Cl$_2$], green, paramagnetic crystals, a nonelectrolyte in MeNO$_2$, prepd. from NiCl$_2$ + AQ in EtOH (121a).

[Ni(AQ)$_2$I$_2$], yellow-green, paramagnetic crystals, a nonelectrolyte in MeNO$_2$, prepd. from NiI$_2$ + AQ in EtOH (121a).

[Ni(AQ)$_3$](ClO$_4$)$_2$·H$_2$O, yellow-orange, paramagnetic crystals, a 1:2 electrolyte in MeNO$_2$, prepd. from Ni(ClO$_4$)$_2$ + AQ in aq. EtOH (121a).

[Pd(AQ)I$_2$], dark brown, diamagnetic crystals, a nonelectrolyte in MeNO$_2$, prepd. from PdI$_2$ + AQ in EtOH (121a).

[Pd(AQ)$_2$](ClO$_4$)$_2$·H$_2$O, yellow, diamagnetic crystals, a 1:2 electrolyte in MeNO$_2$, prepd. from Pd(ClO$_4$)$_2$ + AQ in aq. EtOH (121a).

[Pt(AQ)Cl$_2$], pale-yellow, diamagnetic crystals, a nonelectrolyte in PhNO$_2$, prepd. from PtCl$_2$ + AQ in EtOH (121a).

[Pt(AQ)$_2$][Pt$_2$I$_6$], dark brown, diamagnetic crystals, a 1:2 electrolyte in MeNO$_2$, prepd. from PtI$_2$ + AQ in EtOH (121a).

[Pt(AQ)$_2$Cl$_2$](ClO$_4$)$_2$, buff, diamagnetic crystals, a 1:2 electrolyte in MeNO$_2$, prepd. from [Pt(AQ)$_2$]Cl$_2$ in aq. Me$_2$CO by oxidn. with HNO$_3$, followed by HClO$_4$ (121a).

[Ag(AQ)$_2$]ClO$_4$, colorless, diamagnetic crystals, a 1:1 electrolyte in MeNO$_2$, prepd. from AgClO$_4$ + AQ in EtOH (121a).

[Zn(AQ)$_2$](ClO$_4$)$_2$, colorless, diamagnetic crystals, a 1:2 electrolyte in MeNO$_2$, prepd. from Zn(ClO$_4$)$_2$ + AQ in EtOH (121a).

2,2′-(Phenylarsylene)bis(ethylamine),   AsC$_{10}$H$_{17}$N$_2$,   PhAs(CH$_2$CH$_2$NH$_2$)$_2$
Prop.: IR spectrum, qual. interpretation (1223).

Dibutylethynylarsine,   AsC$_{10}$H$_{19}$,   Bu$_2$AsC≡CH
Synth.: From Bu$_2$AsI in Et$_2$O + CH≡CMgBr (1:1) in THF at ice temp., completed on a steam bath, 41% yield (913).
Prop.: b$_{11}$ 87-88°, n$_D^{20}$ 1.4865, d$_4^{20}$ 1.0470, MR$_D$ 58.82, AR$_{As}$ 11.34, IR spectrum (913).
Rxn. with: EtMgBr, followed by Me$_2$CO → Bu$_2$AsC≡CC(OH)Me$_2$ (913).
(CH$_2$O)$_x$·yH$_2$O + Et$_2$NH in dioxan in the presence of Cu(OAc)$_2$ → Bu$_2$AsC≡CCH$_2$NEt$_2$ (913).

(m-Allylphenyl)dimethylarsine,   AsC$_{11}$H$_{15}$,   3-(Me$_2$As)C$_6$H$_4$CH$_2$CH=CH$_2$
Synth.: From 3-CH$_2$=CHCH$_2$C$_6$H$_4$Br + Mg in Et$_2$O + Me$_2$AsI in C$_6$H$_6$, ~ 60% yield (157).
Prop.: White oil, b$_2$ 70-74°, IR spectrum (157).
Rxn. with: Br in CCl$_4$ → an intractable oil (157).
Coordination Deriv.: [Pt(3-Me$_2$AsC$_6$H$_4$CH$_2$CH=CH$_2$)$_2$Br$_2$], yellow flakes, m. 119-121°, prepd. from PtBr$_2$ + the arsine (1:2) in CHCl$_3$, followed by crystn. from

EtOH. IR spectrum (157).

[Pt(3-Me$_2$AsC$_6$H$_4$CH$_2$CH=CH$_2$)$_2$Br$_4$], orange crystals, m. 135-139°, prepd. from
the preceding dibromo deriv. by treating with Br in CCl$_4$. IR spectrum (157).

(o-Allylphenyl)dimethylarsine,   AsC$_{11}$H$_{15}$,   2-(Me$_2$As)C$_6$H$_4$CH$_2$CH=CH$_2$
Synth.: From 2-CH$_2$=CHCH$_2$C$_6$H$_4$Br + Mg in THF + Me$_2$AsI in C$_6$H$_6$, ~ 60% yield
(157).
Prop.: White oil, b$_1$ 72-77°, IR spectrum (157), NMR spectrum (162).
Rxn. with: Br in CCl$_4$ → an intractable oil (157).
Coordination Deriv.: (In the following complexes (o-AA) stands for o-(Me$_2$As)-
C$_6$H$_4$CH$_2$CH=CH$_2$.)
[Cu(o-AA)$_2$I], m. 98-99°, a nonelectrolyte in PhNO$_2$, monomeric in CHCl$_3$, prepd.
from Cu$_2$I$_2$ + o-AA. IR and NMR spectra; a 3-coordinate Cu complex (162).

[Pd(o-AA)Cl$_2$], yellow, diamagnetic crystals, m. 193-194°, a nonelectrolyte in
PhNO$_2$, monomeric in CHCl$_3$, prepd. from [Pd(PhCN)$_2$Cl$_2$] + o-AA in CHCl$_3$. IR
spectrum - a coordination of o-AA with the Pd atom through the As atom and
the olefinic bond postulated (163).

[Pd(o-AA)Br$_2$], orange, diamagnetic crystals, m. 195-196°, a nonelectrolyte in
PhNO$_2$, prepd. from [Pd(PhCN)$_2$Br$_2$] + o-AA in CHCl$_3$. IR spectrum; a complex
structure analogous to that of the dichloro complex postulated (163).

[Pd(o-AA)I$_2$]$_2$, dark red, diamagnetic crystals, m. 180-182°, a nonelectrolyte
in PhNO$_2$, prepd. from [Pd(o-AA)Cl$_2$] + LiI in Me$_2$CO. UV, visible, and IR
spectra - a dimeric, I-bridged structure with the arsine ligands coordinated
through the As atoms only postulated (163).

[Pd(o-AA)(NCS)$_2$], orange, diamagnetic crystals, m. 132-134°, a nonelectrolyte
in Me$_2$CO, prepd. from [Pd(o-AA)Cl$_2$] + KSCN in Me$_2$CO. IR spectrum indicated
the presence of a noncoordinated olefinic bond (163).

[Pt(o-AA)Cl$_2$], white, diamagnetic crystals, m. 243-244°, a nonelectrolyte in
PhNO$_2$, prepd. from PtCl$_2$ + o-AA in CHCl$_3$ at r.t., IR spectrum - a 4-coordi-
nate Pt complex in which o-AA is coordinated through the As atom and the
olefinic bond (163).

[Pt(o-AA)Br$_2$], cream-colored, diamagnetic crystals, m. 250-251°, a nonelectrolyte in PhNO$_2$, prepd. from PtBr$_2$ + o-AA in CHCl$_3$ and crystd. from EtOH
(157, 163). A recrystn. from C$_6$H$_6$ gave a white hemisolvate, dec. > 225°
(157). IR spectra of both the nonsolvated and solvated complexes were obtd.
A 4-coordinate Pt complex with o-AA chelated through the As atom and the olefinic bond postulated (157).

[Pt(o-AA)I$_2$], yellow, diamagnetic solid, m. 242-243°, a nonelectrolyte in
PhNO$_2$, prepd. from [Pt(o-AA)Cl$_2$] + excess NaI in Me$_2$CO. IR spectrum - a
structure analogous to that of the preceding dibromo complex postulated (163).

[Pt(o-AA)Br$_4$], deep orange solid, m. 174°, prepd. from [Pt(o-AA)Br$_2$] + Br in
C$_6$H$_6$. On standing in org. solvents the complex loses Br and reverts to

(In the following complexes (o-AA) stands for o-$(Me_2As)C_6H_4CH_2CH=CH_2$.)

[Pt(o-AA)Br$_2$]. IR spectrum of the Pt(IV) complex had no bands which could be attributed to the C=C stretching, but it showed bands attributable to the C-Br or Pt-C vibration. A dimeric, Br-bridged octahedral structure with a σ-Pt-C bond was suggested. The position of the Br atom in the aliphatic chain was not ascertained. Rxn. with MeOH → evolution of HBr and probably $[Pt(Me_2AsC_6H_4CH=CHCH_2OMe)Br_2]$ (161).

[Pt(o-AA)$_2$Br$_2$], yellow crystals, m. 146-149°, prepd. from PtBr$_2$ + o-AA (1:2) in CHCl$_3$ and recrystd. from C$_6$H$_6$. IR and UV spectra (157).

[Pt(o-AA)$_2$(NCS)$_2$], yellow, diamagnetic crystals, m. 123-125°, prepd. from $[Pt(o-AA)Cl_2]$ + NaNCS in Me$_2$CO. IR spectrum indicated the presence of free olefinic bonds and the N-bonded NCS groups (163).

[o-Me$_2$AsC$_6$H$_4$CH$_2$CHBrCH$_2$Pt(o-AA)Br$_3$], yellow crystals, m. 140°, prepd. from $[Pt(o-AA)_2Br_2]$ + Br$_2$ in CCl$_4$. IR and UV spectra.

(p-Allylphenyl)dimethylarsine, $AsC_{11}H_{15}$,   4-$(Me_2As)C_6H_4CH_2CH=CH_2$
<u>Synth.</u>: From p-CH$_2$=CHCH$_2$C$_6$H$_4$Br + Mg in Et$_2$O, followed by Me$_2$AsI in C$_6$H$_6$, ~ 60% yield (157).
<u>Prop.</u>: White oil, b$_2$ 76°, IR spectrum (157).
<u>Rxn. with</u>: Br in CCl$_4$ → an intractable oil (157).
<u>Coordination Deriv.</u>: {Pt[4-$(Me_2As)C_6H_4CH_2CH=CH_2$]$_2Br_2$}, yellow needles, m. 131-133°, prepd. from the arsine + PtBr$_2$ (2:1) in CHCl$_3$ and recrystd. from EtOH. IR spectrum (157).

{Pt[4-$(Me_2As)C_6H_4CH_2CH=CH_2$]$_2Br_4$}, orange crystals, m. 129-131°, prepd. from the preceding dibromo complex by treating with Br$_2$ (1:1) in CCl$_4$. IR and UV spectra (157).

[m-(2,3-Dibromopropyl)phenyl]dimethylarsine, $AsC_{11}H_{15}Br_2$,
     3-$(Me_2As)C_6H_4CH_2CHBrCH_2Br$
The arsine was prepared and isolated as a complex with PtBr$_4$ (157).
<u>Coordination Deriv.</u>: {Pt[3-$(Me_2As)C_6H_4CH_2CHBrCH_2Br$]$_2Br_4$}, orange crystals, m. 130-135°, prepd. from {Pt[3-$(Me_2As)C_6H_4CH_2CH=CH_2$]$_2Br_2$} by treating with Br$_2$ (1:3) in C$_6$H$_6$. IR spectrum (157).

[o-(2,3-Dibromopropyl)phenyl]dimethylarsine, $AsC_{11}H_{15}Br_2$,
     2-$(Me_2As)C_6H_4CH_2CHBrCH_2Br$ **(o-AABr$_2$)**
The arsine was prepared and isolated as complexes with Pt salts (157).
<u>Coordination Deriv.</u>: [Pt(o-AABr$_2$)$_2$Br$_2$], yellow solid, m. 171-174°, prepd. from [Pt(o-AABr$_2$)Br$_4$] + SnCl$_2$ in Me$_2$CO or EtOH. IR and UV spectra (157).

[Pt(o-AABr$_2$)$_2$Br$_4$], orange crystals, m. 107°, prepd. from {Pt[o-$(Me_2As)C_6H_4$-CH$_2$CH=CH]$_2$Br$_2$} + Br$_2$ (1:3) in C$_6$H$_4$. IR and UV spectra (157).

[p-(2,3-Dibromopropyl)phenyl]dimethylarsine, $AsC_{11}H_{15}Br_2$,
     4-$(CH_2BrCHBrCH_2)C_6H_4AsMe_2$

The arsine was prepd. and isolated as complexes with PtBr$_4$.
Coordination Deriv.: {Pt[4-(Me$_2$As)C$_6$H$_4$CH$_2$CHBrCH$_2$Br][4-(Me$_2$As)C$_6$H$_4$CH$_2$CH=CH$_2$]-Br$_4$}, orange crystals, m. 118-122°, prepd. from {Pt[4-(Me$_2$As)C$_6$H$_4$CH$_2$CH=CH$_2$]$_2$-Br$_2$} + Br$_2$ (1:2) in CCl$_4$.   IR spectrum (157).

{Pt[4-(Me$_2$As)C$_6$H$_4$CH$_2$CHBrCH$_2$Br]$_2$Br$_4$}, orange solid, m. 159-162°, prepd. from {Pt[4-(Me$_2$As)C$_6$H$_4$CH$_2$CH=CH$_2$]$_2$Br$_2$ + Br$_2$ (1:3) in C$_6$H$_6$-petr.   IR spectrum (157).

Tribromo{2-bromo-3-[2-(dimethylarsino)phenyl]propyl}platinum Complexes,
     AsPtC$_{11}$H$_{15}$Br$_4$·AsC$_{11}$H$_{15}$ 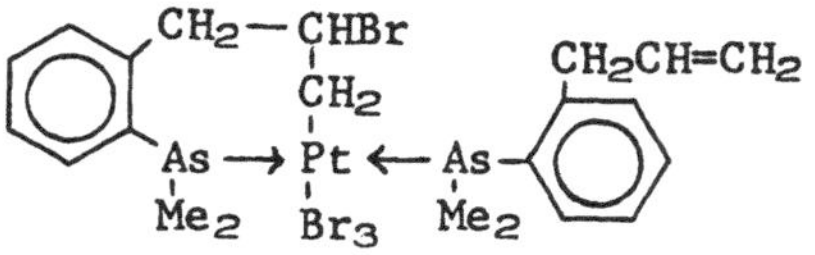
Synth.: From {Pt[2-(Me$_2$As)C$_6$H$_4$CH$_2$CH=CH$_2$]$_2$-Br$_2$} by treating with Br$_2$ (1:1) in CCl$_4$ (157).
Prop.: Yellow crystals, m. 140°, IR and UV spectra (157).
Rxn. with: MeOH → replacement of the Br atom at the β-C atom with the MeO group (157).

     AsPtC$_{11}$H$_{15}$Br$_4$·AsC$_{11}$H$_{15}$Br$_2$ 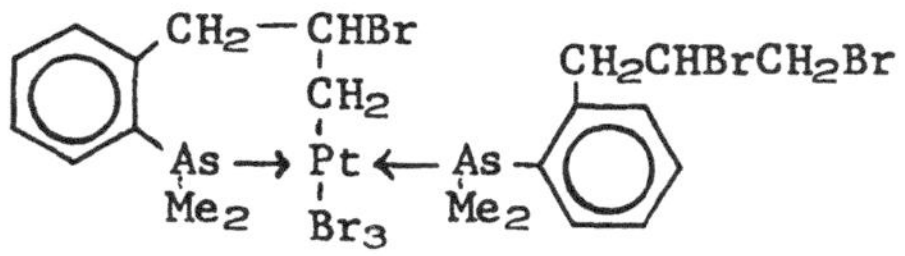
Synth.: From {Pt[2-(Me$_2$As)C$_6$H$_4$CH$_2$CH=CH$_2$]$_2$-Br$_2$} by treating with Br$_2$ (1:2) in C$_6$H$_6$-petr. ether and from {2-(Me$_2$As)C$_6$H$_4$CH$_2$CH-BrCH$_2$Pt[2-(Me$_2$As)C$_6$H$_4$CH$_2$CH=CH$_2$]Br$_3$} by treating with Br$_2$ (1:1) in C$_6$H$_6$-petr. ether (157).
Prop.: Yellow crystals, m. 142-148°, IR spectrum; a structure corr. to the above drawing postulated (157).

[o-(2-Methoxyvinyl)phenyl]dimethylarsine,   AsC$_{11}$H$_{15}$O,   2-Me$_2$AsC$_6$H$_4$CH=CHOMe
The arsine was prepd. and isolated as a complex with PtBr$_2$.
Coordination Deriv.: {Pt[2-(Me$_2$As)C$_6$H$_4$CH=CHOMe]Br$_2$}, white solid, m. 238° (dec.), prepd. from [2-(Me$_2$As)C$_6$H$_4$CHBrCH$_2$PtBr$_3$]$_2$ by heating with MeOH.   NMR spectrum; the mechanism of the complex formation and the position of the MeO group are discussed (161).

Dimethyl(o-propylphenyl)arsine,   AsC$_{11}$H$_{17}$,   2-Me$_2$AsC$_6$H$_6$Pr
Synth.: From 2-PrC$_6$H$_4$Br + Mg in Et$_2$O, followed by Me$_2$AsI in C$_6$H$_6$ (157).
Prop.: White oil, b$_6$ 90-92° (157).
Rxn. with: Br in CCl$_4$ → an intractable oil (157).
Coordination Deriv.: [Pt(o-Me$_2$AsC$_6$H$_4$Pr)$_2$Br$_2$], yellow needles, m. 168-170°, prepd. from PtBr$_2$ + the arsine (1:2) in CHCl$_3$ and recrystd. from EtOH.   UV spectrum (157).

[Pt(o-Me$_2$AsC$_6$H$_4$Pr)$_2$Br$_4$], red solid, m. 192-195°, prepd. from the preceding dibromo complex by treating with Br$_2$ in CCl$_4$ (157).

Dimethyl(p-propylphenyl)arsine,   AsC$_{11}$H$_{17}$,   4-Me$_2$AsC$_6$H$_4$Pr
Synth.: From 4-PrC$_6$H$_4$Br + Mg in Et$_2$O, followed by Me$_2$AsI in C$_6$H$_6$ (157).
Prop.: White oil, b$_6$ 100-102° (157).

Rxn. with: Br in CCl₄ → an intractable oil (157).

Coordination Deriv.: [Pt(4-Me₂AsC₆H₄Pr)₂Br₂], yellow needles, m. 165-167°, prepd. from PtBr₂ + the arsine (1:2) in CHCl₃ and recrystd. from EtOH (157).

[Pt(4-Me₂AsC₆H₄Pr)₂Br₄], red solid, m. 167-170°, prepd. from the preceding dibromo complex by a rxn. with Br in CCl₄ (157).

[o-(3-Methoxypropenyl)phenyl]dimethylarsine,   AsC₁₂H₁₇O,
    2-(Me₂As)C₆H₄CH=CHCH₂OMe
The arsine was prepd. and isolated as a Pt(II) complex.

Coordination Deriv.: [Pt(Me₂AsC₆H₄CH=CHCH₂OMe)Br₂], white solid, m. 205° (dec.), prepd. from [Pt(2-Me₂AsC₆H₄CH₂CH=CH₂)Br₄]₂ by heating with MeOH. NMR spectrum. The rxn. mechanism and complex structure are considered (161).

Dimethyl(o-propylthiophenyl)arsine,   AsC₁₁H₁₇S,   2-Me₂AsC₆H₄SPr
The arsine was prepd. and isolated as a Pd(II) complex.

Coordination Deriv.: [Pd(2-Me₂AsC₆H₄SPr)I₂], red rhombic crystals, prepd. from [2-(Me₂As)C₆H₄S]₂Pd in CHCl₃ by a rxn. with PrI, crystd. with 1:2 CHCl₃. The solvent of crystn. was removed at 135° (971).

Tribromo{2-methoxy-3-[2-(dimethylarsino)phenyl]propyl}platinum Complex,
    AsPtC₁₂H₁₈Br₃O·AsC₁₁H₁₅

Synth.: From {o-(Me₂As)C₆H₄CH₂CHBrCH₂Pt-
[o-(Me₂As)C₆H₄CH₂CH=CH₂]Br₃} + MeOH (157).
Prop.: Yellow solid, m. 166-167°, IR
spectrum (157).

    AsPtC₁₂H₁₈Br₃O·AsC₁₁H₁₅Br₂
Synth.: From {2-(Me₂As)C₆H₄CH₂CHBrCH₂Pt[2-(Me₂As)C₆H₄CH₂CHBrCH₂Br]Br₃} + MeOH (157).
Prop.: Yellow solid, m. 157-161° (157).

Tribromo-2-ethoxy-3-[2-(dimethylarsino)phenyl]propylplatinum Complex,
    AsPtC₁₃H₂₀Br₃O·AsC₁₁H₁₅
Synth.: From {o-(Me₂As)C₆H₄CH₂CHBrCH₂Pt-
[o-(Me₂As)C₆H₄CH₂CH=CH₂]Br₃} + EtOH (157).
Prop.: Yellow solid, m. 160°, UV spectrum
(157).

Methyldiphenylarsine,   AsC₁₃H₁₃,   Ph₂AsMe,   (M-51)
Synth.: From Ph₂AsH + BuLi in petroleum, followed by PhOMe in THF under reflux, 85% yield along with PhOH (1017).
Prop.: b₀.₃₅ 112-115° (1017), b. 302.6° calcd. by the Kinney equation (66).
Spectra: IR (418, 649), Raman (649), and electronic (1165).
Rxn. with: HgO in Me₂CO → MePh₂AsO (1057).
Aq. H₂O₂ in Et₂O → MePh₂AsO (1057).
R₂NCl → [MePh₂AsNR₂]Cl (1420).

Coordination Deriv.: (In the following entries (Ars) stands for Ph_2AsMe.)

$[Cu_2(Ars)_3Cl_3]$, two forms were isolated: blue and brown solids of unknown structure (see M-53, ref. 142). These two forms were shown to be the arsine oxide complexes (R. S. Nyholm, J. Chem. Soc. 1951, 1767). The postulate was made that the blue form has the compn. $[Cu(ArsO)_4][CuCl_2]_2$ and the brown form is essentially the same complex contaminated with 5-10% $CuCl_2$. This reformulated compn. was confirmed by x-ray diffraction analysis (1284).

$[Ir(Ars)(COD)Cl]$*, m. 120-122°, a nonelectrolyte in $ClCH_2CH_2Cl$, monomeric; prepd. from $[Ir(COD)Cl]_2$ + Ars (1:2). IR spectrum indicated the absence of free olefinic bonds. NMR spectrum showed a nonequivalence of the "olefinic" protons. At rising temp. the proton signals broadened and finally coalesced; the rate of coalescence was detd. When the free arsine was present besides the complex in $PhNO_2$, mol. cond. measurements indicated the presence of the equilibrium: $[Ir(Ars)(COD)Cl]$ + Ars $\rightleftharpoons$ $[Ir(Ars)_2(COD)]^+Cl^-$. The concn. of the ionic species increased when the temp. was lowered or the arsine was added. A mechanism of the arsine ligand exchange was suggested (1573).

$[CH_2=C(Me)CH_2Ir(Ars)(CO)Cl_2]$, pale yellow prisms, m. 162-166° (dec.), prepd. from $[CH_2=C(Me)CH_2Ir(CO)(cyclooctene)Cl_2]$ in CH_2Cl_2 + Ars by warming, adding MeOH and removing CH_2Cl_2. IR and NMR spectra were obtd. A structure with a symmetrical π-bonded methallyl group with C-2 atom trans to the Ars ligand more weakly bonded to the Ir atom than C-1 and C-3 (1390).

$[MeIr(Ars)_3I_2]$, yellow prisms, m. 195-203° (dec.), prepd. from fac-$[Me_3Ir-(Ars)_3]$ in $CHCl_3$ by treating with iodine (1:2) at -45°. IR and NMR spectra - a structure with the Me group and one I atom in the apical positions postulated (1391).

$[Me_2Ir(Ars)_3Cl]$, colorless prisms, m. 148-156° (dec.), prepd. from fac-$[Me_3Ir-(Ars)_3]$ in $CHCl_3$ by treating with Cl at -45°; IR and NMR spectra. A structure with one Me group and the Cl atom in the apical positions in C_6H_6 soln. postulated, while in $CHCl_3$ soln. a structure with one Ars and the Cl atom in the apical positions and the remaining Ars and both Me groups cis to each other postulated (1391).

$[Me_2Ir(Ars)_3Br]$, pale yellow prisms, m. 137-143° (dec.), prepd. from fac-$[Me_3Ir(Ars)_3]$ in $CHCl_3$ by treating with Br at -45°. IR and NMR spectra. The structure of the complex in soln. is influenced by the solvent used; in C_6H_6 and $CHCl_3$ the structures analogous to those of the chloro complex in these solvents were postulated (1391).

$[Me_2Ir(Ars)_3I]$, pale yellow prisms, m. 145-148° (dec.), prepd. from fac-$[Me_3-Ir(Ars)_3]$ in $CHCl_3$ by treating with iodine (1:1) at -45°. IR and NMR spectra. A structure with one Me group and the I atom in apical positions was postulated (1391).

---

* COD = 1,5-cyclooctadiene.

(In the following entries (Ars) stands for Ph_2AsMe.)

fac-[$Me_3Ir(Ars)_3$], colorless prisms, m. 179-185° (dec.), prepd. from mer-[Ir-($Ars)_3Cl_3$] + MeMgCl in C_6H_6 under reflux. IR and NMR spectra. A 6-coordinate Ir with 2 Me groups cis to each other and the third in an apical position in $CHCl_3$ was postulated (1391).

[π-$C_5H_5Mo(Ars)_2(CO)I$], red crystals, a nonelectrolyte, prepd. from [π-C_5H_5Mo-$(CO)_3I$] + excess Ars in C_6H_6 under UV irradiation. Soly. data. IR, visible, and NMR spectra (670).

[π-$C_5H_5Mo(Ars)(CO)_2I$], red crystals, m. 138-140°, a nonelectrolyte, prepd. from [π-$C_5H_5Mo(CO)_3I$] + Ars in C_6H_6-cyclohexane under UV irradiation. Soly. data. IR, visible, and NMR spectra (670).

[$Pd(Ars)_2Br_2$] (M-54), m. 168° (1017).

[$Rh(Ars)_3BrCl_2$], prepd. from [$(Ars)_3Cl_2Rh$-$HgBr$] by titration with Br in CH_2Cl_2. Electronic spectrum (1165).

[$Rh(Ars)_3Br_3$] (M-55), electronic spectrum (1165).

[$RhH(Ars)_3Br_2$] (M-55), electronic spectrum (1165).

[$RhH(Ars)_3Cl_2$] (M-56), yellow, diamagnetic crystals, m. 172-175°, prepd. from [$(Ars)_3Cl_2Rh$-$HgCl$] by treating with dry HCl in Et_2O. IR and electronic spectra (1165). Rxn. with $HgCl_2 \rightarrow$ [$(Ars)_3Cl_2Rh$-$HgCl$]. In the same manner the following Rh-Hg complexes were prepared: [$(Ars)_3Rh$-$HgYX_2$], where Y = F, Cl, Br, I, OAc, CN, and SCN and X = Cl and Br (1566).

[$(Ars)_3RhCl_2$-HgF] (M-56), yellow, diamagnetic crystals, m. 195°, a nonelectrolyte in $PhNO_2$, stable in air and to moisture but very sensitive to UV light; prepd. from [$RhH(Ars)_3Cl_2$] and HgF_2, PhHgF, or Hg_2F_2 in hot EtOH. Electronic spectrum - a Rh-Hg bond postulated (1165).

[$(Ars)_3RhCl_2$-$HgCl$] (M-56), yellow, diamagnetic crystals, m. 205°, a nonelectrolyte in $PhNO_2$, stable in air and to moisture but very sensitive to UV light, sol. in org. solvents; prepd. from [$RhH(Ars)_3Cl_2$] and $HgCl_2$, PhHgCl, or Hg_2Cl_2 in hot EtOH. Electronic spectrum - a Rh-Hg bond postulated. Rxn. with $HCl \rightarrow$ [$RhH(Ars)_3Cl_2$] (1165).

[$(Ars)_3RhCl_2$-$HgBr$] (M-56), orange-yellow, diamagnetic crystals, m. 165°, a nonelectrolyte in $PhNO_2$, stable in air and to moisture, but unstable in UV light, sol. in org. solvents; prepd. from [$RhH(Ars)_3Cl_2$] + HgBr, PhHgBr, or Hg_2Br_2 in hot EtOH. Electronic spectrum - a Rh-Hg bond postulated. Rxn. with $Br \rightarrow$ [$Rh(Ars)_3BrCl_2$] (1165).

[$(Ars)_3RhCl_2$-HgI] (M-56), orange, diamagnetic crystals, m. 155°, stable in air, to moisture, and to UV light; prepd. from [$RhH(Ars)_3Cl_2$] and HgI_2, PhHgI, or Hg_2I_2 in hot EtOH. Electronic spectrum - a Rh-Hg bond postulated (1165).

[$(Ars)_3RhBr_2$-$HgBr$] (M-56), yellow-orange, diamagnetic crystals, m. 187°, a nonelectrolyte in $PhNO_2$, stable in air and to moisture but unstable in UV light, sol. in org. solvents; prepd. from $[RhH(Ars)_3Br_2] + HgBr_2$, $PhHgBr$, or Hg_2Br_2 in hot EtOH. Electronic spectrum - a Rh-Hg bond postulated (1165).

$[(Ars)_3RhCl_2-HgOAc]$, yellow, diamagnetic crystals, m. 172°, a nonelectrolyte in $PhNO_2$, stable in air and moisture but dec. in UV light, readily sol. in org. solvents; prepd. from $[RhH(Ars)_3Cl_2] + Hg(OAc)_2$, $PhHgOAc$, or $Hg_2(OAc)_2$ in hot EtOH. IR and electronic spectra - a Rh-Hg bond postulated (1165).

$[(Ars)_3RhBr_2-HgCl]$, yellow, diamagnetic crystals, m. 202°, a nonelectrolyte in $PhNO_2$, stable in air and to moisture but dec. in UV light, sol. in org. solvents; prepd. from $[RhH(Ars)_3Br_2] + HgCl_2$, $PhHgCl$, or Hg_2Cl_2 in hot EtOH. Electronic spectrum - a Rh-Hg bond postulated (1165).

$[(Ars)_3RhBr_2-HgF]$, yellow, diamagnetic crystals, m. 154°, a nonelectrolyte in $PhNO_2$, stable in air and moisture but very sensitive to UV light, sol. in org. solvents; prepd. from $[RhH(Ars)_3Br_2] + HgF_2$, $PhHgF$, or Hg_2F_2 in hot EtOH. Electronic spectrum - a Rh-Hg bond postulated (1165).

$[(Ars)_3RhBr_2-HgI]$, orange, diamagnetic crystals, m. 148°, a nonelectrolyte in $PhNO_2$, stable to air, moisture, and UV light, sol. in org. solvents; prepd. from $[RhH(Ars)_3Br_2] + HgI_2$, $PhHgI$, or Hg_2I_2 in hot EtOH. Electronic spectrum - a Rh-Hg bond postulated (1165).

$[(Ars)_3RhBr_2-HgOAc]$, pale yellow, diamagnetic crystals, m. 190°, a nonelectrolyte in $PhNO_2$, stable to air and moisture, but slowly dec. in UV light, sol. in org. solvents; prepd. from $[RhH(Ars)_3Br_2] + Hg(OAc)_2$, $PhHgOAc$, or $Hg_2(OAc)_2$ in hot EtOH. Electronic spectrum - a Rh-Hg bond postulated (1165).

$[(Ars)_3RhCl_2-HgCN]$, $[(Ars)_3RhCl_2-HgCNS]$, $[(Ars)_3RhBr_2-HgCN]$, and $[(Ars)_3RhBr_2-HgCNS]$, prepd. in the same manner as the preceding complexes, were isolated in crude forms and were not further investigated (1165). The Rh-Hg bonds in the preceding complexes are readily cleaved by chlorine, bromine, iodine, and HCl (1566).

$[MeCH(CN)Rh(Ars)_3Cl_2]$, m. ~ 142°, prepd. from $[RhH(Ars)_3Cl_2] + CH_2=CHCN$ in C_6H_6 under reflux. IR spectrum. Rxn. with Py at 100° → $[MeCH(CN)Rh(Py)_3Cl_2]$ (473).

$[Ru(Ars)_2(NO)Cl_3]$, light orange, paramagnetic solid, m. 209-211°, prepd. from $Ru(NO)Cl_3·3H_2O + Ars$ in EtOH. IR spectrum (543).

$[Ru(Ars)_2(NO)I_2]$, dark brown crystals, m. 168°, prepd. from $Ru(NO)I_2 + Ars$ in C_6H_6 under reflux. IR spectrum (780).

[o-(3-Ethoxypropenyl)phenyl]dimethylarsine, $AsC_{13}H_{19}O$,
    $2-(EtOCH_2CH=CH)C_6H_4AsMe_2$
The arsine was prepd. and isolated as a Pt(II) complex.
Coordination Deriv.: $\{Pt[2-(EtOCH_2CH=CH)C_6H_4AsMe_2]Br_2\}$, white solid, m. 197° (dec.), prepd. from $[Pt(2-CH_2=CHCH_2C_6H_4AsMe_2)Br_3]_2$ by heating with EtOH. NMR spectrum - the structure of the complex is discussed (161).

Dibutyl(3-hydroxy-3-methyl-1-butynyl)arsine, $AsC_{13}H_{25}O$, $Bu_2AsC\equiv CC(OH)Me_2$
Synth.: From $Bu_2AsC\equiv CH$ + EtMgBr in Et_2O at r.t., followed by heating on a steam bath, treating with Me_2CO at r.t., and completing the rxn. on a steam bath, 50% yield (913).
Prop.: $b_{1.5}$ 98.5-100°, n_D^{20} 1.4952, d_4^{20} 1.0550, MR_D 75.18, AR_{As} 12.32, IR spectrum (913).

Ethynyldiphenylarsine, $AsC_{14}H_{11}$, $Ph_2AsC\equiv CH$
Synth.: From Ph_2AsX + $HC\equiv CMgX$ in THF, 85% yield (149).
Prop.: m. 23°, $b_{1.5}$ 140° (149), IR spectrum (149, 1028), NMR spectrum (149, 1412). Elemental analysis by combustion (612).
Rxn. with: Ketones, R_2CO in the presence of KOH → $Ph_2AsC\equiv CC(OH)R_2$ (149).
RMgX, followed by CO_2 → $Ph_2AsC\equiv CCO_2H$ (149).
$p\text{-}MeC_6H_4SO_3Me$ → $[MePh_2(HC\equiv C)As]O_3SC_6H_4Me\text{-}p$ (149).

Di(o-bromophenyl)ethylarsine, $AsC_{14}H_{13}Br_2$, $(2\text{-}BrC_6H_4)_2AsEt$
Synth.: From $(2\text{-}BrC_6H_4)_2AsCl$ in C_6H_6 + EtMgBr in Et_2O, 86% yield (715).
From $EtAsCl_2$ in C_6H_6 + $2\text{-}BrC_6H_4MgI$ in Et_2O, 6% yield (715).
Prop.: m. 97-98°, $b_{0.1}$ 158-162° (715).
Rxn. with: MeBr at 100° → $[(2\text{-}BrC_6H_4)_2EtMeAs]Br$ (715).
n-BuLi (1:2), followed by $EtPCl_2$ → an intractable, probably polymeric prod. (715).
n-BuLi (1:2), followed by CO_2 → $(2\text{-}HO_2CC_6H_4)_2AsEt$ (715).

Diphenyl-1-propynylarsine, $AsC_{15}H_{13}$, $Ph_2AsC\equiv CMe$
Synth.: From Ph_2AsCl + $MeC\equiv CMgCl$ in THF at r.t., 86% yield (149).
From Ph_2AsCl + $MeC\equiv CNa$ in C_6H_{14}, 96% yield (149).
By-prod. from the rxn. of Ph_2AsX + $HC\equiv CCH_2Br$ + Mg in Et_2O, trace amt. (150).
Prop.: b_1 128° (150). IR spectrum (149, 1028), NMR spectrum (149, 1412).
Rxn. with: H_2O_2 in AcOH → $Ph_2(MeC\equiv C)As(OH)_2$ (149).
$p\text{-}MeC_6H_4SO_3Me$ → $[Me(MeC\equiv C)Ph_2As]O_3SC_6H_4Me\text{-}p$ (149).

Diphenyl-2-propynylarsine, $AsC_{15}H_{13}$, $Ph_2AsCH_2C\equiv CH$
Synth.: From Ph_2AsX + $HC\equiv CCH_2Br$ + Mg in Et_2O at r.t., 67% yield along with $Ph_2AsCH=C=CH_2$ (150).
From Ph_2AsCl + $HC\equiv CCH_2AlCl_2$, 47% yield along with $Ph_2AsC=C=CH_2$ and $Ph_2AsC\equiv CMe$ (150).
From $[EtPh_2(HC\equiv CCH_2)As]Br$ by thermal dec. > 200°, 48% yield (150).
Prop.: $b_{0.01}$ 110° (150).

Diphenylpropadienylarsine, $AsC_{15}H_{13}$, $Ph_2AsCH=C=CH_2$
Synth.: From Ph_2AsX + $HC\equiv CCH_2Br$ + Mg in Et_2O at 20°, 79% yield along with $Ph_2AsCH_2C\equiv CH$ (150).
By-prod. from thermal dec. of $[EtPh_2(HC\equiv CCH_2)As]Br$ > 200° (150).

3-(Diphenylarsino)propionic Acid, $AsC_{15}H_{15}O_2$, $Ph_2AsCH_2CH_2CO_2H$, (M-61)
Prop.: Dissocn. const. in 20% aq. dioxan at 20°; a strong tendency to proto-

nation was observed. The acid forms a complex with Ag^+ (1222).

Dimethyl[o-(p-nitrobenzylthio)phenyl]arsine, $AsC_{15}H_{16}NO_2S$,
    $2-(4-O_2NC_6H_4CH_2S)C_6H_4AsMe_2$
The arsine was prepd. and isolated as a complex with $PdBr_2$.
$\{Pd[o-(p-O_2NC_6H_4CH_2S)C_6H_4AsMe_2]Br_2\}$, large yellow crystals, prepd. from
$[Pd(SC_6H_4AsMe_2)_2]$ in $CHCl_3$ by a treatment with $p-O_2NC_6H_4CH_2Br$, crystd. with
1/2 mol. $HCCl_3$ (971).

[o-(Benzylthio)phenyl]dimethylarsine, $AsC_{15}H_{17}S$, $2-PhCH_2SC_6H_4AsMe_2$
The arsine was prepd. and isolated as a complex with $PdCl_2$.
$[Pd(o-PhCH_2SC_6H_4AsMe_2)Cl_2]$, yellow crystals, prepd. from $[Pd(SC_6H_4AsMe_2)_2]$
in $CHCl_3$ by a treatment with $PhCH_2Cl$ (971).

3-(Diphenylarsino)propylamine, $AsC_{15}H_{18}N$, $Ph_2As(CH_2)_3NH_2$
Synth.: From Ph_2AsH by a condensation with $CH_2=CHCN$, followed by a redn. with
$LiAlH_4$ in Et_2O (390).
From $Ph_2AsCH_2CH_2CN$ by a redn. with $LiAlH_4$ in Et_2O, 66% yield (1317).
Prop.: $b_{0.8}$ 161-163° (1317).
Rxn. with: $Ac_2 \rightarrow [Ph_2AsCH_2CH_2CH_2N=C(Me)-]_2$ (390).
$RCHO \rightarrow Ph_2AsCH_2CH_2CH_2N=CHR$ (1317).
Coordination Deriv.: $\{Mo[Ph_2As(CH_2)_3NH_2](CO)_4\}$, prepd. from $Mo(CO)_6$ + the
arsinoamine. Magnetic moment, cond. data, and IR spectrum were obtd. (390).

$\{Ni[Ph_2As(CH_2)_3NH_2]_3\}(ClO_4)_2$, prepd. from $Ni(ClO_4)_2$ + the arsinoamine.
Magnetic moment, cond. data, and IR spectrum were obtd. (390).

Dibutyltolylarsine Isomers, $AsC_{15}H_{25}$, $Bu_2AsC_6H_4Me$
Synth.: From Bu_2AsLi + $4-MeC_6H_4F$ in Et_2O at 25°, a mixt. of $Bu_2AsC_6H_4Me-4$
and $Bu_2AsC_6H_4Me-3$ was obtd. with 44% yield (791).
Prop.: The isomer mixt. b_{1-2} 135-140° (791).

Di-tert-butyl-m-tolylarsine, $AsC_{15}H_{25}$, $t-Bu_2AsC_6H_4Me-3$
Synth.: From $t-Bu_2AsCl$ + $3-MeC_6H_4Li$ in Et_2O under reflux, 72% yield (791).
From $t-Bu_2AsLi$ + $4-MeC_6H_4F$ in Et_2O at 25°, a mixt. of $t-Bu_2AsC_6H_4Me-3$ and
$t-Bu_2AsC_6H_4Me-4$ was obtd. (791).
Prop.: b_{1-2} 116°; the isomer mixt. b_{1-2} 118-122° (791).

Di-tert-butyl-p-tolylarsine, $AsC_{15}H_{25}$, $t-Bu_2AsC_6H_4Me-4$
Synth.: From $t-Bu_2AsCl$ + $4-MeC_6H_4Li$ in Et_2O under reflux, 80% yield (791).
From $t-Bu_2AsLi$ + $4-MeC_6H_4F$ in Et_2O at 25°, a mixt. of $t-Bu_2AsC_6H_4Me-4$ and
$t-Bu_2AsC_6H_4Me-3$ was obtd. with a yield of 44% (791).
Prop.: b_{1-2} 116°, m. 26°; the p- and m-isomer mixt. b_{1-2} 118-122° (791).

(3-Chloropropyl)dicyclohexylarsine, $AsC_{15}H_{28}Cl$, $(C_6H_{11})_2As(CH_2)_3Cl$
Synth.: From $Cl(CH_2)_3Cl$ in dioxan by adding dropwise to $(C_6H_{11})_2AsLi$ (1:1)
in the same solvent under Ar, 52% yield (1533).
Prop.: Colorless liquid, b_5 202-205°, very sensitive to air (1533).

1-(Dicyclohexylarsino)-N-ethylthiocarbamide, $AsC_{15}H_{28}NS$,
   $(C_6H_{11})_2AsC(S)NHEt$
Synth.: From $(C_6H_{11})_2AsLi$ in Et_2O by adding dropwise EtNCS in Et_2O at r.t.,
followed by heating under reflux and dec. with H_2O, 69% yield (1537).
Prop.: Light yellow crystals, m. 52°, sol. in C_6H_6, THF, dioxan, Me_2CO,
sparingly sol. in petr. ether (1537).

3-(Dicyclohexylarsino)propanol, $AsC_{15}H_{29}O$, $(C_6H_{11})_2As(CH_2)_3OH$
Synth.: From $(C_6H_{11})_2AsLi$ + $\overline{CH_2CH_2CH_2O}$ in dioxan or Et_2O under reflux, fol-
lowed by a treatment with H_2O, 66% yield (1531).
Prop.: Colorless liquid, b_1 161-163° (1531).

1-(Dicyclohexylarsino)-2-propanol, $AsC_{15}H_{29}O$, $(C_6H_{11})_2AsCH_2CH(Me)OH$
Synth.: From $(C_6H_{11})_2AsLi$ + $\overline{MeCHCH_2O}$ in dioxan or Et_2O under reflux, fol-
lowed by a treatment with H_2O, 56% yield (1531).
Prop.: Colorless liquid, b_8 185-187° (1531).

Dibutyl(diethylaminopropynyl)arsine, $AsC_{15}H_{30}N$, $Bu_2AsC{\equiv}CCH_2NEt_2$
Synth.: From $Bu_2AsC{\equiv}CH$ + $(HCHO)_x \cdot xH_2O$ + Et_2NH in dioxan in the presence of
$Cu(OAc)_2$ under reflux, 18% yield (913).
Prop.: b_{11} 142.5-144°, n_D^{20} 1.4889, d_4^{20} 0.9908, MR_D 87.24, AR_{As} 12.73, IR
spectrum (913).

Di-4-pentenylphenylarsine, $AsC_{16}H_{23}$, $(CH_2{=}CHCH_2CH_2CH_2)_2AsPh$
Synth.: From $PhAsI_2$ + $CH_2{=}CHCH_2CH_2CH_2MgBr$ in $Et_2O-C_6H_6$ (164).
Prop.: Yellowish oil, b_1 128-130° (164).
Coordination Deriv.: $\{Hg[(CH_2{=}CHCH_2CH_2CH_2)_2AsPh]Cl_2\}_2$, white crystals, m.
92°, prepd. from $HgCl_2$ and the arsine in 80% aq. EtOH. IR spectrum (164).

$\{Pt[(CH_2{=}CHCH_2CH_2CH_2)_2AsPh]Cl_2\} \cdot 1/2\ C_6H_6$, yellow crystals, m. 98-100°, prepd.
from $PtCl_2$ and the arsine in C_6H_6. IR spectrum - a square planar structure
postulated (164).

$\{Pt[(CH_2{=}CHCH_2CH_2CH_2)_2AsPh]Cl_2\}$, yellow crystals, m. 98°, prepd. from $PtCl_2$
and the arsine in Et_2O. IR spectrum (164).

Dipentylphenylarsine, $AsC_{16}H_{27}$, $(C_5H_{11})_2AsPh$
Synth.: From $PhAsCl_2$ + $C_5H_{11}MgBr$ (1:2) in Et_2O, 65% yield (828).
Prop.: b_3 154.5-156.5°, d_4^{20} 1.0513, n_D^{20} 1.5212, MR_D 85.27, AR_D 11.68 (828).
Rxn. with: S at 100-120° → $Ph(C_5H_{11})_2AsS$ (1549).
$BrCH_2CO_2Et$ on a steam bath → $[Ph(C_5H_{11})_2(EtO_2CCH_2)As]Br$ (1549).
Biol. Prop.: Fungistatic and fungicidal effect on Alternaria radicina, Helmin-
thosporium sativum, Bortrytis cinerea, Fusicladium dendriticum, Trichophyton
gypseum, and Epidermophyton; toxic to mice (1133).

Pentafluorophenyldiphenylarsine, $AsC_{18}H_{10}F_5$, $Ph_2AsC_6F_5$
Synth.: From C_6F_5Li in Et_2O at -78° by adding dropwise Ph_2AsCl in Et_2O and
warming to r.t., 55% yield (842).

Prop.: m. 63-64°, IR and NMR spectra (842).
Coordination Deriv.: [Pt(Ph$_2$AsC$_6$F$_5$)$_2$Br$_2$], light orange, m. 203-210° (dec.),
prepd. from K$_2$PtBr$_4$ in H$_2$O + the arsine in warm alc. IR spectrum - a trans
configuration postulated (842).

[Pt(Ph$_2$AsC$_6$F$_5$)$_2$Cl$_2$], pale yellow, m. 201-206° (dec.), prepd. from K$_2$PtCl$_4$ in
H$_2$O + the arsine in warm alc. IR and NMR spectra - a trans configuration
postulated (842).

[Pt(Ph$_2$AsC$_6$F$_5$)$_2$I$_2$], pink crystals, m. 216-230° (dec.), prepd. from PtI$_2$ + the
arsine in xylene under reflux. IR spectrum - a trans configuration postulated
(842).

[Rh(Ph$_2$AsC$_6$F$_5$)$_2$(CO)Cl], yellow crystals, m. 138-144° (dec.), prepd. from
[Rh$_2$(CO)$_4$Cl$_2$] + the arsine in C$_6$H$_6$. IR spectrum (842).

o-Bromophenyldiphenylarsine, AsC$_{18}$H$_{14}$Br, 2-BrC$_6$H$_4$AsPh$_2$, (M-69)
Synth.: From Ph$_2$AsCl in C$_6$H$_6$ by a rxn. with 2-BrC$_6$H$_4$MgBr in Et$_2$O, 95% yield
(715).
From 2-BrC$_6$H$_4$AsCl$_2$ by a rxn. with PhMgBr in Et$_2$O, 52% yield (292, 1151).
Prop.: Crystals, m. 103° (292), m. 100.5-101.5° (1151), m. 112-113° (715).
Rxn. with: BuLi in Et$_2$O, followed by Ph$_2$PCl → 2-(Ph$_2$As)C$_6$H$_4$PPh$_2$ (1151, 1153).
BuLi in Et$_2$O, followed by Ph$_2$P(S)Cl → 2-(Ph$_2$As)C$_6$H$_4$P(S)Ph$_2$ (1151, 1153).
BuLi in Et$_2$O, followed by S → 2-HSC$_6$H$_4$AsPh$_2$ (292).

2-(Diphenylarsinomethyl)pyridine, AsC$_{18}$H$_{16}$N,
Synth.: From Ph$_2$AsK·2 Dioxan in ketylated ether
by adding dropwise α-chloropicoline in the same
solvent and stirring at r.t., 25% yield (1543).
Prop.: m. 73-74°, sol. in org. solvents, stable in air (1543).

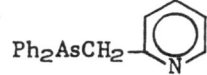

Rxn. with: MeI → [MePh$_2$AsCH$_2$C=CHCH=CHCH=NMe]I$_2$ (1543).
Coordination Deriv.: (In the following complexes DPAP stands for 2-(diphenylarsinomethyl)pyridine.)

[Co(DPAP)$_2$(NCS)$_2$], brown-violet, paramagnetic crystals, m. 183°, a weak electrolyte in Me$_2$CO, prepd. from Co(NCS)$_2$ + the arsinopyridine in EtOH under
reflux. Reflexion and IR spectra. A 6-coordinate octahedral structure with
DPAP ligands as bidentates postulated. On heating at 130°, a dark green melt
was formed which, after cooling, solidifed to a blue-green, paramagnetic substance, m. 69-70°. A weak electrolyte in Me$_2$CO. Reflexion and IR spectra.
A 4-coordinate tetrahedral structure with the DPAP coordinated through the N
atom only (1543).

[Cu(DPAP)$_2$]CuCl$_2$, fine, ebony-colored, diamagnetic crystals, m. 178-180°, an
electrolyte in DMF, prepd. from Cu$_2$Cl$_2$ + the arsinopyridine (1:1) in C$_6$H$_6$
under reflux. A tetrahedral cation structure postulated (1543).

[Ni(DPAP)$_2$(NCS)$_2$], blue-violet, paramagnetic crystals, m. 253°, prepd. from
Ni(NCS)$_2$ + the arsinopyridine in EtOH under reflux. Reflexion and IR spectra.
A 6-coordinate complex with the DPAP ligands chelated through both the As and
N atoms (1543).

(In the following complexes DPAP stands for 2-(diphenylarsinomethyl)pyridine.)

[Ag(DPAP)NO$_3$], colorless crystals, m. 226°, sensitive to light, prepd. from AgNO$_3$ + DPAP in warm EtOH (1543).

1,1'-(Butylarsylene)bis(formanilide),   AsC$_{18}$H$_{21}$N$_2$O$_2$,   BuAs(CONHPh)$_2$
Synth.: From BuAsH$_2$ + PhNCO (1:2) in dioxan in the presence of Bu$_2$Sn(OAc)$_2$ catalyst under reflux, followed by distn. of the solvent and crystn. through addn. of petr. ether, 76% yield (1536).
Prop.: Yellowish oil, IR spectrum (1536).

1,1'-(Butylarsylene)bis(thioformanilide),   AsC$_{18}$H$_{21}$N$_2$S$_2$,   BuAs(CSNHPh)$_2$
Synth.: From BuAsLi$_2$ in Et$_2$O by adding dropwise PhNCS in Et$_2$O at r.t., then heating under reflux, and hydrolyzing with H$_2$O, 38% yield (1537).
Prop.: Orange crystals, m. 78-80°, sol. in Et$_2$O and THF, sparingly sol. in petr. ether (1537).

Dihexylphenylarsine,   AsC$_{18}$H$_{31}$,   (C$_6$H$_{13}$)$_2$AsPh
Synth.: From PhAsCl$_2$ + C$_6$H$_{13}$MgBr (1:2) in Et$_2$O, 52% yield (828).
Prop.: b$_{0.32}$ 156-157°, d$_4^{20}$ 1.0287, n$_D^{20}$ 1.5182, MR$_D$ 94.88, AR$_{As}$ 12.06 (828).
Rxn. with: BrCH$_2$CO$_2$Et on a steam bath → [Ph(C$_6$H$_{13}$)$_2$(EtO$_2$CCH$_2$)As]Br (1549).
PhCH$_2$CH$_2$Br in a sealed tube → [Ph(PhCH$_2$CH$_2$)(C$_6$H$_{13}$)$_2$As]Br (1549).
Biol. Prop.: Fungistatic and fungicidal effect against Alternaria radicina, Helminthosporium sativum, Bortrytis cinerea, Fusicladium dendriticum, Trichophyton gypseum, and Epidermophyton. Toxic to mice (1133).

N,N'-Dicyclohexyl-1,1'-(butylarsylene)bis(formamide),   AsC$_{18}$H$_{33}$N$_2$O$_2$,
    BuAs(CONHC$_6$H$_{11}$)$_2$
Synth.: From BuAsLi$_2$ + C$_6$H$_{11}$NCO (1:2) in Et$_2$O at r.t., followed by heating and dec. with H$_2$O. The prod. was recovered from the Et$_2$O phase by distn. and crystn. from Et$_2$O-petr. ether, 69% yield (1536).
Prop.: Colorless crystals, m. 81°, sol. in C$_6$H$_6$, THF, and EtOH, sparingly sol. in petr. ether and H$_2$O, IR spectrum (1536).

1-(Diphenylarsino)formanilide Lithium Salt,   AsLiC$_{19}$H$_{15}$NO,   Ph$_2$AsCON(Li)Ph
Synth.: From Ph$_2$AsCONHPh in Et$_2$O by treating with PhLi (1:1) and pptg. with petr. ether, 70% yield. Addn. of dioxan to the Et$_2$O soln. gave the dioxan complex, Ph$_2$AsCON(Li)Ph·Dioxan (1536).
Prop.: A solid substance sol. in dioxan and THF, sparingly sol. in petr. ether (1536).

1-(Diphenylarsino)formanilide,   AsC$_{19}$H$_{16}$NO,   Ph$_2$AsCONHPh
Synth.: From Ph$_2$AsH + PhNCO (1:1) in dioxan under reflux in the presence of Bu$_2$Sn(OAc)$_2$, followed by distn. of the solvent and crystn. by addn. of petr. ether, 89% yield (1536).
Prop.: Colorless needles (from EtOH), m. 110° (dec.), sol. in C$_6$H$_6$, THF, and EtOH, sparingly sol. in petr. ether and H$_2$O. IR spectrum (1536).
Rxn. with: PhLi in Et$_2$O → Ph$_2$AsCON(Li)Ph (1536).

1-(Diphenylarsino)thioformanilide, $AsC_{19}H_{16}NS$, $Ph_2AsCSNHPh$
Synth.: From Ph_2AsLi in Et_2O by adding dropwise PhNCS dissolved in Et_2O at
r.t., then heating under reflux and dec. with H_2O, 65% yield (1537).
Prop.: Yellow crystals, m. 44°, sol. in C_6H_6 and THF, sparingly sol. in
petr. ether (1537).
Rxn. with: $PhNH_2$ on a steam bath, followed by H_2O-EtOH → $(PhNH)_2CO$ + Ph_2AsH
(1537).

Benzyldiphenylarsine, $AsC_{19}H_{17}$, $PhCH_2AsPh_2$
Synth.: From Ph_2AsH + BuLi in petroleum, followed by $PhCH_2OPh$ or $PhCH_2SPh$ in
THF under reflux, 74% and 71% yield along with PhOH and PhSH, resp. (1017).
Prop.: m. 63-65°, $b_{0.3}$ 176-178° (1017).
Coordination Deriv.: $[Pd(Ph_2AsCH_2Ph)_2Br_2]$, orange crystals, m. 200° (dec.),
prepd. from $PdBr_2$ and the arsine in DMF (1017).

2-[2-(Diphenylarsino)ethyl]pyridine, $AsC_{19}H_{18}N$,
Synth.: From Ph_2AsH + 2-vinylpyridine in AcOH
under reflux at 150°, 81% yield (1542a).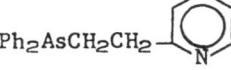
From $Ph_2AsK\cdot2$ Dioxan + 2-(2-chloroethyl)pyridine
(1542a).
Prop.: Crystals, m. 43°, b_3 240°, insol. in H_2O, sol. in org. solvents,
stable in air (1542a).
Coordination Deriv.: (In the following complexes AEP stands for $Ph_2AsCH_2CH_2$-
$\overset{\frown}{C=CHCH=CHCH=N}$.)

$[Co(AEP)_2I_2]$, blue-green, paramagnetic crystals, m. 131°, a nonelectrolyte in
Me_2CO, prepd. from CoI_2 + the arsinopyridine in $MeNO_2$ by heating to 60°. A
tetrahedral structure with AEP coordinated through the N atoms postulated
(1542a).

$[Cu(AEP)_2]CuCl_2$, prepd. from Cu_2Cl_2 and the arsinopyridine in C_6H_6 under
reflux. A tetrahedral structure of the cation with AEP coordinated through
both the As and N atoms postulated (1542a).

$[Cu(AEP)_2]CuBr_2$, crystals, m. 155°, prepd. from Cu_2Br_2 in the same manner as
the preceding complex (1542a).

$[Ni(AEP)Br_2]$, brown, paramagnetic crystals, m. 222°, a nonelectrolyte in
Me_2CO, prepd. from $NiBr_2\cdot2$ THF in Me_2CO + the arsinopyridine under reflux.
Electronic spectrum - a distorted tetrahedral structure postulated. The complex in the solid state might be a coordination polymer with Br bridges
(1542a).

$[Ni(AEP)I_2]$, dark green, diamagnetic crystals, m. 124°, a nonelectrolyte in
Me_2CO, prepd. from NiI_2 in the same manner as the preceding dibromo complex
(1542a).

$[Ni(AEP)_2(NCS)_2]$, dark red, diamagnetic crystals, m. > 154°, prepd. from
$NiCl_2\cdot6$ H_2O in EtOH by treating with KNCS and filtering the soln. into an alc.
soln. of AEP. Reflexion spectrum - a tetrahedral structure with the AEP
ligands coordinated through the N atoms only postulated (1542a).

(In the following complexes AEP stands for $Ph_2AsCH_2CH_2C=CHCH=CHCH=N.$)

[Ni(AEP)$_2$(NCS)$_2$], light green, paramagnetic crystals, m. 175°, prepd. from the preceding dark red complex by crystn. from Me$_2$CO. Reflexion spectrum — an octahedral polymeric structure bridged through the NCS groups suggested (1542a).

[Ni(AEP)$_2$](ClO$_4$)$_2$, orange, diamagnetic crystals, m. 208°, a 1:2 electrolyte in Me$_2$CO, prepd. from Ni(ClO$_4$)$_2$·6 H$_2$O + the arsinopyridine in acetone diethyl-acetate. A square planar Ni complex with the AEP ligands coordinated through the As and N atoms postulated (1542a).

[Zn(AEP)$_2$Br$_2$], colorless, diamagnetic crystals, m. 83°, a nonelectrolyte in Me$_2$CO, prepd. from ZnBr$_2$ in EtOH by heating with AEP. A tetrahedral struc-ture with the AEP ligands coordinated through the N atoms only postulated (1542a).

N-Cyclohexyl-1-(diphenylarsino)formamide,   AsC$_{19}$H$_{22}$NO,   Ph$_2$AsCONHC$_6$H$_{11}$
Synth.: From Ph$_2$AsLi + C$_6$H$_{11}$NCO (1:1) in Et$_2$O at r.t., followed by heating and, after cooling, dec. with H$_2$O, 71% yield (1536).
Prop.: Colorless crystals (from Et$_2$O-petr. ether), m. 78°, sol. in C$_6$H$_6$, THF, and EtOH, sparingly sol. in petr. ether and H$_2$O. IR spectrum (1536).

1-(Dicyclohexylarsino)formanilide,   AsC$_{19}$H$_{28}$NO,   (C$_6$H$_{11}$)$_2$AsCONHPh
Synth.: From (C$_6$H$_{11}$)$_2$AsH + PhNCO (1:1) in dioxan under reflux in the pre-sence of Bu$_2$Sn(OAc)$_2$, followed by distn. of the solvent and crystn. by addn. of petr. ether, 72% yield (1536).
From (C$_6$H$_{11}$)$_2$AsLi + PhNCO (1:1) in Et$_2$O at r.t., followed by heating and, after cooling, dec. with H$_2$O, 66% yield (1536).
Prop.: Colorless crystals (from EtOH), m. 139°, sol. in C$_6$H$_6$, THF, and EtOH, sparingly sol. in petr. ether and H$_2$O. IR spectrum (1536).
Rxn. with: PhLi in Et$_2$O → (C$_6$H$_{11}$)$_2$AsCON(Li)Ph (1536).

1-(Dicyclohexylarsino)thioformanilide,   AsC$_{19}$H$_{28}$NS,   (C$_6$H$_{11}$)$_2$AsCSNHPh
Synth.: From (C$_6$H$_{11}$)$_2$AsLi + PhNCS in Et$_2$O at r.t., followed by heating under reflux and dec. with H$_2$O, 70% yield (1537).
Prop.: Yellow needles (from EtOH-petr. ether), m. 68°, sol. in C$_6$H$_6$ and EtOH, sparingly sol. in Et$_2$O and petr. ether (1537).
Rxn. with: MeI in Et$_2$O → [Me$_2$(C$_6$H$_{11}$)$_2$As]I (1537).

N-Cyclohexyl-1-(dicyclohexylarsino)formamide,   AsC$_{19}$H$_{34}$NO,
  (C$_6$H$_{11}$)$_2$AsCONHC$_6$H$_{11}$
Synth.: From (C$_6$H$_{11}$)$_2$AsLi + C$_6$H$_{11}$NCO (1:1) in Et$_2$O at r.t., followed by heating and dec. with H$_2$O, 71% yield (1536).
Prop.: Colorless needles (from EtOH), m. 122°, sol. in C$_6$H$_6$, THF, and EtOH, sparingly sol. in petr. ether and H$_2$O. IR spectrum (1536).
Rxn. with: MeI in C$_6$H$_6$ → [Me(C$_6$H$_{11}$)$_2$(C$_6$H$_{11}$NHCO)As]I (1536).
PhLi in Et$_2$O → (C$_6$H$_{11}$)$_2$AsCON(Li)C$_6$H$_{11}$ (1536).

Diphenyl-cis-styrylarsine, $AsC_{20}H_{17}$, $Ph_2AsCH=CHPh$
Synth.: From Ph_2AsLi + cis-$PhCH=CHBr$ in THF, 63% yield (43, 88).
From Ph_2AsLi + $PhC\equiv CH$ in THF in the presence of primary or secondary amines, 67% yield (44).
Prop.: m. 91-92°, NMR spectrum. The compd. could not be isomerized by heating with 15% BuLi in THF or with PCl_3. When heated with PCl_5, the compd. isomerized to the trans form (43). Vapor phase chromatography at 200° in a 3% SE-30 silicone column in the presence of Et_2NH or $BuNH_2$ afforded 5% isomerization to the trans form (44).

Diphenyl-trans-styrylarsine, $AsC_{20}H_{17}$, $Ph_2AsCH=CHPh$
Synth.: From Ph_2AsLi + trans-$PhCH=CHBr$ in THF, 71% yield (43, 88).
From cis-$PhCH=CHAsPh_2$ by heating with PCl_5 in THF (43).
Prop.: $b_{0.25}$ 187-189°, NMR spectrum, vapor phase chromatography (43).

Phenylarsylenebis(formanilide), $AsC_{20}H_{17}N_2O_2$, $PhAs(CONHPh)_2$
Synth.: From $PhAsH_2$ + PhNCO (1:2) in dioxan under reflux in the presence of $Bu_2Sn(OAc)_2$, followed by distn. of the solvent and crystn. by adding petr. ether, 60% yield (1536).
Prop.: Colorless crystals (from EtOH-petr. ether), m. 132°, sol. in C_6H_6, THF, and EtOH, sparingly sol. in petr. ether and H_2O. IR spectrum (1536).

Bis(o-methylthiophenyl)phenylarsine, $AsC_{20}H_{19}S_2$, $(2-MeSC_6H_4)_2AsPh$
Synth.: From $2-MeSC_6H_4Br$ in Et_2O by adding BuLi in C_6H_{14} at 0°, warming to r.t., adding $PhAsCl_2$ in Et_2O, and hydrolyzing the rxn. mixt., 84% yield (1526).
Prop.: White solid, m. 120° (1526).
Coordination Deriv.: (In the following complexes DAS stands for $(2-MeSC_6H_4)_2-$AsPh.)

$[Mo(DSA)(CO)_2Br_2]$, diamagnetic crystals, m. 185° (dec.), sparingly sol. in org. solvents, prepd. from $NH_4[Mo(CO)_4Br_3]$ + the arsine (1:1) in CH_2Cl_2 at 25°. IR spectrum (1526).

$[Mo(DSA)(CO)_2I_2]\cdot MeOH$, diamagnetic crystals, m. 156° (dec.), a nonelectrolyte in $MeNO_2$, prepd. from $NH_4[Mo(CO)_4I_3]$ + the arsine in CH_2Cl_2 at 25°, filtering off the salt, concg. the filtrate and adding MeOH. IR spectrum (1526).

(1-Hydroxycyclohexylethynyl)diphenylarsine, $AsC_{20}H_{21}O$,
$Ph_2AsC\equiv C\overline{C(OH)(CH_2)_4CH_2}$
Synth.: From $Ph_2AsC\equiv CH$ by forming a Mg deriv. and coupling with cyclohexanone, 87% yield (149).
Prop.: m. 57° (169).

Diheptylphenylarsine, $AsC_{20}H_{35}$, $(C_7H_{15})_2AsPh$
Synth.: From $PhAsCl_2$ + $C_7H_{15}MgBr$ (1:2) in Et_2O, 70% yield (828).
Prop.: $b_{0.015}$ 134°, d_4^{20} 1.0143, n_D^{20} 1.5131, MR_D 103.73, AR_{As} 11.67 (828).
Rxn. with: S at 100-120° → $Ph(C_7H_{15})_2AsS$ (1549).
$BrCH_2CO_2Et$ on a steam bath → $[Ph(C_7H_{15})_2(EtO_2CCH_2)As]Br$ (1549).

<u>Biol. Prop.</u>: Fungistatic and fungicidal effect against Alternaria radicina, Helminthosporium sativum, Bortrytis cinerea, Fusicladium dendriticum, Trichophyton gypseum, and Epidermophyton. Toxic to mice. Minimum activity doses were detd. (1133).

[o-(Diphenylmethyl)phenyl]dimethylarsine, $AsC_{21}H_{21}$, [o-(Ph$_2$CH)C$_6$H$_4$]AsMe$_2$
<u>Prop.</u>: m. 56-57° (1163).
<u>Rxn. with</u>: BuLi → a highly colored substance which readily reacted with PtBr$_2$ to form a typical arsine complex (1163).

1-(Dicyclohexylarsino)-N-phenylthioformimidic Acid Ethyl Ester,
    $AsC_{21}H_{32}NS$, (C$_6$H$_{11}$)$_2$AsC(SEt)=NPh
<u>Synth.</u>: From (C$_6$H$_{11}$)$_2$AsLi (prepd. in situ) in Et$_2$O by treating dropwise with PhNCS, then heating under reflux, replacing the solvent with C$_6$H$_6$ and heating again with EtBr, 78% yield (1537).
<u>Prop.</u>: Orange-colored oil, sol. in C$_6$H$_6$, sparingly sol. in petr. ether. IR spectrum (1537).

Ferrocenyldiphenylarsine, $AsFeC_{22}H_{19}$, Ph$_2$AsFc
<u>Synth.</u>: From Ph$_2$AsCl by the rxn. with a mixt. of FcLi and FcdLi$_2$ in Et$_2$O under reflux, a mixt. of Ph$_2$AsFc with Fcd(AsPh$_2$)$_2$ was obtd., with the yields of 4 and 2%, resp. (1437).
<u>Prop.</u>: Yellow powder, m. 119.5-121.5°, IR spectrum (1437).

Dioctylphenylarsine, $AsC_{22}H_{39}$, (C$_8$H$_{17}$)$_2$AsPh
<u>Synth.</u>: From PhAsCl$_2$ in Et$_2$O + C$_8$H$_{17}$MgBr, 58% yield (1549).
<u>Prop.</u>: b$_{0.0048}$ 148-150°, d$_4^{20}$ 1.0031, n$_D^{20}$ 1.5119 (1549).
<u>Rxn. with</u>: S at 100-120° → (C$_8$H$_{17}$)$_2$PhAsS (1549).
BrCH$_2$CO$_2$Et on a steam bath → [(C$_8$H$_{17}$)$_2$Ph(EtO$_2$CCH$_2$)As]Br (1549).
MeI on a steam bath → [MePh(C$_8$H$_{17}$)$_2$As]Br (1549).

3-(Diphenylarsino)-N-(o-methoxybenzylidene)propylamine, $AsC_{23}H_{24}NO$,
    2-MeOC$_6$H$_4$CH=N(CH$_2$)$_3$AsPh$_2$
<u>Synth.</u>: From Ph$_2$As(CH$_2$)$_3$NH$_2$ + 2-MeOC$_6$H$_4$CHO on a steam bath, drying with CaSO$_4$, dissolving in BuOH, and crystg. at -20° (1317).
<u>Prop.</u>: Yellowish white crystals, m. 58-60° (1317).
<u>Coordination Deriv.</u>: [Ni(AsC$_{23}$H$_{24}$NO)Br$_2$], dark purple, paramagnetic crystals, m. 184-189°, a nonelectrolyte in ClCH$_2$CH$_2$Cl, prepd. from NiBr$_2$ and the arsinoamine in hot BuOH. Electronic absorption and reflectance spectra - a 5-coordinate Ni complex postulated (1317).

[Ni(AsC$_{23}$H$_{24}$NO)I$_2$], brown, paramagnetic crystals, m. 174-178°, a nonelectrolyte in ClCH$_2$CH$_2$Cl, prepd. from NiI$_2$ and the arsinoamine in hot BuOH. Electronic absorption and reflectance spectra - a 5-coordinate Ni complex postulated (1317).

3-(Diphenylarsino)-N-[o-(methylthio)benzylidene]propylamine, $AsC_{23}H_{24}NS$,
    2-MeSC$_6$H$_4$CH=N(CH$_2$)$_3$AsPh$_2$

Synth.: From $Ph_2As(CH_3)_2NH_2$ + 2-$MeSC_6H_4CHO$ on a steam bath, drying with $CaSO_4$, dissolving in BuOH, and crystg. at -20° (1317).
Prop.: Yellowish white crystals, m. 79-80° (1317).
Coordination Deriv.: $[Ni(AsC_{23}H_{24}NS)Br_2]$, mustard brown, paramagnetic crystals, m. 202-205°, a nonelectrolyte in $ClCH_2CH_2Cl$, prepd. from $NiBr_2$ + the arsinoamine in hot BuOH. Electronic absorption and reflectance spectra - a 5-coordinate Ni complex postulated (1317).

$[Ni(AsC_{23}H_{24}NS)I_2]$, brown, paramagnetic crystals, m. 190-193°, a nonelectrolyte in $ClCH_2CH_2Cl$, prepd. from $NiCl_2$ + the arsinoamine in hot BuOH. Electronic absorption and reflectance spectra - a 5-coordinate Ni complex postulated (1317).

o-{N-[3-(Diphenylarsino)propyl]forimidoyl}-N-methylaniline, $AsC_{23}H_{25}N_2$,
    2-$MeNHC_6H_4CH=N(CH_2)_3AsPh_2$
Synth.: From $Ph_2As(CH_2)_3NH_2$ + 2-$MeNHC_6H_4CHO$ on a steam bath, drying with $CaSO_4$, dissolving in BuOH, and crystg. at -20° (1317).
Prop.: Yellowish white crystals, m. 84-86° (1317).
Coordination Deriv.: $[Ni(AsC_{23}H_{25}N_2)Br_2]$, olive green, paramagnetic crystals, m. 217°, a nonelectrolyte in $ClCH_2CH_2Cl$, prepd. from $NiBr_2$ and the arsino-amine in hot BuOH. Absorption and reflectance spectra - a 5-coordinate Ni complex postulated (1317).

$[Ni(AsC_{23}H_{25}N_2)I_2]$, brown, paramagnetic crystals, m. 220°, a nonelectrolyte in $ClCH_2CH_2Cl$, prepd. from NiI_2 and the arsinoamine in hot BuOH. Absorption and reflectance spectra - a 5-coordinate Ni complex postulated (1317).

Dinonylphenylarsine, $AsC_{24}H_{23}$, $(C_9H_{19})_2AsPh$
Synth.: From $PhAsCl_2$ + $C_9H_{19}MgBr$ (1:2) in Et_2O, 48% yield (828).
Prop.: Odorless liquid, $b_{0.008}$ 187-190°, d_4^{20} 0.9899, n_D^{20} 1.5066, MR_D 121.21, AR_{As} 11.68 (828).
Rxn. with: S at 100-120° → $Ph(C_9H_{19})_2AsS$ (1549).
$BrCH_2CO_2Et$ on a steam bath → $[Ph(C_9H_{19})_2(EtO_2CCH_2)As]Br$ (1549).
Biol. Prop.: Fungistatic and fungicidal effect against Alternaria radicina, Helminthosporium sativum, Bortrytis cinerea, Fuscicladium dendriticum, Tricho-phyton gypseum, and Epidermophyton. Toxic to mice. Minimum efficacy doses were detd. (1133).

1-(Diphenylarsino)-N,N'-diphenylformamidine Lithium Salt, $AsLiC_{25}H_{20}ON_2$,
    $Ph_2AsC(=NPh)N(Li)Ph$
Synth.: From Ph_2AsLi (prepd. in situ) in Et_2O by adding PhN=C=NPh (1:1) and heating under reflux, 88% yield (1538).
From $Ph_2AsC(=NPh)NHPh$ in Et_2O by treating with PhLi (1:1) (1538).
Prop.: Crystals cont. 1 mol Et_2O, readily sol. in THF and dioxan, sparingly sol. in Et_2O, petr. ether, and C_6H_6 (1538).
Rxn. with: H_2O → $Ph_2AsC(=NPh)NHPh$ (1538).

1-(Diphenylarsino)-N,N'-diphenylformamidine, $AsC_{25}H_{21}N_2$, $Ph_2AsC(=NPh)NHPh$

<u>Synth.</u>: From $Ph_2AsC(=NPh)N(Li)Ph$ in Et_2O by treating with H_2O, 78% yield (1538).

<u>Prop.</u>: Yellow amorphous substance, IR spectrum (1538).

<u>Rxn. with</u>: PhLi (1:1) in $Et_2O \rightarrow Ph_2AsC(=NPh)N(Li)Ph$ (1538).

N,N'-Dicyclohexyl-1-(diphenylarsino)formamidine Lithium Salt,
$AsLiC_{25}H_{32}N_2$, $Ph_2AsC(=NC_6H_{11})N(Li)C_6H_{11}$

<u>Synth.</u>: From Ph_2AsLi (prepd. in situ) in Et_2O by adding $C_6H_{11}N=C=NC_6H_{11}$ (1:1) and heating under reflux, 96% yield (1538).

From $Ph_2AsC(=NC_6H_{11})NHC_6H_{11}$ in Et_2O by treating with PhLi (1:1) (1538).

<u>Prop.</u>: Crystals readily sol. in THF and dioxan, sparingly sol. in Et_2O, petr. ether, and C_6H_6 (1538).

<u>Rxn. with</u>: $H_2O \rightarrow Ph_2AsC(=NC_6H_{11})NHC_6H_{11}$ (1538).

RBr in C_6H_6 under reflux $\rightarrow Ph_2AsC(=NC_6H_{11})N(R)C_6H_{11}$ (1538).

1-(Dicyclohexylarsino)-N,N'-diphenylformamidine Lithium Salt,
$AsLiC_{25}H_{32}N_2$, $(C_6H_{11})_2AsC(=NPh)N(Li)Ph$

<u>Synth.</u>: From $(C_6H_{11})_2AsLi$ (prepd. in situ) in Et_2O by adding $PhN=C=NPh$ (1:1) and heating under reflux, 80% yield (1538).

From $(C_6H_{11})_2AsC(=NPh)NHPh$ in Et_2O by treating with PhLi (1:1) (1538).

<u>Prop.</u>: Crystals readily sol. in THF and dioxan, sparingly sol. in Et_2O, petr. ether, and C_6H_6 (1538).

<u>Rxn. with</u>: $H_2O \rightarrow (C_6H_{11})_2AsC(=NPh)NHPh$ (1538).

1-(Dicyclohexylarsino)-N,N'-diphenylformamidine, $AsC_{25}H_{33}N_2$,
$(C_6H_{11})_2AsC(=NPh)NHPh$

<u>Synth.</u>: From $(C_6H_{11})_2AsC(=NPh)N(Li)Ph$ in Et_2O by treating with H_2O, 62% yield (1538).

<u>Prop.</u>: Light yellow crystals, m. 82-83°, IR spectrum (1538).

<u>Rxn. with</u>: PhLi (1:1) in $Et_2O \rightarrow (C_6H_{11})_2AsC(=NPh)N(Li)Ph$ (1538).

MeI in C_6H_6 under reflux $\rightarrow [Me(C_6H_{11})_2AsC(=NPh)NHPh]I$ (1538).

N,N'-Dicyclohexyl-1-(diphenylarsino)formamidine, $AsC_{25}H_{33}N_2$,
$Ph_2AsC(=NC_6H_{11})NHC_6H_{11}$

<u>Synth.</u>: From $Ph_2AsC(=NC_6H_{11})N(Li)C_6H_{11}$ in Et_2O by treating with H_2O, 80% yield (1538).

<u>Prop.</u>:Colorless crystals, m. 89°, IR spectrum (1538).

<u>Rxn. with</u>: PhLi (1:1) in $Et_2O \rightarrow Ph_2AsC(=NC_6H_{11})N(Li)C_6H_{11}$ (1538).

N,N'-Dicyclohexyl-1-(dicyclohexylarsino)formamidine Lithium Salt,
$AsLiC_{25}H_{44}N_2$, $(C_6H_{11})_2AsC(=NC_6H_{11})N(Li)C_6H_{11}$

<u>Synth.</u>: From $(C_6H_{11})_2AsLi$ (prepd. in situ) in Et_2O by adding $C_6H_{11}N=C=NC_6H_{11}$ (1:1) and heating under reflux, 85% yield (1538).

From $(C_6H_{11})_2AsC(=NC_6H_{11})NHC_6H_{11}$ in Et_2O by treating with PhLi (1:1) (1538).

<u>Prop.</u>: Crystals readily sol. in THF and dioxan, sparingly sol. in Et_2O, petr. ether, and C_6H_6 (1538).

Rxn. with: H_2O → $(C_6H_{11})_2AsC(=NC_6H_{11})NHC_6H_{11}$ (1538).
RBr in Et_2O → $(C_6H_{11})_2AsC(=NC_6H_{11})NRC_6H_{11}$ (1538).

N,N'-Dicyclohexyl-1-(dicyclohexylarsino)formamidine, $AsC_{25}H_{45}N_2$,
    $(C_6H_{11})_2AsC(=NC_6H_{11})NHC_6H_{11}$
Synth.: From $(C_6H_{11})_2AsC(=NC_6H_{11})N(Li)C_6H_{11}$ in Et_2O by treating with H_2O,
81% yield (1538).
Prop.: Colorless crystals, m. 96°, IR spectrum (1538).
Rxn. with: PhLi (1:1) in Et_2O → $(C_6H_{11})_2AsC(=NC_6H_{11})N(Li)C_6H_{11}$ (1538).

1-Anthryldiphenylarsine, $AsC_{26}H_{19}$, $Ph_2AsC_{14}H_9$
Synth.: From Ph_2AsK + 1-anthryl-SO_3K in THF at 67° or in $(EtOCH_2CH_2)_2O$ at
130-180°, 4 and 30% yield, resp. (1347).
Prop.: m. 194-196° (1347).
Rxn. with: 30% aq. H_2O_2 in Me_2CO under reflux → the arsine oxide (1347).
S in C_6H_6 under reflux → the arsine sulfide (1347).

2-Anthryldiphenylarsine, $AsC_{26}H_{19}$, $Ph_2AsC_{14}H_9$
Synth.: From Ph_2AsK + 2-anthryl-SO_3Na in $(EtOCH_2CH_2)_2O$ at 130-180°, 16-18%
yield; in THF at 67° no rxn. occurred (1347).
Prop.: m. 163-165° (1347).
Rxn. with: 30% aq. H_2O_2 in Me_2CO under reflux → $Ph_2(2-C_{14}H_9)AsO$ (1347).
S in C_6H_6 under reflux → $Ph_2(2-C_{14}H_9)AsS$ (1347).

cis-(1,2-Diphenylvinyl)diphenylarsine, $AsC_{26}H_{21}$, $PhCH=C(Ph)AsPh_2$
Synth.: From Ph_2AsLi + $PhC≡CPh$ in THF in the presence of a secondary amine,
62% yield (44, 88).
From Ph_2AsLi + trans $PhCH=CBrPh$ (44).
From the trans isomer by heating in EtOH or during vapor phase chromatography
with 3% silicone at 200° (44).
Prop.: m. 105-106°, dipole moment, NMR spectrum (44).
Rxn. with: $PhCH_2Br$ → $\{Ph_2(PhCH_2)[PhCH=C(Ph)]As\}Br$ (44).

trans-(1,2-Diphenylvinyl)diphenylarsine, $AsC_{26}H_{21}$, $PhCH=C(Ph)AsPh_2$
Synth.: From Ph_2AsLi + $PhC≡CPh$ in THF in the presence of primary amines,
62% yield (44, 88).
Prop.: m. 128-130°, dipole moment, NMR spectrum. On heating in EtOH and
during vapor phase chromatography at 200° the arsine isomerizes to the cis
form (44).

2,6-Di-tert-butyl-4-(diphenylarsino)phenoxy Radical, $AsC_{26}H_{30}O$,
Synth.: The formation of the radical was detected by
ESR spectroscopy during dehydrogenation of 4,3,5-HO-
$(t-Bu)_2C_6H_2AsPh_2$ by alk. soln. of $K_3[Fe(CN)_6]$ in C_6H_6
(1105).

Prop.: Yellow-brown soln. in C_6H_6 which rapidly changes to brown, diamagnetic
3,5,3',5'-tetra-tert-butyldiphenoquinone. ESR data of the radical were obtd.
A mesomeric resonance structure of the radical was suggested (1105).

(3,5-Di-tert-butyl-4-hydroxyphenyl)diphenylarsine, $AsC_{26}H_{31}O$,
 4,3,5-HO(t-Bu)$_2$C$_6$H$_2$AsPh$_2$

Synth.: From Ph$_2$AsK in dioxan + 4,3,5-HO(t-Bu)$_2$C$_6$H$_2$I at 50°, followed by
hydrolysis with H$_2$O free of O, and acidifying with concd. HCl, purified by
chromatography, 5% yield. When Ph$_2$AsLi was used instead of Ph$_2$AsK, the title
prod. was not formed; the rxn. gave Ph$_2$AsH, (Ph$_2$As)$_2$O, and 3,5,3',5'-tetra-
tert-butyldiphenoquinone (1105).
From Ph$_2$AsNa + 4,3,5-HO(t-Bu)$_2$C$_6$H$_2$Br in dioxan, a trace amt. of the title
compd. was obtd. (1105).
From 4,3,5-HO(t-Bu)$_2$C$_6$H$_2$Br + Na in THF, followed by Ph$_2$AsCl (1:1), 29% yield
(1105).
Prop.: Colorless needles, m. 97-98°, IR, NMR, and mass spectra (1105).
Rxn. with: H$_2$O$_2$ → the arsine oxide (1105).
S at 110-130° → the arsine sulfide (1105).
Alkaline soln. of K$_3$[Fe(CN)$_6$] in C$_6$H$_6$ → brown, diamagnetic soln. contg.
3,5,3',5'-tetra-tert-butyldiphenoquinone via a yellow brown 2,6-di-tert-
butyl-4-(diphenylarsino)phenoxy radical (1105).

Bis[2-(diethylamino)ethyl][2-(diphenylarsino)ethyl]amine, $AsC_{26}H_{42}N_3$,
 (Et$_2$NCH$_2$CH$_2$)$_2$NCH$_2$CH$_2$AsPh$_2$

Synth.: From (Et$_2$NCH$_2$CH$_2$)$_2$NCH$_2$CH$_2$Cl + Ph$_2$AsK·2 Dioxan in THF (1316).
Coordination Deriv.: [Co(AsC$_{26}$H$_{42}$N$_3$)X]BPh$_4$, where X = Cl, Br, or I, are
paramagnetic 1:1 electrolytes, 5-coordinate Co complexes. Electronic spectra;
a trigonal bipyramidal structure was suggested (1316).

[Ni(AsC$_{26}$H$_{42}$N$_3$)I]BPh$_4$, a paramagnetic 1:1 electrolyte with 5-coordinate Ni.
Absorption spectrum - a trigonal bipyramidal structure suggested (1316).

[Ni(AsC$_{26}$H$_{42}$N$_3$)(NCS)$_2$], a paramagnetic nonelectrolyte, monomeric. Absorption
spectrum - a 5-coordinate Ni complex with a trigonal bipyramidal structure
postulated (1316).

Didecylphenylarsine, $AsC_{26}H_{47}$, (C$_{10}$H$_{21}$)$_2$AsPh
Rxn. with: S at 100-120° → (C$_{10}$H$_{21}$)$_2$PhAsS (1549).
BrCH$_2$CO$_2$Et on a steam bath → [Ph(C$_{10}$H$_{21}$)$_2$(EtO$_2$CCH$_2$)As]Br (1549).
MeI → [MePh(C$_{10}$H$_{21}$)$_2$As]I (1549).
Biol. Prop.: Fungistatic and fungicidal effect against Alternaria radicina,
Helminthosporium sativum, Bortrytis cinerea, Fusicladium dendriticum, Tricho-
phyton gypseum, and Epidermophyton; toxic to mice. Minimum efficacy doses
were detd. (1133).

(3-Hydroxy-3,3-diphenyl-1-propynyl)diphenylarsine, $AsC_{27}H_{21}O$,
 Ph$_2$AsC≡CC(OH)Ph$_2$

Synth.: From Ph$_2$AsC≡CH via a Grignard derivative and its condensation with
benzophenone, 60% yield (149).
From Ph$_2$AsC≡CH + Ph$_2$CO in KOH soln., 87% yield (149).
Prop.: m. 136° (149).
Rxn. with: H$_2$O$_2$ in AcOH → Ph$_2$As(O)C≡CC(OH)Ph$_2$ (149).

MeI → [MePh$_2$AsC≡CC(OH)Ph$_2$]I (149).

[o-(Diphenylarsino)phenyl]diphenylphosphine, AsC$_{30}$H$_{24}$P,
   2-(Ph$_2$As)C$_6$H$_4$PPh$_2$

Synth.: From 2-BrC$_6$H$_4$AsPh$_2$ by treating with BuLi in Et$_2$O, followed by Ph$_2$PCl,
50% yield (1151, 1153).
Prop.: White crystals, m. 190.5-192.5° (1151, 1153).
Rxn. with: KNCSe in BuOH, no rxn. (1151).
Coordination Deriv.: (In the following complexes AP stands for 2-(Ph$_2$As)-
C$_6$H$_4$PPh$_2$.)

[Co(AP)$_2$Cl]ClO$_4$, red, paramagnetic crystals, a 1:1 electrolyte in MeNO$_2$,
prepd. by adding a soln. of CoCl$_2$·6H$_2$O and Co(ClO$_4$)$_2$·6H$_2$O in EtOH to AP
(1:1:2) in CH$_2$Cl$_2$ and concg. the soln. Absorption spectrum (510, 510a).

[Co(AP)$_2$Br]ClO$_4$, red, paramagnetic crystals, a 1:1 electrolyte in MeNO$_2$,
prepd. by adding the arsinophosphine in CH$_2$Cl$_2$ to a soln. of CoBr$_2$·6H$_2$O and
Co(ClO$_4$)$_2$·6H$_2$O in Me$_2$CO. Absorption spectrum (510, 510a).

[Co(AP)$_2$I]ClO$_4$, maroon, paramagnetic crystals, a 1:1 electrolyte in MeNO$_2$,
prepd. from CoI$_2$ and Co(ClO$_4$)$_2$·6H$_2$O in BuOH by adding to a soln. of the
arsinophosphine in PhCl. Absorption spectrum - a 5-coordinate Co(II) complex (510, 510a).

[Co(AP)$_2$Cl]SnCl$_3$, red, paramagnetic crystals, a 1:1 electrolyte in MeNO$_2$,
prepd. from CoCl$_2$ + SnCl$_2$ in boiling BuOH and adding the arsinophosphine.
Electronic and IR spectra - a 5-coordinate Co(II) complex (510, 510a).

[Co(AP)$_2$Br]SnBr$_3$, paramagnetic crystals, a 1:1 electrolyte in MeNO$_2$, prepd.
from CoBr$_2$ + SnBr$_2$ in boiling BuOH and adding the arsinophosphine. Electronic
spectrum (510).

[Mo(AP)(CO)$_3$Br$_2$], diamagnetic crystals, m. 194°, slightly ionized in MeNO$_2$,
prepd. from NH$_4$[Mo(CO)$_4$Br$_3$] by treating with the arsinophosphine (1:1) in
CH$_2$Cl$_2$ at 25°. IR spectrum (1526).

[Mo(AP)(CO)$_3$I$_2$], diamagnetic crystals, m. 160° (dec.), slightly ionized in
MeNO$_2$, prepd. in the same manner as the preceding dibromo complex. IR spectrum (1526).

[Ni(AP)$_2$](ClO$_4$)$_2$, yellow, diamagnetic crystals, prepd. from Ni(ClO$_4$)$_2$·6H$_2$O +
the arsinophosphine (1:2) in hot PhCl-BuOH soln. Elec. cond. data. Electronic spectrum - a planar 4-coordinate cation postulated (500).

[Ni(AP)$_2$X]ClO$_4$, where X = Cl, Br, I, NO$_2$, NCS, and NCSe, all were obtd. as
purple crystals (with the exception of the nitro complex which is deep blue)
by heating the arsinophosphine with the corr. NiXClO$_4$ soln. in BuOH-PhCl.
The complexes are diamagnetic, 1:1 electrolytes. Electronic absorption spectra were obtd. - a 5-coordinate structure of the Ni complex ions was postulated (500, 510a).

(In the following complexes AP stands for $2\text{-}(Ph_2As)C_6H_4PPh_2$.)

$[Ni(AP)X_2]$, where X = NCS, Cl, Br, and I, were obtd. as red-brown, red, brick red, and violet crystals, resp., from hydrated NiX_2 salts by adding to the arsinophosphine in BuOH-PhCl mixture. All the four complexes are diamagnetic nonelectrolytes in $MeNO_2$. Magnetic susceptibility data and absorption spectra were obtd. (500).

$[Ni(AP)_2(NCS)_2]$, yellow-green, paramagnetic crystals, prepd. from $Ni(SCN)_2$ in EtOH + the arsinophosphine in CH_2Cl_2. Elec. cond. data in $MeNO_2$ indicated an extensive dissocn. to the 5-coordinate $[Ni(AP)_2NCS]^+$ cation. Electronic absorption spectrum indicated a tetragonal distortion from octahedral symmetry. IR spectrum indicated only the N-bonded thiocyanate groups (500).

$\{Ni(AP)[(o\text{-}MeSC_6H_4)_2PPh]\}(ClO_4)_2$, brown, diamagnetic crystals, a 2:1 electrolyte in $MeNO_2$ and MeCN, prepd. from $(o\text{-}MeSC_6H_4)_2PPh$ + AP (1:1) in Me_2CO by adding $Ni(ClO_4)_2$. Electronic absorption spectrum - a Ni complex coordinated with the bi- and tri-dentate ligands, probably with a form of a distorted square pyramide (1625).

$[Pd(AP)Cl_2]$, pale yellow crystals, insol. in MeCN, prepd. by adding AP in BuOH to $Na_2(PdCl_4)$ in EtOH. Electronic spectrum (1151, 1153).

$[Pd(AP)Br_2]$, light yellow crystals, prepd. by adding AP in BuOH to a rxn. mixt. of $Na_2[PdCl_4]$ with LiBr in EtOH. Electronic spectrum (1151, 1153).

$[Pd(AP)I_2]$, orange crystals, a nonelectrolyte in MeCN, prepd. by treating $Na_2[PdCl_4]$ with LiI in EtOH, followed by AP in BuOH (1151, 1153).

$[Pd(AP)(SCN)(NCS)]$, yellow crystals, m. 253-253.5°, a nonelectrolyte in MeCN, prepd. by adding AP in BuOH to a rxn. mixt. of $Na_2(PdCl_4)$ with NaSCN in EtOH. IR and electronic spectra indicated the presence of the thiocyanate groups bonded through the S and N atoms (1153).

$[Pd(AP)(SeCN)_2]$, orange crystals, a nonelectrolyte in MeCN, prepd. by treating $Na_2[PdCl_4]$ with KSeCN in EtOH, followed by AP dissolved in BuOH (1151, 1153). IR and electronic spectra (1153).

$[Pd(AP)_2](NO_3)_2$, pale yellow crystals, m. 224°, a 1:2 electrolyte in MeCN, prepd. from AP + $Pd(NO_3)_2$ in DMF and pptg. the prod. with Et_2O (1151, 1153). Electronic absorption spectrum (1153).

$[Pd(AP)_2][Pd(SCN)_4]$, orange crystals, turned yellow at 190-220°, m. 253.5-254.5°, insol. in MeCN, prepd. by mixing AP with $Na_2[Pd(SCN)_4]$ in EtOH (1151, 1153). IR and electronic absorption spectra indicated the dimeric palladium palladate compn. The complex salt dissolves in hot DMF to give a yellow soln., from which bright yellow crystals, m. 253-253.5° (1151, 1153), can be isolated. The latter is a nonelectrolyte in MeCN and monomeric in $CHCl_3$: $[Pd(AP)(SCN)(NCS)]$ (1153).

$[Pt(AP)Cl_2]$, colorless crystals, insol. in MeCN, prepd. by adding the arsino-phosphine in BuOH to $Na_2[PtCl_4]$ in EtOH (1151, 1153).

[W(AP)(CO)$_3$I$_2$], diamagnetic crystals, m. 242° (dec.), a nonelectrolyte in MeNO$_2$, prepd. from NH$_4$[W(CO)$_4$I$_3$] + the arsenophosphine (1:1) in CH$_2$Cl$_2$ at 25°. IR spectrum - a 7-coordinate W complex (1526).

[o-(Diphenylarsino)phenyl]diphenylphosphine Sulfide, AsC$_{30}$H$_{24}$PS, 2-(Ph$_2$As)C$_6$H$_4$P(S)Ph$_2$

Synth.: From 2-BrC$_6$H$_4$AsPh$_2$ by treating with BuLi in Et$_2$O, followed by Ph$_2$P(S)Cl, 46% yield (1151, 1153).
Prop.: Crystals, m. 190-192°, IR spectrum (1151, 1153). The compd. does not form any stable complexes with the divalent metal ions of the first row (1151).
Coordination Deriv.: (In the following complexes APS stands for 2-(Ph$_2$As)-C$_6$H$_4$P(S)Ph$_2$.)

[Au(APS)Cl], colorless crystals, m. 202-203° (dec.), a nonelectrolyte in MeCN, prepd. from Na[AuCl$_4$] in EtOH by adding APS in BuOH. IR spectrum (1151, 1154). The APS ligand is a monodentate (1151).

[Cu(APS)$_2$]ClO$_4$, colorless to pale green crystals, m. 227-230° (dec.), a 1:1 electrolyte in MeCN, prepd. from Cu(ClO$_4$)$_2$ in EtOH by adding APS in BuOH (1151, 1153). IR spectrum (1151, 1153). Electronic spectrum (1151).

[Pd(APS)Cl$_2$], yellow crystals, a nonelectrolyte in MeCN, prepd. by adding APS in BuOH to Na$_2$[PdCl$_4$] in EtOH. IR and electronic spectra - a 4-coordinate Pd complex (1151, 1153).

[Pd(APS)Br$_2$], orange crystals, a nonelectrolyte in MeCN, prepd. by adding APS in BuOH to a rxn. mixt. of Na$_2$[PdCl$_4$] with LiBr in EtOH. IR and electronic spectra - a 4-coordinate Pd complex (1151, 1153).

[Pd(APS)I$_2$], purple crystals, a nonelectrolyte in MeCN, prepd. by adding APS in BuOH to a rxn. mixt. of Na$_2$[PdCl$_4$] with LiI in EtOH. IR and electronic spectra - a 4-coordinate Pd complex (1151, 1153).

[Pd(APS)(SCN)$_2$], orange crystals, a nonelectrolyte in MeCN, prepd. by adding APS in BuOH to a rxn. mixt. of Na$_2$[PdCl$_4$] in NaSCN in EtOH. IR and electronic spectra. A 4-coordinate Pd complex with the SCN group bonded through the S atom (1151, 1153).

[Pd(APS)(SeCN)$_2$], red-orange crystals, m. 233° (dec.), a nonelectrolyte in MeCN, prepd. by adding APS in BuOH to a rxn. mixt. of Na$_2$[PdCl$_4$] and NaSeCN. IR and electronic spectra. A SeCN bonding through the Se atom postulated (1151, 1153).

[Pd(APS)$_2$](NO$_3$)$_2$, yellow crystals, a 1:2 electrolyte in MeCN, prepd. by adding APS in DMF to Pd(NO$_3$)$_2$ in DMF and pptg. the prod. with Et$_2$O. IR spectrum indicated that the P=S groups are trans to each other (1151, 1153).

[Pt(APS)Cl$_2$], yellow crystals, a nonelectrolyte in MeCN, prepd. by adding APS in BuOH to Na$_2$[PtCl$_4$] in EtOH. Prior to recrystn. from DMF-EtOH mixt. the complex is probably a dimer. IR spectrum (1151, 1153).

(In the following complexes APS stands for $2\text{-}(Ph_2As)C_6H_4P(S)Ph_2$.)

$[Pt(APS)Br_2]$, yellow crystals, a nonelectrolyte in MeCN, prepd. by treating $Na_2[PdCl_4]$ with LiBr in EtOH and then adding APS in BuOH. Prior to recrystn. from DMF-EtOH mixt., the complex is probably a dimer: $[Pt(APS)_2][PtBr_4]$. IR spectrum (1151, 1153).

Butylarsylenebis[N,N'-dicyclohexylformamidine Lithium Salt),
$\qquad$ $AsLi_2C_{30}H_{53}N_4$, $\qquad$ $BuAs[C(=NC_6H_{11})N(Li)C_6H_{11}]_2$
<u>Synth.</u>: From $BuAsLi_2$ (prepd. in situ) in Et_2O by adding $C_6H_{11}N=C=NC_6H_{11}$ (1:2) and heating under reflux, 96% yield (1538).
From $BuAs[C(=NC_6H_{11})NHC_6H_{11}]_2$ in Et_2O by treating with PhLi (1:2) (1538).
<u>Prop.</u>: Crystals, readily sol. in THF and dioxan, sparingly sol. in Et_2O, petr. ether, and C_6H_6 (1538).
<u>Rxn. with:</u> $H_2O \rightarrow BuAs[C(=NC_6H_{11})NHC_6H_{11}]_2$ (1538).

Butylarsylenebis(N,N'-dicyclohexylformamidine), $\qquad$ $AsC_{39}H_{55}N_4$,
$\qquad$ $BuAs[C(=NC_6H_{11})NHC_6H_{11}]_2$
<u>Synth.</u>: From $BuAs[C(=NC_6H_{11})N(Li)C_6H_{11}]_2$ in Et_2O by treating with H_2O, 92% yield (1538).
<u>Prop.</u>: Yellowish oil, IR spectrum (1538).
<u>Rxn. with:</u> PhLi (1:2) in $Et_2O \rightarrow BuAs[C(=NC_6H_{11})N(Li)C_6H_{11}]_2$ (1538).
MeI in C_6H_6 under reflux $\rightarrow \{MeBuAs[C(=NC_6H_{11})NHC_6H_{11}]_2\}I$ (1538).

Table 1. Unsymmetric Tertiary Arsines $R_2R'As$

$R_2R'As$	Prepd. from	Yield %	Prop. and Remarks	Ref.
C_4				
$Me_2(CH_2=CH)As$ (M-64)	$R_2AsX + R'MgBr$ or $R'Li$ in THF		b_{760} 79-80°, UV spectra (a)	1594
Me_2AcAs (M-35)	$R_2AsH + CH_2=C=O$ in Et_2O at -15°	71	Colorless liquid, b_{17} 42-43°, instantaneous oxidn. by air	883
Me_2EtAs (M-64)			b. 85.6° (b)	66
C_5				
$Me_2(CF_3CF=CF)As$ (M-64)	$R_2AsMgBr + CF_3CF=CF_2$ in Et_2O at 20° (c)	75	b. 108-110°, IR spectrum, rxn. with 10% NaOH → $CF_3CF=CHF$	421
$Me_2[HOC(CF_3)_2]As$	$R_2AsH + (CF_3)_2CO$ at 20°		b. 124 $\pm$ 0.5° calcd., IR and NMR spectra	426
$(CH_2=CH)_2MeAs$	$R_2AsI + R'MgI$		b_{760} 120°, UV spectra (a)	1594
$Me_2(CH_2=CHCH_2)As$			b. 108.4° (b)	66
C_6				
$(CF_3)_2[CF_3CH=C(CF_3)]As$	$R_2AsH + CF_3C{\equiv}CCF_3$ at 210°		b. 99-110°, IR and NMR spectra	416
$Me_2[\underline{CF_2C(CF_3)=C}]As$	$R_2AsC{\equiv}CCF_3 + Me_3SnCF_3$ at 140°		b. ~120° (dec.), NMR spectrum	425
$Me_2[CF_3CF=C(CF_3)]As$	$R_2AsCl + CF_3C{\equiv}CCF_3$ at 140° in the presence of $AlCl_3$	50	b. 148-156°	415
$(CH_2=CH)_2EtAs$	$R_2AsI + R'MgBr$		b_{40} 54°, UV spectra (a)	1594
Me_2BuAs (M-64)			Rxn. with HgO in Me_2CO → Me_2BuAsO	1057
$Me_2[Me_3SiSC(S)]As$	$R_2AsSiMe_3 + CS_2$		Dec. 90°	1309
C_7				
$(CH_2=CH)_2(CF_3CF_2CF_2)As$	$R_2AsI + R'I$ in the presence of Hg		b_{40} 44°, UV spectra (a)	1594
$Me_2[\underline{CF_2CF_2CF_2CH=C}]As$	$R_2AsH + R'Cl$ at 100° (d)		b_{100} 95°, IR and NMR spectrum	420

(a) Detd. in i-C_8H_{18}, MeOH, and in the vapor phase (1594). (b) Calcd. by the Kinney equation (66).
(c) $Me_2AsCF_2CHFCF_3$ is formed as a by-prod. (421). (d) Me_2AsCl was formed as a by-prod. (420).

77

Table 1 Continued

$R_2R'As$	Prepd. from	Yield %	Prop. and Remarks	Ref.
C_7 (Cont.)				
$Me_2[\overline{CF_3CHCF_2C(CF_3)}]As$	$R_2AsC(CF_3)=CHCF_3 + Me_3SnCF_3$ at $100°$	10	NMR spectrum	425
$Et_2(CH_2=C=CH)As$	By-prod. from $R_2AsX + CH\equiv CCH_2Br + Mg$ in Et_2O at r.t.	8		150
$Et_2(CH\equiv CCH_2)As$	$R_2AsX + R'Br + Mg$ in Et_2O at r.t.	47	b_{14} $56°$	150
C_8				
$Me_2(3\text{-}BrC_6H_4)As$			b_1 $87\text{-}89°$, sepn. from the o- and p-isomers by gas-liq. chromatography	1303
$Me_2(4\text{-}BrC_6H_4)As$ (M-39)			b_1 $90\text{-}92°$, sepn. from the o- and p-isomers by gas-liq. chromatography	1303
$Me_2(2\text{-}IC_6H_4)As$	$2\text{-}LiC_6H_4AsMe_2 + o\text{-}C_6H_4I_2$ in petr. under reflux		Dark oil	1016
$Me_2(2\text{-}LiC_6H_4)As$	$2\text{-}BrC_6H_4AsMe_2 + BuLi$ in petr.		Rxn. with $o\text{-}C_6H_4X_2$ (X = I or Br) $\rightarrow$ $2\text{-}XC_6H_4AsMe_2$	1016
$Me_2[\overline{CF_2CF_2C(Et)=C}]As$	$R_2AsH + R'Cl$ at $140°$	83	b_{59} $102°$, IR and NMR spectra	420
$Et_2[CH_2=C=C(Me)]As$	$R_2AsX + CH\equiv CCHBrMe + Mg$ in Et_2O at r.t.	33(e)	b_{14} $64°$	150
$Et_2(MeCH=C=CH)As$	$R_2AsX + MeC\equiv CCH_2Br + Mg$ in Et_2O at r.t.	35(f)	b_{14} $62°$	150
$Et_2[HC\equiv CCH(Me)]As$	By-prod. from $R_2AsX + R'Br + Mg$ in Et_2O at r.t.	12		150
$Et_2(MeC\equiv CCH_2)As$	By-prod. from $R_2AsX + R'Br + Mg$ in Et_2O at r.t.	15		150
$(CH_2=CH)_2BuAs$ (M-65)	$R_2AsI + R'MgBr$		b_{18} $64°$, UV spectra in $i\text{-}C_8H_{18}$ and MeOH (g)	1594

78

Compound	Preparation	Yield %	Properties	Ref.
Pr$_2$(CH≡C)As	R$_2$AsI in Et$_2$O + R'MgBr in THF	41	b$_{17}$ 66.5-67.5°, n$_D^{20}$ 1.4848, d$_4^{20}$ 1.0850, MR$_D$ 49.39, AR$_{As}$ 11.15, IR spectrum	913
Et$_2$BuAs			Rxn. with HgO in Me$_2$CO → Et$_2$BuAsO	1057
C$_9$				
Me$_2$(o-HO$_2$CC$_6$H$_4$)As (M-66)			IR spectrum, qual. interpretation	1223
Et$_2$(MeC≡C)As	R$_2$AsX + R'MgX in THF	83	b$_{0.5}$ 70°, IR and NMR spectra; NMR spectrum	149; 1412,1412,1413
Me$_2$(PhCH$_2$)As (M-66)		50	b$_{0.5}$ 55°	1163
Me$_2$(3-MeC$_6$H$_4$)As (M-66)	R$_2$AsNa + R'Br		b$_{19}$ 102-107°	1229
Me$_2$(4-MeC$_6$H$_4$)As			b. 219.7° (b)	66
Et$_2$[MeC≡CCH(Me)]As	R$_2$AsX + MeC≡CCHBrMe + Mg in Et$_2$O at r.t.	48	b$_{14}$ 72°	150
Et$_2$[MeCH=C=C(Me)]As	By-prod. from the preceding rxn.			150
Bu$_2$MeAs (M-66)			Rxn. with HgO in Me$_2$CO and with aq. H$_2$O$_2$ in Et$_2$O → Bu$_2$MeAsO	1057
C$_{10}$				
Et$_2$(4-BrC$_6$H$_4$)As	R$_2$AsCl + R'MgBr	80	b$_{17}$ 148°, d$_4^{20}$ 1.4840, n$_D^{20}$ 1.5895 (h)	340
Et$_2$(4-ClC$_6$H$_4$)As	R$_2$AsCl + R'MgBr	90	b$_{20}$ 142°, d$_4^{20}$ 1.2933, n$_D^{20}$ 1.5658 (h)	340
Et$_2$(4-O$_2$NC$_6$H$_4$)As	R'AsCl$_2$ + R$_4$Pb in C$_6$H$_6$; R'AsCl$_2$ + RMgBr + CdCl$_2$ in Et$_2$O		b$_{5-6}$ 136-138°, d$_4^{20}$ 1.3561, n$_D^{20}$ 1.5889 (h)	340
Bu$_2$(CH$_2$=CH)As (M-66)	R$_2$AsX + R'MgBr or R'Li in THF		b$_{10}$ 86-87°, UV spectrum	1594
Pr$_2$BuAs (M-66)			Rxn. with HgO in Me$_2$CO → Pr$_2$BuAsO	1057
C$_{11}$				
(HO$_2$CCH$_2$)$_2$(2-MeC$_6$H$_4$)As			IR spectrum, qual. interpretation	1223
(HO$_2$CCH$_2$)$_2$(3-MeC$_6$H$_4$)As			IR spectrum, qual. interpretation	1223
(HO$_2$CCH$_2$)$_2$(4-MeC$_6$H$_4$)As			IR spectrum, qual. interpretation	1223

(e) Et$_2$AsCH(Me)C≡CH was also formed. (f) Et$_2$AsCH$_2$C≡CMe by-prod. (g) Rxn. with MnO$_2$ in C$_6$H$_{14}$ or dioxan at 50° → Bu(CH$_2$=CH)$_2$AsO (1594). (h) Rxn. with H$_2$O$_2$ in Me$_2$CO → R$_2$R'AsO (340).

Table 1 Continued

R₂R′As	Prepd. from	Yield %	Prop. and Remarks	Ref.
C₁₁ (Cont.)				
Et₂(4-MeC₆H₄)As (M-67)	R₂AsCl + R′MgCl	63	b_{10} 119-120°, d_4^{20} 1.1418, n_D^{20} 1.5504 (h)	340
Et₂(4-MeOC₆H₄)As	R₂AsCl + R′MgBr in MePh at 65°	51	b. 256.9° (b); b_{18} 146°, d_4^{20} 1.2110, n_D^{20} 1.5614 (h)	66; 340
C₁₂				
(MeC≡C)₂PhAs	R′AsCl₂ + RMgCl in THF	85	$b_{0.01}$ 75°, IR spectrum	149
Me₂(1-C₁₀H₇)As (M-50)	By-prod. from R₂AsNa + 1,8-C₁₀H₆Cl₂ in THF		b_{3-5} mμ 98-100°	145
(HO₂CCH₂CH₂)₂(4-ClC₆H₄)As (M-67)			Dissocn. const. in 20% aq. dioxan at 20° (i)	1222
(CH₂=CHCH₂)₂PhAs (M-67)	RMgCl + R′AsCl₂ (2:1)	90	Rxn. with MeBr → [MeR₂R′As]Br	752
	RMgBr + R′AsCl₂ in Et₂O	45	b_{20} 144°, d_4^{20} 1.1577, n_D^{20} 1.5769, MR_D 66.19, AR_D 11.71	822
	RR′AsCl + RMgBr in Et₂O	54	b_{14} 121-121.5°, d_4^{20} 1.1810, n_D^{20} 1.5689 (j,k)	608
Et₂(PhC≡C)As	R₂AsO₂H + R′H (1:2) in C₆H₆ or MePh (l)	31	$b_{0.8}$ 104-105°, n_D^{20} 1.5906, d_{20} 1.1676, MR_D 67.77, AR_{As} 13.27, IR spectrum	911
(MeO₂CCH₂)₂PhAs	R₂Hg + R′AsS in xylene under reflux			901
	R′AsCl₂ + Et₃SnR (1:2) by heating		b_1 160-161°, n_D^{20} 1.5573, d_4^{20} 1.2914, IR spectrum	901
Pr₂PhAs	RR′AsCl + RMgBr in Et₂O	58	b_{8-9} 123-124.5°, d_4^{20} 1.144, n_D^{20} 1.5418; b_3 99-100°, d_4^{20} 1.1183, n_D^{20} 1.5431 (m)	1548; 605,607

Compound	Preparation	Yield (%)	Properties	Ref.
$Me_2[2-(MeOCH_2CH_2CH_2)C_6H_4]As$	$R_2AsI + R'MgBr$ in C_6H_6 under reflux	66	b_{18} 142-146°, rxn. with HBr → cleavage of the phenyl-As bond	1513
$Et_2(4-Me_2NC_6H_4)As$	$R_2AsCl + R'MgBr$ in THF	56	b_{20} 175-178°, d_4^{20} 1.1625, n_D^{20} 1.5896 (h)	340
$(n-C_5H_{11})_2(CH{\equiv}C)As$	R_2AsI in $Et_2O + R'MgBr$ in THF	12.7	b_{16} 125-127°, n_D^{20} 1.4826, d_4^{20} 1.0180, MR$_D$ 68.16, AR$_{As}$ 30.97, IR spectrum	913
$Pr_2(Et_2NCOCH_2)As$	$R_2AsNEt_2 + CH_2{=}C{=}O$		$b_{1.5}$ 119-120°, n_D^{20} 1.5007, d_4^{20} 1.0915, IR spectrum	901
C₁₃				
$(cyclo-C_6H_{11})_2MeAs$ (M-68)			Complexed with Ru carbonyl catalyzes the oxonation of olefins to aldehydes and alcs.	1424
C₁₄				
$Ph_2(CF_3CO)As$	$R_2AsK\cdot 2$ Dioxan + R'Cl or R'$_2$O in THF at -78° and warming to r.t.	10	Orange-red, $b_{0.001}$ 118-120°, IR spectrum	967
Ph_2AcAs	R_2AsX in $Et_2O + CH_2{=}C{=}O$	69	Light yellow viscous liquid, $b_{0.9}$ 70°, readily oxidized by air	883
Ph_2EtAs	R_2AsH + BuLi in petr. + PhOEt in THF under reflux By-prod. from thermal dec. of $[EtPh_2AsC{\equiv}CR'']Br$	6	b. 105-140° (n); b. 318.2° (b)	1017
$Ph_2(HOCH_2CH_2)As$ (M-68)	$R_2AsK\cdot 2$ Dioxan + $\overline{CH_2CH_2O}$ in dioxan or Et_2O under reflux and dec. with H_2O	66	Colorless liquid, b_2 193-196°, soly. data, rxn. with BzCl → $BzOCH_2CH_2AsPh_2$	1531

(i) The diacid forms very weak chelate complexes with transition metals. but strong 1:1 and protonated 1:1 complexes with Ag^+ (1222). (j) On heating at 180-220° in the presence of Bz_2O_2 polymerizes and copolymerizes with Me methacrylate (608). (k) Rxn. with S in C_6H_6 → $R_2R'AsS$ (603). (l) By heating with azeotropic removal of H_2O (911). (m) Rxn. with S → $R_2R'AsS$ (605, 607, 1548). (n) Rxn. with O → $EtPh_2AsO$ (1017).

Table 1 Continued

$R_2R'As$	Prepd. from	Yield %	Prop. and Remarks	Ref.
C_{14} (Cont.)				
$Et_2(1\text{-}C_{10}H_7)As$	By-prod. from the rxn. of $RR'PhAsS + MeI$ at 50°		b_{10} 140°, d_4^{20} 1.2315, n_D^{20} 1.6312	604
$Pr_2(PhC\equiv C)As$	$R_2AsO_2H + R'H$ (1:2) in C_6H_6 or MePh (1)	36	$b_{0.8}$ 120°, n_D^{20} 1.5710, d_{20}^{20} 1.1206, MR_D 76.92, AR_{As} 13.19, IR spectrum	911
$(EtCOCH_2)_2PhAs$	$R_2Hg + R'AsS$ in xylene under reflux		$b_{0.07}$ 130–132°, n_D^{20} 1.5628, d_4^{20} 1.2473, IR spectrum	901
$(EtO_2CCH_2)_2PhAs$ (M-68)	$R_2Hg + R'AsS$ in xylene under reflux		$b_{0.5}$ 143–144°, n_D^{20} 1.5417, d_4^{20} 1.2842, IR spectrum, (o)	901
Bu_2PhAs (M-68)	$RMgBr + R'AsCl_2$ in Et_2O	90	b_1 150°, d_4^{20} 1.0862, n_D^{20} 1.5341, (p)	609
$(n\text{-}C_6H_{13})_2HC\equiv CAs$	R_2AsI in $Et_2O + R'MgBr$ in THF at ice temp., completed on a steam bath	27	b_2 107–108°, n_D^{20} 1.4827, d_4^{20} 0.9949, MR_D 77.49, AR_{As} 11.54, IR spectrum	913
$(cyclo\text{-}C_6H_{11})_2(HOCH_2CH_2)As$	$R_2AsLi + \underline{CH_2CH_2O}$ in dioxan or ether under reflux and dec. with H_2O	63	Colorless liquid, b_3 172–173°, (q)	1531
$Me_2(n\text{-}C_{12}H_{25})As$ (M-68)	$R_2AsCl + R'MgBr$ in Et_2O		Rxn. with $H_2O_2 \rightarrow R_2R'AsO$	939
C_{15}				
$Ph_2(CF_3CF_2CO)As$	$Ph_2AsK\cdot 2$ Dioxan $+ R'Cl$ or R'_2O in THF at -78° and warming to r.t.	2.5	Red brown liquid, $b_{0.001}$ 128–130°, IR spectrum	967
$Ph_2(HO_2CC\equiv C)As$	$R_2AsC\equiv CH$ via its Mg deriv. and treatment with CO_2	58	m. 130°	149
$(o\text{-}HO_2CC_6H_4)_2MeAs$			IR spectrum	1223
$Ph_2(NCCH_2CH_2)As$ (M-61)	$R_2AsH + CH_2=CHCN$		Rxn. with $LiAlH_4 \rightarrow Ph_2As(CH_2)_3NH_2$	390, 1317

	Preparation	Yield (%)	Properties	Ref.
Ph₂(HOCH₂CH₂CH₂)As	R₂AsK·2 Dioxan + $\overline{CH_2CH_2CH_2O}$ in Et₂O under reflux and dec. with H₂O	52	Colorless crystals, m. 42–43°, b₁ 185–187°, (r)	1531
Ph₂(MeCHOHCH₂)As	R₂AsK·2 Dioxan + $\overline{MeCHCH_2O}$ in Et₂O or dioxan under reflux followed by dec. with H₂O	69	Colorless liquid, b₃ 197–199°, (s)	1531
C₁₈				
Ph₂(HC≡CC≡C)As	R₂AsX + R′MgX in THF at −5°	48	m. 55°, IR spectrum (t, u)	149
	R₂AsX + R′H in liq. NH₃	25		149
Ph₂[CH₂=C=C(Me)]As	R₂AsNa + HC≡CCH(Me)Br in liq. NH₃ at −40°, (v)	22		150
	By-prod. from R₂AsX + HC≡CCHBrMe + Mg in Et₂O, (w)	23		150
Ph₂(MeCH=C=CH)As	By-prod. from R₂AsX + MeC≡CCH₂Br + Mg in Et₂O at r.t.	12		150
Ph₂(EtC≡C)As	R₂AsX + R′MgX in THF at −5°	92	b₀.₀₁ 110°, IR spectrum, rxn. with H₂O₂ in Me₂CO → R₂R′As(OH)₂	149
Ph₂[HC≡CCH(Me)]As	R₂AsX + R′Br + Mg in Et₂O at r.t., (x)	42	b₀.₀₁ 120°	150
	R₂AsNa + R′Br in liq. NH₃ at −40°, (y)	18		150
Ph₂(MeC≡CCH₂)As	R₂AsX + R′Br + Mg in Et₂O at r.t., (z)	58	b₀.₀₁ 120°	150
	[EtR₂R′As]Br above 200°, (aa)	49		150

(o) Elemental analysis involving a rxn. with S and thermal dec. (1502, 1503). (p) Rxn. with S in C₆H₆ at 100° → R₂R′AsO (609). (q) Rxn. with AcCl in CHCl₃ → (C₆H₁₁)₂AsCH₂CH₂OAc (1531). (r) Rxn. with AcCl in CHCl₃ → Ph₂As-(CH₂)₃OAc (1531). (s) Rxn. with RCOCl in CHCl₃ → Ph₂AsCH₂CH(Me)O₂CR (1531). (t) R₂As(C≡C)₂AsR₂ was the major prod. (149). (u) NMR spectrum (149, 1412). (v) Ph₂AsCH(Me)C≡CH and Ph₂AsO₂H were formed as by-prods.(150). (w) Ph₂AsCH(Me)C≡CH was the main prod. (150). (x) Ph₂AsC(Me)=C=CH₂ was also formed (150). (y) Ph₂AsC(Me)=C=CH₂ and Ph₂AsO₂H were also formed (150). (z) Ph₂AsCH=C=CMe was formed as by-prod. (150). (aa) EtPh₂As was also formed (150).

Table 1 Continued

$R_2R'As$	Prepd. from	Yield %	Prop. and Remarks	Ref.
C_{18} (Cont.)				
$(2\text{-}HO_2CC_6H_4)_2EtAs$	$(2\text{-}BrC_6H_4)_2EtAs$ in Et_2O + n-BuLi in petr. + CO_2	58	m. 226-228°	715
$Ph_2[HO(CH_2)_4]As$	R_2AsLi in THF under reflux, followed by dec. with H_2O By-prod. from R_2AsH + BuLi followed by PhOEt in THF under reflux	55	Colorless liquid, b_{3-4} 195-197°, rxn. with MeI → $[MeR_2R'As]I$ $b_{0.4}$ 172-174°, IR spectrum	1531 1017
$Bu_2(PhC{\equiv}C)As$	R_2AsO_2H + R'H (1:2) in C_6H_6 or MePh under reflux (bb)	41	b_1 145-147°, n_D^{20} 1.5611, d_{20}^{20} 1.0904, MR_D 86.21, AR_{As} 13.24, IR spectrum	911
$(C_6H_{11})_2(AcOCH_2CH_2)As$	$R_2(HOCH_2CH_2)As$ + AcCl in $CHCl_3$ under reflux	63	Colorless liquid, b_9 182-184°	1531
$(C_6H_{11})_2[Cl(CH_2)_4]As$	R_2AsLi + R'Cl in dioxan, (cc)	55	Oily liquid, dec. during distn., at 170° → $[R_2AsCH_2CH_2CH_2CH_2]Cl$	1533
$(C_6H_{11})_2[HO(CH_2)_4]As$	R_2AsLi in THF under reflux, followed by dec. with H_2O	47	Colorless liquid, b_1 180-181°	1531
$Et_2(C_{12}H_{25})As$	R_2AsCl + R'MgBr in Et_2O		Rxn. with H_2O_2 → $R_2R'AsO$	939
C_{17}				
$Ph_2(MeC{\equiv}CC{\equiv}C)As$	R_2AsX + R'MgX in THF at -5°	76	m. 60°, IR spectrum NMR spectrum	149, 149, 1412
$Ph_2[MeC{\equiv}CCH(Me)]As$	R_2AsX + R'Br + Mg in Et_2O at r.t.	70	$b_{0.01}$ 125°	150
$Ph_2[AcO(CH_2)_3]As$	$R_2As(CH_2)_3OH$ + AcCl in $CHCl_3$ under reflux	64	Colorless liquid, b_3 187-190°	1531
$Ph_2[AcOCH(Me)CH_2]As$	$R_2AsCH_2CH(Me)OH$ + AcCl in $CHCl_3$ under reflux	65	Colorless liquid, b_4 178-180°	1531

Compound	Method of preparation	Yield %	Properties	Ref.
$(C_6H_{11})_2[Cl(CH_2)_5]As$	$R_2AsLi + R'Cl$ (1:1) in dioxan	78	Colorless liquid, $b_{0.0005}$ 142°, sensitive to air	1533
$(C_6H_{11})_2[EtN=C(SEt)]As$	$R_2AsLi + EtNCS$ in Et_2O under reflux, replacing the solvent with C_6H_6 and heating with EtBr	81	Yellowish oil, IR spectrum	1537
C_18				
$(C_6F_5)_2PhAs$	$RLi + R'AsCl_2$ in Et_2O at -78° and warming to r.t.	50	m. 73-75°	842
$(4-ClC_6H_4)_2PhAs$ (M-69)	$RMgI + R'AsCl_2$		m. 61.5°, rxn. with K-Na in decalin at 180° $\rightarrow \{C_6H_4C_6H_4As(Ph)\}_n$	739
$(2-BrC_6H_{11})_2PhAs$ (M-69)	R_2AsCl in $C_6H_6 + R'MgBr$ in Et_2O	95	Crystals, m. 112-113°	715
$Ph_2(2-HSC_6H_4)As$	$R_2(2-BrC_6H_4)As + BuLi$ in Et_2O at 0°, followed by S + 4% NaOH and acidification		White crystals, m. 80-81°, rxn. with Na in EtOH + $BrCH_2CH_2Br \rightarrow (2-Ph_2AsC_6H_4SCH_2-)_2$	292
$Ph_2(2-HOC_6H_{10})As$	$R_2AsK \cdot 2$ Dioxan + cyclohexene oxide in dioxan or ether, (dd)	59	Colorless crystals, m. 130°, sol. in dioxan and ether	1531
$(C_6H_{11})_2(2-HOC_6H_{10})As$	R_2AsLi + cyclohexene oxide in dioxan or Et_2O, (dd)	43	Colorless crystals, m. 53-54°, b_{10} 233-236°	1531
$(C_6H_{11})_2[Cl(CH_2)_6]As$	$R_2AsLi + R'Cl$ in dioxan	60	Colorless liquid, $b_{4.4}$ 165°, sensitive to air	1533
$Me_2(n-C_{18}H_{33})As$	$R_2AsCl + R'MgBr$ in Et_2O		Rxn. with $H_2O_2 \rightarrow R_2RAsO$	939
C_19				
$Ph_2(MeC\equiv CC\equiv CC\equiv C)As$	$R_2AsX + R'MgX$ in THF at -20°	60	m. 125°, IR spectrum, (ee)	149
$Ph_2(PrC\equiv CC\equiv C)As$	$R_2AsX + R'MgX$ in THF at -5°	75	m. 40°, IR spectrum	149
$(C_6H_{11})_2[PhN(Li)CO]As$	$R_2(PhNHCO)As + PhLi$ (1:1) in Et_2O, (ff)	82	Solid, sol. in dioxan and THF, sparingly sol. in petr. ether	1536

(bb) With azeotropic removal of H_2O (911). (cc) $R_2As(CH_2)_4AsR_2$ was also formed (1533). (dd) Under reflux followed by a treatment with H_2O. (ee) NMR spectrum (149, 1412). (ff) The prod. was pptd. with petr. ether.

Table 1 Continued

$R_2R'As$	Prepd. from	Yield %	Prop. and Remarks	Ref.
C_{19} (Cont.)				
$(C_6H_{11})_2[PhN(Li)CS]As$	R_2AsLi in Et_2O + PhNCS under reflux + dioxan	72	Light yellow crystals contg. 1 mol dioxan, soly. data	1537
$(C_6H_{11})_2[C_6H_{11}N(Li)CO]As$	$R_2(C_6H_{11}NHCO)As$ + PhLi (1:1) in Et_2O, pptd. with petr. ether	81	Crystd. with 1 mol dioxan	1536
C_{20}				
$Ph_2(\overline{CH_2CH_2CH_2CH_2CH{=}CC{\equiv}C})As$	R_2AsX + $R'MgX$ in THF	70	m. 42°, IR spectrum	149
$Ph_2[PhCH(OH)CH_2]As$	$R_2AsK \cdot 2$ Dioxan + $PhCHCH_2O$ in dioxan or Et_2O under reflux and dec. with H_2O	76	Pale yellow oil, nondistillable in vacuum	1531
C_{21}				
$Ph_2(PhC{\equiv}CCH_2)As$	$[R_2R'EtAs]Br$ by heating > 200°	63	Colorless liquid, b_2 247-250°, sol. in dioxan, $CHCl_3$, and Et_2O	150
$Ph_2(BzOCH_2CH_2)As$	$R_2AsCH_2CH_2OH$ + BzCl in $CHCl_3$ under reflux	50		1531
C_{22}				
$Ph_2(PhC{\equiv}CC{\equiv}C)As$	R_2AsX + $R'MgX$ in THF at -5°	55	m. 63°, IR spectrum	149
$(PhC{\equiv}C)_2PhAs$	$R'AsX_2$ + $RMgX$ in THF at r.t.	63	m. 63°, IR spectrum	149
$Ph_2(1\text{-}C_{10}H_7)As$ (M-71)	R_2AsK + $R'SO_3Na$ in $(EtOCH_2CH_2)_2O$ at 180° or in THF at 67°		m. 111°, (gg, hh)	1347
$Ph_2(2\text{-}C_{10}H_7)As$ (M-71)	R_2AsK + $R'SO_3Na$ in $(EtOCH_2CH_2)_2O$ at 180°	21	(gg, hh)	1347
$Ph_2[BzOCH(Me)CH_2]As$	$R_2AsCH_2CHOHMe$ + BzCl in $CHCl_3$, (ii)	46	Colorless liquid, b_1 245-247°	1531
C_{27}				
$Ph_2[Et(C_6H_{11})NC({=}NC_6H_{11})]As$	$R_2AsC({=}NC_6H_{11})N(Li)C_6H_{11}$ in C_6H_6 + EtBr, (ii)	84	Colorless oil, sol. in org. solvents	1538

$(C_6H_{11})_2[Et(C_6H_{11})NC(=NC_6H_{11})]As$ $R_2AsC(=NC_6H_{11})N(Li)C_6H_{11}$ in   Viscous oil, sol. in C_6H_6 1538

C_6H_6 + EtBr, (ii)   and THF

94

(gg) Rxn. with 30% H_2O_2 in $Me_2CO \rightarrow R_2R'AsO$ (1347).   (hh) Rxn. with S in C_6H_6 under reflux $\rightarrow R_2R'AsS$ (1347).

(ii) Under reflux.

1.1.3 UNSYMMETRIC AND ASYMMETRIC RR'R"As DERIVATIVES

The methods of preparation of the unsymmetric and asymmetric arsines, RR'R"As, employed through 1964 were summarized in the main volume (M-72). Since then a few new synthesis methods have been reported which are based on secondary arsines and arsinolithiums containing two different organic groups. A condensation of secondary arsines with formaldehyde yields arsinomethanols, with isocyanates yields arsinoformamides, with acetylene derivatives yields alkenylarsines, and with perhalogenated cyclobutene yields cyclobutenylarsines:

$$RR'AsH + HCHO \xrightarrow{HCl} RR'AsCH_2OH$$

$$RR'AsH + R''NCO \xrightarrow{Bu_2Sn(OAc)_2} RR'AsCONHR''$$

$$RR'AsH + HC\equiv CR'' \xrightarrow{20°} R'R'AsCH=CHR''$$

$$RR'AsH + \overline{CF_2CF_2CCl=CCl} \xrightarrow{20°} RR'AsC=CClCF_2CF_2$$

A reaction of secondary arsinolithiums with isocyanates and isothiocyanates, followed by hydrolysis, affords arsinoformamides and arsinothioformamides, respectively.

$$RR'AsLi + R''NCZ \rightarrow +H_2O \rightarrow RR'AsCZNHR'', \quad Z = O \text{ or } S$$

A reduction of tertiary arsine oxides, RR'R"AsO, with sodium sulfite in alcoholic HCl in the presence of trace amounts of potassium iodide affords the corresponding arsines, RR'R"As.

Arsinic acids react with acetylene derivatives in benzene under reflux, with azeotropic removal of water, to form tertiary arsines with one alkynyl group.

$$RR'AsO_2H + 2 HC\equiv CR'' \rightarrow RR'AsC\equiv CR''$$

Optically active tertiary arsines are formed from quaternary arsonium salts containing one or two allyl groups and the other groups different from each other by cyanolysis with 10% aqueous potassium cyanide under reflux.

$$[RR'R''(CH_2=CHCH_2)As]X \xrightarrow{KCN} RR'R''As + CH_2=C(Me)CN + KX$$

Some arsines of the RR'R"As type are described hereafter individually, while others are compiled in Table 2.

(+)-Methylphenylpropylarsine, $AsC_{10}H_{15}$, MePrPhAs, (M-75)
Prop.: The optical activity remained unchanged during heating at 200° for 10 hrs. A methanolic soln. of the arsine contg. mineral acids rapidly racemized at 20°, while such a soln. contg. NaCl remained unchanged (751). The racemization occurred in MeOH, MePh, and MeC_6H_{11} in UV light and in visible light in the presence of sensitizers such as ketones (753).

(-)-Methylphenylpropylarsine, $AsC_{10}H_{15}$, MePrPhAs, (M-75)
Prop.: $[\alpha]_D^{20}$ -10.5° (752).
Rxn. with: $CH_2=CHCH_2Br \rightarrow [(-)-MePrPh(CH_2=CHCH_2)As]Br$ (752).

Allylethylphenylarsine, $AsC_{11}H_{15}$, $(CH_2=CHCH_2)EtPhAs$
Synth.: From $EtPhAsCl + CH_2=CHCH_2MgBr$ in Et_2O, 45-53% yield (608, 822).
Prop.: b_{20} 143-144°, d_4^{20} 1.1577, n_D^{20} 1.5769 (608), b_{14} 121.5°, d_4^{20} 1.1810, n_D^{20} 1.5689, MR_D 61.08, AR_D 11.14 (822). At 180-200° in the presence of Bz_2O_2 the arsine polymerizes and copolymerizes with Me methacrylate (608).
Rxn. with: S in $C_6H_6 \rightarrow (CH_2=CHCH_2)EtPhAsS$ (603).

p-(Ethylpropylarsino)benzoic Acid, $AsC_{12}H_{17}O_2$, $4-(EtPrAs)C_6H_4CO_2H$
Synth.: From $EtPrAsC_6H_4Me-p + KMnO_4$ in H_2O at 90°, followed by a redn. of the arsine oxide deriv. with SO_2 in HCl soln., 91% yield (601).
Prop.: A racemic mixture, m. 59°, the quinine salt m. 190-193°, $[\alpha]_D^{20}$ -108.18°; dec. of the latter salt with H_2SO_4 gave the (+)-isomer, m. 59.3°, $[\alpha]_D^{20}$ 3.01°. The D-α-phenylethylamine salt, m. 132-134°, $[\alpha]_D^{20}$ + 6.56°; dec. of the amine salt gave the (-)-isomer, m. 57°, $[\alpha]_D^{20}$ -3.04° (601).

p-(Butylethylarsino)benzoic Acid, $AsC_{13}H_{19}O_2$, $4-(EtBuAs)C_6H_4CO_2H$
Synth.: From $4-(EtBuAs)C_6H_4Me + KMnO_4$ in aq. soln. at 90°, followed by a redn. with SO_2 in HCl soln., 79% yield (601).
Prop.: A racemic mixt., m. 36-37°; the quinine salt, m. 164-165°, $[\alpha]_D^{20}$ -113.1°, after fractional crystn. and dec. with H_2SO_4 gave the (-)-isomer, m. 36.4°, $[\alpha]_D^{20}$ -3.96° (601).

p-(Ethylphenylarsino)benzoic Acid, $AsC_{15}H_{15}O_2$, $4-(EtPhAs)C_6H_4CO_4H$
Synth.: From $EtPh(4-MeC_6H_4)As + KMnO_4$ in H_2O at 95°, followed by redn. with SO_2 in HCl soln. (606).
Prop.: A racemic mixture, m. 123°; the quinine salt, m. 174-181°, $[\alpha]_D^{20}$ -103.3°. Fractional crystn. of the salt gave a fraction m. 179-181°, $[\alpha]_D^{20}$ -111.1°; dec. of the latter fraction gave the (+)-isomer, m. 119.5°, $[\alpha]_D^{20}$ + 11.64° (602, 606). The (-)-isomer, m. 121°, $[\alpha]_D^{20}$ -5.35° was obtd. from the mother liquor which remained after isolation of the (+)-isomer by dec. with H_2SO_4 (602, 606).
Rxn. with: S in C_6H_6 at 90° → both isomers gave the arsine sulfide (602).

1-(Cyclohexylphenylarsino)-N-ethylthioformamide, $AsC_{15}H_{22}NS$,
  $Ph(C_6H_{11})AsCSNHEt$
Synth.: From $Ph(C_6H_{11})AsLi$ in Et_2O by adding dropwise EtNCS in Et_2O at r.t.,

heating under reflux, and hydrolyzing with H_2O, 50% yield (1537).

Prop.: Light yellow crystals, m. 58°, sol. in Et_2O and EtOH, sparingly sol. in petr. ether (1537).

1-(Cyclohexylphenylarsino)formanilide, $AsC_{19}H_{22}NO$, $Ph(C_6H_{11})AsCONHPh$

Synth.: From $Ph(C_6H_{11})AsH$ + PhNCO (1:1) in dioxan in the presence of Bu_2Sn-$(OAc)_2$ under reflux, followed by distn. of dioxan and crystn. by addn. of petr. ether, 75% yield (1536).

From $Ph(C_6H_{11})AsLi$ + PhNCO (1:1) in Et_2O at r.t., followed by heating and, after cooling, dec. with H_2O, 40% yield (1536).

Prop.: Colorless crystals (from Et_2O), m. 112°, sol. in C_6H_6, THF, and EtOH, sparingly sol. in petr. ether and H_2O; IR spectrum (1536).

(Cyclohexylphenylarsino)thioformanilide, $AsC_{19}H_{22}NS$, $Ph(C_6H_{11})AsCSNHPh$

Synth.: From $Ph(C_6H_{11})AsLi$ in Et_2O by adding dropwise PhNCS in Et_2O at r.t., heating under reflux, and dec. with H_2O, 51% yield (1537).

Prop.: Yellow crystals, m. 55°, sol. in Et_2O and EtOH, sparingly sol. in petr. ether (1537).

o-(2-Naphthylphenylarsino)benzoic Acid, $AsC_{23}H_{27}O_2$,
    $2-[Ph(2-C_{10}H_7)As]C_6H_4CO_2H$

Synth.: From $Ph(2-C_{10}H_7)(2-HO_2CC_6H_4)AsO$ in alc. HCl + Na_2SO_3 and a trace of KI in H_2O, 64% yield (601).

Prop.: A racemic mixt., m. 129-130°. The quinine salt, m. 214-217°, $[\alpha]_D^{20}$ -126.38°. A fractional crystn. and dec. of the salt with 3% HCl gave the (+)-isomer, m. 131.4°, $[\alpha]_D^{20}$ 6.04°, and the (-)-isomer, m. 126-127°, $[\alpha]_D^{20}$ -4.93° (601).

Table 2. Unsymmetric and Asymmetric RR'R"As Derivatives

RR'R"As	Prepd. from	Yield %	Prop. and Remarks	Ref.
C_4				
Me(HOCH$_2$)(HCF$_2$CF$_2$)As	RR"AsH + 36% aq. HCHO in the presence of HCl	38	b$_{10}$ 68°, d$_4^{20}$ 1.7231, n$_D^{20}$ 1.4287	1266
C_9				
(+)-EtMePhAs			The optical rotation remained unchanged at 200° for 10 hrs., (a)	751
C_{10}				
Me(CH$_2$=CHCH$_2$)PhAs (M-75)	[RR'$_2$R"As]Br + 10% aq. KCN under reflux	85	Rxn. with PhCH$_2$Cl, followed by ClO$_4^-$ → [RR'R"(PhCH$_2$)As]ClO$_4$	752
EtMe(4-MeC$_8$H$_4$)As (M-75)			(a)	603
C_{11}				
MePh($\overline{\text{CF}_2\text{CF}_2\text{CCl}=\text{C}}$)As	RR'AsH + R"Cl at 20°	86	b$_{21}$ 133°, IR spectrum	421
MePh($\overline{\text{CF}_2\text{CF}_2\text{CF}=\text{C}}$)As	RR'AsH + R"F at 100°	30	b$_{15}$ 104°, IR and NMR spectra	421
MePh[CF$_3$CH=C(CF$_3$)As	RR'AsH + CF$_3$C≡CCF$_3$ at 20°		b$_{0.001}$ 54°, IR and NMR spectra	416
(+)-MeBuPhAs (M-75)			The optical rotation remained unchanged at 200° for 10 hrs.	751
EtPrPhAs (M-76)			(a)	603
C_{12}				
EtBuPhAs (M-76)			(a)	603
C_{13}				
Et(n-C$_5$H$_{11}$)PhAs (M-76)			(a)	603
C_{14}				
(+)-MePh(PhCH$_2$)As (M-76)	(+)-[RR'R"(CH$_2$=CHCH$_2$)As]Br + 10% aq. KCN under reflux		[α]$_D^{20}$ + 56°	752
Et(n-C$_8$H$_{13}$)PhAs (M-77)			(a)	603

C_{15}				
EtPh(4-MeC_6H_4)As (M-77)	RR'AsCl + R''MgBr in Et_2O	70	b_2 135°, d_4^{20} 1.2189, n_D^{20} 1.6111, (b)	606
EtPh(PhCH_2)As	[RR'R''_2As]Br by cathodic cleavage		Rxn. with BuBr → [RR'R''BuAs]Br	750
PrBu(PhC≡C)As	RR'AsO_2H + R''H (1:2) in MePh under reflux, (c)	26	b_1 141-143°, n_D^{20} 1.5648, d_{20}^{20} 1.1043, MR_D 81.44, AR_{As} 13.09, IR spectrum	911
Et(C_7H_{15})PhAs (M-77)			(a)	603
C_{16}				
EtPh[4-(Et_2N)C_6H_4]	RR'AsCl + R''MgBr in THF under reflux	46	b_3 186-187°, n_D^{20} 1.6296, d^{20} 1.2473	601
Et(n-C_8H_{17})PhAs (M-77)			(a)	603
C_{17}				
Bu(C_5H_{11})(PhC≡C)As	RR'AsO_2H + R''H (1:2) in C_6H_6 or MePh under reflux, (c)	34	b_1 164-165°, n_D^{20} 1.5541, d_{20}^{20} 1.0730, MR_D 90.92, AR_{As} 13.34, IR spectrum	911
Et(n-C_9H_{19})PhAs (M-77)			(a)	603
Et(n-C_8H_{17})(4-MeC_6H_4)As (M-77)			(a)	603
C_{18}				
Et(1-C_{10}H_7)PhAs	RR''AsCl + R'MgBr in Et_2O or without solvent at 100°	70	m. 58°, (a)	604, 605
Bu(C_6H_{13})(PhC≡C)As	RR'AsO_2H + R''H (1:2) in C_6H_6 or MePh under reflux, (c)	31.5	$b_{0.8}$ 151-152°, n_D^{20} 1.5480, d_{20}^{20} 1.0601, MR_D 95.36, AR_{As} 13.15, IR spectrum	911
C_{19}				
Et(1-C_{10}H_7)(4-MeC_6H_4)As (M-78)	RR'AsCl + R''MgBr in THF	49	b_2 203-204°, n_D^{20} 1.6575, d^{20} 1.2464, (b)	601
C_{23}				
Ph(2-C_{10}H_7)(2-MeC_6H_4)As	RR'AsCl + R''MgBr in Et_2O	80	A glassy mass, (b)	601

(a) Rxn. with S in C_6H_6 → RR'R''AsS (603, 604, 605).   (b) Rxn. with KMnO_2 in H_2O, followed by redn. with SO_2 → RR'AsC_6H_4CO_2H (601, 606).   (c) With azeotropic removal of H_2O (911).

1.1.4 DITERTIARY ARSINES

The diarsines and their methods of synthesis were described in the main volume (M-79 to M-104). In addition to the previously reported methods, ditertiary arsines have been prepared as follows:

From secondary arsine halides by a condensation with butadiyne in liquid ammonia and with butadiynylenebis(trialkyltin) compounds, respectively,

$$2\ R_2AsX + HC\equiv CC\equiv CH \xrightarrow{\text{liq. } NH_3} R_2AsC\equiv CC\equiv CAsR_2$$

$$2\ R_2AsX + R'_3SnC\equiv CC\equiv CSnR'_3 \rightarrow R_2AsC\equiv CC\equiv CAsR_2\ \ ;$$

From secondary arsinoalkali metal compounds by a condensation with halo-arenesulfonic acid salts and with arylenedisulfonic acid salts, respectively,

$$2\ R_2AsM + XC_nH_mSO_3Na \rightarrow R_2AsC_nH_mAsR_2$$

$$2\ R_2AsM + NaO_3SC_xH_ySO_3Na \rightarrow R_2AsC_xH_yAsR_2\ \ ;$$

From secondary arsines and halo-substituted tertiary arsines by dehydro-halogenation coupling at elevated temperatures,

$$R_2AsH + ClC_nH_yAsR'_2 \xrightarrow[-HCl]{140°} R_2AsC_nH_yAsR'_2\ \ ;$$

From tertiary arsines containing at least one reactive group by a condensation with appropriate bifunctional reagents such as α-diketones,

$$2\ R_2AsC_xH_yNH_2 + Ac_2 \rightarrow R_2AsC_xH_yN=C(Me)C(Me)=NC_xH_yAsR_2\ \ ;$$

From dihalo-palladium and -platinum complexes with [o-(alkylthio)phenyl]di-methylarsines by a dealkylation during heating in DMF or THF, dimeric dimethyl-arsinophenylthio-palladium and -platinum halides, respectively, bridged through the S atoms, are formed. When the dealkylation is carried out in the presence of (o-alkylthiophenyl)dimethylarsines, bis[(dimethylarsino -phenylthio]-palladiums and -platinums, respectively, are formed. The dimeric arsinophenylthio-palladium halides and -platinum halides are also formed by a thermal dealkylation of bis[(o-alkylthiophenyl)dimethylarsine]-palladium tetrahalopalladates and -platinum tetrahaloplatinates, respectively,

$$[M(o\text{-}RSC_6H_4AsMe_2)X_2] \xrightarrow{\Delta} [Me_2AsC_6H_4SMX]_2 \quad M = Pd \text{ or } Pt$$
$$X = Cl, Br, \text{ or } I$$

$$[M(2\text{-}RSC_6H_4AsMe_2)_2][MX_4] \xrightarrow{\Delta} [Me_2AsC_6H_4SMX]_2$$

$$[M(2\text{-}RSC_6H_4AsMe_2)X_2] \xrightarrow{2\text{-}RSC_6H_4AsMe_2} [2\text{-}(Me_2As)C_6H_4S]_2M\ \ .$$

Moreover, trialkylphosphonium bis(arsino)methylides are formed from tri-alkylphosphonium arsinomethylides by pyrolysis:

$$R_3P=CHAsR'_2 \xrightarrow{\Delta} R_3P=C(AsR'_2)_2\ \ .$$

The following entries are arranged in the order of the increasing numbers of carbon and hydrogen atoms.

Vinylenebis(dimethylarsine), $As_2C_6H_{14}$, $Me_2AsCH=CHAsMe_2$

Synth.: From Me_2AsI + Na in THF, followed by cis-ClCH=CHCl, 30-40% yield contg. ~ 10% cis and 90% trans isomers (559).

From preformed Me_2AsNa in THF and cis-ClCH=CHCl, a prod. contaminated with $(Me_2As-)_2$ was obtd. When trans-ClCH=CHCl was used, a mixt. of the title compd. with cacodyl was obtd. (1229).

Prop.: A mixt. of the cis and trans isomers (1:9), b_{24} 90-93° (559), b_{19} ~ 80° (1229), n_D^{25} 1.5567, extremely flammable and malodorous. The cis isomer was isolated by complexing with $FeCl_3 \rightarrow [FeCl_2(Me_2AsCH=CHAsMe_2)]FeCl_4$ and displacing the diarsine with Ph_3P at 150° (559). NMR spectrum (559, 1229).

Rxn. with: $PdCl_2 \rightarrow$ an extremely insol. complex (1229).

Coordination Deriv.: (VDA in the following complexes denotes $Me_2AsCH=CHAsMe_2$.)

trans-$[Co(cis-VDA)_2Br_2]Br$, green crystals, prepd. from $CoBr_2$ in $MeOH-Et_2O$ (4:1) and VDA by heating to 55° and bubbling air through the rxn. mixt. Electronic spectrum (559).

$[Cu(cis-VDA)_2]ClO_4$, white ppt., prepd. from $Cu(ClO_4)_2 \cdot 6H_2O$ in MeOH by adding the diarsine. Insol. in org. solvents, dec. in DMSO (559).

$[Fe(cis-VDA)_3](ClO_4)_2$, orange solid, prepd. from $Fe(ClO_4)_2 \cdot 6H_2O$ and the diarsine in EtOH on a steam bath. Electronic spectrum (559).

$[Fe(cis-VDA)Cl_2]FeCl_4$, bright crimson ppt., prepd. from $FeCl_3$ and a cis-, trans-VDA mixt. in EtOH (the complex with trans-VDA, if formed, is sol. in EtOH). cis-VDA is readily displaced from the complex when heated with Ph_3P at 150°/0.1 mm (559).

trans-$[Fe(cis-VDA)_2Br_2]ClO_4$, green ppt., electronic absorption spectrum (559).

trans-$[Fe(cis-VDA)_2Cl_2]FeCl_4$, red ppt., electronic absorption spectrum (559).

$[Fe(cis-VDA)_3](ClO_4)_3$, dark green ppt., prepd. from $[Fe(cis-VDA)_3](ClO_4)_2$ by dissolving in concd. HNO_3 and adding $NaClO_4$ soln. Electronic absorption spectrum (559).

$[Pd(cis-VDA)_2Cl]ClO_4$, brown powder, prepd. from K_2PdCl_4 dissolved in aq. EtOH by adding the diarsine, filtering, and boiling with 70% $HClO_4$ (559).

trans-VDA coordinates with Pt(II) in the trans configuration, which isomerizes under irradiation to a chelated complex in a cis configuration (422).

$[Ti(VDA)Br_4]$: two complexes of this empirical formula were isolated; a bright red cryst. matl. (I) was obtd. by adding $TiBr_4$ to excess VDA in cyclohexane. Upon further addition of $TiBr_4$, a deep red matl. (II) was formed. (I) is sol. in C_6H_6 and in other org. solvents, while (II) is almost insol. in C_6H_6. (I) is a nonelectrolyte in $PhNO_2$; it is monomeric in C_6H_6 and is considered to be a 6-coordinate Ti complex with cis-VDA, whereas (II) may be a 6-coordinate Ti complex with trans-VDA in a polymeric form. IR and diffuse reflectance spectra of both complexes were obtd. (363).

1,2-Bis(trifluoromethyl)vinylenebis[bis(trifluoromethyl)arsine],
     $As_2C_8F_{18}$, $(CF_3)_2AsC(CF_3)=C(CF_3)As(CF_3)_2$
Synth.: From $[(CF_3)_2As-]_2$ + $CF_3C\equiv CCF_3$ under UV irradiation (415).
Prop.: b. 156° $\pm$ 2° (calcd.), IR spectrum (415).
Rxn. with: a base → trans-$CF_3CH=CHCF_3$ (415).

(3,3,4,4-Tetrafluoro-1-cyclobuten-1,2-ylene)bis(dimethylarsine),
     $As_2C_8H_{12}F_4$, $\overline{CF_2CF_2C(AsMe_2)=CAsMe_2}$
Synth.: From Me_2AsH by heating with $\overline{CF_2CF_2CCl=CAsMe_2}$ at 140°, 52% yield
(421).
Prop.: b_{47} 120°, sensitive to air, IR spectrum (421).
Coordination Deriv.:
  (In the following complexes TFDA denotes $\overline{CF_2CF_2C(AsMe_2)=CAsMe_2}$.)

{TFDA[Fe(CO)$_3$]$_2$}, red prisms, crystallographic data. The following structure
was postulated: one Fe atom is octahedrally coordinated to 3 CO groups, 2 As
atoms, and the sixth position forms a bond with the other Fe atom. The other
Fe atom has a trigonal bipyramidal coordination with 2 CO and a π-bonded olefinic group in the equatorial positions, the third CO in one apical position,
while the other apex is filled by the Fe-Fe bond (520, 522).

[Fe(TFDA)(CO)$_3$], a by-prod. formed from TFDA + $Fe_3(CO)_{12}$ in UV light (424).

[Fe$_3$(TFDA)(CO)$_{10}$], black cryst. solid, m. ~ 160° (dec.), prepd. from TFDA and
$Fe_3(CO)_{12}$ in UV light. IR, Mössbauer, and mass spectra - an interpretation
of the structure and bonding (424).

[Hg(TFDA)Cl$_2$], white solid, m. 123°, prepd. from TFDA + $HgCl_2$ in Me_2CO (422).

[Mo(TFDA)(CO)$_4$], pale blue crystals, m. 147°, prepd. from TFDA + $Mo(CO)_6$ in
THF under reflux. IR and NMR spectra (422).

[Pd(TFDA)Cl$_2$], yellow solid, m. 220°, prepd. from TFDA in Me_2CO + aq. $(NH_4)_2$-
$PdCl_4$ and from TFDA and $PdCl_4$ in Me_2CO under reflux (422).

[Rh(TFDA)$_2$Cl$_3$], yellow solid which becomes white at 200° and dec. > 200°,
prepd. from TFDA + $RhCl_3$·hydrate in EtOH under reflux. NMR spectrum (422).

[Bis(trifluoromethyl)vinylene]bis(dimethylarsine), $As_2C_8H_{12}F_6$,
     $Me_2AsC(CF_3)=C(CF_3)AsMe_2$
Synth.: From $Me_2AsAsMe_2$ or $[Me_2AsAsMe_2·(CF_3)_2CO]$ by the rxn. with $CF_3C\equiv CCF_3$
at 20°, a mixt. of cis- and trans-isomers was obtd. (426).
Prop.: A mixt. of the cis- and trans-isomers, b_{15} 98°, NMR spectra (426).

Trimethylphosphonium Bis(dimethylarsino)methylide, $As_2C_8H_{21}P$,
     $Me_3P=C(AsMe_2)_2$
Synth.: From $Me_3P=CHAsMe_2$ by a thermal dec. at 120°, the title compd. was
detected by NMR spectroscopy (1355).

[o-(Dimethylarsino)phenyl]methylarsine, $As_2C_9H_{14}$,   2-$(Me_2As)C_6H_4AsHMe$
Prop.: $b_{0.05}$ 75-77° (1163).

3,4,5,6-Tetrafluoro-o-phenylenebis(dimethylarsine), $As_2C_{10}H_{12}F_4$,
     $2\text{-}(Me_2As)C_6F_4AsMe_2$

Synth.: From 2-bromo-3,4,5,6-tetrafluorophenylmagnesium bromide + Me_2AsI,
followed by a treatment with Mg and then with Me_2AsI (502).
Prop.: NMR spectrum (502).
Coordination Deriv.: (In the following entries FDA denotes $2\text{-}(Me_2As)C_6F_4AsMe_2$.)

$[Mo(FDA)_2(CO)_2]$, prepd. from $Mo(CO)_6$ + FDA under drastic conditions (502).

$[Mo(FDA)(CO)_4]$, prepd. from $Mo(CO)_6$ + FDA under mild rxn. conditions (502).
IR spectrum (422, 502).

$[Mo(FDA)_2(CO)_2I]I$, prepd. from $[Mo(FDA)_2(CO)_2]$ by oxidn. with iodine (502).

$[Ni(FDA)_2](ClO_4)_2$, orange brown, diamagnetic solid (502).

$[Ni(FDA)_2Cl]ClO_4$, prepd. from the preceding diperchlorate by treating it with
one equiv. of Cl^- (502).

$[Ni(FDA)_2Cl]Cl$, dark brown, diamagnetic solid was formed from $[Ni(FDA)_2Cl_2]$
by a slow dissocn. in $MeNO_2$ (502).

$[Ni(FDA)_2I]I$, brown diamagnetic solid, a 1:1 electrolyte in $MeNO_2$ (502).

$[Ni(FDA)_2Cl_2]$, pale green, paramagnetic solid, sol. in C_6H_6 without dissocn.,
prepd. from $[Ni(FDA)_2](ClO_4)_2$ by treating with excess Cl^- ions (502).

$[Ni(FDA)_2Br_2]$, green, paramagnetic solid, sol. in C_6H_6 without dissocn. (502).

The following FDA complexes were listed, without any prepn. and/or prop. data:
$[Co(FDA)_2Cl_2]^+$, $[Co(FDA)_2Cl_2]$, $[Fe(FDA)_2Cl_2]^{2+}$, $[Fe(FDA)_2Cl_2]^+$, $[Fe(FDA)_2Cl_2]$,
$[Ni(FDA)(CO)_2]$, $[Pd(FDA)Cl_2]$, $[Pd(FDA)Br_2]$, $[Pd(FDA)I_2]$, $[Pd(FDA)_2Cl]Cl$,
$[Pd(FDA)_2I]I$, $[Pt(FDA)Cl_2]$, and $[Pt(FDA)_2I_2]$ (502).

o-Phenylenebis(dimethylarsine), $As_2C_{10}H_{16}$, $2\text{-}(Me_2As)C_6H_4AsMe_2$,   (M-80)
Synth.: From Me_2AsI + excess Na in THF, followed by $o\text{-}C_6H_4Cl_2$, 44% yield
(558, 560).
From Me_2AsH + Na in THF, followed by $o\text{-}C_6H_4Cl_2$, 48% yield (558, 560).
From $o\text{-}C_6H_4(AsCl_2)_2$ and CD_3I by a Grignard rxn., $2\text{-}[(CD_3)_2As]C_6H_4As(CD_3)_2$ was
obtd. (1285).
Prop.: b_1 100-101°, n_D^{25} 1.6156 (558), $b_{1.1}$ 97-101° (560), d_{25} 1.3992 (560).
Spectra: NMR (560), IR and Raman (649). IR spectrum of the hexadeuterio
deriv. (1285). The monomethiodide of the diarsine formed colorless plates
m. 226-230° (558).
Rxn. with: CH_2X_2 (X = Br or I) → 2,3-dihydro-1,1,3,3-tetramethyl-1H-1,3-
benzodiarsolium salts (372).
CX_4 (X = Br or I) → 1,1′,3,3′-tetramethyl-1,1,1′,1′,3,3,3′,3′-octamethyl-2,2′-
spirobi(2H-1,3-benzodiarsolium)X_2 (372).
$Ph_2P(S)N_3$ → $Ph_2P(S)N=As(Me)_2C_6H_4As(Me)_2=NP(S)Me_2$ (1583).
Coordination Deriv.: (In the following complexes DA denotes $2\text{-}(Me_2As)C_6H_4AsMe_2$.)

$[Cd(DA)Cl_2]$ (M-81), IR spectrum - an octahedral polymeric structure suggested
(453).

(In the following complexes DA denotes 2-$(Me_2As)C_6H_4AsMe_2$.)

[Cd(DA)Br$_2$] (M-81), IR spectrum - an octahedral polymeric structure suggested (453).

[Cr(DA)$_2$Br$_2$]ClO$_4$ (M-81), absorption and reflectance spectra (561).

[Cr(DA)$_2$Cl$_2$]ClO$_4$ (M-81), absorption and reflectance spectra (561). Low frequency IR spectrum. X-ray powder analysis - a trans configuration postulated (963).

[Cr(DA)$_2$I$_2$]ClO$_4$ (M-81), electronic absorption spectrum (561).

[Cr(DA)(CO)$_4$] (M-82), crystallographic data - monoclinic crystals (264).

Et$_4$N[Cr(DA)(NCS)$_4$], blue, paramagnetic crystals, prepd. from K$_3$[Cr(NCS)$_6$] + DA in EtOH under reflux, followed by pptn. with Et$_4$NCl. IR and absorption spectra (159).

[Co(DA)$_2$](ClO$_4$)$_2$, the complex was obtd. in two forms, golden-brown monoclinic and orthorhombic crystals. The monoclinic crystals, extended along the b axis, were prepd. from Co(OAc)$_2$ by adding DA, followed by NaClO$_4$. The orthorhombic crystals, elongated along the c axis, were prepd. from Co(ClO$_4$)$_2$·6H$_2$O by adding DA in aq. EtOH. Crystallographic data and IR spectra of both forms were obtd. These two forms represent a type of bonding isomerism, where the Co-ClO$_4$ bond is essentially covalent in the monoclinic form and essentially ionic in the orthorhombic form. X-ray analysis indicated a planar arrangement of the bidentate ligands about the Co atom (521). IR and absorption spectra of the monoclinic form and IR spectrum of the other form were obtd. The former is paramagnetic and the latter diamagnetic. In MeNO$_2$ soln. the orthorhombic form is a 2:1 electrolyte (1285).

{Co[2-[(CD$_3$)$_2$As]C$_6$H$_4$As(CD$_3$)$_2$]$_2$}(ClO$_4$)$_2$, prepd. in the same manner as the preceding nondeuterated complex. IR spectrum (1285).

[Co(DA)$_2$(NO$_3$)$_2$], paramagnetic substance, prepd. from Co(NO$_3$)$_2$ and DA in aq. EtOH. IR and electronic spectra and x-ray powder data indicated the presence of coordinated NO$_3$ ions (1285).

[Co(DA)$_2$](CNS)$_2$, light brown, paramagnetic substance, prepd. from CoCl$_2$·6H$_2$O by treating with NH$_4$SCN in aq. alc., followed by DA. Electronic spectrum - a square planar structure of the 4-coordinate Co was postulated (1150). See (M-82) for the [Co(DA)$_2$](SCN)$_2$ complex.

trans-[Co(DA)$_2$Br$_2$]ClO$_4$ (M-82), electronic spectrum (561).

[Co(DA)$_2$Cl$_2$]Cl (M-83), low-frequency IR spectrum and x-ray powder analysis indicated a trans configuration (963).

cis-[Co(DA)$_2$Cl$_2$]ClO$_4$ (M-83), brown-violet substance (1209). The cis configuration was confirmed by low frequency IR spectrum and by an x-ray powder study (963). In methanolic soln. at 54.9° the color slowly changes to green, while the complex isomerizes to the trans form, as established spectrophotometrically from the changes of the absorption peaks. The rate of the isomerization depends on the light intensity and the previous history of the complex but is independent of the Cl^- ion concentration added to the soln. as LiCl. The isomerization is catalyzed by $[Co(DA)_2]Cl_2$. A rxn. of the cis-perchlorate with KSCN or $[Me_3PhN]SCN$ in MeOH at $54.9°$ gave $[Co(DA)_2(SCN)_2]^+$. The rate of rxn. was followed spectrophotometrically. The ligand exchange apparently proceeds stepwise through the $[Co(DA)_2Cl(NCS)]^+$ intermediate. When the rxn. with the SCN^- ions was carried out in the presence of excess Cl^- ions, the trans-dichloro complex was formed simultaneously, but much faster than the di(thiocyanato) complex. The mechanism of the isomerization and the Cl displacement are considered (1209).

trans-$[Co(DA)_2Cl_2]ClO_4$ (M-83), green, diamagnetic substance, formed from the cis isomer in MeOH soln. at $54.9°$. Absorption and reflectance spectra (561).

trans-$[Co(DA)_2{}^{36}Cl_2]ClO_4$, prepd. from trans-$[Co(DA)_2Cl_2]Cl$ in MeOH + $Li^{36}Cl$ in a sealed tube kept at $55-75°$ for a few days and then treated with $HClO_4$. The rate of the Cl exchange was influenced by the pH of the rxn. mixt. The final prod. was free of ionic ^{36}Cl and retained the trans configuration. (488).

trans-$[Co(DA)_2I_2]ClO_4$, electronic spectrum (561).

trans-$[Co(DA)_2Cl_2]NO_3$, prepd. from the corr. chloride by a rxn. with $NaNO_3$ in MeOH. The displacement rate of the chloro ligands in the nitrate for SCN was measured conductometrically. Likewise, the rate of the exchange of the chloro ligands for ^{36}Cl was detd. The competitive exchange rxn. of the chloro ligands for SCN^- in the presence of Cl^- ions was studied (488).

trans-$[Co(DA)_2(SCN)_2]ClO_4$, (M-83), prepd. from trans-$[Co(DA)_2Cl_2]ClO_4$ + NaSCN in MeOH. The rate of the exchange was detd. (488).

$[Co(DA)_3](ClO_4)_3$ (M-83), yellow, diamagnetic crystals, dec. in storage to form trans-$[Co(DA)_2Cl_2]^+$ (561).

$[Ga(DA)_2Br_2]GaBr_4$, white, diamagnetic substance, m. $95°$ (dec.), a 1:1 electrolyte in $PhNO_2$, sol. in C_6H_6, Et_2O, THF, and $PhNO_2$, dec. in H_2O, EtOH, and Me_2CO, prepd. from $GaBr_3$ and the diarsine (1:1) in C_6H_6. IR spectrum - an octahedral cation structure with the Br ligands probably trans to each other suggested (1164).

$[Ga(DA)_2I_2]GaI_4$, diamagnetic solid, m. $127-130°$, prepd. from GaI_3 and the diarsine in Et_2O. A 1:1 electrolyte in $PhNO_2$, less sol. than the preceding dibromo analog. IR spectrum - an octahedral cation structure analogous to that of the dibromo complex suggested (1164).

$[Ga(DA)_2Cl_2][Ga_3Cl_{10}]$, white to pink, diamagnetic complex, dec. $\sim 190°$, sol. in $PhNO_2$, C_6H_6, Et_2O, and THF, dec. in H_2O, EtOH, and Me_2CO, a 1:1 electrolyte in $PhNO_2$, prepd. from $GaCl_3$ and the diarsine (2:1) in C_6H_6. IR spectrum - an octahedral cation structure postulated (1164).

$[Au(DA)_2I_2]I$ (M-85), dark red, monoclinic crystals, prepd. by adding the diarsine and NaI to $NaAuCl_4$ soln. in EtOH. Crystallographic data; the 4 As

(In the following complexes DA denotes 2-(Me$_2$As)C$_6$H$_4$AsMe$_2$.)

atoms coordinated with the Au atom in a square-planar arrangement and the I atoms positioned near the axial sites produce a distorted octahedron. A different monoclinic form can be obtd. by recrystn. from PhNO$_2$ (501).

[Hf(DA)$_2$Cl$_4$] (M-85), prepd. from HfCl$_4$ and the diarsine in THF (358). IR spectrum (355).

[In(DA)$_2$Cl$_2$]InCl$_4$, a diamagnetic solid, dec. $\sim$ 175°, a 1:1 electrolyte in PhNO$_2$, prepd. from InCl$_3$ and the diarsine in Et$_2$O at r.t. The complex is sol. in PhNO$_2$, sparingly sol. in Et$_2$O and C$_6$H$_6$ and is dec. by H$_2$O, Me$_2$CO, and alcohols. IR spectrum - an octahedral cation with the Cl ligands probably trans to each other was suggested (1164).

[In(DA)$_2$Br$_2$]InBr$_4$, white, diamagnetic powder, m. 296-298°(dec.), a 1:1 electrolyte in PhNO$_2$, prepd. from InBr$_3$ and the diarsine in C$_6$H$_6$. Soly. similar to that of the chloro complex. IR spectrum - an octahedral structure of the cation was suggested (1164). Rxn. with HClO$_4$ $\rightarrow$ a complex approximating the compn. of [In(DA)$_2$Br$_2$]ClO$_4$ (1162).

[In(DA)$_2$I$_2$][In(DA)I$_4$], diamagnetic ppt., m. 135-138° (dec.), a 1:1 electrolyte in PhNO$_2$, prepd. from InI$_3$ and the diarsine in C$_6$H$_6$. IR spectrum. Rxn. with [Ph$_4$As]I $\rightarrow$ [Ph$_4$As][In(DA)I$_4$] (crude) (1164).

[Ir(DA)$_2$(CO)]Br, ivory white, diamagnetic crystals, a 1:1 electrolyte, prepd. from [Ir(CO)$_3$]Br suspended in C$_6$H$_6$ by heating with the diarsine at 50-60°. Soly. data, IR spectrum (295).

[Ir(DA)$_2$]Br, yellow-maroon crystals, prepd. from [Ir(DA)$_2$(CO)]Br by refluxing in MePh. The complex in soln. readily reacts with O and H to form [IrO$_2$(DA)$_2$]Br and [IrH$_2$(DA)$_2$]Br (295).

[Ir(DA)$_2$Br$_2$]Br, yellow crystals, a 1:1 electrolyte in MeOH, does not melt at 350°, prepd. from IrBr$_3$ dissolved in H$_2$O by adding the diarsine in EtOH and heating under reflux. The complex is sol. in MeOH, EtOH, and Me$_2$CO, but insol. in H$_2$O and MeNO$_2$. IR and absorption spectra (1562).

[Ir(DA)$_2$Br$_2$]NO$_3$, yellow crystals, a 1:1 electrolyte in MeNO$_2$, prepd. from [Ir(DA)$_2$Br$_2$]Br by heating with HNO$_3$. IR spectrum (1562).

[Ir(DA)$_2$Br$_2$]BPh$_4$, yellow ppt., prepd. from [Ir(DA)$_2$Br$_2$]X (X = Br or NO$_3$) in MeOH by treating with NaBPh$_4$ (1562).

[IrO$_2$(DA)$_2$]Br, ivory white, diamagnetic crystals, a 1:1 electrolyte, prepd. from [Ir(DA)$_2$]Br in soln. by treating with O. IR spectrum (295).

[IrH$_2$(DA)$_2$]Br, white, diamagnetic solid, a 1:1 electrolyte, prepd. from [Ir(DA)$_2$]Br in soln. by treating with H. IR spectrum (295).

[Ir(DA)$_2$Cl$_2$]Cl, the complex was obtd. in two forms: yellow and pink crystals, both very stable and nonmelting up to 350°, 1:1 electrolytes. Both were prepd. from K$_3$IrCl$_6$ in H$_2$O by mixing with the diarsine dissolved in EtOH and heating under reflux. The pink isomer formed a ppt., while the yellow isomer was recovered from the supernatant soln. by concn. and pptn. with Et_2O. The pink isomer is converted into the yellow form, which is thermodynamically more stable, by dissolving the former in $MeNO_2$. UV, visible, and low-frequency IR spectra were obtd. The yellow form was identified as a trans- and the pink form as a cis-isomer. In both complexes the 6-coordinate Ir(III) affords a univalent cations (1562). Low-frequency IR spectrum and x-ray powder analysis of the trans isomer were reported (963).

$[Ir(DA)_2Cl_2]ClO_4$, bright yellow crystals, prepd. from the yellow $[Ir(DA)_2Cl_2]^+$ complex salts by a rxn. with 70% $HClO_4$ dissolved in aq. EtOH. The complex is sol. in MeOH and EtOH, insol. in H_2O and $MeNO_2$. UV, visible, and IR spectra (1562). Low-frequency IR spectrum and x-ray powder study - a trans configuration postulated (963).

$[Ir(DA)_2Cl_2]BPh_4$, bright yellow crystals, prepd. from the yellow $[Ir(DA)_2Cl_2]^+$ salts dissolved in MeOH by adding $NaBPh_4$ (1562).

$[Ir(DA)_2Cl_2]NO_3$, bright yellow crystals, a 1:1 electrolyte in MeOH, prepd. from the pink $[Ir(DA)_2Cl_2]Cl$ isomer by a rxn. with warm HNO_3. UV, visible, and IR spectra - a 6-coordinate Ir(III) complex (1562).

$[Fe(DA)_2(CO)_3]$ (M-85), crystal structure by x-ray analysis (245).

$[Fe(DA)_2Cl_2]$, electronic spectrum (561), low-frequency IR spectrum and x-ray powder analysis (963). A trans configuration postulated (561, 963).

$[Fe(DA)_2Br_2]$ (M-85), yellow, diamagnetic compd., electronic spectrum - a tetragonal trans structure postulated (561).

$[Fe(DA)_2I_2]$ (M-85), yellow, diamagnetic crystals, electronic spectrum - a tetragonal trans structure postulated (561).

$[Fe(DA)_2Cl_2]ClO_4$ (M-86), low-frequency IR spectrum and x-ray powder analysis; a trans configuration postulated (963).

$[Fe(DA)_2Cl_2]BF_4$, pink crystals, m. 184° (dec.), a 1:1 electrolyte in $MeNO_2$, prepd. from $[Fe(DA)_2Cl_2]FeCl_4$ + HBF_4 in Me_2CO. Visible and far-IR spectra (714).

$[Fe(DA)_2Cl_2][BF_4]_2$, black, hydroscopic, paramagnetic powder, m. 190° (dec.), a 1:2 electrolyte in $MeNO_2$, stable in vacuum in the dark, prepd. from $[Fe(DA)_2Cl_2]FeCl_4$ or from $[Fe(DA)_2Cl_2]BF_4$ + 42% aq. HBF_4 in 15 N HNO_3. Far-IR spectrum, magnetic susceptibility data. An octahedral structure with the Cl atoms trans to each other postulated (714).

$[Fe(DA)_2Cl_2]FeCl_4$ (M-86), a 1:1 electrolyte in $MeNO_2$; absorption spectrum in the visible range (714).

$[Fe(DA)_2Cl_2][ReO_4]_2$, black powder, m. 130° (dec.), nonhydroscopic but dec. in moist air, can be stored for a short time in a dark, evacuated container, prepd. from $[Fe(DA)_2Cl_2]FeCl_4$ + 4.7 N $HReO_4$ in 15 N HNO_3. Visible and far-IR spectra (714).

(In the following complexes DA denotes $2-(Me_2As)C_6H_4AsMe_2$.)

$[Fe(DA)_2Br_2]Br$ (M-86), green solid, m. 120° (dec.), prepd. from $[Fe(DA)_2Br_2]$-$FeBr_4$ by suspending in H_2O and stirring for 2 min. Visible and far-IR spectra (714).

$[Fe(DA)_2Br_2]BF_4$, green crystals, m. 195° (dec.), a 1:1 electrolyte in $MeNO_2$, prepd. from $[Fe(DA)_2Br_2]FeBr_4$ or from $[Fe(DA)_2Br_2]Br$ by treating with HBF_4 in Me_2CO. Visible and far-IR spectra (714).

$[Fe(DA)_2Br_2][BF_4]_2$, dark green, hydroscopic, paramagnetic powder, m. 170° (dec.), a 1:2 electrolyte in $MeNO_2$, prepd. from $[Fe(DA)_2Br_2]FeBr_4$ or from $[Fe(DA)_2Br_2]BF_4$ + 42% aq. HBF_4 in 15 N HNO_3. Visible and far-IR spectra, magnetic susceptibility data. An octahedral cation structure with the Br ligands in trans positions postulated (714).

$[Fe(DA)_2Br_2]FeBr_4$, a 1:1 electrolyte in $MeNO_2$; absorption spectrum in the visible range (714).

$[Fe(DA)_2Br_2][ReO_4]_2$, crude prod. contaminated with N, prepd. from $[Fe(DA)_2Br_2]$-$FeBr_4$ + 4.7 N $HReO_4$ in 15 N HNO_3. Far-IR spectrum (714).

$[Fe(DA)_3](ClO_4)_2$, orange crystals, prepd. from $Fe(ClO_4)_2$ in MeOH by treating with the diarsine and heating under reflux. Electronic absorption spectrum - a tetragonal Fe(II) complex (561).

trans-$[Fe(DA)_2(NO)Br]Br$, green, paramagnetic crystals, prepd. from anhyd. $FeBr_2$ in MeOH by treating with NO, adding dropwise 1/2 of the diarsine soln. and equilibrating with NO until no more NO was absorbed, then pouring the remaining diarsine soln. at once, and recrystg. from CH_2Cl_2-C_6H_{14}. IR spectrum; a 6-coordinate Fe complex postulated (1411).

trans-$[Fe(DA)_2(NO)Br]ClO_4$, dark green, paramagnetic crystals, a 1:1 electrolyte in MeOH, prepd. from $[Fe(DA)_2(NO)](ClO_4)_2$ suspended in MeOH by treating with LiBr and heating to boiling. IR spectrum - a 6-coordinate Fe complex postulated (1411).

$[Fe(DA)_2(NO)Br][Fe(NO)_2Br_2]$, wine-red, paramagnetic crystals, fairly stable in the solid state under N and in soln. in the absence of air, prepd. from anhyd. $FeBr_2$ dissolved in MeOH or in THF by treating with NO, adding dropwise the diarsine (1:1) in MeOH or THF, and recrystg. from CH_2Cl_2-C_6H_{14}. IR spectrum - a 6-coordinate Fe of the cation postulated (1411).

trans-$[Fe(DA)_2(NO)Cl]ClO_4$, prepd. from $[Fe(DA)_2(NO)](ClO_4)_2$ suspended in MeOH by treating with LiCl and heating to boiling. IR and reflectance spectra - a 6-coordinate Fe postulated (1411).

$[Fe(DA)_2(NO)](ClO_4)_2$, dark blue, paramagnetic crystals, a 2:1 electrolyte in MeOH, prepd. from $Fe(ClO_4)_2 \cdot 6H_2O$ in MeOH by treating with NO, adding dropwise the diarsine, and recrystg. from CH_2Cl_2-C_6H_{14}. IR spectrum - a 5-coordinate Fe complex (1411).

trans-$[Fe(DA)_2(NO)I]I$, green, paramagnetic crystals, prepd. from FeI_2 in the same manner as the bromo analog. IR spectrum (1411).

trans-[Fe(DA)$_2$(NO)I][Fe(NO)$_2$I$_2$], purple, paramagnetic crystals, prepd. from FeI$_2$ dissolved in MeOH by treating with NO and slowly adding the diarsine (1:1) and from [Fe(NO)$_2$I]$_2$ in C$_6$H$_{14}$ by adding dropwise the diarsine in excess and recrystg. from CH$_2$Cl$_2$-C$_6$H$_{14}$. IR spectrum - a 6-coordinate Fe cation postulated (1411).

trans-[Fe(DA)$_2$(NO)(CN)]ClO$_4$, yellow crystals, prepd. from [Fe(DA)$_2$(NO)](ClO$_4$)$_2$ suspended in MeOH by adding LiCN, heating to boiling, and recrystg. from CH$_2$Cl$_2$-C$_6$H$_{14}$ (1411).

{[π-(MeC$_5$H$_4$)Mn(CO)$_2$]$_2$(DA)} (M-87), crystal and mol. structures (165).

[Hg(DA)X$_2$], where X = Cl or Br (M-87), IR spectra - a tetrahedral or distorted tetrahedral structure suggested (453).

[Hg(DA)$_2$](ClO$_4$)$_2$ (M-87), IR spectrum (453).

[Mo(DA)(CO)$_4$] (M-87), crystallographic data (264).

[Mo(DA)$_2$Cl$_2$] (M-88), low-frequency IR spectrum and x-ray powder analysis - a trans configuration established (963).

[Mo(DA)$_2$I$_2$] (M-88), brown, paramagnetic solid, a nonelectrolyte in PhNO$_2$, dec. in air, soly. data, prepd. from MoI$_3$ and excess of the diarsine in an evacuated tube at 160° (480).

[Mo(DA)$_2$Cl$_2$]ClO$_4$, low-frequency IR spectrum and x-ray powder analysis - a trans configuration was established (963).

[π-C$_5$H$_5$Mo(DA)(CO)I], red crystals, m. 126-128°, unstable in air, a nonelectrolyte, prepd. from [π-C$_5$H$_5$Mo(CO)$_3$I] and the diarsine in C$_6$H$_6$ under UV irradiation. Soly. data, IR spectrum (670).

[π-C$_5$H$_5$Mo(DA)(CO)$_2$I], yellow crystals, a nonelectrolyte, prepd. from [π-C$_5$H$_5$-Mo(CO)$_3$I] and the diarsine in C$_6$H$_6$ under UV irradiation. Soly. data, IR spectrum (670).

[Mo$_6$Cl$_8$(DA)$_2$Cl$_2$]Cl$_2$, a 1:2 electrolyte in DMF and DMSO, prepd. from [Mo$_6$Cl$_8$]Cl$_4$ and the diarsine in THF under reflux. Electronic absorption spectrum - a 6-coordination of the Mo$_6$Cl$_8$ cluster postulated (563). IR spectrum (362).

[Ni(DA)$_2$Br$_2$] (M-88), pale red, diamagnetic solid, prepd. from [Ni(DA)(CO)$_2$] + HBr in C$_6$H$_{14}$, 57% yield. An unstable complex; in Me$_2$CO showed a slight elec. cond. due to dec. A square-planar structure suggested (768).

[Ni(DA)Cl$_2$], pale red, paramagnetic crystals, prepd. from [Ni(DA)(CO)$_2$] + HCl gas in C$_6$H$_{14}$, 63% yield. The complex is unstable; in Me$_2$CO soln. showed some elec. cond. due to dec. Far-IR spectrum - a square planar structure suggested (768).

[Ni(DA)I$_2$] (M-88), mauve colored, diamagnetic solid, prepd. from [Ni(DA)(CO)$_2$] + iodine in C$_6$H$_{14}$. An unstable complex; in Me$_2$CO soln. showed slight elec. cond. due to dec. A square-planar structure suggested (768, see also 165).

(In the following complexes DA denotes 2-$(Me_2As)C_6H_4AsMe_2$.)

$[Ni(DA)_2](ClO_4)_2$ (M-88), a diamagnetic ppt., a 1:2 electrolyte in $MeNO_2$, prepd. from $Ni(ClO_4)_2$ by treating with the diarsine in aq. EtOH (1285). IR and electronic absorption spectra (479, 1285). The complex is isomorphous with the monoclinic form of the corr. Co complex and, therefore, essentially a covalent bonding of the ClO_4 group was postulated (1285).

$[Ni(DA)_2(NO_3)_2]$ (M-89), a diamagnetic substance, prepd. from $Ni(NO_3)_2$ and the diarsine in aq. EtOH. IR and electronic absorption spectra indicated a co-valent bonding of the NO_3 groups (1285).

$[Ni(DA)_2Br]Br$ (M-89), stability const. of the 5-coordinate Ni complex and thermodynamic data (534). Electronic absorption spectrum (479).

$[Ni(DA)_2Cl]Cl$ (M-89), stability const. of the 5-coordinate Ni complex and thermodynamic data (534). Electronic absorption spectrum (479).

$[Ni(DA)_2I]I$ (M-89), electronic absorption spectrum (479).

$[Ni(DA)_2SCN]SCN$ (M-89), stability const. of the 5-coordinate Ni complex and thermodynamic data (534). Electronic absorption spectrum (479).

$[Ni(DA)_2N_3]^+$, stability const. of the 5-coordinate Ni complex and thermodyna-mic data (534).

$[Ni(DA)_2Cl_2]Cl$ (M-90), low-frequency IR spectrum (963), x-ray powder analysis (898, 963), ESR spectrum (898), a trans configuration was established (963).

$[Ni(DA)_2Cl_2]ClO_4$ (M-90), low-frequency IR spectrum and x-ray powder analysis - a trans configuration was established (963). ESR spectrum of a powder sample and of methanolic soln. - interpretation of the hyperfine structure (391).

$[Ni(DA)_3]Cl_2$, the complex is formed as by-prod. in the rxn. of $NiCl_2·6H_2O$ with the diarsine (1:3) in diethylene glycol under reflux. The $[Ni(DA)_3]^{++}$ cation was established by a study of its visible absorption spectrum (211).

$[Ni(DA)_3](ClO_4)_2$ (M-89), deep maroon, diamagnetic crystals, prepd. from $NiCl_2·6H_2O$ and the diarsine in diethylene glycol by heating under reflux, pptg. the prod. with $NaBPh_4$, and digesting with $HClO_4$ in Me_2CO. Sol. in DMF and in concd. mineral acids, sparingly sol. in H_2O. Stable in acid soln. but is dec. by alkali. NMR and electronic absorption spectra - an octahedral structure for the $[Ni(DA)_3]^{++}$ cation postulated (211). ESR spectrum (898).

$[Ni(DA)_2CS(NH_2)_2]^{++}$, the formation of the 5-coordinate Ni cation by equilibra-tion of $[Ni(DA)_2](NO_3)_2$ with thiourea in H_2O and in MeOH was studied spectro-photometrically. The stability const. of the cation and thermodynamic data were detd. (534).

$[Nb(DA)F_5]$, white, diamagnetic ppt., prepd. from NbF_5 and the diarsine in CCl_4. A nonelectrolyte; a 7-coordinate Nb complex (361).

$[Nb(DA)Cl_5]$, red, diamagnetic ppt., prepd. from $NbCl_5$ and the diarsine (1:1.3) in CCl_4. Soly. and elec. cond. data. Diffuse reflectance spectrum - a 7-

coordinate Nb complex postulated (361).

[Nb(DA)Br$_5$], brown, diamagnetic ppt., prepd. from NbCl$_5$ and the diarsine in hot Et$_2$O. Soly. and elec. cond. data. Electronic absorption and diffuse reflectance spectra. A 7-coordinate Nb complex postulated (361).

[NbO(DA)Cl$_3$], red, diamagnetic crystals, prepd. from NbCl$_5$ and the diarsine in MeNO$_2$ by boiling for several hours. Elec. cond. data. Electronic absorption and IR spectra (361).

[NbO(DA)Br$_3$], diamagnetic ppt., prepd. from NbBr$_5$ and the diarsine in MeCN. Diffuse reflectance spectrum (361).

[Nb(DA)Cl$_4$]$_2$O, diamagnetic crystals, isolated from the mother liquor remaining after pptn. of [NbO(DA)Cl$_3$]. Elec. cond. data. Electronic absorption and IR spectra. A 7-coordinate Nb complex (361).

[Nb(DA)$_2$Cl$_4$], pale green, paramagnetic ppt., prepd. from the diarsine by heating with NbCl$_4$ at 150°, or with NbCl$_5$ at 200-250°, or with NbOCl$_3$ at 200° in a sealed tube. Diffuse reflectance spectrum (360). IR spectrum (355). An 8-coordinate, dodecahedral structure postulated (360).

[Nb(DA)$_2$Br$_4$], pale green, paramagnetic ppt., prepd. from the diarsine and NbBr$_5$ in a sealed tube at 100°. Diffuse reflectance spectrum - an 8-coordinate, dodecahedral structure postulated (360).

[Os(DA)(CO)$_2$]$_3$, brownish yellow ppt., prepd. from Os$_3$(CO)$_{12}$ by heating with the diarsine in xylene. IR spectrum (216, 217).

[Os(DA)$_2$Cl$_2$] (M-90), low-frequency, IR spectrum and x-ray powder analysis - a trans configuration was established (963).

[Os(DA)(CO)$_2$I$_2$], white crystals, prepd. from Os(CO)$_4$I$_2$ and the diarsine in C$_6$H$_6$ under reflux. IR spectrum (674).

[Pd(DA)Br$_2$] (M-91), white, diamagnetic substance, prepd. from [Pd(PhCN)$_2$Br$_2$] and the diarsine in CHCl$_3$, 75% yield. A nonelectrolyte in MeNO$_2$. A square-planar structure suggested (768).

[Pd(DA)Cl$_2$] (M-91), white, diamagnetic solid, prepd. from [Pd(PhCN)$_2$Cl$_2$] and the diarsine in CHCl$_3$, 40% yield. A nonelectrolyte in MeNO$_2$. Far-IR spectrum, a square-planar structure suggested (768).

[Pd(DA)I$_2$], white, diamagnetic substance, prepd. from [Pd(PhCN)$_2$I$_2$] and the diarsine in CHCl$_3$, 60% yield. A nonelectrolyte in MeNO$_2$. A square-planar structure suggested (768).

[Pd(DA)(SCN)$_2$], pale yellow, diamagnetic substance, prepd. from [Pd(PhCN)$_2$-(SCN)$_2$] and the diarsine in CHCl$_3$, 55% yield. A nonelectrolyte in MeNO$_2$. IR spectrum. A square-planar structure suggested (768).

[Pd(DA)$_2$](NO$_3$)$_2$, prepd. from [Pd(DA)$_2$](ClO$_4$)$_2$ by mixing with LiNO$_3$ in DMF and pptg. with Et$_2$O. UV spectroscopic study of the equilibria: [Pd(DA)$_2$]$^{++}$ + X^n $\rightleftarrows$ [Pd(DA)$_2$X]$^{n+2}$, where X^n = Cl$^-$, Br$^-$, I$^-$, N$_3$$^-$, SCN$^-$, SeCN$^-$, S$_2O_3$$^{2-}$, and thio-

(In the following complexes DA denotes 2-$(Me_2As)C_6H_4AsMe_2$.)

urea in H_2O. The stability consts. of the 5-coordinate complexes $[Pd(DA)_2X]^{n+2}$ increase in the following order: $CS(NH_2)_2 \simeq N_3^- < Cl^- < Br^- \simeq S_2O_3^{2-} < SCN^- < SeCN^- < I^-$. The heats of rxn. favor the formation of the complexes as follows: $Br^- \simeq S_2O_3^{2-} < SCN^- \simeq SeCN^- \simeq CS(NH_2)_2 < I^-$. Thermodynamic data of the formation of the complexes reported (535, see also 534).

$[Pt(DA)Cl_2]$, white, diamagnetic substance, prepd. from $[Pt(PhCN)_2Cl_2]$ by treating with the diarsine in $CHCl_3$, 18% yield. A nonelectrolyte in $MeNO_2$. Far-IR spectrum. A square-planar structure suggested (768).

$[Pt(DA)_2Cl_2]$ (M-93), crystallographic data detd. from 3-dimensional Fourier functions. An octahedral configuration with all 4 As atoms in a planar square and both Cl atoms at the apical positions postulated (1451).

$[Pt(DA)_2]^{++}$ salts (M-93), rxn. with X^- in MeOH (where X^- is Cl^-, Br^-, I^-, N_3^-, SCN^-, and $CS(NH_2)_2$, leads to the equilibrium mixtures: $[Pt(DA)_2]^{++} + X^- \rightleftharpoons [Pt(DA)_2X]^+$, i.e. the formation of 5-coordinate Pt complexes. Their stability consts. detd. spectrophotometrically, increase as follows: $N_3^- < Cl^- < SCN^- < Br^- < I^-$, while the ΔH of their formation increase: $N_3^- < Cl^- < Br^- < SCN^- < I^- < CS(NH_2)_2$ (490, 534). The stability consts., entropy, and enthalpy of $[Pt(DA)_2Br]^+$ and $[Pt(DA)_2I]^+$ were detd. (535).

$[Pt(DA)_2](NO_3)_2$ (M-93), UV spectrum (491). A rxn. of the complex with trans-$[Pt(PEt_3)_2Cl_4]$ in the presence of NaCl in MeOH $\rightarrow [Pt(DA)_2Cl_2]^{++}$ + trans-$[Pt(PEt_3)_2Cl_2]$. When the rxn. was carried out in the presence of NaBr, a mixt. of $[Pt(DA)_2Br_2]^{++}$ with trans-$[Pt(PEt_3)_2Cl_2]$ and trans-$[Pt(PEt_3)_2Br_2]$ was obtd. The rxn. in the presence of NaI gave trans-$[Pt(PEt_3)_2I_2]$ and I_3^-. Analogous rxns. with trans-$[Pt(PEt_3)_2Br_4]$ were carried out in MeOH in the presence of NaCl and NaBr, resp. In the presence of NaCl the rxn. gave, at first, trans-$[Pt(PEt_3)_2Br_2]$ and $[Pt(DA)_2ClBr]^{++}$, followed by a conversion of the latter to $[Pt(DA)_2Cl_2]^{++}$. In the presence of NaBr, $[Pt(DA)_2Br_2]^{++}$ and $[Pt(PEt_3)_2Br_2]$ were obtd. With NaI trans-$[Pt(PtEt_3)_2Br_4]$ was reduced to trans-$[Pt(PEt_3)_2I_2]$, while the Pt of the diarsine complex was not oxidized. The rates of these rxns. were studied spectrophotometrically. A mechanism of these rxns. was suggested (1207).

$[Pt(DA)_2SCN]SCN$ (M-93), in equilibrium with $[Pt(DA)_2](SCN)_2$ was detected spectrophotometrically during a redn. of $[Pt(DA)_2Cl_2]Cl_2$ with NaSCN in MeOH. The rate of the rxn. was detd. (491). The stability const. of the 5-coordinate Pt complex and thermodynamic data were detd. (534).

$[Pt(DA)_2Br_2](NO_3)_2$ was reduced by NaSCN, $NaNO_2$, and NaI in MeOH to $[Pt(DA)_2]X_2$ (X = SCN, NO_2, and I, resp.). The rates of the redn. were followed spectrophotometrically. A mechanism of the redn. was suggested (489).

$[Pt(DA)_2Br_2](ClO_4)_2$ (M-93), was reduced by NaSCN, $NaNO_2$, and NaI in the same manner as the preceding dinitrate to $[Pt(DA)_2]X_2$. The rates of the redn. were followed spectrophotometrically. A mechanism of the rxn. was suggested (489).

$[Pt(DA)_2Cl_2]Cl_2$, pale yellow powder, prepd. from $[Pt(DA)_2](ClO_4)_2$ in DMF on Dowex IX anion exchange resin in the Cl^- form. Redn. of the complex with NaSCN in MeOH gave $[Pt(DA)_2]^{++} \rightleftharpoons [Pt(DA)_2SCN]^+$ equil. mixt. The rate of the redn. was detd. spectrophotometrically (491).

$[Pt(DA)_2Cl_2](NO_3)_2$, yellowish white ppt., prepd. from $[Pt(DA)_2Cl_2](ClO_4)_2$ dissolved in DMF by treating with concd. HNO_3, followed by Me_2CO and Et_2O. A soln. in MeOH when treated with 2.26 N $H^{36}Cl$ at 30° underwent an isotope exchange affording $[Pt(DA)_2{}^{36}Cl_2](NO_3)_2$, which was isolated as a yellowish white ppt. by adding Et_2O. The exchange kinetics was detd. and correlated with the entropy and enthalpy of activation for a single bimol. process (1206). A redn. of $[Pt(DA)_2Cl_2](NO_3)_2$ with NaI in MeOH and the rxn. rate were followed spectrophotometrically. A mechanism of the rxn. was suggested (489).

$[Re(DA)_2(CO)Br]$, white, diamagnetic crystals, stable in air, insol. in org. solvents, prepd. from $[Re(DA)(CO)_3Br]$ and the diarsine at 270° in a sealed tube. IR spectrum (855).

$[Re(DA)_2(CO)Cl]$, white, diamagnetic crystals, with the same prop. as the preceding bromo complex, prepd. in analogous manner from $[Re(DA)(CO)_3Cl]$. IR spectrum (855).

$[Re(DA)_2(CO)I]$, white, diamagnetic crystals, with the same prop. as the preceding chloro and bromo complexes and prepd. in exactly analogous manner from $[Pt(DA)(CO)_3I]$. IR spectrum (855).

$[Re(DA)(CO)_3Br]$, white, diamagnetic crystals, stable in air, sol. in polar and nonpolar solvents, a nonelectrolyte in $PhNO_2$, prepd. from $Re(CO)_5Br$ and the diarsine in EtOH under reflux. IR spectrum (855).

$[Re(DA)(CO)_3Cl]$, white, diamagnetic crystals, stable in air, sol. in polar and nonpolar solvents, a nonelectrolyte in $PhNO_2$, prepd. from $Re(CO)_5Cl$ and the diarsine in EtOH under reflux. IR spectrum (855).

$[Re(DA)(CO)_3I]$, white, diamagnetic crystals, stable in air, sol. in polar and nonpolar solvents, a nonelectrolyte in $PhNO_2$, prepd. from $Re(CO)_5I$ and the diarsine in EtOH under reflux. IR spectrum (855).

$[Re(DA)_2Cl_2]$ (M-93), low-frequency IR spectrum and x-ray powder analysis; a trans configuration was established (963).

$[Re(DA)_2Cl_2]Cl$, low-frequency IR spectrum and x-ray powder analysis; a trans configuration was established (963).

$[Re(DA)_2Cl_2]ClO_4$ (M-93), low-frequency IR spectrum and x-ray powder analysis; a trans configuration was established (963).

$[Re(DA)Br_4]$, deep red, paramagnetic crystals, sol. in C_6H_6, $CHCl_3$ and $PhNO_2$, a nonelectrolyte in $PhNO_2$, prepd. from $[Re(DA)(CO)_3Br]$ in $CHCl_3$ and excess Br under reflux (855).

(In the following complexes DA denotes 2-(Me$_2$As)C$_6$H$_4$AsMe$_2$.)

[Re(DA)Cl$_4$], mustard yellow, paramagnetic crystals, a nonelectrolyte in PhNO$_2$, sol. in C$_6$H$_6$, CHCl$_3$, and PhNO$_2$, prepd. from [Re(DA)(CO)$_3$Cl] in CHCl$_3$ by treating with chlorine (855).

[ReO(DA)Cl$_3$], green, paramagnetic crystals, sol. in most org. solvents, with the exception of light petr., a nonelectrolyte in PhNO$_2$, prepd. from [Re(DA)-(CO)$_3$Cl] in EtOH by treating with Cl (855).

[Re(DA)$_2$(CO)Br$_2$]Br$_3$, deep pink, diamagnetic crystals, stable in air, insol. in polar and nonpolar solvents, with the exception of PhNO$_2$, a 1:1 electrolyte, prepd. from [Re(DA)$_2$(CO)Br] in CHCl$_3$ by treating with Br in CCl$_4$. IR spectrum (855).

[Re(DA)$_2$(CO)I$_2$]I$_3$, black, diamagnetic crystals, stable in air, with other phys. prop. similar to those of the tribromide analog, prepd. from [Re(DA)$_2$-(CO)I] in CHCl$_3$ by warming gently with iodine. IR spectrum (855).

[Re(DA)$_2$(CO)I$_2$]ClO$_4$, mauve, diamagnetic crystals, more sol. than the triiodide but unstable in soln., a 1:1 electrolyte in PhNO$_2$, prepd. from the preceding triiodide and LiClO$_4$ in EtOH. IR spectrum (855).

[Rh(DA)$_2$Cl$_2$]Cl (M-94), low-frequency IR spectrum and x-ray powder analysis; a trans configuration was established (963).

[Rh(DA)$_2$Cl$_2$]ClO$_4$, low-frequency IR spectrum and x-ray powder analysis; a trans configuration was established (963).

trans-[Ru(DA)$_2$Cl$_2$] (M-94), low-frequency IR spectrum and x-ray analysis; a trans configuration was established (963). Absorption spectrum - a tetragonally distorted octahedral structure has been suggested (858).

[Ru(DA)$_2$Cl$_2$]ClO$_4$ (M-95), low-frequency IR spectrum and x-ray powder analysis; a trans configuration was established (963).

[Ru(DA)(NO)Cl$_3$] (in M-95 erroneously entered as [Ru(DA)$_2$(NO)Cl$_3$]), orange, paramagnetic solid, stable up to 320°, prepd. from Ru(NO)Cl$_3$·3H$_2$O and the diarsine in EtOH at 75°. IR spectrum (543).

[Ta(DA)Br$_5$], bright yellow ppt., a nonelectrolyte in PhNO$_2$, prepd. from TaCl$_5$ and the diarsine in Et$_2$O. Diffuse reflectance spectrum. A 7-coordinate Ta complex (361).

[Ta(DA)Cl$_5$], yellow, diamagnetic ppt., a nonelectrolyte in MeNO$_2$, prepd. from TaCl$_5$ and the diarsine in CCl$_4$. Diffuse reflectance spectrum. A 7-coordinate Ta complex (361).

[Tc(DA)$_2$Cl$_2$] (M-96), low-frequency IR spectrum and x-ray powder analysis; a trans configuration was established (963).

[Tc(DA)$_2$Cl$_2$]Cl·hydrate (M-96), low-frequency IR spectrum and x-ray powder analysis; a trans configuration was established (963).

[Tc(DA)$_2$Cl$_2$]ClO$_4$, low-frequency IR spectrum and x-ray powder analysis; a trans configuration was established (963).

[(C$_6$F$_5$)$_2$Tl(DA)Br], cream-colored microcrystals, m. ~ 136°, prepd. from (C$_6$F$_5$)$_2$TlBr and the diarsine in Et$_2$O and pptd. with petroleum; weakly ionized in Me$_2$CO (456).

[Ti(DA)$_2$Cl$_4$] (M-96), IR spectrum - an 8-coordinate Ti complex (355).

[Ti(DA)$_2$I$_2$]I$_2$, yellow ppt., m. > 300°, insol. in common solvents, prepd. from TiI$_4$ and the diarsine in C$_6$H$_6$. IR spectrum (358).

[DA(TiF$_4$)$_2$], white, diamagnetic ppt., chars but does not melt below 300°, insol. in most org. solvents but is dec. by highly polar solvents; prepd. from TiF$_4$ and the diarsine in THF. IR spectrum - a structure of two octahedra linked by F bridges suggested. The phys. prop. may indicate a polymeric structure in the solid state (538).

[V(DA)$_2$Cl$_4$] (M-97), IR spectrum (355).

[V(DA)(CO)$_4$]$_2$ (M-97), light yellow, diamagnetic solid, prepd. from [V(CO)$_6$]$_2$ and the diarsine in light petroleum at r.t. IR spectrum (833).

[V(DA)(CO)$_4$]I, prepd. from [V(DA)(CO)$_4$]$_2$ by titration with iodine (833).

[W(DA)(CO)$_4$] (M-97), crystallographic data, isomorphous with the Mo analog (264).

[W(DA)$_2$I$_2$], brown, paramagnetic powder, dec. in air, a nonelectrolyte in PhNO$_2$, soly. data, prepd. from WI$_3$ and excess of the diarsine at 165° in an evacuated tube (480).

[(W$_6$Cl$_8$)(DA)$_2$Cl$_2$]Cl$_2$, IR spectrum (362).

[Zn(DA)Br$_2$] (M-97), IR spectrum - a tetrahedral or distorted tetrahedral structure was suggested (453).

[Zn(DA)Cl$_2$], IR spectrum - the same structure as that of the dibromo complex was suggested (453).

[Zn(DA)I$_2$] (M-97), IR spectrum - the same structural features as those of the preceding two complexes was suggested (453).

[Zr(DA)$_2$Cl$_4$] (M-98), prepd. from ZrCl$_4$ and the diarsine in THF. When a mixt. of ZrCl$_4$ with HfCl$_4$ is used, the Zr complex is pptd. first so effectively that the complex formation may be a useful method for the sepn. of ZrCl$_4$ from HfCl$_4$ (358). IR spectrum (355).

[Zr(DA)$_2$Br$_4$], white ppt., prepd. from ZrBr$_4$ and the diarsine in THF (358).

[3-(Dimethylarsino)propyl]phenylarsine,   As$_2$C$_{11}$H$_{18}$,   Me$_2$As(CH$_2$)$_3$AsHPh
The arsine was mentioned in ref. (1163).

1,8-Naphthylenebis(dimethylarsine), $As_2C_{14}H_{18}$,

Me$_2$As AsMe$_2$

Synth.: From Me$_2$AsNa and 1.8-dichloronaphthalene in THF
at r.t. (1415).

Prop.: Viscous oil, b$_{5m\mu}$ 125°, which solidified to colorless crystals, m. 86-87° (from EtOH), sol. in THF, Et$_2$O, sparingly sol. in
EtOH, insol. in H$_2$O (1415).

Coordination Deriv.: [In the following complexes NDA denotes 1,8-naphthylene-
bis(dimethylarsine).]

[Ni(NDA)$_2$](ClO$_4$)$_2$, orange crystals, dec. 200°, a 1:1 electrolyte in MeNO$_2$,
prepd. from Ni(ClO$_4$)$_2$·6H$_2$O and the diarsine in Me$_2$CO under reflux. When the
rxn. was carried out in MeNO$_2$, a blue prod. was obtd. which gradually changed
to the orange form. Electronic absorption spectrum. A 4-coordinate Ni complex in 2 isomeric forms is postulated (479).

[Ni(NDA)$_2$Cl]Cl, blue crystals, dec. 158-160°, a 1:1 electrolyte in MeNO$_2$,
prepd. from NiCl$_2$·6H$_2$O and the diarsine in boiling EtOH. Electronic absorption spectrum. A 5-coordinate Ni complex with a square-pyramidal structure
is suggested (479).

[Ni(NDA)$_2$Br]Br, blue crystals, dec. 194°, a 1:1 electrolyte in MeNO$_2$, prepd.
from NiBr$_2$·3H$_2$O and the diarsine in EtOH under reflux. Electronic absorption
spectrum. A 5-coordinate Ni complex with a square pyramidal structure is
suggested (479).

[Ni(NDA)$_2$I]I, dark blue crystals, dec. > 240°, a 1:1 electrolyte in MeNO$_2$,
prepd. from NiI$_2$·6H$_2$O and the diarsine in EtOH under reflux. Electronic
absorption spectrum. A 5-coordinate Ni complex - a square pyramidal structure
is suggested (479).

[Ni(NDA)$_2$SCN]SCN, violet crystals, m. 196-198°, a 1:1 electrolyte in MeNO$_2$,
prepd. from NiCl$_2$·6H$_2$O by treating with NH$_4$SCN in EtOH and filtering the soln.
into the diarsine in boiling EtOH. Electronic absorption spectrum. A 5-
coordinate Ni complex for which a square pyramidal structure is suggested
(479).

o-Phenylenebis(diethylarsine), $As_2C_{14}H_{24}$, 2-Et$_2$AsC$_6$H$_4$AsEt$_2$, (M-99)

Prop.: IR and NMR spectra (356).

Coordination Deriv.: (In the following entries DA denotes 2-Et$_2$AsC$_6$H$_4$AsEt$_2$.)

[Sn(DA)Cl$_4$], white, diamagnetic crystals, m. 203-206° (dec.), a nonelectrolyte
in PhNO$_2$, prepd. from SnCl$_4$ and the diarsine in CCl$_4$ (356). IR spectrum; a
6-coordinate Sn complex postulated (356, 357).

[Sn(DA)Br$_4$], pale yellow, diamagnetic crystals, m. 174-180° (dec.), prepd.
from SnBr$_4$ and the diarsine in CCl$_4$ at 40° (356). IR spectrum; a 6-coordinate
Sn complex is postulated (356, 357).

[Ti(DA)Cl$_4$], red, diamagnetic crystals, m. 160-165°, a nonelectrolyte in
PhNO$_2$, prepd. from TiCl$_4$ and the diarsine in CCl$_4$. IR, NMR, and electronic
spectra; a 6-coordinate Ti structure is suggested (356, see also 357).

[Ti(DA)Br$_4$], maroon, diamagnetic crystals, m. 165-170°, a nonelectrolyte in PhNO$_2$, prepd. from TiBr$_4$ and the diarsine in CCl$_4$ (356). IR (356, 357) and electronic (356) spectra. A 6-coordinate Ti complex.

[V(DA)Cl$_4$], brown black, paramagnetic crystals, a nonelectrolyte in PhNO$_2$, prepd. from VCl$_4$ and the diarsine in CCl$_4$ (356). IR (356, 357) and electronic (356) spectra. A 6-coordinate V complex.

Ethynylenebis(dipropylarsine), As$_2$C$_{14}$H$_{28}$, Pr$_2$AsC≡CAsPr$_2$
Synth.: From Pr$_2$AsI + BrMgC≡CMgBr (2:1) in Et$_2$O under reflux, 42.5% yield (912).
Prop.: b$_2$ 118.5-119.5°, n$_D^{20}$ 1.5268, d$_4^{20}$ 1.1570, MR$_D$ 91.89, AR$_{As}$ 12.42 (912).

Ethynylenebis(diisopropylarsine), As$_2$C$_{14}$H$_{28}$, i-Pr$_2$AsC≡CAsPr-i$_2$
Synth.: From i-Pr$_2$AsI + BrMgC≡CMgBr in Et$_2$O under reflux, 30% yield (912).
Prop.: b$_2$ 134-136°, n$_D^{20}$ 1.5122, d$_4^{20}$ 1.0986, MR$_D$ 110.0, AR$_{As}$ 12.24 (912).

p-Phenylenebis[bis(dimethylamino)arsine], As$_2$C$_{14}$H$_{28}$N$_4$,
   (Me$_2$N)$_2$AsC$_6$H$_4$As(NMe$_2$)$_2$
Rxn. with: 1,4-C$_6$H$_4$[P(S)(Me)N$_3$]$_2$ (1:1) → [=NP(Me)(S)C$_6$H$_4$P(Me)(S)N=As(NMe$_2$)$_2$-C$_6$H$_4$As(NMe$_2$)$_2$=]$_2$ (1583).

Bis[o-(dimethylarsino)phenylthio]palladium, As$_2$PdC$_{16}$H$_{20}$S$_2$,
   [2-(Me$_2$As)C$_6$H$_4$S]$_2$Pd
Synth.: From [Pd(o-MeSC$_6$H$_4$AsMe$_2$)X$_2$] (X = Cl or Br) in DMF by treating with o-MeSC$_6$H$_4$AsMe$_2$ and heating under reflux. When [Pd(o-MeSC$_6$H$_4$AsMe$_2$)I$_2$] was heated in DMF, a mixt. of prods. with mono- and di-demethylated ligands was formed (971).
Prop.: Yellowish orange flakes, nonelectrolyte in MeNO$_2$; chelation of both As atoms with Pd is postulated (971).
Rxn. with: MeI in CHCl$_3$ in a sealed tube → [Pd(o-MeSC$_6$H$_4$AsMe$_2$)I$_2$] (971).

Bis[o-(dimethylarsino)phenylthio]platinum, As$_2$PtC$_{16}$H$_{20}$S$_2$,
   [2-(Me$_2$As)C$_6$H$_4$S]$_2$Pt
Synth.: From [Pt(2-MeSC$_6$H$_4$AsMe$_2$)$_2$][PtCl$_4$] by treating with 2-MeSC$_6$H$_4$AsMe$_2$ in DMF and heating under reflux (971).
From [Pt(2-MeSC$_6$H$_4$AsMe$_2$)$_2$I]I·2H$_2$O by heating in DMF under reflux (971).
Prop.: Yellow crystals, nonelectrolyte in Me$_2$NO$_2$; chelation of both As atoms with Pt is postulated (971).

Dibromobis[o-(dimethylarsino)phenyl-μ-thiolo]dipalladium, As$_2$Pd$_2$C$_{16}$H$_{20}$Br$_2$S$_2$,
   [2-(Me$_2$As)C$_6$H$_4$SPdBr]$_2$
Synth.: From [Pd(o-MeSC$_6$H$_4$AsMe$_2$)Br$_2$] by heating in THF, DMF, or cyclohexanone under reflux (971, see also 969).
Prop.: Orange flakes, nonelectrolyte in MeNO$_2$. An S-bridged dimeric structure postulated (971, see also 969).

Dichlorobis[o-(dimethylarsino)phenyl-μ-thiolo]dipalladium, $As_2Pd_2C_{16}H_{20}Cl_2S_2$,
     $[2-(Me_2As)C_6H_4SPdCl]_2$

Synth.:   From $[Pd(MeSC_6H_4AsMe_2)Cl_2]$ and from $\{Pd[o-(PhCH_2S)C_6H_4AsMe_2]Cl_2\}$ by heating in THF under reflux (971, see also 969).

From $[Pd(o-MeSC_6H_4AsMe_2)_2]^{++}$ and $K_2[PdCl_4]$ by heating in 90% aq. Me_2CO for 40 min. (971).

Prop.:   Orange flakes, nonelectrolyte in $MeNO_2$.   A S-bridged dimeric structure is postulated (971, see also 969).

Diiodobis[o-(dimethylarsino)phenyl-μ-thiolo]dipalladium, $As_2Pd_2C_{16}H_{20}I_2S_2$,
     $[2-(Me_2As)C_6H_4SPdI]_2$

Synth.:   From $[Pd(o-MeSC_6H_4AsMe_2)I_2]$ in THF by heating under reflux (969).

Prop.:   Reddish orange crystals, insol. in org. solvents.   An S-bridged dimeric structure is postulated (971).

Dibromobis[o-(dimethylarsino)phenyl-μ-thiolo]diplatinum, $As_2Pt_2C_{16}H_{20}Br_2S_2$,
     $[2-(Me_2As)C_6H_4SPtBr]_2$

Synth.:   From $[Pt(2-MeSC_6H_4AsMe_2)_2][PtBr_4]$ in DMF by heating under reflux (971).

Prop.:   Yellow crystals, a nonelectrolyte in $MeNO_2$; a S-bridged dimeric structure is postulated (971).

Dichlorobis[o-(dimethylarsino)phenyl-μ-thiolo]diplatinum, $As_2Pt_2C_{16}H_{20}Cl_2S_2$,
     $[2-(Me_2As)C_6H_4SPtCl]_2$

Synth.:   From $[Pt(2-MeSC_6H_4AsMe_2)_2][PtCl_4]$ in DMF by heating under reflux (971).

From $[Pt(2-MeSC_6H_4AsMe_2)_2]^{++} + K_2[PtCl_4]$ in 80% aq. Me_2CO by heating for 40 min. (971).

Prop.:   Yellow crystals, a nonelectrolyte in $MeNO_2$; a S-bridged dimeric structure is postulated (971).

Diiodobis[o-(dimethylarsino)phenyl-μ-thiolo]diplatinum, $As_2Pt_2C_{16}H_{20}I_2S_2$,
     $[2-(Me_2As)C_6H_4SPtI]_2$

Synth.:   From $[Pt(2-MeSC_6H_4AsMe_2)I_2]$ in DMF by heating under reflux (971).

Prop.:   Yellow crystals, insol. in $MeNO_2$; a S-bridged dimeric structure is postulated (971).

Bis[o-(dimethylarsino)phenyl]phosphine, $As_2C_{16}H_{21}P$, $[2-(Me_2As)C_6H_4]_2PH$

Prop.: $b_{1.2}$ 184-188° (1163).

Bis[3-(dimethylarsino)propyl]phenylphosphine, $As_2C_{16}H_{29}P$,
     $PhP[(CH_2)_3AsMe_2]_2$

Coordination Deriv.:   (In the following entries DAP denotes the above diarsinophosphine.)

The compound forms 5-coordinate complexes with Ni(II) salts which undergo solvolysis in polar solvents and afford the $[Ni(DAP)X]^+$ species in soln. (1052).

[Co(DAP)$_2$](ClO$_4$)$_2$, prepd. from Co(ClO$_4$)$_2$ and DAP (1052).

[Ni(DAP)$_2$](ClO$_4$)$_2$, prepd. from Ni(ClO$_4$)$_2$ and DAP (1052).

[Pd(DAP)X]X, square-planar complexes, prepd. from PdX$_2$ and DAP (1052).

[Pt(DAP)X]X, square-planar complexes, prepd. from PtX$_2$ and DAP (1052).

Ethynylenebis(dibutylarsine),   As$_2$C$_{18}$H$_{36}$,   Bu$_2$AsC≡CAsBu$_2$
Synth.: From Bu$_2$AsI + BrMgC≡CMgBr in Et$_2$O under reflux for 1 hr., 43% yield (912).
Prop.: b$_{1.5}$ 159-160°, n$_D^{20}$ 1.5168, d$_4^{20}$ 1.1154, MR$_D$ 110.45, AR$_{As}$ 12.46 (912).

2,2'-Biphenylenebis(diethylarsine),   As$_2$C$_{20}$H$_{28}$,   [2-(Et$_2$As)C$_6$H$_4$-]$_2$,   (M-100)
Rxn. with: BrCH$_2$CH$_2$Br → 5,5,8,8-tetraethyl-5,6,7,8-tetrahydrodibenzo[e,g]-
[1,4]diarsocinium dibromide (57).
o-Xylylene dibromide → 9,9,16,16-tetraethyl-9,10,15,16-tetrahydrotribenzo-
[b,d,h][1,6]diarsecinium dibromide (57).

o-Terphenyl-2,2″-ylenebis(dimethylarsine),   As$_2$C$_{22}$H$_{24}$,
    o-C$_6$H$_4$(C$_6$H$_4$AsMe$_2$-o)$_2$,   (M-101)
Synth.: From o-BrC$_6$H$_4$I + Mg in Et$_2$O, followed by Me$_2$AsI, a conversion to the
dimethiodide through a rxn. with excess MeI in MeOH, and a thermal dec. of
the dimethiodide at 285-290°/0.05 mm (1016).
Prop.: Crystals (from EtOH), m. 92-93° (1016).
Rxn. with: MeI (excess) in MeOH → the dimethiodide (1016).
MeBr in MeOH at 60° in a sealed tube → the dimethobromide (1016).
Dild. HNO$_3$ → the bis(hydroxy nitrate) (1016).
H$_2$O$_2$ soln. at 35-40° → the dioxide·H$_2$O$_2$·2H$_2$O adduct (1016).
CH$_2$Br$_2$ in a sealed tube at 100° → 5,6-dihydro-5,5,7,7-tetramethyl-7H-tribenzo-
[d,f,h][1,3]diarsoninium dibromide (1016).
Br(CH$_2$)$_n$Br → quaternization of both As groups with a ring closure (1016).

Ethynylenebis(dipentylarsine),   As$_2$C$_{22}$H$_{44}$,   (C$_5$H$_{11}$)$_2$AsC≡CAs(C$_5$H$_{11}$)$_2$
Synth.: From (C$_5$H$_{11}$)$_2$AsI + BrMgC≡CMgBr in Et$_2$O under reflux for 1 hr., 17%
yield (912).
Prop.: b$_3$ 196-198°, n$_D^{20}$ 1.5103, d$_4^{20}$ 1.0600, MR$_D$ 129.54, AR$_{As}$ 12.77 (912).

o,o'-Benzylidenebis(phenyldimethylarsine),   As$_2$C$_{23}$H$_{26}$,   PhCH(C$_6$H$_4$AsMe$_2$)$_2$
Prop.: m. 96-98° (1163).

2,2-Bis[p-(diethylarsino)phenyl]propane,   As$_2$C$_{23}$H$_{34}$,
    [4-(Et$_2$As)C$_6$H$_4$]$_2$CMe$_2$,   (M-102)
Rxn. with: Bu$_3$GeN$_3$ (1:2) → [Bu$_3$GeN=As(Et)$_2$C$_6$H$_4$]$_2$CMe$_2$ (1584-1586).

Methylenebis(diphenylarsine),   As$_2$C$_{25}$H$_{22}$,   Ph$_2$AsCH$_2$AsPh$_2$,   (M-102)
Coordination Deriv.: (In the following entries DA denotes Ph$_2$AsCH$_2$AsPh$_2$.)
[Ir(DA)$_2$Cl], prepd. from Ir(CO)$_4$Cl and the diarsine in C$_6$H$_6$. The complex
catalyzes hydrogenation of olefins and acetylenes (776).

(In the following entries DA denotes $Ph_2AsCH_2AsPh_2$.)

$[Mo(DA)_2(CO)_2Br_2]$, orange, diamagnetic crystals, a nonelectrolyte in CH_2Cl_2, prepd. from $Mo(CO)_4Br_2$ and the diarsine in CH_2Cl_2. IR spectrum - a 7-coordination of the Mo atom postulated, with one DA as a mono- and the other as a bi-dentate (78).

$[Mo(DA)_2(CO)_2Cl_2]$, yellow, diamagnetic crystals, a nonelectrolyte, prepd. from $Mo(CO)_4Cl_2$ and the diarsine in CH_2Cl_2. IR spectrum - a 7-coordination of the Mo atom postulated, with a bonding analogous to that of the preceding dibromo complex (78).

$[Mo(DA)_2(CO)_3Br_2] \cdot Me_2CO$, yellow, diamagnetic crystals, prepd. from $Mo(CO)_4Br_2$ and the diarsine in Me_2CO. IR spectrum - a 7-coordinate Mo bonding suggested (78).

$[Mo(DA)_2(CO)_3Cl_2] \cdot Me_2CO$, yellow, diamagnetic crystals, prepd. from $Mo(CO)_4Cl_2$ and the diarsine in Me_2CO. IR spectrum - a 7-coordinate Mo bonding suggested (78).

$[W(DA)_2(CO)_2Br_2]$, orange, diamagnetic crystals, a nonelectrolyte, prepd. from $[W(DA)_2(CO)_3Br_2]$ by boiling in CH_2Cl_2. IR spectrum - a 7-coordinate W complex with one diarsine as a mono- and the other as a bi-dentate is postulated (78).

$[W(DA)_2(CO)_2Cl_2]$, yellow, diamagnetic crystals, a nonelectrolyte in CH_2Cl_2, prepd. from $[W(DA)_2(CO)_3Cl_2]$ by boiling in CH_2Cl_2. IR spectrum - a 7-coordinate W complex, with a bonding analogous to that of the preceding dibromo complex is postulated (78).

$[W(DA)_2(CO)_3Br_2] \cdot CH_2Cl_2$, yellow, diamagnetic crystals, a nonelectrolyte, prepd. from $W(CO)_4Br_2$ and the diarsine in CH_2Cl_2. IR spectrum - a 7-coordinate W bonding is suggested. On boiling in CH_2Cl_2 the complex loses 1 CO group (78).

$[W(DA)_2(CO)_3Cl_2] \cdot Me_2CO$, yellow, diamagnetic crystals, a nonelectrolyte, prepd. from $W(CO)_4Cl_2$ and the diarsine in Me_2CO. IR and NMR spectra - a 7-coordinate W bonding is suggested. On boiling in CH_2Cl_2 the complex loses 1 CO group and Me_2CO (78).

cis-Vinylenebis(diphenylarsine), $As_2C_{26}H_{22}$, $Ph_2AsCH=CHAsPh_2$
Synth.: From Ph_2AsLi and cis-$ClCH=CHCl_2$ (2:1) in THF, 61% yield (45, 88).
Prop.: m. 112-113°, IR and NMR spectra (45).
Coordination Deriv.: (In the following entries VDA denotes cis-$Ph_2AsCH=CHAsPh_2$.)

$[Rh(VDA)_2]BF_4 \cdot MeOH$, orange crystals, m. 238° (dec.), a 1:1 electrolyte in $MeNO_2$, prepd. from $[Rh(CO)_2Cl]_2$ and the diarsine in MeOH by heating under reflux, followed by a treatment with $NaBF_4$. IR spectrum - a square planar configuration of the VDA ligands is suggested (45).

$[Rh(VDA)_2]BPh_4$, a 1:1 electrolyte in $MeNO_2$, prepd. from $[Rh(VDA)_2H_2]Cl$ by dehydrogenation in CH_2Cl_2 by a stream of N, followed by pptn. with $NaBPh_4$.

IR spectrum (1005).

[Rh(VDA)$_2$]Cl, bright yellow substance, prepd. from [RhH$_2$(VDA)$_2$]Cl by dehydrogenation in CH$_2$Cl$_2$ with a stream of N. The complex reversibly activates H$_2$,
readily reacts with halogens, allyl chloride, methyl iodide, SO$_2$, NO, CO, and
PF$_3$ to give simple 5- and 6-coordinate complexes (1005).

[Rh(VDA)$_2$(CO)]BPh$_4$, prepd. from [Rh(VDA)$_2$]BPh$_4$ by treating with CO. IR spectrum (1005).

[RhH$_2$(VDA)$_2$]Cl, buff colored crystals, a 1:1 electrolyte in MeNO$_2$, stable to
air and moisture in the solid state, prepd. from [RhH$_2$(PPh$_3$)$_2$]Cl and the diarsine in CH$_2$Cl$_2$-C$_6$H$_6$ under hydrogen. The complex in soln. readily loses H.
The addn. and removal of H can be followed by IR and NMR spectra (1005).

trans-Vinylenebis(diphenylarsine),   As$_2$C$_{26}$H$_{22}$,   Ph$_2$AsCH=CHAsPh$_2$
Synth.: From Ph$_2$AsLi and trans-ClCH=CHCl (2:1) in THF, 10% yield (45).
Prop.: m. 103-104°, NMR and IR spectra (45).
Coordination Deriv.: [Rh(trans-Ph$_2$AsCH=CHAsPh$_2$)$_2$Cl$_2$]Cl, prepd. from [Rh(cis-
Ph$_2$AsCH=CHAsPh$_2$)$_2$]Cl by treating with Cl (1005).

[RhH(trans-Ph$_2$AsCH=CHAsPh$_2$)$_2$X]X, where X = Cl, Br, or I, are formed from
[Rh(cis-Ph$_2$AsCH=CHAsPh$_2$)$_2$]$^+$ by adding HX. IR and NMR spectra (1005).

[Rh(trans-Ph$_2$AsCH=CHAsPh$_2$)(CO)Cl], pale yellow microcrystals, extremely insol.,
prepd. from [Rh(CO)$_2$Cl]$_2$ and the diarsine (1:2) in C$_6$H$_6$. IR spectrum - a
trans configuration is postulated. A polymeric structure is suggested (45).

Ethylenebis(diphenylarsine),   As$_2$C$_{26}$H$_{24}$,   Ph$_2$AsCH$_2$CH$_2$AsPh$_2$,   (M-102)
Synth.: From Ph$_2$AsK·2 Dioxan and ClCH$_2$CH$_2$Cl in THF under reflux (1315).
By-prod. from the condensation of Ph$_2$AsK with alkali metal arylsulfonates in
(EtOCH$_2$CH$_2$)$_2$O by heating to 180° (1347).
Prop.: Crystals, m. 100-103° (1347). IR spectrum (1268). NMR spectrum
(1526).
Rxn. with: S in C$_6$H$_6$ under reflux → Ph$_2$As(S)CH$_2$CH$_2$As(S)Ph$_2$ (1347).
UCl$_4$ in THF in the presence of O → {U[Ph$_2$As(S)CH$_2$CH$_2$As(S)Ph$_2$]$_2$Cl$_4$} (1383).
Coordination Deriv.: (In the following entries DA denotes Ph$_2$AsCH$_2$CH$_2$AsPh$_2$.)

[Cr(DA)$_3$], bright to dark yellow, diamagnetic crystals, a nonelectrolyte in
Me$_2$CO, very sensitive to air, prepd. from K$_6$[Cr(CN)$_6$] and the diarsine in liq.
NH$_3$ (147).

[Hf(DA)Cl$_4$], prepd. from HfCl$_4$ and the diarsine in C$_6$H$_6$ in a Schlenk tube. IR
spectrum (1268).

[Mo(DA)(CO)$_3$I$_2$], diamagnetic crystals, m. 140°, slightly ionized in PhNO$_2$,
prepd. from NH$_4$[Mo(CO)$_4$I$_3$] and the diarsine (1:1) in CH$_2$Cl$_2$ at 25°. IR and
NMR spectra (1526).

[Ni(DA)$_2$], orange yellow, diamagnetic crystals, prepd. from K$_4$[Ni(CN)$_4$] and
the diarsine in liq. NH$_3$ at 40° (147).

(In the following entries DA denotes $Ph_2AsCH_2CH_2AsPh_2$.)

[Ni(DA)Br$_2$], purple, diamagnetic crystals, m. 215-220°, a nonelectrolyte in ClCH$_2$CH$_2$Cl, prepd. by adding the diarsine dissolved in BuOH to a boiling soln. of NiBr$_2$ in the same solvent. Absorption and reflectance spectra were obtd. A square-planar structure is suggested (1315).

[Ni(DA)I$_2$], amethyst-colored, diamagnetic crystals, m. 200° (dec.), a nonelectrolyte in ClCH$_2$CH$_2$Cl and in EtNO$_2$, prepd. from NiI$_2$ in the same manner as the preceding dibromo complex. Absorption spectrum - a square-planar structure is suggested (1315).

[Pd(DA)Br$_2$] (M-102), yellow crystals, m. 217-217.5°, prepd. from Na[PdCl$_4$] by a rxn. with excess LiBr in EtOH and filtering the soln. into a soln. of the diarsine in CH$_2$Cl$_2$ (1151).

[Pd(DA)Cl$_2$] (M-102), light yellow crystals, m. > 340° (1602), prepd. from Na$_2$[PdCl$_4$] in EtOH by a rxn. with the diarsine in CH$_2$Cl$_2$ (1151) and from K$_2$[PdCl$_4$] in DMF and the diarsine by heating (1602).

[Pd(DA)I$_2$] (M-102), red-orange crystals, prepd. from Na$_2$[PdCl$_4$] by a treatment with excess NaI in EtOH and filtering the soln. into the diarsine dissolved in CH$_2$Cl$_2$ (1151).

[Pd(DA)(SCN)$_2$], yellow crystals, m. 217-217.5°, prepd. from Na$_2$[PdCl$_4$] by treating with excess NaSCN in EtOH and filtering the soln. into the diarsine dissolved in CH$_2$Cl$_2$ and also by adding the diarsine dissolved in boiling BuOH to a boiling soln. of [Pd(SCN)$_4$]$^{--}$ in EtOH and quenching the rxn. mixt. in ice-H$_2$O bath (1151).

[Pd(DA)(SeCN)$_2$], yellow-orange crystals, prepd. from Na$_2$[PdCl$_4$] and excess KSeCN in EtOH by filtering the soln. into the diarsine dissolved in CH$_2$Cl$_2$ (1151).

[Pd(DA)$_2$](ClO$_4$)$_2$ (M-103), dec. 338.5°, a 1:2 electrolyte in MeNO$_2$ - conducto-metric titration with [Et$_4$N]I; prepd. from [Pd(DA)$_2$Cl]Cl in EtOH by treating with excess dild. HClO$_4$ (1602).

[Pd(DA)$_2$Br]Br (?), yellow substance, m. 210-225° (dec.), molar cond. and conductometric titrations in nonaq. solvent indicated a partial assocn. corr. to an equil. compn. [Pd(DA)$_2$]Br$_2$ $\rightleftharpoons$ [Pd(DA)$_2$Br]Br. The complex was prepd. from [Pd(DA)Cl$_2$] by heating with BaBr$_2$ in DMF, pptg. the prod. with Et$_2$O, washing the ppt. with H$_2$O to remove BaCl$_2$, dissolving the residue with DMF, and treating the soln. with the diarsine (1602).

[Pd(DA)$_2$Cl]Cl ( ), yellow substance, m. 231-250° (dec.), molar cond. data in MeNO$_2$ and in MeOH similar to the preceding bromo analog; partial assocn. of the Cl was suggested. The complex was prepd. from [Pd(DA)Cl$_2$] in hot DMF and the diarsine in CH$_2$Cl$_2$ (1602).

[Pt(DA)Cl$_2$], white glistening platelets, m. 331-335° (dec), prepd. from Na$_2$[PtCl$_4$] in EtOH by a rxn. with the diarsine in CH$_2$Cl$_2$, washing the resulting ppt. with H$_2$O and heating it with EtOH-HCl under reflux (1602).

[Pt(DA)$_2$]Br$_2$, white crystals, m. 210.5-212°, molar cond. in MeOH is significantly lower than that of the corr. dinitrate and in MeNO$_2$ the value is close to that of a 1:1 electrolyte. The complex was prepd. from [Pt(DA)Cl$_2$] in warm DMF by treating with BaBr$_2$, pptg. the prod. with Et$_2$O, washing the ppt. with H$_2$O to remove BaCl$_2$, dissolving the crystals in DMF, adding the diarsine and pptg. with Et$_2$O (1602).

[Pt(DA)$_2$]Cl$_2$, white ppt., m. 200-200.5°, its molar cond. in MeOH is lower than that of the corr. dinitrate and in MeNO$_2$ the value is close to that of a 1:1 electrolyte. The complex was prepd. from [Pt(DA)Cl$_2$] in DMF by adding the diarsine in CHCl$_3$, followed by an equal vol. of EtOH and light petroleum (1602).

[Pt(DA)$_2$](ClO$_4$)$_2$, white powder, m. 314.5°, insol. in MeOH, a 2:1 electrolyte in MeNO$_2$, prepd. from [Pt(DA)$_2$]Cl$_2$ in EtOH by adding to an excess of dild. HClO$_4$ (1602).

[Pt(DA)$_2$](NO$_3$)$_2$, white powder, m. 292-295°, a 2:1 electrolyte in MeNO$_2$, prepd. from the corr. dichloride in hot H$_2$O by treating with KNO$_3$ (1602).

[Pt(DA)Br$_4$], prepd. from PtBr$_4$ and the diarsine in C$_6$H$_6$ soln. in a Schlenk tube. IR spectrum (1268).

[Pt(DA)Cl$_4$], prepd. from PtCl$_4$ in the same manner as the tetrabromo complex. IR spectrum (1268).

[Ti$_2$(DA)$_3$Cl$_8$], red-orange substance, m. 145-149°, prepd. from TiCl$_4$ and the diarsine (2:3) in C$_6$H$_6$. IR spectrum (1603).

[W(DA)(CO)$_3$I$_2$]·CH$_2$Cl$_2$, diamagnetic crystals, m. 200° (dec.), a nonelectrolyte in MeNO$_2$, prepd. from NH$_4$[W(CO)$_4$I$_3$] and the diarsine (1:1) in CH$_2$Cl$_2$ at 25°. IR and NMR spectra (1526).

[Zr(DA)Br$_4$], prepd. from ZrBr$_4$ and the diarsine in C$_6$H$_6$ soln. in a Schlenk tube. IR spectrum (1268).

[Zr(DA)Cl$_4$], prepd. from ZrCl$_4$ in the same manner as the preceding tetrabromo complex. IR spectrum (1268).

Ethynylenebis(di-n-hexylarsine), As$_2$C$_{26}$H$_{52}$, (C$_6$H$_{13}$)$_2$AsC≡CAs(C$_6$H$_{13}$)$_2$
Synth.: From (C$_6$H$_{13}$)$_2$AsI and BrMgC≡CMgBr (2:1) in Et$_2$O under reflux for 1 hr., 16% yield (912).
Prop.: Crystals, m.< 35°, b$_{1.5}$ 199-201°, n$_D^{20}$ 1.4963, d$_4^{20}$ 1.0222, MR$_D$ 146.99, AR$_{As}$ 12.26 (912)

Butadiynylenebis(diphenylarsine), As$_2$C$_{28}$H$_{20}$, Ph$_2$AsC≡CC≡CAsPh$_2$
Synth.: From Ph$_2$As halides and HC≡CC≡CH in liq. NH$_3$, 40% yield along with Ph$_2$AsC≡CC≡CH (149).
From Ph$_2$AsCl + Me$_3$SnC≡CC≡CSnMe$_3$ in C$_6$H$_6$ at 100°, 40% yield (694).
Prop.: m. 115° (149, 694), m. 114° (dec.) (695). IR spectrum (694, 695).

Tetramethylenebis(diphenylarsine), $As_2C_{28}H_{28}$, $Ph_2As(CH_2)_4AsPh_2$, (M-103)
Synth.: From $Ph_2AsK \cdot 2$ Dioxan and $Cl(CH_2)_4Cl$ (2:1) in THF under reflux (1315).
Prop.: Crystals (1315).
Coordination Deriv.: {$Ni[Ph_2As(CH_2)_4AsPh_2]I_2$}, brown, paramagnetic crystals,
m. 199-203°, prepd. from NiI_2 and the diarsine in boiling BuOH. Absorption
and reflectance spectra - a distorted tetrahedral structure is postulated
(1315).

(Oxydiethylene)bis(diphenylarsine), $As_2C_{28}H_{28}O$, $(Ph_2AsCH_2CH_2)_2O$
Synth.: From $Ph_2AsK \cdot 2$ Dioxan and $(ClCH_2CH_2)_2O$ (2:1) in THF under reflux
(1315).
Prop.: Oily liquid (1315).
Coordination Deriv.: {$Ni[(Ph_2AsCH_2CH_2)_2O]I_2$}, dark brown, paramagnetic crystals, m. 218-223°, a nonelectrolyte in $ClCH_2CH_2Cl$ and in $EtNO_2$, prepd. from
NiI_2 and the diarsine in boiling BuOH. Absorption and reflectance spectra -
a distorted tetrahedral structure is proposed (1315).

(Thiodiethyl)bis(diphenylarsine), $As_2C_{28}H_{28}S$, $(Ph_2AsCH_2CH_2)_2S$
Synth.: From $Ph_2AsK \cdot 2$ Dioxan and $(ClCH_2CH_2)_2S$ in THF under reflux (1315).
Prop.: Crystals, m. 125-127° (1315).
Coordination Deriv.: {$Ni[Ph_2AsCH_2CH_2)_2S]_2Br_2$}, light green, paramagnetic
crystals, m. 130-131°, a nonelectrolyte in $C_2H_4Cl_2$ and in $EtNO_2$, prepd. from
$NiBr_2$ and the diarsine in boiling BuOH. Absorption and reflectance spectra -
a distorted octahedral structure is suggested (1315).

{$Ni[(Ph_2AsCH_2CH_2)_2S]I_2$}, green, diamagnetic crystals, m. 216-217°, a nonelectrolyte in CH_2Cl_2 and in $EtNO_2$, prepd. from NiI_2 and the diarsine in boiling
BuOH. Absorption and reflectance spectra - a 5-coordinate Ni bonding is
postulated (1315).

Bis[2-(diphenylarsino)ethyl]amine, $As_2C_{28}H_{29}N$, $(Ph_2AsCH_2CH_2)_2NH$
Synth.: From $Ph_2AsK \cdot 2$ Dioxan and $(ClCH_2CH_2)_2NH$ (2:1) in THF under reflux
(1315).
Prop.: Oily liquid (1315).
Coordination Deriv.: {$Ni[(Ph_2AsCH_2CH_2)_2NH]Br_2$}, crimson, paramagnetic crystals, m. 192-194°, a nonelectrolyte in $C_2H_4Cl_2$ but partly ionized in $EtNO_2$,
prepd. from $NiBr_2$ and the diarsine in boiling BuOH. Absorption and reflectance spectra; a possibility of a dimeric mol. in the solid state is discussed
(1315).

{$Ni[(Ph_2AsCH_2CH_2)_2NH]I_2$}, dark gray, diamagnetic crystals, m. 224-226°, a nonelectrolyte in $C_2H_4Cl_2$ but partly ionized in $EtNO_2$, prepd. from NiI_2 and the
diarsine in boiling BuOH. Absorption and reflectance spectra - a distorted
square-pyramidal structure is suggested (1315).

o-Phenylenebis[bis(pentafluorophenyl)arsine], $As_2C_{30}H_4F_{20}$,
    o-[$(C_6F_5)_2As]_2C_6H_4$
The diarsine is mentioned (1163).

116

m-Phenylenebis(diphenylarsine), $As_2C_{30}H_{24}$, $3\text{-}(Ph_2As)C_6H_4AsPh_2$, (M-103)

Rxn. with: $Ph_2Ge(N_3)_2 \rightarrow \left[\begin{array}{c}Ph \quad Ph \quad\quad Ph \\ | \quad\quad | \quad\quad\quad | \\ Ge\text{-}N{=}As\text{-}C_6H_4\text{-}As{=}N \\ | \quad\quad | \quad\quad\quad | \\ Ph \quad Ph \quad\quad Ph \end{array}\right]_n$ (1584-1586).

p-Phenylenebis(diphenylarsine), $As_2C_{30}H_{24}$, $4\text{-}(Ph_2As)C_6H_4AsPh_2$, (M-104)

Synth.: From Ph_2AsK and $p\text{-}ClC_6H_4SO_3Na$ in $(EtOCH_2CH_2)_2O$ by heating up to 180°, 75% yield (1347).
From Ph_2AsH by adding dropwise to a K suspension in boiling THF, cooling, treating the mixt. with $p\text{-}C_6H_4Br_2$ in THF, and completing the rxn. under reflux, 75% yield (1658).
Prop.: White crystals, m. 130-133°, after recrystn. m. 150° (1347), m. 150°, sol. in hydrocarbons, $CHCl_3$, Me_2CO, and Et_2O, sparingly sol. in EtOH and petr. ether. Dipole moment 1.51 D in C_6H_6 at 120°. UV and IR spectra (1658).
Rxn. with: Br → the tetrabromide (1658).
H_2O_2 in Me_2CO → $Ph_2As(O)C_6H_4As(O)Ph_2$ (1658).

1,5-Naphthylenebis(diphenylarsine), $As_2C_{34}H_{26}$,
Synth.: From Ph_2AsK and 1,5-naphthalenedisulfonic acid di-Na salt in $(EtOCH_2CH_2)_2O$ by heating up to 180°, 7% yield (1347).
Prop.: m. 240-243° (1347).
Rxn. with: 30% aq. H_2O_2 in Me_2CO under reflux → $Ph_2As(S)C_{10}H_6As(S)Ph_2$ (1347).

2,6-Naphthylenebis(diphenylarsine), $As_2C_{34}H_{26}$,
Synth.: From Ph_2AsK and 2,6-naphthalenedisulfonic acid di-Na salt in $(EtOCH_2CH_2)_2O$ by heating to 180°, 5% yield (1347).
Prop.: m. 202-204° (1347).
Rxn. with: S at 140° → $Ph_2As(S)C_{10}H_6As(S)Ph_2$ (1347).
30% aq. H_2O_2 in Me_2CO under reflux → $Ph_2As(O)C_{10}H_6As(O)Ph_2$ (1347).

2,7-Naphthylenebis(diphenylarsine), $As_2C_{34}H_{26}$,
Synth.: From Ph_2AsK and 2,7-naphthalenedisulfonic acid di-Na salt in $(EtOCH_2CH_2)_2O$ by heating to 180°, 17% yield (1347).
Rxn. with: S at 140° → $Ph_2As(S)C_{10}H_6As(S)Ph_2$ (1347).
30% aq. H_2O_2 in Me_2CO under reflux → $Ph_2As(O)C_{10}H_6As(O)Ph_2$ (1347).

1,1′-Bis(diphenylarsino)ferrocene, $As_2FeC_{34}H_{28}$,
Synth.: From Ph_2AsCl and a mixt. of $FcdLi_2$ and FcLi in Et_2O under reflux, a mixt. of $Fcd(AsPh_2)_2$ with $FcAsPh_2$ was obtd. with yields of 2 and 4%, resp.
Prop.: Yellow powder, m. 145-146.5°, IR spectrum (1437).

N,N′-(1,2-Dimethylethanediylidene)bis[3-(diphenylarsino)propylamine], $As_2C_{34}H_{38}N_2$, $Ph_2AsCH_2CH_2CH_2N{=}C(Me)C(Me){=}NCH_2CH_2CH_2AsPh_2$

Synth.: From Ph$_2$AsCH$_2$CH$_2$CH$_2$NH$_2$ by condensation with biacetyl (390).

1,5-Anthrylenebis(diphenylarsine), As$_2$C$_{38}$H$_{28}$, Ph$_2$AsC$_{14}$H$_8$AsPh$_2$

Synth.: From Ph$_2$AsK and 1,5-anthracenedisulfonic acid di-Na salt in (EtOCH$_2$CH$_2$)$_2$O by heating to 180°, 2% yield (1347).

Prop.: m. 346-348° (1347).

[Ethylenebis(oxy-o-phenylene)]bis(diphenylarsine), As$_2$C$_{38}$H$_{32}$O$_2$,
  2-(Ph$_2$As)C$_6$H$_4$OCH$_2$CH$_2$OC$_6$H$_4$(AsPh$_2$)-2

Synth.: From [2-(Cl$_2$As)C$_6$H$_4$OCH$_2$-]$_2$ and PhMgBr in Et$_2$O, followed by dild. HCl (292).

Prop.: White crystals, m. 143° (292).

Coordination Deriv.: (In the following entries EPA denotes 2-(Ph$_2$As)C$_6$H$_4$O-CH$_2$CH$_2$OC$_6$H$_4$(AsPh$_2$)-2.)

[Pd(EPA)Br$_2$], yellow-orange crystals, m. 153-154°, prepd. from K$_2$[PdBr$_4$] and the diarsine in EtOH. UV and visible spectra. A trans coordination of the diarsine is suggested (292).

[Pd(EPA)Cl$_2$], yellow crystals, m. 167-168°, prepd. from K$_2$[PdCl$_4$] and the diarsine in EtOH. UV and visible spectra. A trans coordination of the diarsine is suggested (292).

[Pd(EPA)I$_2$], red crystals, m. 79-80°, prepd. from K$_2$[Pd(NO$_3$)$_4$] in aq.-EtOH NaI soln. by adding the diarsine. UV and visible spectra. A trans coordination of the diarsine is suggested (292).

[Pd(EPA)(NCS)$_2$], orange crystals, m. 154-155°, prepd. from K$_2$[Pd(NO$_3$)$_4$] in aq.-EtOH soln. of NaNCS by adding the diarsine. UV and visible spectra. A trans coordination of the diarsine is suggested (292).

[Ethylenebis(thio-o-phenylene)]bis(diphenylarsine), As$_2$C$_{38}$H$_{32}$S$_2$,
  2-(Ph$_2$As)C$_6$H$_4$SCH$_2$CH$_2$SC$_6$H$_4$(AsPh$_2$)-2

Synth.: From 2-(Ph$_2$As)C$_6$H$_4$SH by treating with Na in EtOH under reflux, followed by BrCH$_2$CH$_2$Br (292).

Prop.: Pink ppt., m. 182° (292).

Coordination Deriv.: In the following entries ESPA denotes the diarsine. A trans coordination of ESPA with the Pd atom in all four complexes is suggested.

[Pd(ESPA)Br$_2$], orange crystals, m. 154-155°, prepd. from K$_2$[PdBr$_4$] and the diarsine in EtOH. UV and visible spectra (292).

[Pd(ESPA)Cl$_2$], yellow crystals, m. 174-175°, a nonelectrolyte in PhNO$_2$, prepd. from K$_2$[PdCl$_4$] and the diarsine in EtOH. UV and visible spectra (292).

[Pd(ESPA)I$_2$], red crystals, m. 102-103°, a nonelectrolyte in PhNO$_2$, prepd. from K$_2$[Pd(NO$_3$)$_4$] in aq.-EtOH NaI soln. by a rxn. with the diarsine. UV and visible spectra (292).

[Pd(ESPA)(NCS)$_2$], orange crystals, m. 157-158°, a nonelectrolyte in PhNO$_2$, prepd. from K$_2$[Pd(NO$_3$)$_4$] in aq.-EtOH NaNCS soln. by adding the diarsine. UV and visible spectra (292).

Butadiynylenebis(di-α-naphthylarsine), $\qquad$ As$_2$C$_{44}$H$_{22}$,
$\qquad$ (C$_{10}$H$_7$)$_2$AsC≡CC≡CAs(C$_{10}$H$_7$)$_2$

Synth.: From (α-C$_{10}$H$_7$)$_2$AsCl and Me$_3$SnC≡CC≡CSnMe$_3$ in C$_6$H$_6$ at 100°, 50% yield (694).

Prop.: m. 220° (694), m. 218° (dec.) (695). IR spectrum (694, 695).

1.1.5 TRITERTIARY ARSINES

Besides the methods described in the main volume (M-104), tritertiary arsines have been synthesized from secondary arsino-alkali metals by a reaction with trifunctional organic compounds such as 1,3-dichloro-2-(chloromethyl)-2-methylpropane and tris(chloroalkyl)amines.

$$3\ R_2AsM + (ClCH_2)_3CMe \rightarrow (R_2AsCH_2)_3CMe$$

$$3\ R_2AsM + [X(CH_2)_n]_3N \rightarrow [R_2As(CH_2)_n]_3N$$

Methylbis[3-(dimethylarsino)propyl]arsine, $\qquad$ As$_3$C$_{11}$H$_{27}$,
$\qquad$ MeAs(CH$_2$CH$_2$CH$_2$AsMe$_2$)$_2$, $\qquad$ (M-105)

Coordination Deriv.: (In the following complexes TA denotes MeAs(CH$_2$CH$_2$CH$_2$-AsMe$_2$)$_2$.)

[Cr(TA)(CO)$_3$I]BPh$_4$, brown, paramagnetic plates, prepd. from [Cr(TA)(CO)$_3$] (M-105) in MePh at -70° by adding iodine in C$_6$H$_6$, isolating the resulting solid prod., dissolving it in MeOH and treating with Na[BPh$_4$]. The compd. dec. on standing and turned black. Its magnetic moment increased with time. IR spectrum (1162).

[Mo(TA)(CO)$_2$Br$_2$], yellow powder, a nonelectrolyte in PhNO$_2$, prepd. from [Mo(TA)(CO)$_3$] (M-105) in MePh at -70° by adding Br in C$_6$H$_6$ and warming the rxn. mixt. to 40°. On exposure to light the complex turned brown. IR spectrum (1162).

[Mo(TA)(CO)$_2$I$_2$], orange crystals, a nonelectrolyte in PhNO$_2$, prepd. from [Mo(TA)(CO)$_3$I]I by heating in PrOH at 100°. On exposure to air the complex turned yellow. IR spectrum (1162).

[Mo(TA)(CO)$_3$Br]BPh$_4$, yellow crystals, m. 168°, a 1:1 electrolyte in PhNO$_2$, prepd. from [Mo(TA)(CO)$_3$] (M-105) in MePh at -70° by adding Br in C$_6$H$_6$ and then treating the oily prod. dissolved in MeOH with Na[BPh$_4$]. IR spectrum (1162).

[Mo(TA)(CO)$_3$I]I, yellow ppt., a 1:1 electrolyte in PhNO$_2$, prepd. from [Mo(TA)(CO)$_3$] in MePh at -70° by treating with I dissolved in C$_6$H$_6$. On exposure to light the complex turned green. IR spectrum (1162).

[Ti(TA)Br$_4$] (M-107), crimson solid, a nonelectrolyte in PhNO$_2$, prepd. from TiBr$_4$ and TA in C$_6$H$_6$. IR spectrum; magnetic data - a 7-coordination of Ti is suggested (121).

(In the following complexes TA denotes MeAs(CH$_2$CH$_2$CH$_2$AsMe$_2$)$_2$.)

[(TA)(TiCl$_4$)$_2$] (M-107), crimson ppt., a nonelectrolyte in PhNO$_2$, prepd. by condensing TiCl$_4$ onto a frozen soln. of TA in C$_6$H$_6$. IR spectrum and magnetic data (121).

[(TA)(TiF$_4$)$_3$], white ppt., a nonelectrolyte in PhNO$_2$, prepd. from TA and TiF$_4$ in MeCN. IR spectrum and magnetic data (121).

[Ti(TA)I$_4$] (M-107), pale yellow ppt., a nonelectrolyte in PhNO$_2$, prepd. from TA and TiI$_4$ in CCl$_4$. IR spectrum and magnetic data (121).

[Zr(TA)Br$_4$], white ppt., prepd. from TA and ZrBr$_4$ in MeCN, crystd. with 1/2 MeCN as a cream colored substance. IR spectrum and magnetic data. A 7-coordination of Zr is postulated (121).

{2,2-Bis[(dimethylarsino)methyl]propyl}dimethylarsine,   As$_3$C$_{11}$H$_{27}$,
    MeC(CH$_2$AsMe$_2$)$_3$,   (M-107)

Synth.: From Me$_2$AsI or Me$_2$AsH by treating with Na in THF followed by MeC(CH$_2$Br)$_3$, 47-55% yield (558).
From preformed Me$_2$AsNa and MeC(CH$_2$Cl)$_3$ in THF at 10°, completed under reflux, 70% yield (1229).
From Me$_2$AsLi and MeC(CH$_2$Cl)$_3$ in Et$_2$O (1229).
Prop.: b$_{0.7}$ 110-111°, n$_D^{20}$ 1.5623, d$_{20}$ 1.3571 (558), b$_{0.08}$ 78-79° (1229). IR spectrum (359, 1161), NMR spectrum (558, 559, 1161, 1162, 1229).
Coordination Deriv.: (In the following complexes PTA denotes MeC(CH$_2$AsMe$_2$)$_3$.)

[Cr(PTA)(CO)$_3$], yellow, paramagnetic crystals, a nonelectrolyte in PhNO$_2$, prepd. from Cr(CO)$_6$ and the triarsine in mesitylene under reflux. IR and NMR spectra. A vicinal coordination of the triarsine and a cis arrangement of the CO ligands is postulated (1161).

[Cr(PTA)Cl$_3$], blue ppt., a nonelectrolyte in PhNO$_2$. IR and electronic absorption spectra (359).

[Cr(PTA)(CO)$_2$I]BPh$_4$, red, paramagnetic cryst. plates, m. 100-106°, a 1:1 electrolyte in PhNO$_2$, stable in air, but sensitive to light and slowly becomes green, unstable in Me$_2$CO, soly. data in other solvents, prepd. from [Cr(PTA)-(CO)$_3$] in MeOH at -70° by adding I in C$_6$H$_6$, recovering the red powder, dissolving it in MeOH and adding Na[BPh$_4$]. IR, UV, and NMR spectra (1162).

[(PTA)Cu-Mn(CO)$_5$] (M-107), light brown, diamagnetic ppt., m. 220°, a nonelectrolyte in PhNO$_2$, prepd. from [Cu(PTA)Br] (M-107) suspended in THF by adding Na[Mn(CO)$_5$]. IR spectrum. The complex is fairly stable in dry and moist air; soly. data (832a).

[Mn(PTA)(CO)$_3$]ClO$_4$, pale cream, paramagnetic crystals, sol. in Me$_2$CO and EtOH, insol. in CHCl$_3$ and Et$_2$O, prepd. from [Mn(CO)$_3$(1,3,5-Me$_3$C$_6$H$_3$)]ClO$_4$ and the triarsine in cyclohexanone under reflux. IR and NMR spectra (1161).

[Mo(PTA)(CO)$_2$Br$_2$], yellow, slightly paramagnetic substance, a nonelectrolyte in PhNO$_2$, sensitive to light, prepd. from [Mo(PTA)(CO)$_3$Br]BPh$_4$ and LiBr in

EtOH under reflux. IR, UV, and NMR spectra. A 7-coordination of Mo is postulated (1162).

[Mo(PTA)(CO)$_2$Cl$_2$], yellow, paramagnetic ppt., a nonelectrolyte in PhNO$_2$, with other prop. similar to those of the preceding dibromo complex, prepd. from [Mo(PTA)(CO)$_3$Br]BPh$_4$ and excess LiCl in EtOH under reflux. IR spectrum (1162).

[Mo(PTA)(CO)$_2$I$_2$], red, paramagnetic crystals, a nonelectrolyte in PhNO$_2$, prepd. from [Mo(PTA)(CO)$_3$I]I by heating under reflux in MeOH. Soly. data. IR, UV, and NMR spectra (1162).

[Mo(PTA)(CO)$_3$], colorless, paramagnetic crystals, a nonelectrolyte in PhNO$_2$, prepd. from Mo(CO)$_6$ and the triarsine in mesitylene under reflux. IR and NMR spectra. A vicinal coordination of the triarsine and a cis arrangement of the CO ligands is postulated. A rxn. with halogens (X) in C$_6$H$_6$ → [Mo(PTA)-(CO)$_3$X]X (1162).

[Mo(PTA)(CO)$_3$Br]Br, yellow, paramagnetic ppt., a 1:1 electrolyte in PhNO$_2$, sensitive to light, prepd. from [Mo(PTA)(CO)$_3$] in MePh at -70° by adding Br in C$_6$H$_6$. IR and UV spectra. A 7-coordination of Mo is suggested (1162).

[Mo(PTA)(CO)$_3$Br]BPh$_4$, yellow, diamagnetic plates, m. 148-149°, a 1:1 electrolyte in PhNO$_2$, stable in air and light, soly. data, prepd. from [Mo(PTA)(CO)$_3$-Br]Br in MeOH by adding Na[BPh$_4$]. IR and NMR spectra. A 7-coordination of Mo is suggested (1162).

[Mo(PTA)(CO)$_3$I]I, yellow, paramagnetic ppt., a 1:1 electrolyte in PhNO$_2$, prepd. from [Mo(PTA)(CO)$_3$] by treating with I in C$_6$H$_6$. IR and UV spectra. A 7-coordination of Mo is suggested (1162).

[Re(PTA)(CO)$_3$Cl], diamagnetic crystals, m. 121-122°, resolidifies at 200°, loses CO at 270°, sol. in org. solvents, prepd. from Re(CO)$_5$Cl and the triarsine in boiling xylene. IR and NMR spectra. A cis arrangement of the CO ligands is postulated (1161).

[Re(PTA)(CO)$_3$]ClO$_4$, colorless, paramagnetic crystals, prepd. from [Re(CO)$_6$]-ClO$_4$ and the triarsine in boiling cyclohexanone. Molar cond. data. IR and NMR spectra (1161).

[Re(PTA)(CO)$_3$]BPh$_4$, white plates, prepd. from the preceding perchlorate in EtOH by adding Na[BPh$_4$]. On heating with [Et$_4$N]Cl to 270°, one CO ligand is set free (1161).

[Ti(PTA)Cl$_4$], orange-red, diamagnetic ppt., a nonelectrolyte in PhNO$_2$, prepd. from TiCl$_4$ and the triarsine in CCl$_4$ or MeCN. IR, NMR, and absorption spectra. A 7-coordination of Ti is suggested (359).

[V(PTA)Cl$_3$], red-brown, paramagnetic ppt., a nonelectrolyte in PhNO$_2$, prepd. from VCl$_3$ and the triarsine in THF. IR and absorption spectra. A 6-coordination ov V is suggested (359).

[V(PTA)Cl$_4$], brown, paramagnetic ppt., a nonelectrolyte in MeNO$_2$, prepd. from VCl$_4$ and the triarsine in CCl$_4$. IR and absorption spectra. A 6-coordination of V is suggested (359).

(In the following complexes PTA denotes $MeC(CH_2AsMe_2)_3$.)

$[VO(PTA)Cl_2]$, green, paramagnetic ppt., a nonelectrolyte in $PhNO_2$, was formed from $[V(PTA)Cl_4]$ in a moist solvent. IR and absorption spectra. A 6-coordination of V is suggested (359).

$[W(PTA)(CO)_3]$, paramagnetic crystals, m. 330°, a nonelectrolyte in $PhNO_2$, prepd. from $W(CO)_6$ and the triarsine under reflux in mesitylene. IR and NMR spectra. A vicinal coordination of the triarsine and a cis arrangement of the CO ligands are postulated. Rxn. with halogens (X) in C_6H_6 at r.t. → $[W(PTA)(CO)_3X]X$ (1162).

$[W(PTA)(CO)_4]$, diamagnetic crystals, m. 93.5-94.5°, a nonelectrolyte in $PhNO_2$, prepd. as a by-prod. in the rxn. of $W(CO)_6$ with the triarsine in mesitylene under reflux. On heating to 220° under N, CO is set free and the complex is converted to the tricarbonyl deriv. IR and NMR spectra. A coordination of the triarsine as a bidentate is suggested (1161).

$[W(PTA)(CO)_2Br_2]$, yellow, paramagnetic substance, a nonelectrolyte in $PhNO_2$, prepd. from $[W(PTA)(CO)_3Br]Br$ by heating in PrOH at 100°, IR and NMR spectra (1162).

$[W(PTA)(CO)_2Cl_2]$, yellow, diamagnetic crystals, a nonelectrolyte in $PhNO_2$, prepd. from $[W(PTA)(CO)_3Br]BPh_4$ by heating in PrOH with excess LiCl at 100°. IR spectrum (1162).

$[W(PTA)(CO)_2I_2]$, brick-red, diamagnetic crystals, a nonelectrolyte in $PhNO_2$, stable in air and in light, sol. in highly polar solvents, prepd. from $[W(PTA)(CO)_3I]I$ by heating in $MeOCH_2CH_2OH$ at 200°. IR, UV, and NMR spectra. A 7-coordination of W is suggested (1162).

$[W(PTA)(CO)_3Br]Br$, yellow, paramagnetic crystals, a 1:1 electrolyte in $PhNO_2$, stable in air but slowly dec. when exposed to light, soly. data. IR, UV, and NMR spectra. A 7-coordination of W is suggested (1162).

$[W(PTA)(CO)_3Br]BPh_4$, yellow, diamagnetic platelets, m. 170-172°, a 1:1 electrolyte in $PhNO_2$, stable in air and in light, soly. data, prepd. from $[W(PTA)(CO)_3Br]Br$ in MeOH by treating with $Na[BPh_4]$. IR spectrum. A 7-coordination of W is suggested (1162).

$[W(PTA)(CO)_3Br]ClO_4$, yellow, diamagnetic crystals, a 1:1 electrolyte in MeOH, prepd. from $[W(PTA)(CO)_3Br]Br$ by treating with $Mg(ClO_4)_2$ in a Me_2CO-H_2O mixt. IR spectrum (1162).

$[W(PTA)(CO)_3I]I$, yellow, paramagnetic crystals, a 1:1 electrolyte in $PhNO_2$, soly. data, stable in air but dec. slowly on exposure to light, prepd. from $[W(PTA)(CO)_3]$ by treating with I in C_6H_6 at r.t. IR, UV, and NMR spectra. A 7-coordination of W is suggested (1162).

$[W(PTA)(CO)_3I]ClO_4$, yellow, diamagnetic crystals, a 1:1 electrolyte in $MeNO_2$, stable in air and in light, prepd. from $[W(PTA)(CO)_3I]I$ by treating with $Mg(ClO_4)_2$ in a Me_2CO-H_2O soln. IR spectrum. A 7-coordination of W is suggested (1162).

{$W(CO)_4[(Me_2AsCH_2)_2C(Me)CH_2AsMe_3]$}I, white, diamagnetic crystals, mol. cond. data, stable and sol. in H_2O, prepd. from [$W(PTA)(CO)_4$] by heating with excess MeI in MeOH. IR and NMR spectra (1161).

Tris[3-(dimethylarsino)propyl]phosphine, $As_3C_{15}H_{36}P$,
   ($Me_2AsCH_2CH_2CH_2)_3P$,   (M-107)

<u>Coordination Deriv.</u>: (In the following formulae TAP denotes $(Me_2AsCH_2CH_2CH_2)_3P$.)

[$Co(TAP)Br_2$]ClO_4, prepd. from $CoBr_2$ by a rxn. with $LiClO_4$ in EtOH, followed by the triarsinophosphine. A 6-coordinate Co complex with TAP as a tetradentate is suggested (155, see also 1051).

[$Co(TAP)Cl_2$]ClO_4, prepd. from $CoCl_2$ and $LiClO_4$ in EtOH, followed by TAP soln. A 6-coordinate Co complex is postulated (155, see also 1051).

[$Co(TAP)I_2$]ClO_4, prepd. from CoI_2 and $LiClO_4$ in EtOH, followed by TAP soln. A 6-coordinate Co complex is postulated (155, see also 1051).

[$Co(TAP)(NCS)_2$]ClO_4, diamagnetic substance, a 1:1 electrolyte in $MeNO_2$, prepd. from [$Co(TAP)(H_2O)OH$]$^{++}$ by a rxn. with NaSCN in MeOH. Electronic absorption spectrum. Formulated as a 6-coordinate Co complex. Rxn. with $AgClO_4$ in DMF $\rightarrow$ [$Co(TAP)(DMF)(NCS)$]$(ClO_4)_2$ (155, see also 1051).

[$Co(TAP)(DMF)Cl$]$(ClO_4)_2$, prepd. from [$Co(TAP)Cl_2$]ClO_4 by a rxn. with $AgClO_4$ in DMF. A 6-coordination of Co is postulated (155, see also 1051).

[$Co(TAP)(DMF)(NCS)$]$(ClO_4)_2$, prepd. from [$Co(TAP)(NCS)_2$]ClO_4 by a rxn. with $AgClO_4$ in DMF. Electronic absorption spectrum (155).

[$Ni(TAP)Br$]ClO_4 (M-108), diamagnetic crystals. Electronic absorption spectrum; a trigonal-bipyramidal structure for the 5-coordinate Ni complex is suggested (155, 156, see also 1051).

[$Ni(TAP)Cl$]ClO_4 (M-108), diamagnetic crystals. Electronic absorption spectrum; the same structure as that of the preceding bromo deriv. is postulated (155, 156, see also 1051).

[$Ni(TAP)I$]ClO_4 (M-108), diamagnetic crystals. Electronic absorption spectrum; the same structure as those of the preceding two complexes is postulated (155, 156, see also 1051).

[$Ni(TAP)CN$]ClO_4 (M-108), deep red, diamagnetic crystals, prepd. from [$Ni(TAP)$-Cl]ClO_4 by treating with NaCN. Electronic absorption spectrum (155, 156). Crystallographic data by x-ray analysis. A trigonal-bipyramidal structure is postulated in which the central Ni atom is surrounded by 3 As atoms at the equatorial sites and the P atom and CN group in the apical positions. The absolute conformation of the cation in the crystals was detd. from anomalous dispersion effects (1457, see also 155, 156, and 1051).

[$Ni(TAP)NCS$]ClO_4 (M-108), prepd. from [$Ni(TAP)Cl$]ClO_4 and NaSCN (155). The complex is resistant to oxidation by air or by chemical oxidn. agents (155, see also 1051).

(In the following formulae TAP denotes $(Me_2AsCH_2CH_2CH_2)_3P$.)

[Ni(TAP)NCSe]ClO$_4$, prepd. from [Ni(TAP)Cl]ClO$_4$ by treating with NaSeCN (155, see also 1051).

[Ni(TAP)NO$_2$]ClO$_4$ (M-108), prepd. from [Ni(TAP)Cl]ClO$_4$ by treating with NaNO$_2$ (155, see also 1051).

[Ni(TAP)SEt]ClO$_4$ (M-108), prepd. from [Ni(TAP)Cl]ClO$_4$ by treating with NaSEt (155, see also 1051).

[Ni(TAP)(H$_2$O)](ClO$_4$)$_2$ (M-108), prepd. from [Ni(TAP)Cl]ClO$_4$ by treating with H$_2$O (155, see also 1051).

Bis[o-(dimethylarsino)phenyl]methylarsine, As$_3$C$_{17}$H$_{23}$, [2-(Me$_2$As)C$_6$H$_4$]$_2$AsMe
Synth.: From 2-BrC$_6$H$_4$AsMe$_2$ by treating with BuLi in Et$_2$O, followed by MeAsCl$_2$ in C$_6$H$_6$, 61% yield (428).
From 2-LiC$_6$H$_4$AsMe$_2$ + MeAsCl$_2$ (2:1) in petrol under reflux, 45-55% yield (1659).
Prop.: Viscous oil, b$_{0.5}$ 184-190°, slowly crystallizes to a white solid, m. 64-68° (428). White needles, m. 65° (1659). IR spectrum (359, 428). NMR spectrum (359).
Coordination Deriv.: (In the following formulae TtA denotes [2-(Me$_2$As)C$_6$H$_4$]$_2$-AsMe.)

[Cr(TtA)(CO)$_3$], yellow, diamagnetic needles, a nonelectrolyte in Me$_2$CO, prepd. from Cr(CO)$_6$ and the triarsine in mesitylene at 190-200°. IR spectrum (380).

[Cr(TtA)Br$_3$], deep blue, paramagnetic crystals, a nonelectrolyte in Me$_2$CO, prepd. from [Cr(TtA)(CO)$_3$] by treating with Br in CHCl$_3$. UV and visible spectra (380).

[Cr(TtA)Cl$_3$], blue, paramagnetic crystals, a nonelectrolyte in Me$_2$CO, prepd. from [Cr(THF)$_3$Cl$_3$] by treating with the triarsine in THF soln. UV and visible spectra (359, 380).

[Cr(TtA)I$_3$], deep green, paramagnetic crystals, a nonelectrolyte in Me$_2$CO, prepd. from [Cr(TtA)(CO)$_3$] in CHCl$_3$ by treating with iodine soln. UV and visible spectra (380).

[Cr(TtA)(CO)$_3$I]I$_3$, brown, diamagnetic crystals a 1:1 electrolyte, prepd. from [Cr(TtA)(CO)$_3$] by treating with iodine in C$_6$H$_6$. IR spectrum (380).

[Co(TtA)Br$_3$], dark purple ppt., a nonelectrolyte, prepd. from CoBr$_2$·6H$_2$O in EtOH by treating with the triarsine followed by 46% HBr and from [Co(TtA)Cl$_3$] in EtOH by treating with aq. KBr soln. Soly. data. Electronic absorption and diffuse reflectance spectra (428).

[Co(TtA)Cl$_3$], red-purple, diamagnetic crystals, prepd. from CoCl$_2$·6H$_2$O in EtOH by treating with the triarsine followed by concd. HCl. Soly. data. Electronic absorption and reflectance spectra (428). Kinetics of the Cl substitution with Br and NCS, resp., in MeOH-CH$_2$Cl$_2$ and in MeOH-MeNO$_2$ soln. is influenced by the concn. of MeOH, not by the nature of the anion (1123).

[Co(TtA)I₃], black crystals, a nonelectrolyte, prepd. from Co(NO₃)₂·6H₂O and NaI in EtOH by treating with the triarsine. Soly. data. Electronic absorption and reflectance spectra (428).

[Co(TtA)(NCS)₃], red crystals, a nonelectrolyte, prepd. from [Co(TtA)Cl₃] in MeOH by treating with KCNS in EtOH. Soly. data. Electronic absorption and diffuse reflectance spectra (428).

[Co(TtA)(NO₂)₃], orange-brown crystals, a nonelectrolyte, prepd. from [Co(TtA)Cl₃] and NaNO₂ in MeOH and from Na₃[Co(NO₂)₆] by treating with the triarsine in hot MeOH. Soly. data. Electronic absorption and diffuse reflectance spectra (428).

[Co(TtA)(NO₃)₃], bright magenta-colored ppt., prepd. from [Co(TtA)Cl₃] in 6 N HNO₃ by adding aq. AgNO₃. Electronic absorption spectrum (428).

[Co(TtA)₂](CoCl₄), light green, paramagnetic ppt., prepd. from CoCl₂·6H₂O in EtOH by treating with the triarsine. Absorption and diffuse reflectance spectra (428).

[Co(TtA)₂](ClO₄)₂, yellow-green, diamagnetic needles, prepd. from Co(ClO₄)₂· 6H₂O in aq. MeOH by adding the triarsine in Me₂CO (428).

[Cu(TtA)Br], prepd. from [Cu(PPh₃)₂]₂ by treating with the triarsine and converting the resulting [Cu(TtA)₂]₂ with Br in CHCl₃. A 4-coordinate Cu complex is suggested (941).

[Cu(TtA)]I, prepd. from [(TtA)₂CuV(CO)₆] by oxidn. with iodine (833).

{[Cu(TtA)]₂Fe(CO)₄}, yellow brown, diamagnetic crystals, m. 107°, a nonelectrolyte in PhNO₂, prepd. from [Cu(TtA)Br] in THF by adding to a rxn. mixt. of Fe(CO)₅ and NaBH₄ in MeOCH₂CH₂OMe. IR spectrum (832a).

[Cu(TtA)V(CO)₆], light yellow solid, prepd. from [Cu(TtA)Br] and [Na(diglyme)₂]- [V(CO)₆] in THF. IR spectrum (833).

[Cu(TtA)Co(CO)₄], pale yellow, diamagnetic crystals, m. 171°, a nonelectrolyte in PhNO₂, prepd. from [Cu(TtA)Br] in THF by adding Na[Co(CO)₄] in the same solvent. IR spectrum (832a).

[Cu(TtA)Mn(CO)₅], light brown, diamagnetic crystals, m. 140°, a nonelectrolyte in PhNO₂, prepd. from [Cu(TtA)Br] and Na[Mn(CO)₅] in THF (832a). IR spectrum (215, 832a). Crystallographic data - a tetrahedral Cu complex coordinated with all 3 As atoms (849).

[Cu(TtA)Mo(C₅H₅)(CO)₃], yellow crystals, dec. 187°, a nonelectrolyte in PhNO₂, prepd. from [Cu(TtA)Br] and π-C₅H₅Mo(CO)₃H in THF. IR and NMR spectra (671).

[Cu(TtA)W(C₅H₅)(CO)₃], yellow crystals, dec. 185°, a nonelectrolyte in PhNO₂, prepd. from [Cu(TtA)Br] and π-C₅H₅W(CO)₃H in THF. IR and NMR spectra (671).

[Fe(TtA)₂](BPh₄)₂, pale blue, diamagnetic crystals, dec. > 270°, a 2:1 electrolyte in MeNO₂, prepd. from FeI₃ in EtOH soln. by adding the triarsine and treating the soln. with NaBPh₄. Visible spectrum (1659).

(In the following formulae TtA denotes $[2-(Me_2As)C_6H_4]_2AsMe$.)

$[Fe(TtA)_2](ClO_4)_2$, the complex occurs in two forms, viz. yellow and blue crystals. The yellow, diamagnetic crystals explode at 265°; a 2:1 electrolyte in $MeNO_2$, sol. in MeOH, EtOH, and $MeNO_2$, prepd. from the corr. dinitrate by treating with $LiClO_4$, both dissolved in Me_2CO, and from hydrated $Fe(ClO_4)_2$ by treating with the triarsine in EtOH and recrystg. from $MeNO_2$ by adding Et_2O. Visible and IR spectra and x-ray powder patterns are different from those of the blue complex. The blue, diamagnetic crystals explode at 215°; a 2:1 electrolyte in $MeNO_2$, sol. in CH_2Cl_2 and $CHCl_2$, prepd. from hydrated $Fe(ClO_4)_3$ and the triarsine in EtOH. The yellow and blue forms may have different structures (1659).

$[Fe(TtA)_2](FeCl_4)_2$, yellow, paramagnetic crystals, dec. > 270°, a 2:1 electrolyte in $MeNO_2$, prepd. from anhyd. $FeCl_3$ and the triarsine in $CHCl_3$. Visible spectrum (1659).

$[Fe(TtA)_2](NO_3)_2$, yellow, diamagnetic crystals, explodes at 265°, a 2:1 electrolyte in $MeNO_2$, prepd. from the corr. tetrachloroferrate in MeOH by adding satd. $AgNO_3$ soln. Visible spectrum (1659).

$[Fe(TtA)_2](SCN)_2$, yellow-brown, diamagnetic crystals, m. 250-253° (dec.), a 2:1 electrolyte in $MeNO_2$, prepd. from hydrated $Fe(NO_3)_3$ by treating with NaSCN in EtOH, filtering off and adding the triarsine in EtOH to the Fe salt soln. Visible spectrum (1659).

$[Fe(TtA)_2](ClO_4)_3$, dark green, paramagnetic crystals, explodes at 180°, a 3:1 electrolyte in $MeNO_2$, prepd. from $[Fe(TtA)_2]^{++}$ by oxidn. with 15 N HNO_3, followed by 72% $HClO_4$. Visible spectrum (1659).

$[Fe(TtA)Cl_3]$, reddish brown, paramagnetic crystals, dec. > 270°, a nonelectrolyte in $MeNO_2$, prepd. from anhyd. $FeCl_3$ and the triarsine in EtOH. Visible spectrum (1659).

$[Fe(TtA)(CO)_3]$, yellow, diamagnetic crystals, m. 170-174° (dec.), a nonelectrolyte in $MeNO_2$, prepd. from $Fe(CO)_5$ and the triarsine in a sealed tube at 100°. IR and NMR spectra (1659).

$[Mo(TtA)(CO)_3]$, white, diamagnetic needles, a nonelectrolyte in Me_2CO, prepd. from $Mo(CO)_6$ and the triarsine by heating in mesitylene at 190-200°. IR spectrum (380).

$[Mo(TtA)(CO)_2Br_2]$, orange, diamagnetic crystals, a nonelectrolyte in $MeNO_2$, prepd. from $[Mo(TtA)(CO)_3Br]Br$ by heating in MeOH under reflux. IR spectrum (380).

$[Mo(TtA)(CO)_2I_2]$, orange, diamagnetic crystals, a nonelectrolyte in $MeNO_2$, prepd. from $[Mo(TtA)(CO)_3I]I$ by heating in MeOH under reflux. IR spectrum (380).

$[Mo(TtA)(CO)_3Br]Br$, yellow, diamagnetic crystals, on exposure to air and light turned deep maroon, a 1:1 electrolyte, prepd. from $[Mo(TtA)(CO)_3]$ in

$CHCl_3$ by treating with bromine. IR spectrum (380).

[Mo(TtA)(CO)$_3$I]I, orange, diamagnetic crystals, a 1:1 electrolyte, prepd. from [Mo(TtA)(CO)$_3$] in $CHCl_3$ by treating with iodine. IR spectrum (380).

[Ag(TtA)Co(CO)$_4$], pale yellow, diamagnetic crystals, m. 105°, a nonelectrolyte in $PhNO_2$, prepd. from [Ag(TtA)]Br in THF by adding Na[Co(CO)$_4$]. IR spectrum (832a).

{[Ag(TtA)]$_2$Fe(CO)$_4$}, yellow, diamagnetic crystals, m. 95°, a nonelectrolyte in $PhNO_2$, prepd. by adding [Ag(TtA)]Br in THF to Na$_2$[Fe(CO)$_4$]. IR spectrum (832a).

[Ag(TtA)Mn(CO)$_5$], light brown, diamagnetic crystals, m. 84°, a nonelectrolyte in $PhNO_2$, prepd. from [Ag(TtA)]Br and Na[Mn(CO)$_5$] in THF (832a). IR spectrum (215, 832a).

[Ti(TtA)Cl$_4$], orange-red, diamagnetic ppt., a nonelectrolyte in $MeNO_2$, prepd. from TiCl$_4$ and the triarsine in CCl$_4$. IR, absorption, and NMR spectra - a 7-coordination of Ti is suggested (359).

[V(TtA)Cl$_3$], brown, paramagnetic crystals, a nonelectrolyte in $PhNO_2$, prepd. from VCl$_3$ and the triarsine in THF. IR and absorption spectra - a 6-coordination of V is suggested (359).

[V(TtA)Cl$_4$], red-brown, paramagnetic ppt., a nonelectrolyte in $MeNO_2$, prepd. from VCl$_4$ and the triarsine in CCl$_4$. IR and electronic absorption spectra - a 7-coordination of V is suggested (359).

[VO(TtA)Cl$_2$], green, paramagnetic ppt., a nonelectrolyte in $MeNO_2$, prepd. from [V(TtA)Cl$_4$] in a moist solvent and from [V(TtA)Cl$_3$] by exposure to moist air. IR and electronic absorption spectra (359).

[W(TtA)(CO)$_3$Br]Br, yellow, diamagnetic crystals, darkened when exposed to air and light, prepd. from [W(TtA)(CO)$_3$] in $CHCl_3$ by treating with bromine. IR spectrum (380).

[W(TtA)(CO)$_3$I]I, diamagnetic, shiny orange plates, a 1:1 electrolyte, prepd. from [W(TtA)(CO)$_3$] in $CHCl_3$ by treating with iodine. IR spectrum (380).

Bis[o-(dimethylarsino)phenyl]phenylarsine, As$_3$C$_{22}$H$_{25}$, [2-(Me$_2$As)C$_6$H$_4$]$_2$AsPh
Prop.: m. 124-125° (1163).

Bis[o-(diphenylarsino)phenyl]phenylarsine, As$_3$C$_{42}$H$_{33}$, [2-(Ph$_2$As)C$_6$H$_4$]$_2$AsPh,
(M-108)
Coordination Deriv.: {Cr[2-(Ph$_2$As)C$_6$H$_4$]$_2$AsPh(CO)$_3$}, diamagnetic crystals, m. 265-275°, a nonelectrolyte in $PhNO_2$, prepd. from Cr(CO)$_6$ and the triarsine in Bu$_2$O under reflux. IR spectrum (760).

{Pt[[2-(Ph$_2$As)$_2$C$_6$H$_4$]$_2$AsPh]Br}ClO$_4$, kinetics of the exchange of the Br ligand for CN, CS(NH$_2$)$_2$, SCN, I, and N$_3$ in MeOH was detd. (1202).

Tris[(diphenylarsino)ethyl]amine, $AsC_{42}H_{42}N$, $(Ph_2AsCH_2CH_2)_3N$

Synth.: From $Ph_2AsK\cdot2$ Dioxan and $(ClCH_2CH_2)_3N$ (3:1) in THF under reflux (1315).

Prop.: White crystals, m. 95-96° (1315).

Coordination Deriv.: (In the following formulae TAN denotes $(Ph_2AsCH_2CH_2)_3N$.)

[Ni(TAN)Br]BPh_4, green, diamagnetic crystals, m. 207-208°, a 1:1 electrolyte in $C_2H_4Cl_2$ and in $EtNO_2$, prepd. from anhyd. $NiBr_2$ in BuOH and the solid tri-arsinoamine by heating under reflux and pptg. with Na[BPh_4]. Electronic absorption spectrum - a trigonal bipyramidal structure is postulated (1315).

[Ni(TAN)I]BPh_4, blue, diamagnetic crystals, m. 183-188°, a 1:1 electrolyte in $C_2H_4Cl_2$ and in $EtNO_2$, prepd. by adding the triarsinoamine dissolved in Et_2O to a boiling soln. of anhyd. NiI_2 and pptg. with Na[BPh_4]. Electronic absorption spectrum - a trigonal bipyramidal structure is postulated (1315).

Tris[o-(diphenylarsino)phenyl]phosphine, $As_3C_{54}H_{42}P$, $[2-(Ph_2As)C_6H_4]_3P$,
     (M-109)

Coordination Deriv.: {Mo[$2-(Ph_2AsC_6H_4)_3P$](CO)$_3$}, yellow, diamagnetic ppt., m. 198° (dec.), prepd. by adding dropwise [Mo(CO)$_3$(MeCN)$_3$] to a suspension of the triarsinophosphine in MeCN and heating under reflux. IR spectrum (1526).

1.1.6 TETRATERTIARY ARSINES

Through 1964 two tetratertiary arsines were prepared by the conversion of monofunctionally substituted tertiary arsines to the corresponding Grignard reagents or lithium derivatives and their reaction with arsenic trichloride (M-109). Since then the arsines were also prepared from secondary arsino-sodium derivatives by a metathetic reaction with tetrafunctional haloorganic compounds.

1,2,4,5-Benzenetetrayltetrakis(dimethylarsine),
     $As_4C_{14}H_{26}$,

Synth.: From Me_2AsNa in THF by treating with $1,2,4,5-C_6H_2Cl_4$ at r.t. (1415).

Prop.: Colorless cryst. solid, m. 148-150° (from EtOH), sol. in THF, Et_2O, $CHCl_3$, and CH_2Cl_2 at r.t. and in MeOH and EtOH under reflux. UV spectrum (1415).

Tris[3-(dimethylarsino)propyl]arsine, $As_4C_{15}H_{36}$, $(Me_2AsCH_2CH_2CH_2)_3As$,
     (M-109)

Prop.: A greasy solid, m. 23° (156).

Rxn. with: S → the tetrasulfide (156).

Coordination Deriv.:(In the following formulae TtA denotes $(Me_2AsCH_2CH_2CH_2)_3As$.)

[Co(TtA)X_2]ClO_4 (where X = a halogen) (M-109 - M-110), diamagnetic substances, 1:1 electrolytes in $MeNO_2$, prepd. from CoX_2 and $LiClO_4$ in EtOH by adding the tetrasine soln. A 6-coordination of Co in the complexes is postulated (155).

[Co(TtA)(DMF)Cl](ClO$_4$)$_2$, prepd. from the preceding complexes by a rxn. with AgClO$_4$ in DMF (155).

[Ni(TtA)Br]ClO$_4$, blue crystals, prepd. from Ni(ClO$_4$)$_2$·6H$_2$O and NiBr$_2$·6H$_2$O in EtOH by treating with the tetraarsine. Electronic absorption spectrum - a trigonal-bipyramidal structure is postulated (156).

[Ni(TtA)Cl]ClO$_4$, blue crystals, prepd. from Ni(ClO$_4$)$_2$·6H$_2$O and NiCl$_2$·6H$_2$O in EtOH by treating with the tetraarsine. Electronic absorption spectrum - a trigonal-bipyramidal structure is postulated (156).

[Ni(TtA)I]ClO$_4$, blue-green crystals, prepd. from Ni(ClO$_4$)$_2$·6H$_2$O and NiI$_2$·6H$_2$O mixt. in EtOH by treating with the tetraarsine. Electronic absorption spectrum - the same structure as that of the bromo and chloro analogs is postulated (156).

[Ni(TtA)CN]ClO$_4$, red crystals, prepd. from [Ni(TtA)Cl]ClO$_4$ by treating with NaCN in MeOH. Electronic absorption spectrum - a structure analogous to those of the preceding halo complexes is postulated (156).

[Ni(TtA)H$_2$O](ClO$_4$)$_2$, purple powder, prepd. from Ni(ClO$_4$)$_2$·6H$_2$O in EtOH-Me$_2$C-(OMe)$_2$ by treating with Ph$_3$P and the tetraarsine in EtOH (156).

o-Phenylenebis{[3-(dimethylarsino)propyl]methylarsine},
   As$_4$C$_{18}$H$_{34}$,
Prop.: b$_{0.01}$ 190° (1163).
Coordination Deriv.: The tetraarsine forms crystalline complexes with Ni(II), Pd(II), and Pt(II), in which the central metal atoms are either 4-coordinate [ML](ClO$_4$)$_2$ or 5-coordinate [MLX]Y species, with X = Cl, Br, or I and Y = Cl, Br, I, or ClO$_4$ groups. Electronic spectra of the [MLX]Y complexes indicate a 5-coordinate square-pyramidal structure. The tetraarsine complexes with Fe(III) and Co(II) of the [MLX$_2$]Y type, where X = Cl and Y = Cl or ClO$_4$, were prepd. and were assigned an octahedral trans configuration. Co(III) formed a cis complex, too. Oxidn. of the Fe(III) complex gave a Fe(IV) complex of the [FeLCl$_2$](ClO$_4$)$_2$ formula. The stability of the Fe(IV) complex is similar to that of {Fe[o-(Me$_2$As$_2$)C$_6$H$_4$]$_2$-Cl$_2$}Y$_2$, with Y = ClO$_4$, ReO$_4$, or BF$_4$. A rxn. of the tetraarsine with Ni(CO)$_4$ gave a binuclear complex [Ni$_2$L(CO)$_4$] (1163).

Tris[o-(dimethylarsino)phenyl]arsine,   As$_4$C$_{24}$H$_{30}$,   [2-(Me$_2$As)C$_6$H$_4$]$_3$As
Synth.: From AsCl$_3$ in dry petrol by adding slowly to a suspension of 2-LiC$_6$H$_4$-AsMe$_2$ in the same solvent, 42% yield (1659).
Prop.: Crystals, m. 115° (1163, 1659).
Coordination Deriv.: (In the following complexes Qas denotes [2-(Me$_2$As)C$_6$H$_4$]$_3$As.)

[Fe(Qas)Cl$_2$]FeCl$_4$, red, paramagnetic crystals, dec. 180-185°, a 1:1 electrolyte in MeNO$_2$, prepd. from FeCl$_3$ and the tetraarsine in EtOH. Visible spectrum; an octahedral structure is postulated (1659, see also 1163).

(In the following complexes Qas denotes $[2-(Me_2As)C_6H_4]_3As$.)

$[Fe(Qas)Cl_2]ClO_4$, red, paramagnetic crystals, explodes at 230°, a 1:1 electrolyte in $MeNO_2$, prepd. from $FeCl_2$ in EtOH by treating with the tetraarsine dissolved in a CH_2Cl_2-EtOH mixt. and adding $NaClO_4$ in EtOH. Visible spectrum (1659, see also 1163).

$[Fe(Qas)Br_2]FeBr_4$, reddish brown, paramagnetic crystals, dec. 200-204°, a 1:1 electrolyte in $MeNO_2$, prepd. from $FeBr_3$ and the tetraarsine in EtOH. Visible spectrum (1659, see also 1163).

$[Fe(Qas)Br_2]ClO_4$, dark reddish brown, paramagnetic crystals, explodes at 220°, a 1:1 electrolyte in $MeNO_2$, prepd. from hydrated $Fe(ClO_4)_3$ by grinding with NaBr (1:3), extracting it with EtOH and mixing the soln. with the tetraarsine dissolved in CH_2Cl_2-EtOH mixt. Visible spectrum. An octahedral structure of the complex cation is postulated (1659, see also 1163).

$[Fe(Qas)(SCN)_2]$, reddish-purple, diamagnetic crystals, m. 245-249°, a non-electrolyte in $MeNO_2$, prepd. from $Fe(SCN)_3$ and the tetraarsine in $CHCl_3$-EtOH mixt. Visible and IR spectra. An octahedral structure with the thiocyanate ion bonded to Fe through the N atom (1659, see also 1163).

$[Fe(Qas)I]I$, blue, diamagnetic crystals, dec. 230-235°, a 1:1 electrolyte in $MeNO_2$, prepd. from FeI_3 and the tetraarsine in EtOH. Visible spectrum. A 5-coordinate Fe complex with a trigonal bipyramidal structure is postulated (1659, see also 1163).

$[Fe(Qas)]Cl_2$, purple crystals, prepd. from anhyd. $FeCl_2$ in EtOH soln. by treating with the tetraarsine dissolved in a CH_2Cl_2-EtOH mixt. Visible spectrum (1659, see also 1163).

$[Ni(Qas)Cl]BF_4$, dark blue, diamagnetic crystals, dec. > 360°, a 1:1 electrolyte in $MeNO_2$, prepd. from hydrated $NiCl_2$ by mixing with the tetraarsine in EtOH and adding HBF_4. Visible spectrum. A trigonal bipyramidal structure of the complex cation is postulated (1659, see also 1163).

$[Ni(Qas)Br]Br$, dark blue, diamagnetic crystals, dec. 310-314°, a 1:1 electrolyte in $MeNO_2$, prepd. from hydrated $Ni(NO_3)_2$ by metathesis with NaBr in EtOH and treating the soln. with the tetraarsine dissolved in CH_2Cl_2-EtOH soln. Visible spectrum. A trigonal bipyramidal structure is postulated (1659, see also 1163).

$[Ni(Qas)Br]BF_4$, dark blue, diamagnetic crystals, dec. > 360°, a 1:1 electrolyte in $MeNO_2$, prepd. from hydrated $NiBr_2$ by mixing with the tetraarsine in EtOH and treating the soln. with HBF_4. Visible spectrum. A trigonal bipyramidal structure of the complex cation is postulated (1659, see also 1163).

$[Ni(Qas)Br]BPh_4$, dark blue, diamagnetic crystals, dec. 245-250°, a 1:1 electrolyte in $MeNO_2$, prepd. from the corr. bromo bromide soln. in EtOH contg. CH_2Cl_2 by treating with $NaBPh_4$ in EtOH. Visible spectrum. A trigonal bipyramidal structure of the complex cation is postulated (1659, see also 1163).

[Ni(Qas)I]I, green, diamagnetic crystals, dec. slowly > 330°, a 1:1 electrolyte in MeNO$_2$, prepd. from hydrated Ni(NO$_3$)$_2$ by metathesis with NaI in EtOH and treating the soln. with the tetraarsine dissolved in CH$_2$Cl$_2$-EtOH mixt. Visible spectrum. A trigonal bipyramidal structure of the complex cation in soln. and a distorted octahedral structure in the solid state are considered (1659, see also 1163).

[Ni(Qas)I]BF$_4$, dark blue, diamagnetic crystals, dec. > 360°, a 1:1 electrolyte in MeNO$_2$, prepd. from the corr. iodo iodide complex by treating with 42% HBF$_4$. Visible spectrum. A trigonal bipyramidal structure of the complex cation is postulated (1659, see also 1163).

[Ni(Qas)NCS]SCN, purple-blue, diamagnetic crystals, dec. 310-315°, a 1:1 electrolyte in MeNO$_2$, prepd. from hydrated Ni(NO$_3$)$_2$ by metathesis with NaSCN in EtOH and treating the soln. with the tetraarsine dissolved in CH$_2$Cl$_2$-EtOH mixt. Visible and IR spectra. A trigonal bipyramidal structure with one thiocyanate group N-bonded to the Ni atom is postulated (1659, see also 1163).

[Ni(Qas)NCS]BPh$_4$, purple-blue, diamagnetic crystals, dec. 258-262°, a 1:1 electrolyte in MeNO$_2$, prepd. from Ni(NO$_3$)$_2$ by metathesis with NaSCN in EtOH, followed by a treatment with the tetraarsine dissolved in CH$_2$Cl$_2$-EtOH mixt. followed by NaBPh$_4$ in EtOH. Visible spectrum. A trigonal bipyramidal structure is postulated (1659, see also 1163).

[Ni(Qas)N$_3$]BF$_4$, blue-black, diamagnetic crystals, dec. 265-270°, a 1:1 electrolyte in MeNO$_2$, prepd. from Ni(NO$_3$)$_2$·6H$_2$O in EtOH by metathesis with NaN$_3$ dissolved in H$_2$O, followed by a treatment with the tetraarsine dissolved in CH$_2$Cl$_2$-EtOH mixt., heating under reflux, and adding NaBF$_4$ dissolved in H$_2$O. Visible and IR spectra. A trigonal bipyramidal structure is postulated (1659, see also 1163).

[Ni(Qas)NO$_2$]NO$_2$, reddish-purple, diamagnetic crystals. dec. slowly > 320°, a 1:1 electrolyte in MeNO$_2$, prepd. from Ni(NO$_3$)$_2$·6H$_2$O in MeOH by treating with NaNO$_2$ in aq. MeOH, followed by the tetraarsine dissolved in boiling MeOH and pptg. the prod. with Et$_2$O. Visible and IR spectra. A trigonal bipyramidal structure is postulated (1659, see also 1163).

[Ni(Qas)NO$_2$]BPh$_4$, reddish-purple, diamagnetic crystals, dec. slowly > 250°, a 1:1 electrolyte in MeNO$_2$, prepd. from Ni(NO$_3$)$_2$·6H$_2$O in MeOH by treating with NaNO$_2$ in aq. MeOH, followed by the tetraarsine dissolved in boiling MeOH and then by NaBPh$_4$ in EtOH. Visible and IR spectra. A trigonal bipyramidal structure is postulated. The rxn. of the complex with various tetraethylammonium halides was carried out in Me$_2$CO at 25° (1659).

[Ni(Qas)NO$_3$]BPh$_4$, dark blue, diamagnetic crystals. dec. 260-264°, a 1:1 electrolyte in MeNO$_2$, prepd. from hydrated Ni(NO$_3$)$_2$ in EtOH by treating with the tetraarsine dissolved in CH$_2$Cl$_2$-EtOH mixt., followed by NaBPh$_4$. Visible and IR spectra. A trigonal bipyramidal structure is postulated (1659, see also 1163).

(In the following complexes Qas denotes $[2-(Me_2As)C_6H_4]_3As.$)

[Ni(Qas)OAc]BPh$_4$, purple-blue, diamagnetic crystals, dec. 238-243°, a 1:1 electrolyte in MeNO$_2$, prepd. from hydrated Ni(NO$_3$)$_2$ by metathesis with NaOAc in EtOH, followed by a treatment with the tetraarsine and then with NaBPh$_4$ soln. Visible spectrum. A trigonal bipyramidal structure is postulated (1659, see also 1163).

[Ni(Qas)CN]CN, attempted prepn. (1659).

[Pd(Qas)Cl]BF$_4$, orange-red, diamagnetic crystals. dec. > 330°, a 1:1 electrolyte in MeNO$_2$, prepd. from anhyd. PdCl$_2$ suspended in EtOH by treating with the tetraarsine dissolved in CH$_2$Cl$_2$-EtOH mixt. and heating under reflux, filtering and adding 42% HBF$_4$ to the filtrate. Visible and IR spectra. A trigonal bipyramidal structure is postulated (1659, see also 1163).

[Pd(Qas)Br]Br, red, diamagnetic crystals, dec. 290-294°, a 1:1 electrolyte in MeNO$_2$, prepd. from anhyd. PdBr$_2$ and the tetraarsine in EtOH. Visible and IR spectra. A trigonal bipyramidal structure is postulated (1659, see also 1163).

[Pd(Qas)Br]BF$_4$, dark red, diamagnetic crystals dec. > 330°, a 1:1 electrolyte in MeNO$_2$, prepd. from anhyd. PdBr$_2$ dissolved in DMF by treating with the tetraarsine dissolved in EtOH, heating under reflux, and adding 42% HBF$_4$. Visible and IR spectra. A trigonal bipyramidal structure is postulated (1659, see also 1163).

[Pd(Qas)I]I, dark red, diamagnetic crystals, dec. 290-295°, a 1:1 electrolyte in MeNO$_2$, prepd. from anhyd. PdBr$_2$ by grinding with NaI, extracting the powder mixt. with hot EtOH and adding the tetraarsine dissolved in CH$_2$Cl$_2$-EtOH mixt. to the Pd salt soln. Visible and IR spectra. A trigonal bipyramidal structure is postulated (1659, see also 1163).

[Pd(Qas)(NCS)]SCN, dark red, diamagnetic crystals, dec. 240-244°, a 1:1 electrolyte in MeNO$_2$, prepd. from anhyd. PdBr$_2$ and NaSCN and the tetraarsine in the same manner as the iodo iodide. Visible and IR spectra. A trigonal bipyramidal structure with one thiocyanate N-bonded to the Pd atom is postulated (1659, see also 1163).

[Pt(Qas)Cl]BF$_4$, yellow, diamagnetic crystals, dec. > 330°, a 1:1 electrolyte in MeNO$_2$, prepd. from PtCl$_4$ in EtOH by treating with the tetraarsine dissolved in CH$_2$Cl$_2$-EtOH mixt., isolating the ppt., dissolving it in DMF and treating with 42% HBF$_4$. Visible and IR spectra. A trigonal bipyramidal structure is postulated (1659, see also 1163).

[Pt(Qas)Br]Br, crude prod., yellow, diamagnetic substance, dec. 312-315°, a 1:1 electrolyte, prepd. from a PtBr$_2$ mixt. with NaBr by a rxn. with the tetraarsine dissolved in a CH$_2$Cl$_2$-EtOH mixt. Visible spectrum (1659, see also 1163).

[Pt(Qas)Br]BF$_4$, a crude prod., yellow, diamagnetic crystals, dec. 312-315°, a 1:1 electrolyte in MeNO$_2$, prepd. from anhyd. PtBr$_2$ mixed with NaBr by suspending in EtOH, adding the tetraarsine dissolved in EtOH, heating, filtering, and treating the filtrate with 42% HBF$_4$. Visible and IR spectra. A trigonal bipyramidal structure is postulated (1659, see also 1163).

[Pt(Qas)I]I, yellow-brown, diamagnetic crystals, dec. 320-323°, a 1:1 electrolyte in MeNO$_2$, prepd. from PtBr$_2$ mixed with NaI by boiling in EtOH and treating the soln. with the tetraarsine dissolved in a CH$_2$Cl$_2$-EtOH mixt. Visible and IR spectra. A trigonal bipyramidal structure is postulated (1659, see also 1163).

[Pt(Qas)](SCN)$_2$, yellow, diamagnetic crystals, dec. 221-223°, a 2:1 electrolyte in MeNO$_2$, prepd. from PtBr$_2$ and NaSCN in the same manner as the iodo iodide complex. Visible and IR spectra (1659, see also 1163).

o-Phenylenebis{[(dimethylarsino)phenyl]methylarsine},
    As$_4$C$_{24}$H$_{30}$,
<u>Prop.</u>: m. ~ 25° (1163).
<u>Coordination Deriv.</u>: The tetraarsine forms crystalline complexes with Ni(II), Pd(II), and Pt(II) in which the central metal atoms are either 4-coordinate [ML](ClO$_4$)$_2$ or 5-coordinate [MLX]Y species, with X = Cl, Br, or I and Y = Cl, Br, I, or ClO$_4$ (1163). The crystal and mol. structures of [PdLCl]ClO$_4$ were detd. and a square-pyramidal structure of the 5-coordinate Pd complex was postulated (196). An x-ray diffraction analysis of [PdLCl]ClO$_4$ indicated that one As atom occupies the apical position and Cl a basal position with 3 other As atoms in the square pyramide (1163). Electronic absorption spectra of the [MLX]Y complexes indicated a square-pyramidal structure of the cations (1163). The Fe(III) and Co(III) complexes with the tetraarsine, of the general composition [MLCl$_2$]Y, where Y = Cl or ClO$_4$, were isolated and were assigned an octahedral structure with Cl ligands in trans positions. Oxidn. of the Fe(III) complex gave a Fe(IV) derivative, having a formula [FeLCl$_2$](ClO$_4$)$_2$. Its stability is similar to that of {Fe[o-(Me$_2$As)C$_6$H$_4$]$_2$-Cl$_2$}Y$_2$, where Y = ClO$_4$, ReO$_4$, or BF$_4$ (1163).

Ethylenebis{[o-(dimethylarsino)phenyl]phenylarsine},
    As$_4$C$_{30}$H$_{34}$
<u>Prop.</u>: m. 117-120° (1163).

Neopentanetetrayltetrakis(diphenylarsine),   As$_4$C$_{53}$H$_{48}$,   (Ph$_2$AsCH$_2$)$_4$C
<u>Synth.</u>: From Ph$_2$AsNa and (BrCH$_2$)$_4$C (4:1) in THF, 40% yield (525).
<u>Prop.</u>: Colorless needles, m. 173°, soly. data. IR and NMR spectra. The tetraarsine functions as a bivalent bidentate in complexes with transition metals (525).
<u>Rxn. with</u>: H$_2$O$_2$ → C[CH$_2$As(O)Ph$_2$]$_4$·2H$_2$O$_2$ (527).
<u>Coordination Deriv.</u>: (See next page.)

$$(OC)_4Cr \left\langle \begin{matrix} Ph_2 & & Ph_2 \\ AsCH_2 & CH_2As \\ & C & \\ AsCH_2 & CH_2As \\ Ph_2 & & Ph_2 \end{matrix} \right\rangle Cr(CO)_4,$$

colorless crystals, dec. > 230°, prepd. from $Cr(CO)_6$ and the tetraarsine in xylene in a sealed tube at 150-160°. Soly. data. IR spectrum. Mol. structure and bonding are discussed (526).

$$\begin{matrix} OC \\ ON \end{matrix} Co \left\langle \begin{matrix} Ph_2 & & Ph_2 \\ AsCH_2 & CH_2As \\ & C & \\ AsCH_2 & CH_2As \\ Ph_2 & & Ph_2 \end{matrix} \right\rangle Co \begin{matrix} CO \\ NO \end{matrix},$$

orange-red crystals, dec. > 154°, prepd. from $Co(NO)(CO)_3$ and the tetraarsine in C_6H_6 in a sealed tube at 47°, 90% yield. Soly. data. IR spectrum. Mol. structure and bonding are discussed (526).

$[(ON)_2Fe(Ph_2AsCH_2)_2]_2C$, red, diamagnetic needles, dec. 240°, soly. data; prepd. from $[Fe(NO)_2(CO)_2]$ and the tetraarsine in C_6H_6 (90% yield). IR spectrum. Mol. structure and bonding are discussed (526).

$[(OC)_4Mo(Ph_2AsCH_2)_2]_2C$, colorless crystals, dec. 250°, soly. data; prepd. from $Mo(CO)_6$ and the tetraarsine in xylene at 150-160° in a sealed tube. IR spectrum. Mol. structure and bonding are discussed (526).

$[(OC)_2Ni(Ph_2AsCH_2)_2]_2C$, colorless, diamagnetic crystals, dec. 150°, soly. data; prepd. from $Ni(CO)_4$ and the tetraarsine in C_6H_6 in a Schlenk tube at r.t. (80% yield). IR spectrum. Mol. structure and bonding are discussed (526).

$[(Ph_2AsCH_2)_2C(CH_2AsPh_2)_2W(CO)_4]$, colorless, diamagnetic crystals, m. 220° (dec.), soly. data; prepd. from $W(CO)_6$ and the tetraarsine in xylene or benzene at 150-160° in a sealed tube. IR spectrum. Mol. structure and bonding are discussed (526).

Tris[o-(diphenylarsino)phenyl]arsine, $As_4C_{54}H_{42}$, $[2-(Ph_2As)C_6H_4]_3As$, (M-110)

Coordination Deriv.: (In the following formulae TTA denotes $[2-(Ph_2As)C_6H_4]_3As$.)

$[Co(TTA)(CO)]BPh_4$, red-orange, diamagnetic ppt., dec. 232-234°, a 1:1 electrolyte in $PhNO_2$. UV, visible, and IR spectra. A trigonal-bipyramidal structure of the cation is postulated (691).

$[Ir(TTA)SnCl_3]$, orange, diamagnetic crystals, m. > 330°, a nonelectrolyte in $PhNO_2$, prepd. from $[Ir(cyclooctadiene)_2SnCl_3]$ and the tetraarsine in PhCl under reflux. A 5-coordinate Ir complex is postulated (1245).

$[Ir(TTA)(SnCl_3)Cl]Cl$, pale yellow, diamagnetic solid, m. > 330°, a 1:1 electrolyte in $PhNO_2$, prepd. from $[Ir(TTA)SnCl_3]$ suspension in CH_2Cl_2 by treating with Cl gas. An octahedral structure of the cation is postulated (1245).

$[Hg(TTA)Br_2]$, colorless crystals, dec. 255-258°, a nonelectrolyte in $PhNO_2$, prepd. from $HgBr_2$ and the tetraarsine in EtOH under reflux, crystd. from CH_2Cl_2 with 1 mol. of the solvent held as a clathrate. Crystallographic data by x-ray analysis; a tetrahedral coordination of the Hg atom with 2 terminal As and both Br atoms is postulated (509).

[Hg(TTA)Cl$_2$], colorless crystals, dec. 273-274°, a nonelectrolyte in PhNO$_2$, prepd. from HgCl$_2$ and the tetraarsine in EtOH under reflux, crystd. from CH$_2$Cl$_2$ with 1 mol. of the solvent held as a clathrate (509).

[Hg(TTA)I$_2$], very pale yellow crystals, dec. 224-225°, a nonelectrolyte in PhNO$_2$, prepd. from HgI$_2$ and the tetraarsine in EtOH under reflux, crystd. from CH$_2$Cl$_2$ with 1 mol. of the solvent (509).

[Hg(TTA)(SCN)$_2$], colorless crystals, dec. 238-240°, a nonelectrolyte in PhNO$_2$, prepd. from Hg(SCN)$_2$ and the tetraarsine in EtOH under reflux, crystd. from CH$_2$Cl$_2$ with 1 mol. of the solvent (509).

[Hg(TTA)](ClO$_4$)$_2$, yellow crystals, dec. 198-201°, a 1:2 electrolyte in PhNO$_2$, prepd. from Hg(ClO$_4$)$_2\cdot$6H$_2$O and the tetraarsine in THF by evapn., extraction of the residue with CH$_2$Cl$_2$ and pptn. with CCl$_4$. A trigonal pyramidal structure of the 4-coordinated Hg is postulated (509).

[Ni(TTA)Br]Br, blue-black crystals, dec. 321-333°, a 1:1 electrolyte in PhNO$_2$, prepd. from NiBr$_2\cdot$3H$_2$O and the tetraarsine in EtOH under reflux (511). Electronic spectrum (511, 513). A trigonal bipyramidal structure of the 5-coordinated Ni is postulated (511).

[Ni(TTA)Br]BPh$_4$, dark blue crystals, dec. 225-228°, a 1:1 electrolyte, prepd. from [Ni(TTA)Br]Br by metathesis with NaBPh$_4$ in EtOH. Electronic spectrum - a trigonal bipyramidal structure of the 5-coordinate Ni complex is postulated (511).

[Ni(TTA)Br]ClO$_4$, dark blue crystals, dec. 321-322°, a 1:1 electrolyte, prepd. from [Ni(TTA)Br]Br by metathesis with NaClO$_4$ in EtOH. Electronic spectrum - a trigonal bipyramidal structure of the 5-coordinate Ni complex is postulated (511).

[Ni(TTA)Cl]ClO$_4$, dark blue crystals, dec. 321-322°, a 1:1 electrolyte in PhNO$_2$, prepd. from a mixt. of NiCl$_2\cdot$6H$_2$O with Ni(ClO$_4$)$_2\cdot$6H$_2$O in EtOH by a rxn. with the tetraarsine in CH$_2$Cl$_2$ under reflux. Electronic spectrum. A trigonal bipyramidal structure of the 5-coordinate Ni is postulated (511).

[Ni(TTA)I]I, blue-black crystals, dec. 330-331°, a 1:1 electrolyte in PhNO$_2$, prepd. from Ni(NO$_3$)$_2\cdot$6H$_2$O by metathesis with NaI in EtOH, followed by a rxn. with the tetraarsine under reflux. Electronic spectrum. A trigonal bipyramidal structure of the 5-coordinate Ni complex is postulated (511).

[Ni(TTA)I]ClO$_4$, blue-black crystals, dec. 338-339°, a 1:1 electrolyte in PhNO$_2$, prepd. from [Ni(TTA)I]I by metathesis with NaClO$_4$ in boiling EtOH. Electronic spectrum. A trigonal bipyramidal structure of the 5-coordinate Ni complex is postulated (511).

[Ni(TTA)(CN)]ClO$_4$, brown crystals, dec. 338-339°, a 1:1 electrolyte in PhNO$_2$, prepd. from [Ni(TTA)I]ClO$_4$ in CH$_2$Cl$_2$ by metathesis with NaCN. Electronic and IR spectra. A trigonal bipyramidal structure of the 5-coordinate Ni complex is postulated (511).

(In the following formulae TTA denotes [2-(Ph$_2$As)C$_6$H$_4$]$_3$As.)

[Ni(TTA)(NCS)]ClO$_4$, black crystals, dec. 316-318°, a 1:1 electrolyte in PhNO$_2$, prepd. from Ni(NO$_3$)$_2$·6H$_2$O by a rxn. with NaSCN in EtOH, filtering the soln. into the tetraarsine dissolved in CH$_2$Cl$_2$ and, in turn, filtering the latter soln. into NaClO$_4$ dissolved in EtOH. Electronic and IR spectra. A trigonal-bipyramidal structure of the 5-coordinate Ni complex is postulated (511).

[Ni(TTA)(ClO$_4$)]ClO$_4$, dark blue crystals, dec. > 250°, a 1:1 electrolyte in PhNO$_2$, prepd. from Ni(ClO$_4$)$_2$·6H$_2$O in EtOH by adding to the tetraarsine in boiling C$_6$H$_6$. Electronic spectrum. A trigonal-bipyramidal structure of the 5-coordinate Ni complex is postulated (511).

[Ni(TTA)(NO$_3$)]ClO$_4$, dark blue crystals, dec. 293-296°, a 1:1 electrolyte in PhNO$_2$, prepd. from a mixt. of Ni(NO$_3$)$_2$·6H$_2$O and Ni(ClO$_4$)$_2$·6H$_2$O in EtOH by adding to the tetraarsine in boiling PhCl. Electronic spectrum. A trigonal-bipyramidal structure of the 5-coordinate Ni complex is postulated (511).

[Pd(TTA)Br]Br (M-110), kinetics of the exchange of the Br ligand for CN, I, SCN, N$_3$, Ph$_3$P, and CS(NH$_2$), resp., was detd. (1202).

[Pd(TTA)Br]ClO$_4$, dark red-purple crystals, dec. 318-320°, a 1:1 electrolyte in PhNO$_2$, prepd. from Na$_2$[PdCl$_4$] by metathesis with NaBr in EtOH, followed by refluxing with the tetraarsine and treating the prod. with NaClO$_4$ in EtOH. UV and visible spectra. A trigonal-bipyramidal structure of the 5-coordinate Pd complex is postulated (514).

[Pd(TTA)I]ClO$_4$, dark purple crystals, dec. 325-328°, a 1:1 electrolyte in PhNO$_2$, prepd. from Na$_2$[PdCl$_4$] by metathesis with NaI in EtOH, followed by re-fluxing with the tetraarsine and pptn. with NaClO$_4$ in EtOH. UV and visible spectra. A trigonal-bipyramidal structure of the 5-coordinate Pd complex is postulated (514).

[Pd(TTA)(CN)]ClO$_4$, bright yellow crystals, dec. 300-303°, a 1:1 electrolyte in PhNO$_2$, prepd. from the corr. chloro complex (M-111) by a rxn. with NaCN in MeOH. Electronic spectrum. A trigonal-bipyramidal structure is postulated (514).

[Pd(TTA)(NO$_2$)]ClO$_4$, orange crystals, dec. 299-300°, a 1:1 electrolyte in PhNO$_2$, prepd. from the corr. chloro complex (M-111) by a rxn. of NaNO$_2$ in MeOH under reflux. Electronic spectrum (514).

[Pd(TTA)(NO$_3$)]BPh$_4$, deep red crystals, dec. 243-246°, elec. cond. in PhNO$_2$ showed a partial dissocn. The complex was prepd. from [Pd(SMe$_2$)$_2$Cl$_2$] by a rxn. with Me$_2$S in H$_2$O, followed by AgNO$_3$ in H$_2$O and the tetraarsine dissolved in EtOH by heating under reflux and pptn. with NaBPh$_4$ dissolved in EtOH. Electronic spectrum (514).

[Pd(TTA)(NO$_3$)]ClO$_4$, deep red crystals, dec. 278-282°, a 1:1 electrolyte in PhNO$_2$, prepd. from Pd(NO$_3$)$_2$·2H$_2$O and the tetraarsine in THF under reflux, followed by pptn. with NaClO$_4$. Electronic spectrum (514).

{Pd(TTA)[CS(NH$_2$)$_2$]}(ClO$_4$)$_2$, deep red crystals, dec. 310-311°, a 1:2 electrolyte in PhNO$_2$, prepd. from [Pd(TTA)Cl]ClO$_4$ by treating with urea followed by NaClO$_4$. Electronic spectrum (514).

[Pd(TTA)(SMe$_2$)](ClO$_4$)$_2$, red crystals, dec. 300-302°, a 1:2 electrolyte in PhNO$_2$, prepd. from [Pd(SMe$_2$)$_2$Cl$_2$] and Me$_2$S in H$_2$O by treating the resulting soln. with the tetraarsine in EtOH under reflux, filtering the rxn. prod., and treating the filtrate with NaClO$_4$. Electronic spectrum (514).

[O$_2$NCH$_2$Pd(TTA)]ClO$_4$, scarlet crystals, dec. 267-270°, a 1:1 electrolyte in PhNO$_2$, prepd. from [Pd(TTA)Cl]ClO$_4$ and MeNO$_2$ in EtOH by shaking the mixt. with moist Ag$_2$O. Electronic spectrum (514).

[Pt(TTA)Br]Br (M-111), kinetics of the Br ligand exchange in MeOH for SCN, N$_3$, CN, I, Ph$_3$P, NO$_2$, and CS(NH$_2$)$_2$, resp. was detd. spectrophotometrically and the rxn. mechanism is proposed. The activation energy and entropy of the exchange with I were detd. (1202).

[Pt(TTA)Br]ClO$_4$, orange crystals, dec. 359-360°, a 1:1 electrolyte in PhNO$_2$, prepd. from Na$_2$[PtCl$_4$] by metathesis with NaBr in EtOH, followed by a rxn. with the tetraarsine under reflux and pptn. with NaClO$_4$ soln. in EtOH. UV and visible spectra. A trigonal-bipyramidal structure of the 5-coordinate Pt complex is postulated (514). Kinetics of the Br exchange in MeOH for CN, SCN, N$_3$, CS(NH$_2$)$_2$, I, Ph$_3$P, and NO$_2$, resp., was detd. spectrophotometrically. The mechanism of the exchange is proposed. The activation energy and entropy of the exchange for I were detd. (1202).

[Pt(TTA)Cl]ClO$_4$ (M-111), kinetics of the Cl exchange for I, N$_3$, and NO$_2$, resp., in MeOH was detd. spectrophotometrically. A rxn. mechanism is proposed. The activation energy and entropy of the exchange for I were detd. (1202).

[Pt(TTA)I]ClO$_4$ (M-111), elec. cond. of the complex in MeOH by itself and in the presence of KI was detd. The ion pairing was investigated (1202).

[Pt(TTA)(CN)]ClO$_4$, very pale yellow crystals, dec. 368-369°, a 1:1 electrolyte in PhNO$_2$, prepd. from [Pt(TTA)Br]ClO$_4$ by metathesis with NaCN in MeOH. UV and visible spectra. A trigonal-bipyramidal structure of the 5-coordinate Pt complex is postulated (514).

[Pt(TTA)(NCS)]ClO$_4$, orange crystals, dec. 356-358°, a 1:1 electrolyte in PhNO$_2$, prepd. from Na$_2$[PtCl$_4$] by a rxn. with NaCNS in EtOH, followed by the tetraarsine under reflux and pptn. with NaClO$_4$ soln. in EtOH. UV and visible spectra. A trigonal-bipyramidal structure of the 5-coordinate Pt complex postulated (514).

[Pt(TTA)(N$_3$)]BPh$_4$, yellow crystals, prepd. from [Pt(TTA)Cl]Cl (M-110) by treating with NaN$_3$ in MeOH, followed by NaBPh$_4$ in the same solvent (1202).

[Pt(TTA)(NO$_2$)]ClO$_4$, yellow crystals, dec. 372-373°, a 1:1 electrolyte in PhNO$_2$, prepd. from [Pt(TTA)Br]ClO$_4$ by metathesis with NaNO$_2$ in MeOH under reflux. UV and visible spectra. A trigonal-bipyramidal structure of the 5-coordinate Pt complex is postulated (514).

(In the following formulae TTA denotes [2-(Ph$_2$As)C$_6$H$_4$]$_3$As.)

[Pt(TTA)(SMe$_2$)](ClO$_4$)$_2$, yellow-orange crystals, dec. 349-351°, a 1:2 electrolyte in PhNO$_2$, prepd. from [Pt(SMe$_2$)$_2$Cl$_2$] and Me$_2$S in H$_2$O by shaking with AgNO$_3$ in aq. soln., filtering the mixt., refluxing the filtrate with the tetraarsine in EtOH, and pptg. the prod. with NaClO$_4$. UV and visible spectra. A 5-coordination of Pt in the complex is suggested (514).

{Pt(TTA)[CS(NH$_2$)$_2$]}(ClO$_4$)$_2$, yellow-orange crystals, dec. 344-345°, a 1:2 electrolyte in PhNO$_2$, prepd. from [Pt(TTA)Br]ClO$_4$ by treating with urea and NaClO$_4$ in MeOH under reflux. UV and visible spectra. A 5-coordination of Pt is postulated (514).

[O$_2$NCH$_2$Pt(TTA)]ClO$_4$, yellow crystals, dec. 338-341°, a 1:1 electrolyte in PhNO$_2$, prepd. from [Pt(TTA)Br]ClO$_4$ in MeNO$_2$ and EtOH by shaking with moist Ag$_2$O. UV and visible spectra. A 5-coordination of Pt is postulated (514).

[Ru(TTA)Br$_2$], diamagnetic crystals, a nonelectrolyte, monomeric; a 3-dimensional crystal analysis showed that the 6-coordinate Ru(II) has a slightly distorted octahedral structure (1009).

1.1.7 POLYMERS WITH TERTIARY ARSINE GROUPS

Poly[(phenylarsylene)-p-phenylene], Poly[p-phenylene(phenylarsine)],
     (AsC$_{12}$H$_9$)$_n$,   Ph   (M-113)
                              ⟨C$_6$H$_4$As⟩$_n$,

Synth.: From p-phenylenedilithium and PhAsCl$_2$ in Et$_2$O at -25°, followed by heating under reflux (739).
Prop.: Amber-colored brittle resin at r.t., softens > 40°, liquifies at 100°, sol. in Et$_2$O, C$_6$H$_6$, and CHCl$_3$. A cryst. fraction isolated by adding Et$_2$O to a dilute C$_6$H$_6$ soln. is infusible and sol. in concd. H$_2$SO$_4$ only. The latter fraction softens at 400° while loses 9% of its wt.; its mol. wt. ~ 6,000 (739). Semiconductor prop. of the infusible fraction were detd. (524).

Poly[phenylarsylene(5-bromo-2-methoxy-m-phenylene)],
     (AsC$_{13}$H$_{10}$BrO)$_n$,
Synth.: From 5-bromo-2-methoxy-1,3-phenylenedilithium and PhAsCl$_2$ in Et$_2$O at -25°, followed by heating under reflux (739).
Prop.: A mixt. of C$_6$H$_6$-sol. and -insol. polymers. A fraction with a mol. wt. ~ 5,000, sol. in CHCl$_3$ but insol. in C$_6$H$_6$, softens at 300°, m. 320°, dec. > 320° (739).

Poly[phenylarsylene(4,6-dimethoxy-m-phenylene)],
     (AsC$_{14}$H$_{13}$O$_2$)$_n$,   (M-113),
Synth.: From 4,6-dimethoxy-m-phenylenedilithium and PhAsCl$_2$ in Et$_2$O at -25°, followed by heating under reflux (739).
Prop.: A mixt. of C$_6$H$_6$-sol. and -insol. polymers. A frac-

tion with a mol. wt. ~ 1,900, insol. in C_6H_6 but sol. in $CHCl_3$ and concd. H_2SO_4, softens at 220°, m. 250°, dec. > 250° (739).

Poly[phenylarsylene(p-biphenylylene)], $(AsC_{18}H_{13})_n$, $[C_6H_4C_6H_4\overset{Ph}{As}]_n$, (M-113)

<u>Synth.</u>: From $(4-ClC_6H_4)_2AsPh$ by treating with K-Na alloy in decalin at 180° (739).
<u>Prop.</u>: A polymer from which a high mol. wt. fraction insol. in common org. solvents and concd. H_2SO_4 was isolated. This insol. fraction is a brown solid, darkens > 280° but does not melt < 400° (739).

Poly{phenylarsylene[oxybis(p-phenylene)]}, $(AsC_{18}H_{13}O)_n$, $[C_6H_4OC_6H_4\overset{Ph}{As}]_n$ (M-113)

<u>Synth.</u>: From 4,4'-oxybis(phenylenelithium) and $PhAsCl_2$ in Et_2O at -25° followed by heating under reflux (739).
<u>Prop.</u>: A polymer from which a fraction with a mol. wt. > 2000 was isolated. This fraction is sol. in $CHCl_3$ but insol. in C_6H_6, m. 180° to a yellow liquid; at 400° loses 25% of its wt. (739).

Poly[phenylarsylene(9,10-anthrylene)], $(AsC_{20}H_{13})_n$, (M-114)

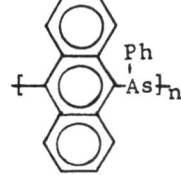

<u>Synth.</u>: From 9,10-anthrylenedilithium and $PhAsCl_2$ in THF at -35 to -45° followed by heating under reflux (739).
<u>Prop.</u>: A fraction with a mol. wt. of ~ 8000, softens at 270°, is sol. in $CHCl_3$ but insol. in C_6H_6 (739).

1.2 SECONDARY ARSINES

As in the main volume (M-114 to M-118), this section contains mono- and di-secondary arsines. Besides the methods of synthesis reported in the main volume, secondary arsines have been prepared as follows:

From arsine, AsH_3, by a condensation with reactive olefins at 75°. When the condensation is carried out in the presence of mercury oxide and a carboxylic acid, arsylenebis(alkylenemercury carboxylates) are formed.

$$AsH_3 + \overset{\backslash}{C}{=}\overset{/}{C} \text{ (1:2) at } 75° \rightarrow (H\overset{|}{C}{-}\overset{|}{C})_2AsH$$

$$AsH_3 + \overset{\backslash}{C}{=}\overset{/}{C} + HgO + RCO_2H \rightarrow HAs(\overset{|}{C}{-}\overset{|}{C}HgO_2CR)_2 \text{ ;}$$

From primary arsines, $RAsH_2$, by a condensation with reactive olefins and with ketones, respectively:

$$RAsH_2 + \overset{\backslash}{C}{=}\overset{/}{C} \text{ (1:1)} \rightarrow R(H\overset{|}{C}{-}\overset{|}{C})AsH$$

$$RAsH_2 + R'_2CO \text{ (1:1)} \rightarrow R[R'_2C(OH)]AsH \text{ ;}$$

From tertiary arsines containing at least one aromatic group by a reaction with sodium in liquid ammonia, to form a secondary arsinosodium derivative, followed by hydrolysis of the latter:

$$R_3As + Na \text{ in } NH_3 \longrightarrow [R_2AsNa] \xrightarrow{H_2O} R_2AsH \text{ ;}$$

From secondary arsinosilanes by hydrolysis with water or alcohols:

$$R_2AsSiR'_3 + HOR'' \rightarrow R_2AsH \quad (R'' = H \text{ or alkyl}) \text{ ;}$$

From heterocyclic compounds of a trivalent arsenic by a reduction with lithium aluminum hydride:

$$R\overline{As(CH_2)_n}AsR + LiAlH_4 \rightarrow R[H(CH_2)_n]AsH \text{ .}$$

A few secondary arsines are described individually hereafter, while others including disecondary arsines are compiled in Table 3.

Bis(trifluoromethyl)arsine, AsC_2HF_6, $(CF_3)_2AsH$, (M-115)
Synth.: From $[(CF_3)_2As-]_2$ by a rxn. with HI in the presence of Hg, 97% yield (310).
From $(CF_3)_2AsI$ by a redn. with Hg in HI, 88% yield (310).
Prop.: A gaseous substance (at r.t.), relatively stable up to 200° (414). Mass spectrum (485).
Rxn. with: $(CF_3)_2AsCl$ at 100° → $[(CF_3)_2As-]_2$ (414).
$CF_3C\equiv CCF_3$ at 200° → $(CF_3)_2AsC(CF_3)C=CHCF_3$ (416).
Me_2AsCl at 100° → no $(CF_3)_3AsAsMe_2$ was obtd. (414).
B_2H_6 at 50-60° → $[(CF_3)_2AsBH_2]_3$ and probably $[(CF_3)_2AsBH_4]_4$ (930).

Dimethylarsine, AsC_2H_7, Me_2AsH, (M-115)
Synth.: From $MeAsH_2$ by treating with Na in NH_3 at -63°, followed by MeCl (185).
From Me_2AsCl by a redn. with Zn and HCl in EtOH, 65% yield (883).
From Me_2AsO_2H by a redn. with Zn and HCl, 90-95% yield (558, 560).
From Me_2AsO_2Na by a redn. with Zn in H_2O-EtOH in the presence of a small amt. of $HgCl_2$ by adding dropwise concd. HCl, 65% yield (627).
From $Me_2AsSiMe_3$ by hydrolysis with H_2O or alcohols (1309).
Prop.: b_{760} 35-37° (558), b. 36-37° (883), b. 40.4° calcd. by the Kinney equation (66), stable up to 100° (414), spontaneously flammable (560). Relative acidity detd. by NMR spectroscopy (185). IR and NMR spectra (627).
Rxn. with: $HCo(CO)_4$ → $Me_2AsCo(CO)_4$ (99a).
Me_3M (M = Al, Ga, In) → $(Me_2AsMMe_2)_3$ (130).
Me_2AsCl at 100° → $Me_2AsAsMe_2$ (414).
HCl at 104° → $[Me_2AsH_2]Cl$ (414).
$[(CF_3)_2As]_2$ → $(CF_3)_2AsAsMe_2$ + $(CF_3)_2AsH$ (311).
$[(CF_3)_2P-]_2$ → $Me_2AsP(CF_3)_2$ (311).
$CF_3C\equiv CCF_3$ at 20° → $Me_2AsC(CF_3)=CHCF_3$ (416).
$CF_3C\equiv CH$ at 20° → $Me_2AsCH=CHCF_3$ (416).
RSSR at 100° → Me_2AsSR and $Me_2AsAsMe_2$ (419).
Me_2AsNMe_2 at 20° → $Me_2AsAsMe_2$ and Me_2NH (419).
CF_3SCl at 20° → Me_2AsCl, Me_2AsCF_3, and CF_3SH (419).
$ClC=CHCF_2\overline{CF_2}$ at 150° → $Me_2As\overline{C=CHCF_2CF_2}$ and Me_2AsCl (420).
$ClC=C(CF_2)_2\overline{CF_2}$ at 100° → $Me_2As\overline{C=CH(CF_2)_2CF_2}$ and Me_2AsCl (420).

ClC=C(Et)CF$_2$CF$_2$ at 140° → Me$_2$AsC=C(Et)CF$_2$CF$_2$ (420).

ClC=C(OMe)CF$_2$CF$_2$ at 130° → Me$_2$AsCl and unidentified As derivs. (420).

CF$_3$CF=CF$_2$ at 100° → CF$_3$CFHCF$_2$AsMe$_2$ (421).

CF$_2$(CF$_2$)$_n$CX=CCl (n = 1 or 2, X = Cl or F) → Me$_2$AsC=CX(CF$_2$)$_n$CF$_2$ (421).

CF$_2$CF$_2$CCl=CAsMe$_2$ at 140° → CF$_2$CF$_2$C(AsMe$_2$)=CAsMe$_2$ (421).

(CF$_3$)$_2$CO at 20° → Me$_2$AsC(CF$_3$)$_2$OH (426).

CF$_3$COCl at 20° → Me$_2$AsCOCF$_3$ and Me$_3$AsCl (427).

Na in THF followed by MeC(CH$_2$Br)$_3$ → MeC(CH$_2$AsMe$_2$)$_3$ (558).

ClCF$_2$CCl=CF$_2$ at r.t. → Me$_2$AsCF$_2$CCl=CF$_2$ (627).

CH$_2$=C=O → Me$_2$AsAc (883).

Me$_3$SiCl → Me$_2$AsSiMe$_3$ (999).

Use: Complexes formed from the arsine and halides or oxyhalides of Ti, V, or Cr and SiO$_2$ or Al$_2$O$_3$ catalyze the polymerization of olefins (42).

Methylphenylarsine,   AsC$_7$H$_9$,   MePhAsH,   (M-117)

Rxn. with: CF$_3$C≡CCF$_3$ at 20° → MePhAsC(CF$_3$)=CHCF$_3$ (416).

CF$_2$CF$_2$C(OMe)=CCl at 130° → MePhAsCl and some unidentified prods. (420).

CF$_2$CF$_2$CF=CF at 100° → CF$_2$CF$_2$CF=CAsMePh (421).

CF$_2$CF$_2$CCl=CCl at 20° → CF$_2$CF$_2$CCl=CAsMePh (421).

Diphenylarsine,   AsC$_{12}$H$_{11}$,   Ph$_2$AsH,   (M-116)

Synth.: From Ph$_2$AsCl + LiAlH$_4$ in Et$_2$O, 58% yield (883).

From Ph$_3$As + Na in liq. NH$_3$ at -65°, replacing the solvent with Et$_2$O and treating the soln. with cold H$_2$O, 75% yield (1017).

Prop.: b$_{1.5}$ 119-120° (883), b$_{0.25}$ 87-90° (1017), NMR spectrum (1312), pK$_a$ 20.3 (787). Photolysis at -196° → Ph$_2$As radical (1358).

Rxn. with: CH$_2$=CHCN → Ph$_2$AsCH$_2$CH$_2$CN (390).

2-Vinylpyridine → Ph$_2$AsCH$_2$CH$_2$C=CHCH=CHCH=N (1542a).

BuLi followed by PhOR → Ph$_2$AsR + PhOH (1017).

BuLi followed by PhSR → Ph$_2$AsR + PhSH (1017).

K in (EtOCH$_2$CH$_2$)$_2$O → Ph$_2$AsK (1347).

PhLi in Et$_2$O → Ph$_2$AsLi (1530).

Et$_2$Zn → (Ph$_2$As)$_2$Zn (1534).

Et$_2$Mg → (Ph$_2$As)$_2$Mg (1534).

Me$_2$NSnMe$_3$ → Ph$_2$AsSnMe$_3$ (811).

CH$_2$=C=O → Ph$_2$AsAc (883).

PhNCO in the presence of Bu$_2$Sn(OAc)$_2$ → Ph$_2$AsCONHPh (1536).

PhNCO at 120° → (Ph$_2$As-)$_2$ + PhNHCONHPh + CO (1536).

Use: The prods. formed from Ph$_2$AsH and halides or oxyhalides of Ti, V, or Cr deposited on SiO$_2$ or Al$_2$O$_3$ catalyze the polymerization of olefins (42).

Table 3. Secondary Arsines

R_2AsH or $RR'AsH$	Prepd. from	Yield %	Prop., Rxn., and Remarks	Ref.
C_3				
$Me(HCF_2CF_2)AsH$	$MeAsH_2 + CF_2=CF_2$ at 75°, (a)	53	b_{756} 74°, d_4^{20} 1.6226, n_D^{20} 1.3990, MR_D 28.56, (b)	1266
$MeEtAsH$			b. 72.1°, (c)	66
C_4				
$(HCF_2CF_2)_2AsH$	By-prod. from AsH_3 and $CF_2=CF_2$ at 75°, along with $HCF_2CF_2AsH_2$	14	b_{739} 94°, d_4^{20} 1.8320, n_D^{20} 1.3610, MR_D 33.57, (d)	1266, 1423
$Me[(CF_3)_2COH]AsH$	$MeAsH_2 + (CF_3)_2CO$ at 20°		IR and NMR spectra, (e)	426
$MePrAsH$			b. 100.8°, (c)	66
Et_2AsH (M-117)			b. 100.8°, (c)	66
C_8				
$Ph(C_2ClF_3H)AsH$	$PhAsH_2 + ClCF=CF_2$ at 100°	34	b_2 88-90°, d_4^{20} 1.6708, n_D^{20} 1.5062, MR_D	1264
$EtPhAsH$ (M-117)			pKa 23.5	787
$Ph(H_2NCH_2CH_2)AsH$			Rxn. with $RCOR' \rightarrow RR'\overline{CNHCH_2CH_2AsPh}$	1532
Bu_2AsH (M-117)			pKa 31.0, (f)	787
$t\text{-}Bu_2AsH$	$t\text{-}Bu_2AsCl + LiAsH_4$ in Et_2O at -10° to 36°	41	b_{18} 49-50°, sensitive to air, rxn. with $PhLi \rightarrow t\text{-}Bu_2AsLi$, (g)	1530
C_9				
$PhPrAsH$ (M-117)	$PhAs\overline{(CH_2)_3AsPh} + LiAlH_4$ in dioxan under reflux, (h)	43	b_5 122-126°, rxn. with BuLi in $Et_2O \rightarrow PhPrAsLi$	790
C_{10}				
$[AcOHgCH_2CH(CH_2Cl)]_2AsH$	$AsH_3 + CH_2=CHCH_2Cl + HgO$ in AcOH		A fungicide and bactericide	1447
C_{12}				
$Ph(cyclo\text{-}C_6H_{11})AsH$ (M-118)	$PhAsH_2 + K$ in dioxan followed by $C_6H_{11}Cl$	30	b_2 111-113°, (i)	1536

(cyclo-C_6H_{11})$_2$AsH (M-118)		Rxn. with PhLi $\rightarrow$ (C_6H_{11})$_2$AsLi, (j)	1530
C_{15}			
PhAsH(CH$_2$)$_3$AsHPh (M-118)	PhAsHK + Br(CH$_2$)$_3$Br in C_6H_6 at r.t. then heating under reflux	A mixt. with PhAs(CH$_2$)$_3$AsPh, b$_1$ 176-178°, (k)	1535
C_{16}			
PhAsH(CH$_2$)$_4$AsHPh (M-118)	PhAsHK + Cl(CH$_2$)$_4$Cl in C_6H_6	A mixt. with PhAs(CH$_2$)$_4$AsPh, b$_2$ 187-189°, (k)	1535
(C_8H_{17})$_2$AsH		Rxn. with NaBH$_4$ + PhOBO at 140° $\rightarrow$ (C_8H_{17})$_2$AsBH$_2$	97
C_{17}			
PhAsH(CH$_2$)$_5$AsPh (M-118)		(k)	1535
C_{18}			
PhAsH(CH$_2$)$_6$AsHPh (M-118)		(k)	1535
C_{22}			
[Me(CH$_2$)$_7$CO$_2$HgCH$_2$CCl$_2$]$_2$AsH	AsH$_3$ + CH$_2$=CCl$_2$ + HgO + Me(CH$_2$)$_7$CO$_2$H in AcOH	A fungicide and bactericide	1447
C_{24}			
[NaO$_2$CC$_6$H$_4$CO$_2$HgCH$_2$-CH(OCH$_2$CH$_2$F)]$_2$AsH	AsH$_3$ + CH$_2$=CHOCH$_2$CH$_2$F + Hg + Na$_2$ terephthalate in AcOH	A fungicide and bactericide	1447

(a) MeAs(CF$_2$CF$_2$H)$_2$ is formed as by-prod.   (b) Rxn. with HCHO in the presence of HCl $\rightarrow$ Me(HCF$_2$CF$_2$)AsCH$_2$OH (1266). (c) Calcd. by the Kinney equation.   (d) Rxn. with Cl $\rightarrow$ (HCF$_2$CF$_2$)$_2$AsCl (1266).   (e) Rxn. with (CF$_3$)$_2$CO at 20° $\rightarrow$ (CF$_3$)$_2$CHOH and an unidentified solid substance (426).   (f) Rxn. with PhLi in Et$_2$O $\rightarrow$ Bu$_2$AsLi (1530).   (g) pK$_a$ 32.0 (787).   (h) PhAsH$_2$ is formed as by-prod. (790).   (i) Rxn. with PhNCO in the presence of Bu$_2$Sn(OAc)$_2$ $\rightarrow$ Ph(C$_6$H$_{11}$)AsCONHPh (1536).   (j) Rxn. with PhNCO in the presence of Bu$_2$Sn(OAc)$_2$ $\rightarrow$ (C$_6$H$_{11}$)$_2$AsCONHPh (1536). (k) Rxn. with BuLi in Et$_2$O followed by dioxan $\rightarrow$ PhAs(Li)(CH$_2$)$_n$As(Li)Ph, n = 3, 4, 5, and 6, resp. (1535).

1.3 PRIMARY ARSINES

Several methods of synthesis of primary arsines were reported in the main volume (M-119). Since then two new methods were developed based on arsine:

$$AsH_3 + \,>\!C\!=\!C\!< \text{ under pressure} \rightarrow HC\text{-}CAsH_2, \text{ along with } (HC\text{-}C)_2AsH$$

$$AsH_3 + R_2CO \rightarrow R_2C(OH)AsH_2 \; .$$

Methylarsine, $AsCH_5$, $MeAsH_2$,   (M-120)

Synth.: From $NaAsH_2$ by a rxn. with MeCl (185).

Prop.: b. 2.3° calcd. by the Kinney equation (66). IR spectrum (697). IR spectra of $MeAsH_2$, $MeAsD_2$, CD_3AsH_2, and CD_3AsD_2 in the vapor and solid states and Raman spectra in the liquid phase were detd. (698). Coriolis coupling (697). The effect of the nonbonded electrons on the barriers to the internal rotation (696). Rel. acidity in liq. NH_3 detd. by NMR spectroscopy (185).

Rxn. with: Me_3M (M = Al, Ga, In) at elevated temp. → colorless polymeric matl. contg. M and As in various ratios (130).

$(CF_3)_2CO$ at 20° → $Me[(CF_3)_2COH]AsH$ (426).

Na in NH_3 → MeAsHNa (185).

$CF_2\!=\!CF_2$ at 75° → $Me(HCF_2CF_2)AsH$ and $Me(HCF_2CF_2)_2As$ (1266).

Uses: The rxn. prods. formed from $MeAsH_2$ and halides or oxyhalides of Ti, V, or Cr deposited on SiO_2 or Al_2O_3 catalyze the polymerization of olefins (42). A chromotropic additive for liquified petroleum gas of not more than 4 C atoms, which produces a blue-colored flame (1313).

1,1,2,2-Tetrafluoroethylarsine, $AsC_2H_3F_4$, $HCF_2CF_2AsH_2$

Synth.: From AsH_3 and $CF_2\!=\!CF_2$ in an autoclave at 75°, 44% yield, along with $(HCF_2CF_2)_2AsH$ (1266, 1432).

Prop.: A bad smelling liquid, b_{739} 34°, d_4^{20} 1.6805, n_D^{20} 1.3660, MR_D 23.37 (1266, 1432).

Rxn. with: Cl in petr. ether at -20° → $HCF_2CF_2AsCl_2$ (1266).

HCHO in the presence of HCl → $HCF_2CF_2As(CH_2OH)_2$ (1266).

Phenylarsine, AsC_6H_7, $PhAsH_2$,   (M-122)

Prop.: pK_a 22.2 (787).

Rxn. with: Me_3M (M = Ga, In) → colored polymeric matl. of variable compn. (130).

$RC\!\equiv\!CC\!\equiv\!CR$ in the presence of BuLi → $Ph\overline{AsC(R)\!=\!CHCH\!=\!C}R$ (1002).

$CF_2\!=\!CF_2$ at 75° → $HCF_2CF_2(Ph)AsH$ (1264).

$CF_2\!=\!CFCl$ at 100° → $Ph(C_2ClF_3H)AsH$ (1264).

Et_2Zn in MePh → $(PhAsH)_2Zn$ as well as EtZnAsHPh and $(PhAsZn)_n$, depending on the order of mixing (1534).

Et_2Mg (2:1) → yellow amorphous solid, probably a mixt. of $(PhAsH)_2Mg$ with $(PhAsMg)_n$ (1534).

Table 4. Primary Arsines

RAsH$_2$	Prepd. from	Yield %	Prop., Rxn. and Remarks	Ref.
C$_1$				
CF$_3$AsH$_2$ (M-120)	CF$_3$AsI$_2$ + Hg to form (CF$_3$As)$_n$ and reducing the latter with Hg + HI	92	(a)	310
	CF$_3$AsCl$_2$ + HCl + Hg at 20°			413
C$_2$				
EtAsH$_2$			b. 38.1°, (b)	66
C$_3$				
[(CF$_3$)$_2$C(OH)]AsH$_2$	AsH$_3$ + (CF$_3$)$_2$CO in a steel bomb at 50–60°	84	b$_{100}$ 41°, n$_D^{20}$ 1.3702, d$_{20}$ 1.8475	251
i-PrAsH$_2$			A conversion with halides or oxy-halides of Ti, V, or Cr and SiO$_2$ or Al$_2$O$_3$ → a catalyst for the polymerization of olefins	42
C$_4$				
BuAsH$_2$	BuAsO$_3$H$_2$ + Zn-Hg in Et$_2$O by adding concd. HCl	41	b. 101°, (c, d)	1536
t-BuAsH$_2$	t-BuAsCl$_2$ in Et$_2$O + LiAlH$_4$ at -10° and then heating on a steam bath	65	b. 65–67°, very sensitive to air, (e, f)	1530
C$_5$				
n-C$_5$H$_{11}$AsH$_2$ (M-120)			b. 125.4°, (b)	66
C$_6$				
2-ClC$_6$H$_4$AsH$_2$			b. 205.1°, (b)	66

(a) Mass spectrum (485). (b) Calcd. by the Kinney equation (66). (c) Rxn. with BuLi in petr. ether → BuAsLi$_2$ (1536). (d) Rxn. with PhNCO (1:2) in the presence of Bu$_2$Sn(OAc)$_2$ → BuAs(CONHPh)$_2$ (1536). (e) Rxn. with PhLi (1:1) in Et$_2$O, followed by dioxan → t-BuAsHLi·Dioxan (1530). (f) Rxn. with PhLi (1:2) in Et$_2$O → t-BuAsLi$_2$ (1530).

1.4.1.1 DIARSINES

The methods of synthesis of diarsines employed through 1964 were summarized in the main volume (M-122). Since then the following new methods were reported:

Reduction of secondary arsine halides with tertiary phosphines and with lithium, respectively,

$$R_2AsX + R'_3P \ (2:1) \rightarrow R_2AsAsR_2$$

$$R_2AsX + Li \rightarrow R_2AsAsR_2 \ ;$$

Reduction of oxybis- and thiobis(secondary arsines) by phosphines and by mercury metal, respectively,

$$(R_2As)_2O + R'_2PX \ (1:1) \rightarrow R_2AsAsR_2$$

$$(R_2As)_2S + Hg \rightarrow R_2AsAsR_2 \ ;$$

Metathesis of secondary arsines with secondary arsinoamines and with diarsines, respectively. (The latter reaction is suitable for the preparation of unsymmetric diarsines.)

$$R_2AsH + R_2AsNR_2 \rightarrow R_2AsAsR'_2$$

$$R_2AsH + R'_2AsAsR'_2 \rightarrow R_2AsAsR'_2 \ ;$$

Decarbonylation of isocyanurate derivatives with secondary arsinolithiums.

$$+ \ 6 \ R_2AsLi \rightarrow 3R_2AsAsR_2 + 3 \ CO + 3 \ R'NH_2 \ .$$

Tetrakis(trifluoromethyl)diarsine, $As_2C_4F_{12}$, $(CF_3)_2AsAs(CF_3)_2$,   (M-123)
<u>Synth.</u>: From $(CF_3)_2AsCl$ by shaking with Hg at 20°, 99% yield (413, 414).
From $[(CF_3)_2As]_2O$ by treating with $(CF_3)_2PCl$ (1:1) at 25° (1416).
From $(CF_3)_2AsCl + CF_3I$ by shaking with Hg at 20°, 75% yield, along with $(CF_3)_3As$ (414).
From $(CF_3)_2AsCl + (CF_3)_2AsH$ by heating at 100°, 51% yield (414).
By-prod. from $(CF_3)_2AsI$ and CF_3I by shaking with Hg at 20°, 32% yield along with $(CF_3)_3As$, 62% yield (414).
By-prod. from the rxn. of CF_3AsCl_2 with CF_3I by shaking with Hg at 200° (414).
<u>Prop.</u>: m. 0.0-0.5° (414). IR and Raman spectra (409). Mass spectrum (485).
<u>Rxn. with</u>: Hg + HI → $(CF_3)_2AsH$ (310).
Me_2AsH → $(CF_3)_2AsAsMe_2$ (311).
Me_2PH → $(CF_3)_2AsPMe_2$ (311).
$[(CF_3)_2P]_2$ → $(CF_3)_2AsP(CF_3)_2$ (311).
$CF_3I + Hg$ → trace amt. of $(CF_3)_3As$ (414).
$CF_3C \equiv CCF_3$ in UV light → $(CF_3)_2AsC(CF_3)=C(CF_3)As(CF_3)_2$ (415).

1,1-Dimethyl-2,2-bis(trifluoromethyl)diarsine, $As_2C_4H_6F_6$, $(CF_3)_2AsAsMe_2$
Synth.: From Me_2AsH and $[(CF_3)_2As-]_2$ at $-63°$ by warming to r.t. and maintaining at $24°$ for 24 hrs., 94% yield along with $(CF_3)_2AsH$ (311).
Prop.: b. $135°$, vapor pressure, thermodynamic data, IR, NMR and mass spectra (311).

Tetramethyldiarsine, $As_2C_4H_{12}$, $Me_2AsAsMe_2$, (M-124)
Synth.: From Me_2AsI by treating with Li in Et_2O, 80% yield (1229).
From Me_2AsCl and Me_2AsH at $100°$, 26% yield (414).
From Me_2AsI by shaking with Hg at $20°$ (optionally in AcOH), 17% (10%) yield (414).
From Me_2AsH and Me_2NAsMe_2 at $20°$ (419).
By-prod. from the rxn. of Me_2AsH with RSSR or Me_2AsSEt (419).
Prop.: m. -1 to $0°$, b. $155°$, NMR spectrum (1229).
Rxn. with: MeI $\rightarrow$ $[Me_4As]I$ and $[Me_4As]I_3$ (414).
HCl $\rightarrow$ Me_2AsH and Me_2AsCl (414).
CF_3SCl at $20°$ $\rightarrow$ Me_2AsCl and Me_2AsSCF_3 (419).
$\overline{CF_2CF_2C(OMe)=C}Cl$ at $120°$ $\rightarrow$ dec. (420).
$\overline{CF_2(CF_2)_nCCl=C}Cl$ at $20°$ $\rightarrow$ $\overline{CF_2(CF_2)_nCCl=C}AsMe_2$ and Me_2AsCl (n = 1 or 2) (421).
$CF_3C\equiv CH$ at $20°$ $\rightarrow$ $CF_3C\equiv CAsMe_2$ and trans-$CF_3CH=CHAsMe_2$ (425).
$(CF_3)_2CO$ $\rightarrow$ $Me_2AsAsMe_2\cdot(CF_3)_2CO$ (426).
$CF_3C\equiv CCF_3$ at $20°$ $\rightarrow$ $Me_2AsC(CF_3)=C(CF_3)AsMe_2$ (426).
CF_3COCl at $20°$ $\rightarrow$ $Me_2AsCOCF_3$ and Me_2AsCF_3 (427).
$(CF_3CO)_2O$ at $20°$ $\rightarrow$ $Me_2AsOCOCF_3$ and Me_2AsCF_3 (427).
MeI $\rightarrow$ Me_2AsI and $[Me_4As]I$ (1229).
Coordination Deriv.: $[Me_2AsAsMe_2\cdot(CF_3)_2CO]$, nonvolatile liquid, unstable in air. IR and NMR spectra. Rxn. with H_2O $\rightarrow$ $Me_2AsAsMe_2$ and $(CF_3)_2CO$; rxn. with $CF_3C\equiv CCF_3$ $\rightarrow$ $Me_2AsC(CF_3)=C(CF_3)AsMe_2$ (426).

$[Me_2AsAsMe_2\cdot PF_5]$, prepd. from $Me_2AsSiMe_3 + PF_5$ (1309).

Complexes with transition metal carbonyls in mixt. with N-chloro or N-bromo amides, imides, or sulfonamides catalyze a radical polymerization of vinyl monomers (773).

$$\begin{array}{c} \pi\text{-}C_5H_5 \diagdown \quad Me\ Me \quad \diagup CO \\ OC-Mn\leftarrow As\text{-}As\rightarrow Mn-CO \\ CO \diagup \quad Me\ Me \quad \diagdown C_5H_5\text{-}\pi \end{array}$$

orange crystals, m. $210\text{-}212°$ (dec.), $subl_{0.2}$ $170°$, prepd. from $Me_2AsAsMe_2$ and $\pi\text{-}C_5H_5Mn\text{-}(CO)_5$ in methylcyclohexane under UV light. IR and NMR spectra (713).

1,2-Dimethyl-1,2-diphenyldiarsine, $As_2C_{14}H_{16}$, $MePhAsAsPhMe$, (M-126)
Prop.: Stereomutation of the meso form to the dl form and vice versa at $150\text{-}210°$ was established by NMR spectroscopy (928).

Tetrakis(pentafluorophenyl)diarsine, $As_{20}C_{24}F_{20}$, $(C_6F_5)_2AsAs(C_6F_5)_2$
Synth.: From $(C_6F_5)_2AsCl$ by shaking with Hg or with Me_3SiH, a mixt. of two rotamer isomers was obtd. (650, 651).
From $[(C_6F_5)_2As]_2S$ by shaking with Hg, a mixt. of two rotamer isomers was obtd. (650).

Prop.: One isomer subl. 140° and the other at 230°; NMR spectra of both isomers were obtd. (650, 651).
Rxn. with: $[\pi\text{-}C_5H_5Fe(CO)_2]_2 \rightarrow \pi\text{-}C_5H_5FeAs(C_6F_5)_2(CO)_2$ (381).
$[\pi\text{-}C_5H_5Mo(CO)_3]_2 \rightarrow \pi\text{-}C_5H_5MoAs(C_6F_5)_2(CO)_3$ (381).
$Fe_3(CO)_{12} \rightarrow [(C_6F_5)_2AsFe(CO)_3]_2$ (381).
$Ru_3(CO)_{12} \rightarrow [(C_6F_5)_2AsRu(CO)_3]_2$ (381).

Tetraphenyldiarsine, $As_2C_{24}H_{20}$, $Ph_2AsAsPh_2$, (M-125)
Synth.: From Ph_2AsBr by electrolytic redn. (467).
From Ph_2AsBr by interaction with the Ph_2As^- anion generated electrolytically from $Ph_2AsAsPh_2 + 2e^-$ (467).
From Ph_2AsCl or Ph_2AsI by treating with Bu_3P, 43% yield (798).
From Ph_2AsH and $(PhNCO)_3$ at 120°, 76% yield (1536).
From $Ph_2AsLi + (PhNCO)_3$ in Et_2O, followed by hydrolysis, 48% yield (1536).
Prop.: m. 128-129° (798), m. 129° (1536). Polarography (467, 472, 1029).
Rxn. with: $2 e^-$ (electrolytic redn.) $\rightarrow 2 Ph_2As^-$ (467, 472).
$M(CO)_nX$ (M = Tc or Re) (1:2) $\rightarrow \{(Ph_2As)_2[M(CO)_{n-1}X]_2\}$ (737, 738).
$M(CO)_nX$ (M = Tc or Re) (1:1) $\rightarrow [Ph_2AsAsPh_2M(CO)_{n-1}X]$ (737, 738).
$[Fe(NO)_2X]$ (1:1) $\rightarrow [Ph_2AsFe(NO)_2]_2$ (735).
$[Co(NO)_2Cl]_2$ (1:1) $\rightarrow [Ph_2AsCo(NO)_2Cl]_2$ (735).
$[Ni(NO)X]_4$ (2:1) $\rightarrow [Ph_2AsNi(NO)X]_2$ (735).
$[Rh(CO)_2Cl]_2$ (4:1) $\rightarrow [Rh(Ph_2AsAsPh_2)_2(CO)Cl]$ (736).
$[Ir(CO)_3Cl]_n \rightarrow [Ph_2AsIr(CO)Cl]_2$ (736).
$V(CO)_6$ (1:2) in $C_6H_6 \rightarrow [Ph_2AsV(CO)_4]_2$ (738).
PhLi in Et_2O followed by Et_2O-dioxan $\rightarrow Ph_2AsLi \cdot C_4H_8O_2 + Ph_3As$ (786).
$LiAlH_4$ in dioxan under reflux $\rightarrow Li[AlH(AsPh_2)_3]$ (790).
Polarographic oxidn. $\rightarrow (Ph_2As)_2O$ (1029).
Coordination Deriv.: $[Re(Ph_2AsAsPh_2)(CO)_4I]$, colorless viscous oil, prepd. from $Re(CO)_5I$ and the diarsine in C_6H_6 under reflux. IR spectrum (738).

$[Rh(Ph_2AsAsPh_2)_2(CO)Cl]$, yellow brown solid, m. 175° (dec.), very stable in soln., prepd. from $[Rh(CO)_2Cl]_2$ in petr. ether and the diarsine (1:4) in C_6H_6. IR spectrum (736).

Complexes with transition metal carbonyls in mixt. with N-chloro or N-bromo amides, imides, or sulfonamides catalyze a radical polymerization of vinyl monomers (773).

Tetracyclohexyldiarsine, $As_2C_{24}H_{44}$, $(C_6H_{11})_2AsAs(C_6H_{11})_2$
Prop.: Polarographic oxidn. - 1/2 wave potential (1029).

1.4.1.2 TRIARSINES

The first compound containing three tertiary arsenic atoms bonded to each other is described hereafter.

Pentaphenyltriarsine, $As_3C_{30}H_{25}$, $Ph_2AsAsAsPh_2$
$\phantom{Pentaphenyltriarsine, As_3C_{30}H_{25}, Ph_2AsAs}\overset{|}{\underset{}{Ph}}$

Synth.: From Ph₂AsCl and (Me₃Sn)₂AsPh in Et₂O at r.t., removing the solvent, and treating the residue with C_5H_{12}, 65% yield (1373).
Prop.: Colorless crystals, m. 185-190°, sol. in org. solvents, sensitive to oxygen (1373).

1.4.1.3 TETRAARSINES

The first arsenic compound containing four tertiary arsenic atoms bonded to each other is described hereafter.

2-(Diphenylarsino)-1,1,3,3-tetraphenyltriarsine, $As_4H_{36}H_{30}$,   (Ph₂As)₃As
Synth.: From Ph₂AsCl and (Me₃Sn)₃As in Et₂O at r.t., removing the solvent, and treating the residue with C_5H_{12}, 51% yield (1373).
Prop.: Colorless crystals, m. 143-147°, very sensitive to O, sol. in org. solvents; IR spectrum (1373).

1.4.2 ARSENO DERIVATIVES

The research of arseno compounds has enormously subsided since the discovery of antibiotics and their application in treating syphilis.

1.4.2.1 SYMMETRIC ARSENO COMPOUNDS

Besides the methods of synthesis summarized in the main volume (M-128 to M-129), symmetric arseno compounds have been prepared as follows: (a) from primary arsines by a reduction with carbodiimides, (b) from primary arsylene-dilithiums by a reaction with α,ω-dihaloalkanes, (c) from primary dihaloar-sines by a reduction with mercury, (d) from arsenoso derivatives by a reaction with sodium amalgam in ethanol and by heating with phenylhydrazine, respectively, and (e) from oxybis(organohaloarsines) by a reduction with zinc and hydrochloric acid in ethanol.

Arsenotrifluoromethane Tetramer, $As_4C_4F_{12}$,   (CF₃As<)₄
Synth.: The tetramer was obtd. as by-prod. (11% yield) from the redn. of CF₃AsI₂ with Hg, followed by the removal of excess Hg with (CF₃)₂AsCl, and by the sepn. of [(CF₃)₂As]₂ and excess (CF₃)₂AsCl from HgCl₂, (CF₃As<)₅ and the less volatile arseno derivatives with a higher mol. wt. (1598).
Prop.: Large colorless crystals, m. 98.5° (1598).
Rxn. with: Hg + HI → CF₃AsH₂ (310).

Arsenotrifluoromethane Pentamer, $As_5C_5F_{15}$,   (CF₃As<)₅
Synth.: The pentamer was prepd. in a mixt. with the tetramer from CF₃AsI₂ by dehalogenation with Hg and was isolated from excess Hg, [(CF₃)₂As]₂, and other by-prod. as described in the procedure given for the synth. of the tetramer (1598).
Prop.: Dense liquid; NMR spectrum and its changes in the temp. range of -55° to +160° (neat) and between -100° and +42° (in CFCl₃ soln.) are interpreted as

an indication of a puckered 5-membered ring with an effective plane symmetry
in the liquid phase below 160° on a time scale of the order of 10^{-2} sec. The
symmetry is attributed to a fast "butterfly" motion of the ring puckering
around the ring plane and a simultaneous inversion of the puckering atom. A
mechanism of the pseudo-rotation is suggested (1598).

Rxn. with: $Hg + HI \rightarrow CF_3AsH_2$ (310).

Arsenomethane, $As_5C_5H_{15}$, $(MeAs{<})_5$,   (M-130)

Synth.: From $MeAsO(OH)(ONa)$ by a redn. with H_3PO_2 (1598).

Prop.: Yellow oil which, after extensive purification and standing at r.t.
or at 0°, formed a red solid reverting to the oily state at 180°. NMR spectrum and its changes over the temp. range from 42 to 202° is interpreted as an
indication of a puckered 5-membered ring with an effective plane of symmetry
in the liquid phase < 160° on a time scale of 10^{-2} sec. This symmetry is
attributed to a fast "butterfly" motion of the ring puckering through the
ring plane and simultaneous inversion of the puckering atom. The collaps of
the NMR spectrum above 170° is ascribed to intramolecular pseudo-rotation of
the puckering and at still higher temp. to intermolecular equilibrium caused
by ring opening (1598). Mass spectrum (528).

Rxn. with: $CH_2=C(Me)C(Me)=CH_2$ under irradiation with a Hg lamp → 1,2,3,6-
tetrahydro-1,2,4,5-tetramethyl-1,2-diarsenin (1357).

$CH_2=C(Me)CH=CH_2$ under irradiation with a Hg lamp → 1,2,3,6-tetrahydro-1,2,4-
trimethyl-1,2-diarsenin (1357).

Metal carbonyls → complex compds. (528).

Coordination Deriv.: $[(MeAs{<})_5 \cdot M(CO)_3]$, where M = Cr, Mo, or W, red cryst.
prods., prepd. from $(MeAs{<})_5$ and $M(CO)_6$ at 170° in C_6H_6 soln. IR and NMR
spectra. The 5-membered As ring functions as a tridentate (528).

$\{[(MeAs{<})_5[M(CO)_5]_2\}$, where M = Cr, Mo, or W, yellow complexes, prepd. from
$(MeAs{<})_5$ and $M(CO)_5 \cdot EtOH$ at r.t. IR spectra showed the carbonyl pattern expected for the monosubstituted Group VI A carbonyls, thus $(MeAs{<})_5$ functions
as a bidentate (528).

Arsenobenzene, $(AsC_6H_5)_n$, $(PhAs{<})_n$,   (M-130)

Synth.: From $PhAsH_2$ and $C_6H_{11}N=C=NC_6H_{11}$ by heating on a steam bath (80%
yield) - probably a hexamer (1538).

From $PhAsLi_2$ in THF by treating with $BrCH_2CH_2Br$, a tetramer was obtd. (583).

Prop.: Hexamer, m. 196° (1538). IR spectrum - qual. interpretation (1223).

Biol. Prop.: A weak inhibitory effect on the growth of Lactobacillus leich-
mannii, L. arabinosus, Streptococcus faecalis, and Saccharomyces carlsbergensis
(1071). A fuel-sol. biocide effective against Cladosporium resinae and
Pseudomonas species (1295).

Uses: Complexed with $RAlX_2$ and halides of Ti, V, Zr, Cr, or Mo, the prods.
catalyze the polymerization of α-olefins (384).

Coordination Deriv.: $[(PhAs{<})_6Mo(CO)_4]$, yellow needles, m. 200°, prepd. from
$(PhAs{<})_6$ and $Mo(CO)_6$ (1:1) in C_6H_6/diglyme at 130° (74).

$[(PhAs{<})_4Mo(CO)_4]$, prepd. from $(PhAs{<})_4$ and $Mo(CO)_6$ in C_6H_6 under reflux (583).

{(PhAs⊂)$_4$[Mo(CO)$_4$]$_2$}, prepd. from (PhAs⊂)$_4$ and Mo(CO)$_6$ in MePh under reflux (583).

[(PhAs⊂)$_6$W(CO)$_5$], yellow crystals, m. 163°, prepd. from (PhAs⊂)$_6$ and W(CO)$_6$ in MePh at 116°. IR spectrum (74).

3,3′-Diamino-4,4′-dihydroxyarsenobenzene, Salvarsan, Arsphenamine,
    (AsC$_6$H$_6$NO)$_n$,   [3,4-H$_2$N(HO)C$_6$H$_3$As⊂]$_n$,   (M-131)

Biol. Prop.: A potential carcinogen in prolonged medicinal applications (769). A combined use of salvarsan and sulfanilamide in rabbits causes myocardial fibrosis and regressive changes of the heart muscle (904). A dose of 5 ml 2-5% Na arsphenamine soln. injected intravenously repeatedly after 30 min. into a cat produced each time the same type of shock as Congo Red. When added to blood plasma, the soln. produced a rapidly disappearing turbidity (1466). Its fungicidal and bactericidal activity is compared to that of phenarsazine and phenoxarsine derivatives (1473).

3,3′-Diamino-4,4′-dihydroxyarsenobenzene-N,N′-bis(methylenesulfonic Acid) Disodium Salt, Sulfarsenamine, Myarsenol,   (AsNaC$_7$H$_7$NO$_4$S)$_n$,
    [4,3-HO(NaOSO$_2$CH$_2$NH)C$_6$H$_3$As⊂]$_n$,   (M-136)

Prop.: Anal. procedure - detn. of As, using HNO$_3$ vapor (1435).

p-[N,N-Bis(2-chloroethyl)amino]arsenobenzene,   (AsC$_{10}$H$_{12}$Cl$_2$N)$_n$,
    {4-[(ClCH$_2$CH$_2$)$_2$N]C$_6$H$_4$As⊂}$_n$

Synth.: From (ClCH$_2$CH$_2$)$_2$NC$_6$H$_4$AsO$_3$H$_2$ in MeOH by redn. with SnCl$_2$ and concd. HCl contg. KI, 59% yield (122, 441).

Prop.: Yellow powder, m. 120-121° (dec.), sol. in THF and dioxan but insol. in EtOH and in H$_2$O (122, 441). IR spectrum (441).

Rxn. with: 30% H$_2$O$_2$ in aq. NaOH → 4-[(ClCH$_2$CH$_2$)$_2$N]C$_6$H$_4$AsO$_3$H$_2$ (441).

Biol. Prop.: Inhibitory effect on Ehrlich's ascites tumors (441).

p-(N,N-Diethylamino)arsenobenzene,   (AsC$_{10}$H$_{14}$N)$_n$,   [4-(Et$_2$N)C$_6$H$_4$As⊂]$_n$

Synth.: From 4-(Et$_2$N)C$_6$H$_4$AsO in EtOH by a redn. with Na-Hg, 78% yield (441).

Prop.: Yellow powder, m. 180°, IR and NMR spectra (122, 441).

Biol. Prop.: Antitumor activity in rats (122).

Arsenoferrocene,   (AsFeC$_{10}$H$_9$)$_n$,   (FcAs⊂)$_n$

Synth.: From FcAsCl$_2$ or (FcAsCl)$_2$O in 95% EtOH by reducing with Zn powder and HCl (1436).

From (FcAsO)$_2$ in 95% EtOH by heating with excess H$_3$PO$_2$ (1436).

From (FcAsO)$_2$ or (FcAsO)$_n$ dissolved in PhNHNH$_2$ by heating almost to boiling, 10-15% yield (1436).

Prop.: Bright yellow powder, crystd. from Py, darkened > 290°, m. 328-331°, IR spectrum (1436).

1.4.2.2 UNSYMMETRIC ARSENO COMPOUNDS

The methods of synthesis of the arseno compounds were summarized in the main volume (M-143).

3-Amino-4,4'-dihydroxy-3'-(sulfinomethylamino)arsenobenzene Sodium Salt,
     Neosalvarsan, Neoarsphenamine, $(As_2NaC_{13}H_{13}N_2O_4S)_n$,   (M-144)

Biol. Prop.: Tests in mucous membrane leishmaniasis
(321). Potential carcinogen in prolonged medicinal
applications (769). A sensitizing effect on Micrococcus
sodonensis to x-radiation in the presence of O (859).
In a combination with $MeSO_3Et$ and with Hg_2Cl_2 a synergistic effect on the chromosome aberrations in Hordeum
sativum was observed (1091, 1092, 1094). A toxic and muta-
genic effect on Schizosaccharomyces pombe (1093). A protective action against
latent hemobartonelosis in splenectomized rats caused by $PhNHNH_2$ (63). Sensi-
tization of guinea pigs and induction of a hemorrhagic reactivation of the
inflammatory lesions by repeated Neosalvarsan injections which resemble the
localized Schwartzman phenomenon to bacterial endotoxins (1590). An injection
of 2-5% Neosalvarsan soln. into a cat. in 5 ml doses repeated in 30 min. inter-
vals produced, each time, the same type of shock as that after Congo Red.
Addn. of the soln. to blood plasma produced a transient turbidity. Neosalvarsan prevented the shock induced by Congo Red (1466).

3-{5-{3-[[Bis(2,3-dihydroxypropyl)amino]-4-hydroxyphenyl]arseno}-2-benzoxazo-
     lylthio}propionic Acid, $(As_2C_{22}H_{26}N_2O_8S)_n$,   (M-151)

Biol. Prop.: Chemotherapeutic effect against extra-
cellular stages of Trypanosoma cruzi (1062). In a
combination with $MeSO_3Et$ a synergistic effect on
the chromosome aberrations in Hordeum sativum was
observed (1092, 1094).

1.5   COMPOUNDS WITH ARSENIC-HALOGEN AND ARSENIC-PSEUDOHALOGEN BONDS

1.5.1   SECONDARY HALOARSINES, R_2AsX AND $RR'AsX$

The methods of synthesis of the haloarsines employed through 1964 were
summarized in the main volume (M-156 to M-158). Since then several new methods
and modifications of the known methods have been reported.

Dihaloarsinoamines have been extensively used as the starting materials for
the synthesis of the haloarsines through the reaction with Grignard reagents
and with alkyllithiums, respectively, followed by hydrolysis with hydrohalic
acids.

$$R'_2NAsX_2 + 2\ RMgX' \longrightarrow R'_2NAsR_2 \xrightarrow{HX''} R_2AsX''$$

$$R'_2NAsX_2 + 2\ RLi \longrightarrow R'_2NAsR_2 \xrightarrow{HX''} R_2AsX'' \quad .$$

Starting from primary dihaloarsines, secondary haloarsines have been prepared by a condensation with secondary amines to alkylhaloarsinoamines, followed by a conversion of the latter with Grignard reagents and hydrolysis with hydrohalic acids as follows:

$$RAsX_2 + R'_2NH \rightarrow RAs(NR_2)X \rightarrow + R''MgX \rightarrow RR''AsNR'_2 \xrightarrow{HX'} RR''AsX' \ .$$

The latter method is useful in the preparation of the haloarsines with two different organic radicals.

Secondary arsines have been converted to the corresponding haloarsines by a reaction with halogens and with alkanesulfenyl halides, respectively.

$$R_2AsH + X_2 \text{ or } R'SX \rightarrow R_2AsX$$

Secondary fluoroarsines have been prepared by heating oxybis(secondary arsines) with phosphonic difluorides.

$$(R_2As)_2O + R'P(O)F_2 \rightarrow R_2AsF$$

Fluoroarsines have been obtained from alkoxydialkylarsines by a cleavage with boron trifluoride. They have also been prepared from chloroarsines by heating with ammonium fluoride in the presence of boron trifluoride.

$$R_2AsOR' + BF_3 \rightarrow R_2AsF$$

$$R_2AsCl + NH_4F + BF_3 \rightarrow R_2AsF$$

Metathesis of thiobis(secondary arsines) with mercury(II) chloride in ether yields the corresponding chloroarsines.

$$(R_2As)_2S + HgCl_2 \rightarrow R_2AsCl$$

Secondary iodoarsines exchange the iodine for chlorine during shaking with mercury(II) chloride.

$$R_2AsI + HgCl_2 \rightarrow R_2AsCl$$

Unsymmetric tertiary arsines have been converted to secondary haloarsines by heating with halogens or with alkyl halides at 100° in a sealed tube.

Metathesis of dialkyl(alkylthio)arsines with alkyl halides and with alkanesulfenyl halides, respectively, affords secondary haloarsines.

The haloarsines have been prepared from tertiary arsine sulfides by heating with alkyl or acyl halides in a sealed tube.

Some compounds of this group are described individually hereafter, while others are compiled in Table 5.

Chlorobis(trifluoromethyl)arsine, AsC_2ClF_6, $(CF_3)_2AsCl$, (M-158)
Synth.: From $(CF_3)_2AsI$ by repeated treatments with $HgCl_2$ (266).
From $[(CF_3)_2As]_2O$ by treating with dry HCl (266) or with $(CF_3)_2PCl$ (1416).
By-prod. from the rxn. of CF_3AsCl_2 with CF_3I and Hg at 20° (414).
Prop.: IR and Raman spectra (409). Mass spectrum (485).

Rxn. with: MeOH + Me$_3$N → (CF$_3$)$_2$AsOMe (266).

Hg → [(CF$_3$)$_2$As-]$_2$ (413, 414).

CF$_3$I + Hg → [(CF$_3$)$_2$As-]$_2$ and (CF$_3$)$_3$As (414).

(CF$_3$)$_2$AsH at 100° → [(CF$_3$)$_2$As-]$_2$ (414).

Me$_2$AsH at 100° → CF$_3$Cl, (Me$_2$As-)$_2$, (CF$_3$)$_2$AsH, Me$_2$AsCl, and SiF$_4$ (414).

Me$_2$NNH$_2$ in Et$_2$O → (CF$_3$)$_2$AsNHNMe$_2$ (1221).

(CF$_3$)$_2$PNHR in the presence of Me$_3$N → (CF$_3$)$_2$AsN(R)P(CF$_3$)$_2$ (1416).

[(CF$_3$)$_2$P]$_2$NNa in THF → (CF$_3$)$_2$AsN[P(CF$_3$)$_2$]$_2$ (1416).

Iodobis(trifluoromethyl)arsine, AsC$_2$F$_6$I, (CF$_3$)$_2$AsI, (M-159)

Prop.: IR and Raman spectra (409). Dec. at 210° → (CF$_3$)$_3$As (414).

Rxn. with: HgCl$_2$ → (CF$_3$)$_2$AsCl (266).

Hg + HI → (CF$_3$)$_2$AsH (310).

CF$_3$I + Hg → (CF$_3$)$_3$As and [(CF$_3$)$_2$As-]$_2$ (414).

Fe(CO)$_5$ → [Fe(CO)$_3$As(CF$_3$)$_2$I]$_2$ (657).

Bromodimethylarsine, AsC$_2$H$_6$Br, Me$_2$AsBr, (M-159)

Prop.: b. 129.0° calcd. by the Kinney equation (66). Dipole moment (349).
Bond and mol. polarizabilities were calcd. (1118).

Rxn. with: Ph$_3$P=NPh in Et$_2$O → [Ph$_3$PN(Ph)AsMe$_2$]Br (1008).

(RO)$_3$P + R'CHO → Me$_2$AsOCH(R')P(O)(OR)$_2$ (1274).

Chlorodimethylarsine, AsC$_2$H$_6$Cl, Me$_2$AsCl, (M-160)

Synth.: From (Me$_2$As-)$_2$, Me$_2$AsH, or Me$_2$AsSEt by treating with CF$_3$SCl at 20°,
a mixt. of the title compd. with Me$_2$AsSCF$_3$ was obtd. (419).
From Me$_2$AsO$_2$H and PCl$_3$ at 25-27° (883, 939).

Prop.: b$_{37}$ 77-78° (883), b. 105.5° calcd. by the Kinney equation (66). Force
const. calcd. (349b). Dipole moment (349). Bond and mol. polarizabilities
calcd. (1118). NMR spectrum (798).

Rxn. with: Me$_3$SiCo(CO)$_4$ → Me$_2$AsCo(CO)$_3$ (99a).

Me$_2$AsH at 100° → (Me$_2$As-)$_2$ (414).

CF$_3$C≡CCF$_3$ in UV light → Me$_2$AsC(CF$_3$)=CClCF$_3$ (415).

CF$_3$C≡CCF$_3$ at 140° in the presence of AlCl$_3$ → Me$_2$AsC(CF$_3$)=CFCF$_3$ (415).

CF$_3$SCl at 20° → MeAsCl$_2$, Me$_2$AsCl, and several by-prods. (419).

CF$_3$C≡CH at 155° → CF$_3$C≡CAsMe$_2$, trans-CF$_3$CH=CHAsMe$_2$, and CF$_3$C(=CHCl)AsMe$_2$ (425).

(CF$_3$)$_2$CO at 20° → no rxn. (426).

RMgBr → Me$_2$AsR (939).

Zn + concd. HCl in EtOH → Me$_2$AsH (883).

LiN(Me)SiMe$_3$ → Me$_2$AsN(Me)SiMe$_3$ (1343).

LiN=C(Ph)N(Me)SiMe$_3$ in Et$_2$O → PhC(=NAsMe$_2$)N(Me)SiMe$_3$ and PhC(=NSiMe$_3$)N(Me)-
AsMe$_2$ (1343).

MeNCH$_2$CH$_2$N(Me)PN(Me)Li → MeNCH$_2$CH$_2$N(Me)PN(Me)AsMe$_2$ (1346).

LiN=PMe$_3$ → Me$_3$P=NAsMe$_2$ (1352).

(RO)$_3$P + R'CHO → Me$_2$AsOCH(R')P(O)(OR)$_2$ (1274).

Iododimethylarsine, AsC$_2$H$_6$I, Me$_2$AsI, (M-161)

Synth.: From Me$_2$AsO$_2$H in H$_2$O contg. KI by adding simultaneously HCl and SO$_2$,

90-95% yield (558).

Prop.: b. 150.3° calcd. by the Kinney equation (66). Dipole moment (349).
NMR spectrum (798). Bond and mol. polarizabilities were calcd. (1118).

Rxn. with: RMgX → Me$_2$AsR (157, 164, 425, 1513).
Hg at 20° → (Me$_2$As-) (414).
Na in THF under reflux → Me$_2$AsNa (1229).
Na in THF followed by o-C$_6$H$_4$Cl$_2$ → o-C$_6$H$_4$(AsMe$_2$)$_2$ (558).
Na in THF followed by MeC(CH$_2$Br)$_3$ → MeC(CH$_2$AsMe$_2$)$_3$ (558).
Na in THF followed by RBr → Me$_2$AsR (1414).
K in THF under reflux → Me$_2$AsK (1229).
Li in Et$_2$O → (Me$_2$As-)$_2$ (1229).
RLi → Me$_2$AsR (1414).
Bu$_3$P → [Bu$_3$PAsMe$_2$]I (798).
Tert. arsines → quaternary arsonium iodides (1016).
Me$_3$SnMn(CO)$_5$ (1:1) → [Me$_2$AsMn$_2$(CO)$_6$I]$_2$ (24).
Me$_3$SnCo(CO)$_4$ → [Me$_2$AsCo(CO)$_3$]$_n$ (24).

Chlorodipropylarsine,   AsC$_6$H$_{14}$Cl,   Pr$_2$AsCl,   (M-163)
Rxn. with: RR'C=NOH + base → Pr$_2$AsON=CRR' + base·HCl (825).
KSC(S)SR → Pr$_2$AsSC(S)SR (960).

Iododipropylarsine,   AsC$_6$H$_{14}$I,   Pr$_2$AsI,   (M-170)
Rxn. with: BrMgC≡CMgBr → Pr$_2$AsC≡CAsPr$_2$ (912).
HC≡CMgBr → Pr$_2$AsC≡CH (913).

Dibutylchloroarsine,   AsC$_8$H$_{18}$Cl,   Bu$_2$AsCl,   (M-171)
Synth.: From Et$_2$NAsCl$_2$ + BuMgBr in Et$_2$O-C$_6$H$_6$, followed by HCl, 74% yield (194).
Prop.: b$_{17}$ 100-102°, n$_D^{20}$ 1.5028, d$_4^{20}$ 1.1539 (194).
Rxn. with: RSH → Bu$_2$AsR (191).
p- and m-MeC$_6$H$_4$Li → p- and m-MeC$_6$H$_4$AsBu$_2$, resp. (791).

Di-tert-butylchloroarsine,   AsC$_8$H$_{18}$Cl,   t-Bu$_2$AsCl
Synth.: From AsCl$_3$ in Et$_2$O by adding dropwise t-BuMgCl (1:2) soln. in Et$_2$O
at -10° in the dark and completing the rxn. on a steam bath, 55% yield (1530).
Prop.: Colorless liquid, b$_{10}$ 73-77° (1530).
Rxn. with: LiAlH$_4$ → t-Bu$_2$AsH (1530).

Dibutyliodoarsine,   AsC$_8$H$_{18}$I,   Bu$_2$AsI,   (M-164)
Rxn. with: BrMgC≡CMgBr → Bu$_2$AsC≡CAsBu$_2$ (912).
HC≡CMgBr → Bu$_2$AsC≡CH (913).

Iododipentylarsine,   AsC$_{10}$H$_{22}$I,   (C$_5$H$_{11}$)$_2$AsI
Rxn. with: BrMgC≡CMgBr → (C$_5$H$_{11}$)$_2$AsC≡CAs(C$_5$H$_{11}$)$_2$ (912).
HC≡CMgBr → (C$_5$H$_{11}$)$_2$AsC≡CH (913).

Chlorobis(pentafluorophenyl)arsine, $AsC_{12}ClF_{10}$, $(C_6F_5)_2AsCl$

Synth.: From H_2NAsCl_2 and C_6F_5MgBr (1:2) in Et_2O under reflux, followed by hydrolysis with HCl, 77% yield (651, see also 650).
From [$(C_6F_5)_2As]_2S$ + $HgCl_2$ in Et_2O under reflux (651).
Prop.: Colorless liquid, $b_{0.1}$ 82°, IR and NMR spectra (651).
Rxn. with: Hg → [$(C_6F_5)_2As]_2$ (650, 651).
Me_3SiH → [$(C_6F_5)_2As]_2$ (650).
H_2O → [$(C_6F_5)_2As]_2O$ (651).
Aq. NaOH → C_6F_5H (651).
Ag_2S in C_6H_6 → [$(C_6F_5)_2As]_2S$ (651).
[$\pi-C_5H_5Fe(CO)_2]^-$ → [$(C_6F_5)_2AsFe(CO)_2-\pi-C_5H_5$] (381).
[$\pi-C_5H_5Mo(CO)_3]^-$ → [$(C_6F_5)_2AsMo(CO)_3-\pi-C_5H_5$] (381).

Bromodiphenylarsine, $AsC_{12}H_{10}Br$, Ph_2AsBr, (M-173)

Synth.: From Et_2NAsCl_2 in C_6H_6 by adding to PhMgCl and treating the rxn. mixt. with HBr, 57% yield (194).
Prop.: m. 55°, b_2 175-177° (194). 1/2-wave potential (467, 471).
Rxn. with: Ph_2As^- → $Ph_2AsAsPh_2$ (467).
Electrolytic redn. (1e⁻) → $Ph_2AsAsPh_2$ (467).
Ph_3Sn^- → $Ph_3SnAsPh_2$ (467).
Ph_2AsO^- → $(Ph_2As)_2O$ (467).
Polarographic redn. with 2e⁻ → Ph_2As^-, reactivity of the anion with i-PrBr was detd. (471).

Chlorodiphenylarsine, Clark I, $AsC_{12}H_{10}Cl$, Ph_2AsCl, (M-165)

Synth.: From Et_2NAsCl_2 + PhMgBr in $Et_2O-C_6H_6$, followed by hydrolysis with HCl, 83% yield (194).
Ph_2AsO_2H by reducing with SO_2 in HCl (1184).
Prop.: m. 38°, b_8 165° (194), b_8 168-169° (1184), b. 329.0° calcd. by the Kinney equation (66), m. 39° (1421). Safety precautions and disposal (1368).
Rxn. with: $(RO)_2P(S)SH$ → $Ph_2AsSP(S)(OR)_2$ (192).
RSNa → Ph_2AsSR (316).
$Me_3SnC\equiv CC\equiv CSnMe_3$ → $Ph_2AsC\equiv CC\equiv CAsPh_2$ (694, 695).
Bu_3P → $Ph_2AsAsPh_2$ and Bu_3PCl_2 (798).
RLi → Ph_2AsR (842).
KSC(S)OR → $Ph_2AsSC(S)OR$ (957).
KSC(S)SR → $Ph_2AsSC(S)SR$ (960).
K in dioxan → $Ph_2AsK\cdot 2$ Dioxan (1104).
4,3,5-HO(t-Bu)$_2C_6H_2Br$ + Na in THF → 4,3,5-HO(t-Bu)$_2C_6H_2AsPh_2$ (1105).
4,3,5-LiO(t-Bu)$_2C_6H_2SLi$ in THF, followed by hydrolysis → 4,3,5-HO(t-Bu)$_2C_6H_2$-$SAsPh_2$ (1105).
PhSH at 60° → Ph_2AsSPh (1105).
Cl_2 → Ph_2AsCl_3 (1421).
NH_2Cl → [$Ph_2As(NH_2)Cl]Cl$ (1421).
NH_2Cl-NH_3, followed by recrystn. from $CHCl_3$ → [$Ph_2AsAs(NH_2)-N-As(NH_2)Ph_2]Cl\cdot$-$CHCl_3$ (1421).
RSH → Ph_2AsSR (1184).

LiSnR$_3$ → Ph$_2$AsSnR$_3$ (1371).

(Me$_3$Sn)$_3$X (X = P or As) → (Ph$_2$As)$_3$X (1373).

(Me$_2$Sn)$_2$XPh (X = P or As) → (Ph$_2$As)$_2$XPh (1373).

FcH + AlCl$_3$, no ferrocenyl-As deriv. (1436).

Mixt. of FcdLi$_2$ with FcLi → Fcd(AsPh$_2$)$_2$ + FcAsPh$_2$ (1437).

Chlorodicyclohexylarsine,   AsC$_{12}$H$_{22}$Cl,   (C$_6$H$_{11}$)$_2$AsCl,   (M-173)

<u>Synth.</u>: From AlCl$_3$ in cyclohexane by neutraon irradiation, a mixt. of radio-active (C$_6$H$_{11}$)$_2$AsCl and C$_6$H$_{11}$AsCl$_2$ was obtd. The rxn. in the liquid phase gave a higher yield than in the gas phase (1286).

Dihexyliodoarsine,   AsC$_{12}$H$_{26}$I,   (C$_6$H$_{13}$)$_2$AsI

<u>Rxn. with</u>: BrMgC≡CMgBr → (C$_6$H$_{13}$)$_2$AsC≡CAs(C$_6$H$_{13}$)$_2$ (912).

HC≡CMgBr → (C$_6$H$_{13}$)$_2$AsC≡CH (913).

KSC(S)SR → (C$_6$H$_{13}$)AsSC(S)SR (960).

Chlorodi-α-naphthylarsine,   AsC$_{20}$H$_{14}$Cl,   (α-C$_{10}$H$_7$)$_2$AsCl,   (M-175)

<u>Rxn. with</u>: Me$_3$SnC≡CC≡CSnMe$_3$ → (α-C$_{10}$H$_7$)$_2$AsC CC CAs(C$_{10}$H$_7$-α)$_2$ (694, 695).

Electrolytic redn. (2e⁻) → (α-C$_{10}$H$_7$)$_2$As⁻ (467).

Table 5. Monohaloarsines, R_2AsX and $RR'AsX$

R_2AsX or $RR'AsX$	Prepd. from	Yield %	Prop., Rxn., and Remarks	Ref.
C$_2$				
$(CF_3)_2AsF$ (M-168)	R_2AsOMe or R_2AsOBu-t + BF_3 at 24°		Vapor pressure data and IR spectrum	266
Me_2AsF (M-168)	R_2AsCl + NH_4F in the presence of SbF_3 at 85°	40	b. 71.5-72°, dipole moment	349
	$(R_2As)_2O$ + $MeP(O)F_2$ at 40-95°		b_{750} 80-81°, n_D^{20} 1.3968, d_4^{20} 1.4550	620
C$_3$				
$(O_2NCH_2)_2AsCl \cdot MeNO_2$	$AsCl_3$ in $MeNO_2$ under reflux		Light grey solid, m. > 225°	1199
$MeEtAsBr$ (M-168)			b. 153.3° calcd. by the Kinney equation	66
C$_4$				
$(HCF_2CF_2)_2AsCl$	R_2AsH + Cl in petr. ether	53	b_{747} 120°, n_D^{20} 1.3964, d_4^{20} 1.7921	1266
$(CH_2=CH)_2AsI$ (M-168)	AsI_3 + $CH_2=CHSnBu_3$ at 130°		Rxn. with $RMgBr$ → $(CH_2=CH)_2AsR$, rxn. with $CF_3CF_2CF_2I$ + Hg → $(CH_2=CH)_2AsCF_2CF_2CF_3$	1594
Et_2AsBr (M-169)	Et_2NAsCl_2 + $RMgBr$ in Et_2O-C_6H_6 followed by HBr	45	b_5 47°, n_D^{20} 1.5472, d_4^{20} 1.6120, (a)	194
Et_2AsCl (M-162)	R_2AsO_2H + PCl_3 at 25-30°		Rxn. with $RMgX$ → Et_2AsR, b. 155.6°, (b), (c)	939
Et_2AsI (M-162)			b. 194.6°, (b)	66
			Rxn. with $HC\equiv CMgBr$ → $Et_2AsC\equiv CH$	913
C$_5$				
$Me[CF_3CCl=C(CF_3)]AsBr$	$R_2R'As$ + Br in CCl_4 and heating at 100°		b_{70} 106-115°, IR and NMR spectra	415
$EtPrAsCl$ (M-169)			b. 178.1°, (b)	66
C$_6$				
Pr_2AsBr (M-170)	Et_2NAsCl_2 in C_6H_6 + $RMgX$ in Et_2O-C_6H_6, (g)	80	b_{20} 98°, n_D^{20} 1.5339, d_4^{20} 1.4531, (d)	194

Compound	Yield (%)	Preparation	Properties	Ref.
C₇				
MePhAsBr			b. 249.9°, (b)	66
MePhAsCl (M-163)			b. 233.3°, (b)	66
MePhAsI (M-163)	57	R'AsI₂ + Et₂NH, followed by RMgBr in Et₂O-C₆H₆ and HI	b₂ 100-104°, d_4^{20} 1.9568	194
C₈				
Ph(HO₂CCH₂)AsCl			IR spectrum interpretation	1223
EtPhAsBr	21	RR'AsSEt + EtBr in C₆H₆	b₂ 74°	609
EtPhAsCl (M-170)	66	R'AsCl₂ + Et₂NH, followed by RMgBr in Et₂O-C₆H₆, (f)	b₁₄ 130-132°, n_D^{14} 1.6051, d_4^{14} 1.4610 b₁ 251.7°, (b), (e)	194
EtPhAsI (M-164)	16	R₂R'AsS + EtI in a sealed tube at 100°	b₉ 137-140°, n_D^{20} 1.6566, d_4^{20} 1.6951	1549
Bu₂AsBr (M-170)	78	Et₂NAsCl₂ + RLi in petr. ether, (g)	b₂ 85-86°, n_D^{28} 1.5330, d_4^{28} 1.3545	194
i-Bu₂AsI		Rxn. with BrMgC≡CMgBr → i-Bu₂AsC≡C-As-i-Bu₂		912
C₉				
CH₂=CHCH₂(Ph)AsCl		Rxn. with RMgBr → CH₂=CHCH₂(Ph)AsR		608
MeC₆H₄(HO₂CCH₂)AsCl		IR spectrum interpretation		1223
PrPhAsCl	70	R₂R'AsS + AcCl at 80-90° in a sealed tube	b₈₋₉ 124-126°, n_D^{20} 1.5784, d_4^{20} 1.3273, AR_As 11.45, IR spectrum, (h)	1548
Et(4-MeC₆H₄)AsCl (M-171)		Rxn. with RMgBr → Et(4-MeC₆H₄)AsR		601
C₁₀				
BuPhAsCl (M-171)	61	R'AsCl₂ + Et₂NH followed by RMgBr in Et₂O-C₆H₆, (f)	b₁₀ 135-137°, n_D^{20} 1.5725, d_4^{20} 1.3043	194
(n-C₅H₁₁)₂AsBr	81	Et₂NAsCl₂ + MgX in Et₂O-C₆H₆, (g)	b₂ 114°, n_D^{22} 1.5072, d_4^{22} 1.2624	194
(i-C₅H₁₁)₂AsCl			b. 265.0°, (b)	66

(a) Rxn. with RR'C=NOH + base → Et₂AsON=CRR' + base·HBr (827). (b) Calcd. by the Kinney equation (66). (c) Rxn. with $\overline{CH_2CH_2S}$ → Et₂AsSCH₂CH₂Cl (961). (d) RR'C=NOH → Pr₂AsON=CRR' (827). (e) Rxn. with RMgBr → EtPhAsR (601, 605, 606, 608, 822). (f) Followed by hydrolysis with HCl. (g) Followed by hydrolysis with HBr. (h) Rxn. with RMgX → PhPrAsR (1548).

Table 5 Continued

R_2AsX or $RR'AsX$	Prepd. from	Yield %	Prop., Rxn., and Remarks	Ref.
C_{11}				
$Ph(C_5H_{11})AsI$	RR'_2As + EtI in a sealed tube at 100°	27	$b_{0.0021}$ 100–103°, n_D^{20} 1.6249, d_4^{20} 1.5930	1549
C_{12}				
$(2\text{-}BrC_6H_4)_2AsCl$	$RAsCl_2$ + RAsO at 260–270° under N, (i)	71	Crystals, m. 108–109°, $b_{0.2}$ 166°, (j, k)	715
$(2\text{-}BrC_6H_4)_2AsI$	R_2AsCl + NaI in Me_2CO		Yellow crystals, m. 103–104°	715
$(4\text{-}FC_6H_4)_2AsCl$	Et_2NAsCl_2 + RMgBr in $Et_2O\text{-}C_6H_6$, followed by HCl	65	m. 37°, b_3 151–153°	194
$(4\text{-}ClC_6H_4)_2AsCl$ (M-172)	Et_2NAsCl_2 + RMgBr in $Et_2O\text{-}C_6H_6$, followed by HCl	65	m. 50°, b_3 195–19 °	194
$Ph(2,4\text{-}Cl_2C_6H_3)AsCl$	$RR'AsO_2H$ + SO_2 in HCl		b. 154–156°, (l)	1184
$Ph(4\text{-}O_2NC_6H_4)AsCl$			Rxn. with RSH + Et_3N → $Ph(4\text{-}O_2NC_6H_4)$-AsSR + $Et_3N\cdot HCl$	820
Ph_2AsClO_4			1/2-wave potential, electrolytic redn. → Ph_2As^-	467
Ph_2AsI (M-173)			1/2-wave potential, (m)	467
$Ph(C_6H_{13})AsCl$	$RR'AsS$ + AcCl at 80–90° in a sealed tube	48	$b_{0.02}$ 106–108°, n_D^{20} 1.5339, d_4^{20} 1.2170, AR_{As} 11.76	1548
$(n\text{-}C_6H_{13})_2AsBr$	Et_2NAsCl_2 + RMgX in $Et_2O\text{-}C_6H_6$, (g)	53	b_2 125–126°, n_D^{22} 1.5025, d_4^{22} 1.2014	194
C_{13}				
$Ph(2\text{-}MeC_6H_4)AsCl$	$RR'AsO_2H$ + SO_2 in HCl		$b_{0.2}$ 94–96°, rxn. with RSH → $Ph(2\text{-}Me\text{-}C_6H_4)AsSR$	1184
$Ph(4\text{-}MeC_6H_4)AsCl$ (M-174)	$RR'AsO_2H$ + SO_2 in HCl		b_9 187–188°, rxn. with RSH → $Ph(4\text{-}Me\text{-}C_6H_4)AsSR$, b. 343.7°, (b)	1184
$Ph(4\text{-}MeOC_6H_4)AsCl$	$RR'AsO_2H$ + SO_2 in HCl		$b_{0.3}$ 78–80°, rxn. with RSH → $Ph(4\text{-}MeO\text{-}C_6H_4)AsSR$	1184

	Yield (%)	Preparation	Properties	Ref.
C$_{14}$				
(2-MeC$_6$H$_4$)$_2$AsBr	61	Et$_2$NAsCl$_2$ in C$_6$H$_6$ + RMgX in Et$_2$O-C$_6$H$_6$, (g)	m. 63°, b$_3$ 195-197°	194
(n-C$_7$H$_{15}$)$_2$AsBr	58	Et$_2$NAsCl$_2$ in C$_6$H$_6$ + RMgX in Et$_2$O-C$_6$H$_6$, (g)	b$_3$ 153°, n$_D^{22}$ 1.4990, d$_4^{22}$ 1.1492	194
Ph(n-C$_8$H$_{17}$)AsBr	69	RAsBr$_2$ + Et$_2$NH → [Ph(Et$_2$N)AsBr] → + R'MgX in Et$_2$O-C$_6$H$_6$, (g)	b$_1$ 174-176°, n$_D^{16}$ 1.5632, d$_4^{16}$ 1.3082	194
(n-C$_7$H$_{15}$)$_2$AsCl			Rxn. with KSC(S)OR → (C$_7$H$_{15}$)$_2$AsSC(S)OR	957
C$_{16}$				
Ph(2-C$_{10}$H$_7$)AsCl	22	RR'$_2$AsS + EtI at 100° in a sealed tube	Rxn. with RMgBr → Ph(2-C$_{10}$H$_7$)AsR	601
Ph(C$_{10}$H$_{21}$)AsI			b$_{0.0025}$ 124-127°, n$_D^{20}$ 1.5985, d$_4^{20}$ 1.4933	1549
(n-C$_8$H$_{17}$)$_2$AsBr	81	Et$_2$NAsCl$_2$ + RMgX in Et$_2$O-C$_6$H$_6$, (g)	b$_3$ 175-177°, n$_D^{20}$ 1.4965, d$_4^{20}$ 1.1270	194
C$_{18}$				
(n-C$_9$H$_{19}$)$_2$AsBr	92	Et$_2$NAsCl$_2$ + RMgX in Et$_2$O-C$_6$H$_6$, (g)	b$_4$ 196-198°, n$_D^{23}$ 1.4960, d$_4^{23}$ 1.1191	194
C$_{20}$				
(n-C$_{10}$H$_{21}$)$_2$AsCl	93	Et$_2$NAsCl$_2$ + RMgBr in Et$_2$O-C$_6$H$_6$, (f)	n$_D^{22}$ 1.4745, d$_4^{22}$ 1.4745	194
C$_{24}$				
(2-PhC$_6$H$_4$)$_2$AsCl (M-167)	100	(R$_2$As)$_2$O by boiling with concd. HCl	m. 110-112°, m. 115-116° (from EtOH), (n)	722

(i) Isolated by extraction with C$_6$H$_6$ (715). (j) Rxn. with NaI in Me$_2$CO → (2-BrC$_6$H$_4$)$_2$AsI (715). (k) Rxn. with RMgBr → (2-BrC$_6$H$_4$)$_2$AsR (715). (l) RSH → Ph(2,4-Cl$_2$C$_6$H$_3$)AsSR (1184). (m) Rxn. with Bu$_3$P in Et$_2$O → Ph$_2$AsAsPh$_2$ and Bu$_3$PI$_2$ (798). (n) Rxn. with Cl$_2$ in CCl$_4$ → (2-PhC$_6$H$_4$)$_2$AsCl$_3$ (722).

1.5.2 PRIMARY DIHALOARSINES

Besides the methods of synthesis summarized in the main volume (M-176 - M-177), primary dihaloarsines have recently been prepared from thioarsenoso compounds by a treatment with mercury(II) chloride and from arsenic trichloride and cyclohexane by neutron irradiation. The latter method affords mixtures of dichlorocyclohexyl- and chlorodicyclohexyl-arsine. The conversion of arsenic trichloride with acetylene to dichloro(2-chlorovinyl)arsine is catalyzed by aluminum chloride.

Some compounds of this group are described individually hereafter, while others are compiled in Table 6.

Dichloro(trifluoromethyl)arsine, $AsCCl_2F_3$, CF_3AsCl_2,   (M-177)
Synth.: From $(CF_3)_3As$ and $AsCl_3$ by heating at 210° for 20 hrs., a mixt. of the title compd. with $(CF_3)_2AsCl$ was obtd. (413).
Prop.: IR and Raman spectra (409).
Rxn. with: HCl in the presence of Hg $\rightarrow$ CF_3AsH_2 (413).
CF_3I + Hg $\rightarrow$ $(CF_3)_3As$, $(CF_3)_2AsCl$, and $[(CF_3)_2As-]_2$ (414).

Diiodo(trifluoromethyl)arsine, $AsCF_3I_2$, CF_3AsI_2,   (M-178)
Rxn. with: CF_4 $\rightarrow$ dec. (323).
Hg $\rightarrow$ $(CF_3As)_4$ (408).
Hg, followed by Hg + HI $\rightarrow$ $(CF_3As)_4$ and $(CF_3As)_5$ $\rightarrow$ CF_3AsH_2 (310).

Dibromomethylarsine, $AsCH_3Br_2$, $MeAsBr_2$,   (M-188)
Prop.: b. 180.4° calcd. by the Kinney equation (66). Dipole moment (349). Force const. calcn. (349b).
Rxn. with: $Ph_3P=NPh$ in Et_2O $\rightarrow$ $[Ph_3PN(Ph)As(Me)N(Ph)PPh_3]Br_2$ (1008).
$Ph_3P=NC_6H_4N=PPh_3$ in dioxan $\rightarrow$ $\{N(PPh_3)C_6H_4N(PPh_3)As(Me)\}_n \cdot 2nBr$ (1008).
$(RO)_3P$ + R'CHO $\rightarrow$ $MeAs[OCH(R')P(O)(OR)_2]_2$ (1274).

Dichloromethylarsine, $AsCH_3Cl_2$, $MeAsCl_2$,   (M-178)
Synth.: From As_2O_3 dissolved in aq. NaOH by adding MeOH followed by MeI, recovering the solid $MeAsO(ONa)_2$, redissolving in H_2O, acidifying with H_2SO_4 to pH 3, adding excess concd. HCl, and reducing with SO_2, 32% yield (1659).
Prop.: Oily liquid with extremely unpleasant odor (1659). b. 126.3° calcd. by the Kinney equation (66). Dipole moment (349). Force const. calcn. (349b). Indirect chronopotentiometric detn. by inhibition of xanthine oxidase (894). IR spectrum, qual. interpretation (1223). NMR spectrum (798).
Rxn. with: RMgX $\rightarrow$ R_2AsMe (425).
RLi $\rightarrow$ R_2AsMe (428).
Bu_3P $\rightarrow$ $[Bu_3PAs(Me)Cl]Cl$ (798).
$2-HOC_6H_4CONHNHCS_2K$ $\rightarrow$ $(2-HOC_6H_4CONHNHCS_2)_2AsMe$ (876).
$LiN(Me)SiMe_3$ $\rightarrow$ $MeAs[N(Me)SiMe_3]_2$ (1343).
$\overline{SCH_2CH_2S}AsSCH_2CH_2S\overline{AsSCH_2CH_2S}$ $\rightarrow$ $Me\overline{AsSCH_2CH_2S}$ (1438).
$HSCH_2CH_2NH_2$ in the presence of Et_3N $\rightarrow$ $Me\overline{AsNHCH_2CH_2S}$ along with $\overline{SCH_2CH_2N}As(Me)As$-$(Me)\overline{NCH_2CH_2S}AsMe$ (1438).

Biol. Prop.: Toxicity to barley (1122).

Coordination Deriv.: A formula is given for the calcn. of the solvent effect on the wave number of the CO vibration applicable to the following complexes: [M(CO)$_5$(MeAsCl$_2$)], where M = Cr, W, Mo; [Fe(CO)$_4$(MeAsCl$_2$)], [π-C$_5$H$_5$V(CO)$_3$-(MeAsCl$_2$)], [π-C$_5$H$_5$Mn(CO)$_2$(MeAsCl$_2$)], {π-[(AcO)$_2$C$_6$H$_4$]Cr(CO)$_2$(MeAsCl$_2$)}, [π-C$_6$H$_6$Cr(CO)$_2$(MeAsCl$_2$)], [π-(1,3,5-Me$_3$C$_6$H$_3$)Cr(CO)$_2$(MeAsCl$_2$)], [π-(Me$_6$C$_6$)Cr-(CO)$_2$(MeAsCl$_2$)], and [Ni(CO)$_3$(MeAsCl$_2$)] (1459).

Difluoromethylarsine, AsCH$_3$F$_2$, MeAsF$_2$, (M-188)
Prop.: b. 70.7° calcd. by the Kinney equation (66). Dipole moment (349). Torsional energy level data (463). Polarizability data from the σ-function potential model (973).

Diidomethylarsine, AsCH$_3$I$_2$, MeAsI$_2$, (M-179)
Synth.: From MeAsO(ONa)$_2$ and KI (1:2) in aq. HCl by passing SO$_2$ and acidifying with HCl, 60% yield (349).
Prop.: m. 30°, dipole moment (349). NMR spectrum (798). The crystal and mol. structures by x-ray diffraction (287).
Rxn. with: Bu$_3$P → [Bu$_3$PAs(Me)I]I (798).
HSCRR′CR″R‴SH → $\overline{SCRR'CR''R'''S}$AsMe (1608).
HOCRR′CR″R‴OH → $\overline{OCRR'CR''R'''O}$AsMe (1608).

Dichloro(2-chlorovinyl)arsine, Lewisite, AsC$_2$H$_2$Cl$_3$, ClCH=CHAsCl$_2$, (M-179).
Synth.: From AsCl$_3$ and C$_2$H$_2$ in the presence of AlCl$_3$ at 40-60°, followed by hydrolysis with 20% HCl, 17.5% yield (850).
Prop.: Detection of Lewisite with paper impregnated with 0.01% fluorescein, 0.1% fuchsin, 0.25% Au chloride, or 0.5% mercurochrome (850).
Biol. Prop.: Toxicity to mice and its inhibition by thiol agents (837). Inhibition of xanthine oxidase, an indirect chronopotentiometric detn. of Lewisite (894).

Dichloroethylarsine, AsC$_2$H$_5$Cl$_2$, EtAsCl$_2$, (M-180)
Prop.: b. 150.8° calcd. by the Kinney equation (66).
Rxn. with: NaSP(S)(OR)$_2$ in C$_6$H$_6$ → EtAs[SP(S)(OR)$_2$]$_2$ (315).
Me$_2$C=NOH in the presence of Py → EtAs(ON=CMe$_2$)$_2$ + Py·HCl (826).
Biol. Prop.: Inhibition of cholinesterase from human plasma (307). Inhibition of xanthine oxidase, an indirect chronopotentiometric detn. of EtAsCl$_2$ (894).

Dichloro(pentafluorophenyl)arsine, AsC$_6$Cl$_2$F$_5$, C$_6$F$_5$AsCl$_2$
Synth.: From AsCl$_3$ by adding dropwise C$_6$F$_5$MgCl (4:1) in Et$_2$O and heating under reflux, 66% yield (651, see also 650).
From (C$_6$F$_5$AsS)$_4$ and HgCl$_2$ in Et$_2$O under reflux (651).
Prop.: b$_{0.1}$ 52° (650, 651). IR and NMR spectra (651).
Rxn. with: Hg → (C$_6$F$_5$As)$_4$ (650, 651).
H$_2$O → (C$_6$F$_5$AsO)$_n$ (651).

Aq. NaOH → C_6F_5H (651).

Ag_2S in C_6H_6 under reflux → $(C_6F_5AsS)_4$ (651).

(o-Bromophenyl)dichloroarsine, $AsC_6H_4BrCl_2$, $2\text{-}BrC_6H_4AsCl_2$,   (M-191)

Synth.:   From $2\text{-}BrC_6H_4NH_2$ in $AcOH\text{-}EtCO_2H\text{-}H_2SO_4$ mixt. contg. $AsCl_3$ at 0-5° by diazotization ($NaNO_2$), followed by adding Cu_2Br_2, heating on a steam bath and redn. with SO_2 in H_2SO_4 (1151) or 10 N HCl contg. KI (292), 52% yield (1151). From $2\text{-}BrC_6H_4AsO_3H_2$ by redn. with SO_2 in concd. HCl contg. KI, 61-64% yield (1659).

Prop.: $b_{0.6}$ 94-96° (292), $b_{<1}$ 120-130° (1151), $b_{0.1}$ 125-130°, m. 63-64° (715), b_2 138-142°, crystd. during cooling (1659).

Rxn. with: RMgX (1:2) → $2\text{-}BrC_6H_4AsR_2$ (1151, 1659).

$2\text{-}BrC_6H_4AsO$ at 260-270° → $(2\text{-}BrC_6H_4)_2AsCl$ (715).

Dichlorophenylarsine, $AsC_6H_5Cl_2$, $PhAsCl_2$,   (M-182)

Prop.:   b. 248.1° calcd. by the Kinney equation (66).   Half-wave potential (467).   IR spectrum, qual. interpretation (1223).

Rxn. with: RSNa → $PhAs(SR)_2$ (316).

RMgBr → $PhAsR_2$ (609, 752, 822, 828, 1549).

RLi (1:2) → $PhAsR_2$ (842).

Li in THF at -30 to 10° → $PhAsLi_2$ (1514).

Phenylenedilithium → $\{C_6H_4As(Ph)\}_n$ (739).

$LiSnR_3$ → $PhAs(SnR_3)_2$ (1371).

Hexamethylcyclotrisilthiane (3:1) → $(PhAsS)_4$ (20).

2,2′-Dilithio-4,5,4′,5′-bis(methylenedioxy)biphenyl → 2,3,7,8-bis(methylene-dioxy)-5-phenyldibenzarsole (434).

Bu_3P → $[Bu_3PAs(Ph)Cl]Cl$ (798).

FcH + $AlCl_3$, no ferrocenyl-As deriv. is formed (1436).

$Ph(CH_2)_3Br$ in aq. NaOH soln. under reflux → $Ph(PhCH_2CH_2CH_2)AsO_2H$ (1513).

$Me_2C=NOH$ + Py → $PhAs(ON=CMe_2)_2$ + $Py\cdot HCl$ (826).

$\overline{SCH_2CH_2SAsSCH_2CH_2SAsSCH_2CH_2S}$ → $Ph\overline{AsSCH_2CH_2S}$ (1438).

$[Fe_2(CO)_6(PhC{\equiv}CPh)_2]$ → pentaphenylarsole (767).

$[Fe_2(CO)_5(PPh_3)(PhC{\equiv}CPh)_2]$ → pentaphenylarsole (767).

Biol. Prop.:   Toxicity to barley (1122).   Inhibition of xanthine oxidase, an indirect chronopotentiometric method for the detn. of $PhAsCl_2$ (894).

Diiodophenylarsine, $AsC_6H_5I_2$, $PhAsI_2$,   (M-184)

Rxn. with: RMgX → $PhAsR_2$ (164).

Bu_3P → $[Bu_3PAs(Ph)I]I$ (798).

$NCN=C(SK)_2$ in DMF → $Ph\overline{AsSC(=NCN)S}$ (1089).

$NCC(SNa)=C(SNa)CN$ in $CH_2(OMe)_2$ → $Ph\overline{AsSC(CN)=C(CN)S}$ (1090).

p-$(LiC{\equiv}C)C_6H_4C{\equiv}CLi$ in Et_2O under reflux, followed by oxidative coupling in Py in the presence of $CuCl_2$ → a polymer sol. in $Cl_2CHCHCl_2$, $PhNO_2$, and PhCl (613).

PhBr + $HC{\equiv}CC_6H_4C{\equiv}CH$ + Li in Et_2O under reflux, followed by oxidative coupling in Py in the presence of Cu_2Cl_2 → a polymer which formed a flexible film (613).

2-Amino-4-(dichloroarsino)phenol, $AsC_6H_6Cl_2NO$, $3,4-H_2N(HO)C_6H_3AsCl_2$, (M-184)

Biol. Prop.: Philaricidal effect on Setaria digitata in vitro (875). In a combination with $MeSO_3Et$ a synergistic effect on the chromosome aberrations of Hordeum sativum (1092, 1094).

Dichloro-p-tolylarsine, $AsC_7H_7Cl_2$, $4-MeC_6H_4AsCl_2$, (M-185)

Prop.: b. 265.8° calcd. by the Kinney equation (66). IR spectrum, qual. interpretation (1223).

Rxn. with: ROH in the presence of Py → $4-MeC_6H_4As(OR)_2$ and Py·HCl (819). RSH → $2-MeC_6H_4As(SR)_2$ (1184).

Dichloroferrocenylarsine, $AsFeC_{10}H_9Cl_2$, $FcAsCl_2$

Synth.: From $(FcAsO)_2$ by titrating with concd. HCl, filtering off the solid, extracting with hot C_7H_{16}, and evapg. to dryness, 95% yield (1436, 1437). From $(FcAsCl)_2O$ or from $(FcAsO)_n$ by the same procedure as described above (1436, 1437).

Prop.: Yellow powder or golden leaflets, m. 63.5-64°, IR spectrum. The compd. turned brown in storage under N. The brown prod. was insol. in C_7H_{14} (1436).

Rxn. with: Atm. moisture or with H_2O → $(FcAsO)_2$ (1436, 1437).
Aq. alkali followed by neutralization → $(FcAsO)_2$ (1436).
Air during evapn. of a hot C_6H_6 or petr. ether soln. → $(FcAsCl)_2O$ (1436, 1437).
Zn + HCl in 95% EtOH → $(FcAs\leftrightharpoons)_n$ (1436).

1,2-Bis[o-(dichloroarsino)phenoxy]ethane, $As_2C_{14}H_{12}Cl_4O_2$, $[2-(Cl_2As)C_6H_4OCH_2-]_2$

Synth.: From $[2-(H_2O_3As)C_6H_4OCH_2-]_2$ by a treatment with SO_2 in 10 N HCl contg. KI (292).

Prop.: Clear oil (292).

Rxn. with: PhMgBr (1:4) → $[2-(Ph_2As)C_6H_4OCH_2-]_2$ (292).

Table 6. Primary Dihaloarsines

RAsX₂	Prepd. from	Yield %	Prop., Rxn., and Remarks	Ref.
C₂				
$HCF_2CF_2AsCl_2$	$RAsH_2$ in petr. ether + Cl at $-20°$	70	b_{747} 106-108°, d_4^{20} 1.5051, n_D^{20} 1.4537	1266
$EtAsBr_2$ (M-189)			b. 201.5°, (a)	66
$EtAsF_2$ (M-189)			b. 99.5°, (a)	66
$EtAsI_2$ (M-189)			IR spectrum interpretation, (b)	1223
C₃				
$PrAsCl_2$ (M-189)			b. 173.7°, (a, c)	66
$i\text{-}PrAsCl_2$ (M-190)			b. 167.7°, (a)	66
C₄				
$CF_3CCl=C(CF_3)AsBr_2$	$RAsMe_2 + Br_2$ (1:2) in CCl_4		Colorless ppt.	415
	$RMeAsBr + Br_2$ (1:1) in CCl_4			415
$BuAsCl_2$ (M-190)			b. 194.1°, (a, d)	66
$s\text{-}BuAsCl_2$			IR spectrum interpretation	1223
$t\text{-}BuAsCl_2$	$AsCl_3$ in Et_2O + $RMgCl$ in Et_2O at -30 to -35° in the dark	52	b. 189.5°, (a)	66
$BuAsI_2$ (M-190)			Crystals sol. in common org. solvents, (e)	1530
			IR spectrum interpretation	1223
C₅				
$n\text{-}C_5H_{11}AsCl_2$ (M-190)			b. 215.5°, (a)	66
C₆				
$C_6F_5AsBr_2$	$RHgMe + AsBr_3$		The prod. contaminated with MeHgBr	322a
$2,4\text{-}Cl_2C_6H_3AsCl_2$ (M-191)			Rxn. with RSH → $2,4\text{-}Cl_2C_6H_3As(SR)_2$	1184
$4\text{-}O_2NC_6H_4AsCl_2$ (M-182)			Rxn. with NaSR → $2\text{-}O_2NC_6H_4As(SR)_2$	316
$2\text{-}ClC_6H_4AsCl_2$ (M-192)			IR spectrum interpretation	1223

3-ClC$_6$H$_4$AsCl$_2$	IR spectrum interpretation	1223
PhAsBr$_2$ (M-192)	b. 288.0°, (a)	66
C$_6$H$_{11}$AsCl$_2$ (M-193)	The prod. contd. C$_6$H$_{11}$AsCl$_2$ along with (C$_6$H$_{11}$)$_2$AsCl	1286
AsCl$_3$ + cyclohexane by neutron irradiation		
C$_7$		
2-MeC$_6$H$_4$AsCl$_2$ (M-194)	b. 265.8°, (a)	66
3-MeC$_6$H$_4$AsCl$_2$	b. 265.8°, (a)	66
	IR spectrum, qual. interpretation	1223
C$_8$		
2,4-Me$_2$C$_6$H$_3$AsCl$_2$ (M-195)	b. 282.9°, (a)	66
2,5-Me$_2$C$_6$H$_3$AsCl$_2$	b. 282.9°, (a)	66

(a) Calcd. by the Kinney equation. (b) Rxn. with 2,2'-biphenylylenedilithium → 5-ethyldibenzarsole (57). (c) Rxn. with Me$_2$C=NOH + Py → PrAs(ON=CMe$_2$)$_2$ + Py·HCl (826). (d) Rxn. with Me$_2$C=NOH + Py → BuAs(ON=CMe$_2$)$_2$ + Py·HCl (825, 826). (e) Rxn. with LiAlH$_4$ in Et$_2$O → t-BuAsH$_2$ (1530).

1.5.5 SECONDARY PSEUDOHALOARSINES, R_2AsX

Table 7. Pseudohaloarsines, R_2AsX

R_2AsX*	Prop. and Rxn.	Ref.
C_3		
Me_2AsCN (M-202)	b. 161.5°, (a)	66
	Crystal and mol. structures by x-ray diffraction	287
Me_2AsSCN	Rxn. with $(RO)_3P \rightarrow Me_2AsOP(OR)_2$	871
	Rxn. with $\overline{CH_2CH_2O} \rightarrow Me_2AsOCH_2CH_2SCN$	872
	Rxn. with $\overline{CH_2CH_2S} \rightarrow Me_2AsSCH_2CH_2SCN$	873
C_5		
Et_2AsSCN	Rxn. with $\overline{CH_2CH_2S} \rightarrow Et_2AsSCH_2CH_2SCN$	872
C_9		
Bu_2AsSCN	Rxn. with $\overline{CH_2CH_2O} \rightarrow Bu_2AsOCH_2CH_2SCN$	872
C_{13}		
Ph_2AsSCN	Rxn. with $\overline{CH_2CH_2O} \rightarrow Ph_2AsOCH_2CH_2SCN$	872

* For the methods of synthesis see the main volume (M-201). (a) Calcd. by the Kinney equation.

1.5.6 PRIMARY DIPSEUDOHALOARSINES, $RAsX_2$

This type of arsenic derivatives was described in the main volume on page 206.

Dicyanomethylarsine, $AsC_3H_3N_2$, $MeAs(CN)_2$
Synth.: From $MeAsCl_2$ and $AgCN$ in C_6H_6 under reflux (1349).
Prop.: Flat needles, m. 126-127.5°, IR spectrum, crystal structure data (1349).

1.6 COMPOUNDS OF ARSENIC BONDED TO GROUP VIA ELEMENTS

1.6.1 COMPOUNDS WITH ARSENIC-OXYGEN BONDS

1.6.1.1 ARSINOUS ACIDS

For the nomenclature and methods of synthesis of the so-called arsinous acids see the main volume, page 206.

Hydroxyphenylarsinoacetic Acid, $AsC_8H_9O_3$, $PhAs(OH)CH_2CO_2H$
Rxn. with: $BuOH$ in C_6H_6 under reflux $\rightarrow MePhAs(O)OBu$ (958).

1.6.1.2 ARSINOUS ACID ESTERS

The esters, R_2AsOR' and $RR'AsOR''$, and their methods of preparation employed through 1964 were described in the main volume, pp. 208-211. Since then a new type of arsinous acid esters has been prepared from secondary haloarsines by a condensation with oximes in the presence of a tertiary amine or pyridine as a dehydrohalogenating agent. The oxime esters undergo transesterification

$$R_2AsX + HON=CR'_2 + Base \rightarrow R_2AsON=CR'_2 + Base \cdot HCl$$

when heated with ketones boiling at a higher temperature than that formed by the displacement of the alkylidene moiety. Likewise, the alkylidenamino-oxy moiety can be replaced by an alkoxy group through heating the oxime esters with a higher boiling alcohol.

2-Thiocyanatoethoxyarsines have been prepared from secondary thiocyanato-arsines by heating with ethylene oxide.

Methoxydimethylarsine, AsC_3H_9O, Me_2AsOMe, (M-209)
Coordination Deriv.: cis-[Pt(Me$_2$AsOMe)(PPh$_3$)Cl$_2$], prepd. from cis-[Pt(Me$_2$AsCl)-(PPh$_3$)Cl$_2$] + NaOMe in MeOH at 20°. IR spectrum (327).

α-[(Dimethylarsino)oxy]furfurylphosphonic Acid Diethyl Ester,
$AsC_{11}H_{20}O_5P$,
Synth.: From Me$_2$AsBr by adding to a mixt. of (EtO)$_3$P
and furfuraldehyde at 10° and then warming to 60° (1274).
Prop.: Colorless liquid, b$_{0.15}$ 132-136° (1274).

$Me_2AsOCHP(O)(OEt)_2$

Methoxydiphenylarsine, $AsC_{13}H_{13}O$, Ph_2AsOMe, (M-211)
Rxn. with: RSH $\rightarrow$ Ph$_2$AsSR (317).
Coordination Deriv.: cis-[Pt(Ph$_2$AsOMe)(PEt$_3$)Cl$_2$], prepd. from cis-[Pt(Ph$_2$AsCl)-(PEt$_3$)Cl$_2$] and NaOMe in MeOH at 20°. IR spectrum (327).

Table 8. Arsinous Acid Esters

R_2AsOR' or $RR'AsOR''$	Prepd. from	Yield %	Prop. Rxn., and Remarks	Ref.
C_3				
$(CF_3)_2AsOMe$	$R_2AsCl + MeOH + Me_3N$ at $-30°$ to $+24°$ or at $70°$	85	b. $70.7°$ (calcd.), vapor pressure data, mol. wt. detn. in the vapor phase, (a, b, c)	266
C_5				
$Me_2AsOCH_2CH_2SCN$	$R_2AsSCN + \overline{CH_2CH_2O}$ at 40-$50°$	80	b_3 $86°$, d_4^{20} 1.3738, n_D^{20} 1.5345, (d, e)	872
$Me_2AsOSiMe_3$ (M-209)			Rxn. with $Me_3P{=}CHSiMe_3$ in $Et_2O \rightarrow Me_3P{=}CHAsMe_2$	1355
C_6				
$(CF_3)_2AsO$-t-Bu	$R_2AsCl + t$-$BuOH$ (1:1) $+ Me_3N$ at $-20°$ to $+20°$	95	b. $116.1°$ (calcd.), vapor pressure data, IR spectrum, (c, f)	266
C_7				
$Et_2AsON{=}CMe_2$	$R_2AsBr + HON{=}CMe_2$ in $Et_2O + Et_3N$, (h)	40	b_{10} $70°$, d_4^{20} 1.1447, n_D^{20} 1.4847, (g)	827
C_8				
$Et_2AsON{=}CEtMe$	$R_2AsBr + HON{=}CEtMe$ in $Et_2O + Et_3N$, (h)	52	b_{15} $91°$, d_4^{20} 1.1233, n_D^{20} 1.4839, (i)	827
$Me_2AsOCH(Me)P(O)(OEt)_2$	$R_2AsCl + (EtO)_3P + MeCHO$			1274
C_9				
$Et_2AsOCH_2C(CH_2O)_3As$			IR spectrum	1387
$Pr_2AsON{=}CMe_2$	$R_2AsBr + HON{=}CMe_2$ in $Et_2O + Et_3N$, (h)	32	b_8 $92°$, d_4^{20} 1.0901, n_D^{20} 1.4810	827
	$R_2AsCl + HON{=}CMe_2$ in $Et_2O + Py$ at 15-$20°$	26	b_{20} 120-$122°$, d_4^{20} 1.1601, n_D^{20} 1.4610, MR_D 62.60	825
C_{10}				
$Et_2AsON{=}CBuMe$	$R_2AsBr + HON{=}CBuMe$ in $Et_2O + Et_3N$, (h)	46	b_{11} $113°$, b_{17} 125-$126°$, d_4^{20} 1.0795, n_D^{20} 1.4814	827
	$R_2AsON{=}CMe_2 + MeCOBu$ in CO_2 at $80°$	57		827
$Pr_2AsON{=}CEtMe$	$R_2AsBr + HON{=}CEtMe$ in $Et_2O + Et_3N$, (h)	64	b_{13} $109°$, d_4^{20} 1.0747, n_D^{20} 1.4800, (j)	827

(Et$_2$AsOCH$_2$-)$_2$	R$_2$AsON=CEtMe + HOCH$_2$CH$_2$OH at 80°	24	b$_{18}$ 148-150°, d$_4^{20}$ 1.2732, n$_D^{20}$ 1.5038	827
C$_{11}$				
EtPhAsOCH$_2$CH=CH$_2$	(RR′As)$_2$O + R″OH in the presence of CuSO$_4$ on a steam bath	87	b$_9$ 114.5°, d$_4^{20}$ 1.1883, n$_D^{20}$ 1.5468, MR$_D$ 63.20, AR$_{As}$ 11.53, (k)	822
Bu$_2$AsOCH$_2$CH$_2$SCN	R$_2$AsSCN + $\overline{\text{CH}_2\text{CH}_2\text{O}}$ at 40-50° in a sealed tube	77	b$_3$ 125-127°, d$_4^{20}$ 1.1312, n$_D^{20}$ 1.5172, (e)	872
Et$_2$AsON=CPr$_2$	R$_2$AsBr + HON=CPr$_2$ in Et$_2$O + Et$_3$N, (h)	66	b$_8$ 112°, d$_4^{20}$ 1.0656, n$_D^{20}$ 1.4788, (1)	827
C$_{12}$				
Pr$_2$AsON=CBuMe	R$_2$AsBr + HON=CBuMe in Et$_2$O + Et$_3$N, (h)	38	b$_{13}$ 140°, d$_4^{20}$ 1.0426, n$_D^{20}$ 1.4779, (m)	827
C$_{13}$				
Pr$_2$AsON=CPr$_2$	R$_2$AsBr + HON=CPr$_2$ in Et$_2$O + Et$_3$N, (h)	28	b$_5$ 129°, d$_4^{20}$ 1.0275, n$_D^{20}$ 1.4763	827
C$_{14}$				
Ph$_2$AsOEt (M-211)	Rxn. with RSH → Ph$_2$AsSR			318
Pr$_2$AsOC$_8$H$_{17}$	R$_2$AsON=CBuMe + n-C$_8$H$_{17}$OH in CO$_2$ at 80°	24	b$_{15}$ 160-161°, d$_4^{20}$ 0.9941, n$_D^{20}$ 1.4639	827
C$_{15}$				
Ph$_2$AsOCH$_2$CH$_2$SCN	R$_2$AsSCN + $\overline{\text{CH}_2\text{CH}_2\text{O}}$ at 40-50° in a sealed tube	97	b$_1$ 200-225°, d$_4^{20}$ 1.3667, n$_D^{20}$ 1.6325, (e)	872
C$_{16}$				
Pr$_2$AsOCH$_2$CH$_2$CH(Me)OAsPr$_2$	R$_2$AsON=CEtMe + HOCH$_2$CH$_2$CH(Me)OH at 80°	27	b$_{23}$ 194-195°, d$_4^{20}$ 1.1455, n$_D^{20}$ 1.4894	827
C$_{17}$				
Ph$_2$AsOCH$_2$C(CH$_2$O)$_3$As	IR spectrum			1387

(a) IR spectrum (266, 326). (b) Rxn. with HCl (1:2) → (CF$_3$)$_2$AsCl (266). (c) Rxn. with BF$_3$ → (CF$_3$)$_2$AsF (266). (d) Rxn. with RCOCl → Me$_2$AsCl + RCO$_2$CH$_2$CH$_2$SCN (874). (e) Useful as a plasticizer and corrosion inhibitor (872). (f) Rxn. with MeI at 85°, no Arbuzov rearrangement (266). (g) Rxn. with MeCOBu at 80° → Et$_2$AsON=CMeBu (827). (h) At 0-5° and completed by heating (827). (i) Rxn. with HOCH$_2$CH$_2$OH at 80° → (Et$_2$AsOCH$_2$-)$_2$ + EtMeC=NOH (827). (j) Rxn. with HOCH$_2$CH$_2$CH(Me)OH → Pr$_2$AsOCH$_2$CH$_2$CH(Me)OAsPr$_2$ (827). (k) On heating with Me methacrylate (1:1 and 1:4) at 180° and 240°, resp., in the presence of Bz$_2$O$_2$, copolymers were obtd. (608). (l) Rxn. with AcCl at 70° → Et$_2$AsCl + Pr$_2$C=NOAc (827). (m) Rxn. with n-C$_8$H$_{17}$OH at 80° → Pr$_2$AsOC$_8$H$_{17}$ + BuMeC=NOH (827).

1.6.1.3 ARSINOUS ACID ANHYDRIDES WITH CARBOXYLIC AND PHOSPHORUS ACIDS

The main volume contains the anhydrides with carboxylic acids only, which appeared on page 212.

(Fluoromethylphosphinyloxy)dimethylarsine, $AsC_3H_9FO_2P$, $Me_2AsOP(O)(F)Me$
Synth.: By-prod. from the rxn. of $(Me_2As)_2O$ with $MeP(O)F_2$ (620).
Prop.: b_4 46-49°, n_D^{20} 1.4562, d_4^{20} 1.5138 (620).

Dimethyl(trifluoroacetoxy)arsine, $AsC_4H_6F_3O_2$, $Me_2AsO_2CCF_3$, (M-212)
Synth.: From $Me_2AsAsMe_2$ by treating with $(CF_3CO)_2O$ at 20°, 78% yield (427).
Prop.: b. 136° (427).

1.6.1.4 OXYBISARSINES, $(R_2As)_2O$ AND $(RR'As)_2O$

These arsenic derivatives and their methods of preparation were described in the main volume, pp. 212-218. Besides the methods described there, oxybis-arsines have been prepared from diarsines by anodic oxidation and from arsenic trioxide by the reaction with organolithium compounds. While numerous oxybis-arsines were prepared from secondary haloarsines by alkaline hydrolysis and by metathesis with mercuric oxide, oxybis[bis(pentafluorophenyl)arsine] was pre-pared in a high yield from chlorobis(pentafluorophenyl)arsine by shaking with water.

Oxybis[bis(trifluoromethyl)arsine], $As_2C_4F_{12}O$, $[(CF_3)_2As]_2O$, (M-216)
Rxn. with: Dry HCl → $(CF_3)_2AsCl$ (266).
$(CF_3)_2PCl$ (1:1) at 25° → $(CF_3)_2AsCl$ and $[(CF_3)_2As-]_2$ (1416).
$(CF_3)_2PCl$ (excess) at 25° → $(CF_3)_2AsCl$ and $[(CF_3)_2P]_2O$ (1416).

Oxybis(dimethylarsine), Cacodyl Oxide, $As_2C_4H_{12}O$, $(Me_2As)_2O$, (M-213)
Rxn. with: $CF_3C\equiv CH$ at 20° → $Me_2AsC\equiv CCF_3$ (425).
$MeP(O)F_2$ → Me_2AsF and $Me_2AsOP(O)(F)Me$ (620).

Oxybis(ethylphenylarsine), $As_2C_{16}H_{20}O$, $(EtPhAs)_2O$, (M-216)
Rxn. with: ROH → $EtPhAsOR$ (822).

Oxybis[bis(pentafluorophenyl)arsine], $As_2C_{24}F_{20}O$, $[(C_6F_5)_2As]_2O$
Synth.: From $(C_6F_5)_2AsCl$ by shaking with H_2O, 95% yield (651).
Prop.: White ppt., m. 108-108.5°, IR and NMR spectra (651).

Oxybis[(p-chlorophenyl)phenylarsine], $As_2C_{24}H_{18}Cl_2O$, $[Ph(4-ClC_6H_4)As]_2O$, (M-217)
Biol. Prop.: Highly active against the germination of Cochliobolus miyabeanus spores (1493).

Oxybis(diphenylarsine), $As_2C_{24}H_{20}O$, $(Ph_2As)_2O$, (M-214)
Synth.: From $Ph_2AsAsPh_2$ by anodic oxidn. (1029). Sepn. from mixts. by paper chromatography and paper electrophoresis (1422).

Prop.: Half-wave potential (467).
Rxn. with: $RCO_2H + HOCH_2CH_2SH \rightarrow Ph_2AsSCH_2CH_2O_2CR$ (191).
Polarographic redn. in 0.1 molar $NaClO_4$ in MeOH in 2 steps $\rightarrow Ph_2As^-$ (1029).
Electrolytic redn. (2e$^-$) $\rightarrow Ph_2As^-$ and Ph_2AsO^- (467).
Biol. Prop.: Fungistatic and fungicidal effect against Alternaria radicina, Bortrytis cinerea, Fusicladium dendriticum, Trichophyton gypseum, and Epidermophyton; lethal effect on mice (1132, 1133).

Oxybis(di-2-biphenylylarsine), $As_2C_{48}H_{36}O$, $[(2\text{-}PhC_6H_4)_2As]_2O$
Synth.: From $2\text{-}PhC_6H_4Li + As_2O_3$ in Et_2O-petr. ether mixt., 80% yield (722).
Prop.: Colorless, voluminous ppt., m. 140-141° (722).
Rxn. with: Concd. HCl $\rightarrow (2\text{-}PhC_6H_4)_2AsCl$ (722).

1.6.1.5 ARSONOUS ACIDS AND MONOTHIO ANALOGS

The compounds of this class were described in the main volume, pp. 219-226. In the period covered by this supplement, only one arsonous acid was tested for its biological activity.

Dihydroxy(p-nitrophenyl)arsine, $AsC_6H_6NO_4$, $4\text{-}O_2NC_6H_4As(OH)_2$, (M-221)
Biol. Prop.: Effective against the growth of Xanthomonas oryzae (1493).

1.6.1.6 ARSENOSO COMPOUNDS

Arsenoso derivatives have been prepared by the methods reported in the main volume, pp. 226-227. Arsenosoferrocene dimer and polymers were prepared by a condensation of arsenic trichloride with ferrocene in the presence of aluminum chloride, followed by hydrolysis.

Several arsenoso compounds have been tested for their biological properties. Some of them have shown fungistatic, fungicidal, antiviral, antitumor, and enzyme-inhibiting activity. However, like other arsenicals, arsenoso compounds are potential carcinogens.

Besides the arsenoso compounds described individually hereafter, several compounds of this class are compiled in Table 9.

(2-Chlorovinyl)oxoarsine, AsC_2H_2ClO, $ClCH=CHAsO$, (M-228)
Biol. Prop.: Inhibition of uridine diphosphate transglucuronylase activity of a mouse-liver homogenate in the synthesis of o-aminophenol and p-nitrophenol glucuronides (1458).

p-Phenylenebis(oxoarsine), $As_2C_6H_4O_2$, $4\text{-}OAsC_6H_4AsO$, (M-236)
Synth.: From p-arsanilic acid by the Bart reaction, followed by a redn. of the diarsonic acid with SO_2 in HCl contg. KI (739)
Rxn. with: $p\text{-}C_6H_4(CH_2Br)_2$ + NaOH in aq. MeOH $\rightarrow \{C_6H_4As(O)(OH)CH_2C_6H_4CH_2As(O)$-$(OH)\}_n$ (739).
$BrCH_2CH_2Br$ + NaOH $\rightarrow \{C_6H_4As(O)(OH)CH_2CH_2As(O)(OH)\}_n$ (739).

Oxophenylarsine, Arsenosobenzene, $(AsC_6H_5O)_4$, $(PhAsO)_4$, (M-228)

Prop.: IR spectrum, qual. interpretation (1223). Sepn. from mixtures by paper chromatography and paper electrophoresis (1422).

Rxn. with: ROH in C_6H_6 in the presence of acid or basic catalysts, with azeotropic removal of $H_2O \rightarrow PhAs(OR)_2$ (741).

Pyrocatechol in C_6H_6 in the presence of acid or basic catalysts, with azeotropic removal of $H_2O \rightarrow Ph\overline{AsOC_6H_4O}$ (741).

2,3-Dihydroxynaphthalene in C_6H_6 in the presence of acid or basic catalysts, with azeotropic removal of $H_2O \rightarrow Ph\overline{AsOC_{10}H_6O}$ (741).

$RSH \rightarrow PhAs(SR)_2$ (596a).

Me N-acetylcysteinate $\rightarrow$ binding of the SH group (1639).

Biol. Prop.: Inhibition of a pyruvate oxidn. in suspensions of Aerobacter aerogenes cells and of mitochondria cells (89). Inhibition of the synth. of 2-aminophenol and 4-nitrophenol glucuronides by the uridine diphosphate transglucoronylase of mouse-liver homogenates (1458). Fungistatic and fungicidal effect on Alternaria radicina, Helmintosporium sativum, Fusicladium dendriticum, Trichophyton gypseum, and Trichophyton rubrum; toxic to mice - minimum effective doses were detd. (1133). Effective against nematodes in soil (1378). Effective against Alternaria solani spores and against Botrylis cinerea spores and micelia (645). Inactivation tests with Adenovirus type gave a negative result (1639).

Uses: Feed additive, dosage (1088). A fungicide and bactericide for polyvinyl resins used in plasticizers (1635).

p-Arsenosoaniline, AsC_6H_6NO, $4-OAsC_6H_4NH_2$, (M-229)

Synth.: From $4-H_2NC_6H_4AsO_3H_2$ by reducing with $PhNHNH_2$ (441).

From $4-H_2NC_6H_4AsO_3H_2$ in EtOAc by treating with PCl_3, pptg. with Et_2O, dissolving the ppt. in aq. NaOH and satg. the soln. with NH_4Cl, 98% yield (441). From $4-[(HOCH_2CH_2)_2N]C_6H_4AsO_3H_2$ in $CHCl_3$ by reducing with $SOCl_2$ at r.t., removing the solvent and other volatile compds., dissolving the residue in 95% EtOH, passing through Amberlite Type I (OH form), eluting with 95% EtOH, concg., and treating with H_2O and Me_2CO (441).

Prop.: White ppt., m. 80-85°, IR spectrum (441), m. 105° after softening at 70° (441). Identification of the ^{74}As-labeled compd. by paper chromatographic and electrophoretic methods (1185).

Rxn. with: $HSCH_2CH_2SH$ in Et_2O under reflux $\rightarrow 4-H_2NC_6H_4\overline{AsSCH_2CH_2S}$ (122, 441).

2-Amino-4-arsenosophenol, Oxophenarsine, $AsC_6H_6NO_2$, $3,4-H_2N(HO)C_6H_3AsO$, (M-229)

Biol. Prop.: The sensitivity of human cancers to the arsenoso compd. (864). A tumor regression induced by the arsenoso deriv. alone and in combination with other anticancer drugs (865).

2-Amino-4-arsenosophenol Hydrochloride, Oxophenarsine Hydrochloride, Mapharsen, $AsC_6H_7ClNO_2$, $2,5-HO(OAs)C_6H_3NH_2 \cdot HCl$, (M-230)

Biol. Prop.: In combination with dimethylcysteine effective against Trypanosoma equiperdum in man (531). Potential carcinogen in a prolonged medicinal application (769). Philaricidal action on adult and microfilariae of Setaria digitata in vitro (875). Eradication of Eperythrozoon coccoides in normal and drug-resistant lines of Plasmodium berghei in mice (1510). Normal to enhanced effect on some drug-resistant strains of Plasmodium berghei in mice (1511). Inhibition of glutamine synthetases from Neurospora and mammalian tissues (1626). The effect of various thiol agents on the LD_{50} dose of Mapharsen in mice (837).

4-Arsenosophenylurea, $AsC_7H_7N_2O_2$, $4-(H_2NCONH)C_6H_4AsO$, (M-238)
Biol. Prop.: Detn. of the As distribution in the bowel and liver after intravenous and intragastrical administration of the arsenoso deriv. to rabbits and rats as an amoebicide (1617).

Ferrocenyloxoarsine Dimer, Arsenosoferrocene Dimer, $(AsFeC_{10}H_9O)_2$, $(FcAsO)_2$
Synth.: From ferrocene (3 moles) by adding dropwise $AsCl_3$ (1 mole) dissolved in $n-C_7H_{16}$ in the presence of $AlCl_3$ (1 mole) in the same solvent and heating under reflux. The solid fraction was extracted with C_6H_6 and the soln. combined with the $n-C_7H_{16}$ soln. was evapd. to dryness. The orange-brown solid residue, after boiling with 2 N NaOH, gave an orange solid, $(FcAsO)_n$ polymer. The latter was extd. with $n-C_7H_{14}$ to remove ferrocene, then was dissolved in alc. NaOH or NaOEt and acidified. The title compd. was recrystd. from Py; 22% yield (1436).
From $FcAsCl_2$ by hydrolysis with H_2O (1437).
Prop.: Pale yellow crystals, m. 261-262° (darkened > 250°), instantaneous m. 275.5-276.0°. IR spectrum (1436, 1437).
Rxn. with: Concd. HCl → $FcAsCl_2$ (1436, 1437).
H_3PO_2 in 95% EtOH → $(FcAs<)_n$ (1436).
Atm. moisture during concn. of $(FcAsO)_2$ soln. in $CHCl_3$ or C_6H_6 → $[FcAs(OH)]_2O$ (1437).

Ferrocenyloxoarsine Polymer, Arsenosoferrocene Polymer, $(AsFeC_{10}H_9O)_n$, $(FcAsO)_n$
Synth.: From $(FcAsO)_2$ by boiling in $CHCl_3$, C_6H_6, EtOH, or 2 N NaOH soln. The polymer was obtd. in a pure form by freezing a soln. of the dimer or of the polymer in Py and allowing it to thaw and stand at 0° (1436).
Prop.: Fine, pale yellow powder, infusible in a normal melting procedure, darkened > 260°, instantaneous m. 266-267°. IR spectrum. Recrystn. from Py at -10° → $(FcAsO)_2$ (1436).
Rxn. with: NaOEt or alc. NaOH, followed by dilution with H_2O and acidification → $(FcAsO)_n$ (1436).
$PhNHNH_2$ → $(FcAs<)_n$ (1436).

4-Arsenoso-N,N-bis(2-chloroethyl)aniline, $AsC_{10}H_{12}Cl_2NO$,
$4-[(ClCH_2CH_2)_2N]C_6H_4AsO$
Synth.: From $4-[(ClCH_2CH_2)_2N]C_6H_4AsO_3H_2$ in EtOH contg. aq. HI by reducing with SO_2, ~ 100% yield (122, 441).

<u>Prop.</u>: m. 30-50°, IR spectrum (122, 441).
<u>Rxn. with</u>: H_2O_2 in alk. soln. → 4-[(ClCH$_2$CH$_2$)$_2$N]C$_6$H$_4$AsO$_3$H$_2$ (122, 441).
HSCH$_2$CH$_2$SH in EtOH under reflux → 4-[(ClCH$_2$CH$_2$)$_2$N]C$_6$H$_4$AsSCH$_2$CH$_2$S (122, 441).
<u>Biol. Prop.</u>: Inhibitory effect on the Ehrlich ascites tumors (122, 441).

p-Arsenoso-N,N-diethylaniline,   AsC$_{10}$H$_{14}$NO,   4-(Et$_2$N)C$_6$H$_4$AsO
<u>Synth.</u>: From PhNEt$_2$ by heating with AsCl$_3$ on a steam bath, then dec. the rxn.
prod. with ice-water, pouring the soln. into excess concd. NaOH soln. and
mixing with solid NH$_4$Cl, 70% yield (122, 441).
<u>Prop.</u>: m. 57-58° (122, 441). IR and NMR spectra (441).
<u>Rxn. with</u>: H_2O_2 in alk. soln. → 4-(Et$_2$N)C$_6$H$_4$AsO$_3$H$_2$ (122, 441).
HSCH$_2$CH$_2$SH in EtOH under reflux → 4-(Et$_2$N)C$_6$H$_4$AsSCH$_2$CH$_2$S (441).
Na-Hg in EtOH → [4-(Et$_2$N)C$_6$H$_4$As=]$_n$ (441).
<u>Biol. Prop.</u>: Antitumor activity in rats (122). Inhibition of the synth. of
o-aminophenol and p-nitrophenol glucuronides catalyzed by the uridine diphosphate transglucuronylase of mouse-liver homogenates (1458).

p-Arsenoso-N,N-di(2-hydroxyethyl)aniline,   AsC$_{10}$H$_{14}$NO$_3$,
    4-[(HOCH$_2$CH$_2$)$_2$N]C$_6$H$_4$AsO
<u>Synth.</u>: From 4-[(HOCH$_2$CH$_2$)$_2$N]C$_6$H$_4$AsO$_3$H$_2$ by reducing with PhNHNH$_2$ in MeOH
under reflux and treating the prod. with aq. NaOH, followed by NH$_4$Cl, 52%
yield (122, 441).
<u>Prop.</u>: m. 245-250° (dec.), IR spectrum (122, 441).
<u>Rxn. with</u>: HSCH$_2$CH$_2$SH in EtOH under reflux → 4-[(HOCH$_2$CH$_2$)$_2$N]C$_6$H$_4$AsSCH$_2$CH$_2$S
(122, 441).

Table 9. Arsenoso Derivatives

RAsO	Prepd. from	Yield %	Prop. and Rxn.	Ref.
C$_1$				
MeAsO (M-227)			IR spectrum, (a, b)	1057
C$_2$				
EtAsO (M-228)			(a)	1378
C$_3$				
PrAsO (M-234)			(a)	1378
C$_4$				
BuAsO (M-234)			(a)	1378
C$_5$				
C$_5$H$_{11}$AsO			(a)	1378

(a) Effective against nemathodes in the soil (1378).

Table 9 Continued

RAsO	Prepd. from	Yield %	Prop. and Rxn.	Ref.
C_6				
C_6F_5AsO	$RAsCl_2$ + H_2O	79	m. ~ 217° (dec.), IR and NMR spectra	651
$2,3-Cl_2C_6H_3AsO$			(c, d, e, f)	645
$2-BrC_6H_4AsO$			Rxn. with $2-BrC_6H_4AsCl_2$ at $260-270°$ → $(2-BrC_6H_4)_2AsCl$	715
$2-ClC_6H_4AsO$ (M-235)			(e)	645
$4-ClC_6H_4AsO$ (M-235)			(d, e, f, g)	645
$2-O_2NC_6H_4AsO$ (M-235)			(c)	645
$4-O_2NC_6H_4AsO$ (M-236)			(c)	645
$C_6H_{13}AsO$			(a)	1378
C_7				
$4-HO_2CC_6H_4AsO$			Inhibition of glutamine synthetase from Neurospora and from mammalian tissues	1626
$4-MeOC_6H_4AsO$			Rxn. with pyrocatechol → $4-MeOC_6H_4\overline{AsOC_6H_4O}$	741
C_8				
$4-EtC_6H_4AsO$			Rxn. with pyrocatechol → $4-EtC_6H_4\overline{AsOC_6H_4O}$	741
$4-(Me_2N)C_6H_4AsO$ (M-240)			Rxn. with pyrocatechol → $4-(Me_2N)C_6H_4\overline{AsOC_6H_4O}$	741
C_{10}				
$4-[HO_2C(CH_2)_3]C_6H_4AsO$			Inactivation of aldehyde hydrogenase	461

(b) Effective in controlling soil-borne Pythium fungi (1379). Effective against Alternaria solani (c) spores and (d) micelia and against Botrylis cinerea (e) spores and (f) micelia (645). (g) Highly active against the germination of numerous fungi spores including Xanthomonas oryzae (1493)

1.6.1.7 ARSONOUS ACID ESTERS

The esters, $RAs(OR')_2$ and $RAs(OR')(OR'')$, and their methods of synthesis were reported in the main volume, pp. 250-255. This class of arsenic compounds includes oxime esters $RAs(ON=CR'_2)_2$, which have been prepared from primary dihaloarsines by a condensation with oximes in alcohol in the presence of a tertiary amine or pyridine as a dehydrohalogenating agent. A reaction of primary dihaloarsines with trialkyl phosphites and aldehydes leads to the formation of arsonous acid diesters with α-hydroxyalkylphosphonates:

$$RAsX_2 + 2 (R'O)_3P + 2 R''CHO \rightarrow RAs[OCH(R'')P(O)(OR')_2]_2 \quad .$$

Mixed arsonous acid esters, RAs(OR′)(OR″), have been prepared from the dioxime esters by heating with alcohol and from α-(alkylchloroarsinooxy)alkylphos-phonates by a condensation with alkylene oxides.

$$RAs(ON=CR′_2)_2 + R″OH \rightarrow RAs(OR″)(ON=CR′_2)$$

$$RAs \begin{smallmatrix} Cl \\ R′ \\ OCHP(O)(OR″)_2 \end{smallmatrix} + R‴\overline{CHCH_2O} \rightarrow RAs \begin{smallmatrix} OCH_2CHClR‴ \\ \\ OCHP(O)(OR″)_2 \\ R′ \end{smallmatrix}$$

(All the esters of this class are compiled in Table 10 on page 179.)

1.6.1.9 OXYBIS(HYDROXYARSINES), [RAs(OH)]$_2$O

Several compounds of this type were described in the main volume on page 257.

Oxybis(ferrocenylhydroxyarsine),   As$_2$Fe$_2$C$_{20}$H$_{20}$O$_3$,   (FcAsOH)$_2$O
Synth.: From (FcAsO)$_2$ dissolved in CHCl$_3$ or C$_6$H$_6$ by evapn. in the presence of atm. moisture (1437).
Prop.: Light yellow powder, m. 266-267° (dec.) (1437).
Rxn. with: Concd. HCl → FcAsCl$_2$ (1437)

Table 10. Arsonous Acid Esters

RAs(OR')$_2$ or RAs(OR')(OR")	Yield %	Prepd. from	Prop. and Rxn.	Ref.
C$_4$				
EtAs(OMe)$_2$ (M-251)			Rxn. with RSH → EtAs(SR)$_2$	317
C$_6$				
EtAs(OEt)$_2$ (M-251)			Raman spectrum	1388
			Rxn. with RSH → EtAs(SR)$_2$	318
			Rxn. with PhOH → EtAs(OPh)$_2$	824
C$_7$				
PrAs(OEt)$_2$ (M-251)			Rxn. with ROH → PrAs(OR)$_2$	824
C$_8$				
PhAs(OMe)$_2$ (M-252)			Rxn. with RSH → PhAs(SR)$_2$	317
EtAs(ON=CMe$_2$)$_2$		EtAsCl$_2$ + Me$_2$C=NOH + Py in EtOH	b$_3$ 86-87°, d$_4^{20}$ 1.2040, n$_D^{20}$ 1.4940, MR$_D$ 60.01, IR spectrum, (a)	826
C$_9$				
PrAs(ON=CMe$_2$)$_2$		PrAsCl$_2$ + Me$_2$C=NOH + Py in EtOH	b$_6$ 116°, d$_4^{20}$ 1.1833, n$_D^{20}$ 1.4934, MR$_D$ 64.40, IR spectrum, (b)	826
C$_{10}$				
4-O$_2$NC$_6$H$_4$As(OEt)$_2$			Rxn. with RSH → 4-O$_2$NC$_6$H$_4$As(SR)$_2$	318
PhAs(OEt)$_2$ (M-252)			Raman spectrum	1388
			Rxn. with RSH → PhAs(SR)$_2$	318
BuAs(ON=CMe$_2$)$_2$	58	BuAsCl$_2$ + Me$_2$C=NOH + Py in Et$_2$O	b$_4$ 110°, n$_D^{20}$ 1.4900, d$_4^{20}$ 1.1530, MR$_D$ 69.28, (c)	825, 826
PrAs(OBu)(ON=CMe$_2$)		PrAs(ON=CMe$_2$)$_2$ + BuOH by heating	b$_{19}$ 122-124°, d$_4^{20}$ 1.1305, n$_D^{20}$ 1.4739, MR$_D$ 65.42, IR spectrum	826
C$_{11}$				
4-MeC$_6$H$_4$As(OEt)$_2$	64	4-MeC$_6$H$_4$AsCl$_2$ + EtOH in Et$_2$O + Py under reflux	b$_9$ 123-124°, n$_D^{20}$ 1.5375, d$_4^{20}$ 1.2280, MR$_D$ 64.75, (d, e)	819
MeAs(OCH$_2$CHClEt)[OCH(Me)P(O)(OEt)$_2$]		MeAs(Cl)OCH(Me)P(O)(OEt)$_2$ + EtCH$_2$CH$_2$O		1274

(a) Rxn. with n-C$_7$H$_{15}$OH → EtAs(OC$_7$H$_{15}$)$_2$ (824). (b) Rxn. with BuOH → PrAs(OBu)(ON=CMe$_2$) (826). (c) IR spectrum (826). (d) Rxn. with MeCH=CHCH(R)OH in CO$_2$ at 150-180° → 4-MeC$_6$H$_4$As[OCH(R)CH=CHMe]$_2$ (819). (e) Useful as insecticide (819).

Table 10 Continued

RAs(OR')$_2$ or RAs(OR')(OR'')	Prepd. from	Yield %	Prop. and Rxn.	Ref.
C$_{11}$ (Cont.)				
PrAs(OCH$_2$CH$_2$NMe$_2$)$_2$	PrAs(OEt)$_2$ + Me$_2$NCH$_2$CH$_2$OH (1:2) at 150-180°	70	b$_4$ 118-119°, n$_D^{20}$ 1.4732, d$_4^{20}$ 1.0950, (f)	824
C$_{12}$				
PhAs(ON=CMe$_2$)$_2$	PhAsCl$_2$ + Me$_2$C=NOH + Py in Et$_2$O		b$_4$ 142°, n$_D^{20}$ 1.5556, d$_4^{20}$ 1.2735, MR$_D$ 74.73, IR spectrum	826
C$_{13}$				
4-MeC$_6$H$_4$As(OCH$_2$CH=CH$_2$)$_2$	4-MeC$_6$H$_4$AsCl$_2$ + CH$_2$=CHCH$_2$OH in petr. ether + Py under reflux	77	b$_{11.5}$ 155-156°, n$_D^{20}$ 1.5468, d$_4^{20}$ 1.2320, MR$_D$ 72.40, (e, g)	819
C$_{14}$				
EtAs(OPh)$_2$	EtAs(OEt)$_2$ + PhOH (1:2) at 150-180°	54	b$_5$ 154-155°, n$_D^{20}$ 1.5922, d$_4^{20}$ 1.3095	824
PhAs(OBu)$_2$ (M-253)	PhAsO + BuOH in C$_6$H$_6$, (h)			741
C$_{15}$				
PrAs(OPh)$_2$	PrAs(OEt)$_2$ + PhOH (1:2) at 150-180°	47	b$_3$ 157-158°, n$_D^{20}$ 1.5805, d$_4^{20}$ 1.2671	824
MeAs[OCH(Et)P(O)(OEt)$_2$]$_2$	MeAsBr$_2$ + (EtO)$_3$P + EtCHO (1:2:2), (i)	96		1274
C$_{16}$				
EtAs(OC$_7$H$_{15}$)$_2$ (M-254)	EtAs(ON=CMe$_2$)$_2$ + n-C$_7$H$_{15}$OH at 150-180°	76	b$_4$ 149-150°, n$_D^{20}$ 1.4580, d$_4^{20}$ 1.0041	824
CH$_2$=CHAs[OCH(Et)P(O)(OEt)$_2$]$_2$	CH$_2$=CHAsBr$_2$ + (EtO)$_3$P + EtCHO, (j)			1274
C$_{17}$				
PrAs(OC$_6$H$_4$Me-4)$_2$	PrAs(OCH$_2$CH$_2$NMe$_2$) + 4-MeC$_6$H$_4$OH (1:2) at 150-180°	70	b$_4$ 178-179°, n$_D^{20}$ 1.5668, d$_4^{20}$ 1.2218	824
4-MeC$_6$H$_4$As[OCH(Et)CH=CH$_2$]$_2$	4-MeC$_6$H$_4$As(OEt)$_2$ + CH=CHCH(Et)OH under CO$_2$ at 150-180°	64	b$_{14}$ 170-171°, n$_D^{20}$ 1.5278, d$_4^{20}$ 1.1545, MR$_D$ 89.69, (e)	819
C$_{18}$				
PhAs(OC$_6$H$_{11}$)$_2$ (M-254)	PhAsO + C$_6$H$_{11}$OH in C$_6$H$_6$, (h)			741

C_{19}				
$4\text{-MeC}_6\text{H}_4\text{As}[\text{OCH(Et)CH=CHMe}]_2$	$4\text{-MeC}_6\text{H}_4\text{As(OEt)}_2$ + MeCH=CHCH(Et)OH under CO_2 at 150-180°	88	b_9 173-174°, n_D^{20} 1.5210, d_4^{20} 1.1070, MR_D 100.72, (e)	819
C_{21}				
$4\text{-MeC}_6\text{H}_4\text{As}[\text{OCH(Pr)CH=CHMe}]_2$	$4\text{-MeC}_6\text{H}_4\text{As(OEt)}_2$ + MeCH=CHCH(Pr)OH under CO_2 at 150-180°	96	$b_{1.5}$ 162-163°, n_D^{20} 1.5185, d^{20} 1.0911, MR_D 109.04, (e)	819
C_{22}				
$\text{PhAs}[\text{OCH(Me)C}_6\text{H}_{13}]_2$	PhAsO + $\text{C}_6\text{H}_{13}\text{CH(Me)OH}$ in C_6H_6, (h)			741
C_{23}				
$4\text{-MeC}_6\text{H}_4\text{As}[\text{OCH(Bu)CH=CHMe}]_2$	$4\text{-MeC}_6\text{H}_4\text{As(OEt)}_2$ + MeCH=CHCH(Bu)OH under CO_2 at 150-180°	42	b_{11} 206-207°, n_D^{20} 1.5154, d^{20} 1.0771, MR_D 118.60, (e)	819
$4\text{-MeC}_6\text{H}_4\text{As}[\text{OCH(Me)C}_6\text{H}_{13}]_2$	$4\text{-MeC}_6\text{H}_4\text{AsO}$ + $\text{C}_6\text{H}_{13}\text{CH(Me)OH}$ in C_6H_6, (h)			741
$4\text{-MeOC}_6\text{H}_4\text{As}[\text{OCH(Me)C}_6\text{H}_{13}]_2$	$4\text{-MeOC}_6\text{H}_4\text{AsO}$ + $\text{C}_6\text{H}_{13}\text{CH(Me)OH}$ in C_6H_6, (h)			741
C_{24}				
$4\text{-EtC}_6\text{H}_4\text{As}[\text{OCH(Me)C}_6\text{H}_{13}]_2$	$4\text{-EtC}_6\text{H}_4\text{AsO}$ + $\text{C}_6\text{H}_{13}\text{CH(Me)OH}$ in . C_6H_6, (h)			741
C_{25}				
$4\text{-MeC}_6\text{H}_4\text{As}[\text{OCH(Ph)CH=CH}_2]_2$ (M-255)	$4\text{-MeC}_6\text{H}_4\text{As(OEt)}_2$ + CH=CHCH(Ph)OH under CO_2 at 150-180°	67	b_3 213-214°, n_D^{20} 1.5889, d^{20} 1.2015, MR_D 120.25, (e)	819

(f) Rxn. with $4\text{-MeC}_6\text{H}_4\text{OH} \rightarrow \text{PrAs(OC}_6\text{H}_4\text{Me-4})_2$ (824). (g) Rxn. with SO_2 in C_6H_6 contg. $AgNO_3 \rightarrow$ a polysulfone (819). (h) In the presence of catalytic amt. of $4\text{-MeC}_6\text{H}_4\text{SO}_3\text{H}$, Py, or Bu_3N with azeotropic removal of H_2O (741). (i) By combining the reagents < 10°, gradually warming to 15-18°, and completing at 70°. (j) By combining the reagents at a low temp. and completing at 60°.

1.6.2   COMPOUNDS WITH ARSENIC-SULFUR BONDS

1.6.2.2   THIOARSINOUS ACID ESTERS WITH THIOLS

The thioesters and their methods of synthesis were described in the main volume pp. 258-260. Since 1964 several new methods for the preparation of these esters were reported.

Thus, tertiary arsine sulfides have been converted to the thioesters by a ligand rearrangement at 240° in a sealed tube and by a reaction with alkyl halides at elevated temperatures. The latter reaction with optically active tertiary arsine sulfides yields the corresponding thioesters with an optical rotation opposite to that of the arsine sulfide.

A condensation of secondary halo- and pseudohalo-arsines with ethylene sulfide proceeds at room temperature and affords the corresponding 2-halo- and 2-pseudohalo-ethylthio esters, respectively.

$$R_2AsX + \overline{CH_2CH_2S} \rightarrow R_2AsSCH_2CH_2X$$

Starting from secondary arsines, the thioesters are formed at room temperature by a reaction with alkylsulfenyl halides and with dialkyl disulfides, respectively. Likewise, a reaction of alkylsulfenyl chloride with tetraalkyl-diarsines at room temperature affords the corresponding alkylthio ethers.

A metathesis of arsinous acid esters, R_2AsOR', with thiols, $R''SH$, at moderately elevated temperatures yields arsinous acid thioesters, R_2AsSR''.

Oxybis(secondary arsines) react with 2-merceptoethanol in the presence of carboxylic acids in a hydrocarbon solvent under reflux, with continuous removal of water, to form 2-[(diorganoarsino)thio]ethyl esters.

$$(R_2As)_2O + HSCH_2CH_2OH + R'CO_2H \rightarrow R_2AsSCH_2CH_2OCOR'$$

2-Hydroxyethylthio esters are readily esterified with phosphorochloridic esters in the presence of tertiayy amines to form the corresponding phosphates.

$$R_2AsSCH_2CH_2OH + (R'O)_2P(O)Cl \xrightarrow{R''_3N} R_2AsSCH_2CH_2OP(O)(OR')_2 \quad .$$

Many of the thioesters have shown fungistatic and fungicidal properties.

The compounds of this class are compiled in Table 11, with the exception of two individual entries which follow the table.

Table 11. Thioarsinous Acid Esters with Thiols

R_2AsSR' or $RR'AsSR''$	Prepd. from	Yield %	Prop. and Rxn.	Ref.
C_3				
Me_2AsSCF_3 (M-258)	$R_2AsH + R'SSR'$ at 20°			419
	$R_2AsH + R'SCl$ at 20°			419
	$R_2AsAsR_2 + R'SCl$ at 20°			419
C_4				
Me_2AsSEt (M-258)	$R_2AsH + R'SSR'$ at 100°		Rxn. with Me_2AsH at 100° → $(Me_2As)_2$, Me_2AsH, and EtSH, (a, b)	419
C_5				
$Me_2AsSCH_2CH_2SCN$	$R_2AsSCN + \overline{CH_2CH_2S}$ at 25°	92	b_1 104-105°, n_D^{24} 1.5990, d_4^{24} 1.5298, (c)	873
C_6				
$Et_2AsSCH_2CH_2Cl$	$R_2AsCl + \overline{CH_2CH_2S}$ at 20-30°, (d)	86	$b_{1.5}$ 73-76°, n_D^{24} 1.5499, d_4^{24} 1.3265	961
C_7				
$Et_2AsSCH_2CH_2SCN$	$R_2AsSCN + \overline{CH_2CH_2S}$ at 25°	92	b_2 119-120°, d_4^{20} 1.3143, n_D^{20} 1.5773, (c)	873
C_8				
Me_2AsSPh			Rxn. with CF_3SCl → Me_2AsCl + $PhSSCF_3$	419
$Pr_2AsSCH_2CH_2OH$			Rxn. with $(RO)_2P(O)Cl$ → $Pr_2AsSCH_2CH_2OP-(O)(OR)_2$	193
C_{10}				
$EtPhAsSEt$	$R_2R'AsS$ at 240° in a sealed tube	6	$b_{1.5}$ 98-101°, n_D^{20} 1.5955, d^{20} 1.2571	607
	$R_2R'AsS$ at 200°/3 mm	9	b_1 96-98°, n_D^{20} 1.5949, d^{20} 1.2585	607
	$R_2R'AsS$ + EtBr at 240°	45	(e)	605, 607, 609
	$RR'AsCl$ + EtSH in Et_2O + Et_3N at 20°	48		605, 607

(a) Rxn. with CF_3SCl at 20° → Me_2AsCl and $EtSSCF_3$ (419). (b) Rxn. with $CF_3C\equiv CH$ at 125° → $CF_3CH=CAsMe_2$ (425).
(c) Useful as a fungicide (873). (d) A yield of 92% was obtd. when Py was added to the rxn. mixt. (961).
(e) Rxn. with EtBr in C_6H_6 → $EtPhAsBr$ + Et_3SBr (609).

Table 11 Continued

R$_2$AsSR' or RR'AsSR"	Prepd. from	Yield %	Prop. and Rxn.	Ref.
C$_{10}$ (Cont.)				
Bu$_2$AsSCH$_2$CH$_2$OH			Rxn. with (RO)$_2$P(O)Cl → Bu$_2$AsSCH$_2$CH$_2$OP-(O)(OR)$_2$	193
C$_{11}$				
EtPhAsSPr	R$_2$R'AsS + PrBr at 100°	41	b$_{1.5}$ 112-113°, d$_4^{20}$ 1.2203, n$_D^{20}$ 1.5858, MR$_D$ and AR$_D$ data	609
PhPrAsSEt	RR'$_2$AsS + EtBr at 100°	60	b$_{1.5}$ 111-113°, d$_4^{20}$ 1.2209, n$_D^{20}$ 1.5861, (f)	605, 607, 609
C$_{12}$				
BuPhAsSEt	R$_2$R'AsS + EtBr at 100°	52	b$_2$ 123-124°, d$_4^{20}$ 1.2963, n$_D^{20}$ 1.5801, (f)	609
EtPhAsSBu	R$_2$R'AsS + BuBr at 100°	43	b$_{1.5}$ 121-122°, d$_4^{20}$ 1.1959, n$_D^{20}$ 1.5795, (f)	609
PhPrAsSPr	RR'$_2$AsS + PrBr at 100°	41	b$_{1.5}$ 129-130°, d$_4^{20}$ 1.1951, n$_D^{20}$ 1.5787, (f)	605, 607, 609
Bu$_2$AsSCH$_2$CH$_2$OCOCCl$_3$	R$_2$AsCl + R'SH		Liquid, n$_D^{20}$ 1.5235, d$_4^{20}$ 1.3929, (g)	191
Bu$_2$AsSCH$_2$CH$_2$OCOCH$_2$Cl	R$_2$AsCl + R'SH in MePh under reflux		Liquid, n$_D^{20}$ 1.5350, d$_4^{20}$ 1.2771, (g)	191
Bu$_2$AsSCH$_2$CH$_2$OAc	R$_2$AsCl + R'SH in MePh under reflux		Liquid, n$_D^{20}$ 1.5049, d$_4^{20}$ 1.1291, MR$_D$ 80.8, (g)	191
Pr$_2$AsSCH$_2$CH$_2$OP(O)-(OEt)$_2$	R$_2$AsSCH$_2$CH$_2$OH + (EtO)$_2$P(O)Cl in C$_6$H$_6$ + Et$_3$N at 50-80°	87	b$_2$ 123-125°, n$_D^{20}$ 1.4890, d$_4^{20}$ 1.2016	193
C$_{13}$				
BuPhAsSPr	R$_2$R'AsS + PrBr at 100°	44	b$_2$ 134-135°, n$_D^{20}$ 1.5733, d$_4^{20}$ 1.1773, (g)	609
PhPrAsSBu	RR'$_2$AsS + BuBr at 100°	38	b$_7$ 153-154°, n$_D^{20}$ 1.5760, d$_4^{20}$ 1.1797, (f)	609
C$_{14}$				
Ph(4-O$_2$NC$_6$H$_4$)AsSEt	RR'AsCl + R"SH in C$_6$H$_6$ + Et$_3$N on a steam bath	62	b$_{0.2}$ 185-187°, n$_D^{20}$ 1.6622, d$_4^{20}$ 1.3890	820
Ph$_2$AsSCH$_2$CH$_2$OH			Rxn. with (RO)$_2$P(O)Cl → Ph$_2$AsSCH$_2$CH$_2$OP-(O)(OR)$_2$	193
BuPhAsSBu	R$_2$R'AsS + BuBr at 100°	42	b$_5$ 155-156°, n$_D^{20}$ 1.5701, d$_4^{20}$ 1.1628, (f)	609

	Yield (%)	Properties	Preparation	Ref.
C_{15}				
$Ph(2,4\text{-}Cl_2C_6H_3)AsSPr$	37	$b_{0.07}$ 100-102°, n_D^{20} 1.6288, d_4^{20} 1.3939, MR_D 95.1, AR_{As} 10.94	$RR'AsCl + PrSH$ at r.t., (h)	1184
$Ph_2AsSCH_2CH(OH)CH_2Cl$	98	Crude oil, unstable > 120° and in H_2O under reflux, (i)	$R_2AsOEt + R'SH$ at 80-120°	318
Ph_2AsSPr	44	$b_{0.14}$ 112-116°, n_D^{20} 1.6360, d_4^{20} 1.2952, MR_D 84.19, AR_{As} 9.7, (j, k)	$R_2AsCl + PrSH$ at r.t., (h)	1184
C_{16}				
$Ph_2AsSCH_2CH_2OCOCl_3$		Liquid, n_D^{20} 1.6180, (g)	$(R_2As)_2O + Cl_3CCO_2H + HOCH_2CH_2SH$, (l)	191
$Ph_2AsSCH_2CH_2OCOCH_2Cl$		n_D^{20} 1.6210, (g)	$(R_2As)_2O + ClCH_2CO_2H + HOCH_2CH_2SH$, (l)	191
$Ph_2AsSCH_2CH_2OAc$		m. 168°, (g)	$(R_2As)_2O + AcOH + HOCH_2CH_2SH$, (l)	191
$Ph(4\text{-}HO_2CC_6H_4)AsSPr$		(-)-Isomer $[\alpha]_D^{20}$ -3.18° / (+)-Isomer $[\alpha]_D^{20}$ +2.88°	(+)- and (-)-$RR'EtAsS + PrBr$ in C_6H_6 at 90°, the esters were obtd. with inversed optical activity	602
$Ph(2,4\text{-}Cl_2C_6H_3)AsSBu$	42	b_2 182°, n_D^{20} 1.6190, d_4^{20} 1.3666, MR_D 99.35, AR_{As} 11.07	$RR'AsCl + R''SH$ at r.t., (h)	1184
Ph_2AsSBu	61.5	$b_{1.5}$ 152-154°, n_D^{20} 1.6252, d_4^{20} 1.2534, MR_D 89.79, AR_{As} 10.75, (m)	$R_2AsCl + R'SH$ at r.t., (h)	1184
$Ph(4\text{-}MeC_6H_4)AsSPr$	58	$b_{0.1}$ 116-118°, n_D^{20} 1.6320, d_4^{20} 1.2701, MR_D 89.2, AR_{As} 10.16	$RR'AsCl + R''SH$ at r.t., (h)	1184
$Ph(2\text{-}MeC_6H_4)AsSPr$	75	$b_{0.09}$ 96-98°, n_D^{20} 1.6160, d_4^{20} 1.2411, MR_D 89.15, AR_{As} 10.11	$RR'AsCl + R''SH$ at r.t., (h)	1184
$Ph(4\text{-}MeOC_6H_4)AsSPr$	76	$b_{0.2}$ 99-101°, n_D^{20} 1.6220, d_4^{20} 1.2922, MR_D 91.06, AR_{As} 10.37	$RR'AsCl + R''SH$ at r.t., (h)	1184

(f) MR_D and AR_D data (609). (g) Fungicidal props. (191). (h) The rxn. was completed on a water bath. (i) Fungicidal activity against Alternaria radicina, Botrytis cinerea, Fusicladium dendriticum, Trichophyton gypseum, and Epidermophyton; lethal to mice (1132, 1133). (j) Rxn. with 30% H_2O_2 → Ph_2AsO_2H (1184). (k) Rxn. with Cu_2Cl_2 → $[Ph_2AsPr \cdot CuCl]$, crystals with a greenish tinge, m. 164-166° (1184). (l) The rxn. was carried out in MePh in Dean-Stark apparatus. (m) Rxn. with H_2O on a steam bath → Ph_2AsO_2H (1184).

Table 11 Continued

R_2AsSR' or $RR'AsSR''$	Prepd. from	Yield %	Prop. and Rxn.	Ref.
C_{17}				
$Ph(1-C_{10}H_7)AsSMe$	$RR'EtAsS + MeI$ in C_6H_6 at 80°	50	b_{10} 202-203°, m. 54°	604, 605
$Ph_2AsSCH_2CH(OAc)CH_2Cl$	$R_2AsOEt + R'SH$ at 80-120°	95	Crude oil, unstable > 120° and in boiling water, (i)	318
$Ph(4-HO_2CC_6H_4)AsSBu$	(+)- and (-)-$RR'EtAsS + BuBr$ in C_6H_6 at 90°, the esters were obtd. with inversed optical rotation		(-)-Isomer $[\alpha]_D^{20}$ -3.09° (+)-Isomer $[\alpha]_D^{20}$ +2.12	602
$Ph_2AsSCH_2CH_2SPr-i$	$R_2AsOMe + R'SH$ by heating and distg. MeOH	53	Yellowish liquid, $b_{0.002}$ 182°	317
C_{18}				
$Ph(4-O_2NC_6H_4)AsSPh$	$RR'AsCl + R''SH$ in $C_6H_6 + Et_3N$ on a steam bath	90	Oil dec. during distn., d_4^{20} 1.3449, n_D^{20} 1.6781, (n, o)	820
$Ph(4-O_2NC_6H_4)AsSC_6H_4-NH_2-2$	$RR'AsCl + R''SH$ in C_6H_6 on a steam bath and neutralized	83	m. 92-93°, HCl salt m. 182-183°, (i)	820
Ph_2AsSPh (M-260)	$R_2AsCl + R'SH$ at 60°	62	b_{13} 168-170°	1105
$Ph_2AsSC=NC(Me)=NC(Me)=N$	$R_2AsCl + R'SNa$ in C_6H_6	47	m. 87.5°	316
$Ph(1-C_{10}H_7)AsSEt$	$RR'EtAsS + MeI$ at 50°		m. 57° (i)	604
$Ph_2AsSCH_2CH_2NEt_2·HCl$ (M-260)				1132, 1133
$Ph_2AsSCH_2CH_2OP(O)(OEt)_2$	$R_2AsSCH_2CH_2OH + (EtO)_2P(O)Cl + Et_3N$ in C_6H_6 at 50-80°	81	n_D^{20} 1.5670, d_4^{20} 1.3010	193
$Bu_2AsSCH_2CH_2OCOCH_2O-C_6H_3Cl_2-2,4$	$R_2AsCl + R'SH$ in MePh under reflux		n_D^{20} 1.5515, d_4^{20} 1.2968, (g)	191
$Bu_2AsSCH_2CH_2OP(O)(OBu)_2$	$R_2AsSCH_2CH_2OH + (BuO)_2P(O)Cl + Et_3N$ in C_6H_6 at 50-80°	84	n_D^{20} 1.5853, d_4^{20} 1.1172	193
C_{19}				
$Ph(1-C_{10}H_7)AsSPr$	$RR'EtAsS + PrI$ at 50°	59	m. 47.6°	604

$Ph_2AsSCH_2CH_2OP(O)$-$(OBu$-$i)Me$	$R_2AsSCH_2CH_2OH + Me(i$-$BuO)P(O)Cl$ $+ Et_3N$ in C_6H_6 at 50-80°	83	n_D^{20} 1.5630, d_4^{20} 1.2551	193
C_{20}				
$Ph(4$-$O_2NC_6H_4)AsSCH_2$-$CONHPh$	$RR'AsCl + R''SH$ in C_6H_6 on a steam bath	71	m. 97-98°, HCl salt m. 104-106°	820
$Ph(1$-$C_{10}H_7)AsSBu$	$RR'EtAsS + BuI$ at 50°	46	m. 55-56°	604
$Bu_2AsSCH_2CH_2OCOCH_2O$-$C_6H_3Me_2$-3,5	$R_2AsCl + R'SH$ in MePh under reflux		n_D^{20} 1.5260, (g)	191
C_{22}				
$Ph_2AsSCH_2CH_2OCOCH_2O$-$C_6H_3Cl_2$-2,4	$(R_2As)_2O + 2,4$-$Cl_2C_6H_3OCH_2CO_2H$ $+ HSCH_2CH_2OH$, (1)		Glassy solid, (g)	191
$Ph_2AsSCH_2CH_2OCOCH_2OPh$	$(R_2As)_2O + PhOCH_2CO_2H +$ $HSCH_2CH_2OH$, (1)		Liquid, n_D^{20} 1.6260, d_4^{20} 1.3271, MR_D 1.3271	191
$Ph_2AsSCH_2CH_2OP(O)(OBu)_2$	$R_2AsSCH_2CH_2SH + (BuO)_2P(O)Cl$ in $C_6H_6 + Et_3N$ at 50-80°	81	n_D^{20} 1.5536, d_4^{20} 1.2267	193
C_{24}				
$Ph_2AsSCH_2CH_2OCOCH_2O$-$C_6H_3Me_2$-3,5	$(R_2As)_2O + 3,5$-$Me_2C_6H_3OCH_2CO_2H$ $+ HSCH_2CH_2OH$, (1)		Glassy solid, (g)	191

(n) Rxn. with $H_2O_2 \rightarrow Ph(4$-$O_2NC_6H_4)AsO_2H$ (820). (o) Rxn. with $Cu_2Cl_2 \rightarrow [Ph(4$-$O_2NC_6H_4)AsSPh \cdot CuCl]$, crystals, m. 106-108° (820).

(3,5-Di-tert-butyl-4-oxyphenylmercapto)diphenylarsine Radical, AsC$_{26}$H$_{30}$OS,

Synth.: The formation of the radical was detected by ESR spectroscopy during dehydrogenation of 4,3,5-HO-(t-Bu)C$_6$H$_2$SAsPh$_2$ by alk. K$_3$[Fe(CN)$_6$] soln. in C$_6$H$_6$ (1105).

Prop.: Yellow-red soln. rapidly changing to brown, diamagnetic 3,5,3',5'-tetra-tert-butyldiphenylquinone. ESR spectrum of the radical was recorded; a mesomeric resonance structure was suggested (1105).

(3,5-Di-tert-butyl-4-hydroxyphenylmercapto)diphenylarsine, AsC$_{26}$H$_{31}$OS, 4,3,5-HO(t-Bu)$_2$C$_6$H$_2$SAsPh$_2$

Synth.: From 4,3,5-HO(t-Bu)$_2$C$_6$H$_2$SH in THF + BuLi in petr. ether followed by Ph$_2$AsCl in THF at 80° and hydrolysis, 91% yield (1105).

Prop.: Colorless needles, m. 63-64°, IR, NMR, and mass spectra (1105).

Rxn. with: H$_2$O$_2$ → Ph$_2$AsO$_2$H and 3,5,3',5'-tetra-tert-butyldiphenoquinone (1105).

PhLi in THF-Et$_2$O → 4,3,5-HO(t-Bu)$_2$C$_6$H$_2$SH, Ph$_2$, and a residue which upon shaking with H$_2$O$_2$ gave Ph$_2$AsO$_2$H (1105).

Alkaline soln. of K$_3$[Fe(CN)$_6$] in C$_6$H$_6$ → a brown soln. contg. a yellow-red

radical (1105).

1.6.2.3 THIOARSINOUS ACID ANHYDROSULFIDES WITH THIOACIDS

The anhydrosulfides described in the main volume, pp. 261-262, include those of carbothioic, xanthogenic, phosphorodithioic, and dithiocarbamic acids. In the four years covered in this supplement volume no anhydrosulfides of carbothioic acids were reported. The other anhydrosulfides have been prepared from secondary haloarsines by the reaction with appropriate acid salt derivatives. Numerous arsenic derivatives of this class have fungicidal properties.

Table 12. Thioarsinous Acid Anhydrosulfides with Thioacids

Anhydrosulfide	Prepd. from	Prop. and Rxn.	Ref.
$R_2AsSC(S)OR'$			
C_5			
$Me_2AsSC(S)OEt$ (M-261)		(a)	645
C_6			
$Me_2AsSC(S)O\text{-}i\text{-}Pr$ (M-261)		(a)	645
C_7			
$Me_2AsSC(S)OBu$		(a)	645
$Et_2AsSC(S)OEt$		(a)	645
C_8			
$Et_2AsSC(S)O\text{-}i\text{-}Pr$		(a)	645
C_9			
$Et_2AsSC(S)OBu$		(a)	645
$Pr_2AsSC(S)OEt$		(a)	645
C_{10}			
$Et_2AsSC(S)O\text{-}i\text{-}C_5H_{11}$		(b)	645
$Pr_2AsSC(S)O\text{-}i\text{-}Pr$		(c)	645
C_{11}			
$Pr_2AsSC(S)OBu$		(c)	645
$Bu_2AsSC(S)OEt$		(c)	645

(a) A fungicide effective against the Alternaria solani spores. (b) A fungicide effective against the spores and micelia of Alternaria solani and Botrytis cinerea. (c) A fungicide effective against the spores of Alternaria solani and Botrytis cinerea.

Table 12 Continued

Anhydrosulfide	Prepd. from	Prop. and Rxn.	Ref.
R$_2$AsSC(S)OR′ (Cont.)			
C$_{12}$			
Pr$_2$AsSC(S)O-i-C$_5$H$_{11}$		(d)	645
Bu$_2$AsSC(S)O-i-Pr		(c)	645
C$_{13}$			
Bu$_2$AsSC(S)OBu		(c)	645
(C$_5$H$_{11}$)$_2$AsSC(S)OEt		(a)	645
C$_{14}$			
Bu$_2$AsSC(S)O-i-C$_5$H$_{11}$		(d)	645
(C$_5$H$_{11}$)$_2$AsSC(S)O-i-Pr		(b)	645
C$_{15}$			
Ph$_2$AsSC(S)OEt (M-261)		(b)	645
(C$_5$H$_{11}$)$_2$AsSC(S)OBu		(a)	645
(C$_6$H$_{13}$)$_2$AsSC(S)OEt		(a)	645
C$_{16}$			
Ph$_2$AsSC(S)O-i-Pr (M-261)		(c)	645
(C$_5$H$_{11}$)$_2$AsSC(S)O-i-C$_5$H$_{11}$		(b)	645
(C$_6$H$_{13}$)$_2$AsSC(S)O-i-Pr		(d)	645
C$_{17}$			
Ph$_2$AsSC(S)OBu		(c)	645
(2-MeC$_6$H$_4$)$_2$AsSC(S)OEt		(b)	645
(4-MeC$_6$H$_4$)$_2$AsSC(S)OEt		(b)	645
(C$_6$H$_{13}$)$_2$AsSC(S)OBu		(d)	645
(C$_7$H$_{15}$)$_2$AsSC(S)OEt		(c)	645

Compound	Preparation	Properties	Ref.
C_{18}			
$Ph_2AsSC(S)O-i-C_5H_{11}$		(c)	645
$(2-MeC_6H_4)_2AsSC(S)O-i-Pr$		(e)	645
$(4-MeC_6H_4)_2AsSC(S)O-i-Pr$		(b)	645
$(C_6H_{13})_2AsSC(S)O-i-C_5H_{11}$		(d)	645
C_{19}			
$(2-MeC_6H_4)_2AsSC(S)OBu$		(b)	645
$(4-MeC_6H_4)_2AsSC(S)OBu$		(c)	645
$(C_7H_{15})_2AsSC(S)OBu$		(c)	645
$(C_8H_{17})_2AsSC(S)OEt$		(d)	645
C_{20}			
$(2-MeC_6H_4)_2AsSC(S)O-i-C_5H_{11}$		(b)	645
$(4-MeC_6H_4)_2AsSC(S)O-i-C_5H_{11}$		(b)	645
$(C_7H_{15})_2AsSC(S)O-i-C_5H_{11}$		(c)	645
$(i-C_7H_{15})_2AsSC(S)O-i-C_5H_{11}$		(b)	645
C_{21}			
$Ph_2AsSC(S)OCH_2CH_2SC_6Cl_5$	$R_2AsCl + R'OC(S)SK$ in Me_2CO at 50-60°, 86% yield	m. 129-130°, a fungicide	957
$Ph_2AsSC(S)OCH_2CH_2SPh$	$R_2AsCl + R'OC(S)SK$ in aq. Me_2CO, 75% yield	d_4^{20} 1.2858, n_D^{20} 1.6463, a fungicide	957
$(C_9H_{19})_2AsSC(S)OEt$		(a)	645
C_{23}			
$(C_7H_{15})_2AsSC(S)OCH_2CH_2SC_6Cl_5$	$R_2AsCl + R'OC(S)SK$ in aq. Me_2CO, 93% yield	d_4^{20} 1.2865, n_D^{20} 1.5697, a fungicide	957
$(C_7H_{15})_2AsSC(S)OCH_2CH_2SC_6H_3Cl_2-2,4$	$R_2AsCl + R'OC(S)SK$ in aq. Me_2CO, 95% yield	d_4^{20} 1.1861, n_D^{20} 1.5520, a fungicide	957

(d) A fungicide effective against the spores and micelia of Alternaria solani and against the spores of Botrytis cinerea. (e) A fungicide effective against the spores and micelia of Alternaria solani.

Table 12 Continued

Anhydrosulfide	Prepd. from	Prop. and Rxn.	Ref.
$R_2AsSC(S)OR'$ (Cont.)			
C_{23} (Cont.)			
$(C_7H_{15})_2AsSC(S)OCH_2CH_2SPh$	$R_2AsCl + R'OC(S)SK$ in aq. Me_2CO at 50-60°, 82% yield	d_4^{20} 1.0083, n_D^{20} 1.5370, a fungicide	957
$(C_9H_{19})_2AsSC(S)OBu$		(b)	645
$(C_{10}H_{21})_2AsSC(S)OEt$		(a)	645
C_{24}			
$(C_9H_{19})_2AsSC(S)O-i-C_5H_{11}$		(f)	645
$R_2AsSC(S)SR'$			
C_{16}			
$Pr_2AsSC(S)SCH_2CH_2SC_6H_4Me-2$	$R_2AsCl + R'SC(S)SK$ in Me_2CO at 20-60°, 61% yield	Viscous liquid, d_4^{20} 1.1999, n_D^{20} 1.5925, a fungicide	960
C_{17}			
$Ph_2AsSC(S)SBu$	$R_2AsCl + R'SC(S)SK$ in Me_2CO at 20-60°, 88% yield	b. 105-107°, a fungicide	960
C_{19}			
$Ph_2AsSC(S)SPh$	$R_2AsCl + R'SC(S)SK$ in Me_2CO at 20-60°, 78% yield	Viscous liquid, d_4^{20} 1.2725, n_D^{20} 1.6762, a fungicide	960
C_{22}			
$(C_6H_{13})_2AsSC(S)SCH_2CH_2SC_6H_4Me-2$	$R_2AsCl + R'SC(S)SK$ in Me_2CO at 20-60°, 74% yield	Viscous liquid, d_4^{20} 1.1342, n_D^{20} 1.5620, a fungicide	960
$R_2AsSP(S)(OR')_2$			
C_{24}			
$Ph_2AsSP(S)(OC_6H_3Cl_2-2,4)_2$	$R_2AsCl + [Me_3NH]SP(S)(OR')_2$ in C_6H_8, 97% yield	Amorphous mass, translucent at 80°, a fungicide	192
$Ph_2AsSP(S)(OPh)_2$	$R_2AsCl + (R'O)_2P(S)SH$ in C_6H_8 under reflux, 67% yield	m. 76-78°, a fungicide	192

C_{26}			
$Ph_2AsSP(S)(OC_6H_4Me-2)_2$	$R_2AsCl + [Me_3NH]SP(S)(OR')_2$ in Me_2CO under reflux, 87% yield	Amorphous solid, a fungicide	192
$Ph_2AsSP(S)(OC_6H_4Me-3)_2$	$R_2AsCl + KSP(S)(OR')_2$ in Me_2CO under reflux, 62% yield	Viscous oil, a fungicide	192
$(C_7H_{15})_2AsSP(S)(OPh)_2$	$R_2AsBr + (R'O)_2P(S)SH$ in C_6H_6 under reflux, 89% yield	n_D^{20} 1.5505, d_4^{20} 1.1730, a fungicide	192
$R_2AsSC(S)NR'_2$			
C_9			
$Et_2AsSC(S)NEt_2$		(c)	645
C_{11}			
$Me_2AsSC(S)NBu_2$		(a)	645
$Pr_2AsSC(S)NEt_2$		(c)	645
C_{13}			
$Et_2AsSC(S)NBu_2$		(c)	645
C_{15}			
$(C_5H_{11})_2AsSC(S)NEt_2$		(c)	645
C_{17}			
$Bu_2AsSC(S)NBu_2$		(c)	645
$(C_6H_{13})_2AsSC(S)NEt_2$		(a)	645
C_{23}			
$(C_9H_{19})_2AsSC(S)NEt_2$		(b)	645

(f) A fungicide effective against the spores of Botrytis cinerea.

1.6.2.4 THIOBISARSINES, $(R_2As)_2S$ AND $R_2AsSAs(S)R_2$

The methods of synthesis of thiobisarsines were described in the main volume, pp. 262-263. Many compounds of this class have shown fungicidal properties, while $(Me_2As)_2S$ was found to be toxic to cauliflower.

[(Dimethylarsino)thio]dimethylarsine Sulfide, $\quad As_2C_4H_{12}S_2$,
    $Me_2AsSAs(S)Me_2$, $\quad$ (M-262)
Prop.: X-ray diffraction data indicate a trigonal-pyramidal and tetrahedral configurations for the As(III) and As(V), respectively (288).
Rxn. with: $NaN(SiMe_3)_2 \rightarrow Me_2AsN(SiMe_3)_2$ (1345).

Table 13. Thiobis(secondary Arsines)

$(R_2As)_2S$	Prepd. from	Prop. and Rxn.	Ref.
C_4			
$(Me_2As)_2S$ (M-262)		Lethal to cauliflower seedlings in pre-emergence damping-off control	641
		(a)	645
C_8			
$(Et_2As)_2S$		(a)	645
C_{12}			
$(i\text{-}Pr_2As)_2S$		(a)	645
C_{20}			
$[(C_5H_{11})_2As]_2S$		(a)	645
C_{24}			
$[(C_6F_5)_2As]_2S$	$R_2AsCl + Ag_2S$ in dry C_6H_6 under reflux, 86% yield	White solid, m. 85-85.5°, IR and NMR spectra, (b,c)	651
$[(4\text{-}ClC_6H_4)_2As]_2S$		(d)	645
$(Ph_2As)_2S$ (M-263)		(e)	645
C_{28}			
$[(2\text{-}MeC_6H_4)_2As]_2S$		(d)	645
$[(4\text{-}MeC_6H_4)_2As]_2S$		(d)	645
C_{32}			
$[(C_8H_{17})_2As]_2S$		(d)	645
C_{40}			
$[(C_{10}H_{21})_2As]_2S$		(a)	645
C_{64}			
$[(C_{16}H_{33})_2As]_2S$		(a)	645

(a) A fungicide effective against the Botrytis cinerea spores. (b) Rxn. with $Hg \rightarrow [(C_6F_5)_2As]_2$ (650). (c) Rxn. with $HgCl_2 \rightarrow (C_6F_5)_2AsCl$ (651). (d) A fungicide effective against the spores of Alternaria solani and Botrytis cinerea.

(e) A fungicide effective against the spores and micelia of Alternaria solani and Botrytis cinerea.

1.6.2.6 THIOARSENOSO COMPOUNDS

Several thioarsenoso compounds and their methods of preparation were reported in the main volume, pp. 263-265.

Methylthiooxoarsine, $AsCH_3S$,   MeAsS,   (M-264)
Biol. Prop.: A highly protective and curative effect in the control of rice sheath blight caused by Pellicularia sasakii (1493). Effective control of Pythium aphanidermatum in tomatoes (640).

Phenylthiooxoarsine Tetramer, $(AsC_6H_5S)_4$, $(PhAsS)_4$,   (M-264)
Synth.: From $PhAsCl_2$ by heating with $(Me_2SiS)_3$ (3:1) in MePh (20).
Prop.: m. 175-176°, mol. wt. detn. by osmometric method in C_6H_6 (20).
Rxn. with: $HgR_2 \rightarrow PhAsR_2 + HgS$ (901).

1.6.2.7 DITHIOARSONOUS ACID ESTERS WITH THIOLS

The esters, $RAs(SR')_2$, and their methods of synthesis were compiled in the main volume, pp. 265-278. Besides the methods of synthesis reported there, the esters have recently been prepared as follows: (a) from arsonous acid esters, $RAs(OR')_2$, by transesterification with mercaptans at elevated temperatures, (b) from primary dihaloarsines by a condensation with ethylene sulfide in the presence of a base, and (c) from arsenoso compounds by a condensation with 2-mercaptoethanol and a carboxylic acid in a hydrocarbon solvent, with azeotropic distn. of the water formed during the reaction. The latter reaction leads to arsylenebis(2-thioethyl carboxylates).

$$RAs(OR')_2 + 2\ R''SH \xrightarrow{\text{reflux}} RAs(SR'')_2$$

$$RAsX_2 + 2\ \overline{CH_2CH_2S} \xrightarrow{\text{base}} RAs(SCH_2CH_2X)_2$$

$$RAsO + 2\ HSCH_2CH_2OH + 2\ R'CO_2H \xrightarrow[-H_2O]{} RAs(SCH_2CH_2O_2CR')_2$$

Moreover, the dithioester, $RAs(SR')_2$, undergo transesterification through heating with higher boiling mercaptans than those of the starting dithioesters.

$$RAs(SR')_2 + 2\ R''SH \rightarrow RAs(SR'')_2$$

Many compounds of this class have shown fungicidal properties and are compiled in Table 14.

1.6.2.8 DITHIOARSONOUS ACID ANHYDROSULFIDES WITH THIOACIDS

The anhydrosulfides of this class were compiled in the main volume on pages 278-284. The compounds are biologically active as fungicides and herbicides and are compiled in Table 15, page 199.

Table 14. Dithioarsonous Acid Esters with Thiols

$RAs(SR')_2$	Yield %	Prop. and Rxn.	Prepd. from	Ref.
C_8				
$BuAs(SCH_2CH_2Cl)_2$	~100	b_1 134–136°, d_4^{23} 1.3790, n_D^{23} 1.5838	$RAsCl_2 + \overline{CH_2CH_2S}$ (2:5) in the presence of 1–2% Py	961
C_{10}				
$PhAs(SCH_2CH_2Cl)_2$	~100	d_4^{20} 1.4723, n_D^{20} 1.6543	$RAsCl_2 + \overline{CH_2CH_2S}$ (2:5) in the presence of 1–2% Py at 0°	961
C_{12}				
$4,3\text{-}HO(AcNH)C_6H_4As(SCH_2CO_2H)_2$		A chemotherapeutic for infections caused by Trichomonas vaginalis		885
$4\text{-}(H_2NCOCH_2NH)C_6H_4As(SCH_2CO_2H)_2$		In vitro tests on Trypanosoma rhodesiense		701, 703
$2,4\text{-}Cl_2C_6H_4As(SPr)_2$	67	$b_{0.25}$ 110–112°, d_4^{20} 1.3796, n_D^{20} 1.6144, MR_D 93.9, AR_{As} 10.93	$RAsCl_2$ + R'SH at 80–120°	1184
$PhAs(SCH_2CHOHCH_2Cl)_2$	97	Crude oil, unstable > 120° and in boiling H_2O, (a) n_D^{20} 1.5377, d_4^{20} 1.4686	$RAs(SEt)_2$ + R'SH at 80–120°	318
$BuAs(SCH_2CH_2OCOCCl_3)_2$			$RAsO$ + R'SH (1:2) in C_6H_6, (b)	191
$EtAs[SCH_2CH(OAc)CH_2Cl]_2$	91	Crude oil, unstable > 120° and in boiling H_2O, (a) n_D^{20} 1.5452, d_4^{20} 1.3166	$RAs(OEt)_2$ + R'SH at 80–120°	318
$BuAs(SCH_2CH_2OCOCH_2Cl)_2$			$RAsO$ + R'SH (1:2) in C_6H_6, (b)	191
$BuAs(SCH_2CH_2OAc)_2$		n_D^{20} 1.5267, d_4^{20} 1.2238	$RAsO$ + R'SH (1:2) in C_6H_6, (b)	191
$EtAs(SCH_2CH_2S\text{-}i\text{-}Pr)_2$	41	Yellowish liquid, $b_{0.9}$ 151°	$RAs(OMe)_2$ + R'SH (1:2) under reflux and distg. MeOH	317
C_{13}				
$4\text{-}MeC_6H_4As(SPr)_2$	68	b_1 184–186°, n_D^{20} 1.6112, d_4^{20} 1.2386, MR_D 89.17, AR_{As} 11.32	$RAsCl_2$ + R'SH at r.t. and blowing out HCl with N	1184
C_{14}				
$PhAs(SCH_2CH_2OCOCCl_3)_2$		oil, n_D^{20} 1.5850	$RAsO$ + $HSCH_2CH_2OH$ + CCl_3CO_2H in MePh, (b)	191

Compound	Method	Yield	Properties	Ref.
PhAs(SCH$_2$CH$_2$OCOCH$_2$Cl)$_2$	RAsO + HSCH$_2$CH$_2$OH + ClCH$_2$CO$_2$H in MePh, (b)		n_D^{20} 1.5950, d_4^{20} 1.4922	191
EtAs(SC$_6$H$_4$NH$_2$-2·HCl)$_2$ (M-269)			(a)	1132
PhAs(SCH$_2$CH$_2$OAc)$_2$	RAsO + HSCH$_2$CH$_2$OAc in C$_6$H$_6$, (b)		n_D^{20} 1.5900, d_4^{20} 1.3744, MR$_D$ 95.8	191
2,4-Cl$_2$C$_6$H$_4$As(SBu)$_2$	RAsCl$_2$ + R'SH at r.t. and blowing out HCl with N	72	b_2 175-177°, n_D^{20} 1.5850, d_4^{20} 1.2963, MR$_D$ 103.2, AR$_{As}$ 11.00	1184
PhAs(S-t-Bu)$_2$	RAsO + R'SH in C$_6$H$_6$ under reflux, with distn. of H$_2$O		m. 38-40°, a fungicide	596a
EtAs(SCH$_2$CH$_2$NEt$_2$)$_2$			(a)	1132
EtAs(SCH$_2$CH$_2$NEt$_2$·HCl)$_2$			(a)	1132
C$_{16}$				
4-O$_2$NC$_6$H$_4$As[SCH$_2$CH(OAc)CH$_2$Cl]$_2$	RAs(OEt)$_2$ + R'SH at 80-120°	90	Crude oil, unstable > 120° and in boiling H$_2$O, (a)	318
PhAs[SCH$_2$CH(OAc)CH$_2$Cl]$_2$	RAs(OEt)$_2$ + R'SH at 80-120°	96	Crude oil, unstable > 120° and in boiling H$_2$O, (a)	318
PhAs(SCH$_2$CH$_2$S-i-Pr)$_2$	RAs(OMe)$_2$ + R'SH by heating and distg. MeOH	52	$b_{0.003}$ 204°, soly. data	317
C$_{18}$				
PhAs(SC$_6$H$_4$Cl-4)$_2$	RAsO + R'SH in C$_6$H$_6$ under reflux with distn. of H$_2$O		m. 61-63°, useful as a fungicide	596a
4-O$_2$NC$_6$H$_4$As(SPh)$_2$	RAs(OEt)$_2$ + R'SH at 80-120°	67	Crude prod. m. 79.5-80.5°, unstable > 120° and in boiling H$_2$O	318
PhAs(SC$_6$H$_4$NH$_2$-2)$_2$ (M-271)	R'SH·HCl + NaOMe in MeOH at 50°, followed by RAsCl$_2$ in C$_6$H$_6$		(a)	1133
PhAs(SC$_6$H$_4$NH$_2$-2·HCl)$_2$ (M-271)			(a)	1132
4-O$_2$NC$_6$H$_4$As(S–[pyrimidinyl-Me$_2$])$_2$		56	m. 184-185°	316

(a) A fungicide effective against Alternaria radicina, Botrytis cinerea, Fusicladium dendriticum, Trichophyton gypseum, and Epidermophyton; toxic to mice (1132, 1133). (b) The rxn. is carried out in a Dean-Stark apparatus.

Table 14 Continued

RAs(SR')₂	Prepd. from	Yield %	Prop. and Rxn.	Ref.
C_{18} (Cont.)				
PhAs(S–[pyrimidine ring: N, Me, N, Me]Me)₂	RAsCl₂ + R'SH in the same manner as the preceding compd.	86	White crystals, m. 150–152°	316
4-O₂NC₆H₄As(SCH₂CH₂NEt₂·HCl)₂			(a)	1132
PhAs(SCH₂CH₂NEt₂·HCl)₂			(a)	1132, 1133
C_{20}				
PhAs(SCH₂Ph)₂ (M-272)	RAsO + R'SH in C₆H₆ under reflux, with distn. of H₂O		m. 62–64°, a fungicide	596a
	RAs(OEt)₂ + R'SH at 80–120°	97	Crude prod. m. 53–55°, unstable >120° and in boiling H₂O	318
C_{22}				
PhAs(SC₈H₁₇-n)₂	RAsO + R'SH in C₆H₆ under reflux, with distn. of H₂O		Oil, a fungicide	596a
C_{25}				
MeAs(SC₁₂H₂₅)₂ (M-275)			Fungicidal effect against Hypochnus sasakii	1406
C_{28}				
PhAs(SC₁₀H₇)₂ (M-275)	RAsO + 2-C₁₀H₇SH in C₆H₆ under reflux, with distn. of H₂O		m. 155–157°, a fungicide	596a
PhAs(SCH₂CH₂OCOCH₂OC₆H₃Cl₂-2,4)₂	RAsO + HSCH₂CH₂OH + 2,4-Cl₂C₆H₃-OCH₂CO₂H in MePh, (b)		Glassy solid	191
C_{30}				
PhAs(SCH₂CH₂OCOCH₂OC₆H₃Me₂-3,5)₂	RAsO + HSCH₂CH₂O + 3,5-Me₂C₆H₃-OCH₂CO₂H in MePh, (b)		Glassy solid	191

Table 15. Dithioarsonous Acid Anhydrosulfides with Thioacids

Anhydrosulfides	Prepd. from	Prop.	Ref.
RAs[SC(S)R']			
C_7			
MeAs[SC(S)Et]$_2$		Fungicidal effect against Hypochnus sasakii	1406
C_{17}			
MeAs[SC(S)CH$_2$Ph]$_2$		Fungicidal effect against Hypochnus sasakii	1406
RAs[SC(S)OR']$_2$			
C_{12}			
PhAs[SC(S)OEt]$_2$ (M-280)		Phytotoxic barley	1122
C_{14}			
PhAs[SC(S)OPr]$_2$ (M-280)		Phytotoxic to barley	1122
RAs[SP(S)(OR')$_2$]$_2$			
C_{14}			
EtAs[SP(S)(O-i-Pr)$_2$]$_2$	RAsCl$_2$ + NaSP(S)(O-i-Pr)$_2$ in C$_6$H$_6$, 97% yield	Oily prod.	315
RAs[SC(S)NR'$_2$]$_2$			
C_7			
MeAs[SC(S)NMe$_2$]$_2$ (M-282)		Phytotoxic to barley	1122
		In mixt. with Tuzet and Ziram has a fungicidal effect on Mycosphaerella musicola in banana	895
		Protective and curative effects in the control of rice sheath blight caused by Pellicularia sasakki	1493

Table 15 Continued

Anhydrosulfides	Prepd. from	Prop.	Ref.
RAs[SC(S)NR$'_2$]$_2$ (Cont.)			
C$_7$(Cont.)			
MeAs[SC(S)NMe$_2$]$_2$ (M-282) (Cont.)		Growth inhibition of Alternaria kikuchiana	1520
C$_{12}$			
3,4-Cl$_2$C$_6$H$_3$As[SC(S)NMe$_2$]$_2$		Moderately active against Trichophyton asteroides	1473
4-ClC$_6$H$_4$As[SC(S)NMe$_2$]$_2$ (M-283)		Inhibition of the grown of Xanthomonas oryzae	1493
C$_{16}$			
PhAs[SC(S)NEt$_2$]$_2$ (M-283)		Crystallographic data	113
C$_{17}$			
MeAs[SC(S)NHNHCOC$_6$H$_4$OH-4]$_2$	RAsCl$_2$ + KSC(S)NHR' (1:2) in H$_2$O	m. 78-80° (dec.), a fungicide	876
MeAs[SC(S)NHC$_6$H$_4$Me-4]$_2$		Fungicidal effect against Hypochnus sasakii	1406

1.7   COMPOUNDS OF ARSENIC BONDED TO GROUP VA ELEMENTS

1.7.1   COMPOUNDS WITH ARSENIC-NITROGEN BONDS

1.7.1.1 AMINOARSINES, $R_2AsNR'_2$

The amino derivatives of secondary arsines were compiled in the main volume on pages 285-287. Besides the methods of synthesis described in the main volume, the aminoarsines have recently been prepared as follows:

(a) From secondary haloarsines by a reaction with phosphinoamines, phosphonium imides, lithium amides, and imidinolithium derivatives, respectively;

(b) From diarsinoamines by a reaction with secondary halophosphines; and

(c) From tertiary arsines by a reaction with secondary haloamines.

$$R_2AsX + R'_2PNH_2 \rightarrow R_2AsNHPR'_2$$

$$R_2AsX + R'_3P=NR'' \rightarrow [R_2AsN(R'')PR'_3]X$$

$$R_2AsX + LiNR'_2 \rightarrow R_2AsNR'_2$$

$$R_2AsX + LiN=PR'_3 \rightarrow R_2AsN=PR'_3$$

$$R_2AsX + PC(=NLi)N(R')SiR''_3 \longrightarrow \begin{array}{c} R_2AsN(R')C(Ph)=NSiR''_3 \\ + \\ R_2AsN=C(Ph)N(R')SiR''_3 \end{array}$$

$$(R_2As)_2NH + R'_2PCl \rightarrow R_2AsNHPR'_2$$

$$R_3As + R'_2NX \rightarrow R_2AsNR'_2$$

Some compounds of this class are described individually hereafter, while others are compiled in Table 16.

(Dichloroboryl)[bis(trifluoromethyl)arsino]amine, $AsBC_2HCl_2F_6N$,
    $(CF_3)_2AsNHBCl_2$
The formation of this species was detected by IR spectroscopy during the reaction of $[(CF_3)_2As]_2NH$ with BCl_3 at 25° (1416).

[Bis(trifluoromethyl)arsino][bis(trifluoromethyl)phosphino]amine,
    $AsC_4HF_{12}NP$, $(CF_3)_2AsNHP(CF_3)_2$
Synth.: From $(CF_3)_2AsCl$ by a rxn. with $(CF_3)_2PNH_2$ in the presence of Me_3N at 25° (1416).
From $[(CF_3)_2As]_2NH$ by a rxn. with $(CF_3)_2PCl$ (1:3) in the presence of Me_3N at 25°, a mixt. of the title compd. and $[(CF_3)_2P]_2NH$ was obtd.; at 80° the rxn. was nearly complete (1416).
Prop.: m. -44.8 to -44.4°, vapor pressure data, IR spectrum (1416).
Rxn. with: Alkali → HCF_3 (1416).
Dry HCl → $(CF_3)_2AsCl$, $(CF_3)_2PCl$, and NH_4Cl (1416).
NH_3 in the vapor phase, no rxn. (1416).
NH_3 in the liquid phase → $(CF_3)_2PNH_2$, $(CF_3)_2AsNH_2$, and $[(CF_3)_2As]_2NH$ (1416).
Me_3N at -78 to -30° → $(CF_3)_2AsNHP(CF_3)_2 \cdot Me_3N$ (1416).

N,N-Dimethyl-N'-[bis(trifluoromethyl)arsino]hydrazine, $AsC_4H_7F_6N_2$,
     $(CF_3)_2AsNHNMe_2$

Synth.: From $(CF_3)_2AsCl$ and Me_2NNH_2 in Et_2O at a low temp., 80% yield (1221).
Prop.: White solid, m. 3° to a colorless liquid; a supercooled liquid is intensely yellow. IR and NMR spectra. The compd. dec. slowly at r.t. and more rapidly at 80° to CHF_3, N_2, and a yellow prod. (1221).
Rxn. with: HCl gas → $(CF_3)_2AsCl$ and $Me_2NNH_2 \cdot HCl$ (1221).

[Bis(trifluoromethyl)arsino][bis(trifluoromethyl)phosphino]methylamine,
     $AsC_5H_3F_{12}NP$, $(CF_3)_2AsN(Me)P(CF_3)_2$

Synth.: From $(CF_3)_2AsCl$ by the rxn. with $(CF_3)_2PNHMe$ in the presence of Me_3N at 25°, followed by heating to 80° (1416).
Prop.: m. -30.7 to -30.0°, vapor pressure data, IR spectrum (1416).
Rxn. with: Alkali → HCF_3 (1416).
Dry HCl → $(CF_3)_2AsCl$, $(CF_3)_2PCl$, and $MeNH_3Cl$ (1416).
BCl_3 at 25° → traces of BF_3 and $(CF_3)_2AsCl$ (1416).
NH_3 in the vapor phase → $(CF_3)_2AsNH_2$, $(CF_3)_2PNHMe$, and $[(CF_3)_2As]_2NH$ (1416).

[Bis(trifluoromethyl)arsino]bis[bis(trifluoromethyl)phosphino]amine,
     $AsC_6F_{18}NP_2$, $(CF_3)_2AsN[P(CF_3)_2]_2$

Synth.: From $[(CF_3)_2P]_2NH$ by treating with Na in THF at 25°, followed by $(CF_3)_2AsCl$ at 0°, without any solvent (1416).
Prop.: m. 25.2-25.9°, vapor pressure data, IR spectrum (1416).
Rxn. with: Alkali → HCF_3 (1416).
HCl at 25° → $(CF_3)_2AsCl$ and $[(CF_3)_2P]_2NH$ (1416).
NH_3 → $(CF_3)_2AsNH_2$, $[(CF_3)_2As]_2NH$, and $[(CF_3)_2P]_2NH$ (1416).

N-(or N'-)-(Dimethylarsino)-N-methyl-N'-(or N-)-(trimethylsilyl)benzamidine
     Isomers, $AsC_{13}H_{23}N_2Si$, $PhC(=NAsMe_2)N(Me)SiMe_3$ and
                                             $PhC(=NSiMe_3)N(Me)AsMe_2$

Synth.: From Me_2AsCl and $LiN=C[N(Me)SiMe_3]Ph$ in Et_2O at r.t., a mixt. of the isomers, formed by intramolecular rearrangement, was obtd. with 72% yield (1344).
Prop.: Colorless oil, $b_{0.1}$ 88-92°, consists of the isomers mixt., extremely sensitive to air and moisture, dec. within a few days in storage in a closed flask. IR and NMR spectra (1344).

[N-(Dimethylarsino)anilino]triphenylphosphonium Salts, $AsC_{26}H_{26}NPX$,
     $[Ph_3PN(Ph)AsMe_2]X$ where X = Br, $HgBr_3$, BPh_4, $[Cr(SCN)_4(NH_3)_2]$

Synth.: From $Ph_3P=NPh$ in Et_2O by adding dropwise Me_2AsBr soln. in Et_2O, the bromide was obtd. in 71% yield. A conversion of the bromide with $HgBr_2$, $K[BPh_4]$, and $NH_4[Cr(SCN)_4(NH_3)_2]$ in Et_2O gave the $[HgBr_3]^-$, BPh_4^-, and $[Cr(SCN)_4(NH_3)_2]^-$ salts, resp. (1008).
Prop.: The bromide forms white crystals, m. 194-198°, sinters at 183-188°. The tribromomercurate, m. 147-148°; the tetraphenylborate and the reineckate form white crystals (1008).

Table 16. Aminoarsines, $R_2AsNR'_2$

$R_2AsNR'_2$	Prepd. from	Prop. and Rxn.	Ref.
C_2			
$(CF_3)_2AsNH_2$ (M-286)	$(CF_3)_2AsNHP(CF_3)_2$ + NH_3		1416
	$(CF_3)_2AsN(Me)P(CF_3)_2$ + NH_3		1416
C_4			
$(CF_3)_2AsN(CF_3)_2$	$(CF_3)_3As + (CF_3)_2NX$ (1:1), X = Cl or Br, (a)	b. 70° (extrapolated)	71, 72
		Thermodynamic and kinetic data, IR and NMR spectra	72
Me_2AsNMe_2 (M-286)		Rxn. with $Me_2NH \rightarrow (Me_2As)_2$ and Me_2NH	419
C_5			
$Me_2AsN=PMe_3$	$Me_2AsCl + LiN=PMe_3$ in Et_2O, 83% yield	b_{12} 96-99°, IR and NMR spectra, (b)	1353
C_6			
$Me_2AsN(Me)SiMe_3$ (M-286)	Me_2AsCl in C_5H_{12} + $LiN(Me)SiMe_3$ in C_6H_{14} at -70° and warming to r.t., 44% yield	Rxn. with MeI $\rightarrow [Me_3PNAsMe_3]I$	1352
		Colorless liquid, b_{11} 44-46°, m. -50°, NMR spectrum	1343
C_7			
$Me_2AsN(Me)\overline{PN(Me)CH_2CH_2}NMe$	$MeNHPN(Me)CH_2CH_2NMe$ in C_6H_{14} + BuLi, followed by Me_2AsCl, 80% yield	$b_{0.1}$ 76-78°, m. -38 to -37°, NMR spectrum	1346
C_8			
$Me_2AsN=PEt_3$	$Me_2AsCl + LiN=PEt_3$ in Et_2O, 69% yield	b_{12} 114-115°, IR and NMR spectrum	1353
$Me_2AsN(SiMe_3)_2$ (M-287)	$Me_2AsSAs(S)Me_2 + NaN(SiMe_3)_2$ in C_6H_6 at 20°, 68% yield	Colorless liquid, b_1 45-46°, m. -46°, NMR spectrum	1345
C_{10}			
$Bu_2As\overline{NCH_2CH_2}$ (M-287)		IR and Raman spectra	1387a
Pr_2AsNEt_2 (M-287)		Rxn. with $CH_2=C=O \rightarrow Pr_2AsCH_2CONEt_2$	901

(a) $[(CF_3)_2N]_2AsCF_3$ is also formed. (b) See also ref. 1352.

1.7.1.4 DIARSINOAMINES, (R$_2$As)$_2$NR'

Bis[bis(trifluoromethyl)arsino]amine,   As$_2$C$_4$HF$_{12}$N,   [(CF$_3$)$_2$As]$_2$NH,   (M-288)
Synth.: From (CF$_3$)$_2$AsNHP(CF$_3$)$_2$ by ammonolysis in the liquid phase and from
(CF$_3$)$_2$AsN(Me)P(CF$_3$)$_2$ by ammonolysis in the gas phase (1416).
Rxn. with: BF$_3$ at 85° → (CF$_3$)$_2$AsF (1416).
BCl$_3$ at 25° → extensive conversion to (CF$_3$)$_2$AsCl, a polymeric (HNBCl)$_n$ prod.,
and an intermediate corr. to the formula (CF$_3$)AsNHBCl$_2$ was detected by IR
spectroscopy (1416).
(CF$_3$)$_2$PCl in the presence of Me$_3$N → (CF$_3$)$_2$AsNHP(CF$_3$)$_2$ and [(CF$_3$)$_2$P]$_2$NH (1416).

1.7.1.6 DIAMINOARSINES, (R$_2$N)$_2$AsR

A few diaminoarsines and their methods of synthesis were described in the
main volume on pages 289-90. Besides the methods reported there, the compounds
have been prepared as follows:

(a) From tertiary arsines by a reaction with secondary haloamimes,

(b) From triaminoarsines by treating with trifluoroiodomethane at elevated
temperatures under pressure, and

(c) From primary dihaloarsines by a reaction with lithium amides and imino-
phosphorane derivatives (ylids), respectively.

Bis[bis(trifluoromethyl)amino](trifluoromethyl)arsine,   AsC$_5$H$_{15}$N$_2$,
    [(CF$_3$)$_2$N]$_2$AsCF$_3$
Synth.: From (CF$_3$)$_3$As by reacting with (CF$_3$)$_2$NX (where X = Cl or Br), a mixt.
of the title compd. with (CF$_3$)$_2$NAs(CF$_3$)$_2$ was obtd. (71, 72).
Prop.: b. 109° (extrapolated) (71, 72); vapor pressure and thermodynamic data,
IR and NMR spectra (72).
Rxn. with: (CF$_3$)$_2$NI (1:1) → an unstable 1:1 adduct (71).

Bis(dimethylamino)(trifluoromethyl)arsine,   AsC$_5$H$_{12}$F$_3$N$_2$,   (Me$_2$N)$_2$AsCF$_3$
Synth.: From (Me$_2$N)$_3$As by treating with CF$_3$I (2:1) at 145° in a sealed tube
(73).
Prop.: b. 138° (extrapolated), thermodynamic data, IR and NMR spectra (73).

Methylbis[methyl(trimethylsilyl)amino]arsine, Methylarsylenebis[methyl(tri-
    methylsilyl)amine],   AsSi$_2$C$_9$H$_{27}$N$_2$,   MeAs[N(Me)SiMe$_3$]$_2$
Synth.: From MeAsCl$_2$ in Et$_2$O by adding dropwise to LiN(Me)SiMe$_3$ in C$_6$H$_{14}$ at
-70° and completing the rxn. under reflux, 52% yield (1343).
Prop.: Crude prod. b$_{0.1}$ 55-59°, NMR spectrum (1343).

Bis(1-aziridinyl)phenylarsine,   AsC$_{10}$H$_{13}$N$_2$,   ($\overline{\text{CH}_2\text{CH}_2\text{N}}$)$_2$AsPh,   (M-290)
Prop.: IR spectrum (1387a).

Bis[N-(bromotriphenylphosphoranyl)anilino]methylarsine, AsC$_{49}$H$_{43}$Br$_2$N$_2$P$_2$,
{[Ph$_3$PN(Ph)]AsMe}Br$_2$

Synth.: From Ph$_3$P=NPh in Et$_2$O by adding dropwise MeAsBr$_2$ soln. in Et$_2$O, 79%
yield (1008).

Prop.: Crystals, m. 198-200° (1008).

N,N'-(Methylarsylene)-p-phenylenebis(aminotriphenylphosphonium Bromide),
(AsC$_{43}$H$_{37}$Br$_2$N$_2$P$_2$)$_n$

Synth.: From Ph$_3$P=NC$_6$H$_4$N=PPh$_3$ in dioxan by adding
dropwise a soln. of MeAsBr$_2$ in the same solvent,
11% yield (1008).

Prop.: Crystals, m. 272-280° (1008).

1.7.2   COMPOUNDS WITH ARSENIC-PHOSPHORUS BONDS

1.7.2.1 PHOSPHINOARSINES, R$_2$PAsR'$_2$

The following four compounds constitute a new class of phosphino-arsenic
derivatives.

[Bis(trifluoromethyl)arsino]bis(trifluoromethyl)phosphine, AsC$_4$F$_{12}$P,
(CF$_3$)$_2$AsP(CF$_3$)$_2$

Synth.: From (CF$_3$)$_2$AsH by a reaction with [(CF$_3$)$_2$P-]$_2$ at r.t. (311).
From [(CF$_3$)$_2$As-]$_2$ by metathesis with [(CF$_3$)$_2$P-]$_2$ in CCl$_3$F at r.t. (311).

Prop.: NMR and mass spectra (311).

[Bis(trifluoromethyl)arsino]dimethylphosphine, AsC$_4$H$_6$F$_6$P, (CF$_3$)$_2$AsPMe$_2$

Synth.: From [(CF$_3$)$_2$As-]$_2$ by mixing with Me$_2$PH at -63°, warming to r.t. and
maintaining at 24° for 24 hrs., 90% yield (311).

Prop.: b. 123°, vapor pressure and thermodynamic data; IR, NMR and mass
spectra (311).

(Dimethylarsino)bis(trifluoromethyl)phosphine, AsC$_4$H$_6$F$_6$P, Me$_2$AsP(CF$_3$)$_2$

Synth.: From (CF$_3$)$_2$P-]$_2$ by mixing with Me$_2$AsH at -63°, warming up to r.t.,
and maintaining at 24° for 24 hrs., 96% yield (311).

Prop.: b. 125°, vapor pressure and thermodynamic data, IR, NMR, and mass
spectra (311).

(Diphenylarsino)diphenylphosphine, AsC$_{24}$H$_{20}$P, Ph$_2$AsPPh$_2$

Synth.: From Ph$_2$AsSnMe$_3$ by mixing with Ph$_2$PCl in Et$_2$O at r.t., removing the
solvent, and treating the residue with C$_5$H$_{12}$, 71% yield (1373).

Prop.: Colorless crystals, m. 115-117°, sol. in org. solvents, sensitive to
oxygen; IR spectrum (1373).

1.7.2.3 ARSINOPHOSPHONIUM DERIVATIVES

Several arsinophosphonium derivatives and their methods of synthesis were reported in the main volume on pages 290-91. As previously, the following compounds were prepared from tertiary phosphines through the coordination with primary and secondary haloarsines, respectively.

Table 17. Arsinophosphonium Derivatives

$[R_2AsPR'_3]X$	Prepd. from	Prop.	Ref.
C_{13}			
$[MeAs(Cl)PBu_3]Cl$	$Bu_3P + MeAsCl_2$	White cryst. solid, m. 69-70°, (a)	798
$[MeAs(I)PBu_3]I$	$Bu_3P + MeAsI_2$	Yellow cryst. solid, m. 89-90°, (a)	798
C_{14}			
$[Me_2AsPBu_3]I$	$Bu_3P + Me_2AsI$	White cryst. solid, m. 81-82°, (a)	798
C_{18}			
$[PhAs(Cl)PBu_3]Cl$	$Bu_3P + PhAsCl_2$	White cryst. solid, m. 80-81°, (a)	798
$[PhAs(I)PBu_3]I$	$Bu_3P + PhAsI_2$	Orange cryst. solid, m. 94-96°, (a)	798

(a) IR and NMR spectra (798).

1.7.2.4 DIARSINOPHOSPHINES, $(R_2As)_2PR'$

The following diarsinophosphine is a new type compound.

Bis(diphenylarsino)phenylphosphine, $As_2C_{30}H_{25}P$, $(Ph_2As)_2PPh$
Synth.: From Ph_2AsCl by mixing with $(Me_3Sn)_2PPh$ in Et_2O at r.t., removing the solvent, and treating the residue with C_5H_{12}, 90% yield (1373).
Prop.: Colorless crystals, m. 155-158°, sol. in org. solvents, sensitive to oxygen, IR spectrum (1373).

1.7.2.5 TRIARSINOPHOSPHINES, $(R_2As)_3P$

The following triarsinophosphine represents a new class of arsenic-phosphorus derivatives.

Tris(diphenylarsino)phosphine, $As_3C_{36}H_{30}P$, $(Ph_2As)_3P$
Synth.: From Ph_2AsCl by mixing with $(Me_3Sn)_3P$ in Et_2O at r.t., removing the solvent, and treating the residue with C_5H_{12}, 60% yield (1373).
Prop.: Colorless crystals, m. 169-172°, sol. in org. solvents, sensitive to air, IR spectrum (1373).

1.7.2.6 DIPHOSPHINOARSINES, $(R_2P)_2AsR'$

The following diphosphinoarsine is the first representative of organoarsenic bonded to two phosphorus atoms.

Phenylarsylenebis(diphenylphosphine), $AsC_{30}H_{25}P_2$, $(Ph_2P)_2AsPh$
Synth.: From Ph_2PCl by mixing with $(Me_3Sn)_2AsPh$ in Et_2O at r.t., removing the solvent, and treating the residue with C_5H_{12}, 66% yield (1373).
Prop.: Colorless crystals, m. 125-129°, sol. in org. solvents, sensitive to oxygen; IR spectrum (1373).

1.7.2.7 ARSINOPHOSPHONIC ACID DERIVATIVES

Numerous compounds of this class were compiled in the main volume on pages 292-93.

Dimethylarsinophosphonic Acid Dimethyl Ester, $AsC_4H_{12}O_3P$,
    $Me_2AsP(O)(OMe)_2$, (M-292)
Synth.: From Me_2AsSCN by mixing with $(MeO)_3P$ (1:1) in C_6H_6 at 7-10° and warming the mixt. to r.t., 92% yield (871).
Prop.: b_1 59-60°, d_4^{20} 1.4248, n_D^{20} 1.5085 (871).

1.8 COMPOUNDS OF ARSENIC BONDED TO METALS

1.8.1 COMPOUNDS WITH ELEMENTS OF GROUPS IA - IVA

1.8.1.1 COMPOUNDS WITH ALKALI METALS

Numerous arsinoalkali metal derivatives and their methods of synthesis were compiled in the main volume on pages 294-96. They include the following types: R_2AsM, $RAsHM$, $RAsM_2$, $RAs(M)(CH_2)_nAs(M)R$, and $R_2As(CH_2)_nAs(M)K$, in which M is Li, Na, or K. Besides the methods described in the main volume, the arsino-alkali metal derivatives have also been prepared from primary and secondary haloarsines by a reaction with alkali metals, from secondary arsines, and from tetraorganodiarsines by a reaction with butyllithium, and from 1,2-diarsolane and from 1,2-diarsenane through a ring cleavage with phenyllithium or butyllithium.

Some compounds of this class are described individually hereafter, while others are compiled in Table 18.

Diphenylarsinolithium, $AsLiC_{12}H_{10}$, Ph_2AsLi, (M-295)
Synth.: From Ph_3As + Li (1:2) in THF, followed by the elimination of PhLi with t-BuCl, the compd. was obtd. as a soln. in THF (43, 88).
From Ph_2AsH + PhLi in Et_2O, 82% yield (1531).
From $(Ph_2As-)_2$ + PhLi in Et_2O, followed by Et_2O-dioxan, 92% yield in the form of $Ph_2AsLi \cdot C_4H_8O_2$ adduct (786).
Prop.: Yellow solid, sol. in Et_2O and THF, insol. in C_6H_6 and petr. ether

(1531). The $Ph_2AsLi \cdot C_4H_8O_2$ adduct is a yellow solid (M-295); its molar cond. was detd. in DMSO (789).

Rxn. with: cis- and trans-$PhCH=CH_2Br$ → cis- and trans-$PhCH=CHAsPh_2$ (43, 88).
$PhC \equiv CPh$ in THF in the presence of R_2NH or RNH_2 → cis- and trans-$PhCH=C(Ph)AsPh_2$ (44, 88).
trans-$PhCH=CBrPh$ → cis-$PhCH=C(Ph)AsPh_2$ (44).
$PhC \equiv CH$ in THF + RNH_2 or R_2NH → cis-$PhCH=CHAsPh_2$ (44, 88).
cis- and trans-$ClCH=CHCl$ → cis- and trans-$Ph_2AsCH=CHAsPh_2$, resp. (45, 88).
RNCS in Et_2O, followed by hydrolysis → $Ph_2AsC(S)NHR$ (1537).
Indene in THF $\rightleftharpoons$ Ph_2AsH + indenyl-Li, 31% exchange (787).
t-BuOH in THF $\rightleftharpoons$ Ph_2AsH + t-BuOLi, 82% exchange (787).
THF by heating, followed by a treatment with H_2O → $Ph_2As(CH_2)_4OH$ (1531).
RNCO in Et_2O at r.t., followed by heating and dec. with H_2O → $Ph_2AsCONHR$ (1536).
Triphenylisocyanuric acid in Et_2O, followed by dec. with H_2O → $(Ph_2As)_2$ + $PhNH_2$ (1536).
RNCO in Et_2O, followed by hydrolysis → $(Ph_2As)_2$ + $PhNHCONHPh$ (1536).
$RN=C=NR$ → $Ph_2AsC(=NR)N(Li)R$ (1538).
$R\overline{CHCH_2O}$ (using $Ph_2AsLi \cdot C_4H_8O_2$) in Et_2O or dioxan, followed by hydrolysis → $Ph_2As\overline{CH_2CH(R)O}H$ (1530).
Phenylene oxide (using $Ph_2AsLi \cdot C_4H_8O_2$) in Et_2O or dioxan, followed by hydrolysis → o-$HOC_6H_4AsPh_2$ (1530).

Dicyclohexylarsinolithium, $AsLiC_{12}H_{22}$, $(C_6H_{11})_2AsLi$, (M-295)
Synth.: From $(C_6H_{11})_2AsH$ in Et_2O by treating with PhLi, 92% yield (1531).
Prop.: Pale yellow solid, sol. in THF and dioxan, insol. in C_6H_6 and petr. ether (1531).
Rxn. with: $R\overline{CHCH_2O}$ in Et_2O, followed by hydrolysis → $(C_6H_{11})_2AsCH_2CH(R)OH$ (1531).
$Cl(CH_2)_nCl$ (1:1) → $(C_6H_{11})_2As(CH_2)_nCl$ (1533).
$Cl(CH_2)_nCl$ (2:1) → $(C_6H_{11})_2As(CH_2)_nAs(C_6H_{11})_2$ (1533).
$MgBr_2 \cdot 4$ THF in THF → $[(C_6H_{11})_2As]_2Mg$ (1534).
PhNCO in Et_2O and dec. with H_2O → $(C_6H_{11})_2AsCONHPh$ (1536).
RNCS in Et_2O and dec. in H_2O → $(C_6H_{11})_2AsCSNHR$ (1537).
RNCS in Et_2O, followed by dioxan → $(C_6H_{11})_2AsCSN(Li)R$ (1537).
RNCS in Et_2O under reflux, replacing the solvent by C_6H_6, and treating with EtBr → $(C_6H_{11})_2AsC(SEt)=NR$ (1537).
$RN=C=NR$ → $(C_6H_{11})_2AsC(=NR)N(Li)R$ (1538).

Phenylarsylenedilithium, $AsLi_2C_6H_5$, $PhAsLi_2$
Synth.: From $PhAsCl_2$ by adding dropwise to a soln. of Li in THF at -25°, maintaining the temp. at -30 to -15° and then gradually warming to +10° (1514).
Prop.: Red soln. in THF (1514).
Rxn. with: 2-$(BrCH_2)C_6H_4CH_2CH_2Br$ → 2-phenyl-1,2,3,4-tetrahydroisoarsinoline (1514).
EtBr → $PhAsEt_2$ (1514).
$BrCH_2CH_2Br$ → $(PhAs \lessdot)_n$ (1514).

Dimethylarsinosodium,　　　$AsNaC_2H_6$,　　　Me_2AsNa

Synth.: From Me_2AsI + Na wire in THF at r.t. and completing the rxn. under reflux (1229).

Prop.: Green soln. in THF (1229).

Rxn. with: 8-Chloroquinoline → 8-(Me_2As)-quinoline (121a).

$MeC(CH_2Cl)_3$ in THF at 10° → $MeC(CH_2AsMe_2)_3$ (1229).

RBr → Me_2AsR (1229, 1415).

$ClCH_2CH_2Cl$ → $Me_2AsCH_2CH_2AsMe_2$ (1229).

Phenylarsinopotassium,　　　$AsKC_6H_6$,　　　PhAsHK,　　　(M-296)

Rxn. with: Indene in THF ⇌ $PhAsH_2$ + indenyl-K, 80% exchange (787).

$X(CH_2)_nX$ (X = Br or Cl, n = 3 or 4) in C_6H_6 or in THF on a steam bath → $PhAsH(CH_2)_nAsHPh$ and $PhAs\overline{(CH_2)_n}AsPh$ (1535).

Diphenylarsinopotassium,　　　$AsKC_{12}H_{10}$,　　　Ph_2AsK,　　　(M-296)

Synth.: From Ph_2AsH + K in $(EtOCH_2CH_2)_2O$ (1347).

From Ph_3As and from Ph_2AsCl by treating with K in dioxan under reflux, the $Ph_2AsK\cdot2$ Dioxan complex was obtd. (1104, 1105).

Prop.: Red soln. in $(EtOCH_2CH_2)_2O$ (1347). Molar conductance of $Ph_2AsK\cdot2$ Dioxan in DMSO was detd. (789).

Rxn. with: RSO_3Na → Ph_2AsR (1347).

$4\text{-}ClC_6H_4AsSO_3Na$ → $Ph_2AsC_6H_4AsPh_2$ (1347).

$p\text{-}Br_2C_6H_4$ in THF → $Ph_2AsC_6H_4AsPh_2$ (1658).

RCOX → Ph_2AsOCR (967).

$4,3,5\text{-}HO(t\text{-}Bu)_2C_6H_2I$ → $4,3,5\text{-}KO(t\text{-}Bu)_2C_6H_2AsPh_2$ + Ph_2AsH + KI (1104, 1105).

α-Chloropicoline → $Ph_2AsCH_2\overline{C=CHCH=CHCH=N}$ (1543).

$N(CH_2CH_2Cl)_3$ → $N(CH_2CH_2AsPh_2)_3$ (1315).

2-(2-Chloroethyl)pyridine → $Ph_2AsCH_2CH_2\overline{C=CHCH=CHCH=N}$ (1542a).

$HN(CH_2CH_2Cl)_2$ → $HN(CH_2CH_2AsPh_2)_2$ (1315).

$O(CH_2CH_2Cl)_2$ → $O(CH_2CH_2AsPh_2)_2$ (1315).

$S(CH_2CH_2Cl)_2$ → $S(CH_2CH_2AsPh_2)_2$ (1315).

$Cl(CH_2)_nCl$ → $Ph_2As(CH_2)_nAsPh_2$ (1315).

RCl → Ph_2AsR (1316, 1543, 1542a).

$\overline{CH_2CH_2CH_2O}$ in Et_2O or dioxan, followed by H_2O → $Ph_2AsCH_2CH_2CH_2OH$ (1531).

$R\overline{CHCH_2O}$ in Et_2O or dioxan, followed by H_2O → $Ph_2AsCH_2CH(R)OH$ (1531).

Phenylene oxide in Et_2O or dioxan, followed by H_2O → $2\text{-}HOC_6H_4AsPh_2$ (1531).

Table 18. Arsino-Alkali Metal Compounds

Arsino-Alkali Metal	Prepd. from	Yield %	Prop. and Rxn.	Ref.
R_2AsLi				
C_2				
Me_2AsLi	$Me_2AsH + BuLi$		(a)	999
	$(Me_2As-)_2 + BuLi$ in C_6H_{12}		Rxn. with $MeC(CH_2Cl)_3$ in $Et_2O \rightarrow MeC(CH_2AsMe_2)_3$	1229
C_8				
$EtPhAsLi$ (M-295)			Rxn. with fluorene $\rightleftarrows$ EtPhAsH + fluorenyl-Li	787
			Rxn. with $Ph_2NH \rightleftarrows EtPhAsH + Ph_2NLi$	787
Bu_2AsLi	$Bu_2AsH + PhLi$ in Et_2O		Colorless powder, sensitive to air, sol. in Et_2O, dioxan, and THF, (b, c)	1530
$t-Bu_2AsLi$	$t-Bu_2AsH + PhLi$ in Et_2O, isolated with petr. ether	85	Yellowish substance, sol. in Et_2O and THF, insol. in petr. ether, (d)	1530
C_9				
$PhPrAsLi$	$PhPrAsH + BuLi$ in Et_2O, (h)			790
C_{12}				
$Ph(C_6H_{11})AsLi$ (M-295)			Rxn. with RNCO, followed by hydrolysis $\rightarrow Ph(C_6H_{11})AsCONPh$, (e)	1536
C_{15}				
$PhAs(Li)(CH_2)_3As(Li)Ph\cdot 2$ Dioxan	$PhAsH(CH_2)_3AsHPh + BuLi$ in Et_2O, (h)	85	A solid, sensitive to air and moisture, sol. in THF to a yellow soln., insol. in C_6H_6, Et_2O, and petr. ether, (f)	1535

C₁₆ PhAs(Li)(CH₂)₄As(Li)Ph·2 Dioxan	PhAsH(CH₂)₄AsHPh + BuLi in Et₂O, (h)	86	A solid, sensitive to air and moisture, sol. in THF to a yellow soln., insol. in C₆H₆, Et₂O, and petr. ether	1535
C₁₇ PhAs(Li)(CH₂)₅As(Li)Ph·2 Dioxan (M-295)	PhAsH(CH₂)₅AsHPh + BuLi in Et₂O, (h)	89	A solid, sensitive to air and moisture, sol. in THF to a yellow soln., insol. in C₆H₆, Et₂O, and petr. ether	1535
C₁₈ PhAs(Li)(CH₂)₆As(Li)Ph·2 Dioxan	PhAsH(CH₂)₆AsPh + PhLi in Et₂O, (h)	88	A solid, sensitive to air and moisture, sol. in THF to a yellow soln., insol. in C₆H₆, Et₂O, and petr. ether	1535
C₁₉ BuPhAs(CH₂)₃As(Li)Ph·1 Dioxan	PhAs(CH₂)₃AsPh + BuLi in Et₂O under reflux		Crystals, very sensitive to air and moisture, sol. in THF, insol. in petr. ether	1535
C₂₀ BuPhAs(CH₂)₄AsLiPh·1 Dioxan	PhAs(CH₂)₄AsPh + BuLi in Et₂O under reflux, (i)	68	Pale yellow solid	1535
C₂₁ Ph₂As(CH₂)₃As(Li)Ph·1 Dioxan	PhAs(CH₂)₃AsPh + PhLi in Et₂O under reflux, (h)	70	Yellow solid, sensitive to air, sol. in THF, insol. in C₆H₆ and Et₂O	1535
C₂₂ Ph₂As(CH₂)₄As(Li)Ph·1 Dioxan	PhAs(CH₂)₄AsPh + PhLi in Et₂O under reflux, (h)	74	Solid substance, very sensitive to air	1535

(a) Rxn. with $Me_3SiCl \rightarrow Me_2AsSiMe_3$ (999, 1309). (b) Rxn. with Ph_3CH in THF $\rightleftarrows R_2AsH + Ph_3CLi$, 10% exchange (787). (c) $4\text{-}MeC_6H_4F$ in Et_2O at 25° $\rightarrow R_2AsC_6H_4Me\text{-}4 + R_2AsC_6H_4Me\text{-}3$ (791). (d) $t\text{-}Bu_2AsLi\cdot1\text{-}1/2$ Dioxan complex, colorless crystals prepd. from $t\text{-}Bu_2AsLi$ + Dioxan in Et_2O (1530). (e) Rxn. with RNCS, followed by hydrolysis $\rightarrow$ $Ph(C_6H_{11}AsCSNHR$ (1537). (f) Rxn. with $BrCH_2CH_2Br \rightarrow PhAs(CH_2)_3AsPh$ (1535). (g) Rxn. with $BrCH_2CH_2Br \rightarrow$ $PhAs(CH_2)_4AsPh$ (1535). (h) The rxn. mixt. is treated with dioxan. (i) The rxn. mixt. is treated with dioxan and, after distn. of the solvent, is treated with petr. ether.

211

Table 18 Continued

Arsino-Alkali Metal	Prepd. from	Yield %	Prop. and Rxn.	Ref.
RAsHLi				
C$_4$				
t-BuAsHLi·1 Dioxan	t-BuAsH$_2$ in Et$_2$O + PhLi in Et$_2$O, (h)	70	Colorless crystals	1530
RAsLi$_2$				
C$_4$				
BuAsLi$_2$	BuAsH$_2$ + BuLi in petr. ether	75	Pale yellow ppt., insol. in org. solvents, (j, k)	1536
t-BuAsLi$_2$	t-BuAsH$_2$ + PhLi (1:2) in Et$_2$O		Yellow ppt., very sensitive to air and moisture, sol. in THF to orange soln., (l)	1530
R$_2$AsNa				
C$_{12}$				
Ph$_2$AsNa	Ph$_3$As + Na (1:2) in liq. NH$_3$ followed by NH$_4$Br		Orange-red soln. in liq. NH$_3$, (m)	525
Ph$_2$AsNa·Dioxan (M-295)			Molar conductance in DMSO	789
R$_2$AsK				
C$_2$				
Me$_2$AsK	Me$_2$AsI + K in THF under reflux		Rxn. with PhBr → Me$_2$AsPh	1229

(j) Rxn. with RNCY (Y = O or S) in Et$_2$O, followed by H$_2$O → BuAs(CYNHR)$_2$ (1536).   (k) Rxn. with RN=C=NR (1:2) in Et$_2$O → BuAs[C(=NR)NLiR]$_2$ (1538).   (l) Rxn. with RN=C=NR → t-BuAs[C(=NR)NLiR]$_2$ (1538).   (m) RBr → Ph$_2$AsR (150).

1.8.1.2 COMPOUNDS WITH GROUP IIA METALS

Through 1964 only two arsinomagnesium halides were prepared (M-297). Now this class also includes bis(secondary arsino)magnesium compounds.

Two arsinomagnesium derivatives were described in the main volume on page 297.

Dimethylarsinomagnesium Bromide, $AsMgC_2H_6Br$, $Me_2AsMgBr$, (M-297)
Synth.: From Me_2AsH + MeMgBr in Et_2O at 20° (421).
Rxn. with: $CF_3CF=CF_2$ at 20° → $CF_3CF=CFAsMe_2$ + $CF_3CHFCF_2AsMe_2$ (421).
$CF_2CF_2CF=CF$ at 20° → $CF_2CF_2CF=CAsMe_2$ (421).

Bis(diphenylarsino)magnesium, $As_2MgC_{24}H_{20}$, $(Ph_2As)_2Mg$
Synth.: From Ph_2AsH by adding dropwise to a boiling suspension of Et_2Mg in MePh and heating under reflux, 64% yield (1534).
Prop.: Yellowish powder, sensitive to air, sol. in THF and DMSO, sparingly sol. or insol. in Et_2O, C_6H_6, and petr. ether. A polymeric structure is considered (1534).
Rxn. with: MeI → $[Me_2Ph_2As]I$ (1534).

Bis(dicyclohexylarsino)magnesium, $As_2MgC_{24}H_{44}$, $[(C_6H_{11})_2As]_2Mg$
Synth.: From $(C_6H_{11})_2AsLi$ + $MgBr_2 \cdot 4$ THF in THF soln. at r.t., 51% yield (1534).
Prop.: White powder, sensitive to air, sol. in THF, insol. in MePh, C_6H_6, Et_2O, and dioxan (1534).

1.8.1.3 COMPOUNDS WITH GROUP IIIA ELEMENTS

As in the main volume (M-297 to M-298), this class comprises compounds of arsenic linked with boron, aluminum, gallium, and indium. Since 1964, a few arsinoboranes, an arsinoaluminum derivative, two arsino-aluminates, an arsino-gallium, and an arsinoindium were synthesized.

Boryldioctylarsine, Dioctylarsinoborane, $AsBC_{16}H_{36}$, $(C_8H_{17})_2AsBH_2$
Synth.: From $(C_8H_{17})_2AsH$ + $NaBH_4$ + $(PhO)_3B$ in aromatic naphtha at 140° (97).

Fluoroborylenebis(dimethylarsine), Bis(dimethylarsino)fluoroborane, $As_2BC_4H_{12}F$, $(Me_2As)_2BF$
Synth.: From $Me_2AsSiMe_3$ + BF_3 at -50 to 90° (1309).

Borylbis(trifluoromethyl)arsine Trimer, $As_3B_3C_6H_6F_{18}$, $[(CF_3)_2AsBH_2]_3$
Synth.: From $(CF_3)_2AsH$ + B_2H_6 in a sealed tube under H by heating one end to 50-60° and maintaining the other at 0°, 85% yield, along with a by-prod., probably $[(CF_3)_2AsBH_2]_4$ (930).
Prop.: m. 3.0-3.5°, dec. > 40° → $(CF_3)_2AsH$ + BF_3, log P = 11.0454 + 1.75 log T-0.0100T-3754/T, Trouton const., IR spectrum (930).
Rxn. with: HCl in MeOH at 98° → $(CF_3)_2AsH$ + CH_3Cl (930).
Cl at 70-400° → BF_3 (930).

The formation of polymers containing phenylarsylene groups linked to two decaborane(12) moieties by the condensation of (PhAsO)$_2$ with (H$_3$O)$_2$B$_{10}$H$_{10}$ is claimed (499).

(Dimethylarsino)dimethylaluminum Trimer,   As$_3$Al$_3$C$_{12}$H$_{36}$,   (Me$_2$AsAlMe$_2$)$_3$
Synth.: From Me$_3$Al + Me$_2$AsH (1:1) by heating at 200-215° in a sealed tube (130).
Prop.: Glassy substance, sensitive to air and moisture, subl. 100-110°, IR spectrum (130).

Lithium Tris(diphenylarsino)aluminate,   As$_3$AlLiC$_{36}$H$_{31}$,   Li[AlH(AsPh$_2$)$_3$]
Synth.: From Ph$_2$AsAsPh$_2$ + LiAlH$_4$ in dioxan under reflux, 85% yield (790).
Prop.: Yellow amorphous solid, with a diphenylarsine-like odor, sensitive to air, sol. in C$_6$H$_6$, dioxan, THF, and Pr$_2$O; dec. in contact with H$_2$O or alc. (790).
Rxn. with: H$_2$O → Ph$_2$AsH (790).

Lithium Tris(dicyclohexylarsino)aluminate,   As$_3$AlLiC$_{36}$H$_{67}$,
    Li{AlH[As(C$_6$H$_{11}$)$_2$]$_3$}
Synth.: From (C$_6$H$_{11}$)$_2$AsAs(C$_6$H$_{11}$)$_2$ + LiAlH$_4$ in dioxan under reflux, 70% yield (790).
Prop.: Yellow amorphous solid, with an arsine-like odor, sensitive to air, sol. in C$_6$H$_6$, dioxan, THF, and Pr$_2$O, dec. in contact with H$_2$O and alc. (790).
Rxn. with: H$_2$O → (C$_6$H$_{11}$)$_2$AsH (790).

(Dimethylarsino)dimethylgallium Trimer,   As$_3$Ga$_3$C$_{12}$H$_{36}$,   (Me$_2$AsGaMe$_2$)$_3$
Synth.: From Me$_3$Ga + Me$_2$AsH (1:1) at 30-50° in a sealed tube (130).
Prop.: Glassy solid, m. 199-201° (dec.), subl. 100-110°, sensitive to air and moisture, IR spectrum (130).

(Dimethylarsino)dimethylindium Trimer,   As$_3$In$_3$C$_{12}$H$_{36}$,   (Me$_2$AsInMe$_2$)$_3$
Synth.: From Me$_3$In + Me$_2$AsH (1:1) at 15-25° in a sealed tube (130).
Prop.: Glassy solid, m. 84° (dec.), sensitive to air and moisture, IR spectrum (130).

1.8.1.4 COMPOUNDS WITH GROUP IVA ELEMENTS

The compounds of arsenic bonded to Group IVA elements compiled in the main volume (M-298 - M-299) comprised only silyl- and stannyl-arsines of the general formulae: R$_3$SiAsR$'_2$, (R$_3$Si)$_2$AsR$'$, and R$_3$SnAsR$_2$. Since 1964, several new arsenic derivatives including germyl- and plumbyl-arsines have been synthesized, with structures corresponding to the formulae: R$_3$SiAsR$'_2$, R$_2$Si(AsR$'_2$)$_2$, RSi(AsR$'_2$)$_3$, (R$_3$Si)$_2$AsR$'$, R$_3$GeAsR$'_2$, R$_3$SnAsR$_2$, (R$_3$Sn)$_2$AsR$'$, R$_2$Sn(AsR$'_2$)$_2$, RSn(AsR$'_2$)$_3$, Sn(AsR$_2$)$_4$, and R$_3$PbAsR$'_2$. These compounds have been prepared by the following reactions:

$$R_2AsM + R'_{4-n}M'X_n \rightarrow (R_2As)_nM'R'_{4-n}$$ M = alkali metal
    M$'$ = Group IVA element

$$R_2AsH + R'_3M'NR''_2 \rightarrow R_2AsM'R'_3$$

$$R_2AsH + R'_3M'X \rightarrow R_2AsM'R'_3$$

$$R_2AsX + R'_3M'M \rightarrow R_2AsM'R'_3$$

$$R_2AsX + R'_3M'^- \rightarrow R_2AsM'R'_3; \qquad R'_3M'^- = \text{electrolytically generated anion}$$

$$RAsX_2 + R'_3M'M \rightarrow RAs(M'R'_3)_2$$

$$R_2AsSiR'_3 + R''_{4-n}M'X_n \rightarrow (R_2As)_nM'R'_{4-n}$$

Dimethyl(trimethylsilyl)arsine, $AsSiC_5H_{15}$, $Me_3SiAsMe_2$

Synth.: From $Me_2AsLi + Me_3SiCl$, 76% yield (999, 1309).
From Me_2AsH in Et_2O by adding BuLi in C_6H_{14} at -80°, stirring the yellow suspension into Me_3SiCl in Et_2O at -80°, and heating under reflux, 88% yield (1660).

Prop.: m. -89°, b. 136° (999), m. -87.6°, b. 136.4° (1309), b. 139°, sensitive to oxygen and H_2O, IR spectrum (1660). NMR spectrum (1660, 1661).

Rxn. with: $H_2O \rightarrow Me_2AsH$ and $(Me_3Si)_2O$ (1309).
$ROH \rightarrow Me_2AsH$ and Me_3SiOR (1309).
HBr at -90° $\rightarrow [Me_3SiAsHMe_2]Br$ (1309).
MeI at r.t. $\rightarrow Me_3As$, $[Me_4As]I$, and Me_3SiI (1309).
BF_3 at -50 to 90° $\rightarrow (Me_2As)_2BF$ and Me_3SiF (1309).
$PF_5 > -50° \rightarrow Me_2AsAsMe_2 \cdot PF_5$, Me_3SiF, and PF_3 (1309).
$CS_2 \rightarrow Me_3SiSC(S)AsMe_2$ (1309).
CS_2 (excess) $\rightarrow Me_3SiAsMe_2 \cdot CS_2$ (1660).
$Me_2SnCl_2 \rightarrow Me_2Sn(AsMe_2)_2$ (1661).

Coordination Deriv.: $Me_3SiAsMe_2 \cdot CS_2$, obtd. from $Me_3SiAsMe_2$ and excess CS_2, stable at 20°/200 mm, at 0.02 mm dec. to $Me_3SiAsMe_2$ and CS_2 (1660).

$[Me_3SiAsMe_2 \cdot Ni(CO)_3]$, colorless crystals, prepd. from $Ni(CO)_4 + Me_3SiAsMe_2$; IR and NMR spectra (1661).

Dimethylsilylenebis(dimethylarsine), $As_2SiC_6H_{18}$, $Me_2Si(AsMe_2)_2$

Synth.: From Me_2AsLi in C_6H_6-Et_2O by adding to Me_2SiCl_2 in Et_2O at -80° with stirring, 61% yield (1660).

Prop.: $b_{0.02}$ 46-48°, sensitive to dec. by oxidn. or hydrolysis, IR and NMR spectra (1660).

Coordination Deriv.: $[Me_2Si(AsMe_2)_2 \cdot Ni(CO)_2]$, yellow crystals, prepd. from $Ni(CO)_4$ and $Me_2Si(AsMe_2)_2$ at 30° (1661). IR spectrum (22, 1661).

$[Me_2Si(AsMe_2)_2 \cdot Mo(CO)_4]$, yellow crystals, prepd. from $Me_2Si(AsMe_2)_2$ by adding to $Mo(CO)_6$ in MeCN and heating under reflux (1661). IR spectrum (22, 1661).

$[Me_2Si(AsMe_2)_2 \cdot Cr(CO)_4]$, yellow crystals, prepd. from $Me_2Si(AsMe_2)_2$ by adding to norbornadiene-$Cr(CO)_4$ in methylcyclohexane and heating under reflux (1661). IR spectrum (22, 1661).

Methylsilylidynetris(dimethylarsine), $As_3SiC_7H_{21}$, $MeSi(AsMe_2)_3$

Synth.: From Me_2AsLi in C_6H_6-Et_2O at $-80°$ by adding to $MeSiCl_3$ in Et_2O at $-80°$ with stirring, 66% yield (1660).

Prop.: $b_{0.05}$ 82°, sensitive to dec. by oxidn. or hydrolysis, IR and NMR spectra (1660).

Coordination Deriv.: $[MeSi(AsMe_2)_3 \cdot Cr(CO)_3]$, orange-yellow crystals, prepd. from $MeSi(AsMe_2)_3$ + cycloheptatriene-$Cr(CO)_3$ in cyclohexane under reflux (1661). IR spectrum (22, 1661).

$[MeSi(AsMe_2)_3 \cdot Mo(CO)_3]$, orange-yellow crystals, prepd. from $MeSi(AsMe_2)_3$ + cycloheptatriene-$Mo(CO)_3$ in C_6H_6, pptd. by cyclohexane (1661). IR spectrum (22, 1661).

Dimethyl(trimethylgermyl)arsine, $AsGeC_5H_{15}$, $Me_3GeAsMe_2$

Synth.: From Me_2AsLi in C_6H_{14}-Et_2O at $-80°$ by adding to Me_3GeCl in Et_2O with stirring, 91% yield (1660).

Prop.: b_{99-101} 85-87°, IR and NMR spectra (1660).

Diphenyl(triphenylgermyl)arsine, $AsGeC_{30}H_{25}$, $Ph_3GeAsPh_2$

Synth.: From Ph_3As + Na in liq. NH_3, followed by NH_4Br and Ph_3GeCl, removing the NH_3, dissolving in C_6H_6, and treating, after distn. of C_6H_6, with C_5H_{12}, 64% yield (1374).

Prop.: m. 114°, sensitive to oxygen (1374).

Rxn. with: H_2O_2 in EtOH-H_2O → $Ph_2As(O)OGePh_3$ (1374).

Dimethyl(trimethylstannyl)arsine, $AsSnC_5H_{15}$, $Me_3SnAsMe_2$

Synth.: From Me_2AsH by adding to Me_3SnNMe_2, 80% yield (1660). From Me_2AsH and Me_3SnCl in deoxygenated H_2O by adding slowly NaOH soln. in deoxygenated H_2O, 46% yield of very pure prod. was obtd. by distn. (1660).

Prop.: $b_{7.5-8}$ 52-53°, b_{10} 57°, n_D^{19} 1.5483, d_4^{20} 1.57, IR spectrum (1660). NMR spectrum (21, 1660, 1661).

Rxn. with: CS_2 (excess) → $Me_3SnAsMe_2 \cdot CS_2$ (1660).

Coordination Deriv.: $[Me_3SnAsMe_2 \cdot Ni(CO)_3]$, yellow crystals, prepd. from $Ni(CO)_4$ by treating with the arsine (1661). IR spectrum (22, 1661). NMR spectrum (1661).

Diphenyl(trimethylstannyl)arsine, $AsSnC_{15}H_{19}$, $Me_3SnAsPh_2$, (M-299)

Synth.: From Ph_2AsH + Me_2NSnMe_3 at 20°, 77% yield (811).

Prop.: $b_{0.05}$ 136°, n_D^{20} 1.6438, d_4^{20} 1.4682 (811).

Rxn. with: Ph_2PCl → Ph_2PAsPh_2 (1373).

Diphenyl(triethylstannyl)arsine, $AsSnC_{18}H_{25}$, $Et_3SnAsPh_2$, (M-299)

Rxn. with: Ph_3SnH in PrCN → $Et_3SnSnPh_3$ + Ph_2AsH (410).

Diphenyl(triphenylstannyl)arsine, $AsSnC_{30}H_{25}$, $Ph_3SnAsPh_2$

Synth.: From Ph_3As by treating with Na in liq. NH_3, followed by NH_4Br and Ph_3SnCl, 42% yield, crystd. by treating the oily prod. with C_5H_{12} (1371).

From Ph$_2$AsCl in THF by adding Ph$_3$SnLi in THF and sepg. the prod. with C$_6$H$_6$, 81% yield (1371).

The title compd. is formed from Ph$_2$AsBr by the rxn. with Ph$_3$Sn$^-$ generated electrolytically (467).

Prop.: m. 117-119° (1371).

Rxn. with: 30% H$_2$O$_2$ in EtOH → Ph$_2$As(O)OSnPh$_3$ (1371).

Phenylbis(trimethylstannyl)arsine,   AsSn$_2$C$_{12}$H$_{23}$,   (Me$_3$Sn)$_2$AsPh

Rxn. with: Ph$_2$PCl → (Ph$_2$P)$_2$AsPh (1373).

Phenylbis(triphenylstannyl)arsine,   AsSn$_2$C$_{42}$H$_{35}$,   (Ph$_3$Sn)$_2$AsPh

Synth.: From PhAsCl$_2$ in THF by adding LiSnPh$_3$ in THF and sepg. the prod. with C$_6$H$_6$, 45% yield (1371).

Prop.: m. 112-115° (1371).

Dimethylstannylenebis(dimethylarsine),   As$_2$SnC$_6$H$_{18}$,   Me$_2$Sn(AsMe$_2$)$_2$

Synth.: From Me$_3$SiAsMe$_2$ by adding slowly to Me$_2$SnCl$_2$ at -80°, warming to r.t., and distg. off Me$_3$SiCl (1661).

Prop.: b$_{0.01}$ 55°, d$_4^{20}$ 1.67, dec. during distn., NMR spectrum (1661).

Coordination Deriv.: [Me$_2$Sn(AsMe$_2$)$_2$·Ni(CO)$_2$], yellow crystals, prepd. by stirring Ni(CO)$_4$ with Me$_2$Sn(AsMe$_2$)$_2$ at 20° (1661). IR spectrum (22, 1661).

[Me$_2$Sn(AsMe$_2$)$_2$·Cr(CO)$_4$], yellow crystals, prepd. from Me$_2$Sn(AsMe$_2$)$_2$ + norbornadiene-Cr(CO)$_4$ in methylcyclohexane under reflux (1661). IR spectrum (22, 1661).

Diphenylstannylenebis(diphenylarsine),   As$_2$SnC$_{36}$H$_{30}$,   Ph$_2$Sn(AsPh$_2$)$_2$

Synth.: From Ph$_3$As + Na in liq. NH$_3$, followed by NH$_4$Br and Ph$_2$SnCl$_2$, 33% yield, crystd. by treating the oily prod. with C$_5$H$_{12}$ (1371).

Prop.: m. 80° (dec.) (1371).

Rxn. with: 30% H$_2$O$_2$ in EtOH → [Ph$_2$As(O)O]$_2$SnPh$_2$ (1371).

Phenylstannylidynetris(diphenylarsine),   As$_3$SnC$_{42}$H$_{35}$,   PhSn(AsPh$_2$)$_3$

Synth.: From Ph$_3$As + Na in liq. NH$_3$, followed by NH$_4$Br and PhSnCl$_3$, 80% yield, crystd. by treating the oily prod. with C$_5$H$_{12}$ (1371).

Prop.: m. 85° (dec.) (1371).

Rxn. with: 30% H$_2$O$_2$ in EtOH → [Ph$_2$As(O)O]$_3$SnPh (1371).

Stannanetetrayltetrakis(diphenylarsine),   As$_4$SnC$_{48}$H$_{40}$,   Sn(AsPh$_2$)$_4$

Synth.: From Ph$_3$As + Na in liq. NH$_3$, followed by NH$_4$Br and SnCl$_4$, 37% yield, crystd. by treating the oily prod. with C$_5$H$_{12}$ (1371).

Prop.: m. 68-70° (1371).

Diphenyl(triphenylplumbyl)arsine,   AsPbC$_{30}$H$_{25}$,   Ph$_3$PbAsPh$_2$

Synth.: From Ph$_3$As + Na (1:1) in liq. NH$_3$, followed by NH$_4$Br and Ph$_3$PbCl, removing the NH$_3$, dissolving the prod. in C$_6$H$_6$, and treating it, after distn. of C$_6$H$_6$, with C$_5$H$_{12}$, 42% yield (1374).

Prop.: Dec. 115°, sensitive to oxygen (1374).
Rxn. with: H_2O_2 in EtOH-H_2O → $Ph_2As(O)OPbPh_3$ (1374).

1.8.2 COMPOUNDS WITH TRANSITION METALS

1.8.2.1 COMPOUNDS WITH GROUP IB METALS

Diphenylarsinosilver, $AsAgC_{12}H_{10}$, Ph_2AsAg
Synth.: The compd. is formed from $AgClO_4$ and Ph_2As^- generated electrochemically from Ph_2AsClO_4 (467).
Prop.: A golden-brown soln. (467).
Rxn.: Electrolytic redn. → Ph_2As^- + Ag (467).

1.8.2.2 COMPOUNDS WITH GROUP IIB METALS

Bis(phenylarsino)zinc, $As_2ZnC_{12}H_{12}$, $(PhAsH)_2Zn$
Synth.: From $PhAsH_2$ in MePh soln. by adding dropwise Et_2Zn in the same solvent and heating under reflux, 78% yield. When $PhAsH_2$ was added to Et_2Zn, a prod. was formed with a Zn:As ratio of 1:1.5-1.7, probably a mixt. of $EtZnAsHPh$ and $(PhAsZn)_n$ (1534).
Prop.: Yellow amorphous ppt., insol. in C_6H_6, Et_2O, THF, and petr. ether (1534).

Bis(diphenylarsino)zinc, $As_2ZnC_{24}H_{20}$, $(Ph_2As)_2Zn$
Synth.: From Ph_2AsH by adding dropwise to a boiling soln. of Et_2Zn in petr. ether, 88% yield (1534).
Prop.: White amorphous ppt., sensitive to air, insol. in MePh, petr. ether, dioxan, Et_2O, and THF, sol. in DMSO; the insoly. suggests a polymeric structure (1534).
Rxn. with: MeI in i-Bu_2O under reflux → partial conversion to $[Me_2Ph_2As]I$ (1534).

1.8.2.5 COMPOUNDS WITH GROUP VB METALS

Octacarbonylbis[μ-(diphenylarsino)]divanadium, $As_2V_2C_{32}H_{20}O_8$,
Synth.: From $Ph_2AsAsPh_2$ + $V(CO)_6$ (1:2) in
C_6H_6 (734).

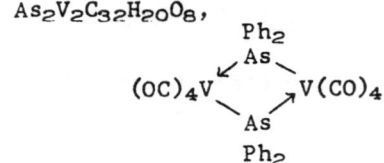

1.8.2.6 COMPOUNDS WITH GROUP VIB METALS

Decacarbonyl-μ-(tetramethyldiarsine)dichromium, $As_2Cr_2C_{14}H_{12}O_4$,

$$[(OC)_5Cr \leftarrow \underset{\underset{Me}{|}}{\overset{\overset{Me}{|}}{As}}-\underset{\underset{Me}{|}}{\overset{\overset{Me}{|}}{As}} \rightarrow Cr(CO)_5]_n ,$$ (M-300)

Rxn.: Polarographic redn. with $2e^-$ (n = 1) → gives a soln. which shows no ESR signal. The As-As bond is ruptured and metalloid anion fragments are formed (468).

Decacarbonyl-μ-(tetramethyldiarsino)dimolybdenum, $As_2Mo_2C_{14}H_{12}O_2$,

$$(OC)_5Mo \leftarrow \overset{\overset{Me}{|}}{\underset{\underset{Me}{|}}{As}} - \overset{\overset{Me}{|}}{\underset{\underset{Me}{|}}{As}} \rightarrow Mo(CO)_5 , \qquad (M\text{-}300)$$

Prop.: IR spectrum (28).

[Bis(pentafluorophenyl)arsino]dicarbonyl-π-cyclopentadienylmolybdenum Polymer, $(AsMoC_{19}H_5F_{10}O_2)_n$, $[(C_6F_5)_2AsMo(CO)_2\text{-}\pi\text{-}C_5H_5]_n$

Synth.: From $(C_6F_5)_2AsMo(CO)_3\text{-}\pi\text{-}C_5H_5$ by UV irradiation in methylcyclohexane (381).
Prop.: Solid, m. > 350°, insol. in org. solvents, IR spectrum (381).

[Bis(pentafluorophenyl)arsino]tricarbonyl-π-cyclopentadienylmolybdenum, $AsMoC_{20}H_5F_{10}O_3$, $(C_6F_5)_2AsMo(CO)_3\text{-}\pi\text{-}C_5H_5$

Synth.: From $(C_6F_5)_2AsCl + [\pi\text{-}C_5H_5Mo(CO)_3]^-$ in THF, 65% yield (381).
From $(C_6F_5)_2AsAs(C_6F_5)_2 + [\pi\text{-}C_5H_5Mo(CO)_3]_2$ in MePh, 15% yield (381).
Prop.: Yellow green crystals, m. 130°, IR and NMR spectra. A soln. in methyl-cyclohexane on exposure to UV irradiation dec., giving off CO and a polymer (381).

Octacarbonylbis[μ-(dimethylarsino)]ditungsten, $As_2W_2C_{12}H_{12}O_8$, (M-301)
Prop.: Half-wave redn. potential (472).
Rxn.: Polarographic redn. ($2e^-$) → a stable dianion (468, 472).

$$(OC)_4W \underset{\overset{As}{Me_2}}{\overset{\overset{Me_2}{As}}{\diamond}} W(CO)_4$$

Decacarbonyl-μ-(tetramethyldiarsino)ditungsten, $As_2W_2C_{14}H_{12}O_{10}$,

$$(OC)_5W \leftarrow \overset{\overset{Me}{|}}{\underset{\underset{Me}{|}}{As}} - \overset{\overset{Me}{|}}{\underset{\underset{Me}{|}}{As}} \rightarrow W(CO)_5 , \qquad (M\text{-}301)$$

Prop.: IR spectrum (28).

1.8.2.7 COMPOUNDS WITH GROUP VIIB METALS

Octacarbonyl-μ-[bis(trifluoromethyl)arsino]-μ'-iododimanganese,

$AsMn_2C_{10}F_6IO_8$,

$$(OC)_4Mn \underset{\underset{I}{\searrow \nearrow}}{\overset{\overset{CF_3 \diagdown \diagup CF_3}{As}}{\swarrow \searrow}} Mn(CO)_4 , \qquad (M\text{-}301)$$

Prop.: IR spectrum interpretation (658).

μ-[Bis(trifluoromethyl)arsino]octacarbonyl-μ'-(trifluoromethylthio)dimanganese,

AsMn₂C₁₁F₉O₈S,

$$CF_3 \quad CF_3$$
$$As$$
$$(OC)_4Mn \qquad Mn(CO)_4$$
$$S$$
$$CF_3$$

<u>Synth.</u>: From {μ-I-μ'-[(CF₃)₂As]Mn₂(CO)₈} by treating with Hg(SCF₃)₂ (658).
<u>Prop.</u>: IR spectrum and its interpretation (658).

μ-[Bis(trifluoromethyl)arsino]octacarbonyl-μ'-(trifluoromethylseleno)diman-
ganese,

AsMn₂C₁₁F₉O₈Se,

$$CF_3 \quad CF_3$$
$$As$$
$$(OC)_4Mn \qquad Mn(CO)_4$$
$$Se$$
$$CF_3$$

<u>Synth.</u>: From {μ-I-μ'-[(CF₃)₂As]Mn₂(CO)₈} by treating with Hg(SeCF₃)₂ (658).
<u>Prop.</u>: IR spectrum and its interpretation (658).

Bis{μ-[bis(trifluoromethyl)arsino]}octacarbonyldimanganese,   As₂Mn₂C₁₂F₁₂O₈,

$$CF_3 \quad CF_3$$
$$As$$
$$(OC)_4Mn \qquad Mn(CO)_4 \quad , \qquad (M-301)$$
$$As$$
$$CF_3 \quad CF_3$$

<u>Prop.</u>: IR spectrum and its interpretation (658).

Octacarbonylbis[μ-(dimethylarsino)]dimanganese,   As₂Mn₂C₁₂H₁₂O₈,

$$Me \quad Me$$
$$As$$
$$(OC)_4Mn \qquad Mn(CO)_4 \quad , \qquad (M-302)$$
$$As$$
$$Me \quad Me$$

<u>Prop.</u>: IR spectrum (23, 658).

Dodecacarbonylbis[μ-(dimethylarsino)]di-μ'-iodotetramanganese,
As₂Mn₄C₁₆H₁₂I₂O₁₂,

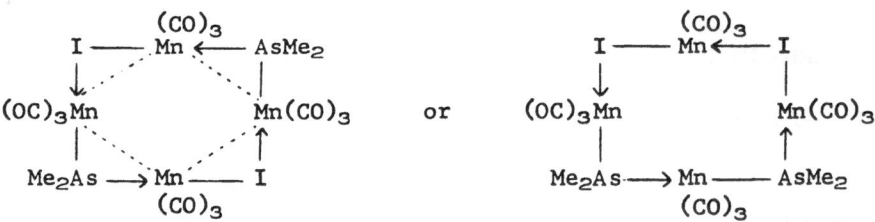

<u>Synth.</u>: From Me₂AsI + Me₃SnMn(CO)₅ (1:1) by heating in C₆H₁₂ at 50° under N,
60% yield (24).
<u>Prop.</u>: Dark orange ppt., recrystd. from CHCl₃-C₅H₁₂, IR spectrum (24).

Octacarbonylbis[μ-(diphenylarsino)]dimanganese, $As_2Mn_2C_{32}H_{20}O_8$,

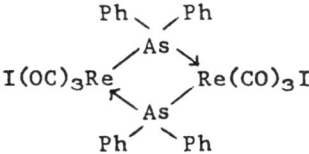

(OC)$_4$Mn   Mn(CO)$_4$,   (M-302)

Synth.: From [Mn(CO)$_5$]$_2$ and Ph$_3$As by heating in a sealed tube at 140° (1181).
Prop.: IR spectrum (23, 1181).

Hexacarbonyldichlorobis[μ-(diphenylarsino)]ditechnetium, $As_2Tc_2C_{30}H_{20}Cl_2O_6$,

Cl(OC)$_3$Tc   Tc(CO)$_3$Cl

Synth.: From Ph$_2$AsAsPh$_2$ + Tc(CO)$_5$Cl (1:2) in C$_6$H$_6$ under reflux (738).
Prop.: Colorless cryst. powder, dec. 248°, IR spectrum (738).

Hexacarbonylbis[μ-(diphenylarsino)]diiododirhenium, $As_2Re_2C_{30}H_{20}I_2O_6$,

I(OC)$_3$Re   Re(CO)$_3$I

Synth.: From Ph$_2$AsAsPh$_2$ + Re(CO)$_5$I (1:2) in C$_6$H$_6$ under reflux (738).
Prop.: Colorless cryst. powder, dec. 276°, IR spectrum (738).

1.8.2.8 COMPOUNDS WITH GROUP VIII METALS

[Bis(pentafluorophenyl)arsino]dicarbonyl-π-cyclopentadienyliron,
    $AsFeC_{19}H_5F_{10}O_2$, $π\text{-}C_5H_5Fe(CO)_2As(C_6F_5)_2$
Synth.: From (C$_6$F$_5$)$_2$AsCl + [π-C$_5$H$_5$Fe(CO)$_2$]$^-$ in THF, 39% yield (381).
From (C$_6$F$_5$)$_2$AsAs(C$_6$F$_5$)$_2$ + [π-C$_5$H$_5$Fe(CO)$_2$]$_2$ in MePh under reflux, 20% yield
(381).
Prop.: Yellow crystals, m. 138°, IR and NMR spectra. A soln. in methylcyclo-
hexane under UV irradiation gives off CO, while a polymer is formed (381).

[Bis(pentafluorophenyl)arsino]carbonyl-π-cyclopentadienyliron Polymer,
    $(AsFeC_{18}H_5F_{10}O)_n$, $[π\text{-}C_5H_5Fe(CO)As(C_6F_5)_2]_n$
Synth.: From π-C$_5$H$_5$Fe(CO)$_2$As(C$_6$F$_5$)$_2$ in methylcyclohexane by UV irradiation
(381).
Prop.: Brown cryst. prod., insol. in org. solvents, m. > 350°, IR spectrum
(381).

Bis{μ-[bis(trifluoromethyl)arsino]}hexacarbonyldiiododiiron, $As_2Fe_2C_{10}F_{12}I_2O_6$,

$$\begin{array}{c} CF_3 \quad CF_3 \\ I \quad\ \diagdown \ / \\ \diagdown As \\ (OC)_3Fe \leftarrow\quad\rightarrow Fe(CO)_3 \ , \qquad (M\text{-}303) \\ \diagup As \diagdown \\ CF_3 \quad CF_3 \quad I \end{array}$$

Synth.: From $Fe(CO)_5$ + $(CF_3)_2AsI$ at 60-70°, 95% yield (657).
Prop.: Dark red, IR and NMR spectra (657).

Hexacarbonylbis[μ-(dimethylarsino)diiron, $As_2Fe_2C_{10}H_{12}O_6$,

$$\begin{array}{c} Me \quad Me \\ \diagdown \ / \\ As \\ (OC)_3Fe \text{———} Fe(CO)_3 \ , \qquad (M\text{-}303) \\ \diagup As \diagdown \\ Me \quad Me \end{array}$$

Prop.: Polarographic redn., 1/2-wave potential (472).
Rxn.: Polarographic redn. by $2e^-$ → a stable dianion (468, 472).

Dibromobis[μ-(diphenylarsino)]tetranitrosyldiiron, $As_2Fe_2C_{24}H_{20}Br_2N_4O_4$,

$$\begin{array}{c} Ph \quad Ph \\ Br \ \diagdown \ / \ NO \\ \diagdown As \diagdown \\ ON\text{-}Fe \qquad Fe\text{-}NO \\ ON \ \diagup As \diagdown \ Br \\ Ph \quad Ph \end{array}$$

Synth.: From $Ph_2AsAsPh_2$ + $[Fe(NO)_2Br]_2$ (1:1) in THF (735).
Prop.: Black crystals, m. 187-189° (dec.), IR spectrum (735).

Bis[μ-(diphenylarsino)]diiodotetranitrosyldiiron, $As_2Fe_2C_{24}H_{20}I_2N_4O_4$,

$$\begin{array}{c} Ph \quad Ph \\ I \ \diagdown \ / \ NO \\ \diagdown As \diagdown \\ ON\text{-}Fe \qquad Fe\text{-}NO \\ ON \ \diagup As \diagup \ I \\ Ph \quad Ph \end{array}$$

Synth.: From $Ph_2AsAsPh_2$ + $[Fe(NO)_2I]_2$ (1:1) in THF (735).
Prop.: Black crystals, m. 125-127° (dec.), IR spectrum, magnetic susceptibility data (735).

Bis[μ-(diphenylarsino)]tetranitrosyldiiron, $As_2Fe_2C_{24}H_{20}N_4O_4$,

$$\begin{array}{c} Ph \quad Ph \\ ON \ \diagdown \ / \ NO \\ \diagdown As \diagdown \\ Fe \qquad Fe \\ ON \ \diagup As \diagup \ NO \\ Ph \quad Ph \end{array}$$

Synth.: From $Ph_2AsAsPh_2$ + $[Fe(NO)_2Br]_2$ (2.5:1) in THF at 50°, 88% yield (735).
Prop.: Purple-red crystals, m. 252° (dec.), sol. in warm THF, IR spectrum, magnetic susceptibility data (735).

Bis{μ-[bis(pentafluorophenyl)arsino]}hexacarbonyldiiron, $As_2Fe_2C_{30}F_{20}O_6$,

$$
\begin{array}{c}
C_6F_5 \quad\quad C_6F_5 \\
\diagdown \quad \diagup \\
As \\
\diagdown \\
(OC)_3Fe \quad\quad\quad Fe(CO)_3 \\
\diagup \\
As \\
\diagup \quad \diagdown \\
C_6F_5 \quad\quad C_6F_5
\end{array}
$$

Synth.: $[(C_6F_5)_2As-]_2$ + $Fe_3(CO)_3$ in MePh under reflux, 29% yield (381).
Prop.: Red crystals, m. 252°, IR and NMR spectra (381).

Hexacarbonyl(tetramethylarsetane)diiron (?), $As_4Fe_2C_{10}H_{12}O_6$,

$$
\begin{array}{cc}
\begin{array}{c}
Me \quad Me \\
As - As \\
\diagup \quad\quad \diagdown \\
(OC)_3Fe \cdots\cdots Fe(CO)_3 \\
\diagdown \quad\quad \diagup \\
As - As \\
Me \quad Me \\
I
\end{array}
&
\begin{array}{c}
Me \quad\quad\quad Me \\
As \text{———} As \\
\diagup \quad\quad\quad\quad | \\
(OC)_3Fe \quad Fe(CO)_3 \quad | \\
\diagdown \quad\quad\quad\quad | \\
As \text{———} As \\
Me \quad\quad\quad Me \\
II
\end{array}
\end{array}
$$

or

Synth.: From $(MeAs \!\!<\!)_5$ by treating with $Fe(CO)_5$ (528).
Prop.: Dark orange crystals, IR, NMR, and mass spectra; structure II is
slightly favored (528).

Tricarbonyl(dimethylarsino)cobalt Polymer, $(AsCoC_5H_6O_3)_n$, $Me_2AsCo(CO)_3$
Synth.: From Me_2AsI + $Me_3SnCo(CO)_4$ (1:1) in C_6H_{12} at 0° (24).
From Me_2AsCl + $Me_3SiCo(CO)_4$ in C_6H_{12} (99a).
From $Me_2AsCo(CO)_4$ by dec. at r.t. (99a).
Prop.: Red-brown ppt., insol. in org. solvents (24, 99a). IR spectrum (24).

Tetracarbonyl(dimethylarsino)cobalt, $AsCoC_6H_6O_4$, $Me_2AsCo(CO)_4$
Synth.: From Me_2AsH + $HCo(CO)_4$ (99a).
Prop.: Unstable compd., dec. at r.t. $\rightarrow$ $Me_2AsCo(CO)_3$ (99a).

Dibromotetranitrosyl-μ-(tetraphenyldiarsine)dicobalt, $As_2Co_2C_{24}H_{20}Br_2N_4O_4$,

$$
\begin{array}{c}
Ph \quad Ph \\
| \quad\quad | \\
Br(ON)_2Co \leftarrow As - As \rightarrow Co(NO)_2Br \\
| \quad\quad | \\
Ph \quad Ph
\end{array}
$$

Synth.: From $Ph_2AsAsPh_2$ + $[Co(NO)_2Br]_2$ (1:1) in THF (735).
Prop.: Black crystals, m. 137-139° (dec.), IR spectrum (735).

Dichlorotetranitrosyl-μ-(tetraphenyldiarsine)dicobalt, $As_2Co_2C_{24}H_{20}Cl_2N_4O_4$,

$$
\begin{array}{c}
Ph \quad Ph \\
| \quad\quad | \\
Br(ON)_2Co \leftarrow As - As \rightarrow Co(NO)_2Br \\
| \quad\quad | \\
Ph \quad Ph
\end{array}
$$

Synth.: From $Ph_2AsAsPh_2$ + $[Co(NO)_2Cl]_2$ (1:1) in THF (735).
Prop.: Black crystals, m. 124-126° (dec.), IR spectrum (735).

Dibromobis[μ-(diphenylarsino)]dinitrosyldinickel, $As_2Ni_2C_{24}H_{20}Br_2N_2O_2$,

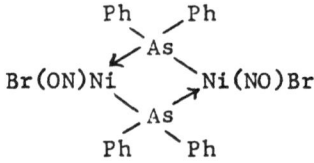

<u>Synth.</u>: From $Ph_2AsAsPh_2$ + Ni(NO)Br (1:2) in THF (735).
<u>Prop.</u>: Blue-green crystals, m. 219° (dec.), subl. > 160° in high vacuum, IR spectrum, magnetic susceptibility data (735).

Diiodobis[μ-(diphenylarsino)dinitrosyldinickel, $As_2Ni_2C_{24}H_{20}I_2Ni_2O_2$,

I(ON)Ni   Ni(CO)I

with Ph As Ph bridges

<u>Synth.</u>: From $Ph_2AsAsPh_2$ + Ni(NO)I (1:2) in THF (735).
<u>Prop.</u>: Blue-green crystals, subl. > 140° in high vacuum, m. 212° (dec.), IR spectrum, magnetic susceptibility data (735).

Bis{μ-[bis(pentafluorophenyl)arsino]}hexacarbonyldiruthenium,
$As_2Ru_2C_{30}F_{20}O_6$, C_6F_5 C_6F_5

(OC)$_3$Ru   Ru(CO)$_3$

C_6F_5 C_6F_5

<u>Synth.</u>: From $(C_6F_5)_2AsAs(C_6F_5)_2$ + $Ru_3(CO)_{12}$ in MePh, 7.6% yield (381).
<u>Prop.</u>: Crystals, m. 231°, IR and NMR spectra (381).

Dicarbonyldichlorobis[μ-(diphenylarsino)]dirhodium, $As_2Rh_2C_{26}H_{20}Cl_2O_2$,

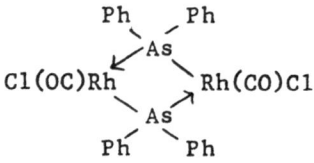

<u>Synth.</u>: From $Ph_2AsAsPh_2$ + $[Rh(CO)_2Cl]_2$ (0.65:0.48) in C_6H_6, a mixt. of cis- and trans-isomers was obtd. (736).
<u>Prop.</u>: The trans-isomer, sol. in C_6H_6, forms brown-black crystals, m. 208° (dec.); soly. data and IR spectrum (736). The cis-isomer, insol. in C_6H_6, black crystals, dec. ~ 235°, IR spectrum (736).

Dicarbonyldichlorobis[μ-(diphenylarsino)]diiridium, $As_2Ir_2C_{26}H_{20}Cl_2O_2$,

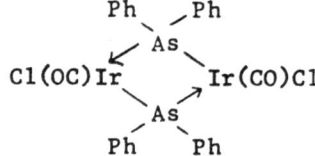

Synth.: From $Ph_2AsAsPh_2$ + $[Ir(CO)_3Cl]_n$ in Py (736).
Prop.: Orange crystals, IR spectrum (736).

1.9 MISCELLANEOUS COMPOUNDS

This section includes various derivatives of trivalent arsenic which do not conform with any of the preceding groups (M-305).

Oxybis(chloroferrocenylarsine), $As_2Fe_2C_{20}H_{18}Cl_2O$, $(FcAsCl)_2O$
Synth.: From $FcAsCl_2$ by dissolving in hot C_6H_6 or petr. ether and evaptg. the solvent in a stream of air, dissolving the residue in hot petr. ether, and crystg. by cooling the soln., 12-34% yield (1436, 1437).
Prop.: Yellow powder, m. 118.5-120°, sensitive to moisture but relatively stable under N. IR spectrum (1436, 1437).
Rxn. with: H_2O → $(FcAsO)_n$ (1436, 1437).
Concd. HCl → $FcAsCl_2$ (1436, 1437).
Aq. alkali followed by acidification → $(FcAsO)_2$ (1436).
Zn + HCl in 95% EtOH → $(FcAs\bar{\ })_n$ (1436).

3-Methyl-2-[(3-methyl-2-benzothiazolinylidene)arsino]benzothiazolium Tetra-
    fluoroborate, $AsBC_{16}H_{14}F_4N_2S_2$,
Synth.: From 2-chloro-3-methylbenzothiazolium tetrafluoroborate and $(Me_3Si)_3As$ in MeCN, 57% yield (1003).
Prop.: m. 143-145°, electronic spectrum (1003).

3-Ethyl-2-[(3-ethyl-2-benzothiazolinylidene)arsino]benzothiazolium Tetra-
    fluoroborate, $AsBC_{18}H_{18}F_4N_2S_2$,
Synth.: From 2-chloro-N-ethylbenzothiazolium tetrafluoroborate + $(Me_3Si)_3As$ in MeCN, 57% yield (1003).
Prop.: m. 194-198°, electronic and IR spectra (1003).

1-Methyl-2-[(1-methyl-2(1H)-quinolinylidene)arsino]quinolinium Tetrafluoro-
    borate, $AsBC_{20}H_{18}F_4N_2$,
Synth.: From 2-chloro-1-methylquinolinium tetrafluoroborate + $(Me_3Si)_3As$ in MeCN, 48% yield (1003).
Prop.: m. 201-203°, electronic spectrum (1003).

3-Ethyl-2-[(3-ethyl-6-methoxy-2-benzothiazolylidene)arsino]-6-methoxybenzothia-
    zolium Tetrafluoroborate, $AsBC_{20}H_{22}F_4N_2O_2S_2$,
Synth.: From 2-chloro-3-ethyl-6-methoxy-benzothiazolium tetrafluoroborate and $(Me_3Si)_3As$ in MeCN, 43% yield (1003).
Prop.: m. 168-172°, electronic spectrum (1003).

1-Ethyl-2-[(1-ethyl-2(1H)-quinolinylidene)arsino]quinolinium Tetrafluoroborate
   AsBC$_{22}$H$_{22}$F$_4$N$_2$,

Synth.: From 2-chloro-1-ethylquinolinium
tetrafluoroborate + (Me$_3$Si)$_3$As in MeCN,
32% yield (1003).

Prop.: m. 217-221°, electronic spectrum (1003).

Arsenic Cyanide Free Radical,   AsCN,   AsCN

Formation: From a mixt. of AsH$_3$ or AsCl$_3$ + (CN)$_2$ + N$_2$ (0.2:40:210) by flash
photolysis at 1000j, detected by electronic absorption spectrum (127).

Diphenylarsino Radical,   AsC$_{12}$H$_{10}$,   Ph$_2$As

Formation: From Ph$_2$AsH by photolysis at 50° and condensation on a cold finger
at -196° (1358).

Prop.: ESR spectrum at -196° showed a characteristic signal; stable at -140°
(1358).

Tris(p-dimethylaminophenyl)arsine Radical Cation,   AsC$_{24}$H$_{30}$N$_3$,
   [4-(Me$_2$N)C$_6$H$_4$]$_3$As$^+$

Formation: From [4-(Me$_2$N)C$_6$H$_4$]$_3$As dissolved in PhNO$_2$ by oxidn. with N/20
AgClO$_4$ soln. in C$_6$H$_6$ (1519).

Prop.: Green soln. which produced an ESR signal (1519).

Rxn. with: Ph$_3$C· in C$_6$H$_6$ → extinction of the green color (1519).

2. COMPOUNDS OF PENTAVALENT ARSENIC

2.1 QUINQUENARY DERIVATIVES

Pentaphenylarsenic was the only representative of this group described in the main volume (M-307). Now two more compounds of this type have been synthesized.

Pentakis(p-chlorophenyl)arsenic, $AsC_{30}H_{20}Cl_5$, $(4-ClC_6H_4)_5As$
Synth.: From $(4-ClC_6H_4)_3As=NO_2SC_6H_4Me-4$ + $4-ClC_6H_4Li$ in Et_2O, followed by hydrolysis, 49% yield (723).
Prop.: Crystals, m. 149°, NMR spectrum (723).

Pentaphenylarsenic, $AsC_{30}H_{25}$, Ph_5As,   (M-307)
Synth.: A tritium-marked prod. is formed when Ph_5As is treated with $[4-^3H]PhLi$ (1:6) in Et_2O-THF mixt., followed by hydrolysis (438).
Prop.: IR spectrum (1001).
Rxn. with: BuLi (1:6) in Et_2O → exchange of the Ph groups in Ph_5As for the Bu groups (438).
$[4-^3H]PhLi$ in Et_2O-THF mixt. → $[Li(AsPh_6)] \xrightarrow{H_2O} Ph_5As$ (438).
PhLi in THF at -70°, followed by bis(2,2'-biphenylylene)arsonium iodide → bis(2,2'-biphenylylene)phenylarsenic (722).

Penta-p-tolylarsenic, $AsC_{35}H_{37}$, $(4-MeC_6H_4)_5As$
Synth.: From $(4-MeC_6H_4)_3As=NO_2SC_6H_4Me-4$ + $4-MeC_6H_4Li$ at r.t., followed by hydrolysis and crystn. from Et_2O and then from diglyme, 25% yield (723).
Prop.: Crystals, dec. 139-140° (723). NMR spectrum: the penta-coordinate As compd. in soln. apparently undergoes a rapid transformation between a trigonal-bipyramidal and tetragonal pyramidal structure (719, see also 723).

2.2 QUATERNARY DERIVATIVES

2.2.1 ARSONIUM YLIDS

Several arsonium ylids and their methods of synthesis were compiled in the main volume, pp. 307-309. More recently new ylids have been prepared as follows:

(a) From tertiary arsine oxides and cyclopentadiene derivatives containing a methylene group by dehydration with acetic anhydride or phosphorus pentoxide in triethylamine;

(b) From quaternary arsonium halides containing at least one α-methylene group by dehydrohalogenation with sodium hydride; and

(c) From tertiary arsines by a condensation with diazoalkanes.

Moreover, a metathetic reaction of trimethylarsonium trimethylsilylmethylide, $Me_3As=CHSiMe_3$, with trimethylsilanol affords trimethylarsonium methylide.

The nomenclature for these compounds used by Chemical Abstracts is inconsistent. For example, in CA $\underline{69}$ Ph₃As=CHBz is named phenacylidenetriphenylarsenic, in CA $\underline{68}$ Ph₃As=CH₂ is named triphenylarsonium methylide, and in CA $\underline{70}$ is named triphenylarsonium tetraphenylcyclopentadienylide.

$$Ph_3As \diagdown \underset{Ph}{\overset{Ph}{\diagup}} \overset{Ph}{\underset{Ph}{\diagup}}$$

Trimethylarsonium Methylide, AsC₄H₁₁, Me₃As⁺-CH₂⁻

Synth.: From Me₃As=CHSiMe₃ + Me₃SiOH (1354).

Prop.: Colorless crystals, m. 33-35°, subl.₀.₁ 30-35°, sensitive to air and moisture; on exposure to air turns brown-violet, dec. > 38°; monomeric in C₆H₆ soln. NMR spectrum in C₆H₆ indicated the ylid form (1354).

Rxn. with: MeI → [Me₃EtAs]I (1354).

HX (X = halogen) → [Me₄As]X (1354).

Me₃P at r.t. → Me₃As + Me₃P=CH₂ (1354).

Trimethylarsonium Trimethylsilylmethylide, AsSiC₇H₁₉, Me₃As⁺-CHSiMe₃⁻

Synth.: From [Me₃(Me₃SiCH₂)As]Cl by treating with BuLi (1:1) in Et₂O, removing the solvent, heating the residue at 120-140°, and isolating the volatile matl., 54% yield (1066).

Prop.: Clear liquid, b₆ 69°, m. -29 to -27.7°, fuming in air and having a Me₃As-like odor. IR and NMR spectra: the following resonance structure is suggested: Me₃As⁺-CHSiMe₃⁻ ⇌ Me₃As=CHSiMe₃ ⇌ Me₃AsCH=SiMe₃⁻ (1066).

Rxn. with: HCl gas → [Me₃(Me₃SiCH₂)As]Cl (1066).

MeI, followed by NH₄PF₆ → [Me₃(Me₃SiCHMe)As]PF₆ (1066).

Me₃SiOH → Me₃As=CH₂ and (Me₃Si)₂O (1354).

Triphenylarsonium Methylide, AsC₁₉H₁₇, Ph₃As⁺-CH₂⁻, (M-309)

Synth.: From [MePh₃As]I + NaH in THF (610).

Rxn. with: PhCH=NNClPh in THF at -5 to r.t. → Ph₃As, 1,3-diphenyl-Δ²-pyrazoline, 1,3,4,5-tetraphenyldihydro-1,2,4,5-tetrazine, and BzNHN(Ph)C(Ph)=NNHPh (610).

PhCNO in DMSO → Ph₃As, 3,5-diphenylisoxazole, 3-phenyl-4-benzoyl-Δ²-isoxazoline anti-oxime, and [PhC(=NOH)]₂CH₂ (610).

PhCNO in Et₂O → Ph₃As, diphenyl-1,2,4-oxadiazole, 3,5-diphenylisoxazole, and [PhC(=NOH)]₂CH₂ (610).

Triphenylarsonium Benzylide, AsC₂₅H₂₁, Ph₃As⁺-CHPh⁻

Synth.: From [Ph₃(PhCH₂)As]Br + BuLi or PhLi in Et₂O (806, 1146).

Prop.: A soln. in C₆H₆-Et₂O, dec. at elevated temp. to Ph₃As and stilbene isomers; a rxn. mechanism is suggested (1146).

Rxn. with: 4-O₂NC₆H₄CHO in Et₂O → Ph₃As, 4-O₂NC₆H₄C̅H̅C̅H̅(Ph)O, Ph₃AsO, and 4-O₂NC₆H₄CH=CHPh (806).

PhNO → PhCH=N(O)Ph and Ph₃As (806).

[Ph₃(PhCH₂)As]X → {Ph₃[PhCH(CH₂Ph)]As}X and Ph₃As; the former dec. to Ph₃As, stilbene, and [Ph₃(PhCH₂)As]X (1146).

Table 19. Arsenic Ylids

Ylid	Yield %	Prepd. from	Prop. and Rxn.	Ref.
C_{21} $Ph_3\overset{+}{A}s-\overset{-}{C}HCO_2Me$ (M-309)		$[Ph_3(MeO_2CCH_2)As]Br$ + NaOMe in MeOH	Rxn. with aromatic aldehydes and alicyclic ketones → the corr. α,β-unsatd. acid Me esters	766
C_{26} $Ph_3\overset{+}{A}s-\overset{-}{C}HCOC_6H_4Br-4$	68	$[Ph_3(4-BrC_6H_4COCH_2)As]Br$ + NaOMe in MeOH	m. 155–156°	1144
$Ph_3\overset{+}{A}s-\overset{-}{C}HCOC_6H_4Cl-4$	50	$[Ph_3(4-ClC_6H_4COCH_2)As]Br$ + NaOMe in MeOH	m. 156–157°	1144
$Ph_3\overset{+}{A}s-\overset{-}{C}HCOC_6H_4NO_2-4$	53	$[Ph_3(4-O_2NC_6H_4COCH_2)As]Br$ + NaOMe in MeOH	m. 110–112°	1144
$Ph_3\overset{+}{A}s-\overset{-}{C}HBz$ (M-309)	70	$[Ph(BzCH_2)As]Br$ + NaOMe in MeOH	m. 167–169°	1144
	80	$\{Ph_3[BzCH(HgCl)]As\}Cl$ + NH_3 or Na_2SO_3 in DMF	IR spectrum, rxn. with $HgCl_2$ in MeOH → $\{Ph_3[BzCH(HgCl)]As\}Cl$	1145
C_{27} $Ph_3\overset{+}{A}s-\overset{-}{C}HCOC_6H_4Me-4$	70	$[Ph_3(4-MeC_6H_4COCH_2)As]Br$ + NaOMe in MeOH	m. 150–151°	1144
$Ph_3\overset{+}{A}s-\overset{-}{C}HCOC_6H_4OMe-4$	82	$[Ph_3(4-MeOC_6H_4COCH_2)As]Br$ + NaOMe in MeOH	m. 145–146°	1144
C_{31} $Ph_3\overset{+}{A}s-\overset{-}{F}luorenylide$ (M-308)			Rxn. with PhNO → N-phenylfluorenone ketoxime	806
C_{37} 	80	Ph_3AsO + 1,2-dibenzoyl- cyclopentadiene in Ac_2O	Almost colorless, m. 233–234° (dec.)	687

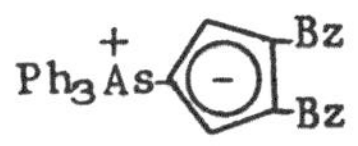

229

Table 19 Continued

Ylid	Prepd. from	Yield %	Prop. and Rxn.	Ref.
C$_{41}$				
[Ph$_3$As$^+$ cyclopentadienylide with Ph, Ph, Ph]	Ph$_3$AsO + 1,2,3-triphenyl-cyclopentadiene + P$_2$O$_5$ in Et$_3$N	50	Pale yellow crystals, m. 198-200° (dec.), (a)	687
C$_{43}$				
[Ph$_3$As$^+$ cyclopentadienylide with Ac, Ph, Ph, Ph]	Ph$_3$AsO + 1,2,3-triphenyl-cyclopentadiene in Ac$_2$O for 5 min. $\overline{Ph_3As{-}C{-}CH{=}C(Ph)C(Ph){=}CPh}$ + Ac$_2$O by heating for 2 min.	67	Yellow solid, m. 225-226°	687 687
C$_{47}$				
[Ph$_3$As$^+$ cyclopentadienylide with Ph, Ph, Ph, Ph]	Ph$_3$As + diazotetraphenyl-cyclopentadiene by fusion		Golden yellow crystals, m. 228-230° (dec.), stable in air, UV spectrum. A resonance cyclopentadienylide ⇌ cyclopentadienyli-dene structure is considered, (b, c)	974

(a) Rxn. with Ac$_2$O → PhC=$\overline{C(Ph)C(Ph)=C(Ac)C}$=AsPh$_3$ (687). (b) Rxn. with HClO$_4$ → [Ph$_3$AsCHC(Ph)=C(Ph)C(Ph)=$\overline{CPh}$]ClO$_4$ (974). (c) Rxn. with BzH → 1,2,3,4,6-pentaphenylfulvene (974).

2.2.2 QUATERNARY ARSONIUM BASES AND SALTS

The methods of synthesis of the arsonium compounds were summarized in the main volume on pages 310-311 and the compounds themselves were compiled in Table 33. Since then a great number of quaternary arsonium salts, particularly tetraphenylarsonium salts including basic and complex anions, have been prepared. Many of them are extractable with organic solvents from aqueous solutions and are useful in the extraction and analytical determination of certain metals.

Besides the methods of preparation described in the main volume, quaternary arsonium hydroxides can be prepared by electrolysis of quaternary arsonium halides as anolytes with aqueous sodium hydroxide as catholyte.

The quaternization of triarylarsines with aryl halides in the presence of equimolar amounts of aluminum chloride is enhanced by the presence of nickel bromide at 180-190° under pressure.

Tetraalkyldiarsines react with alkyl halides at elevated temperatures to form quaternary arsonium halides.

1,2-Diarsolanes and 1,2-diarsenanes are cleaved by methyl iodide to form the corresponding arsinoalkylarsonium iodides. Likewise, 1,2-oxarsolanes are cleaved by hydrogen iodide to hydroxypropylarsonium iodides.

Phosphates, thiophosphates, and sulfonate esters combine with tertiary arsines at room temperature to form quaternary arsonium phosphates, thiophosphates, and sulfonates, respectively.

Quaternary arsonium halides, phosphates, germanates, and stannates react with transition metal carbonyls under irradiation to form complex anions through an insertion of metal carbonyl moieties, with evolution of a part of carbon monoxide.

Tetraphenylarsonium ion and chloride are described individually hereafter. All the other quaternary arsonium salts are compiled in Table 20, where they are arranged in the order of the increasing number of carbon and hydrogen atoms in the arsonium cations.

Tetraphenylarsonium Ion, $AsC_{24}H_{20}$, $[Ph_4As]^+$
Prop.: Sepn. from $[Ph_2I]^+$ by TLC and by paper-chromatography and identification by a salmon-pink color with dithizone (269). Sepn. from $[Ph_2I]^+$ by electrophoresis and identification by a salmon-pink color with dithizone (270). Extraction of Cr(VI) into $HCCl_3$ soln. as $[Ph_4As] \cdot HCrO_4$ (672). A theoretical consideration of the assocn. of ions in the extraction of submicroscopic amts. of certain anions (1310).
Biol. Prop.: Inhibition of the uptake of $[Et_4N]^+$ and N'-methylnicotinamide (A) by slices of dog kidney and of the renal excretion of A by the hen (1430).

Tetraphenylarsonium Chloride, $AsC_{24}H_{20}Cl$, $[Ph_4As]Cl$, (M-330)

Synth.: From Ph_3As + PhCl under γ-irradiation, 2.5% yield (306).

Prop.: White crystals, m. 256-260° (306). Spectra: UV and visible (132), IR (1001, 1049). Mol. cond. data (353). Polarographic redn. (87) and 1/2-wave potential (467). The salt is extracted with $HCCl_3$ from aq. soln. and from NaCl soln. ≤ 0.1 M in HCl (574). Changes in the heat of soln. in aq. EtOH (92) and in aq. t-BuOH (93) with the change of the solvent compn. In HCl_3 soln. oligomerizes to a dimer and tetramer (574).

Rxn. with: $(\pi-C_5H_5)_2TiBr_2$ or $(\pi-C_5H_5)_2Ti(SCN)_2$ in dry Me_2CO → exchange of the anions (931).

Uses: A reagent for conductometric titration of ClO_4^- and of MnO_4^- (104). Gravimetric detn. of ClO_4^- (623, 625, 1172). Gravimetric, amperometric, and spectrophotometric detn. of PF_6^- and BF_4^- (87). Pptn. of W thiocyanate complex extracted into $HCCl_3$ and spectrophotometric detn. of W in steel (1114). Coppptn. of Tc(VII) with ReO_4^- (1254). A corrosion inhibitor for ferrous metals (747). Extraction of Ir from aq. HCl with $CHCl_3$ contg. $[Ph_4As]Cl$ (574). Extraction of Fe(III) and Th from EDTA with i-BuOH-$MeNO_2$ mixt. contg. $[Ph_4As]Cl$ (1655). Extraction of Fe(III), Mo(VI), Ti(IV), Nb(V), V(V), U(VI), and W(VI) as ternary complexes with $[Ph_4As]Cl$ and catechol by $HCCl_3$ from aq. soln. (1565). Extraction of TcO_4^-, ReO_4^-, $FeCl_4^-$, $AuCl_4^-$, $ZnCl_4^{--}$, $AgCl_3^{--}$, and other anions into $HCCl_3$ from 0.5-12 N HCl (1539, see also 1507). Extraction of blue peroxychromic acid with suitable solvents and formation of a stable 1:1 adduct suitable for a photometric detn. of Cr (1463). Extraction of Cr(VI) into $CHCl_3$ as $[Ph_4As]HCrO_4$ (672). A mixt. of $[Ph_4As]Cl$ with thenoyltrifluoroacetone at a ratio of 1:10 has a synergistic effect on the extraction of Ga from a $ClCH_2CO_2H$-NaOAc buffer soln. of pH 2.4-2.6 into o-dichlorobenzene (1256). $[Ph_4As]Cl$ catalyzes an oxidative polymerization of 2,6-xylenol by PbO_2 (1242).

Table 20. Quaternary Arsonium Bases and Salts

R_4As^+	X^-	Prepd. from	Prop. and Rxn.	Ref.
C_4				
Me_4As	Cl (M-312)		IR spectrum	417
	I (M-312)	Me_3As + MeI (1:2) in Et_2O	White crystals, m. 325–326°	90
		$Me_2AsAsMe_2$ + MeI at 100°	Colorless crystals, m. 325° (dec.)	414
		$Me_2AsSiMe_3$ + MeI	(a, b)	1309
	I_3 (M-312)	$Me_2AsAsMe_2$ + MeI at 100°	Violet crystals, m. 133°	414
	HgI_3 (M-312)		(a)	417
	Pc, (c) (M-312)	$[Me_4As]I$ + a picrate	(a)	417
	$2,4\text{-}(O_2N)_2C_6H_4SO_3$	Me_3As + $2,4\text{-}(O_2N)_2C_6H_3SO_3Me$, (d)	m. 308–309°	1012
$Me(HOCH_2)_3As$	I	$(HOCH_2)_3As$ + MeI at 60°	m. 186.5–188°	1012
			Brown viscous substance	1265
C_5				
$EtMe_3As$	I	$Me_3As^+\!\text{-}CH_2^-$ + MeI		1354
C_7				
Et_3MeAs	I		IR spectrum	417
	HgI_3 (M-313)		IR spectrum	417
	$\begin{array}{c}MeS\diagdown\ \ O\\ \quad P\diagup\\ Cl_3C_6H_2O\diagup\ \diagdown O\end{array}$	Et_3As + $Cl\text{-}C_6H_3(Cl)\text{-}OP(S)(OMe)_2$ in petr. ether at 20°	n_D^{20} 1.5667, d_{20} 1.4899	1054, 1055
	$\begin{array}{c}MeNHCOCH_2S\diagdown\ \ O\\ \quad P\diagup\\ MeS\diagup\ \diagdown O\end{array}$		Herbicidal activity against wheat, oat, millet, radish, and vetch	1056

(a) IR spectrum (417). (b) NMR spectrum, solvent and concn. effects (1262). (c) Pc = picrate anion. (d) The rxn. gave a purple soln. and a cryst. reddish-purple deposit which after a prolonged standing became pink. Recrystn. from EtOH removed the color.

Table 20 Continued

R_4As^+	X^-	Prepd. from	Prop. and Rxn.	Ref.
C_7 (Cont.)				
$[Et_3MeAs]_2$	HgI_4 (M-313)		IR spectrum	417
$Me_3(Me_3SiCH_2)As$	Cl	$Me_3As + Me_3SiCH_2Cl$ at 100-130° in a sealed tube	White solid, IR spectrum, rxn. with BuLi → $Me_3\overset{+}{A}s$-$\overset{-}{C}HSiMe_3$	1066
	PF_6	$[Me_3(Me_3SiCH_2)As]Cl + NH_4PF_6$ in H_2O	White solid, IR spectrum	1066
C_8				
$Me_3[CH_2=CH(CH_2)_3]As$	I (M-313)	$Me_2[CH_2=CH(CH_2)_3]As + MeI$	White crystals, m. 153°, IR spectrum	164
Et_4As	OAc		(e, f)	343
	Br		A catalyst for the rxn. of buta-diene + $CHCl_3 + \overline{CH_2CH_2O}$ to form 2-vinyl-1,1-dichlorocyclopropane	857
	Cl		(e, f)	343
	F		(e, f)	343
	OH		(e, f)	343
	I (M-313)		IR spectrum, (e, f, g)	417
	NO_3		(e, f)	343
	ClO_4		(e, f)	343
$[Et_4As]_2$	HgI_4 (M-314)		IR spectrum	417,455
$Me_3AsCH=CHAsMe_3$	I_2	$Me_2AsCH=CHAsMe_2 + MeI$ (1:2)	m. 260°	1229
$Me_3(Me_3SiCHMe)As$	PF_6	$Me_3\overset{+}{A}s$-$\overset{-}{C}HSiMe_3 + MeI$ in Et_2O, followed by NH_4PF_6	White insol. salt, IR spectrum	1066
C_9				
$Me_3[2,4-(O_2N)_2C_6H_3]As$	Pc, (c)	$Me_3As + 2,4-(O_2N)_2C_6H_3Cl$ → cream-colored ppt. which was treated with picric acid	m. 191-191.5°	1012

Cation	Anion	Preparation	Properties	Ref.
$Me_3(2\text{-}BrC_6H_4)As$	I	$Me_2(2\text{-}BrC_6H_4)As + MeI$	m. 286° (dec.)	1016
	Pc, (c)		m. 178°	1016
$Me_3(2\text{-}IC_6H_4)As$	I	$Me_2(2\text{-}IC_6H_4)As + MeI$	m. 240-241° (dec.)	1016
	Pc, (c)		m. 189°	1016
$Me_3(3\text{-}O_2NC_6H_4)As$	NO_3 (M-314)	$[Me_3PhAs]NO_3 + HNO_3\text{-}H_2SO_4$, (h)	UV spectrum	600
Me_3PhAs	I (M-314)	$Me_2PhAs + MeI$ in EtOH	White crystals, m. 248-249°	90
			m. 263°, (i, j)	1229
$[Me_3PhAs]_2$	HgI_3 (M-314)		IR spectrum	418
	HgI_4 (M-314)		IR spectrum	418
$Et_3(CH_2=CHCH_2)As$	ClO_4	$Et_3As + CH_2=CHCH_2Cl$ (1:2) in MePh + $NaClO_4$		464
	$PtCl_3$	$[Et_3(CH_2=CHCH_2)As]ClO_4 + Na_2PtCl_4$ in EtOH under reflux	Yellow crystals, thermodynamic data	464
$Et_3(MeO_2CCH_2)As$	Br	$Et_3As + BrCH_2CO_2Me$ in Et_2O	Crystals, m. 89-90°, $pk_a > 11$	788
	BPh_4		Crystals, m. 167-168°, pk_a 10.9	788
C_{10}				
$Et_2Me(MeC{\equiv}CC{\equiv}C)As$	I	$Et_2(MeC{\equiv}CC{\equiv}C)As + MeI$ in Et_2O or MeCN	m. 137° (dec.), NMR spectrum	149
$Me_3(3\text{-}MeC_6H_4)As$	I	$Me_3(3\text{-}MeC_6H_4)As + MeI$	NMR spectrum, chem. shift and long-range As-H coupling const.	1412
			m. 249°	1229
$Me_3(2\text{-}MeOC_6H_4)As$	I	$Me_2(2\text{-}MeOC_6H_4)As + MeI$, (k)	m. 285°, UV spectrum	1414, 1415
$Me_3(3\text{-}MeOC_6H_4)As$	I (M-316)	$Me_2(3\text{-}MeOC_6H_4)As + MeI$, (k)	m. 250°, UV spectrum	1414
$Me_3(4\text{-}MeOC_6H_4)As$	I	$Me_2(4\text{-}MeOC_6H_4)As + MeI$, (k)	m. 216-217°, UV spectrum	1414
$Et_3(CH_2=CHCH_2CH_2)As$	ClO_4	$Et_3As + CH_2=CHCH_2CH_2Br$ (1:2) in MePh + $NaClO_4$		464
	$PtCl_3$	$[Et_3(CH_2=CHCH_2CH_2)As]ClO_4 + Na_2PtCl_4$ in EtOH under reflux	Thermodynamic data	464

(e) NMR spectrum (343). (f) The effects of solvents and of the anion on the internal chemical shift in NMR spectrum (1027). (g) NMR spectrum, the effect of concn. and of solvents (1262). (h) The 2- and 4-nitro isomers are formed to the extent of 4% (600). (i) UV spectrum (90). (j) IR spectrum (418). (k) In Et_2O.

Table 20 Continued

R_4As^+	X^-	Prepd. from	Prop. and Rxn.	Ref.
C_{10} (Cont.)				
$MePr_3As$	$(MeS)_2P(O)O$	$Pr_3As + (MeO)_2(MeS)PS$, (1)	n_D^{20} 1.5290, d^{20} 1.3836 (m)	1054, 1055
	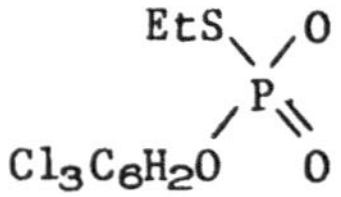 $EtS\!-\!P(=O)(O)\!-\!OC_6H_2Cl_3$	$Pr_3As + EtO(MeO)(2,4,5-Cl_3-C_6H_2O)PS$, (1)	n_D^{20} 1.5528, d^{20} 1.4334 (m, n)	1054, 1055
	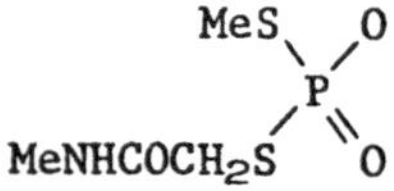 $MeS\!-\!P(=O)(O)\!-\!SCH_2CONHMe$	$Pr_3As + (MeO)_2(MeNHCOCH_2S)PS$ in petr. ether at 20°	n_D^{20} 1.5177, d^{20} 1.4566	1054, 1055
	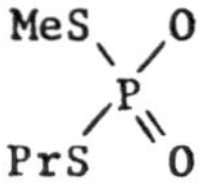 $MeS\!-\!P(=O)(O)\!-\!SPr$	$Pr_3As + (MeO)_2(PrS)PO$, (1)	n_D^{20} 1.5202, d^{20} 1.3337	1054, 1055
	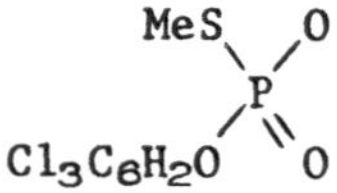 $MeS\!-\!P(=O)(O)\!-\!OC_6H_2Cl_3$	$Pr_3As + (MeO)_2(2,4,5-Cl_3-C_6H_2O)PS$ in petr. ether at 20°	n_D^{20} 1.5550, d^{20} 1.4068 (n)	1054, 1055
$Me_3(2-CH_2=CHC_6H_4)As$	I	$Me_2(2-CH_2=CHC_6H_4)As + MeI$	White, m. 218–220° (dec.)	157
$Me_3(3-CH_2=CHC_6H_4)As$	I	$Me_2(3-CH_2=CHC_6H_4)As + MeI$	White, m. 185–187°	157
$Me_3(4-CH_2=CHC_6H_4)As$	I	$Me_2(4-CH_2=CHC_6H_4)As + MeI$	White, m. 197–200°	157
C_{11}				
$Et_3(MeC≡CC≡C)As$	Br	$Et_3As + MeC≡CC≡CBr$	m. 160°	149
$MeEt_2PhAs$	Pc, (c)	$Et_2PhAs + Me$ picrate	NMR spectrum	149, 1412
	$2,4-(O_2N)_2C_6H_3SO_3$	$Et_2PhAs + 2,4-(O_2N)_2C_6H_3SO_3Me$	m. 85–87°	1012
			m. 124–125°	1012
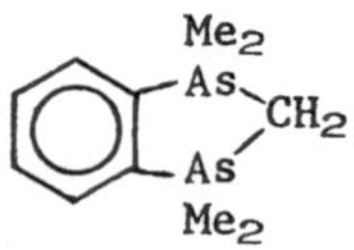	I_2	$2-Me_2AsC_6H_4AsMe_2 + CHI_3$ or CCl_4	NMR spectrum	1163

236

Compound	Anion	Preparation	Properties	Ref.
$2\text{-Me}_2\text{AsC}_6\text{H}_4\text{AsMe}_3$	I	$2\text{-Me}_2\text{AsC}_6\text{H}_4\text{AsMe}_2$ + MeI	Colorless plates, m. 226-230°	558
C₁₂				
$\text{Me}_3(2\text{-CH}_2\text{=CHCH}_2\text{C}_6\text{H}_4)\text{As}$	I	$\text{Me}_2(2\text{-CH}_2\text{=CHCH}_2\text{C}_6\text{H}_4)\text{As}$ + MeI	White, m. 134-136°, (o)	157
$\text{Me}_3(3\text{-CH}_2\text{=CHCH}_2\text{C}_6\text{H}_4)\text{As}$	I	$\text{Me}_2(3\text{-CH}_2\text{=CHCH}_2\text{C}_6\text{H}_4)\text{As}$ + MeI	White, m. 147-151°, (o)	157
$\text{Me}_3(4\text{-CH}_2\text{=CHCH}_2\text{C}_6\text{H}_4)\text{As}$	I	$\text{Me}_2(4\text{-CH}_2\text{=CHCH}_2\text{C}_6\text{H}_4)\text{As}$ + MeI	White, m. 144-146°, (o)	157
$\text{Me}_3(2\text{-PrC}_6\text{H}_4)\text{As}$	I	$\text{Me}_2(2\text{-PrC}_6\text{H}_4)\text{As}$ + MeI	White, m. 120-123°	157
$\text{Me}_3(4\text{-PrC}_6\text{H}_4)\text{As}$	I	$\text{Me}_2(4\text{-PrC}_6\text{H}_4)\text{As}$ + MeI	White, m. 145-148°	157
Et_3PhAs	HgI_3 (M-317)		IR spectrum	418
(benzene ring with AsMe₃, AsMe₃)	$[\text{B}_{12}\text{H}_7\text{Br}_4\text{PMe}_3]_2$	$[o\text{-C}_6\text{H}_4(\text{AsMe}_3)_2]\text{Cl}_2$ + $\text{Na}[\text{B}_{12}\text{H}_7\text{Br}_4\text{PMe}_3]$ in H_2O		1067
C₁₃				
$\text{Me}_3(1\text{-C}_{10}\text{H}_7)\text{As}$	I (M-318)	$\text{Me}_2(1\text{-C}_{10}\text{H}_7)\text{As}$ + MeI	m. 236°	1415
$\text{MePh}(\text{CH}_2\text{=CHCH}_2)_2\text{As}$	Br	$\text{Ph}(\text{CH}_2\text{=CHCH}_2)_2\text{As}$ + MeBr	Rxn. with KCN → $\text{CH}_2\text{=C(Me)-CN}$ + $\text{MePh}(\text{CH}_2\text{=CHCH}_2)\text{As}$	752
$\text{MeEt}(\text{CH}_2\text{=CHCH}_2)(\text{MeC}_6\text{H}_4)\text{As}$	Br	$\text{MeEt}(\text{MeC}_6\text{H}_4)\text{As}$ + $\text{CH}_2\text{=CHCH}_2\text{Br}$ at 130°	m. 85-86°	821
$(-)\text{-MePhPr}(\text{CH}_2\text{=CHCH}_2)\text{As}$	Br	$(-)\text{-MePhPrAs}$ + $\text{CH}_2\text{=CHCH}_2\text{Br}$ at r.t.	Rxn. with 10% aq. KCN → $(-)\text{-MePhPrAs}$	852
$\text{Me}_3\{2\text{-}[\text{MeO(CH}_2)_3]\text{C}_6\text{H}_4\}\text{As}$	Pc, (c)	$\text{Me}_2[2\text{-(MeOCH}_2\text{CH}_2\text{CH}_2)\text{C}_6\text{H}_4]\text{As}$ + MeI, followed by picric acid	m. 108.5-110°	1513
MeBu_3As	MeO(MeS)P(O)O	Bu_3As + $(\text{MeO})_3\text{PS}$, (1)	n_D^{20} 1.4840, d^{20} 1.2585	1054,1055
	EtS–P(=O)–O, $\text{Cl}_3\text{C}_6\text{H}_2\text{O}$	Bu_3As + $(\text{EtO})(\text{MeO})(2,4,5\text{-Cl}_3\text{C}_6\text{H}_2\text{O})\text{PO}$, (1)	n_D^{20} 1.5424, d^{20} 1.3919, (n)	1054,1055

(1) In petr. ether-Et₂O mixt. in a sealed tube at 80-100°. (m) IR spectrum (1054). (n) Herbicidal activity against wheat, oat, millet, radish, and vetch (1056). (o) Rxn. with Br → probably the corr. $[\text{R}_3\text{R}'\text{As}]\text{IBr}_2$ (157).

Table 20 Continued

R_4As^+	X^-	Prepd. from	Prop. and Rxn.	Ref.
C_{13} (Cont.)				
$MeBu_3As$	MeS–P(=O)–O / MeNHCOCH₂S	$Bu_3As + (MeO)_2(MeNHCOCH_2S)PS$	n_D^{20} 1.5227, d^{20} 1.4118, (n)	1054, 1055
	MeS–P(=O)–O / Cl₃C₆H₂O		(n)	1056
C_{14}				
Me_2Ph_2As	I (M-318)	$MePh_2As + MeI$ in EtOH	m. 189-190°, m. 209-211°, (p)	90
			UV spectrum	90
			m. 215-216°	1017
	HgI_3 (M-319)		IR spectrum	418
$Et_2(MeC_6H_4)(CH_2=CHCH_2)As$	Br	$Et_2(MeC_6H_4)As + CH_2=CHCH_2Br$ at 130°	m. 104-106°	821
$Et_3(BzCH_2)As$	Br	$Et_3As + BzCH_2Br$ in Et_2O	Crystals, m. 146-147°, pKa 9.7	788
$Me_2(C_6H_{11})_2As$	I (M-319)	$(C_6H_{11})_2AsCONHPh + MeI$ in Et_2O under reflux	Shiny crystals	1537
C_{15}				
$MePh_2(HC\equiv C)As$	$4\text{-}MeC_6H_4SO_3$	$Ph_2(HC\equiv C)As + p\text{-}MeC_6H_4SO_3Me$ in MeCN	m. 159°, IR spectrum, (q)	149
$MeEt(2\text{-}BrC_6H_4)_2As$	Br	$Et(2\text{-}BrC_6H_4)_2As + MeBr$ at 100°	m. 203-204° (dec.)	715
$MePh_2(HOCH_2CH_2)As$	I (M-319)	$Ph_2(HOCH_2CH_2)As + MeI$ in Et_2O	Colorless crystals, m. 175°, soly. data	1531
$o\text{-}Et_2AsC_6H_4AsEt_2Me$	I	$o\text{-}C_6H_4(AsEt_2)_2 + MeI$	Pale yellow crystals, cond. data	356

C₁₈

Compound	Anion	Preparation	Properties	Ref.
Me(MeC≡C)Ph₂As	4-MeC₆H₄SO₃	Ph₂AsC≡CMe + 4-MeC₆H₄SO₃Me in MeCN	m. 170°, (q)	149
Et₂Ph[2,4-(O₂N)₂C₆H₃]As	Cl	Et₂PhAs + 2,4-(O₂N)₂C₆H₃Cl	Dark gummy solid	1012
	Pc, (c)	{Et₂Ph[2,4-(O₂N)₂C₆H₃]As}Cl + a picrate salt	m. 131-132°	1012
Et₂Ph₂As	HgI₃ (M-320)		IR spectrum	418
(-)-EtMePh(PhCH₂)As	Br		Racemization by heating in aprotic solvents	750
	Cl (M-320)		Racemization by heating in aprotic solvents	750
	I (M-320)		Racemization by heating in aprotic solvents	750
	ClO₄		Soln. in aprotic solvents do not racemize	750
(+)-EtMePh(PhCH₂)As	BCS, (r)	[EtMePh(PhCH₂)As]Cl + Ag[BCS] in H₂O	m. 144-145°, [α]$_D$ +50.4°	823
Et₃(PhC≡CC≡C)As	Br	Et₃As + PhC≡CC≡CBr	m. 170-180° (dec.)	149
EtBu(CH₂=CHCH₂)(MeC₆H₄)As	Br	EtBu(MeC₆H₄)As + CH₂=CHCH₂Br at 130°	m. 44-48°	821
o-Et₂AsC₆H₄AsEt₃	I	o-C₆H₄(AsEt₂)₂ + EtI	Pale yellow crystals, cond. data	356
[W(CO)₄(Me₂AsCH₂)₂C(Me)CH₂AsMe₃]	I	[W(CO)₄(Me₂AsCH₂)₂C(Me)CH₂AsMe] + MeI in MeOH	White diamagnetic crystals, IR and NMR spectra	1161
Bu₄As	N₃		Rxn. with (π-C₅H₅)₂TiBr → exchange of the anions	931

(p) Two forms were obtd., the lower melting one was formed from fresh MePh₂As and the higher melting one was formed from the arsine 8 months old (90). (q) NMR spectrum (149, 1412). (r) BCS = d-π-bromocamphorsulfonate.

Table 20 Continued

R_4As^+	X^-	Prepd. from	Prop. and Rxn.	Ref.
C_{17}				
$EtPh_2(HC\equiv CCH_2)As$	Br	$EtPh_2As + HC\equiv CCH_2Br$ in MeCN	m. 194° (dec.), IR spectrum, (s)	149
	I	$[EtPh_2(HC\equiv CCH_2)As]Br + I$ in H_2O or EtOH	m. 178°, IR spectrum	149
	$4\text{-}MeC_6H_4SO_3$	$EtPh_2As + 4\text{-}MeC_6H_4SO_3CH_2C\equiv CH$	m. 129°, 144°, IR spectrum	149
$MePh(CH_2=CHCH_2)(PhCH_2)As$	ClO_4	$MePh(CH_2=CHCH_2)As + PhCH_2Cl$, followed by ClO_4^-	Resolution into the optical antipodes with (−)-n-dibenzoyl hydrogen tartrate	752
(+)-$MePh(CH_2=CHCH_2)(PhCH_2)As$	Br	The racemic perchlorate was resolved with (−)-n-dibenzoyl hydrogen tartrate and pptd. with Br^-	$[\alpha]_D^{20}$ +10.1°, rxn. with KCN → (+)-MePh(PhCH_2)As	752
$Et_2Ph(PhCH_2)As$	Br	$Et_2PhAs + PhCH_2Br$ at 130°	m. 174−175°	821
(+)-$MePhPr(PhCH_2)As$	BPh_4 (M−321)		The (+)-cation has the absolute S-configuration, (t)	755
(−)-$MePhPr(PhCH_2)As$	Br		Racemization by heating in aprotic solvents	750
	Cl		Racemization by heating in aprotic solvents	750
	I		Racemization by heating in aprotic solvents	750
	ClO_4		Solns. in aprotic solvents do not racemize	750
$PhAs(I)(CH_2)_3AsMe_2Ph$	I	$PhAs(CH_2)_3AsPh + MeI$ in EtOH-Et_2O under reflux	Yellow crystals, m. 151−152°, soly. data	1535
$MePh_2[HO(CH_2)_4]As$	I	$Ph_2As(CH_2)_4OH + MeI$ in Et_2O	Colorless crystals, m. 114−115°, soly. data	1531

EtPh(CH₂=CHCH₂)(C₆H₁₃)As	Br	EtPh(C₆H₁₃)As + CH₂=CHCH₂Br at 130°	m.48–49°	821
MePh(C₅H₁₁)₂As	I	Ph(C₅H₁₁)₂As + MeI at r.t. or on a steam bath	m. 72–74°	828

C_{18}

EtPh₂(MeC≡CCH₂)As	Br	Thermal dec. → EtPh₂As + Ph₂(MeC≡CCH₂)As		150
PhAs(I)(CH₂)₄AsMe₂Ph	I	PhAs(CH₂)₄AsMe₂Ph + MeI in EtOH-Et₂O under reflux	Crystals, m. 98–99°	1535
Et₂(PhCH₂)(MeC₆H₄)As	Br	Et₂(MeC₆H₄)As + PhCH₂Br at 130°	m. 136–138°	821
(+)-EtPhPr(PhCH₂)As	BCS, (r)	[EtPhPr(PhCH₂)As]Cl + Ag[BCS] in H₂O	m. 155–157°, [α]_D +50.9°	823
Et₃(C₁₂H₂₅)As	B₁₂H₁₁AsMe₃	[Et₃(C₁₂H₂₅)As]OH + H₃O[B₁₂H₁₁AsMe₃]	Useful in firework for a green color effect and in resistors made of cellulose	1067
MePh₃As	B(CH=CH₂)₄	[MePh₃As]Cl + Li[B(CH=CH₂)₄]	m. 141.3°	1512
	Br (M-323)	[MePh₃As]I + AgBr in H₂O	m. 198° (dec.)	90
			Cathodic cleavage, (u, v, w)	748
	Cl (M-323)	[MePh₃As]I + AgCl in H₂O	m. 118–120°, dec. 135°, IR spectrum	90
		Ph₃As + MeI under reflux followed by AgCl	(x)	364
		[MePh₃As]I anolyte + HCl catholyte by electrolysis		899

(s) Dec. > 200° → EtPh₂As, Ph₂AsCH₂C≡CH, and Ph₂AsCH=C=CH₂ (150).   (t) Detd. as a quasi-racemate with [(−)-MePhPr-(PhCH₂)P]BPh₄ which formed an eutectic with 60% arsonium component, m. 150° (755).   (u) Cathodic cleavage in aq. soln. → MePh₂As and Ph₃As: = 81.4:18.6 (748).   (v) A catalyst for the polymerization of CF₃CFCF₂O (1080).   (w) Inhibition of the uptake of the [Et₄N]⁺ cation and of N′-methylnicotinamide by the dog kidney and of the excretion of the nicotinamide by the renal tubes of the hen (1430).   (x) K-edge x-ray absorption spectrum (962).

Table 20 Continued

R_4As^+	X^-	Prepd. from	Prop. and Rxn.	Ref.
C_{18} (Cont.)				
$MePh_3As$	$HCrO_6$	$[MePh_3As]I$ in aq. H_2SO_4 + H_2O_2 at $-4°$ + $K_2Cr_2O_7$	Violet diamagnetic ppt., unstable at r.t., IR spectrum	91
	CrO_5Cl, (y)	$[MePh_3As]Cl + (BuO)_3PO·CrO_5$	Royal blue leaflets, dec. 119°, diamagnetic	91
	CrO_3Cl	$[MePh_3As][CrO_3Cl] + H_2O_2$	IR spectrum	91
		$[MePh_3As]Cl + K_2Cr_2O_7$ in 3 N HCl	Orange diamagnetic ppt., dec. 136°, IR spectrum	91
	CrO_5Br	$[MePh_3As]Br + (BuO)_3PO·CrO_5$	Royal blue, diamagnetic leaflets, dec. 110°, IR spectrum	91
	$Co[N(CN)_2]_3$	$Co(NO_3)_2·6 H_2O$ in Et_2O + $K[N(CN)_2]$, followed by $[MePh_3As]N(CN)_2$	Pink-red paramagnetic needles, m. 174°, electronic and IR spectra	866
	$Co(C_7H_8N_2)I_3$, (z)		Near-IR spectrum	633
	$Co(TDT)_2·1/2$ EtOH, (aa) (M-323)		Crystal and mol. structures	523
	OH (M-323)	$[MePh_3As]Cl$ anolyte + aq. NaOH catholyte by electrolysis		899
	I (M-323)	Ph_3As + MeI in EtOH under reflux	White plates, m. 171-172° (dec.), UV spectrum, (bb)	90
			m. 176-177°	1012
			m. 176°, (cc, dd, ee)	1147
	$FeCl_3$		Elec. cond. data	393
	$C(CN)_3$	$[MePh_3As]Cl + K[C(CN)_3]$ in H_2O	m. 126°	866
	$N(CN)_2$	$[MePh_3As]Cl + K[N(CN)_2]$ in H_2O	Colorless, microcryst. ppt., m. 146°	866
	NO_3 (M-323)	$[MePh_3As]I + Ag_2O$ in H_2O under reflux followed by HNO_3	Colorless, microcryst. ppt., m. 133-134°, IR and UV spectra	90

Ni(C$_4$N$_2$S$_2$)$_2$, (ff)	Na$_2$(C$_4$N$_2$S$_2$) + NiCl$_2$·6 H$_2$O + [MePh$_3$As]I in aq. EtOH under reflux	Black needles, m. 208-210° (dec.), elec. resistivity data	166
PaI$_6$	[MePh$_3$As]I + PaI$_5$ in CHCl$_3$ at 50-60° or in MeCN	Dark brown solid, sensitive to H$_2$O and NH$_4$OH	243
Pc, (c) (M-324)	[MePh$_3$As]I + a picrate	m. 145.5-146.5°	1012
SCN (M-324)	[MePh$_3$As]I + AgSCN in hot H$_2$O	Crystals, m. 65-66°	199
p-MeC$_6$H$_4$SO$_3$	Ph$_3$As + p-MeC$_6$H$_4$SO$_3$Me	Oil	1012
2,4-(O$_2$N)$_2$C$_6$H$_3$SO$_3$	Ph$_3$As + 2,4-(O$_2$N)$_2$C$_6$H$_3$SO$_3$Me	m. 191-192.5°	1012
[Q$\cdot$Q], (gg) (M-324)	[MePh$_3$As]I + tetracyanoquino-dimethane (Q) (1:2) in MeCN at 60°	Black prisms, m. 224-227° (dec.), elec. resistivity data, d. 1.397, (hh, ii, jj, kk)	26,507
[D$_4$Q$\cdot$D$_4$Q], (ll)		ESR spectrum, isotope effect	252
[Me-Q$\cdot$Q], (mm) (M-324)	[MePh$_3$As]I + Q + 2-Me-Q (1:1:1) in MeCN	Black rosettes, m. 198-200° (dec.), elec. resistivity data	26

(y) Originally tentatively identified as [MePh$_3$As]$_2$CrO$_6$ (M-324). (z) C$_7$H$_6$N$_2$ = benzimidazole. (aa) TDT = [structure]. (bb) Rxn. with AgX → [MePh$_3$As]X (90). (cc) Rxn. with Ph$_3$P in DMF at 140-150° → Ph$_3$As + [MePh$_3$P]I (1147). (dd) A catalyst for the condensation of bisphenols with dicarboxylic chlorides (377, 378) and with bisphenol A bischloroformate (614). (ee) IR spectrum (418). (ff) C$_4$N$_2$S$_2$ = [structure]. (gg) Q$\cdot$Q = [structure]. (hh) Overhauser effect (624). (ii) Optical consts. by the reflectance technique (1079). (jj) A pressure-induced transition was detd. by the triplet-exciton magnetic resonance spectroscopy (1058). (kk) Useful as a voltage regulator and a circuit amplifying element (507).

(11) D$_4$Q$\cdot$D$_4$Q = [structure]. (mm) Me-Q$\cdot$Q = [structure].

Table 20 Continued

R$_4$As$^+$	X$^-$	Prepd. from	Prop. and Rxn.	Ref.
C$_{19}$ (Cont.)				
MePh$_3$As	[Me-Q·MeQ], (nn)	[MePh$_3$As]I + MeQ (1:2) in MeCN at 70°	Small blue-black, rod-like crystals, m. 171-173°, elec. resistivity data	26
	TFM, (oo)	[MePh$_3$As]I + TFM in hot MeCN	Deep colored, paramagnetic compd., dec. 226°, stable in the solid state, in soln. slow decay of paramagnetism	1112
	TlI$_4$	TlCl + Cl gas in H$_2$O, followed by NaI and [MePh$_3$As]I in Me$_2$CO-H$_2$O	Red-orange crystals, m. 141-142°	395
	VCl$_4$ (M-324)	[MePh$_3$As][VCl$_4$·2 MeCN] at 100° in vacuum	Bright blue-green solid, x-ray diffraction pattern	364
	VCl$_4$·2 MeCN	[MePh$_3$As]Cl + VCl$_3$ in MeCN	Cream-yellow solid, dec. at 80-100° → [MePh$_3$As]VCl$_4$	364
[MePh$_3$As][MePh$_3$P]	[Q·Q]$_2$, (gg) (M-324)	[MePh$_3$As]I + [MePh$_3$P]I + Q (1:1:4) in MeCN	Large black prisms, m. 227-230° (dec.), elec. resistivity data	26
[MePh$_3$As]$_2$	Cr$_2$O$_7$ (M-324)	[MePh$_3$As]Cl + K$_2$Cr$_2$O$_7$ in H$_2$O	Orange, diamagnetic crystals, dec. > 230°, IR spectrum	91
	CoBr$_4$ (M-324)	[MePh$_3$As]Br + CoBr$_2$ in EtOH	Dendritic crystals, crystallographic data	1200
	CoCl$_4$ (M-324)	[MePh$_3$As]Cl + CoCl$_2$ in EtOH	Dendritic crystals, crystallographic data	1200
	CoI$_4$ (M-324)	[MePh$_3$As]I + CoI$_2$ in EtOH	Crystallographic data, the crystals are different from the chloride and bromide	1200
	Co[N(CN)$_2$]$_4$	[MePh$_3$As] [Co[N(CN)$_2$]$_3$] + [MePh$_3$As]N(CN)$_2$ in Me$_2$CO / Co(NO$_3$)$_2$·6 H$_2$O + K[N(CN)$_2$] in Me$_2$CO, followed by [MePh$_3$As]N(CN)$_2$	Pink-red, paramagnetic plates, m. 128°	866
			Elec. cond. data; electronic and IR spectra	866

Compound	Preparation	Description	Ref.
CuBr$_4$ (M-325)	[MePh$_3$As]Br + CuBr$_2$ in EtOH	Orthorhombic crystals, crystallographic data, (pp)	1200
CuCl$_4$ (M-325)	[MePh$_3$As]Cl + CuCl$_2$ in EtOH	Orthorhombic crystals, crystallographic data, (pp)	1200
FeBr$_4$ (M-325)	[MePh$_3$As]Br + FeBr$_2$ in EtOH	Crystal structure data	1200
FeCl$_4$ (M-325)	[MePh$_3$As]Cl + FeCl$_2$ in EtOH	Dendritic crystals, crystallographic data, (qq)	1200
FeI$_4$ (M-325)	[MePh$_3$As]I + FeI$_2$ in EtOH	Crystallographic data	1200
MnBr$_4$ (M-325)	[MePh$_3$As]Br + MnBr$_2$ in EtOH	Dendritic crystals, crystallographic data	1200
MnCl$_4$ (M-325)	[MePh$_3$As]Cl + MnCl$_2$ in EtOH	Dendritic crystals, crystallographic data	1200
MnI$_4$ (M-326)	[MePh$_3$As]I + MnI$_2$ in EtOH	Cryst llographic data	1200
HgI$_4$ (M-326)	[MePh$_3$As]I + HgI$_2$, (rr)	m. 166-167°, (ss)	452
NiBr$_4$ (M-326)	[MePh$_3$As]Br + NiBr$_2$ in EtOH	Dendritic crystals, crystallographic data, (tt)	1200
NiCl$_4$ (M-326)	[MePh$_3$As]Cl + NiCl$_2$ in EtOH	Blue, paramagnetic, dendritic crystals, crystallographic data, (tt, uu, vv)	1200
NiI$_4$ (M-326)	[MePh$_3$As]I + NiI$_2$ in EtOH	Crystallographic data, (tt)	1200
Ni$\left(^S_S{>}C{=}CHNO_2\right)_2$	K$_2$(S$_2$C=CHNO$_2$) + Ni(OAc)$_2$·4 H$_2$O + [MePh$_3$As]I in 95% EtOH	Acicular ruby red crystals, visible and IR spectra	67
		Blue violet, diamagnetic needles, m. 195-196°	406

(nn)

MeQ·MeQ = [structure: two methyl-substituted benzene rings each bearing $\bar{C}(CN)_2$ and $\dot{C}(CN)_2$ / C(CN)$_2$ groups]

(oo)

TFM = [structure: a fluorenyl framework substituted with four NO$_2$ groups and a $\bar{C}(CN)_2$ group]

(pp) Far-IR spectrum (32). (qq) Magnetic data (365). (rr) The salt was also prepd. by boiling (C$_6$F$_5$)$_2$Hg, C$_6$F$_5$HgCl, or C$_6$F$_5$HgBr with NaI in EtOH and adding [MePh$_3$As]I (452). (ss) IR spectrum (418). (tt) Magnetic props. (565a, 634). (uu) Crystal spectrum (112). (vv) Elec. cond. data (393).

Table 20 Continued

R_4As^+	X^-	Prepd. from	Prop. and Rxn.	Ref.
C_{19} (Cont.)				
$[MePh_3As]_2$	Re_2Cl_8	$[MePh_3As]Cl + [Bu_4N]_2[Re_2Cl_8]$ in MeOH contg. HCl	Blue-green crystals, a 2:1 electrolyte in Me_2CO	393
	$Re_2(NCS)_8$	$[Bu_4N]_2[Re_2Cl_8] + NaSCN$ in MeOH contg. AcOH, followed by $[MePh_3As]I$	Dark red crystals, soly. data, a 2:1 electrolyte, IR and electronic spectra	404
	$ZnBr_4$ (M-326)	$[MePh_3As]Br + ZnBr_2$ (2:1) in EtOH by evapn.	Dendritic crystals, crystallographic data	1200
	$ZnBrCl_3$	$[MePh_3As]Br + [MePh_3As]Cl + ZnCl_2$ in EtOH	m. 210°, far-IR spectrum	457
	$ZnBr_2Cl_2$	$[MePh_3As]Cl + ZnBr_2$ (2:1) in EtOH	m. 197°, far-IR spectrum	457
	$ZnBr_3Cl$	$[MePh_3As]Br + [MePh_3As]Cl + ZnBr_2$ (1:1:1) in EtOH	m. 184°, far-IR spectrum	457
	$ZnCl_4$ (M-326)	$[MePh_3As]Cl + ZnCl_2$ (2:1) in EtOH by evapn.	Dendritic crystals, crystallographic data	1200
	ZnI_4 (M-326)	$[MePh_3As]I + ZnI_2$ (2:1) in EtOH by evapn.	Crystallographic data	1200
$[MePh_3As]_3$	Re_3Cl_{12} (M-326)	$[MePh_3As]Cl$ in EtOH + $ReCl_3$ in aq. or alc. HCl	Purple-red plates, molar cond. data, absorption spectrum, effective cation volume 460 Å	1282
	$Re_3(CNS)_9Cl_3$	$[MePh_3As]Cl$ in EtOH + $ReCl_3$ + HSCN in EtOH	Dark brown-green powder, molar cond. data, absorption spectrum, (ww)	1282
	$V(NCS)_6 \cdot 2 \, MeCN$	$[MePh_3As]NCS + [V(NCS)_3(MeCN)_3] \cdot 2 \, MeCN$ in MeCN	Crystals	199

246

Compound	Anion	Synthesis	Properties	Ref.
Et(CH$_2$=CHCH$_2$)(PhCH$_2$)-(MeC$_6$H$_4$)As	Br	Et(PhCH$_2$)(MeC$_6$H$_4$)As + CH$_2$=CHCH$_2$Br at 130°	m. 122–124°	821
Me(2-HOC$_6$H$_{10}$)Ph$_2$As	I	(2-HOC$_6$H$_{10}$)Ph$_2$As + MeI in Et$_2$O	Colorless crystals, m. 206–207°, soly. data	1531
EtBuPh(PhCH$_2$)As	Br	EtPh(PhCH$_2$)As + BuBr	m. 135–136°	750
(-)-EtBuPh(PhCH$_2$)As	Br	EtPh(PhCH$_2$)As + BuBr	m. 111°, [α]$_D^{20}$ -2°, racemized (70%) when heated in CHCl$_3$	750
	Cl (M-328)		Racemized (32.5%) during heating in CHCl$_3$ at 103° for 70 hrs.	750
	HDBT, (xx)	[EtBuPh(PhCH$_2$)As]Br + C$_{18}$H$_{14}$O$_8$, (xx)	m. 118–120°, [α]$_D^{20}$ -63°	750
	I		Racemized during heating in org. aprotic solvents at 103° easier than the bromide or chloride	750
	ClO$_4$	[(-)-EtBuPh(PhCH$_2$)As][C$_{18}$H$_{13}$O$_8$] + ClO$_4$	m. 72–73°, [α]$_D^{20}$ -1.9°, resistant to racemization when heated in CHCl$_3$ at 103°	750
CH$_2$=CHCH$_2$(C$_5$H$_{11}$)$_2$PhAs	Br	Ph(C$_5$H$_{11}$)$_2$As + CH$_2$=CHCH$_2$Br at 80–100°	Yellow oil which crystd. after washing with Et$_2$O, m. 94–96°	828
	(-)-HDBT, (xx)	[CH$_2$=CHCH$_2$(C$_5$H$_{11}$)$_2$PhAs]Br + Ag[-HDBT]		828
Me(C$_6$H$_{13}$)$_2$PhAs	I	Ph(C$_6$H$_{13}$)$_2$As + MeI at r.t. or on a steam bath	m. 72–73°	828
PrPh(C$_5$H$_{11}$)$_2$As	Br		(yy)	1133

(ww) Rxn. with SCN⁻ under reflux → a prod. free of Cl (1282). (xx) HDBT = hydrogen p-(-)-dibenzoyltartrate. (yy) A fungicide and fungistat for Alternaria radicina, Helminthosporium sativum, Fusicladium dendriticum, Trichophyton gypseum, and T. rubrum; toxic to mice. The activity increases from the dipentyl to dihexyl and diheptyl deriv. (1133).

Table 20 Continued

R$_4$As$^+$	X$^-$	Prepd. from	Prop. and Rxn.	Ref.
C$_{19}$ (Cont.)				
Me(C$_6$H$_{11}$)$_3$As	Cl	[Me(C$_6$H$_{11}$)$_3$As]I + AgCl in H$_2$O	White hydroscopic crystals contg. 1 H$_2$O, m. 93-94°	90
	I	(C$_6$H$_{11}$)$_3$As + MeI in EtOH under reflux	White crystals, m. 153-154°	90
C$_{20}$				
Me(2-HOC$_6$H$_{10}$)(C$_6$H$_{11}$)$_2$As	I	(2-HOC$_6$H$_{10}$)(C$_6$H$_{11}$)$_2$As + MeI in CHCl$_3$	Colorless crystals, m. 130-131°, soly. data	1531
EtPh$_3$As	Br (M-327)		Cathodic cleavage in aq. soln. → EtPh$_2$As + Ph$_3$As = 52.5:47.5 m. 156-157°, (zz)	748
	I	Ph$_3$As + EtI at 100°		998
	Q$^-$·Q, (gg) (M-327)	[EtPh$_3$As]I + Q (1:2) in MeCN at 60°	Black crystals, m. 212-219° (dec.), d. 1.342, elec. resistivity data	26,507
MePh$_2$(PhCH$_2$)As	I	Ph$_2$(PhCH$_2$)As + MeI	m. 186-188°	1017
MePh$_2$AsCH$_2$[pyridine]	I$_2$	Ph$_2$AsCH$_2$[N-pyridine] + MeI in EtOH under reflux	m. 155°	1543
(+)-MeEt(PhCH$_2$)(C$_{10}$H$_7$)As	BCS, (r) (M-328)	[MeEt(PhCH$_2$)(C$_{10}$H$_7$)As]Cl + Ag[BCS] in H$_2$O	m. 245-246°, [α]$_D^{20}$ +56.7°	823
(-)-MeEt(PhCH$_2$)(C$_{10}$H$_7$)As	BCS, (r)	[MeEt(PhCH$_2$)(C$_{10}$H$_7$)As]Cl + Ag[BCS] in aq. KBr soln.		823
EtPh(n-C$_5$H$_{11}$)(PhCH$_2$)As	Br (M-328)		Resolution into optical antipodes on NaClO$_3$ in CHCl$_3$	181
(+)-EtPh(n-C$_5$H$_{11}$)(PhCH$_2$)As	BCS, (r)	[EtPh(n-C$_5$H$_{11}$)(PhCH$_2$)As]Cl + Ag[BCS] in H$_2$O	m. 158-159°, [α]$_D$ +53.4°	823

Compound	Anion	Preparation	Properties	Ref.
(−)-EtPh(n-C$_5$H$_{11}$)(PhCH$_2$)As	BCS, (r)	[EtPh(n-C$_5$H$_{11}$)(PhCH$_2$)As]Cl + Ag[BCS] in aq. KBr soln.		823
EtPh(CH$_2$=CHCH$_2$)(C$_9$H$_{17}$)As	Br	EtPh(C$_9$H$_{17}$)As + CH$_2$=CHCH$_2$Br at 130°	, m. 33–34°	821
Ph(EtO$_2$CCH$_2$)(C$_5$H$_{11}$)$_2$As	Br	Ph(C$_5$H$_{11}$)$_2$As + BrCH$_2$CO$_2$Et on a steam bath	Mobile oil, converted to a hydroscopic crystal in a desiccator, n_D^{20} 1.5478, d_4^{20} 1.2268	1549
	NO$_3$	[Ph(EtO$_2$CCH$_2$)(C$_5$H$_{11}$)$_2$As]Br + AgNO$_3$ in MeOH on a steam bath	Mobile oil	1549
Me(C$_6$H$_{11}$NHCO)(C$_6$H$_{11}$)$_2$As	I	(C$_6$H$_{11}$)$_2$AsCONHC$_6$H$_{11}$ + MeI in C$_6$H$_6$ under reflux	Crystals	1536
C$_{21}$				
Ph$_3$(HC≡CCH$_2$)As	Br	Ph$_3$As + HC≡CCH$_2$Br in MeCN	m. 246° (dec.), IR spectrum	149
	I	[Ph$_3$(HC≡CCH$_2$)As]Br + I in H$_2$O or EtOH	m. 196° (dec.), IR spectrum	149
	4-MeC$_6$H$_4$SO$_3$	Ph$_3$As + 4-MeC$_6$H$_4$SO$_3$CH$_2$C≡CH in MeCN	m. 215° (dec.), IR spectrum	149
Ph$_3$(MeO$_2$CCH$_2$)As	Br (M-328)	Ph$_3$As + BrCH$_2$CO$_2$Me	Rxn. with NaOMe → Ph$_3\overset{+}{\text{A}}$s–$\overset{-}{\text{C}}$HCO$_2$Me	766
Ph$_3$(BrCH$_2$CHOHCH$_2$)As	Br		m. 183°	1147
Ph$_3$[HO(CH$_2$)$_3$]As	I	Ph$_3$As + HO(CH$_2$)$_3$I in EtOH under reflux	Needles, m. 170°, (ab)	676
		Ph$_3$AsCH$_2$CH$_2$CH$_2$O + HI		676
Ph(CH$_2$=CHCH$_2$)(C$_6$H$_{13}$)$_2$As	Br	Ph(C$_6$H$_{13}$)$_2$As + CH$_2$=CHCH$_2$Br at 80–100°	m. 90–92°	828
PrPh(C$_6$H$_{13}$)$_2$As	Br		(yy)	1133
MePh(C$_7$H$_{15}$)$_2$As	I	Ph(C$_7$H$_{15}$)$_2$As + MeI at r.t. and completed on a steam bath	m. 62–64°	828

(zz) Rxn. with PhLi in Et$_2$O + 4-MeC$_6$H$_4$CHO under reflux → Ph$_3$As + 4-MeC$_6$H$_4$CH$_2$Ac + trans-MeCH=CHC$_6$H$_4$Me (998).
(ab) Rxn. with NaOH in THF under reflux → Ph$_3$AsCH$_2$CH$_2$CH$_2$O (676).

Table 20 Continued

R_4As^+	X^-	Prepd. from	Prop. and Rxn.	Ref.
C_{22}				
$Ph_3(\overline{HNCH=CHCH=C})As$	Br	Ph_3AsBr_2 in C_6H_6 + 2-pyrrolyl-MgBr in Et_2O under reflux	m. 254-257°	1646
$Me(PhCH_2)(4-MeC_6H_4)_2As$	Cl		A corrosion inhibitor for porous metals	747
$EtPh(PhCH_2)_2As$	Br	$Ph(PhCH_2)_2As$ + EtBr	m. 160-161°, cathodic cleavage → $EtPh(PhCH_2)As$	750
$[2-(Et_2MeAs)C_6H_4-]_2$	I_2 (M-328)	$[2-(Et_2As)C_6H_4-]_2$ + MeI (excess) in C_6H_6 under reflux	m. 199-200°, (ac)	57
$Ph(EtO_2CCH_2)(C_6H_{13})_2As$	Br	$Ph(C_6H_{13})_2As$ + $BrCH_2CO_2Et$ on a steam bath	Mobile oil, formed hydroscopic crystals in a dessicator, n_D^{20} 1.5400, d_4^{20} 1.2295	1549
	NO_3	$[Ph(EtO_2CCH_2)(C_6H_{13})_2As]Br$ + $AgNO_3$ in MeOH on a steam bath	Mobile oil	1549
C_{23}				
$EtPh_2(PhC{\equiv}CCH_2)As$	Br		> 200° → $EtPh_2As$ + $Ph_2AsCH_2C{\equiv}CPh$	150
$Ph(CH_2=CHCH_2)(C_7H_{15})_2As$	Br	$Ph(C_7H_{15})_2As$ + $CH_2=CHCH_2Br$ at 80-100°	m. 97-99°	828
	(-)-HDBT, (xx)	$[Ph(CH_2=CHCH_2)(C_7H_{15})_2As]Br$ + Ag[(-)-HDBT] in MeOH	Dark yellow oil	828
$PrPh(C_7H_{15})_2As$	Br		(yy)	1133
$MePh(C_8H_{17})_2As$	I	$Ph(C_8H_{17})_2As$ + MeI on a steam bath	Hydroscopic crystals	1549
C_{24}				
Ph_4As	$Au(N_3)_2$	$[Ph_4As][Au(N_3)_4]$ in THF by exposure to daylight	Dec. 127°, IR spectrum	131,132
		$[Ph_4As]Cl$ + AuCl + NaN_3 in CH_2Cl_2	Elec. cond. data	132

Anion	Preparation	Properties	Ref.
$Au(N_3)_4$ (ag)	$K[AuCl_4]$ soln. + NaN_3 in H_2O followed by $[Ph_4As]Cl$	Orange, diamagnetic needles, dec. 145°, elec. cond. data, UV, visible, and IR spectra, (ad, ae, af)	132
$AuS_4C_4(CF_3)_4$, (ag)	$Ph_3P \cdot ClAuS_4C_4(CF_3)_4$ in aq. EtOH + $[Ph_4As]Cl$	Green plates, m. 179–181°, elec. cond. data	447, (ah)
$\left[Au\left\langle \begin{smallmatrix} N-N \\ = \\ N-N \end{smallmatrix} \right\rangle_4 C_6H_{11}\right]$	$[Ph_4As][Au(N_3)_4] + C_6H_{11}NC$ (1:4.5) in CH_2Cl_2 at 20°	Shiny colorless leaflets, m. 184° (dec.), NMR and IR spectra	131
$Au(SeCN)_4$	$K[AuCl_4] + KSeCN$ by grinding under CH_2Cl_2, adding $[Ph_4As]Cl$, and pptg. with petr. ether	Purple shiny needles and plates, m. 102°, mol. cond., IR spectrum, (ai, aj)	1361
N_3		Rxn. with $(\pi\text{-}C_5H_5)_2TiX_2$ (X = Cl or Br) in dry Me_2CO → exchange of the anions	931
$NS(O)F_2$	$[Ph_4As]Cl + NH_4[NS(O)F_2]$ in H_2O	Dec. > 240°, m. 285°, IR spectrum	1294
	$[Ph_4As]Cl$ in H_2O + 30% $HNS(O)F_2$ in Et_2O		1294
B_3H_8	$[Ph_4As]Cl + [Me_4N]B_3H_8$ in H_2O	Long colorless needles, sol. in MeOH and Me_2CO, m. 255°, sinters at 250°, IR spectrum	61
BF_4 (M-332)	$[Ph_4As]Cl + BF_3$	Sparingly sol.	39

(ac) Thermal dec. → $[2\text{-}(Et_2As)C_6H_4\text{-}]_2$ (57). (ad) Thermally stable, deflagrates in flame (140). (ae) IR spectrum and elec. cond. (1360). (af) A soln. in THF on exposure to daylight → $[Ph_4As]Au(N_3)_2$ (131, 132).

(ag) $S_4C_4(CF_3)_4 = \begin{pmatrix} -S-CCF_3 \\ \| \\ -S-CCF_3 \end{pmatrix}_2$. (ah) See also ref. 445. (ai) A "soft" Se bonding to Au is postulated (1361).

(aj) Absorption and reflectance spectra in the visible and UV ranges (1359).

Table 20 Continued

R_4As^+	X^-	Prepd. from	Prop. and Rxn.	Ref.
C_{24} (Cont.)				
Ph_4As	BPh_4 (M-332)	$[Ph_4As]Cl + NaBPh_4$	Crystals, dec. 255°, soly. data, (ak, al)	371
	(benzene ring fused to $S\!-\!B\!-\!S$ ring)$_2$	$H[B(O_2C_6H_4)_2]$ in 10% aq. $[Me_4N]OH$ + $[Ph_4As]Cl$	White crystals, m. 195-197°, insol. in H_2O, sol. in CH_2Cl_2, polarographic data	1298
	Br (M-330)	$Ph_3As + PhBr + AlCl_3$ at 180-190° under pressure	m. 281-283°, (am, an)	748
		$Ph_3As + PhBr$ under γ-irradiation	Long, colorless needles, m. 228-230°	306
	HBr_2	$[Ph_4As]Br$ by crystn. from liq. HBr	White crystals, IR spectrum	680
	$HBr_2 \cdot 2\ H_2O$	$[Ph_4As]Br$ by crystn. from 48% HBr	White crystals, IR spectrum	680
	$CdCl_3$	$[Ph_4As]Cl + CdCl_2$ in EtOH	Diamagnetic ppt., a 1:1 electrolyte in Me_2CO, IR spectrum, a dimeric bridged anion suggested	1196
	$\{Cd[(H_2N)_2CS]Cl_3\}$	$[Ph_4As]CdCl_3 + (H_2N)_2CS$ in EtOH	Needle-shaped crystals, m. 177°, IR spectrum	1197
	$[Cd(PhNHCSNH_2)Cl_3]$	$[Ph_4As]CdCl_3 + PhNHCSNH_2$ in EtOH under reflux	A 1:1 electrolyte in Me_2CO, IR spectrum	1198
	HCl_2	$[Ph_4As]Cl$ by crystn. from liq. HCl	White crystals, IR spectrum	680
	$HCl_2 \cdot 2\ H_2O$	$[Ph_4As]Cl$ by crystn. from 37% HCl	White needles, IR spectrum (an)	680
	ClO_3			1310
	ClO_4 (M-332)	$[Ph_4As]Cl + KClO_4$	Crystals, dec. 230°, soly. data, (ao)	371
			m. 347°	748
			UV spectrum, soly. in H_2O, and heat of dissocn., (al, ap)	1172
	CrO_3Br	$K_2Cr_2O_7$ soln. + HBr, followed by $[Ph_4As]Br$, extn. with $CHCl_3$ and pptn. with petr. ether	Yellow solid, IR spectrum	1330

Compound	Preparation	Properties	Ref.
CrO(O₂)₂Br	[Ph₄As][CrO(O₂)₂Br] by thermal dec.	Blue solid, stable at r.t., dec. >110° without explosion, IR spectrum, (aq)	1330
	K₂Cr₂O₇ soln. + HBr + [Ph₄As]Br, followed by extn. with CHCl₃, oxidn. with H₂O₂ and pptn. with petr. ether		1330
CrO₃Cl	K₂Cr₂O₇ soln. + HCl + [Ph₄As]Cl, followed by extn. with CHCl₃, and pptn. with petr. ether	Yellow solid, IR spectrum	1330
	[Ph₄As][CrO(O₂)₂Cl] by thermal dec.		1330
	[Ph₄As]HCl₂ + aq. K₂Cr₂O₇ in 0.05 N H₂SO₄	IR spectrum	1529
CrO(O₂)₂Cl	[Ph₄As][CrO₃Cl] + 30% H₂O₂ in CHCl₃-EtOAc	Royal blue crystals, soly. data, a 1:1 electrolyte in MeNO₂, IR spectrum	1529
	K₂Cr₂O₇ soln. + HCl + [Ph₄As]Cl, followed by extn. with CHCl₃, oxidn. with 3% H₂O₂, and pptn. with petr. ether	Blue solid, relatively stable at r.t., dec. > 80°, without explosion, IR spectrum, (aq)	1330
HCrO₄ / Cr(S-CCF₃ with =S–CCF₃)		Extn. with CHCl₃, (ao)	672
	[Ph₄As]₂[Cr(S-CCF₃)(=S-CCF₃)₃] + Cr(S-CCF₃)(=S-CCF₃)₃ in CH₂Cl₂	Black crystals, m. 127.5-129°	444
[Cr(CO)₅GeCl₃]	[Ph₄As]GeCl₃ in CH₂Cl₂ + Cr(CO)₆ under irradiation	Yellow to orange solid, m. 116-117°, (ar)	1307

(ak) A catalyst for the prepn. of thermoplastic polycarbonates (551). (al) Solvent activity coeff. of the ions at 25° for a transfer from MeOH to H₂O, HCONH₂, HCONMe₂, AcNMe₂, Me₂SO, MeCN, and (Me₂N)₃PO (54). (am) Cathodic cleavage in aq. soln. → Ph₃As + C₆H₆ (748). (an) IR spectrum (1001, 1049). (ao) Extractability with CHCl₃, extractability const., and distribution ratio (1310). (ap) Spectrophotometric and gravimetric detn. (625). (aq) Thermal dec. removes the peroxide O only, leaving Cr(VI) (1330). (ar) Stable in air for a short time, sol. in polar org. solvents, a 1:1 electrolyte in MeNO₂ (1307).

253

Table 20 Continued

R_4As^+	X^-	Prepd. from	Prop. and Rxn.	Ref.
C_{24} (Cont.)				
Ph_4As	$[Cr(CO)_5SP(O)F_2]$	$[Ph_4As]SP(O)F_2 + Cr(CO)_6$ in CH_2Cl_2 under irradiation	Yellow solid, m. 273-275°, dec. at 135°, IR and NMR spectra, molar cond. in $MeNO_2$ detd.	1308
	$CoCl_3 \cdot H_2O$	$[Ph_4As]Cl + CoCl_2 \cdot 6\ H_2O$	Crystals, dec. 295°, soly. data	371
	$Co(PPh_3)I_3$	$[Ph_4As]I + Ph_3P + CoI_2 \cdot 2\ H_2O$ in BuOH	Green crystals, m. 236°, NMR spectrum, (as)	927
	$Co\left[\begin{smallmatrix}S-CCN\\ \| \\ S-CCN\end{smallmatrix}\right]_2$	$NCCSNa + CoCl_2 \cdot 6\ H_2O$ in H_2O + $NCCSNa$ $[Ph_4As]Br$		166
	$\{Co[P(OMe)_3](NO_3)_2\}$	$[Ph_4As]NO_3 + \{Co[P(OMe)_3]_5\}-\{Co[P(OMe)_3](NO_3)_2\}$ in MeCN	Azure blue crystals; UV, visible, and IR spectra; molar cond. data	392
	$Cu\left[\begin{smallmatrix}S-CCF_3\\ \| \\ S-CCF_3\end{smallmatrix}\right]_2$	$[Ph_3PCuI]_4 + CF_3C{=}C(CF_3)SS$ in C_6H_6 + $[Ph_4As]Cl$	Olive green needles, m. 175-175.5°	447
	CN		(at)	1158
	$N(CN)_2$	$[Ph_4As]Cl + K[N(CN)_2]$ in H_2O	Colorless solid, m. 184°	867
	$C_6H_{11}CS_2$		Forms chelates with Cu^{++}, Bi^{3+}, Mo^{5+}, and Pd^{++}; extractive photometric detn.	80
	$\alpha\text{-}C_{10}H_7CS_2$	$[Ph_4As]Cl + Na\ \alpha\text{-dithionaphthoate}$	Soly. data	117
	$Eu[OCH(Bz)CF_3]_4$		Fluorescence spectrum in C_6H_6 at 20°, cation effect	1400, 1401
	F	$[Ph_4As]Cl + HF$ followed by NaOR and heating in vacuum	Useful in organometallic complexes for electrolytic processes	1547
	C_9FN_6, (au)	$[Ph_4As]I + Na[C_9FN_6]$ in H_2O	Large ruby needles, m. 161-163°	302
		$[Ph_4As]^+$ + rxn. prod. of $\overline{CF_2CF_2CF_2CCl}{=}CCl + NaCN$ or KCN	Deep red, surfboard-shaped crystals, x-ray crystal analysis	1183

$C_{11-12}FN_6$, (av)	$[Ph_4As]^+$ + $Na[C_{11-12}FN_6]$ in H_2O	Large needles with metallic blue sheen, m. 208°	302
$Ga(SCH=CHS)_2$		Colorless diamagnetic crystals, m. 205° (dec.), NMR and electronic spectra	762
$GeCl_3$	$[Ph_4As]Cl$ + Ge^{++} in HCl	IR spectrum	810
$GeCl_3 \cdot BF_3$	$[Ph_4As]GeCl_3$ + BF_3 in CH_2Cl_2	IR and NMR spectra, (aw)	810
$GeCl_3 \cdot BCl_3$	$[Ph_4As]GeCl_3$ + BCl_3 in CH_2Cl_2	IR and NMR spectra, (aw)	810
GeF_5	$[Ph_4As]Cl$ + GeO_2 in MeOH-HF	IR spectrum	351
$[In(DiArs)_2I_4]$, (ax)	$[In(DiArs)_2I_2][In(DiArs)I_4]$ + $[Ph_4As]I$	Crude solid	1164
$[In(Py)_2Cl_4]$	$InCl_3$ in MeCN + excess $[Ph_4As]Cl$ + excess Py	White crystals, IR spectrum, a trans-octahedral anion suggested	1581
$[In(2,2'-BiPy)Cl_4]$	$[Ph_4As]Cl$ + $In(2,2'-BiPy)_{1.5}Cl_3$ in MeCN under reflux	Crystals with 1 mole MeCN, IR and far-IR spectra	1580
Me_2InCl_2		White crystals, molar cond. in $MeNO_2$ detd., IR spectrum	353
$In(SCH=CHS)_2$		Beige, diamagnetic crystals, m. 181° (dec.), NMR and electronic spectra	762
I (M-331)		UV spectra in various solvents; the cation effect on the energy and temp. coeff. (al, ao)	189
I_3 (M-331)	Ph_3As + PhI under γ-irradiation	Dark red crystals, (ay)	306
		IR spectrum	1001

(as) Ion pairing and interionic distances calcd. (927). (at) A mixt. with $1,3,5-(O_2N)_3C_6H_3$ under irradiation with 3500-6500 Å light, paramagnetism is induced due to generation of a radical species (1158). (au) C_9FN_6 = $(NC)_2C=C(CN)CF=NC(CN)_2$. (av) $C_{11-12}FN_6$ is a rxn. prod. formed from NaCN + $\overline{CF_2CF_2CF_2CCl}$=CCl (302). (aw) The structure of the anion is considered (810). (ax) DiArs = o-$(Me_2As)_2C_6H_4$. (ay) The effect of the ionic lattice on the electronic structure of I_3 (247).

Table 20 Continued

R_4As^+	X^-	Prepd. from	Prop. and Rxn.	Ref.
C_{24} (Cont.)				
Ph_4As	$Ir(CO)_2Cl_2$	$[Ph_4As]Cl + (OC)_2Ir\overset{O=CMe}{\underset{N-CMe}{\overset{\cdot\cdot CH}{\diagup\diagdown}}}$ MeC_6H_4	White ppt., m. 214-215°, IR spectrum, elec. cond. data	203
	$Ir(SbPh_3)_2Br_4$	$[Ph_4As]Br + K[Ir(SbPh_3)_2Br_4]$ in MeOH	Brown crystals, m. 181°	84
	$Ir(SbPh_3)_2Cl_4$	$[Ph_4As]Cl + Na[Ir(SbPh_3)_2Cl_4]$ in MeOH	Brown crystals, m. 197°	86
	$FeCl_4$ (M-332)	$[Ph_4As]Cl + FeCl_3$ in $MeNO_2$ and adding Et_2O	Yellow needles, m. 183-185°, UV spectrum in MeCN, (az)	1623
	$Fe(NCO)_4$	$[Ph_4As]FeCl_4 + AgCNO$ in Me_2CO	Orange crystals, m. 159°, IR spectrum in soln. and in the solid state	576
	$Fe(CO)_4(HgBr)_2Cl$	$[Ph_4As]Cl + [Fe(CO)_4(HgBr)_2]$	Pale yellow, m. 168-169°, IR and far-IR spectra, molar cond. data, (ba)	322
	$Fe(CO)_4(HgCl)_2Cl$	$[Ph_4As]Cl + [Fe(CO)_4(HgCl)_2]$	Pale yellow, m. 183-184°, IR and far-IR spectra, molar cond. data, (bb)	322
	$Fe(C_8H_8S_2)_2$, (bc)	$Fe^{3+} + Na_2C_8H_8S_2 + [Ph_4As]^+$, (bd)	Black, paramagnetic crystals, m. 240-245° (dec.), 1/2-wave potential, electronic spectrum	761
	$Fe(C_8H_8S_2)_2Py$, (bc)	$Fe^{3+} + Na_2C_8H_8S_2 + [Ph_4As]^+$ in Py, (bd)	Black, paramagnetic crystals, m. 240-245° (dec.), 1/2-wave potential, electronic spectrum	761
	$Fe(CO)_4GeCl_3$	$[Ph_4As]GeCl_3$ in $CH_2Cl_2 + Fe(CO)_5$ under irradiation	Yellow to orange solid, m. 113-114°, (be)	1307

Fe(CO)$_4$SnCl$_3$	[Ph$_4$As]SnCl$_3$ in CH$_2$Cl$_2$ + Fe(CO)$_5$ under irradiation	Yellow to orange solid, m. 96-97°, (be)	1307
Fe(NO)C$_8$F$_{12}$S$_4$, (bf)	[FeC$_8$F$_{12}$S$_4$]$_2$ in Me$_2$CO + NO + [Ph$_4$As]Cl	(be)	1037
Fe(SCH=CHS)$_2$Py		Black, paramagnetic crystals, m. 160-162° (dec.), 1/2-wave potential, electronic spectrum	761
Fe(NO)$_2$(2-HSC$_6$H$_4$NH$_2$)	[Ph$_4$As]Cl + Na[Fe(NO)$_2$(2-HSC$_6$H$_4$NH$_2$)] in THF	Olive green, paramagnetic crystals, m. 128°, stable in air and in H$_2$O, molar cond. data	732
NCO·2 H$_2$O	[Ph$_4$As]Cl in H$_2$O + KNCO	White needles contg. H$_2$O of crystn., loses H$_2$O at 100°, m. 265° (dec.), (bg)	1157
H(NCS)$_2$	[Ph$_4$As]Cl·HCl + KSCN (1:2) in SO$_2$	Pink crystals, m. 136° (dec.), sol. in H$_2$O (acidic pH), IR and NMR spectra	493
MnO$_4$ (M-333)	[Ph$_4$As]Cl + KMnO$_4$	Crystals, dec. < 110°, soly. data CHCl$_3$ soln. is an oxidn. reagent for org. compds. insol. in H$_2$O, (ao)	371 615
Mn(CO)$_5$	[Mn(CO)$_5$]$_2$ + aq. KOH + [Ph$_4$As]$^+$	IR spectrum	808
π-C$_5$H$_5$Mn(CO)$_2$CN	[Ph$_4$As]Cl + Na[π-C$_5$H$_5$Mn(CO)$_2$CN] in H$_2$O	Light yellow ppt., IR and NMR spectra	568

(az) Mössbauer spectrum (517, 1623). (ba) Crystd. from EtOH with 1 mol. solvent (322). (bb) Crystd. from MeOH with 1 mol. solvent (322). (bc)   (bd) The salts may be prepd. from Fe^{2+} under oxidizing

$C_8H_8S_2$ = -S–⟨benzene ring⟩(Me)(Me)–S-   cond. (761). (be) Stable in air for a short time, sol. in polar solvents, a 1:1 electrolyte in MeNO$_2$ (1307).

(bf) $C_8F_{12}S_4$ = $\left[\begin{array}{c} \text{-SCCF}_3 \\ \| \\ \text{-SCCF}_3 \end{array}\right]_2$.   (bg) The dehydrated prod. reabsorbs moisture on exposure to atm.; molar cond. in Me$_2$CO and IR and electronic spectra were detd. (1157).

257

Table 20 Continued

258

R_4As^+	X^-	Prepd. from	Prop. and Rxn.	Ref.
C_{24} (Cont.)				
Ph_4As	$Mn(C_8H_8S_2)_2Py$, (bc)		Brown, paramagnetic crystals, m. ~190° (dec.), 1/2-wave potential, electronic spectrum	761
	$Mn(SCH=CHS)_2Py$		Brown, paramagnetic crystals, m. 134-136° (dec.), 1/2-wave potential, electronic spectrum	761
	$Hg(CNO)_2N_3$	$[Ph_4As]Cl + Hg(CNO)_2 + NaN_3$ in H_2O	Colorless ppt., IR spectrum	132
	$MoCl_6$	$[Ph_4As]Cl + MoCl_5$ in CH_2Cl_2 at 35° in a sealed tube	Dark green solid, sensitive to moisture, electronic spectrum	232
	$MoOBr_4$	$[Ph_4As]Br + MoOBr_3$ in SO_2 in a sealed tube	A 1:1 electrolyte, diffuse reflectance, electronic and NMR spectra, magnetic data, (bh)	232
	$MoOBr_4 \cdot H_2O$		Brown tetragonal crystals, x-ray diffraction analysis, (bi)	1340
	$MoOCl_4$	$[Ph_4As]Cl + MoCl_5$ in concd. HCl by satn. with HCl gas	Green solid, hydrolytically unstable, IR and electronic spectra	1236
		$[Ph_4As]Cl + MoO_3$ in HCl by electrochem. redn.	(bh)	1236
	$MoOCl_4 \cdot H_2O$	$[Ph_4As]Cl + MoOCl_3$ in SO_2 in a sealed tube	A 1:1 electrolyte, diffuse reflectance and electronic spectra, magnetic data	232
			IR spectrum, the effect of the H_2O of crystn.	1341
	$Mo\left[\begin{smallmatrix} S-CCF_3 \\ \parallel \\ S-CCF_3 \end{smallmatrix}\right]_3$	$[Ph_4As]_2Mo\left[\begin{smallmatrix} S-CCF_3 \\ \parallel \\ S-CCF_3 \end{smallmatrix}\right]_3 + Mo\left[\begin{smallmatrix} S-CCF_3 \\ \parallel \\ S-CCF_3 \end{smallmatrix}\right]_3$ in CH_2Cl_2	Very dark crystals, sensitive to moisture, m. 189-192.5°	444
	$Mo(CO)_5C(CN)_3$	$Mo(CO)_6 + [Et_4N]C(CN)_3$ in diglyme + $[Ph_4As]Cl$ in EtOH	Dark yellow crystals, soly. data	138

Compound	Preparation	Properties	Ref.
$Mo(CO)_5GeCl_3$	$[Ph_4As]GeCl_3$ in CH_2Cl_2 + $Mo(CO)_6$ under irradiation	Yellow to orange solid, m. 110-112°, (bj)	1307
$Mo(CO)_5SnCl_3$	$[Ph_4As]SnCl_3$ in CH_2Cl_2 + $Mo(CO)_6$ under irradiation	Yellow to orange solid, m. 120-121°, (bj, bk)	1307
$Mo(CO)_5SP(O)F_2$	$[Ph_4As]SP(O)F_2$ + $Mo(CO)_6$ in CH_2Cl_2 under irradiation	Yellow solid, m. 282-285°, dec. 125°, IR and NMR spectra, molar cond. in $MeNO_2$	1308
$Mo(CO)_4(PPh_3)GeCl_3$	$[Ph_4As][Mo(CO)_5GeCl_3]$ in CH_2Cl_2 + Ph_3P at r.t.	Yellow to orange solid, m. 162-163°, (bj)	1307
$Ni(PPh_3)I_3$	$[Ph_4As]I + Ph_3P$ in BuOH + NiI_2 in H_2O	Dark red crystals, m. 237°, NMR spectrum, ion pairing and interatomic distances calcd.	927
$Ni\left[\begin{array}{c}S-CCF_3\\ \parallel\\ S-CCF_3\end{array}\right]_2$	$[Ph_4As]Cl$ in i-PrOH-H_2O by adding to a soln. of $NiS_4C_4(CF_3)_4$ in DMSO	Red-black, paramagnetic crystals, m. 174-174.5°, mol. cond. data	443
$Ni\left[\begin{array}{c}S-CPh\\ \parallel\\ S-CPh\end{array}\right]_2$	$[Ph_4As]Cl$ in MeOH + $NiS_4C_4Ph_4$ + p-$C_6H_4(NH_2)_2$ in DMSO	Purple, paramagnetic crystals, m. 254-256° (dec.), mol. cond. data	443
$Ni\left[\begin{array}{c}Se-CCF_3\\ \parallel\\ Se-CCF_3\end{array}\right]_2$	$[Ph_4As]Cl$ + $NiSe_4C_4(CF_3)_4$ in DMSO	Light green, paramagnetic crystals, m. 162°, polarographic data, ESR spectrum	448
$NbCl_6$	$[Ph_4As]Cl + NbCl_5$ in $SOCl_2$ at r.t.	Crystals, IR spectrum	756
$NbO_2(NCS)_2$	Nb in HCl + KSCN + $[Ph_4As]Cl$	Ppt. insol. in H_2O, IR spectrum, (bl)	40, 1283
$Nb(NCS)_6$	$[Ph_4As]Cl + K[Nb(NCS)_6]$ in MeCN	IR spectrum	248
NO_3 (M-332)	$[Ph_4As]Br + AgNO_3$ in H_2O	White needles, m. 263-265°, molar cond. data, (ao)	544, 545
$H(NO_3)_2$	$[Ph_4As]NO_3$ by dissolving in HNO_3	White flat needles, m. 137°, molar cond. data	544

(bh) IR spectrum (232, 1341). (bi) IR spectrum (1340, 1341). (bj) Stable in air for a short time, sol. in polar org. solvents, a 1:1 electrolyte in $MeNO_2$ (1307). (bk) Rxn. with Ph_3P in $CH_2Cl_2 \rightarrow [Ph_4As]SnCl_3$ + $Mo(CO)_4(PPh_3)_2$ (1307). (bl) Extractable from H_2O into $CHCl_3$-Me_2CO (9:2); anal. procedure for detn. of Nb in steel and alloys (40, 1283).

Table 20 Continued

R_4As^+	X^-	Prepd. from	Prop. and Rxn.	Ref.
C_{24} (Cont.)				
Ph_4As	$H(NO_3)_2$	[Ph_4As]Cl by dissolving in 8 Ṅ HNO_3	(bm)	544
	$H_n(NO_3)_{n+1}$	By-prod. from [Ph_4As]NO_3 dissolved in HNO_3	Unstable solid, dec. to [Ph_4As]$H(NO_3)_2$	544
	$NO_2 \cdot H_2O$	[Ph_4As]$_2SO_4$ + $Ba(NO_2)_2$ in H_2O, extn. with CH_2Cl_2 and pptn. with petr.	White solid, m. 249-251°, UV spectrum	545
	$H(NO_2)_2$	[Ph_4As]$NO_2 \cdot H_2O$ in CH_2Cl_2 + $NaNO_2$ in NaOH + H_2SO_4	White crystals, dec. > 30° to form [Ph_3As]NO_3, soly. data, IR and UV spectra	545
	OsO_3N	[Ph_4As]Cl + K[OsO_3N] in H_2O	Colorless crystals, sparingly sol., IR spectrum	1103
	$Pd(Me_2SO)Cl_3$	$Pd(OAc)_2$ + Me_2SO + [Ph_4As]Cl, (bn)	Lemon yellow crystals, m. 190° (dec.), IR spectrum	1452
	$PC_{36}H_{24}$, (bo)	[Ph_4As]Cl + Na[$PC_{36}H_{24}$] in DMF-EtOH	Needles contg. 1 Me_2CO, dec. 246-247°, loses solvent at 120° in vacuum	718
	$SP(O)Cl_2$	[Ph_4As]Cl in H_2O + $P(S)Cl_3$	m. 223° (dec.), IR spectrum	1288
	$SP(O)ClF$	[Ph_4As]Cl in H_2O + $P(S)Cl_2F$	m. 248° (dec.), IR spectrum	1288
	$SP(O)F_2$	[Ph_4As]Cl in dild. NaOH soln. + $P(S)F_3$	m. 320°, IR spectrum, (bp)	1287
	$SP(O)(SMe)F$	[Ph_4As]Cl + $MeSP(S)FCl$ in H_2O	m. 164°, IR spectrum	1292
	$SP(O)(SEt)F$	[Ph_4As]Cl + $EtSP(S)FCl$ in H_2O	m. 126°, IR spectrum	1292
	$SP(O)(SPh)F$	$PhSP(S)FCl$ in H_2O-MeCO + 2 N NaOH, followed by [Ph_4As]Cl	Colorless cryst. ppt., m. 171°, IR spectrum	1291
	$SP(S)F_2$	[Ph_4As]Cl + Cs[$SP(S)F_2$]	Dec. 270-275°, IR spectrum	996
	$SP(S)(Me)F$	[$MeP(S)S$]$_2$ + KF in MeCN or NaF in H_2O + [Ph_4As]Cl, (bq)	m. 196°, IR and NMR spectra	1289
	$SP(S)(Et)F$	[$EtP(S)S$]$_2$ + KF in MeCN or NaF in H_2O + [Ph_4As]Cl, (bq)	m. 172°, IR and NMR spectra	1289

SP(S)(N$_3$)Me	[Ph$_4$As]Cl + Na[SP(S)(N$_3$)Me] in H$_2$O	Colorless crystals, m. 123°, dec. ~128° in vacuo, stable in air, IR spectrum	1293
SP(S)(N$_3$)Et	[Ph$_4$As]Cl + Na[SP(S)(N$_3$)Et] in H$_2$O	Colorless crystals, m. 93°, dec. ~98° in vacuo, stable in air, IR spectrum	1293
Pt$\left[\begin{array}{c}S-CCF_3\\\|\\S-CCF_3\end{array}\right]_2$	Pt$\left[\begin{array}{c}S-CCF_3\\\|\|\\S-CCF_3\end{array}\right]_2$ in Me$_2$CO-EtOH + [Ph$_4$As]Cl in EtOH	Brown, paramagnetic needles, m. 168-169°, elec. cond. data	446
	(Ph$_3$P)$_2$PtBr$_2$ + CF$_3$C=C(CF$_3$)SS in C$_6$H$_6$ + [Ph$_4$As]Cl		447
Pb(N$_3$)$_3$	[Ph$_4$As]Cl + (NH$_4$)$_2$PbCl$_4$ + NaN$_3$ in H$_2$SO$_4$	Yellow orange ppt.	132
Pc, (c) (M-333)		(al)	54
Re(O)Br$_4$	KReO$_4$ or K$_2$ReBr$_6$ + Zn in H$_2$SO$_4$, followed by 48% HBr + [Ph$_4$As]Br	Coral red, diamagnetic crystals, soly. data, hydrolytically unstable, (br, bs, bt)	399
	[Ph$_4$As][ReOBr$_4$] + concd. HCl	IR spectrum, (br, bu, bv)	399
Re(O)Br$_4$·MeCN		X-ray diffraction data, (bw, bx)	398,400
Re(O)Cl$_5$	[Ph$_4$As]Cl + Re(O)$_2$Cl$_4$ in CHCl$_3$ or CH$_2$Cl$_2$	Red-brown, paramagnetic crystals, a 1:1 electrolyte, IR and electronic spectra	230,231

(bm) Crystallographic data (546). (bn) By shaking Pd(OAc)$_2$ in C$_6$H$_6$ with Me$_2$SO, dissolving the oily prod. in H$_2$O and pouring it into [Ph$_4$As]Cl dissolved in aq. HCl. (bo) PC$_{36}$H$_{24}$ = tris(2,2'-biphenylylene)phosphate anion. (bp) Rxn. with M(CO)$_6$, where M = Cr, Mo, or M $\rightarrow$ [Ph$_4$As][M(CO)$_5$SP(O)F$_2$] (1308). (bq) The H$_2$O sol. prod. fraction is treated with [Ph$_4$As]Cl. (br) A 1:1 electrolyte, x-ray diffraction data (398). (bs) Rxn. with Py under reflux $\rightarrow$ [Re(Py)$_4$O$_2$]Br·2 H$_2$O (399). (bt) Rxn. with Ph$_3$P in MeCN $\rightarrow$ Re(Ph$_3$P)$_2$(O)Br$_3$ (399). (bu) Rxn. with Py under reflux $\rightarrow$ [Re(Py)$_4$O$_2$]Cl·2 H$_2$O (399). (bv) Rxn. with Ph$_3$P in MeCN $\rightarrow$ Re(Ph$_3$P)$_2$(O)Cl$_3$ (399). (bw) IR spectrum (399). (bx) A square-pyramidal Re(O)Br$_4$ with weakly bonded MeCN postulated (400).

Table 20 Continued

R₄As⁺	X⁻	Prepd. from	Prop. and Rxn.	Ref.
C₂₄ (Cont.)				
Ph₄As	ReO₄ (M-333)	[Ph₄As]Re(O)Br₄ + diaryl sulfoxide in MeCN or hydrolysis with H₂O	Pale yellow tetrahedra, a 1:1 electrolyte, (by)	399
		[Ph₄As]Cl + KReO₄	Crystals, dec. 285°, soly. data (ao, bz, ca, cb, cc)	371 / 404
		[Bu₄N]₂Re₂Cl₈ + H₂O₂ + NaSCN in MeOH, followed by [Ph₄As]Cl [Ph₄As]⁺ + ReO₄⁻ in aq. soln.		1113, 1507
	Re₃Br₁₀	[Ph₄As]Br in EtOH + ReBr₃ in aq. or alc. HBr	Crystals, molar cond. data, absorption spectrum	1282
	Rh(CO)₂Br₂	[Ph₄As]Cl + Rh(CO)₂Cl in EtOH + LiBr under reflux	Deep yellow complex, a 1:1 electrolyte, IR spectrum	853
		[Ph₄As]Br + Rh₂(CO)₄Br₂ or Rh₂(CO)₄Cl₂ in MeOH contg. HBr [Ph₄As]Rh(CO)Cl₂ + HBr in MeOH	Yellow, diamagnetic plates, m. 150°, a 1:1 electrolyte in PhNO₂ IR spectrum, (cd, ce, cf)	1551 1551
	Rh(CO)₂Cl₂	[Rh₂(CO)₄Cl₂] in MeOH contg. HCl + [Ph₄As]Cl	Pale yellow, diamagnetic plates, m. 181°, a 1:1 electrolyte in PhNO₂, (cd, ce, cg, ch)	1551
		[Rh₂(CO)₄Cl₂] + [Ph₄As]Cl in EtOH	Yellow crystals, a 1:1 electrolyte in MeNO₂	853
		[Rh₂(CO)₄Cl₂] + HCl or NaCl in Me₂CO + [Ph₄As]Cl	Yellowish crystals, m. 185-190° (dec.), sparingly sol.	1618
	Rh(CO)₂I₂	[Rh₂(CO)₄Cl₂] + HI + [Ph₄As]I in MeOH	Light brown, diamagnetic leaflets, m. 121°, a 1:1 electrolyte in PhNO₂, IR spectrum, (cd, ce)	1551
		[Ph₄As][Rh(CO)₂X₂] (X = Cl or Br) in MeOH + HI at 0°		1551
	Rh(CO)L₄	[Ph₄As][Rh(CO)₂I₂] or [Ph₄As]₂-[Rh₂(CO)₂I₄] in MeOH contg. HI, (ci)	Dark brown leaflets, sol. in PhNO₂ (dec.), IR spectrum	1551

Compound	Preparation	Properties	No.
Rh(C$_{12}$H$_9$NO)Cl$_4$, (cj)	RhI$_3$ + NaI + CO at 120°/200 atm. in the presence of Cu, extn. with MeOH and treating with [Ph$_4$As]I		1551
	[Ph$_4$As]Cl + RhCl$_3$·3 H$_2$O in EtOH + 2-benzoylpyridine under reflux	Orange substance, molar cond. in MeNO$_2$, IR and diffuse reflectance spectra	1182
Rh(PMe$_2$Ph)$_2$Cl$_4$	[Ph$_4$As]Cl + [Me$_2$PhPH]$^+$-[Rh(PMe$_2$Ph)$_2$Cl$_4$]$^-$ in EtOH-HCl	m. 225-231°, molar cond. data, NMR spectrum	241
Rh(O)(OH)(CN)$_4$	K$_3$[Rh(CN)$_8$] in H$_2$O-HCl + aq. [Ph$_4$As]Cl	Blue purple salt, soly. data, IR spectrum	978
	K$_3$[RhO$_2$(CN)$_4$] in 4 N HCl + [Ph$_4$As]Cl		978
Ru(H$_2$O)$_2$Cl$_4$·H$_2$O	[Ph$_4$As]Cl + HRu(H$_2$O)$_2$Cl$_4$ in aq. HCl	Ruby red prisms, mol. and cryst. structure	742
Ru(CO)(Py)Cl$_4$	[Ru(AsPh$_3$)(Py)$_2$Cl$_3$] in H$_2$O by adding to [Ph$_4$As]Cl in HCl	Orange brown ppt., m. 175°, IR spectrum	1453
SiF$_5$	H$_2$SiF$_6$ + [Ph$_4$As]$^+$ in acid soln.	Ppt. stable up to 245°, sol. in org. solvents, (ck)	146
	[Ph$_4$As]F + SiF$_4$ in CH$_2$Cl$_2$	Crystallographic data, NMR spectrum	146
	[Ph$_4$As]Cl + SiO$_2$ in MeOH-HF	IR spectrum, molar cond. data	351
FSO$_2$NCF$_3$	[Ph$_4$As]Cl in H$_2$O + FSO$_2$NHCF$_3$	White solid, m. 190° (dec.)	1290
TaCl$_6$	[Ph$_4$As]Cl + TaCl$_5$ in SOCl$_2$ at r.t.	Crystals, IR spectrum	756
Ta(NCS)$_6$	[Ph$_4$As]Cl + K[Ta(NCS)$_6$] in MeCN	IR spectrum	248

(by) IR spectrum (399, 404, 1113). (bz) Soly. in H$_2$O detn. by detg. the absorption spectrum and heat of dissocn. (1171). (ca) Extractable with CHCl$_3$ from aq. soln. at pH 10 and sepd. from excess W, Mo, Cu, and Ni (1507). (cb) Redn. of Re(VII) in CHCl$_3$ soln. with α-furildioxime and photometric detn. of Re at 500 mμ (1507). (cc) Co-pptn. of Tc(VII) tracer in the spallation prods. of Mo with high energy protons, after adding NH$_4$ReO$_4$, Fe^{3+}, Nb, Zr, Y, Sr, Rb, and Se, by adding H$_2$O$_2$ and [Ph$_4$As]Cl (1254). (cd) A 4-coordinate Rh with a square-planar structure postulated (1551). (ce) Stable in the solid state and in soln. contg. small amt. HCl (1551). (cf) A soln. in 10% HBr-MeOH on standing for 1 day at r.t. or 1 hr. under reflux → [Rh$_2$(CO)$_2$Br$_4$]$^{--}$ (1551). (cg) Rxn. with R$_3$P and RNC → [Rh(CO)(R$_3$P)Cl] and [Rh(RNC)$_4$]Cl, resp. (1551). (ch) IR spectrum (853, 1551, 1618). (ci) Under reflux for a short time or prolonged standing at r.t. (cj) C$_{12}$H$_9$NO = 2-benzoylpyridine. (ck) Rxn. with H$_2$O → [Ph$_4$As]F, SiO$_2$, and HF (146).

Table 20 Continued

R_4As^+	X^-	Prepd. from	Prop. and Rxn.	Ref.
C_{24} (Cont.)				
Ph_4As	TcO_4 (M-334)	$[Ph_4As]^+ + NH_4[TcO_4]$ in H_2O	Crystals, IR spectrum	1118
	$TlBr_4$		IR spectrum	36,1049
	$TlCl_4$	$TlCl_3 + [Ph_4As]Cl$ in MeCN	White needles, (cl, cm)	36
		$TlCl + Cl$ in MeCN $+ [Ph_4As]Cl$	White needles, m. 146-147°, x-ray diffraction data, (cn)	395,1581
	TlI_4	$[Ph_4As]TlCl_4$ in $Me_2CO + NaI$ in H_2O	Red-orange crystals, m. 142-143°, (cl)	395
	$TlBr_2(N_3)_2$	$TlBr + Br + NaN_3$ in $H_2O + [Ph_4As]Cl$	Yellow needles, dec. 120°, IR spectrum	132,141
	$PhTlBr_3$	$[Ph_4As]Br + PhTlBr_2$ in Me_2CO	m. 148-150°	547
	$PhTlCl_3$	$[Ph_4As]Cl + PhTlCl_2$ in EtOH	m. 165-166°	547
	$PhTlI_3$	$[Ph_4As]I + PhTlI_2$ in Me_2CO	m. 158-159°	547
	Ph_2TlBr_2	$[Ph_4As]Br + Ph_2TlBr$ in Me_2CO	m. 199-200°	547
	Ph_2TlCl_2	$[Ph_4As]Cl + Ph_2TlCl$ in Me_2CO	m. 206-208°	547
	Ph_2TlI_2	$[Ph_4As]I + Ph_2TlI$ in Me_2CO	m. 168-170°	547
	$Me_2Tl(SCN)_2$	$[Ph_4As]SCN + Me_2TlSCN$ in Me_2CO	m. 165-167°	547
	$Ph_2Tl(SCN)_2$	$[Ph_4As]SCN + Ph_2TlSCN$ in Me_2CO	m. 186-188°	547
	$Tl(SCH=CHS)_2$, (co)		Red, diamagnetic crystals, m. 140-142° (dec.), (cp)	762
	$Tl(C_8H_8S_2)_2$, (co)		Yellow, diamagnetic crystals, (cp)	762
	SCN (M-334)		(al, ao, cq)	931
	$SnCl_3$	$[Ph_4As]Cl + Sn^{++}$ in HCl	IR spectrum	810,1408
	$SnCl_3 \cdot BCl_3$	$[Ph_4As]SnCl_3 + BCl_3$ in CH_2Cl_2	IR and NMR spectra	810
	$SnCl_3 \cdot BF_3$	$[Ph_4As]SnCl_3 + BF_3$ in CH_2Cl_2	IR and NMR spectra	810
	$SnCl_5$	$SnCl_4 + [Ph_4As]Cl$, (cr)	(cs)	1471
	$MeSnCl_4$	$MeSnCl_3 + [Ph_4As]Cl$, (cr)	(cs)	1471
	Me_2SnCl_3	$Me_2SnCl_2 + [Ph_4As]Cl$, (cr)	(cs, ct, cu, cv)	1471
	$EtSnCl_4$	$EtSnCl_3 + [Ph_4As]Cl$, (cw)	White solid, m. 180-185°, (cs, ct, cu, cv)	1470

264

Compound	Preparation	Properties	References
Et_2SnCl_3	Et_2SnCl_2 + $[Ph_4As]Cl$, (cr, cw)	White solid, m. 165-166°, (cs, ct, cu, cv)	1470, 1471
Pr_2SnCl_3	Pr_2SnCl_2 + $[Ph_4As]Cl$, (cr)	(cs, cv)	1471
$BuSnCl_4$	$BuSnCl_3$ + $[Ph_4As]Cl$, (cr, cw)	White solid, m. 142-143°, (cs, ct, cu, cv)	1470, 1471
Bu_2SnCl_3	Bu_2SnCl_2 + $[Ph_4As]Cl$, (cr, cw)	White solid, m. 94-96°, (cs)	1470, 1471
$PhSnCl_4$	$PhSnCl_3$ + $[Ph_4As]Cl$, (cr, cw)	White solid, m. 161-162°, (cs, cu, cv)	1470, 1471
Ph_2SnCl_3	Ph_2SnCl_2 + $[Ph_4As]Cl$, (cr, cw)	White solid, m. 225-227°, (cs, cu, cv)	1470, 1471
Ph_3SnCl_2	Ph_3SnCl + $[Ph_4As]Cl$, (cr)	(cs)	1471
$W(O)Cl_4 \cdot H_2O$	$[Ph_4As]Cl$ + $WOCl_3$ in SO_2	A 1:1 electrolyte, diffuse reflectance, electronic and IR spectra, magnetic data	232
$W(CO)_5NCO$	$W(CO)_6$ + KNCO in diglyme at 100° + $[Ph_4As]Cl$ in H_2O-EtOH	Lemon-yellow crystals, m. 157° (dec.), soly. data, IR spectrum, molar cond.	142, 145
	$[Ph_4As]Cl$ + $[Et_4N][W(CO)_5NCO]$ in EtOH		145
$W(CO)_5SP(O)F_2$	$[Ph_4As]SP(O)F_2$ + $W(CO)_6$ in CH_2Cl_2 under irradiation	Yellow solid, m. 298°, dec. 130°, IR and NMR spectra, molar cond. in $MeNO_2$	1308
$W(CO)_5GeCl_3$	$[Ph_4As]GeCl_3$ + $W(CO)_6$ in CH_2Cl_2 under irradiation	Yellow to orange solid, m. 110-111°	1307

(cl) IR spectrum (36, 1049). (cm) Extn. from aq. soln. with i-BuAc (1656). (cn) IR and Raman spectra (1440).
(co)
$C_8H_8S_2 = $ (structure with $-S-$ and Me, Me)
(cp) NMR and electronic spectra (762). (cq) Rxn. with (π-C_5H_5)$_2$TiBr$_2$ or (π-C_5H_5)$_2$TiCl$_2$ in Me_2CO → exchange of the anions (931). (cr) By potentiometric titration of the tin compd. with $[Ph_4As]Cl$ in MeCN at 25°. (cs) A 5-coordinate Sn anion (1471). (ct) Mol. wt. detn. in Me_2CO (1644). (cu) Molar cond. in Me_2CO at 25° was detd. (1644). (cv) Stability const. of the $R_{4-n}SnCl_{n+1}$ anion in MeCN at 35° was detd. (1644). (cw) By adding $[Ph_4As]Cl$ to $R_{4-n}SnCl_n$ in aq. HCl or HCl-NaCl soln. (1470).

Table 20 Continued

R_4As^+	X^-	Prepd. from	Prop. and Rxn.	Ref.
C_{24} (Cont.)				
Ph_4As	$W(CO)_5SnCl_3$	$[Ph_4As]SnCl_3 + W(CO)_6$ in CH_2Cl_2 under irradiation	Yellow to orange solid, m. 129-130°	1307
	$\pi\text{-}C_5H_5W(SCH=CHS)_2$	$\pi\text{-}C_5H_5W(CO)_3Cl + NaSCH=CHSNa$ in MeOH, followed by $[Ph_4As]Cl$ in MeOH	Deep purple flakes, m. 212-214°, IR, UV, visible, and NMR spectra	852
	$W\left[\begin{smallmatrix}S-CCF_3\\ \parallel\\ S-CCF_3\end{smallmatrix}\right]_3$ (M-334)	$[Ph_4As]\left(W\left[\begin{smallmatrix}S-CCF_3\\ \parallel\\ S-CCF_3\end{smallmatrix}\right]_3\right) + W\left[\begin{smallmatrix}S-CCF_3\\ \parallel\\ S-CCF_3\end{smallmatrix}\right]_2$ in CH_2Cl_2	Blue-black crystals, m. 177.5-181°	444
	$UO_2(C_6H_4NO_2)_3 \cdot 3\ H_2O$, (cx)	$[Ph_4As]^+ + H[UO_2(C_6H_4NO_2)_3]$ in H_2O		516
	VCl_4	$[Ph_4As][VCl_4 \cdot 2\ MeCN]$ at 80-100°	Bright blue-green powder, x-ray diffraction analysis, (cy)	364
	$VCl_4 \cdot 2\ MeCN$	$[Ph_4As]Cl + VCl_3$ in MeCN	Yellow salt, electronic spectrum and magnetic data, (cz)	305
	$V\left[\begin{smallmatrix}S-CCF_3\\ \parallel\\ S-CCF_3\end{smallmatrix}\right]_3$	$[Ph_4As]V(CO)_6$ in $CH_2Cl_2 + CF_3C{=}C(CF_3)SS$	Dark purple, m. 175-177°	444
	$Y(CF_3COCHCOMe)_4$, (da)		NMR spectrum	396
	$Y(CF_3COCHCOCF_3)_4$, (da)		NMR spectrum	396
	$Y(CF_3COCHCOMe)_{4-n}\text{-}(CF_3COCHCOCF_3)_n$, (da)		NMR spectrum	396
$[Ph_4As]_2$	Cr_2O_7	$[Ph_4As]Cl + K_2Cr_2O_7$	Crystals, dec. 185°, soly. data	371
	$Cr\left[\begin{smallmatrix}S-CCF_3\\ \parallel\\ S-CCF_3\end{smallmatrix}\right]_3$	$[Ph_4As]Cl + Cr\left[\begin{smallmatrix}S-CCF_3\\ \parallel\\ S-CCF_3\end{smallmatrix}\right]_3 + N_2H_4$ in EtOH	Shiny black needles, m. 193-194° (dec.)	444
	$Co(N_3)_4$	$[Ph_4As]_2CoCl_4 + NaN_3$ in Me_2CO	m. 154°, IR spectrum, (db)	580
	$CoCl_4$	$[Ph_4As]Cl + CoCl_2$ (2:1) in EtOH	Dissocn. in MeOH LiCl soln. $\rightarrow$ $CoCl_2 \cdot 4$ MeOH $\rightleftharpoons$ $CoCl_2 \cdot 2$ MeOH, temp. effect	1339

$Co(NO_3)_4$	$[Ph_4As]Cl + AgNO_3 + CoCl_2$ in MeCN	Crystal and mol. structure, (dc)	171
$Co(CNS)_4$ (M-335)		An indicator (blue color) for the titration of Co(II) with EDTA	1075
$Co(O_2CCF_3)_4$	$[Ph_4As]Cl + AgO_2CCF_3 + CoCl_2$ in MeCN	Blue-violet crystals, mol. structure by x-ray analysis, (dd)	172
$Co(SC_6F_5)_4$	$CoSO_4 + NaSC_6F_5 + [Ph_4As]Cl$ in H_2O	Green, paramagnetic crystals which in air turned yellow and diamagnetic, dec. 240°, (de, df)	143, 144
$Cu[N(CN)_2]_4$	$[Ph_4As][N(CN)_2] + Cu[N(CN)_2]_2$ in Me_2NCN and pptd. with CCl_4	Brown-green crystals, m. 115°, absorption, reflection, and IR spectra, (dg)	867
$Cu_4\left[\begin{smallmatrix}S-\\S-\end{smallmatrix}C=C(CN)_2\right]_3$	$[Ph_4As]Cl + [Pr_4N]\{Cu_4[S_2C=C-(CN)_2]_3\}$	Yellow-orange crystals, m. 293-295°, a 2:1 electrolyte	541
$Cu\left[\begin{smallmatrix}S-CCF_3\\ \| \\ S-CCF_3\end{smallmatrix}\right]_2$	$[Ph_4As]Cl + Cu\left[\begin{smallmatrix}S-CCF_3\\ \| \\ S-CCF_3\end{smallmatrix}\right]_2^{--}$, (dh)	Brick red crystals, m. 198-200°	447
$IrCl_6$ (M-335)		Extn. with $CHCl_3$ from 0.1 N HCl and photometric detn. of Ir, (di)	574
$Fe(N_3)_5$	$Fe(NO_3)_3 \cdot 9\,H_2O + NaN_3$ in 1 N $H_2SO_4 + [Ph_4As]Cl$	Red shiny needles, dec. 155°	132
$Fe(NO)\left[\begin{smallmatrix}S-CCF_3\\ \| \\ S-CCF_3\end{smallmatrix}\right]_2$	$[Ph_4As]\left\{Fe(NO)\left[\begin{smallmatrix}S-CCF_3\\ \| \\ S-CCF_3\end{smallmatrix}\right]_2\right\}$ in $Me_2CO + Na_2SO_3 \cdot 7\,H_2O$ in H_2O	Dark green-brown crystals, m. 168° (dec.), IR spectrum	1037

(cx) $C_6H_4NO_2 = \alpha$-picolinate ion. (cy) Electronic spectrum and magnetic data (305). (cz) Electronic and IR spectra (364). (da) Y = rare earth metals, mostly Eu. (db) UV and visible spectra (132). (dc) A 8-coordinate Co with 4 bidentate NO_3 groups postulated (171). UV, visible, and IR spectra (392). (dd) Visible spectrum and magnetic data; a tetrahedral structure of the anion is postulated (172). (de) Absorption and reflection spectra (133). (df) Molar cond. in Me_2CO (144). (dg) A distorted tetrahedral structure of the anion is postulated (867). (dh) $Cu\left[\begin{smallmatrix}S-CCF_3\\ \| \\ S-CCF_3\end{smallmatrix}\right]_2^{--}$ was prepd. by treating $[Ph_3PCuI]_4$ with $CF_3\overline{C=C(CF_3)S}S$ in C_6H_6 and reducing the liquid phase with N_2H_4. (di) In $CHCl_3 + [Ph_4As]_2IrCl_6 \cdot 2[Ph_4As]Cl \rightarrow [Ph_4As]_2IrCl_6 \cdot 2[Ph_4As]Cl$ (574).

Table 20 Continued

R_4As^+	X^-	Prepd. from	Prop. and Rxn.	Ref.
C_{24} (Cont.)				
$[Ph_4As]_2$	$[Fe(NO)_2SC_6H_4CO_2]_2$	$[Ph_4As]Cl$ in H_2O + $[Fe(NO)_2S-C_6H_4CO_2H]_2$ in THF	Dark brown, granular crystals, m. 178° (dec.), molar cond. data	732
	$Mg(C_6H_4O_2)_2$	$o\text{-}C_6H_4(OH)_2$ in aq. Me_4NOH + $MgSO_4$ + $[Ph_4As]Cl$	Slightly yellow crystals, m. 201-204°, soly. data, polarographic data	1298
	$Hg(CNO)_4$	$Hg(CNO)_2$ + $NaCNO$ + $[Ph_4As]Cl$ in H_2O	White ppt., dec. 175-176°, soly. data, IR spectrum	139
	MoS_4	$[Ph_4As]Cl$ + $[Me_4N]_2MoS_4$ (2:1) in H_2O	IR spectrum	951
	$Mo_6Cl_8(Cl)_6$	Mo_6Cl_{12} in EtOH + $[Ph_4As]Cl$ in dild. HCl	Yellow powder, electronic spectrum	405
	$Mo(CO)_4(GeCl_3)_2$	By-prod. in the synth. of $[Ph_4As]$-$[Mo(CO)_5GeCl_3]$	Yellow to orange solid, m. 180-182°, (dj)	1307
	$Mo(CO)_4(SnCl_3)_2$	$[Ph_4As]SnCl_3$ + $Mo(CO)_6$ in CH_2Cl_2 under irradiation	IR spectrum	1307
	$Mo(NO)_2Cl_4$ (M-336)	$[Ph_4As]Cl$ + $Mo(NO)_2Cl_2$	IR spectrum	807
	$Mo\begin{bmatrix}S-CCF_3 \\ \parallel \\ S-CCF_3\end{bmatrix}_3$ (M-336)	$[Ph_4As]Cl$ + $Mo\begin{bmatrix}S-CCF_3 \\ \parallel \\ S-CCF_3\end{bmatrix}_3$ + N_2H_4 in EtOH	Bright blue needles, m. > 250°	444
	$Np(O)Cl_5$	$[Ph_4As]Cl$ + $Cs_2Np(O)Cl_5$ in EtOH-HCl	Bright yellow, paramagnetic salt, soly. data, IR spectrum	108
	$NiCl_4$ (M-336)	Rxn. with LiCl in ROH → $NiCl_2 \cdot 4$ ROH $\rightleftharpoons NiCl_2 \cdot 2$ ROH, temp. effect on the equil.	1339
	$Ni(NCO)_4$	$[Ph_4As]Cl$ + $NiCl_2$ + $AgOCN$ + $NaOCN$ in $MeNO_2$ and pptd. with Et_2O	Blue, paramagnetic crystals, m. 173-174°, a 2:1 electrolyte, IR and electronic spectra, (dk)	542

Ni(NCS)$_4$	[Ph$_4$As]NCS + Ni(NCS)$_2$ in Me$_2$CO	Olive-yellow salt, at ~155° for 24 hrs. is converted into blue-green salt, m. 219°; both forms are paramagnetic, (dl)	579
Ni$\left[\begin{array}{c}\text{S-CCF}_3\\ \text{S-CCF}_3\end{array}\right]_2$	(Ph$_3$P)$_2$NiBr$_2$ + CF$_3$C=C(CF$_3$)SS in C$_6$H$_6$, redn. of the Ni salt with N$_2$H$_4$ and pptn. with [Ph$_4$As]Cl	Orange needles, m. 234-235°, (dm)	447
Ni$\left[\begin{array}{c}\text{S-CPh}\\ \text{S-CPh}\end{array}\right]_2$	[Ph$_4$As]Ni$\left[\begin{array}{c}\text{S-CPh}\\ \text{S-CPh}\end{array}\right]_2$ + NaHg in THF	Orange, diamagnetic needles	443
	[Ph$_4$As]Cl + Ni$\left[\begin{array}{c}\text{S-CPh}\\ \text{S-CPh}\end{array}\right]_2$ (2:1) in EtOH + N$_2$H$_4$	Bright orange needles, m. 238-239°	444
Ni$\left[\begin{array}{c}\text{Se-CCF}_3\\ \text{Se-CCF}_3\end{array}\right]_2$	[Ph$_4$As]Cl + Ni$\left[\begin{array}{c}\text{Se-CCF}_3\\ \text{Se-CCF}_3\end{array}\right]_2$ in DMSO	Orange brown, diamagnetic crystals, m. 256-257°, polarographic data	448
Ni(S$_2$C=NCN)$_2$		Mol. and crystal structure by x-ray diffraction	394
Ni(CS$_3$)$_2$	Ni(OAc)$_2$ + K$_2$CS$_3$ in EtOH and treating the filtrate with [Ph$_4$As]Cl in H$_2$O	Red, diamagnetic needles, m. 217-218°, a 2:1 electrolyte, electronic spectrum, (dn)	406, 541
Ni(CS$_4$)$_2$ (do)	Ni(OAc)$_2$·4 H$_2$O + K$_2$CS$_4$ in H$_2$O + [Ph$_4$As]Cl in Me$_2$CO·H$_2$O	Brown, diamagnetic crystals, m. 208-209°, a 2:1 electrolyte, x-ray diffraction data	406
Nb$_6$Cl$_{18}$(OH)·H$_2$O	Nb$_6$Cl$_{14}$·8 H$_2$O + HCl + [Ph$_4$As]Cl + O, (dp)	Brown, paramagnetic crystals, visible and near-UV spectra	571

(dj) Rxn. with Ph$_3$P → cis-[Mo(CO)$_4$(PPh$_3$)$_2$] (1307). (dk) A tetrahedral anion with Ni-N bonds is postulated (542). (dl) Electronic and IR spectra; a tetrahedral somewhat distorted geometry for the blue-green salt and a 6-coordinate tetragonal environment of 4 N and 2 S atoms for the olive green salt are postulated (578, 579). (dm) See also ref. 445. (dn) X-ray diffraction data (541, 1044, see also 540). (do) CS$_4$$^{--}$ = perthiocarbonate dianion. (dp) Nb$_6$Cl$_{14}$·8 H$_2$O in MeOH is heated with concd. HCl, [Ph$_4$As]Cl is added, heating is continued and O is bubbled through the soln.

Table 20 Continued

R_4As^+ X^-	Prepd. from	Prop. and Rxn.	Ref.
C_{24} (Cont.)			
$[Ph_4As]_2$ $Nb_6Cl_{17} \cdot H_2O$	Prepn. same as the preceding compd. but with more HCl	Dark brown, paramagnetic crystals. visible and near-UV spectra	571
$ON(SO_3)_2$	$[Ph_4As]Br$ in H_2O + solid $K_2[ON(SO_3)_2]$	Paramagnetic crystals, soly. data, stable in the absence of moisture, ESR spectrum	567
$Pd(N_3)_4$	$[Ph_4As]Cl + Pd(NO_3)_2 + NaN_3$ in H_2O	Red-brown crystals, dec. 187°, (dq)	132
$Pd_2(N_3)_6$	$[Ph_4As]Cl + Pd(NO_3)_2 + NaN_3$ in H_2O	Brown needles, dec. 164-165°, (dq)	132
Pd_2Cl_6		Far-IR spectrum	30
$Pd(CNS)_4$	$PdCl_2 + 4$ KCNS + 2 $[Ph_4As]Cl$	Sol. in $CHCl_3$, spectrophotometric detn. (see also ref. 1257)	1004
$Pd\begin{bmatrix} S-CCF_3 \\ \| \\ S-CCF_3 \end{bmatrix}_2$	$(Ph_3P)_2Pd \overset{S-CCF_3}{\underset{S-CCF_3}{\|}} + CF_3C=C(CF_3)SS$ in aq. EtOH + N_2H_4 + $[Ph_4As]Cl$	Pale green, diamagnetic needles, m. 248-251° (dec.), (dr)	445,446
$Pd(CS_3)_2$	$PdCl_2 + K_2CS_3$ in EtOH + $[Ph_4As]Cl$	Yellow orange crystals, m. 224-225°, a 2:1 electrolyte, (ds)	541
$PdCl_2(SnCl_3)_2$	$Pd + Sn$ (1:5) in 2 N HCl + MeOH (1:1), followed by $[Ph_4As]Cl$	Red ppt., dec. > 50°	847
$Pt(N_3)_4 \cdot H_2O$	$[Ph_4As]Cl + K_2[PtCl_4] + NaN_3$ in H_2O	Orange-red, diamagnetic crystals, dec. 185°, (dt, du, dv)	132
	$H_2[PtCl_6] + NaN_3$ in H_2O at 40-50° + $[Ph_4As]Cl$		1360
$[SP(S)(Me)N=]_2$	$[Ph_4As]SP(S)(N_3)Me$ at ~ 128°	Crude substance, IR spectrum	1295
$Pt(N_3)_6$	$[Ph_4As]Cl + H_2[PtCl_6] + NaN_3$ in H_2O	Orange-yellow, diamagnetic crystals, dec. 205°, molar cond. data, (dw, dx)	132
Pt_2Br_6 (M-336)		Far-IR spectrum	30
$PtCl_4$	$[Ph_4As]Cl$ in H_2O + $K_2[PtCl_4]$	Far-IR spectrum	1157
$PtCl_6$	$[Ph_4As]Cl + K_2[PtCl_6]$	Far-IR and laser Raman spectra	33

Compound	Preparation	Properties	Ref.
$Pt(NCO)_4$	$[Ph_4As]_2PtCl_4 + AgNCO$ in Me_2CO	IR and electronic spectra, molar cond. data	1157
$Pt(CS_3)_2$	$K_2PtCl_4 + K_2CS_3$ in EtOH $+ [Ph_4As]Cl$	Yellow-orange crystals, m. 244-245°, electronic spectrum, x-ray diffraction data, molar cond. data (see also 540)	541
$Pt(CS_4)_2$, (do)		Red, diamagnetic crystals, m. 259-262°, molar cond. data	406
$Pt\left[\begin{smallmatrix}S-CCF_3\\\parallel\\S-CCF_3\end{smallmatrix}\right]_2$	$Pt\left[\begin{smallmatrix}S-CCF_3\\\parallel\\S-CCF_3\end{smallmatrix}\right]_2 + NaHg$ in THF $+ [Ph_4As]Cl$ in H_2O	Diamagnetic salt, m. 240-241.5°, molar cond. data	446
$Pb(N_3)_6$	$[Ph_4As]Cl + NaN_3 + [NH_4]_2PbCl_6$ in CH_2Cl_2	Deep red, diamagnetic needles, dec. 148-149°, molar cond. data, IR, visible and UV spectra	132, 141
$PbCl_6$	$[Ph_4As]Cl + K_2PbCl_6$	Far-IR and laser-Raman spectra	34
$ReBr_6$	$[Ph_4As]Cl + ReBr_3$ in 48% HBr	Golden yellow solid, molar cond. data, IR spectrum	397
	$[Ph_4As]Br + H_2[ReBr_6]$		397
	$[Ph_4As][Re(O)Br_4] +$ aq. HBr		399
Re_2Cl_8	$[Ph_4As]Cl + [Bu_4N]_2[Re_2Cl_8]$ in MeOH contg. HCl	Blue crystals	393
	$[Re(O_2CEt)_2Cl]_2 + HCl + EtCO_2H$ on a steam bath $+ [Ph_4As]Cl$		393
	$ReCl_4$ in HCl-MeOH $+ [Ph_4As]Cl$ in MeOH	Blue-green insol. salt, IR spectrum	403

(dq) Molar cond. data, IR, visible, and UV spectra (132). (dr) Molar cond. data (446). (ds) Electronic spectrum and x-ray diffraction data (541, see also 540). (dt) Thermally stable but deflagrates in flame (140). (du) Molar cond. data (132, 1360). (dv) Spectra: IR (132, 1360, 1506), UV and visible (132). (dw) Thermally stable, but deflagrates when heated with flame (140). (dx) Spectra: IR (132, 1506), UV and visible (132).

Table 20 Continued

R_4As^+ / X^-	Prepd. from	Prop. and Rxn.	Ref.
C_{24} (Cont.)			
$[Ph_4As]_2$ Re_2Cl_9	$ReCl_4$ in MeOH + HCl at 25-30° + $[Ph_4As]Cl$	Violet, insol. salt, IR spectrum, (dy, dz)	403
Re_3Cl_{11} (M-336)	$[Ph_4As]Cl$ in EtOH + $ReCl_3$ in aq. or alc. HCl, (ea)	Crystals, molar cond. data, absorption spectrum (eb, ec)	1282
$Re_3(PPh_3)Cl_{11}$	$[Ph_4As]Re_3Cl_{11}$ by recrystn. from Me_2CO contg. Ph_3P	Crystals, molar cond. data, absorption spectrum	1282
$Re_2O_3Cl_8$	$[Ph_4As]Cl$ + $ReOCl_4$ in $CHCl_3$ in contact with air	Deep mauve, paramagnetic ppt., molar cond. data; UV, visible, and IR spectra, (ef)	321
$Re_2(NCS)_8$	$[Bu_4N]_2[Re_2Cl_8]$ + NaSCN in MeOH contg. HOAc + $[Ph_4As]Cl$	Glistening orange plates, soly. data, molar cond. data, IR and electronic spectra	404
$Re(NCS)_6$	$[Ph_4As]Cl$ in MeOH + $[Bu_4N]_2$-$[Re(NCS)_6]$ in THF	Red-brown ppt., molar cond. and magnetic moment data; IR and electronic spectra	404
$ReO(SCN)_5$	$[Ph_4As]Cl$ + $[ReO(SCN)_5]^{2-}$	Greenish yellow complex, photometric detn. of Re	1170
$Re_2\left[\begin{smallmatrix}S-CCN\\ \| \| \\ S-CCN\end{smallmatrix}\right]_4$	$[Bu_4N]_2[Re_2Cl_8]$ + NaSC(CN)=C(CN)SNa in EtOH + $[Ph_4As]Cl$	Black microcrystals, molar cond. data; visible spectrum	402
$Rh(CO)(SnCl_3)_3$	$Rh(CO)_2Cl$ + $SnCl_2$ (1:3) + $[Ph_4As]Cl$	A 2:2 electrolyte in $MeNO_2$, IR spectrum	853
$Rh_2(CO)_2Br_4$	$[Ph_4As][Rh(CO)_2Cl_2]$ in MeOH contg. 10% aq. HBr under reflux	Deep red, diamagnetic leaflets, m. 250-260°, molar cond. data, (eg, eh)	1551
$Rh_2(CO)_2I_4$	$[Ph_4As][Rh(CO)_2I_2]$ in MeOH contg. concd. HI + 1% H_3PO_2 at r.t.	Reddish brown leaflets, dec. in $PhNO_2$ soln., (eh, ei)	1551
$Ru(CO)_2Br_2(SnBr_3)_2$	$RuCl_2$ in EtOH + CO, followed by $SnBr_2$ under reflux + $[Ph_4As]Cl$	Yellow crystals, molar cond. data, IR spectrum	854

$(Ta_6Cl_{12})Cl_6$	$[Ph_4As]Cl + [Ta_6Cl_{12}]^{++}$ in EtOH satd. with HCl	IR spectrum	1061
$Ta(O_4C_2)_3(OH)$	$[Ph_4As]Cl + $ aq. soln. contg. 1.1 $\times 10^{-3}$ N Ta $+ 1.0 \times 10^{-2}$ N $(CO_2H)_2$	Crystals, m. 206°, molar cond. data, IR spectrum	234
$TeCl_6$ (M-336)		Far-IR spectrum	35
Me_2TlBr_3	$[Ph_4As]Br + Me_2TlBr$ in Me_2CO	m. 194-196°	547
Me_2TlCl_3	$[Ph_4As]Cl + Me_2TlCl$ in Me_2CO	m. 177-181°	547
Me_2TlI_3	$[Ph_4As]I + Me_2TlI$ in Me_2CO	m. 210-211°	547
$S_2C{=}NCN$	$[Ph_4As]Cl + K_2S_2C{=}NCN$ in H_2O	Colorless crystals, m. 270-272°, soly. data, molar cond. data	401
ThI_6	$[Ph_4As]I + ThI_4 \cdot 4$ MeCN in MeCN at 70°	Yellow, paramagnetic crystals, (ej)	106, 107
$Sn(N_3)_6$	$[Ph_4As]Cl + [NH_4]_2SnCl_6 + NaN_3$ in H_2SO_4	Colorless needles, dec. 193°, molar cond. data, IR, visible, and UV spectra	132, 141
$SnCl_6$	$[Ph_4As]Cl + K_2SnCl_6$	Far-IR and laser-Raman spectra	34
$TiOCl_4$	$[Ph_4As]Cl + TiOCl_2$ in MeCN	Bright yellow ppt., molar cond. data, absorption spectrum	584
WS_4	$[Ph_4As]OH + H_2WS_4$	Yellow ppt., UV spectrum	198
	$[Ph_4As]Cl + [Me_4N]_2WS_4$ in H_2O	IR spectrum	951
$WO(NCS)_5$	$[Ph_4As]Cl$ in HCl + a mixt. of $NH_4NCS + [NH_4]_2[WOCl_5]$ in HCl	Polarographic redn.	200

(dy) Shaking in Me_2CO or MeCN at r.t. $\rightarrow [Ph_4As]_2Re_2Cl_8$ (403). (dz) In concd. HCl at 95° $\rightarrow [Ph_4As]_2Re_2Cl_8$ (403). (ea) See also ref. 611. (eb) Effective cation volume 860 Å (1282). (ec) X-ray diffraction data (611, 1210). (ed) Rxn. with SCN⁻ under reflux $\rightarrow$ replacement of the remaining Cl atoms with CNS (1282). (ef) Crystallographic data; an O-bridged dimeric structure is postulated (231). (eg) Stable in the solid state and in soln. contg. HBr (1551). (eh) IR spectrum - a halogen-bridged dimeric structure is postulated (1551). (ei) The salt in soln. is rapidly oxidized to $[Rh(CO)I_4]^-$ unless a reducing agent is present. (ej) Hydrolytically unstable in moist air, moderately stable to heat, dec. >200°; x-ray diffraction analysis (106, 107).

Table 20 Continued

R_4As^+	X^-	Prepd. from	Prop. and Rxn.	Ref.
C_{24} (Cont.)				
$[Ph_4As]_2$	$W(CO)_4(GeCl_3)_2$	$[Ph_4As]GeCl_3 + W(CO)_6$ in CH_2Cl_2 under irradiation	IR spectrum	1307
	$W(CO)_4(SnCl_3)_2$	$[Ph_4As]SnCl_3 + W(CO)_6$ in CH_2Cl_2 under irradiation	IR spectrum	1307
	$W\left[\begin{smallmatrix} S-CCF_3 \\ \parallel \\ S-CCF_3 \end{smallmatrix}\right]_3$ (M-336)	$[Ph_4As]Cl + W[S_2C_2(CF_3)_2]_3 + N_2H_4$ in EtOH	Red-purple needles, m. > 250°	444
	$UO(N_3)_4$	$UO_2(NO_3)_2 + NaN_3$ in 2 N HNO_3 + $[Ph_4As]Cl$	Yellow crystals, dec. 170-171°, molar cond. data; UV, visible, and IR spectra	132,141
	UCl_6	$[Ph_4As]Cl + UCl_4$ in HCl		107
	UI_6	$[Ph_4As]I + UI_4 \cdot 4$ MeCN in MeCN at 70°	Red, paramagnetic crystals, (ej)	106,107
	$VO(N_3)_4$	$VOCl_2$ in 2 N $H_2SO_4 + NaN_3 + Ph_4AsCl$	Green needles, dec. 155°, molar cond. data, UV, visible, and IR spectra	132,141
	$V\left[\begin{smallmatrix} S-CCF_3 \\ \parallel \\ S-CCF_3 \end{smallmatrix}\right]_3$ (M-336)	$[Ph_4As]Cl + V[S_2C_2(CF_3)_2]_3 + N_2H_4$ in EtOH	Purple plates, m. 205-206°	444
	$V\left[\begin{smallmatrix} Se-CCF_3 \\ \parallel \\ Se-CCF_3 \end{smallmatrix}\right]_3$		Purple, paramagnetic salt, m. 190° (dec.), polarographic redn.	448
	$ZnCl_4$	$[Ph_4As]Cl + ZnCl_2$	Crystals, dec. 315°, soly. data	371
	$Zn(o-OC_6H_4O)_2$	$Zn(NO_3)_2 \cdot 6 H_2 + o-C_6H_4(OH)_2$ in $H_2O + NaOH + [Ph_4As]Cl$	Yellow-orange crystals, m. 213-214° (dec.), soly. data, polarographic redn.	1298
	$Zn(S_2C=NCN)_2$	$[Ph_4As]_2ZnCl_4 + K_2S_2C=NCN$ in H_2O-Me_2CO	White needles, m. 242-244°	401

Cation	Anion	Preparation	Properties	Ref.
[Ph4As]3	As(o-OC6H4O)3	AlCl3 + aq. NH4OH + o-C6H4(OH) + [Ph4As]Cl in H2O	Yellow crystals, m. 291-293°, soly. data, polarographic redn.	1298
	Fe(CN)6	[Ph4As]Cl + K3[Fe(CN)6] in H2O	Electron exchange between [Fe(CN)6]3- and [Fe(CN)6]4-	289
		[Fe(CN)6]4- + I in CHCl3 by extn. with CHCl3 contg. [Ph4As]Cl		289
	Nb(O)(O4C2)3	[Ph4As]Cl + aq. 0.36 x 10^-2 N Nb soln. + 0.85 x 10^-2 N (CO2H)2	Crystals, m. 140°, molar cond. data, IR spectrum	234
	Nb6Cl17(OH)	Nb6Cl14·8 H2O in MeOH + concd. HCl by heating, adding [Ph4As]Cl and bubbling O through the soln.	Dark brown, diamagnetic crystals; visible and near-UV spectra	571
	Os(N3)6	KOsCl4 + NaN3 in H2O + [Ph4As]Cl	Orange-yellow ppt., highly explosive at elevated temp., molar cond. data	1360
	Re2O2(SCN)8	[Re2O2(SCN)8]3- + [Ph4As]Cl	Orange-red complex, photometric detn. of Re	1170
	Re2(CO)2(NCS)10	[Bu4N]3[Re2(CO)2(NCS)10] + [Ph4As]Cl in MeCN	Dark green crystals, IR spectrum	404
	Rh(N3)6	Rh(NO3)2·2 H2O + NaN3 in 2 N H2SO4 + [Ph4As]Cl	Red-brown crystals, dec. 186°, molar cond. data; UV, visible, and IR spectra	132
		Na3[RhCl6]·12 H2O + NaN3 in H2O + [Ph4As]Cl	Soly. data, IR spectrum, molar cond. data	1360
	[Rh(SnCl3)4]·SnCl2	[RhCl6]3- or [RhCl6]2- in HCl + SnCl2, extn. with i-C5H11OH and treating with [Ph4As]Cl	Molar cond. data	596
[Ph4As]4	Fe(CN)6	[Ph4As]Cl + K4[Fe(CN)6] in H2O	Rxn. with I in CCl4 → [Fe(CN)6]3-	289

Table 20 Continued

R_4As^+	X^-	Prepd. from	Prop. and Rxn.	Ref.
C$_{24}$ (Cont.)				
Ph$_3$As–[2,5-Me, N–H pyrrole]	Br	Ph$_3$AsBr$_2$ in C$_6$H$_6$ + 2,5-dimethyl-3-pyrryl-MgBr in Et$_2$O	m. 210-211°	1646
Ph$_3$[OH(CH$_2$)$_6$]As	I	Ph$_3$As + HO(CH$_2$)$_6$I in EtOH under reflux	Prisms, m. 138-139°, IR spectrum, (ek)	677
2,2'-C$_6$H$_4$(C$_6$H$_4$AsMe$_3$)$_2$	Br$_2$·H$_2$O	o-Terphenyl-2,2''-ylenebis(dimethylarsine) + MeBr (1:2) in MeOH at 60°	Hydroscopic substance, m. 296°, absorbs moisture from air to form a cryst. trihydrate	1016
	I$_2$ (M-337)	From the preceding diarsine + MeI in MeOH	m. 316° (dec.), absorbs moisture from air to form a dihydrate, m. 296-299°	1016
	Pc$_2$, (c)	From the corr. dibromide + Na picrate	m. 198°	1016
Ph(C$_7$H$_{15}$)$_2$(EtO$_2$CCH$_2$)As	Br	Ph(C$_7$H$_{15}$)$_2$As + BrCH$_2$CO$_2$Et on a steam bath	m. 71-74°	1549
	NO$_3$	From the corr. bromide + AgNO$_3$ in MeOH on a steam bath	Mobile oil	1549
C$_{25}$				
Ph$_3$(PhCH$_2$)As	Br (M-337)		A corrosion inhibitor for Fe in aq. mineral acids, (el)	981,982
	Cl (M-337)		Cathodic cleavage → Ph$_3$As + MePh	748
			Inhibition of Fe dissoln. in acids	533
	I (M-337)		m. 157°, (em)	1147
Ph$_3$(2-MeC$_6$H$_4$)As	Br (M-337)	Ph$_3$As + 2-MeC$_6$H$_4$Br + AlCl$_3$ + NiCl$_2$ at 180-190°	m. 126-127°, cathodic cleavage → Ph$_2$(2-MeC$_6$H$_4$)As and Ph$_3$As (84.3:15.7)	748

	ClO_4		m. 243°	748
$Ph_3(3\text{-}MeC_6H_4)As$	Br	$Ph_3As + 3\text{-}MeC_6H_4Br + AlCl_3 + NiCl_2$ at 180-190°	m. 125-126°, cathodic cleavage → $Ph_2(3\text{-}MeC_6H_4)As + Ph_3As$ (80:20)	748
	ClO_4		m. 260-261°	748
$Ph_3(4\text{-}MeC_6H_4)As$	Br	$Ph_3As + 4\text{-}MeC_6H_4Br + AlCl_3 + NiCl_2$ at 180-190°	m. 197-198°, cathodic cleavage → $Ph_2(4\text{-}MeC_6H_4)As + Ph_3As$ (91.8:8.2)	748
	ClO_4		m. 245°	748
$MePh(C_9H_{19})_2As$	I	$Ph(C_9H_{19})_2As$ + MeI at r.t. or on a steam bath	Paraffin-like substance	828
C_{26}				
$Ph_3[BzCH(HgCl)]As$	Cl	$Ph_3\overset{+}{A}s\text{-}\overset{-}{C}HBz + HgCl_2$ in MeOH	Crystals, m. 189-190°, IR spectrum, (en, eo, ep, eq)	1145
$Ph_3(BzCH_2)As$	Br (M-338)		m. 178°, (er)	1147
$Me[PhNHC(=NPh)](C_6H_{11})_2As$	I	$[PhNHC(=NPh)](C_6H_{11})_2As$ + MeI in C_6H_6 under reflux	Yellowish crystals, m. 175°	1538
$Ph(PhCH_2CH_2)(C_6H_{13})_2As$	Br	$Ph(C_6H_{13})_2As + PhCH_2CH_2Br$ in a sealed tube at 110-120°	Greasy substance	1549
$Ph(EtO_2CCH_2)(C_8H_{17})_2As$	Br	$Ph(C_8H_{17})_2As + BrCH_2CO_2Et$ on a steam bath	Greasy substance	1549
	NO_3	From the corr. bromide + $AgNO_3$ in MeOH on a steam bath	Mobile oil	1549

(ek) Rxn. with NaOH at 140-160° → $CH_2{=}CH(CH_2)_4OH + Ph_3As$ (677). (el) The arsonium salt inhibits the electrode rxn. in the Fe/1 N HCl and Fe/1 N HCl-MeOH systems (1033). (em) Rxn. with Ph_3P in DMF at 60-80° → Ph_3As + $[Ph_3(PhCH_2)P]I$ (1147). (en) Rxn. with NH_3 in DMF → $Ph_3\overset{+}{A}s\text{-}\overset{-}{C}HBz$ (1145). (eo) Rxn. with aq. Na_2SO_3 in THF → $Ph_3\overset{+}{A}s\text{-}\overset{-}{C}HBz$ (1145). (ep) Rxn. with $p\text{-}O_2NC_6H_4CHO$ in DMF → $p\text{-}O_2NC_6H_4CH{=}CHBz + Ph_3AsO$ (1145). (eq) Rxn. with AcCl → $[Ph_3AsCH{=}C(OAc)Ph]HgCl_3$ (1145). (er) Rxn. with Ph_3P in DMF at 100-120° → $Ph_3As + [Ph_3(BzCH_2)P]Br$ (1147).

Table 20 Continued

R_4As^+	X^-	Prepd. from	Prop. and Rxn.	Ref.
C_{27}				
$Ph_3(i-PrC_6H_4)As$	Br	$Ph_3As(i-Pr)C_6H_4Br + AlCl_3$ $+ NiCl_2$ at 180-190°	m. 226-228°, cathodic cleavage $\rightarrow Ph_2(i-PrC_6H_4)As + Ph_3As$ (90.9:9.1)	748
	ClO_4		m. 190°	748
$MePh(C_{10}H_{21})_2As$	I	$Ph(C_{10}H_{21})_2As + MeI$ on a steam bath	m. 50-53°	1549
C_{28}				
$Ph_3[PhC(OAc)=CH]As$	$HgCl_3$	$\{Ph_3[BzCH(HgCl)]As\}Cl + AcOH$ in THF	White crystals, m. 169-170°	1145
$MePh_2[Ph_2C(OH)C\equiv C]As$	I	$Ph_2[Ph_2C(OH)C\equiv C]As + MeI$ in Et_2O	m. 260° (dec.)	149
$(4-MeC_6H_4)_4As$	BF_4	$[(4-MeC_6H_4)_4As]Cl + NaBF_4$	Soly. data	87
	Cl		Spectrophotometric detn., polarographic redn.	87
	I (M-339)	By-prod. from $(4-MeC_6H_4)_3$-$As=NO_2SC_6H_4Me-4 + 4-MeC_6H_4Li$, followed by hydrolysis and dec. with $HCl + I^-$	Dec. 254-256°	723
$Ph(EtO_2CCH_2)(C_9H_{19})_2As$	PF_6	$[(4-MeC_6H_4)_4As]Cl + KPF_6$	Soly. data	87
	Br	$Ph(C_9H_{19})_2As + BrCH_2CO_2Et$ on a steam bath	Mobile oil, formed hydroscopic crystals on standing in dessicator, d_4^{20} 1.1662, n_D^{20} 1.5285	1549
	NO_3	From the corr. bromide + $AgNO_3$ in MeOH on a steam bath	Mobile oil	1549
C_{29}				
$Ph_3[HO(CH_2)_{11}]As$	Br	$Ph_3As + HO(CH_2)_{11}Br$ in EtOH under reflux	Needles, m. 84-86°, IR spectrum, (es)	677

C30				
$Ph_3(4\text{-}PhC_6H_4)As$	Br	$Ph_3As + 4\text{-}PhC_6H_4Br + AlCl_3$ $+ NiCl_2$ at 180-190°	m. 223-224°, cathodic cleavage $\rightarrow Ph_2(PhC_6H_4)As + Ph_3As$ (65.4:34.6)	748
	ClO_4		m. 228°	748
$Ph(EtO_2CCH_2)(C_{10}H_{21})_2As$	Br	$Ph(C_{10}H_{21})_2As + BrCH_2CO_2Et$ on a steam bath	Paraffin-like substance	1549
	NO_3	From the corr. bromide + $AgNO_3$ in MeOH on a steam bath	Mobile oil	1549
C31				
$MeBu[C_6H_{11}NHC(=NC_6H_{11})]_2As$	I	$Bu[C_6H_{11}NHC(=NC_6H_{11})]_2As +$ MeI in C_6H_6 under reflux	Colorless crystals, m. 130-132°	1538
C33				
$Ph_2(PhCH_2)[PhCH=C(Ph)]As$	Br	$Ph_2[PhCH=C(Ph)]As + PhCH_2Br$ in C_6H_6	m. 160-162°, NMR spectrum	44
C42				
$Ph_3As(CH_2)_6AsPh_3$	I_2	By-prod. from $Ph_3As +$ $HO(CH_2)_6I$ in EtOH under reflux	Needles, m. 233-234°	677
C47				
$Ph_3\overset{+}{As}\text{-}\overline{C}CPh=CPhCPh=CPh$ by dissolving in $HClO_4$	ClO_4		Pale yellow crystals, m. 216-217° (dec.)	974

(es) Rxn. with NaOH at 140-160° $\rightarrow Ph_3As + CH_2=CH(CH_2)_9OH$ (677).

2.3   TERTIARY DERIVATIVES

2.3.5   COMPOUNDS WITH ARSENIC-HALOGEN AND -PSEUDOHALOGEN BONDS

2.3.5.2 DIHALO, TETRAHALO, AND DIPSEUDOHALO- DERIVATIVES

The methods of synthesis of the dihalo and tetrahalo derivatives, R_3AsX_2 and R_3AsX_4, were summarized in the main volume, p. 342, and the compounds were compiled in Table 34. More recently the dichloro derivatives have been synthesized from tertiary arsines by a treatment with thionyl chloride, sulfonyl chloride, and phosgene, respectively. The dihalo derivatives containing two different halogen atoms, R_3AsXY, have been prepared by equilibration of two dihalotriorganoarsenic derivatives containing different halogens: $R_3AsX_2 + R_3AsY_2 \rightarrow 2\,R_3AsXY$. Tetrahalotriorganoarsenic compounds have been prepared by the following reactions:

$$R_3AsX_2 + Y_2 \rightarrow R_3AsX_2Y_2$$

$$R_3AsXY + Y_2 \rightarrow R_3AsXY_3$$

$$R_3AsX_2 + XY \rightarrow R_3AsX_3Y$$

Molar conductivity measurements of some dihalotriorganoarsenic compounds indicate that in appropriate solvents they ionize, at least partly, to $[R_3AsX]^+X^-$ type ions, while in the solid state both halogens may be covalently bonded to the arsenic atom.

Diazidotrimethylarsenic can be prepared from dichlorotrimethylarsenic by a metathetic reaction with sodium azide. Diisocyanatotriphenylarsenic is formed from tertiary arsine oxides by fusion with urea.

Several dihalo and pseudodihalo derivatives of this class are described individually hereafter, while others are compiled in Table 21.

Dichlorotrimethylarsenic, $AsC_3H_9Cl_2$, Me_3AsCl_2,   (M-343)
Synth.: From Me_3As by treating with Cl (1167). $(CD_3)_3As$ gave $(CD_3)_3AsCl_2$ (1167).
Prop.: White needles, m. 156-157°. IR and NMR spectra indicated a trigonal bipyramidal geometry for both Me_3AsCl_2 and $(CD_3)_3AsCl_2$ in the solid state (1167). NMR spectrum at -60° (1083).
Rxn. with: $Me_3AsF_2 \rightleftharpoons 2\,Me_3AsClF$ (1083).
$AgF \rightarrow Me_3AsF_2$ (1167).
$NH_4OH \rightarrow Me_3AsO$ (1167).
Me_3AsO (1:1) $\rightarrow [Me_3AsOH]Cl$ (1167).
NaN_3 (1:4) in $ClCH_2CH_2Cl \rightarrow Me_3As(N_3)_2$ (1356).

Dibromotriphenylarsenic, $AsC_{18}H_{15}Br_2$, Ph_3AsBr_2,   (M-345)
Synth.: From Ph_3As + Br in MeCN (177) or in CCl_4 (1167, 1521).
Prop.: White crystals, m. 215° (177). Very sensitive to atm. moisture and readily undergoes hydrolysis to $Ph_3As(OH)Br$ (1167). IR spectrum (1001, 1167).

Molar cond. in MeCN was detd.; a covalent bonding of both Br atoms in the solid state, but partial ionization in soln. to $[Ph_3AsBr]^+Br^-$ is suggested (177, 178, 683). Half-wave potential was detd. (467).
Rxn. with: $H_2O \rightarrow [Ph_3AsOH]Br$ (685, 1167).
Aq. EtOH $\rightarrow [Ph_3AsOH]Br$ (1167).
$AgNO_3$ in MeCN $\rightarrow Ph_3As(NO_3)_2$ (1521).
$N_2O_4 \rightarrow Ph_3As(NO_3)_2$ (1521).

Dichlorotriphenylarsenic, $AsC_{18}H_{15}Cl_2$, Ph_3AsCl_2, (M-345)
Synth.: From $Ph_3As + SO_2Cl_2$ in MePh, 95% yield (116).
From Ph_3As + a Cl-N mixt. in CCl_4 (177).
From Ph_3As + Cl in petr. ether (1167) or in Et_2O at 20° (1421).
From $Ph_3AsO + COCl_2$ or $SOCl_2$ in MeCN (79).
Prop.: Colorless crystals (116), white crystals, m. 205° (177), m. 200°, after a phase change at 155° (1421). Very sensitive to atm. moisture and readily hydrolyzable to $[Ph_3AsOH]Cl$ (1167). IR spectrum (1167). Molar cond. data (177, 178, 683). A weak electrolyte in MeCN due to the formation of the $[Ph_3AsCl]^+Cl^-$ (1167) or perhaps $[Ph_3AsCl]^+[Ph_3AsCl_3]^-$ ions (177) is suggested. A covalent bonding of both Cl atoms in the solid state is postulated (177).
Rxn. with: $MoCl_5 + SO_2 \rightarrow [Ph_3AsCl][MoOCl_4]$ (845).
$MoOCl_3$ in CH_2Cl_2-$CCl_4 \rightarrow [Ph_3AsCl][MoOCl_4]$ (845).
NH_3 in $CHCl_3$ at r.t. $\rightarrow [(Ph_3As)_2N]Cl$ (1421).
$AgClO_4$ in EtOH $\rightarrow (Ph_3AsO)_2HClO_4$ (1167).
$AgF \rightarrow Ph_3AsF_2$ (1167).
$H_2O \rightarrow [Ph_3AsOH]Cl$ (685, 1167).
Aq. EtOH $\rightarrow [Ph_3AsOH]Cl$ (1167).
$AgNO_3$ in aq. EtOH $\rightarrow Ph_3As(OH)ONO_2$ (1167).
$AcC(SNa)=C(SNa)Ac$ in MeCN $\rightarrow Ph_3As\overline{SC(Ac)=C(As)S}$ (1090).
Electrochem. redn. with 2 $e^- \rightarrow Ph_3As + 2 Cl^-$ (467).

Diiodotriphenylarsenic, $AsC_{18}H_{15}I_2$, Ph_3AsI_2, (M-345)
Synth.: From Ph_3As + I in anhyd. petroleum, a mixt. of Ph_3AsI_2 and Ph_3AsI_4 was obtd. (177, see also 179).
Prop.: Solid substance, m. 116°; UV spectrum is interpreted as an indication of an associated outer charge transfer (c.t.) complex which gradually changes to inner $[Ph_3AsI]^+I^-$ complex (179). Kinetics and energy of the $Ph_3As^+I_2^- \rightarrow [Ph_3AsI]^+I^-$ change were detd. (180). Molar cond. in MeCN was detd.; the cond. was attributed to disproportionation: $2 Ph_3AsI_2 \rightarrow Ph_3As + [Ph_3AsI]^+I_3^-$ (177, 683).

(Two more compounds of this class are described following Table 21 on page 284.)

Table 21. Tertiary Dihalo- and Tetrahaloarsenic Derivatives

R_3AsX_2 or R_3AsX_4	Prepd. from	Prop. and Rxn.	Ref.
C_3			
$(CF_3)_3AsCl_2$ (M-343)	$R_3As + Cl$	Rxn. with AgF $\to$ $(CF_3)_3AsF_2$	323
$(CF_3)_3AsF_2$ (M-343)	$R_3AsCl_2 + AgF$	Rxn. with CsF $\to$ Cs[$(CF_3)_3AsF_3$], white solid, IR and NMR spectra	323
Me_3AsBr_2 (M-343)	$R_3As + Br$	White needles, dec. $\sim 170°$, IR spectrum indicated a tetracovalent As and ionic structure: [Me_3AsBr]$^+Br^-$, (a, b, c)	1167
$(CD_3)_3AsBr$	$R_3As + Br$	IR spectrum	1167
Me_3AsClF	$R_3AsCl_2 + R_3AsF_2$ by equilibration, (d)	NMR spectra in TMS and $CFCl_3$ at $-60°$	1083
Me_3AsF_2	R_3AsCl_2 in hot EtOH + aq. AgF	White solid, m. 69-70°, b_{12} 54° (a, e, f)	1167
M_3AsI_2	$R_3As + I$	Light brown powder, dec. $\sim 130°$, IR spectrum, a tetracovalent As and ionic structure postulated	1167
C_{18}			
$(C_6F_5)_3AsCl_2$	$R_3As + Cl$ or ICl	m. 214-216°, a nonelectrolyte in MeCN, rxn. with H_2O $\to$ $(C_6F_5)_3AsO$	684
Ph_3AsBrI	$R_3As + IBr$ in MeCN	Yellow crystals, m. 154-155°, molar cond. data - a strong electrolyte due to disproportionation: 2 Ph_3AsBrI $\to$ Ph_3As + $Ph_3AsBr_2I_2$ $\to$ [Ph_3AsBr]$^+BrI_2^-$	177
Ph_3AsBrI_3	$R_3AsBrI + I$ in CCl_4	Red crystals, m. 115-116°, mol. cond. data, an electrolyte in MeCN: [Ph_3AsBr]$^+I_3^-$	177
$Ph_3AsBr_2I_2$	$R_3As + BrI$ in MeCN, followed by Et_2O	Red crystals, m. 104°, molar cond. data	177
Ph_3AsBr_3I	$R_3AsBr_2 + I$ in MeCN	A strong electrolyte in MeCN: [R_3AsBr]$^+BrI_2^-$	177
	$R_3AsBr_2 + IBr$ in MeCN, followed by Et_2O	Orange crystals, m. 120-121°, molar cond. data, ionizes in MeCN: [Ph_3AsBr]$^+IBr_2^-$	177

Ph$_3$AsBr$_4$ (M-345)	R$_3$As + Br in MeCN, followed by Et$_2$O or by freeze-drying	Hydroscopic orange crystals, m. 89°, molar cond. data, a strong electrolyte in MeCN: [Ph$_3$AsBr]$^+$Br$_3^-$, (g)	177, 683
Ph$_3$AsClF		NMR spectrum in CFCl$_3$ at -60°	1083
Ph$_3$AsF$_2$ (M-345)	R$_3$AsCl$_2$ in EtOH + AgF in H$_2$O	Needle-shaped crystals, m. 139-140°, IR and NMR spectra; a pentacovalent As as in a trigonal bipyramidal configuration is postulated (i)	1167
	R$_3$As + C$_5$F$_{11}$N in C$_6$H$_6$ at r.t., (h)		118
Ph$_3$AsI$_4$ (M-345)	R$_3$As + I in MeCN	Purple crystals, m. 139-140°, molar cond. data, ionizes to [Ph$_3$AsI]$^+$I$_3$ (see also 683)	177
C$_{21}$			
(PhCH$_2$)$_3$AsClF	R$_3$AsCl$_2$ + R$_3$AsF$_2$ by equilibration	NMR spectrum in Me$_4$Si and CFCl$_3$ at -60°	1083
(PhCH$_2$)$_3$AsCl$_2$		NMR spectrum, (j, k)	1083
(PhCH$_2$)$_3$AsF$_2$	R$_3$AsCl$_2$ + AgF in MeCN by heating	Crystals, m. 104°, NMR spectra in Me$_4$Si and CFCl$_3$ at -60°, (k)	1083
C$_{30}$			
p-C$_6$H$_4$(AsPh$_2$Br$_2$)$_2$	p-C$_6$H$_4$(AsPh$_2$)$_2$ + Br in C$_6$H$_6$	Dec. 275-280°, very sensitive to moisture, rxn. with H$_2$O → p-C$_6$H$_4$[As(O)Ph$_2$]$_2$·(H$_2$O)$_n$	1658

(a) NMR spectrum in Me$_4$Si and CFCl$_3$ at -60° (1090). (b) Rxn. with NCC(SNa)=C(SNa)CN in CH$_2$(OMe)$_2$ → Me$_3$AsSC(CN)=C(CN)S (1090). (c) Rxn. with moist air in hot 95% EtOH → [Me$_3$AsOH]Br (1167). (d) Equilibrium const. at -37° was detd. (1083). (e) IR and NMR spectra indicated a pentacovalent As with trigonal bipyramidal structure in the solid state (1167). (f) Rxn. of Me$_3$AsF$_2$ + Me$_3$AsCl$_2$ ⇌ 2 Me$_3$AsClF (1083). (g) Rxn. with H$_2$O in MeCN → [Ph$_3$AsOH]Br$_3$ (685). (h) C$_5$F$_{11}$N = perfluoro-N-fluoropiperidine. (i) NMR spectrum in CFCl$_3$ at -60° (1083). (j) Rxn. with AgF in MeCN → (PhCH$_2$)$_3$AsCl$_2$ + (PhCH$_2$)$_3$AsF$_2$ (1083). (k) (PhCH$_2$)$_3$AsCl$_2$ + (PhCH$_2$)$_3$AsF$_2$ ⇌ 2 (PhCH$_2$)$_3$AsClF (1083).

Diazidotrimethylarsenic, $AsC_3H_9N_6$, $Me_3As(N_3)_2$

Synth.: From Me_3AsCl_2 + NaN_3 (1:4) in $ClCH_2CH_2Cl$ at r.t., 91% yield (1356).
Prop.: Colorless solid, m. 69°, very little sensitive to heat and impact, monomeric in C_6H_6 soln., stable at 0° under exclusion of atm. moisture. IR and NMR spectra - a trigonal bipyramidal structure is postulated (1356).

Diisocyanatotriphenylarsenic, $AsC_{20}H_{15}N_2O_2$, $Ph_3As(NCO)_2$

Synth.: From Ph_3AsO by mixing with powdered urea, heating to fusion at 135°, and removing the H_2O and NH_3 liberated at 140-150° in vacuum, 96% yield (1443, 1445).
Prop.: Amorphous glassy mass, m. ~ 40°, IR spectrum (1443).

2.3.5.3 HALOHYDROXY AND HALOALKOXY DERIVATIVES

The halohydroxy and haloalkoxy derivatives were described in the main volume on page 346. This class now also comprises trihalohydroxytriorganoarsenic compounds, $R_3As(OH)X_3$. The halohydroxy derivatives, $R_3As(OH)X$ or $[R_3\overset{+}{As}OH]X^-$, have been prepared from dihalotriorganoarsenics by partial hydrolysis with aqueous alcohol or acetone and by equilibration with tertiary arsine oxides in acetonitrile. The compounds react with halogens under anhydrous conditions to form triorganohydroxyarsonium trihalides, $[R_3AsOH]X_3$, and mixed mono- and trihalides, $[R_3\overset{+}{As}OH]_2X^-X_3^-$, respectively. Hydrolysis of tetrahalotriorganoarsenics leads to triorganohydroxyarsonium trihalides, $[R_3AsOH]X_3$. Most of the compounds have been formulated as ionic compounds, but for $Me_3As(OH)Cl$ a covalent As-Cl bond has been postulated.

Table 22. Halohydroxytriorganoarsenic Derivatives (M-346)

$R_3As(OH)X$	Prepd. from	Prop. and Rxn.	Ref.
C_3			
$Me_3As(OH)Cl$	Me_3AsCl_2 + Me_3AsO (1:1) in MeCN	White needles, m. 133-134°, IR spectrum - a covalent As-Cl bond and trigonal bipyramidal structure are postulated	1167
$[Me_3AsOH]Br$	Me_3AsBr_2 in hot 95% EtOH + moist Et_2O	m. 149-150°, IR spectrum - an ionic As-Br bond is postulated	1167
$[Me_3AsOD]Br$	$[Me_3AsOH]Br$ soln. in D_2O by evapn. and crystn. from MeCN	Needles, m. 148-149°, IR spectrum - an ionic AsBr bond is postulated	1167
C_{18}			
$[Ph_3AsOH]Br$	Ph_3AsBr_2 in Me_2CO + H_2O	White crystals, m. 165-166°, IR spectrum, (a, b, c)	685
	Ph_3AsBr_2 + aq. EtOH	IR spectrum - no definite structure could be concluded from the spectral data	1167

Table 22 Continued

$R_3As(OH)X$	Prepd. from	Prop. and Rxn.	Ref.
C_{18} (Cont.)			
$[Ph_3AsOH]Br_3$	$[Ph_3AsOH]Br + Br$ in MeCN	Orange crystals, m. 106-108°, UV spectrum	685
	$Ph_3AsBr_4 + H_2O$ in MeCN		685
$Ph_3As(OH)Cl$ (M-346)	$Ph_3AsCl_2 + aq.$ EtOH	IR spectrum - no definite structure could be concluded from the spectral data	1167
$[Ph_3AsOH]Cl_2I$	$Ph_3As(OH)Cl + ICl$ in MeCN, freeze-drying and a treatment with Et_2O	Bright yellow solid, m. 76-78°, IR spectrum, molar cond. data	685
$[Ph_3AsOH]_2Br_4$	$[Ph_3AsOH]Br + Br$ in MeCN by freeze-drying	Orange solid, m. ~ 65°, stable in a sealed tube, IR spectrum, molar cond. data	685

(a) Molar cond. data (685).   (b) Rxn. with Br_2 in MeCN $\rightarrow [Ph_3AsOH]Br_3$ (685).
(c) Rxn. with $AgBF_4 \rightarrow Ph_3AsO \cdot BF_3$ (685).

2.3.5.9   MISCELLANEOUS HALOTRIORGANOARSENIC DERIVATIVES

Chlorotriphenylarsonium Tetrachlorooxymolybdate, $AsMoC_{18}H_{15}Cl_5O$, $[Ph_3AsCl][MoOCl_4]$

Synth.: From $Ph_3AsO + MoCl_5$ in CCl_4-CH_2Cl_2 (845).
From $Ph_3AsCl_2 + MoCl_5 + SO_2$ in a sealed tube at r.t. (845).
From Ph_3AsCl_2 in $CH_2Cl_2 + MoOCl_3$ in CCl_4 at r.t. (845).
Prop.: Large, bright green, paramagnetic crystals, a 1:1 electrolyte in $PhNO_2$. IR and visible spectra. Magnetic susceptibility and magnetic moment data (845). The compd. was previously formulated as a $Ph_3AsO \cdot MoCl_5$ complex (M-354).

2.3.6   COMPOUNDS WITH ARSENIC BONDED TO GROUP VIA ELEMENTS

2.3.6.1   COMPOUNDS WITH ARSENIC-OXYGEN BONDS

2.3.6.1.1   MONOHYDROXY DERIVATIVES

The compounds containing three organic groups, one hydroxy group, and a fifth ligand, probably an anion, were described in the main volume on pages 347-349. In addition to the synthesis methods reported in the main volume, these compounds have been prepared from dihalotriorganoarsenic derivatives and from hydroxytriorganoarsonium halides by metathetic reactions with various inorganic salts in alcoholic or aqueous-alcoholic solutions. Tertiary arsines react with dilute nitric acid to form the corresponding hydroxyarsonium nitrates. Hydroxyarsonium halides combine with mercuric halides to form dimeric hydroxyarsonium tetrahalomercurates.

Table 23. Hydroxyarsonium Salts (M-348)

Table 23. Hydroxyarsonium Salts (M-348)

[R₃As(OH)]X	Prepd. from	Prop. and Rxn.	Ref.
C_3			
[Me₃AsOH]ClO₄	R₃AsCl₂ + AgClO₄ in MeOH, followed by a treatment with moist Et₂O	Crystals, m. 120-121°, IR spectrum, an ionic structure is postulated	1167
[Me₃AsOH]NO₃	R₃AsCl₂ in alc. + aq. AgNO₃	Needles, m. 129-130°, IR spectrum, an ionic structure is postulated	1167
[Me₃AsOH]HSO₄	R₃AsCl₂ + Ag₂SO₂ in aq. soln.	m. 102-104°, IR spectrum - the presence of the HSO₄⁻ ion is postulated	1167
C_{14}			
[EtPh₂AsOH] Picrate		m. 101-102°	1017
C_{18}			
Ph₃As(OH)NO₃ (M-348)	R₃AsCl₂ in alc. + aq.-alc. AgNO₃	IR spectrum indicated a covalently bonded NO₃ group	1167
[Ph₃AsOH]ClO₄	[R₃AsOH]Br + AgClO₄ in EtOH	White crystals, m. 163-165°, IR spectrum, molar cond. data	685
[Ph₃AsOH]₂HgBr₄	[R₃AsOH]Br in EtOH + HgBr₂	White solid, m. 152-156°, IR spectrum, molar cond. data	685
[Ph₃AsOH]₂HgCl₄	[R₃AsOH]Cl in EtOH + HgCl₂ under reflux	White crystals, m. 160-161°, IR spectrum, molar cond. data	685
[(Ph₃AsO)₂H]₂Hg₂Br₆	R₃AsO·H₂O + HCl + HgCl₂		686
	2 R₃AsO·H₂O + 2 HBr + HgBr₂	White crystals, m. 146-147°, crystallographic data, (a)	686
C_{22}			
[o-C₆H₄(C₆H₄AsMe₂OH)₂](NO₃)₂	o-C₆H₄(C₆H₄AsMe₂-o)₂ + dild. HNO₃	m. 207°, rxn. with aq. NaOH → the dioxide	1016

(a) A two-fold symmetry with each Hg atom bonded to 4 Br atoms, two of them form bridges between the Hg atoms (686).

286

2.3.6.1.5 DIHYDROXY DERIVATIVES

The methods of synthesis of the dihydroxy derivatives, $R_3As(OH)_2$, were described in the main volume on page 350. More recently two new compounds of this type were reported which were prepared from tertiary arsines by the oxidation with hydrogen peroxide in acetic acid and acetone, respectively.

Dihydroxydiphenylpropynylarsenic, $AsC_{15}H_{15}O_2$, $Ph_2(MeC\equiv C)As(OH)_2$
<u>Synth.</u>: From $Ph_2(MeC\equiv C)As$ by oxidn. with H_2O_2 in AcOH, 95% yield (149).
<u>Prop.</u>: m. 168°, IR spectrum; loses water during drying in vacuum (149).

1-Butynyldihydroxydiphenylarsenic, $AsC_{16}H_{17}O_2$, $Ph_2(EtC\equiv C)As(OH)_2$
<u>Synth.</u>: From $Ph_2(EtC\equiv C)As$ by oxidn. with H_2O_2 in Me_2CO, 95% yield (149).
<u>Prop.</u>: m. 80°, IR spectrum; loses water during drying in vacuum (149).

Dihydroxytriphenylarsenic, $AsC_{18}H_{17}O_2$, $Ph_3As(OH)_2$, (M-350)
<u>Rxn. with</u>: Carboxylic acid anions, attempted extraction of the anions from aq. soln. into $CHCl_3$ (1376).

2.3.6.1.6 TERTIARY ARSINE OXIDES

The methods of synthesis of tertiary arsine oxides were summarized in the main volume on pages 351-2. Several new compounds of this class described in this supplement were prepared by the methods reported in the main volume. The arsine oxides form coordination derivatives with various compounds, particularly with transition metal salts. A few tertiary arsine oxides with their coordination derivatives are described individually hereafter, while others are compiled in Table 25.

Trimethylarsine Oxide, AsC_3H_9O, Me_3AsO, (M-356)
<u>Synth.</u>: From Me_3As + HgO (1:1) in Me_2CO under reflux, taking up the prod. with H_2O, evapg. to dryness, and purifying by sublimation (1057).
From Me_3As in Et_2O by adding dropwise aq. 30% H_2O_2 (1057).
<u>Prop.</u>: m. 191.2-195.2° (1057). IR spectrum (236, 1057, 1387b).
<u>Rxn. with</u>: Iodine in various solvents → a charge transfer complex (663).
Me_3AsCl_2 (1:1) → $Me_3As(OH)Cl$ (1167).
<u>Coordination Deriv.</u>: [$Co(Me_3AsO)_2Br_2$], paramagnetic solid, prepd. from $CoBr_2$ + Me_3AsO in Me_2CO-EtOH. Electronic and IR spectra, x-ray diffraction data, a 4-coordinate Co complex (235).

[$Co(Me_3AsO)_2Cl_2$], paramagnetic solid, prepd. from $CoCl_2$ + Me_3AsO in Me_2CO-EtOH. Electronic and IR spectra; x-ray diffraction diagram; a 4-coordinate Co is postulated (235).

[$Co(Me_3AsO)_2(CNS)_2$], pink, paramagnetic solid, prepd. from $Co(CNS)_2$ + Me_3AsO in Me_2CO-EtOH. Electronic and IR spectra. In the solid state the Co is probably 6-coordinate by bridging through the S atoms, but in $MeNO_2$ dissolves with a blue color and its spectrum is typical for a tetrahedral structure (235).

[Co(Me₃AsO)₄]I₂, solid substance, prepd. from CoI₂ + Me₃AsO in Me₂CO-EtOH. Electronic and IR spectra. A 4-coordinate Co is postulated (235).

[Co(Me₃AsO)₄](ClO₄)₂, paramagnetic solid, prepd. from Co(ClO₄)₂ + Me₃AsO in Me₂CO-EtOH. Electronic and IR spectra. A 4-coordinate Co with ionic ClO₄ ligands is suggested (235).

[Co(Me₃AsO)₅](ClO₄)₂, mauve, diamagnetic ppt., prepd. from Co(ClO₄)₂ + excess Me₃AsO in anhyd. Me₂CO. Reflectance and IR spectra. A square-pyramidal structure is postulated (236).

[Cu(Me₃AsO)₄(ClO₄)]ClO₄, pale blue, paramagnetic ppt., prepd. from Cu(ClO₄)₂ + Me₃AsO (1:4) in hot EtOH. Electronic reflectance and IR spectra; x-ray diffraction character. A 5-coordinate Cu is suggested (236).

[Me₃Ga(OAsMe₃)], m. 54°, prepd. from Me₃AsO + Me₃Ga in C₆H₆; thermally stable up to 150°, sol. in nonpolar solvents. NMR spectrum (1348).

[Fe(Me₃AsO)₄(ClO₄)]BPh₄, buff solid, prepd. from Fe(ClO₄)₂ + Me₃AsO (1:4) + excess NaBPh₄ in hot EtOH. IR spectrum; a square-pyramidal structure is suggested (236).

[Fe(Me₃AsO)₄(ClO₄)]ClO₄, buff, paramagnetic ppt., prepd. from Fe(ClO₄)₂ + Me₃AsO (1:4) in hot EtOH. Reflectance and IR spectra; a square-pyramidal structure is suggested (236).

[Mn(Me₃AsO)₄(ClO₄)]ClO₄, white ppt., sensitive to moisture, prepd. from Mn(ClO₄)₂ + Me₃AsO (1:4) in hot EtOH. IR spectrum; a square-pyramidal structure is suggested (236).

[Mn(Me₃AsO)₅](ClO₄)₂, white, paramagnetic ppt., sensitive to moisture, prepd. from Mn(ClO₄)₂ + excess Me₃AsO in anhyd. Me₂CO. IR spectrum; a square-pyramidal structure is suggested (236).

[Ni(Me₃AsO)₄(NO₃)]BPh₄, yellow ppt., prepd. from Ni(NO₃)₂ + Me₃AsO (1:4) + NaBPh₄ in hot EtOH. Electronic reflectance spectrum; a square-pyramidal structure is suggested (236).

[Ni(Me₃AsO)₄(ClO₄)]BPh₄, yellow ppt., prepd. from Ni(ClO₄)₂ + Me₃AsO (1:4) + NaBPh₄ in hot EtOH. Electronic reflectance and IR spectra; a square-pyramidal structure is suggested (236).

[Ni(Me₃AsO)₄(ClO₄)]ClO₄, yellow, paramagnetic ppt., prepd. from Ni(ClO₄)₂ + Me₃AsO (1:4) in hot EtOH. Electronic reflectance and IR spectra; a square-pyramidal structure is suggested (236).

[Ni(Me₃AsO)₅](BF₄)₂, orange, paramagnetic ppt., prepd. from Ni(BF₄)₂ + excess Me₃AsO in anhyd. Me₂CO. Electronic reflectance and IR spectra; a square-pyramidal Ni complex is suggested (236).

[Ni(Me₃AsO)₅](NO₃)₂, orange ppt., prepd. from Ni(NO₃)₂ + excess Me₃AsO in anhyd. Me₂CO. Electronic reflectance and IR spectra; a square-pyramidal structure is suggested (236).

$[Ni(Me_3AsO)_5](ClO_4)_2$, orange, paramagnetic ppt., prepd. from $Ni(ClO_4)_2$ + excess Me_3AsO in anhyd. Me_2CO. A 2:1 electrolyte in Me_2CO. Electronic reflectance and IR spectra; a square-pyramidal structure is suggested (236).

$[Zn(Me_3AsO)_2Br_2]$, prepd. from $ZnBr_2$ + Me_3AsO in Me_2CO-EtOH; IR spectrum (235).

$[Zn(Me_3AsO)_2Cl_2]$, prepd. from $ZnCl_2$ + Me_3AsO in Me_2CO-EtOH; IR spectrum and x-ray powder diffraction data (235).

$[Zn(Me_3AsO)_4](ClO_4)_2$, prepd. from $Zn(ClO_4)_2$ + Me_3AsO in Me_2CO-EtOH; IR spectrum (235).

Methyldiphenylarsine Oxide, $AsC_{13}H_{13}O$, $MePh_2AsO$, (M-357)

Synth.: From $MePh_2As$ + HgO (1:1) in Me_2CO under reflux, followed by dissolving the prod. in H_2O, concg. to dryness, and subliming in vacuo (1057). From $MePh_2As$ in Et_2O by adding dropwise aq. 30% H_2O_2 and purifying the prod. as described above (1057).

Prop.: m. 152.7-153.8°, IR spectrum (1057).

Coordination Deriv.: $[Co(MePh_2AsO)_2Br_2]$, prepd. from $CoBr_2$ + $MePh_2AsO$ in EtOH. IR spectrum; a tetrahedrally coordinated Co(II) is postulated (1284).

$[Co(MePh_2AsO)_2Cl_2]$, prepd. from $CoCl_2$ + $MePh_2AsO$ in EtOH. IR spectrum; a tetrahedrally coordinated Co(II) is postulated (1284).

$[Co(MePh_2AsO)_4](NO_3)_2$, IR spectrum indicated ionic NO_3 groups; a reference is made to UV and visible spectra (964). Structural analysis by single-crystal x-ray diffraction technique (1201).

$[Co(MePh_2AsO)_4(NO_3)]ClO_4$, IR spectrum indicated a covalent NO_3 and ionic ClO_4; a reference is made to the UV and visible spectra (964).

$[Co(MePh_2AsO)_4(ClO_4)]ClO_4$, prepd. from $Co(ClO_4)_2$ + $MePh_2AsO$ in EtOH. Magnetic data. IR spectrum indicated a covalent and an ionic ClO_4 (964). X-ray diffraction data (964, 1201). A distorted tetragonal pyramidal structure with 4 equiv. O atoms of the arsine oxide groups and one O atom of the ClO_4 group coordinated with the Co atom is postulated (964). The energy level diagrams for the 5-coordinated Co(II) complex are presented (347).

$[Cu(MePh_2AsO)_4](ClO_4)_2$, paramagnetic solid, prepd. from $Cu(ClO_4)_2$ + $MePh_2AsO$ in EtOH. IR spectrum indicated ionic ClO_4 groups (964).

$[Cu(MePh_2AsO)_4][CuCl_2]_2$ (M-357), the complex occurs in two forms, a blue form, prepd. from $CuCl_2$ + $MePh_2AsO$ in aq. EtOH, and a brown form, prepd. in the same manner as the blue form, but in abs. EtOH. The brown form was shown to be essentially the same compd. as the blue species but contg. 5-10% excess of $CuCl_2$ (Nyholm, J. Chem. Soc. 1951, 1767). The composition of the complex salt was confirmed by x-ray diffraction analysis and by IR spectrum (1284).

$[Fe(MePh_2AsO)_2Br_2]$, prepd. from $FeBr_2$ + $MePh_2AsO$ in EtOH. IR spectrum (1284). Electronic absorption spectrum in the solid state (577). A tetrahedrally coordinated Fe(II) complex is postulated (1284).

[Fe(MePh$_2$AsO)$_4$](ClO$_4$)$_2$, prepd. from MePh$_2$AsO in EtOH + Fe(ClO$_4$)$_2$. IR spectrum indicated ionic ClO$_4$ groups (964).

[Mn(MePh$_2$AsO)$_2$Br$_2$], prepd. from MnBr$_2$ + MePh$_2$AsO in EtOH. IR spectrum; a tetrahedral structure of the Mn(II) complex is postulated (1284).

[Mn(MePh$_2$AsO)$_2$Cl$_2$], prepd. from MnCl$_2$ + MePh$_2$AsO in EtOH. IR spectrum; a tetrahedral structure of the Mn(II) complex is postulated (1284).

[Mn(MePh$_2$AsO)$_4$(ClO$_4$)]ClO$_4$, paramagnetic solid, prepd. from MePh$_2$AsO in EtOH + Mn(ClO$_4$)$_2$. IR spectrum indicated the presence of covalent and ionic ClO$_4$ groups (964).

[Ni(MePh$_2$AsO)$_2$Cl$_2$], prepd. from NiCl$_2$ and MePh$_2$AsO in EtOH. IR spectrum; a tetrahedral structure of the Ni(II) complex is postulated (1284).

[Ni(MePh$_2$AsO)$_4$](NO$_3$)$_2$, IR spectrum indicated ionic NO$_3$ groups; a reference was made to the UV and visible spectra (964).

[Ni(MePh$_2$AsO)$_4$(NO$_3$)]ClO$_4$, IR spectrum indicated a covalent NO$_3$ and ionic ClO$_4$; a reference was made to the UV and visible spectra (964).

[Ni(MePh$_2$AsO)$_4$(ClO$_4$)]ClO$_4$, prepd. from MePh$_2$AsO in EtOH + Ni(ClO$_4$)$_2$. IR spectrum indicated the presence of one covalent and the other ionic ClO$_4$ (964).

[Zn(MePh$_2$AsO)$_2$Br$_2$], prepd. from ZnBr$_2$ + MePh$_2$AsO in EtOH. IR spectrum; a tetrahedral structure of the Zn(II) complex is postulated (1284).

[Zn(MePh$_2$AsO)$_2$Cl$_2$], prepd. from ZnCl$_2$ + MePh$_2$AsO in EtOH. IR spectrum; a tetrahedral structure of the Zn(II) complex is postulated (1284).

[Zn(MePh$_2$AsO)$_4$(ClO$_4$)]ClO$_4$, solid substance, prepd. from MePh$_2$AsO in EtOH + Zn(ClO$_4$)$_2$. IR spectrum indicated one covalent and the other ionic ClO$_4$ (964).

Triphenylarsine Oxide,   AsC$_{18}$H$_{15}$O,   Ph$_3$AsO,   (M-352)
Synth.: From Ph$_3$As + aq. 30% H$_2$O$_2$ in Me$_2$CO under reflux (1347).
Prop.: m. 192-193° (90), m. 189° (1347), m. 194° (1445); crystd. from H$_2$O as a monohydrate which starts melting at 117-118° and is completely molten at 191-192° (90). Soly. data (1235). IR spectrum (389, 803, 1223, 1246) and its interpretation (1223, 1387b). IR spectrum of the monohydrate indicated the hydrate structure rather than Ph$_3$As(OH)$_2$ (90). K-edge x-ray absorption spectrum (962). Basicity data were obtd. in terms of the shifts of the OH and OD bands of MeOH and MeOD, resp., in CCl$_4$ (669). Protonation (861). Ionization const. in AcOH by spectrophotometric titration with HClO$_4$ (668). Dipole moment and molar Kerr const.; the conformation of the Ph groups was studied (94).
Rxn. with: Iodine in CHCl$_3$ → a charge transfer complex [Ph$_3$AsO]$^+$I$_2^-$ (869).
HX (X = Cl or Br) → Ph$_3$AsX$_2$ (389).
HCl + HgCl$_2$ (2:2:1) → [Ph$_3$AsOH]$_2$HgCl$_4$ (686).
HBr + HgBr$_2$ (2:2:1) → [(Ph$_3$AsO)$_2$H]$_2$Hg$_2$Br$_6$ (686).

BuI → $[Ph_3As(OH)OAsPh_3]I$ + $CH_2=CHCH_2CH_3$ (334).

$ClCH_2CO_2H$, Cl_2CHCO_2H, or Cl_3CCO_2H → the corr. adducts (669).

H_2O_2 → $[Ph_3AsO]_2 \cdot H_2O_2$ (389).

Urea → $Ph_3As(NCO)_2$ (1443, 1445).

CS_2 under reflux → Ph_3AsS (1151).

1,2,3-Triphenylcyclopentadiene in Ac_2O → $Ph\overline{C=C(Ph)C(Ph)=C(Ac)C}=AsPh_3$ (687).

1,2,3-Triphenylcyclopentadiene + P_2O_5 in Et_3N → $Ph\overline{C=C(Ph)C(Ph)=CHC}=AsPh_3$ (687).

$Bz\overline{C=C(Bz)CH=CHCH_2}$ in Ac_2O → $H\overline{C=C(Bz)C(Bz)=CHC}=AsPh_3$ (687).

$MoCl_5$ in $CCl_4-CH_2Cl_2$ → $[Ph_3AsCl][Mo(O)Cl_4]$ (845).

$MoCl_5$ in CH_2Cl_2 → $[MoO_2(Ph_3AsO)_2Cl_2]$ (845).

Irradiation with neutrons (350).

Uses: A catalyst for the trimerization of isocyanates (727). Extn. of ReO_4^- and TcO_4^- from aq. mineral acids into org. solvents contg. Ph_3AsO (897). Sepn. of $FeCl_3$ from other metals dissolved in > 8 N HCl by extn. with Ph_3AsO soln. in $CHCl_3$ and re-extn. of $FeCl_3$ with H_2O (1231). Quant. sepn. of U^{8+} by extn. of aq. acid ($\sim$ pH 2) soln. with $CHCl_3$ contg. Ph_3AsO and re-extn. into $(NH_4)_2CO_3$ soln. (1234). Pptn. of Bi in strongly acidic media, probably as $[Ph_3As(OH)O]_3Bi$ (1235). Extn. of UO_2^{2+}, Bi^{3+}, Fe^{3+}, CrO_4^{2-}, Th^{4+}, Ce^{4+}, and VO_4^{3-} from aq. acids with $CHCl_3$ contg. Ph_3AsO (1235). Complexes with $CuCl_2$ (1:1), picric acid, or rare earth metal salts are suitable semiconductors for thermisistors based on organophosphorus compds. (1433).

Coordination Deriv.:* $[Ph_3AsO \cdot H_2NC_6H_4NO_2-3]$, m. 112-113°, prepd. from Ph_3As + Br in dry C_6H_6 at < 7-10°, followed by $3-O_2NC_6H_4NH_2$ in Et_3N. IR spectrum indicated H-bonding (342).

$[Ph_3AsO \cdot H_2NC_6H_4NO_2-4]$, m. 170-171°, prepd. in the same manner as the preceding complex. IR spectrum indicated H-bonding (342).

$[Ph_3AsO \cdot H_2NC_6H_3(NO_2)_2-2,4]$, m. 102-103°, prepd. in the same manner as the preceding two complexes. IR spectrum indicated H-bonding (342).

$[Ph_3AsO \cdot H_2NSO_2Ph]$, prepd. in the same manner as the preceding complexes, IR spectrum indicated H-bonding (342).

$[Ph_3AsO \cdot H_2NSO_2C_6H_4NO_2-3]$, m. 143-143.5°, prepd. from Ph_3As + Br in dry C_6H_6, followed by $3-O_2NC_6H_4SO_2NH_2$ in Et_3N. IR spectrum; H-bonding is suggested (342).

$[Ph_3AsO \cdot (H_2NSO_2C_6H_4Me-4)_3]$, IR spectrum (342).

$[Ph_3AsO \cdot BF_3]$, white crystals, m. 230-232°, prepd. from $[Ph_3AsOH]Br$ + $AgBF_4$ in EtOH. IR spectrum (685).

$[Ph_3AsO \cdot CrO_3]$, orange, diamagnetic crystals, dec. 180°, a non-electrolyte; IR spectrum; rxn. with H_2O_2 → $[Ph_3AsO \cdot CrO_5]$ (91).

---

* The coordination derivatives of Ph_3AsO with yttrium and lanthanide metal nitrates are compiled in Table 24, page 297.

[$Ph_3AsO \cdot CrO_5$], a peroxychromate, bright royal blue, diamagnetic crystals, dec. 110-120°, stable in a cold dark place, a nonelectrolyte, prepd. from Ph_3AsO by combining with a C_6H_6 soln. formed from a mixt. of $K_2Cr_2O_7$ in N H_2SO_4 + C_6H_6 + $(BuO)_3PO$, followed by H_2O_2. UV, IR, and NMR spectra (91).

[$Cr(Ph_3AsO)_4Br_2$], pink, paramagnetic crystals, prepd. from $CrBr_2$ + Ph_3AsO in abs. EtOH. Visible reflectance and near-IR spectra; a distorted tetrahedral structure is postulated (934).

[$Cr(Ph_3AsO)_4Cl_2$], violet-pink, paramagnetic crystals, prepd. from $CrCl_2$ + Ph_3AsO in abs. EtOH. Visible reflectance and near-IR spectra; a distorted octahedral structure is postulated (934).

[$Cr(Ph_3AsO)_4$](ClO_4)$_2$, pale blue, paramagnetic crystals, prepd. from $Cr(ClO_4)_2$ + Ph_3AsO in abs. EtOH. Visible reflectance and near-IR spectra; mol. structure is discussed (934).

[$Co(Ph_3AsO)_2Br_2$] (M-352), IR spectrum (1284).

[$Co(Ph_3AsO)_2Cl_2$] (M-352), the complex occurs in two forms in the solid state which are distinguishable by their IR spectra. Their magnetic moments and electronic spectra indicate a tetrahedral coordination of Co(II) in both forms (1284). Near-IR spectrum (633).

[$Co(Ph_3AsO)_2(NO_3)_2$] (M-352), mauve, paramagnetic crystals, m. 183°, prepd. from [$Co(Ph_3AsO)_2Cl_2$] + $AgNO_3$ in Me_2CO. The complex is monomeric in Me_2CO and $CHCl_3$. IR and electronic spectra of the solid and dissolved samples were obtd. A coordination of the NO_3 groups as chelating bidentates is postulated (635).

[$Cu(OAc)_2 \cdot Ph_3AsO$], pale green, paramagnetic substance, prepd. from $Cu(OAc)_2$ + Ph_3AsO in EtOH by evapn. in vacuum and drying over H_2SO_4. IR and electronic spectra as well as magnetic data suggest that the complex may be a dimer. ESR spectrum in $CHCl_3$ indicates a monomeric structure in soln. (638).

[$Cu(Ph_3AsO)_2Br_2$] (M-353), IR spectrum suggests a dinuclear [$Cu(Ph_3AsO)_4$][$CuBr_4$] structure, but its electronic spectrum is inconclusive (1284).

[$Cu(Ph_3AsO)_2Cl_2$] (M-353), IR and reflectance spectra suggest a dinuclear [$Cu(Ph_3AsO)_4$][$CuCl_4$] structure, but the far-IR and electronic spectra are inconclusive (1284).

[$Ph_3AsO \cdot H_2O_2$], white solid, m. 118°, loses H_2O_2 at 135° and at 60°/10^{-4} mm; prepd. from Ph_3As + H_2O_2 in MeOH (759) and from Ph_3As by shaking with 30% aq. H_2O_2 (1305). IR spectrum; the mol. structure of the complex is considered (759).

[($Ph_3AsO)_2 \cdot H_2O_2$], useful as a catalyst, with improved thermostability, for crosslinking polyesters with styrene (1366).

[$Me_2In(Ph_3AsO)Cl$], colorless, needle-like crystals, m. 97-99° after softening at 63°, prepd. from Ph_3AsO + Me_2InCl in Et_2O-$CHCl_3$-petr. ether (353).

$[Ph_3AsO]^+I_n^-$, a charge transfer complex which was studied by the UV, visible, and IR spectra in different solvents (663). Equil. const. in $CHCl_3$ was detd. (869).

$[Fe(Ph_3AsO)_2Br_2]$, yellow, paramagnetic crystals, m. 185°, prepd. from $FeBr_2$ in Me_2CO + Ph_3AsO (both dehydrated with a molecular sieve); a nonelectrolyte in $PhNO_2$. Electronic spectra in $CHCl_3$ and in the solid state are reported. A tetrahedral structure is postulated (577). IR spectrum band assignment (1284).

$[Fe(Ph_3AsO)_2Cl_2]$, pale yellow, paramagnetic crystals, m. 182°, a nonelectrolyte in $PhNO_2$, prepd. from $FeCl_2 \cdot 4\ H_2O$ in EtOH + Ph_3AsO under reflux. Electronic spectrum in the solid state suggests a tetrahedrally coordinate Fe(II) (577). IR spectrum band assignment (1284).

$[Mn(Ph_3AsO)_2Br_2]$ (M-354), far-IR spectrum (55), IR spectrum band assignment (1284). ESR spectrum; a pseudo-tetrahedral structure is postulated (496).

$[Mn(Ph_3AsO)_2Cl_2]$ (M-353), far-IR spectrum (55). IR spectrum band assignment; a tetrahedral structure of the Mn(II) complex is postulated (1284).

$[Hg(Ph_3AsO)Cl_2]_2$ (M-354), colorless crystals, prepd. from $HgCl_2$ + Ph_3AsO in EtOH (851). Crystallographic data; a dimeric centrosymmetric molecule in which the O atom of each donor is bonded to two Hg atoms and the Hg atoms attain a highly distorted tetrahedral coordination (219, 1047).

$[Hg(Ph_3AsO)_2Cl_2]$ (M-354), crystallographic data; a highly distorted tetrahedral coordination of Hg with 2 Cl and 2 O atoms is postulated (219, 1047).

$[Hg(Ph_3AsO)_4](ClO_4)_2$, white ppt., m. 175-178° (dec.), prepd. from $Hg(ClO_4)_2 \cdot 6\ H_2O$ + Ph_3AsO in EtOH. IR spectrum (1246).

$[Mo(Ph_3AsO)_4Cl_4]$ (M-354), this compn. is rejected. The substance could be a mixt. of $[MoO_2(Ph_3AsO)_2Cl_2]$ and/or $[MoO(Ph_3AsO)Cl_4]$ with $Ph_3As(OH)Cl$ (59).

$[Mo(Ph_3AsO)_2Br_2]$, blue solid, prepd. from $MoBr_3$ + Ph_3AsO in a sealed tube at 200° (300).

$[Mo(Ph_3AsO)_2Cl_2]$, blue solid, prepd. from $MoCl_3$ + Ph_3AsO and from $K_3[MoCl_6]$ + Ph_3AsO in a sealed tube at 200° (300).

$[MoO(Ph_3AsO)_2Br_3]$, ppt. prepd. from $MoOBr_3$ + Ph_3AsO in MeCN; a nonelectrolyte in MeCN and $PhNO_2$. IR and UV spectra (300).

$[MoO_2(Ph_3AsO)_2Br_2]$, ppt. prepd. from MoO_2Br_2 + Ph_3AsO (1:2) in dioxan (124, 125) or in MeCN (301). A nonelectrolyte in MeCN (301). UV, visible, and IR spectra (301). Electronic structure (124, 125).

$[MoO_2(Ph_3AsO)_2Cl_2]$ (M-354), blue-green ppt., a nonelectrolyte in $PhNO_2$ (845), prepd. from MoO_2Cl_2 + Ph_3AsO (1:2) in dioxan (124, 125) or in CH_2Cl_2 (301) and from $MoOCl_3$ + Ph_3AsO in CH_2Cl_2 (845). Electronic spectrum (124, 125). IR spectrum (301, 845). UV and visible spectra (301).

$[MoO_2(Ph_3AsO)(Dioxan)Br_2]$, ppt. prepd. from MoO_2Br_2 + Ph_3AsO (1:1) in dioxan. Electronic spectrum (124, 125).

[$MoO_2(Ph_3AsO)(Dioxan)Cl_2$], ppt. prepd. from MoO_2Cl_2 + Ph_3AsO (1:1) in dioxan. Electronic spectrum (124, 125).

[$Mo(Ph_3AsO)_2(CO)_4$], orange solid, prepd. from [$Mo(CO)_3(MeCN)_3$] + Ph_3AsO by refluxing in C_6H_6. IR spectrum and x-ray powder diagram (324).

[$Ni(Ph_3AsO)_2Br_2$] (M-354), magnetic data (412), magnetic prop. for the $^3T_{1g}$ term (565a); IR spectrum bands assignment, a tetrahedral coordination of Ni(II) is postulated (1284).

[$Ni(Ph_3AsO)_2Cl_2$] (M-354), magnetic data (412), magnetic prop. for the $^3T_{1g}$ term (565a). Far-IR spectrum (55). Near-IR spectrum (633). IR spectrum bands assignment; a tetrahedral coordination of Ni(II) is postulated (1284).

[$Ni(Ph_3AsO)_2(NO_3)_2$] (M-355), magnetic prop. for the $^3T_{1g}$ term (565a).

[$Ni(Ph_3AsO)_4$](ClO_4)$_2$ (M-355), magnetic prop. for the $^3T_{1g}$ term (565a).

[$Ni(Ph_3AsO)_2(meso-PhCH(NH_2)CH(NH_2)Ph)_2$]($ClO_4$)$_2$, yellow, diamagnetic ppt., prepd. from Ph_3AsO by heating with finely powdered [$Ni(meso-PhCH(NH_2)CH(NH_2)-Ph)_2$]($ClO_4$)$_2$ in BuOH. IR and electronic spectra indicate a weak coordination of Ph_3AsO in the 6-coordinate Ni complex (636).

[$Ni(Ph_3AsO)_2(rac-PhCH(NH_2)CH(NH_2)Ph)_2$]($ClO_4$)$_2$, yellow, diamagnetic complex, prepd. in the same manner as the preceding compd. IR and electronic spectra indicate a weak coordination of Ph_3AsO in the 6-coordinate Ni complex (636).

[$Nb_6(Ph_3AsO)_4Cl_{14}$], prepd. from $Nb_6Cl_{14}\cdot 8 H_2O$ + Ph_3AsO in EtOH by evapn. to dryness. IR and visible spectra. A structure with the Nb atoms in an octahedral cluster with 12 bridging Cl atoms above the octahedron edges and one chlorine atom or ligand atom attached to each metal atom in a centrifugal position is postulated (564).

[$NbO(Ph_3AsO)_2Cl_3$], pale yellow solid, prepd. from Ph_3AsO or [Ph_3AsO]$_2\cdot H_2O_2$ + $NbCl_5$. IR spectrum (389).

[$Ph_3AsO\cdot HNO_3$], white cryst. solid, m. 104°, insensitive to moisture, sol. in MeCN and EtOAc, insol. in CCl_4 and petrol, prepd. from $Ph_3As(NO_3)_2$ by dissolving in moist Me_2CO and adding light petrol and from Ph_3As by treating with concd. HNO_3, dilg. with H_2O, dissolving the oily prod. in EtOH, and cooling to -78°. Partial ionization in MeCN was established. IR spectrum (1531).

[Ph_3AsO]$_2HClO_4$, white crystals, m. 210-211°, prepd. from Ph_3AsCl_2 + $AgClO_4$ in EtOH (1167).

[$Re(Ph_3AsO)Cl_3$]$_3$ (M-355), prepd. from $ReCl_3$ + Ph_3AsO; x-ray crystal structure data (611).

[(Ph_3AsO)$_2(ReCl_3)_3$], prepd. from $ReCl_3$ + Ph_3AsO; props. were detd.; x-ray crystal structure data (611).

[$TaO(Ph_3AsO)_2Br_3$], yellow solid, prepd. from [Ph_3AsO]$_2\cdot H_2O_2$ in CH_2Cl_2 + $TaBr_3$; IR spectrum (389).

[(C₆F₅)₂Tl(Ph₃AsO)Br], white needles, m. 146-147°, a nonelectrolyte in Me_2CO, prepd. from $(C_6F_5)_2TlBr$ + Ph_3AsO in Et_2O-MeOH. Soly. data; monomeric in C_6H_6 (458). IR spectrum (454, 458). A tetrahedral structure is postulated (458).

[(C₆F₅)₂Tl(Ph₃AsO)Cl], white tablets, m. 176-177°, a nonelectrolyte in Me_2CO, prepd. from $(C_6F_5)_2TlCl$ + Ph_3AsO in Et_2O-MeOH. Soly. data; monomeric in C_6H_6 (458). IR spectrum (454, 458). A tetrahedral structure is postulated (458).

[(C₆F₅)₂Tl(Ph₃AsO)NO₃], white microcrystals, m. 116-118°, a nonelectrolyte in Me_2CO, prepd. from [(C₆F₅)₂Tl(Ph₃AsO)Br] + $AgNO_3$ in MeOH. Soly. data; monomeric in C_6H_6 (458). IR spectrum (454, 458). A 4-coordinate Tl with a tetrahedral structure is postulated (458).

[(C₆F₅)₂Tl(Ph₃AsO)₂NO₃], small white plates, m. 152.5-153.5°, a nonelectrolyte in Me_2CO, prepd. from $(C_6F_5)_2TlNO_3$ + Ph_3AsO in boiling MeOH by adding H_2O (458). IR spectrum (454, 458). A 5-coordinate Tl complex with a trigonal bipyramidal structure is postulated (458).

[(C₆F₅)₂Tl(Ph₃AsO)(O₂CCF₃)], white microcrystals, m. 173.5-174.5°, a nonelectrolyte in Me_2CO, prepd. from $(C_6F_5)_2TlO_2CCF_3$ + Ph_3AsO in boiling MeOH. Soly. data; monomeric in C_6H_6 (458). IR spectrum (454, 458). A 4-coordinate Tl with a tetrahedral structure is postulated (458).

[Sn(Ph₃AsO)₂Br₄] (M-355), IR spectrum (354).

[Sn(Ph₃AsO)₂Cl₄] (M-355), IR spectrum (354).

[Sn(Ph₃AsO)₂F₄], IR spectrum (354).

[Sn(Ph₃AsO)₂I₄], IR spectrum (354).

[MeSn(Ph₃AsO)₂I₃], IR spectrum (354).

[Me₂Sn(Ph₃AsO)₂Br₂], IR spectrum (354).

[Me₂Sn(Ph₃AsO)₂Cl₂], crystals, dec. 206-208°, prepd. from Me_2SnCl_2 + Ph_3AsO in CH_2Cl_2 or $CHCl_3$ (440). IR spectrum (354, 440). NMR spectrum. A 6-coordinated Sn with an octahedral structure is postulated (440).

[Me₂Sn(Ph₃AsO)₂I₂], IR spectrum (354).

[Ph₂Sn(Ph₃AsO)₂Cl₂], crystals, dec. 183°, prepd. from Ph_2SnCl_2 + Ph_3AsO in CH_2Cl_2 or $CHCl_3$. A 6-coordinate Sn with an octahedral structure is postulated (440).

[Me₃Sn(Ph₃AsO)Br], crystals, dec. 156-158°, prepd. from Me_3SnBr + Ph_3AsO in CH_2Cl_2 or $CHCl_3$. IR and NMR spectra. A 5-coordinate Sn complex with a trigonal bipyramidal structure is postulated (440).

[Ph₃Sn(Ph₃AsO)Cl], crystals, dec. 200-202°, prepd. from Ph_3SnCl + Ph_3AsO in CH_2Cl_2 or $CHCl_3$. IR and NMR spectra. A 5-coordinate Sn complex with a trigonal bipyramidal structure is postulated (440).

[Sn(Ph$_3$AsO)$_4$I$_2$]I$_2$, IR spectrum (354).

[UO$_2$(Ph$_3$AsO)(OAc)$_2$], crystallographic data (115).

[UO$_2$(Ph$_3$AsO)(NO$_3$)$_2$], crystallographic data (648).

[UO$_2$(Ph$_3$AsO)$_2$(OAc)$_2$], crystallographic data (115).

[UO$_2$(Ph$_3$AsO)$_2$Br$_2$], far-IR spectrum (690).

[UO$_2$(Ph$_3$AsO)$_2$Cl$_2$], pale yellow, cryst. solid, m. 297-298°, prepd. from UO$_2$Cl$_2$·3 H$_2$O in EtOAc + Ph$_3$AsO. Soly. data; IR spectrum (689). Far-IR spectrum (690).

[UO$_2$(Ph$_3$AsO)$_2$(NO$_3$)$_2$], pale yellow, cryst. solid, m. 279-280° (265), pale yellow needles, m. 285° (1645), prepd. from UO$_2$(NO$_3$)$_2$·6 H$_2$O in EtOAc + Ph$_3$AsO (689), from UO$_2$(NO$_3$)$_2$·2 H$_2$O + Ph$_3$AsO (1:2) in Me$_2$CO under reflux (265), from UO$_2$(NO$_3$)$_2$·6 H$_2$O + Ph$_3$As by heating in Et$_2$O, dioxan, or cyclohexanol (1191), from UO$_2$(NO$_3$)$_2$·6 H$_2$O in Et$_2$O + excess Ph$_3$As at r.t. in the presence of air (30% yield), and from UO$_2$(NO$_3$)$_2$·6 H$_2$O in dioxan + a deficient amt. of Ph$_3$As under reflux in the presence of air (97% yield) (1645). Soly. data (689). IR spectrum (265, 689). Far-IR spectrum (690). X-ray diffraction data: an 8-coordinate U with 2 bidentate NO$_3$ ligands, 2 oxide oxygens, and 2 O atoms of the Ph$_3$AsO ligands is postulated (1191).

[Zn(Ph$_3$AsO)$_2$Br$_2$], prepd. from ZnBr$_2$ + Ph$_3$AsO in EtOH. IR spectrum bands assignment (1284).

[Zn(Ph$_3$AsO)$_2$Cl$_2$], the complex occurs in 2 solid forms distinguishable by their IR spectra; one form was obtd. by adding a cold Ph$_3$AsO soln. in EtOH to a 25% excess ZnCl$_2$ in cold EtOH, and the other form was prepd. by adding ZnCl$_2$ to excess Ph$_3$AsO in cold EtOH (1284).

Three types of crystalline yttrium and lanthanide metal complexes with Ph$_3$AsO were reported: [M(Ph$_3$AsO)$_2$(EtOH)(NO$_3$)$_3$] (A), [M(Ph$_3$AsO)$_3$(NO$_3$)$_3$] (B), and [M(Ph$_3$AsO)$_4$(NO$_3$)$_2$]NO$_3$ (C). The (A) complexes were prepd. by adding a boiling, 2.5 molar Ph$_3$AsO soln. in EtOH to a boiling, one molar soln. of a lanthanide metal trinitrate·hexahydrate in EtOH. The (B) complexes were prepd. by adding a boiling, 3 molar Ph$_3$AsO soln. in Me$_2$CO to a boiling, one molar soln. of a lanthanide metal trinitrate in Me$_2$CO, and removing the solvent of crystallization by heating at 65-80°/5 mm. The (C) complexes were prepd. by adding one molar soln. of a lanthanide metal trinitrate hexahydrate in Me$_2$CO to a warm, 6 molar Ph$_3$AsO soln. in Me$_2$CO, and removing the solvent of crystallization by heating at 78°/1 mm. Analogous reaction of Lu(NO$_3$)$_3$·6 H$_2$O with Ph$_3$AsO in EtOH gave the corresponding monoethanolate, from which the solvent was removed at 78°/5 mm (407).

Table 24. Lanthanide Metal and Yttrium Complexes with Ph_3AsO (407)

$[M(Ph_3AsO)_2(EtOH)(NO_3)_3]$

M	m.° (dec.)	Color
La	149–153	Colorless*
Ce	152–160	Pale yellow
Pr	160–167	Green
Nd	159–167	Lilac
Sm	165–168	Colorless
Eu	164–170	Colorless*
Gd	163–169	Colorless*
Tb	155–160	Colorless
Dy	148–162	Colorless
Ho	154–159	Pale yellow
Er	147–150	Pink*
Tm	148–150	Colorless
Yb	144–145	Colorless
Lu	131–133	Colorless*
Y	150–156	Colorless

$[M(Ph_3AsO)_3(NO_3)_3]$

M	m.°	Color
La	269–272	Colorless
Ce	256–258	Colorless
Pr	243–246	Pale green
Nd	236–239	Lavender
Sm	220–222	Colorless
Eu	218–219	Colorless
Gd	219–220	Colorless*
Tb	220–221	Colorless
Dy	215–217	Colorless**
Ho	213–215	Colorless
Er	207–210	Pink
Tm	115–127	Colorless***
Yb	108–110	Colorless***
Y	132–137	Colorless

$[M(Ph_3AsO)_4(NO_3)_2]NO_3$

M	m.°	Color
La	239–240	Colorless
Nd	250–260	Lilac
Sm	278	Colorless
Eu	279–280	Colorless*
Dy	284	Colorless
Ho	284	Yellow brown
Er	284–286	Pale rose red
Tm	288	Colorless
Yb	284–285	Colorless
Lu	286	Colorless
Y	287–288	Colorless

*   IR spectrum.

**   On boiling with EtOH for 15 min. → $[Dy(Ph_3AsO)_2(EtOH)(NO_3)_3]$.

***   IR spectrum of the complex with 2 Me_2CO of crystn. was obtd.

Tricyclohexylarsine Oxide, AsC$_{18}$H$_{33}$O, (C$_6$H$_{11}$)$_3$AsO, (M-358)
Synth.: From (C$_6$H$_{11}$)$_3$As + HgO (1:1) in Me$_2$CO under reflux, taking-up the
prod. with H$_2$O, evapg. to dryness, and purifying by sublimation (1057).
From (C$_6$H$_{11}$)$_3$As in Et$_2$O by adding dropwise 30% aq. H$_2$O$_2$ and isolating the
prod. as described in the preceding procedure (1057).
Prop.: m. 210-212°, IR spectrum (1057).
Rxn. with: I$_2$ in CCl$_4$ → [(C$_6$H$_{11}$)$_3$AsO]$^+$I$_2^-$, a charge transfer complex (869).
Coordination Deriv.: {Co[(C$_6$H$_{11}$)$_3$AsO]$_2$Cl$_2$}, bright blue crystals, a nonelec-
trolyte, prepd. from CoCl$_2$·6 H$_2$O + (C$_6$H$_{11}$)$_3$AsO in EtOH. IR spectrum (1057).

{Ni[(C$_6$H$_{11}$)$_3$AsO]$_4$}(ClO$_4$)$_2$, crystals, prepd. from Ni(ClO$_4$)$_2$ + (C$_6$H$_{11}$)$_3$AsO in
EtOH (1057).

(C$_6$H$_{11}$)$_3$AsO + I$_2$, a solid, dec. 88-90°, with uncertain As/I ratio (1057).

(C$_6$H$_{11}$)$_3$AsO + IBr, a solid, m. 137-139°, with uncertain As/halogen ratio (1057).

(C$_6$H$_{11}$)$_3$AsO + ICl, a solid, m. 154-157°, with uncertain As/halogen ratio (1057).

Tri-n-octylarsine Oxide, AsC$_{24}$H$_{51}$O, (n-C$_8$H$_{17}$)$_3$AsO
Synth.: From (n-C$_8$H$_{17}$)$_3$As + HgO (1:1) in Me$_2$CO under reflux, taking-up the
prod. with H$_2$O, evapg. to dryness, and purifying by sublimation (1057).
From (n-C$_8$H$_{17}$)$_3$As in Et$_2$O + aq. 30% H$_2$O$_2$ and isolating the prod. as described
in the preceding method (1057, see also 966).
Prop.: m. 69.5-71° (966), m. 69.6-71.2°, IR spectrum (1057).
Rxn. with: I$_2$ in CCl$_4$ → [(C$_8$H$_{17}$)$_3$AsO]$^+$I$_n^-$, a charge transfer complex (869).
Aq. HClO$_4$, HCl, HBr, HI, and HNO$_3$ → complex compds., sol. in, and extractable
into, C$_6$H$_6$ (965, 966).
Uses: Extn. of Am^{3+}, Cm^{3+}, and Pm^{3+} from aq. soln. into org. phase contg.
(C$_8$H$_{17}$)$_3$AsO (188). Extn. of Fe^{3+} from aq. hydrohalic acid soln. at pH $\geqslant$ 0,
Zn^{++} from aq. hydrohalic acid soln. at pH $\geqslant$ 4.5, Cd^{++} from aq. hydrohalic
acid soln. at pH 2, and Hf^{4+} and Zr^{4+} from citric acid media at pH 2.7 and
4.0, resp., into C$_6$H$_6$ (778, 965).
Coordination Deriv.: (In the following complexes TOAO denotes (C$_8$H$_{17}$)$_3$AsO.)
[TOAO·H]HBr$_2$ is formed when 4.0 to 7.0 molar aq. HBr is extd. with TOAO
dissolved in C$_6$H$_6$; IR spectrum. Aq. HBr at the conc. $\leqslant$ 2 molar is extd. as a
hydrate of the approx. formula: [TOAO·H$_3$O]Br (965, 966).

[TOAO·H]HCl$_2$ is formed when 1 molar aq. HCl is extd. with TOAO dissolved in
C$_6$H$_6$; IR spectrum. The extn. of 0.1 molar HCl gives [TOAO·H$_3$O]Cl (965, 966).

[TOAO·H$_3$O]ClO$_4$ is formed when aq. HClO$_4$ is extd. with TOAO dissolved in C$_6$H$_6$;
IR spectrum (965, 966).

[TOAO·H]HI$_2$ is formed when 4.0 to 7.0 molar aq. HI is extd. with TOAO dissolved in C$_6$H$_6$; IR spectrum (965, 966).

[TOAO·xHNO$_3$·yH$_2$O] is formed when aq. HNO$_3$ is extd. with TOAO dissolved in
C$_6$H$_6$; the x:y ratio depends on the concn. of the acid. IR spectrum (965, 966).

[TOAO·HFeCl$_4$] is formed when FeCl$_3$ soln. in HCl is extd. with TOAO dissolved
in C$_6$H$_6$; IR spectrum (965).

[(TOAO)$_3$(Citric acid)$_2$] is formed when > 2 molar aq. citric acid soln. is extd. with TOAO dissolved in C$_6$H$_6$ (965).

[(TOAO)$_2$CdI$_2$] is formed when CdI$_2$ soln. in aq. HI is extd. with TOAO dissolved in C$_6$H$_6$; IR spectrum (965).

[(TOAO)$_3$·CdX$_2$]$_2$, where X = Cl or Br, is formed when CdX$_2$ soln. in aq. HX is extd. with TOAO dissolved in C$_6$H$_6$; IR spectra (965).

[TOAO·Hf(OH)$_2$]$_2$ is formed when aq. Hf(OH)$_4$ soln. is extd. with TOAO dissolved in C$_6$H$_6$ (965).

[(TOAO)$_2$ZnX$_2$], where X = Cl, Br, or I, is formed when aq. ZnX$_2$ soln. is extd. with TOAO dissolved in C$_6$H$_6$; IR spectra (965).

[TOAO·Zr(OH)$_4$]$_2$ is formed when aq. Zr(OH)$_4$ soln. is extd. with TOAO dissolved in C$_6$H$_6$ (965).

(3,5-Di-t-butyl-4-oxyphenyl)diphenylarsine Oxide Radical,   AsC$_{26}$H$_{30}$O$_2$,

Synth.: The formation of the radical was detected by ESR spectroscopy during dehydrogenation of Ph$_2$[4,3,5-HO(t-Bu)$_2$C$_6$H$_2$]AsO with alk. K$_3$[Fe(CN)$_6$] soln. in C$_6$H$_6$ (1105).
Prop.: Green soln. which rapidly changed to light brown, diamagnetic 3,5,3',5'-tetra-t-butyldiphenoquinone. ESR data of the radical were obtd. (1105).

Neopentyltetrayltetrakis(diphenylarsine Oxide),   As$_4$C$_{53}$H$_{48}$O$_4$,
    C[CH$_2$As(O)Ph$_2$]$_4$
Synth.: From C(CH$_2$AsPh$_2$)$_4$ + H$_2$O$_2$ (1:6) in Me$_2$CO under reflux, the prod. crystd. with 2 H$_2$O$_2$; the H$_2$O$_2$ was removed at 130-140° in a high vacuum (527).
Prop.: Colorless crystals, m. 204-207° (dec.); the complex with 2 H$_2$O$_2$ forms colorless crystals, m. 104° with loss of H$_2$O$_2$; sol. in CH$_2$Cl$_2$, CHCl$_3$, Me$_2$CO, and MeCN; insol. in Et$_2$O, petr. ether, CS$_2$, and C$_6$H$_6$. IR spectrum (527).

Table 25. Tertiary Arsine Oxides

R_3AsO	Prepd. from	Prop. and Rxm.	Ref.
C_5			
Et_2MeAsO	Et_2MeAs + HgO (1:1) in Me_2CO under reflux, (a)	Crystals, m. 97.6-98.6°, IR spectrum	1057
	Et_2MeAs in Et_2O + aq. 30% H_2O_2, (a)		1057
C_6			
$Et_2(CH_2=CH)AsO$	$Et_2AsCH=CH_2$ + MnO_2 in dioxan or C_6H_{14} at 50°	m. 95-96°, insol. in i-C_8H_{18}, (b)	1594
Et_3AsO (M-356)	Et_3As + HgO (1:1) in Me_2CO under reflux, (a)	m. 105-106.5°, IR spectrum, (c, d)	1057
	Et_3As in Et_2O + aq. 30% H_2O_2, (a)		1057
Me_2BuAsO (M-356)	Me_2BuAs + HgO (1:1) in Me_2CO under reflux, (a)	m. 93.2-93.7°, IR spectrum	1057
C_8			
$Bu(CH_2=CH)_2AsO$	$Bu(CH_2=CH)_2As$ + MnO_2 in C_6H_{14} or dioxan at 50°	m. 66-67°, insol. in i-C_8H_{18}	1594
$BuEt_2AsO$	$BuEt_2As$ + HgO (1:1) in Me_2CO under reflux, (a)	m. 36-37°	1057
C_9			
(+)-$EtMePhAsO$		$[\alpha]_D^{20}$ + 2.4° in AcOEt, on contact with H_2O spontaneous racemization	754
Pr_3AsO (M-356)	Pr_3As + HgO (1:1) in Me_2CO under reflux, (a)	m. 96.1-96.8°, IR spectrum, (c)	1057
	Pr_3As in Et_2O + aq. 30% H_2O_2, (a)		1057
$MeBu_2AsO$ (M-356)	$MeBu_2As$ + HgO (1:1) in Me_2CO under reflux, (a)	m. 65.6-66.7°, IR spectrum	1057
	$MeBu_2As$ in Et_2O + aq. 30% H_2O_2, (a)		1057
C_{10}			
$Et_2(4-BrC_6H_4)AsO$	$Et_2(4-BrC_6H_4)As$ + H_2O_2 in Me_2CO	m. 148-149°, (c)	340
$Et_2(4-ClC_6H_4)AsO$	$Et_2(4-ClC_6H_4)As$ + H_2O_2 in Me_2CO	m. 134°, (c)	340
$Et_2(4-O_2NC_6H_4)AsO$	$Et_2(4-O_2NC_6H_4)As$ + H_2O_2 in Me_2CO	m. 156°, (c)	340
Et_2PhAsO	Et_2PhPh + H_2O_2 in Me_2CO	m. 89°, (c)	340
(+)-$MePhPrAsO$		$[\alpha]_D^{20}$ + 6.4° in AcOEt, spontaneous racemization in H_2O	754
$Me_2(C_8H_{17})AsO$		(e)	939
$BuPr_2AsO$	$BuPr_2As$ + HgO (1:1) in Me_2CO under reflux, (a)	m. 86.2-87.2°	1057

C_{11}			
$Et_2(4\text{-}MeC_6H_4)AsO$	$Et_2(4\text{-}MeC_6H_4)As + H_2O_2$ in Me_2CO	m. 128-129°, (c)	340
$Et_2(4\text{-}MeOC_6H_4)AsO$	$Et_2(4\text{-}MeOC_6H_4)As + H_2O_2$ in Me_2CO	m. 71-72°, (c)	340
C_{12}			
$Et_2(4\text{-}Me_2NC_6H_4)AsO$	$Et_2(4\text{-}Me_2NC_6H_4)As + H_2O_2$ in Me_2CO	m. 97°, (c)	340
Bu_3AsO (M-357)	$Bu_3As + HgO$ (1:1) in Me_2CO under reflux, (a)	m. 109.1-110°, IR spectrum, (c)	1057
	Bu_3As in Et_2O + aq. 30% H_2O_2	(f)	1057
$i\text{-}Bu_3AsO$		Rxn. with PhNCO → $i\text{-}Bu_3AsN(Ph)C(O)O$, (d)	223
C_{14}			
$EtPh_2AsO$	$EtPh_2As$ by oxidn.		1017
$Me_2(C_{12}H_{25})AsO$ (M-357)	$Me_2(C_{12}H_{25})As + H_2O_2$	(e)	939
$Me_2[C_{10}H_{21}CH(OH)\text{-}CH_2]AsO$		(e)	939
C_{15}			
$Ph_2(MeC{\equiv}C)AsO$	$Ph_2(MeC{\equiv}C)As(OH)_2$ by dehydration in vacuum	m. 105°, IR and NMR spectra, (g)	149
$EtPr(C_{10}H_{21})AsO$		(e)	939
C_{16}			
$Ph_2(EtC{\equiv}C)AsO$	$Ph_2(EtC{\equiv}C)As(OH)_2$ by dehydration in vacuum	m. 130°, IR spectrum	149
$Et_2(C_{12}H_{25})AsO$	$Et_2(C_{12}H_{25})As + H_2O_2$	(e)	939
$Me_2(C_{14}H_{29})AsO$		(e)	939
$(HOCH_2CH_2)_2\text{-}(C_{12}H_{25})AsO$		(e)	939

(a) The prod. is dissolved in H_2O, evapd. to dryness, and sublimed in vacuo (1057). (b) Polymerization by BuMgI in MePh → a cryst. prod. (56). (c) IR spectra in C_6H_6 soln. and in the solid state (1387b). (d) Useful as a catalyst for trimerization of isocyanates (728). (e) Useful in detergents as a solubilizer for alk. earth soaps and a bacteriostat (939). (f) Useful as a catalyst in the polyurethan foam prodn. from diisocyanates and polyesters (757). (g) NMR spectrum – chem. shift and long-range As-H coupling consts. (1412).

Table 25 Continued

R₃AsO	Prepd. from	Prop. and Rxn.	Ref.
C₁₇			
MeEt($C_{14}H_{29}$)AsO		(e)	939
C₁₈			
(C_6F_5)₃AsO	(C_6F_5)₃AsCl₂ by hydrolysis	m. 198-199°, IR spectrum	648
[4,3-Cl(O_2N)C_6H_3]₃AsO	(4-ClC_6H_4)₃AsO + 90% HNO_3-H_2SO_4 soln. at 3-60°	m. 230-232°, (h)	1392
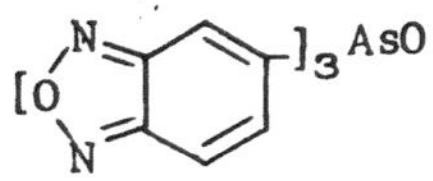 ₃AsO	(4-ClC_6H_4)₃As in Me_2CO + 30% H_2O_2 at 25-30°, nitration of the oxide with HNO_3-H_2SO_4, followed by NaN_3 in DMSO-H_2O and dec. in o-$C_6H_4Cl_2$ at 80-130°	m. 226° (violent dec.), (i)	1392
(4-BrC_6H_4)₃AsO (M-358)		IR spectrum	1057
(4-ClC_6H_4)₃AsO		m. 202-204°, rxn. with HNO_3-H_2SO_4 → [4,3-Cl(O_2N)C_6H_3]₃AsO	1392
(3-H_2NC_6H_4)₃AsO (M-358)		IR spectrum	1057
Me_2($C_{16}H_{33}$)AsO	Me_2($C_{16}H_{33}$)As + H_2O_2	(e)	939
Et_2($C_{14}H_{29}$)AsO		(e)	939
Pr_2($C_{12}H_{25}$)AsO		(e)	939
(HOCH₂CH₂CH₂)₂($C_{12}H_{25}$)AsO		(e)	939
C₂₀			
Ph₂[CH₂(CH₂)₄C(OH)C≡C]AsO	Ph₂[CH₂(CH₂)₄C(OH)C≡C]As + H_2O_2 in AcOH	m. 205°, IR spectrum	149
Me_2[Me(CH₂)₇CH=CH(CH₂)₇CH₂]AsO		(e)	939
C₂₁			
(4-MeC_6H_4)₃AsO (M-359)		Forms the (4-MeOC_6H_4)₃AsO·H_2NSO₂Ph complex, (j)	342

302

(4-MeOC$_6$H$_4$)$_3$AsO		Forms the (4-MeOC$_6$H$_4$)$_3$AsO·H$_2$NSO$_2$Ph complex, m. 154-155°, IR spectrum, (k)	342
C$_{22}$			
Ph$_2$(α-C$_{10}$H$_7$)AsO	Ph$_2$(α-C$_{10}$H$_7$)As + 30% aq. H$_2$O$_2$ in Me$_2$CO under reflux	m. 191-193°	1347
Ph$_2$(β-C$_{10}$H$_7$)AsO	Ph$_2$(β-C$_{10}$H$_7$)As + 30% aq. H$_2$O$_2$ in Me$_2$CO under reflux	m. 173-176°	1347
2,2'-(o-C$_6$H$_4$)[C$_6$H$_4$As(O)Me$_2$]$_2$	2,2'-(o-C$_6$H$_4$)[C$_6$H$_4$As(OH)Me$_2$]$_2$·2 NO$_3$ + aq. NaOH	m. 191-194°, sol. in H$_2$O	1016
	2,2'-(o-C$_6$H$_4$)[C$_6$H$_4$AsMe$_2$]$_2$ in Me$_2$CO + excess H$_2$O$_2$	Crystd. with 1 H$_2$O$_2$·2 H$_2$O, m. 166° (dec.)	1016
C$_{23}$			
Ph(β-C$_{10}$H$_7$)(2-HO$_2$CC$_6$H$_4$)AsO	Ph(β-C$_{10}$H$_7$)(2-MeC$_6$H$_4$)As + aq. KMnO$_4$ at 90°, followed by Na$_2$SO$_3$ in HCl soln.	Ppt.; rxn. with alc. HCl + Na$_2$SO$_3$ + KI → (±)-Ph(β-C$_{10}$H$_7$)AsC$_6$H$_4$CO$_2$H	601
C$_{26}$			
Ph$_2$(1-Anthryl)AsO	Ph$_2$(1-C$_{14}$H$_9$)As + 30% aq. H$_2$O$_2$ in Me$_2$CO under reflux	m. 221-223°	1347
Ph$_2$(2-Anthryl)AsO	Ph$_2$(2-C$_{14}$H$_9$)As + 30% aq. H$_2$O$_2$ in Me$_2$CO under reflux	m. 235-236°	1347
Ph$_2$[4,3,5-HO(t-Bu)$_2$C$_6$H$_2$]AsO	Ph$_2$[4,3,5-HO(t-Bu)$_2$C$_6$H$_2$]As in C$_6$H$_6$ + 30% H$_2$O$_2$	Colorless needles, m. 235-240° (dec.), NMR and mass spectra, (l)	1105

(h) Rxn. with NaN$_3$ in DMSO contg. H$_2$O and dec. of the prod. at 80-130° in o-C$_6$H$_4$Cl$_2$ → tris(5-benzofurazanyl)-arsine oxide (1392). (i) Useful as a depolarizer for dry cells contg. a metallic anode, a C cathode, and metal salt electrolyte (1392). (j) The complex was prepd. from (4-MeC$_6$H$_4$)$_3$As + Br in dry C$_6$H$_6$, followed by PhSO$_2$NH$_2$ in Et$_3$N; IR spectrum was obtd. (342). (k) The complex was prepd. from (4-MeOC$_6$H$_4$)$_3$As + Br in C$_6$H$_6$, followed by PhSO$_2$NH$_2$ in Et$_3$N (342). (l) Rxn. with alk. soln. of K$_3$[Fe(CN)$_6$] in C$_6$H$_6$ → light brown soln. contg. 3,5,3',5'-tetra-tert-butyldiphenoquinone, formed via a green radical (1105).

Table 25 Continued

R_3AsO	Prepd. from	Prop. and Rxn.	Ref.
C_{28} (Cont.) $[-CH_2As(O)Ph_2]_2$		$\{U[Ph_2As(O)CH_2CH_2As(O)Ph_2]_2Cl_4\}$, green solid, m. 203° (dec.), prepd. from $UCl_4 + (-CH_2AsPh_2)_2$ in THF in contact with air, IR spectrum	1383
C_{27} $Ph_2[Ph_2C(OH)C\equiv C]AsO$	$Ph_2[Ph_2C(OH)C\equiv C]As + H_2O_2$ in AcOH	m. 260° (dec.), IR spectrum	149
C_{30} $1,4-C_6H_4[As(O)Ph_2]_2$	$1,4-C_6H_4[As(Br_2)Ph_2]_2$ by hydrolysis $1,4-C_6H_4(AsPh_2)_2 + H_2O_2$ in Me_2CO	m. 330°	1658 1658
C_{34} $1,5-C_{10}H_6[As(O)Ph_2]_2$, (m)	$1,5-C_{10}H_6(AsPh_2)_2 + 30\%$ aq. H_2O_2 in Me_2CO under reflux	m. 340-343°	1347
$2,6-C_{10}H_6[As(O)Ph_2]_2$, (m)	$2,6-C_{10}H_6(AsPh_2)_2 + 30\%$ aq. H_2O_2 in Me_2CO under reflux	m. 296-298°	1347
$2,7-C_{10}H_6[As(O)Ph_2]_2$, (m)	$2,7-C_{10}H_6(AsPh_2)_2 + 30\%$ aq. H_2O_2 in Me_2CO under reflux	m. 193-195°	1347

(m) $C_{10}H_6$ = naphthylene diradical.

2.3.6.1.7 DIALKOXY AND DINITRATO DERIVATIVES

The formation of dialkoxytriorganoarsenics, $R_3As(OR')_2$, as intermediates in the synthesis of tertiary arsine peroxides was indicated in the main volume (M-360). Dinitratotriphenylarsenic, a new compound, can be considered as an analog of the dialkoxy derivatives.

Dinitratotriphenylarsenic, $AsC_{18}H_{15}N_2O_6$, $Ph_3As(ONO_2)_2$
Synth.: From Ph_3AsBr_2 in MeCN by adding dropwise $AgNO_3$ dissolved in MeCN (1521).
From Ph_3AsBr_2 by treating with liquid N_2O_4 (1521).
Prop.: White crystals, m. 161-162°, highly hydroscopic; molar cond. in MeCN too low for an ionic structure. IR spectrum indicated a covalent bonding of the ONO_2 groups and a trigonal bipyramidal structure with equatorial Ph and apical ONO_2 groups (1521).
Rxn. with: Moist Me_2CO, followed by light petrol $\rightarrow Ph_3AsO \cdot HNO_3$ (1521).

2.3.6.1.9 OXYBIS(TERTIARY ARSINE) DERIVATIVES

The oxygen-bridged dinuclear compounds of tertiary arsenic derivatives were described in the main volume on pages 362-3.

Oxybis(triphenylarsenic) As-Hydroxide As'-Iodide, $As_2C_{36}H_{31}IO_2$,
Synth.: From Ph_3AsO + BuI at 55° in various solvents; the highest rate of rxn. was obtd. in $PhNO_2$ (334).
Prop.: Cryst. substance, m. 164° (334).

$$Ph_3As \overset{\nearrow OH}{\underset{\searrow O}{}}$$
$$Ph_3As \overset{\nearrow}{\underset{\searrow I}{}}$$

2.3.6.2 COMPOUNDS WITH ARSENIC-SULFUR BONDS

2.3.6.2.6 TERTIARY ARSINE SULFIDES

The compounds and their methods of synthesis were described in the main volume on pages 363-66. Numerous new tertiary arsine sulfides were prepared from tertiary arsines by the reaction with sulfur in benzene on a steam bath or at 90-100° and without the solvent at 100-120° in a sealed tube. Triphenylarsine sulfide was also prepared from the tertiary arsine by a reaction with a variety of organic disulfides. Several compounds of this class have been used in the preparation of coordination complexes with transition metal salts. The latter and their complexes are described individually hereafter. Many more arsine sulfides are compiled in Table 26.

Trimethylarsine Sulfide, AsC_3H_9S, Me_3AsS,   (M-364)
Coordination Deriv.: $[Co(Me_3AsS)_2Br_2]$, paramagnetic solid, prepd. from $CoBr_2$ + Me_3AsS in Me_2CO-EtOH. Electronic and IR spectra and x-ray powder diagram; a 4-coordinate Co complex is postulated (235).

$[Co(Me_3AsS)_2Cl_2]$, paramagnetic solid, prepd. from $CoCl_2$ + Me_3AsS in Me_2CO-EtOH. Electronic and IR spectra and x-ray powder diagram; a 4-coordinate Co complex is postulated (235).

[Co(Me₃AsS)₂I₂], paramagnetic solid, prepd. from CoI₂ + Me₃AsS in Me₂CO-EtOH. Electronic and IR spectra; a 4-coordinate Co complex is postulated (235).

[Co(Me₃AsS)₂(NCS)₂], paramagnetic solid, prepd. from Co(NCS)₂ + Me₃AsS in Me₂CO-EtOH. Electronic and IR spectra; a 4-coordinate Co complex and Co-N bonds are postulated (235).

[Co(Me₃AsS)₄](ClO₄)₂, paramagnetic solid, prepd. from Co(ClO₄)₂ + Me₃AsS in Me₂CO-EtOH. Electronic and IR spectra; a 4-coordinate Co complex and ionic ClO₄ groups are postulated (235).

[Zn(Me₃AsS)₂Br₂], prepd. from ZnBr₂ + Me₃AsS in Me₂CO-EtOH. IR spectrum (235).

[Zn(Me₃AsS)₂Cl₂], prepd. from ZnCl₂ + Me₃AsS in Me₂CO-EtOH. IR spectrum and x-ray powder diagram (235).

[Zn(Me₃AsS)₄](ClO₄)₂, prepd. from Zn(ClO₄)₂ + Me₂AsS in Me₂CO-EtOH; IR spectrum (235).

Triphenylarsine Sulfide,　　AsC₁₈H₁₅S,　　Ph₃AsS,　　(M-365)
Synth.: From Ph₃As + S in C₆H₆ on a steam bath, 85% yield (603, 1347).
From Ph₃AsBr₂ in ethanolic NH₃ + H₂S (803).
From Ph₃As by heating with PhSSPh in MePh or with RSSR (R = Bz , 4-(Me₂N)C₆H₄-, or Me₂NC(S)-) in xylene under reflux (1115).
From Ph₃As by refluxing in CS₂, 97% yield (1151).
Prop.: White crystals, m. 166° (603), colorless crystals, m. 163.5° (803, 1151), m. 162° (1347), m. 165.5-166.5° (1151). Soly. data (603). IR spectrum (237, 803, 1151). Dipole moment and molar Kerr const. (94). Microdetn. of S (126).
Biol. Prop.: A fungicide (603).
Coordination Deriv.: [Cd(Ph₃AsS)I₂], white crystals, prepd. from CdI₂ + Ph₃AsS in warm Et₂O. IR spectrum (1151).

[Co(Ph₃AsS)₂Br], solid substance, prepd. from CoBr₂ + Ph₃AsS in hot EtOH-HC(OEt)₃. Electronic and IR spectra (237, see also 1151).

[Co(Ph₃AsS)₄](ClO₄)₂, solid substance, prepd. from Co(ClO₄)₂ + Ph₃AsS in hot EtOH-HC(OEt)₃. Electronic and IR spectra; relatively unstable in MeNO₂ soln. (237).

[Cu(Ph₃AsS)Br]ₓ, cryst. substance, stable, prepd. from CuBr₂ in alc. soln. by addn. of Ph₃AsS (1151, 1152). IR spectrum (1151).

[Au(Ph₃AsS)Cl], colorless crystals, m. 155° (dec.), prepd. from Ph₃AsS with freshly reduced NH₄[AuCl₄] soln. (851) and from Ph₃AsS with NaAuCl₄ (1:1) in EtOH or MeCN, a crude prod. was obtd. (1151).

[Hg(Ph₃AsS)Br₂], colorless crystals, m. 238° (dec.) (851), white crystals, prepd. from HgBr₂ + Ph₃AsS in boiling EtOH. IR spectrum (1151).

[Hg(Ph$_3$AsS)Cl$_2$], colorless crystals, m. 228° (dec.) (851), white needles, prepd. from HgCl$_2$ + Ph$_3$AsS in boiling EtOH. IR spectrum (1151).

[Hg(Ph$_3$AsS)I$_2$], yellow crystals, m. 198° (dec.) (851), white yellow crystals, prepd. from HgI$_2$ + Ph$_3$AsS in boiling MeOH. IR spectrum (1151).

[Pd(Ph$_3$AsS)$_2$Br$_2$], orange brown crystals, a nonelectrolyte in Me$_2$CO (1152), prepd. from Na$_2$[PdCl$_4$] + LiBr in EtOH, followed by Ph$_3$AsS in boiling EtOH. Electronic spectrum (1151).

[Pd(Ph$_3$AsS)$_2$Cl$_2$], brown solid, m. 182° (dec.), prepd. from Ph$_3$AsS in Me$_2$CO by adding PdCl$_2$ in 18% HCl (851) and from Ph$_3$AsS and Na$_2$[PdCl$_4$] in boiling EtOH (1151, see also 1152). Electronic spectrum (1151).

[Pd(Ph$_3$AsS)$_2$I$_2$], brown crystals, prepd. from Na$_2$[PdCl$_4$] + NaI in EtOH and filtering the soln. into Ph$_3$AsS in boiling EtOH. IR and electronic spectra (1151).

2 PtCl$_2$·3 Ph$_3$AsS, a solid of undetermined structure, prepd. from aq. K$_2$[PtCl$_4$] soln. and Ph$_3$AsS (1:2) in CHCl$_3$ (851).

[Ag(Ph$_3$AsS)$_2$]ClO$_4$, colorless needles (1151, 1152), a 1:1 electrolyte in MeCN (1152), prepd. from AgClO$_4$ + Ph$_3$AsS in boiling EtOH. IR spectrum (1151).

Ethyl-α-naphthylphenylarsine Sulfide,   AsC$_{18}$H$_{17}$S,   Et(1-C$_{10}$H$_7$)PhAsS
Synth.: From Et(1-C$_{10}$H$_7$)PhAs + S in C$_6$H$_6$ at 100°, 95% yield (604, 605).
Prop.: Crystals, m. 109° (604, 605).
Rxn. with: MeI at 50° → Ph$_2$(1-C$_{10}$H$_7$)AsS, Et$_2$(1-C$_{10}$H$_7$)As, and Ph(1-C$_{10}$H$_7$)AsSMe (604).
MeI in C$_6$H$_6$ at 80° → Et$_2$(1-C$_{10}$H$_7$)As, Ph(1-C$_{10}$H$_7$)AsSMe, and Ph$_2$(1-C$_{10}$H$_7$)AsS (604, 605).
EtI at 50° → Ph(1-C$_{10}$H$_7$)AsSEt (604).

(3,5-Di-t-butyl-4-oxyphenyl)diphenylarsine Sulfide Radical,   AsC$_{26}$H$_{30}$OS,

Synth.: The formation of the radical was detected by ESR spectroscopy during dehydrogenation of Ph$_2$[4,3,5-HO(t-Bu)$_2$C$_6$H$_2$]AsS with PbO$_2$ (1104) or with alk. soln. of K$_3$[Fe(CN)$_6$] (1105) in C$_6$H$_6$.
Prop.: Green soln. which rapidly changed to light brown diamagnetic 3,5,3′,5′-tetra-t-butyldiphenoquinone. ESR data of the radical were obtd. (1105).

(One more compound of this class may be found following Table 26 on page 310.)

Table 26. Tertiary Arsine Sulfides

R_3AsS	Prepd. from	Prop. and Rxn.	Ref.
C_9			
MeEtPhAsS (M-364)	$RR'R''As + S$ in C_6H_6 on steam bath	White crystals, m. 56-57°, soly. data, (a)	603
C_{10}			
Et_2PhAsS	$R_2R'As + S$ in C_6H_6 on steam bath	White crystals, m. 44-45°, soly. data, (a-d)	603
$MeEt(4-MeC_6H_4)AsS$	$RR'R''As + S$ in C_6H_6 on steam bath	White crystals, m. 64-65°, soly. data, (a)	603
C_{11}			
$Et(CH_2=CHCH_2)PhAsS$	$RR'R''As + S$ in C_6H_6 on steam bath	Oil, (a)	603
EtPhPrAsS	$RR'R''As + S$ in C_6H_6 on steam bath	White crystals, m. 53-54°, soly. data, (a)	603
C_{12}			
$Ph(CH_2=CHCH_2)_2AsS$	$RR'_2As + S$ in C_6H_6 on steam bath	Oil, soly. data, (a)	603
$PhPr_2AsS$	$RR'_2As + S$ at 100-120°	m. 92-94°, rxn. with AcCl at 80-90° → PhPrAsCl and AcSPr	1548
BuEtPhAsS	$RR'_2As + S$ in C_6H_6 at 100°	m. 89°, (e)	605,607
	$RR'R''As + S$ in C_6H_6 on steam bath	White crystals, m. 49-50°, soly. data, (a)	603
C_{13}			
$Et(n-C_5H_{11})PhAsS$	$RR'R''As + S$ in C_6H_6 on steam bath	Oil, (a)	603
C_{14}			
$PhBu_2AsS$	$RR'_2As + S$ at 100° in C_6H_6	m. 53°, rxn. with RBr at 100° → BuPhAsSR	609
$Et(n-C_6H_{13})PhAsS$	$RR'R''As + S$ in C_6H_6 on steam bath	Oil, (a)	603
C_{15}			
$(+)-EtPh(4-HO_2CC_6H_4)AsS$	$(+)-RR'R''As + S$ in C_6H_6 at 90°	m. 114°, $[\alpha]_D^{20}$ + 19.12°, (f)	602
$(-)-EtPh(4-HO_2CC_6H_4)AsS$	$(-)-RR'R''As + S$ in C_6H_6 at 90°	m. 112.6°, $[\alpha]_D^{20}$ - 6.93°, (f)	602
$Et(n-C_7H_{15})PhAsS$	$RR'R''As + S$ in C_6H_6 on steam bath	Oil, (a)	603
C_{16}			
$Ph(C_5H_{11})_2AsS$	$RR'_2As + S$ at 100-120°, (g)	White, odorless crystals, m. 50-52°, rxn. with EtI at 100° → $Ph(C_5H_{11})AsI$	1549
$Et(n-C_8H_{17})PhAsS$	$RR'R''As + S$ in C_6H_6 on steam bath	Oil, (a)	603

C_{17}

Compound	Method	Properties	Ref.
$EtPh(n\text{-}C_9H_{19})AsS$ (M-365)	$RR'R''As + S$ in C_6H_6 on steam bath	Oil, (a)	603
$Et(C_8H_{17})(4\text{-}MeC_6H_4)AsS$	$RR'R''As + S$ in C_6H_6 on steam bath	Oil, (a)	603

C_{18}

Compound	Method	Properties	Ref.
$Ph(C_6H_{13})_2AsS$	$RR'_2As + S$ at 100-120°, (g)	White odorless crystals, m. 59-61°, rxn. with AcCl at 80-90° → $Ph(C_6H_{13})AsCl$ and $AcSC_6H_{13}$	1549

C_{20}

Compound	Method	Properties	Ref.
$Ph(C_7H_{15})_2AsS$	$RR'_2As + S$ at 100-120°, (g)	White odorless crystals, m. 41-42°	1549

C_{22}

Compound	Method	Properties	Ref.
$Ph_2(1\text{-}C_{10}H_7)AsS$	$R_2R'As + S$ in C_6H_6 under reflux By-prod. from $EtPh(1\text{-}C_{10}H_7)As + MeI$ at 50° or in C_6H_6 at 80°	m. 145-147° m. 177°	1347 604, 605

C_{24}

Compound	Method	Properties	Ref.
$Ph_2(2\text{-}C_{10}H_7)AsS$	$R_2R'As + S$ in C_6H_6 under reflux	m. 155-157°	1347
$Ph(C_8H_{17})_2AsS$	$RR'_2As + S$ at 100-120°, (g)	White odorless crystals, m. 38-41°	1549
$Ph(C_9H_{19})_2AsS$	$RR'_2As + S$ at 100-120°, (g)	White odorless crystals, m. 50-52°	1549

C_{26}

Compound	Method	Properties	Ref.
$Ph_2(1\text{-}C_{14}H_9)AsS$, (h)	$R_2R'As + S$ in C_6H_6 under reflux	m. 228-230°	1347
$Ph_2(2\text{-}C_{14}H_9)AsS$, (h)	$R_2R'As + S$ in C_6H_6 under reflux	m. 194-196°	1347
$[Ph_2As(S)CH_2\text{-}]_2$ (M-366)	$(Ph_2AsCH_2\text{-})_2 + S$ in C_6H_6 under reflux	m. 200-203°	1347
$Ph(C_{10}H_{21})_2AsS$	$RR'_2As + S$ at 100-120°, (g)	White odorless crystals, m. 54-55°	1549

C_{34}

Compound	Method	Properties	Ref.
$2,6\text{-}C_{10}H_6[As(S)Ph_2]_2$	$2,6\text{-}C_{10}H_6(AsPh_2)_2 + S$ at 140°, (i)	m. 240-242°	1347
$2,7\text{-}C_{10}H_6[As(S)Ph_2]_2$	$2,7\text{-}C_{10}H_6(AsPh_2)_2 + S$ at 140°, (i)	m. 162-164°	1347

(a) A fungicide (603). (b) Intramolec. rearrangement at 240° in a sealed tube or at 200° in vacuum → EtPhAsSEt (607). (c) Rxn. with RBr → EtPhAsSR (605, 607, 609). (d) Rxn. with EtI at 100° → EtPhAsI (1549). (e) Rxn. with RBr at 100° → PhPrAsSR (605, 607). (f) Rxn. with RBr → the $Ph(4\text{-}HO_2CC_6H_4)AsSR$ isomers with inverted rotation (602). (g) In a sealed tube. (h) $C_{14}H_9$ = anthryl. (i) Followed by extn. with Na_2S soln. and recrystn. of the residue from C_6H_6.

(3,5-Di-t-butyl-4-hydroxyphenyl)diphenylarsine Sulfide, $AsC_{26}H_{31}OS$,
   $Ph_2[4,3,5-HO(t-Bu)_2C_6H_2]AsS$

Synth.: From $Ph_2AsK + 4,3,5-HO(t-Bu)_2C_6H_2I$, followed by hydrolysis and a rxn.
with S (1104).
From $4,3,5-HO(t-Bu)_2C_6H_2AsPh_2$ by melting with excess S at 110-130° (1105).
Prop.: Colorless, hexagonal plates, m. 180-181°, IR spectrum (1104). m. 181-
183°, IR, NMR, and mass spectra (1105).

Rxn. with: $PbO_2 \rightarrow$ the ·O—⟨t-Bu ring t-Bu⟩—As(S)Ph₂ radical (1104).

Hg in dioxan under reflux $\rightarrow 4,3,5-HO(t-Bu)_2C_6H_2AsPh_2$ (1105).
PhLi in THF-Et₂O, no rxn. (1105).
Alk. soln. of $K_3[Fe(CN)_6]$ in $C_6H_6 \rightarrow$ a light brown soln. contg. O=⟨...⟩=O

which is formed via the ·O—⟨t-Bu ring t-Bu⟩—AsPh₂ radical (1105).

2.3.6.3   COMPOUNDS WITH ARSENIC-SELENIUM BONDS

2.3.6.3.6 TERTIARY ARSINE SELENIDES

The arsine selenides were reported in the main volume on page 367.

Triethylarsine Selenide, $AsC_6H_{15}Se$, Et_3AsSe,   (M-367)
Coordination Deriv.: The arsine selenide gave a complex with iodine, m. 68°
(dec.), with the As/I ratios which varied from prepn. to prepn. (1057).

Triphenylarsine Selenide, $AsC_{18}H_{15}Se$, Ph_3AsSe,   (M-367)
Synth.: From Ph_3AsCl_2 by treating with aq. NH_3 satd. with H_2Se (803).
Prop.: White crystals, m. 125-130°, IR spectrum (803).

Tricyclohexylarsine Selenide, $AsC_{18}H_{33}Se$, $(C_6H_{11})_3AsSe$,   (M-367)
Coordination Deriv.: The arsine selenide gave a complex with iodine, m. 160-
161° (dec.), with the As/I ratios which varied from prepn. to prepn. (1057).

2.3.7   COMPOUNDS OF ARSENIC BONDED TO GROUP VA ELEMENTS

2.3.7.1 COMPOUNDS WITH ARSENIC-NITROGEN BONDS

Four types of tertiary arsenic bonded to nitrogen were described in the main
volume, pp. 368-70. Several new $R_3As=NR'$ type compounds are described in this
supplement. Besides the methods for their preparation reported in the main
volume, they have been prepared from tertiary arsines by the oxidation with
bromine, followed by the condensation of the resulting dibromotriorganoarsenic
derivatives with sulfonamides. $R_3As + Br_2 \rightarrow Ph_3AsBr_2 + H_2NO_2SR' \rightarrow Ph_3As=NO_2SR'$.

These compounds have been named "arsenioamide" derivatives.

A few new tertiary arsenanylidenammonium salts, $[R_3AsNR'_2]X$, were prepared by a condensation of tertiary arsines with chloramine and substituted chloramines, respectively. Triphenylarsenanylidenammonium chloride was also obtained from dichlorotriphenylarsenic by a condensation with ammonia at $-78°$ in chloroform.

Three new representatives with a structural formula $[R_3As\text{---}N\text{---}MR_3]X$, where M = As or P, are described hereafter.

The other compounds with the As-N bonds discussed above are compiled in Tables 27 and 28.

N-(Dimethylarsino)-P,P,P-trimethylphosphine Imide Methiodide, $\quad AsC_6H_{18}INP$,
$\quad$ $[Me_3As\text{---}N\text{---}PMe_3]I$
Synth.: From $Me_2AsN=PMe_3 + MeI$ in C_6H_6 (1352, 1353).
Prop.: m. $237°$ (dec.), IR and NMR spectra (1352, 1353).

N-(Diphenylarsino)-As-amino-As,As-diphenylarsine-Imide Phenochloride,
$\quad As_2C_{30}H_{27}ClN_2$, $\quad$ $[Ph_3As\text{---}N\text{---}As(NH_2)Ph_2]Cl$ (?)
Synth.: From a mixt. of $[Ph_3AsNH_2]Cl + [Ph_2As(NH_2)_2]Cl$ by heating in DMF at $150°$ (1421).
From a mixt. of $Ph_3AsCl_2 + Ph_2AsCl_3$ by heating and treating with NH_3 and heating again with DMF at $150°$ (1421).
Prop.: White crystals, IR spectrum (1421).

N-(Diphenylarsino)-As,As,As-triphenylarsine Imide Phenochloride,
$\quad As_2C_{36}H_{30}ClN$, $\quad$ $[Ph_3As\text{---}N\text{---}AsPh_3]Cl$
Synth.: From Ph_3AsCl_2 in $CHCl_3 + NH_3$ at r.t. (1421).
From $[Ph_3AsNH_2]Cl$ by heating in DMF at $150°$ the $[Ph_3As\text{---}N\text{---}AsPh_3]Cl\cdot DMF$ complex was obtd. (1421).
From $Ph_3AsCl_2 + NH_3$ in $CHCl_3$ at $-78°$, a mixt. of the title compd. with $[Ph_3AsNH_2]Cl$ was obtd. (1421).
Prop.: White solid, m. $255\text{-}260°$; IR spectrum. The complex with DMF, m. $255\text{-}260°$, loses DMF during heating in vacuum (1421).

Table 27. Tertiary Arsenioamides, $R_3As=NR'$

$R_3As=NR$	Prepd. from	Prop. and Rxn.	Ref.
C_{20}			
$Me_2PhAs=NP(O)Ph_2$	$Me_2PhAs + Ph_2P(O)N_3$ (1:1) in Py under reflux	Useful as an insecticide, lubricating oil additive, UV stabilizer for polymers, and flame retardant	1583
C_{24}			
$Ph_3As=NO_2SC_6H_4NO_2-3$	$Ph_3As + Br_2 + 3-O_2NC_6H_4SO_2NH_2$, (a)	m. 141-142°	341
C_{25}			
$(4-ClC_6H_4)_3As=NO_2SC_6H_4Me-4$	$(4-ClC_6H_4)_3As$ + chloramine T in THF under reflux	m. 238-240°, (b)	723
$Ph_3As=NO_2SC_6H_4Me-4$ (M-369)	$Ph_3As + 4-MeC_6H_4SO_2NSO$ in C_6H_6 under reflux	Crystals, m. 199-201°	1385
C_{27}			
$(4-MeC_6H_4)_3As=NO_2SC_6H_4NO_2-3$	$(4-MeC_6H_4)_3As + Br_2 + 3-O_2NC_6H_4-SO_2NH_2$, (a)	m. 169-170°	341
$(4-MeOC_6H_4)_3As=NO_2SC_6H_4NO_3-3$	$(4-MeOC_6H_4)_3As + Br_2 + 3-O_2NC_6H_4-SO_2NH_2$, (a)	m. 198-199°	341
$(4-MeC_6H_4)_3As=NO_2SPh$	$(4-MeC_6H_4)_3As + Br_2 + PhSO_2NH_2$, (a)	m. 159-160°	341
C_{28}			
$(4-MeC_6H_4)_3As=NO_2SC_6H_4Me-4$ (M-369)	$(4-MeC_6H_4)_3As + Br_2 + 4-MeC_6H_4-SO_2NH_2$, (a)	m. 182-183°	341
	$(4-MeC_6H_4)_3As$ + chloramine T (1:1) in THF	m. 184-186°, (c)	723
$(4-MeOC_6H_4)_3As=NO_2SC_6H_4Me-4$	$(4-MeOC_6H_4)_3As + Br_2 + 4-MeOC_6H_4-SO_2NH_2$, (a)	m. 165-166°	341
C_{30}			
$Ph_3As=NP(O)Ph_2$	$Ph_3As + Ph_2P(O)N_3$ (1:1) in xylene under reflux	(d)	1583

C34

1,4-[Ph$_2$P(S)N=AsMe$_2$]$_2$C$_6$H$_4$	1,4-C$_6$H$_4$(AsMe$_2$)$_2$ + Ph$_2$P(S)N$_3$ (1:2) in MePh under reflux	(d)	1583

C38

(Bu$_3$As=N)$_2$Si(C$_6$H$_4$OMe-4)$_2$ (M-369)	Bu$_3$As + (4-MeOC$_6$H$_4$)$_2$Si(N$_3$)$_2$ (2:1) by heating in MeCN		1586
(Bu$_3$As=N)$_2$Sn(C$_6$H$_4$OMe-4)$_2$	Bu$_3$As + (4-MeOC$_6$H$_4$)$_2$Sn(N$_3$)$_2$ (2:1) in MeCN under reflux	Useful as a UV light stabilizer for polymers	1584

C47

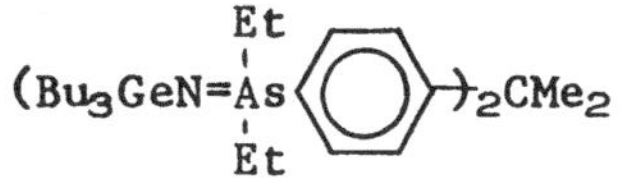 $\left(\text{Bu}_3\text{GeN=As} \overset{\text{Et}}{\underset{\text{Et}}{}} \right)_2\text{CMe}_2$ (M-369)	(4-Et$_2$AsC$_6$H$_4$)$_2$CMe$_2$ + Bu$_3$GeN$_3$ (1:2) in Py under reflux	Useful as a UV light stabilizer for polymers	1584-1586

C90

1,4-[(Ph$_3$As=N)$_2$GePh]$_2$C$_6$H$_4$ (M-369)	Ph$_3$As + NaN$_3$ (1:4) in MePh under reflux and adding slowly 1,4-(PhGeCl$_2$)$_2$C$_6$H$_4$	Useful as a UV light stabilizer for polymers	1584,1585

Polymers

{P(S)(Me)-C$_6$H$_4$-P=N-As-C$_6$H$_4$-As=N}$_n$ with NMe$_2$ substituents	1,4-C$_6$H$_4$[As(NMe$_2$)$_2$]$_2$ + 1,4-C$_6$H$_4$-[P(S)(Me)N$_3$]$_2$ (1:1) in Py under reflux	(d)	1583
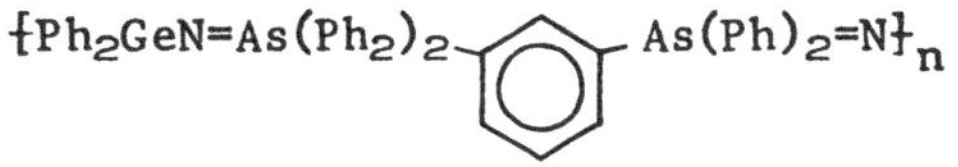 {Ph$_2$GeN=As(Ph)$_2$-C$_6$H$_4$-As(Ph)$_2$=N}$_n$	1,3-C$_6$H$_4$(AsPh$_2$)$_2$ + Ph$_2$Ge(N$_3$)$_2$ (1:1) in Py under reflux	Oxidatively, thermally, and hydrolytically stable polymer, (d)	1584-1586

(a) The tertiary arsine is treated with Br$_2$ in CCl$_4$ to form the corr. R$_3$AsBr$_2$ and the latter is converted with the appropriate arenesulfonamide in CCl$_4$ in the presence of Et$_3$N. (b) Rxn. with 4-ClC$_6$H$_4$Li, followed by hydrolysis → (4-MeC$_6$H$_4$)$_5$As (723). (c) Rxn. with 4-MeC$_6$H$_4$Li, followed by hydrolysis → (4-MeC$_6$H$_4$)$_5$As + (4-MeC$_6$H$_4$)$_4$As$^+$ (723). (d) Useful as an insecticide, lubricating oil additive, UV light stabilizer for polymers, and flame retardant (1583).

313

Table 28. Tertiary Aresenanylidenammonium Salts, $[R_3AsNR'_2]X$

$[R_3AsNR'_2]X$	Prepd. from	Prop.	Ref.
C$_3$			
[Me$_3$AsNH$_2$]Cl	Me$_3$As + H$_2$NCl in Et$_2$O at -78°	m. 136-138°, IR spectrum	1421
C$_5$			
[Me$_3$AsNMe$_2$]Cl		NMR spectrum	1420
C$_8$			
[Me$_2$PhAsNH$_2$]Cl	Me$_2$PhAs + H$_2$NCl in Et$_2$O, (a)	White ppt., m. 175° (dec.), IR and NMR spectra	1420
C$_{10}$			
[Me$_2$PhAsNMe$_2$]Cl	Me$_2$PhAs + Me$_2$NCl (excess) in Et$_2$O at r.t.	White ppt., m. 132-134°, IR and NMR spectra	1420
C$_{13}$			
[MePh$_2$AsNH$_2$]Cl	MePh$_2$As + H$_2$NCl in Et$_2$O, (a)	White ppt., m. 166-170°, (dec.), IR and NMR spectra	1420
C$_{15}$			
[MePh$_2$AsNMe$_2$]Cl	MePh$_2$As + Me$_2$NCl (excess) in Et$_2$O at r.t.	White ppt., m. 131-132°, IR and NMR spectra	1420
C$_{18}$			
[Ph$_3$AsNH$_2$]Cl (M-368)	Ph$_3$As + H$_2$NCl in Et$_2$O at -78° to -10°	White solid, m. 170° (dec.), (c)	1421
	Ph$_3$As + H$_2$NCl-NH$_3$ in C$_6$H$_6$, (b)	(d)	1421
	Ph$_3$AsCl$_2$ + NH$_3$ in CHCl$_3$ at -78°, (e)	IR spectrum	1421

(a) At the temp. of liq. N$_2$ and warming to r.t. under vacuum. (b) [Ph$_2$As-(NH$_2$)$_2$]Cl is formed as by-prod. (c) On heating with DMF at 150° → [Ph$_3$As^{---}N^{---}AsPh$_3$]Cl·DMF; by mixing Ph$_3$AsCl$_2$ with NH$_3$ → [Ph$_3$As^{---}N^{---}AsPh$_3$]Cl (1421). (d) At 150° probably dec. → [Ph$_3$As=N-As(NH$_2$)Ph$_2$]Cl (1421). (e) [Ph$_3$As^{---}N^{---}AsPh$_3$]Cl is formed as by-prod. (1421).

2.3.7.2 COMPOUNDS WITH ARSENIC-PHOSPHORUS BONDS

(Dichlorophosphinyl)triphenylarsonium Chloride, $AsC_{18}H_{15}Cl_3OP$, [Ph$_3$AsP(O)Cl$_2$]Cl

<u>Synth.</u>: From Ph$_3$As + POCl$_3$ under reflux, 60% yield (698).
<u>Prop.</u>: Colorless powder, thermally unstable, relatively stable to moisture, and readily sol. in polar solvents (698).

2.3.8   COMPOUNDS WITH ARSENIC BONDED TO METALS

2.3.8.1.2 COMPOUNDS WITH GROUP IIB METALS

(Methylmercuri)triphenylarsonium Perchlorate, $AsHgC_{19}H_{18}ClO_4$,
    $[Ph_3AsHgMe]ClO_4$
Synth.: From Ph_3As + $MeHgClO_4$ in Me_2CO-H_2O (368).
Prop.: m. 224-225° (368).

2.4   SECONDARY DERIVATIVES

2.4.2 SECONDARY ARSONIUM SALTS

   Two new arsonium salts have been described which contain two organic ligands
and at least one hydrogen atom covalently bonded to the arsenic atom.

Dimethylarsonium Chloride, AsC_2H_8Cl, $[Me_2AsH_2]Cl$
Synth.: From Me_2AsH by the rxn. with HCl at 104° (414).
Prop.: A solid, m. -46°, dissoc. at 20°, vapor pressure data (414).

Dimethyl(trimethylsilyl)arsonium Bromide, $AsSiC_5H_{16}Br$, $[Me_2(Me_3Si)AsH]Br$
Synth.: From $Me_2AsSiMe_3$ + HBr at -96° (1309).
Prop.: Dec. > -50° → Me_2AsH + Me_3SiBr (1309).

2.4.5   COMPOUNDS WITH ARSENIC-HALOGEN BONDS

2.4.5.1 TRIHALO DERIVATIVES

   Secondary trihaloarsenic derivatives were described in the main volume on
page 371. Besides the methods previously reported, the trihaloarsenic derivatives are formed from tertiary arsines by oxidative cleavage with halogens
and from aminohalodiarylarsonium salts by a reaction with hydrogen chloride.

Trifluorobis(trifluoromethyl)arsenic, AsC_2F_9, $(CF_3)_2AsF_3$
Rxn. with: ScF → $Sc[(CF_3)_2AsF_4]$ (323).
Complex Salts: $Cs[(CF_3)_2AsF_4$, white solid, prepd. from $(CF_3)_2AsF_3$ + CsF in
MeCN at 25°. IR and NMR spectra (323).

$Ag[(CF_3)_2AsF_4]$, white solid, turned grey on exposure to light, prepd. from
$Ag[(CF_3)_2AsO_2]$ + SF_4 at 25°. IR spectrum (323).

Trichlorodiphenylarsenic, $AcC_{12}H_{10}Cl_3$, Ph_2AsCl_3,   (M-371)
Synth.: From Ph_2AsCl by treating with Cl in Et_2O at -78° (1421).
From Ph_3As and Cl_2 in C_6H_6 at 20°, a mixt. of Ph_2AsCl_3 and Ph_3AsCl_2 is formed
(1421).
From $[Ph_2As(NH_2)Cl]Cl$ by treating with HCl in Et_2O at 20° (1421).
Prop.: m. 172° (1421).

Rxn. with: NH_3 + NH_4Cl at -35° → $(Ph_2As=N)_4$ (679).
NH_3 → $Ph_3As_2N_2H_3Cl_4$, probably $[Ph_2As(Cl)_2\overset{\cdots}{\,}N\overset{\cdots}{\,}As(Cl)_2Ph]Cl$, $(Ph_2As=N)_3$, and $[Ph_2As(NH_2)\overset{\cdots}{\,}N\overset{\cdots}{\,}As(NH_2)Ph_2]Cl$ (1421).
NH_3 in $CHCl_3$ at -78° → $(Ph_2As=N)_3$ (1421).

Di(2-biphenylyl)trichloroarsenic, $AsC_{24}H_{18}Cl_3$, $(2-PhC_6H_4)_2AsCl_3$, (M-371)
Synth.: From $(2-PhC_6H_4)_2AsCl$ in CCl_4 by bubbling Cl and removing the solvent at 250°/12 mm (722).
Prop.: Gray mass, sol. in H_2O (722).
Rxn. with: Aq. KI → 5,5'-spirobi(dibenzarsolium)iodide (722).

2.4.5.3 HALOARSINE OXIDES

Chlorodiphenylarsine Oxide, $AsC_{12}H_{10}ClO$, $Ph_2As(O)Cl$
Synth.: From $Ph_2As(O)OR$ + $SOCl_2$ at 80°, 92% yield (870).

2.4.5.6 AMINO-HALO-ARSONIUM SALTS AND RELATED COMPOUNDS

Aminochlorodiphenylarsonium Chloride, $AsC_{12}H_{12}Cl_2N$, $[Ph_2(H_2N)ClAs]Cl$
Synth.: From Ph_2AsCl + H_2NCl in Et_2O at -78° (1421).
Prop.: m. 150-180° (dec.), IR spectrum. When heated in DMF to 150° → a substance m. 120°, corr. to the empirical formula $Ph_3As_2N_2H_3Cl_4$, and $[Ph_2As(Cl)\overset{\cdots}{\,}N\overset{\cdots}{\,}As(Cl)Ph_2]Cl$ (1421).
Rxn. with: HCl in Et_2O at 20° → Ph_2AsCl_3 (1421).

As-Chloro-N-(chlorophenylarsino)-As,As-diphenylarsine Imide Phenochloride,
    $As_2C_{24}H_{20}Cl_3N$, $[Ph_2As(Cl)\overset{\cdots}{\,}N\overset{\cdots}{\,}As(Cl)Ph_2]Cl$
Synth.: From $[Ph_3As(NH_2)]Cl$ by heating in DMF at 150° (1421).
From $[Ph_2(H_2N)AsH]Cl$ by heating in DMF at 150° (1421).
Prop.: m. 178-180°, IR spectrum (1421).

2.4.6   COMPOUNDS WITH ARSENIC BONDED TO GROUP VIA ELEMENTS

2.4.6.1   COMPOUNDS WITH ARSENIC-OXYGEN BONDS

2.4.6.1.1 ARSINIC ACIDS

   The methods of synthesis of arsinic acids, presently called hydroxyarsine oxides, were summarized in the main volume, pp. 372-373. In addition to the methods reported previously, the compounds have been prepared from arsinous and thioarsinous acid esters by the oxidation with hydrogen peroxide, from arsinic acid esters by hydrolysis, and from dichloroarsinodimethylamine by the reaction with Grignard reagents at a molar ratio of 1:2, followed by hydrolysis and oxidation with hydrogen peroxide.

Di(fluoroaryl)arsinic acids have been prepared from arenediazonium hexa-fluoroarsenates, with unsubstituted para position of the arene moieties, by thermal decomposition followed by hydrolysis. Di(4-fluorophenyl)arsenic acid was also prepared by a condensation of fluorobenzene with arsenic penta-fluoride, followed by hydrolysis.

In the synthesis of aliphatic arsinic acids from sodium arsenite, Na_3AsO_3, in aqueous solutions by the reaction with alkyl bromides, the reaction is im-proved by maintaining the temperature 20° below the boiling point of the alkyl bromide, but never higher than 100°, while stirring the rxn. mixture, and in the case of long-chain alkyl bromide adding up to 40% methanol (1230).

A decomposition of a diazonium salt of arsenosoaniline with alkali in the presence of $CuSO_4$ gives poly(arsinicophenylene). Moreover, polymeric arsinic acids can be obtained from diarsenoso compounds by a condensation with poly-methylene dihalides and xylylene dihalides, respectively, in the presence of alkali.

Arsinic acids, R_2AsO_2H, suspended in nonpolar solvents react with equimolar quantities of $SOCl_2$ to form addition products $R_2AsO_2H \cdot SOCl_2$ (A). These adducts are quite soluble in organic solvents. They are readily decomposed by water and atmospheric moisture according to the reaction formula: $R_2AsO_2H \cdot SOCl_2 + H_2O \rightarrow R_2AsO_2H + SO_2 + HCl$. The adducts of dialkylarsinic acids decompose at 35-40°, while the diaryl analogs decompose at 80-110°. A thermal decomposition of the dialkylarsinic acid adducts affords $R_2As(O)Cl \cdot HCl$ (B), while that of the diaryl analogs gives $R_2As(O)Cl$ (C). Above 50° (B) decomposes, giving off alkyl chlorides. (B) and (C) react with $SOCl_2$ at 30-35° and 80°, respectively, to form R_2AsCl_3 derivatives (870).

A few arsinic acids and three polymeric derivatives are described indivi-dually hereafter; many more arsinic acids are compiled in Table 29.

Bis(trifluoromethyl)arsinic Acid, Hydroxybis(trifluoromethyl)arsine Oxide,
   $AsC_2HF_6O_2$, $(CF_3)_2AsO_2H$,   (M-373)
Rxn. with: $Ag_2O \rightarrow (CF_3)_2AsO_2Ag$ (323).
Salt: $(CF_3)_2AsO_2Ag$, rxn. with SF_4 at 25° $\rightarrow Ag[(CF_3)_2AsF_4]$ (323).

Dimethylarsinic Acid, Cacodylic Acid, Hydroxydimethylarsine Oxide,
   $AsC_2H_7O_2$, Me_2AsO_2H,   (M-373)
Prop.: Dissocn. consts. in H_2O and in H_2O-MeOH mixt. and thermodynamic data (816). Crystallographic data, a tetrahedral configuration of As was confirmed (1525).
Rxn. with: $SO_2 + HCl$ in the presence of KI $\rightarrow Me_2AsI$ (558).
$Zn + HCl \rightarrow Me_2AsH$ (558, 560).
Ph_3PbCl (using Me_2AsO_2Na) $\rightarrow Ph_3PbOAs(O)Me_2$ (726).
$PCl_3 \rightarrow Me_2AsCl$ (883, 939).
ROH in C_6H_6 under reflux $\rightarrow Me_2As(O)OR$ (959).
Tris(2,4-pentanedionato)chromium at 250° under N $\rightarrow$ a polymer, sol. in $CHCl_3$, along with a solid insol. in $CHCl_3$ but sol. in C_6H_6 (1211).

$Cr(OAc)_2 \cdot H_2O$ in EtOH free of air at r.t., followed by distn. of the solvent and AcOH and oxidn. with atm. O → a viscous polymeric substance contg. a

$$-Cr(O\overset{\overset{\displaystyle Me}{|}}{\underset{\underset{\displaystyle Me}{|}}{As}}O)_2-$$ backbone (1327).

Biol. Prop.: Herbicidal effect on rape and millet (100) and on blackjack oak (218). A defoliant for living cotton and soybean plants (411, 1149). Defoliation of trees after injection of the acid soln. into growing stems; injections of subphytotoxic concn. produce malformed influorescence and other abnormalities at later growth stages (1455). Effective in control of jack pine trees, red maple, and other trees (451). Topkill effect in herbicidal application to sod-planted corn and sorghum (860). Phytotoxic effect on bearing and nonbearing peaches after a postemergent weed control application (1423). Insecticidal effect on bark beetle in trees after injection (325). Activation of a plant invertase from the root nodules of Lupinus luteus (848). Antiprotozoal action, in vitro, on bovine ruminal protozoa for control of bloat (1614).

Uses: A catalyst for the esterification of H_3PO_4 (190). Control of the production of seeds of yellow foxtail, goose grass, and crowfoot grass; the most effective application is at the boot stage (1501). A post-emergence application has no ill effects on 8 cotton varieties (1379a). Herbicidal prepns. in the granular form or liquid sprays contg. Me_2AsO_2H or its Na salt kill weeds and certain grasses, e.g., coarse fescue and Kentucky blue grass (1381). Granular prepn. contg. Me_2AsO_2H or its Na salt after 2 applications in 7 days gave 100% mortality of Nimblewill and Canadian bluegrass; after 7 days the turf is suitable for reseeding with Kentucky bluegrass (1382). A renovation of pastures by killing unproductive sods with the acid and drilling desirable grasses into the dead sod, without tillage (1441). Prepns. contg. the acid, monophosphate esters of ethoxylated alcohols, and a wetting agent in aq. soln. are used as pre- and post-emergent herbicides (1448).

Salts: Me_2AsO_2Na and Me_2AsO_2K exposed to β- or γ-radiation produce a blue color in a modified Fiske-Subba Row test; the color intensity is proportional to the radiation dose and can be used in radiation dosimetry (1275).

The sodium salt is a potential carcinogen in prolonged medicinal applications (769).

$Me_2AsO_2GaMe_2$, prepd. from Me_2AsO_2H in C_6H_6 soln. + Me_3Ga, m. 144-145°, $subl_{0.01}$ 110°; IR spectrum. A dimeric structure with an 8-membered ring is postulated (368a).

$(Me_2AsO_2)_2Zn$, prepd. from Me_2AsO_2H + $ZnSO_4$ in EtOH, polymerized in C_6H_6 soln. (348).

Dipropylarsinic Acid, Hydroxydipropylarsine Oxide, $AsC_6H_{15}O_2$,

Pr_2AsO_2H,   (M-378)

Synth.: From Pr_3As or its soln. by exposure to the atm. oxygen or to excess H_2O_2, 3% yield (1057).

Prop.: Crystals, m. 134-135.5° (1057). IR spectrum interpretation (1223).
Rxn. with: PhC≡CH → Pr$_2$AsC≡CPh (911).
Tris(2,4-pentanedionato)chromium at 250° under N → a polymeric prod. with a
high thermal stability (1211).
Salts: The Fe^{3+}, Zn^{++}, UO$_2$$^{2+}$, and Al^{3+} salts are extractable with CHCl$_3$ from
H$_2$O at various pH values (1233).

Butylpropylarsinic Acid, Hydroxybutylpropylarsine Oxide,   AsC$_7$H$_{17}$O$_2$,
     BuPrAsO$_2$H
Synth.: From BuAsCl$_2$ + PrBr, followed by hydrolysis (1233).
Rxn. with: PhC≡CH → BuPrAsC≡CPh (911).
Salts: The pH values were detd. for the pptn. of Be, Mg, Al, Y, La, Ce, Ti^{4+},
Zr, Th^{4+}, Cr^{3+}, UO$_2$$^{++}$, Mn^{++}, Fe^{++}, Fe^{3+}, Co^{++}, Ni^{++}, Pd^{++}, Cu^{++}, Zn, Cd, In,
Sn^{++}, Pb^{++}, and Bi^{3+}. The salts of Fe^{3+}, Zn, UO$_2$$^{++}$, Al, Cu^{++}, Pb^{++}, Be, and
Bi^{3+} are extractable with CHCl$_3$ (1233).

Dibutylarsinic Acid, Dibutylhydroxyarsine Oxide,   AsC$_8$H$_{19}$AsO,
     Bu$_2$AsO$_2$H,   (M-375)
Synth.: From Bu$_3$As or its soln. by exposure to air or H$_2$O$_2$ (1057).
Prop.: Crystals, m. 139°, IR spectrum (1057).
Rxn. with: PhC≡CH → Bu$_2$AsC≡CPh (911).
Use: A catalyst for the esterification of H$_3$PO$_4$ (190).
Salts: (Bu$_2$AsO$_2$)$_2$Zn, prepd. from Bu$_2$AsO$_2$Na + ZnSO$_4$ (2:1) in H$_2$O, forms a
polymer in C$_6$H$_6$ soln.; viscosity data (348).

Butylpentylarsinic Acid, Butylhydroxypentylarsine Oxide,   AsC$_9$H$_{21}$O$_2$,
     Bu(n-C$_5$H$_{11}$)AsO$_2$H
Synth.: From BuAsCl$_2$ + n-C$_5$H$_{11}$Br, followed by hydrolysis (1233).
Rxn. with: PhC≡CH → Bu(n-C$_5$H$_{11}$)AsC≡CPh (911).
Salts: The pH values were detd. for the pptn. of Be, Mg, Al, Y, La, Ce, Ti^{4+},
Zr, Th^{4+}, Cr^{3+}, UO$_2$$^{++}$, Mn^{++}, Fe^{++}, Fe^{3+}, Co^{++}, Ni^{++}, Pd^{++}, Cu^{++}, Zn, Cd, In,
Sn^{++}, Pb^{++}, Bi^{3+}, and V^{5+}. The Fe^{3+}, Zn, UO$_2$$^{++}$, Al, Cu^{++}, Pb^{++}, Be, Bi^{3+}, Cd,
and Co^{++} salts are extractable with CHCl$_3$ (1233).

Butyl-n-heptylarsinic Acid, Butyl-n-heptylhydroxyarsine Oxide,   AsC$_{11}$H$_{25}$O$_2$,
     Bu(n-C$_7$H$_{15}$)AsO$_2$H
Synth.: From BuAsCl$_2$ + n-C$_7$H$_{11}$Br, followed by hydrolysis (1233).
Salts: The pH values were detd. for the pptn. of Be, Mg, Al, Y, La, Ce, Ti^{4+},
Zr, Th^{4+}, V^{5+}, Cr^{3+}, UO$_2$$^{++}$, Mn^{++}, Fe^{++}, Fe^{3+}, Co^{++}, Ni^{++}, Pd^{++}, Cu^{++}, Zn, In,
Sn^{++}, Pb^{++}, and Bi^{3+}. The following salts are extractable with CHCl$_3$: Fe^{3+},
Zn, UO$_2$$^{++}$, Al, Cu^{++}, Pb^{++}, Be, Bi^{3+}, Cd, and Co^{++} (1233).

Diphenylarsinic Acid, Hydroxydiphenylarsine Oxide,   AsC$_{12}$H$_{11}$O$_2$,
     Ph$_2$AsO$_2$H,   (M-376)
Synth.: From Ph$_3$As or its soln. by atm. oxygen or excess H$_2$O$_2$ (1057).
From Ph$_2$AsSR by hydrolysis with H$_2$O in the presence of O or with H$_2$O$_2$ (1184).
Prop.: Crystals, m. 179-181° (1057), m. 173-174° (1184), IR spectrum inter-

pretation (1233). Sepn. from mixtures by paper chromatography and paper electrophoresis (1422).

Rxn. with: Ph_3MCl (M = Ge or Pb) in C_6H_6 in the presence of Et_3N → $Ph_2As(O)$-$OMPh_3$ (1374).

ROH in C_6H_6 under reflux → $Ph_2As(O)OR$ (959).

SO_2 in HCl → Ph_2AsCl (1184).

Transition metal carbonyls → complex salts, see salts below (877).

Biol. Prop.: A selective herbicide for rape (100). A fungicide effective against Alternaria radicina, Bortrytis cinerea, Fusicladium dendriticum, Trichophyton gypseum, and Epidermophyton; lethal to mice (1132, 1133).

Use: A catalyst for the esterification of H_3PO_4 (190).

Salts: $Ph_2As(O)OGePh_3$, m. 178°, prepd. from $Ph_2AsGePh_3$ by oxidn. with H_2O_2 in $EtOH$-H_2O soln., 97% yield; and from Ph_2AsO_2H + Ph_3GeCl + Et_3N in C_6H_6 under reflux (1374).

$Ph_2As(O)OPbPh_3$, m. 280°, prepd. from $Ph_2AsPbPh_3$ by oxidn. with H_2O_2 in $EtOH$-H_2O_2 soln., 92% yield, and from Ph_2AsO_2H + Ph_3PbCl + Et_3N in C_6H_6 under reflux (1374).

$Ph_2As(O)OSnBu_3$ (M-377), a fungicide and bactericide for paint formulations (617).

$Ph_2As(O)OSnPh_3$ (M-377), colorless ppt., m. 323°, prepd. from $Ph_2AsSnPh_3$ in C_6H_6 by the oxidn. with 30% H_2O_2 in $EtOH$, 80% yield (1371); useful as a fungicide and bactericide for paint formulations (617).

$[Ph_2As(O)O]_2SnPh_2$, colorless ppt., m. 330°, prepd. from $(Ph_2As)_2SnPh_2$ in C_6H_6 by treating with 30% H_2O_2 in $EtOH$, 80% yield (1371).

$[Ph_2As(O)O]_3SnPh$, colorless ppt., m. 180° (dec.), prepd. from $(Ph_2As)_3SnPh$ in C_6H_6 by treating with 30% H_2O_2 in $EtOH$, 70% yield (1371).

light green powder, insol. in org. solvents, dec. > 395°, thermogravimetric curve; prepd. from Ph_2AsO_2H + $Cr(CO)_6$ in THF under reflux in UV light, 86% yield (877).

light yellow powder, insol. in org. solvents, dec. > 360°, thermogravimetric curve; the complex salt was prepd. from Ph_2AsO_2H + $Fe(CO)_5$ in THF under reflux in UV light, 85% yield (877).

n = 3-4, orange powder, insol. in EtOH and H_2O, sol. in THF, C_6H_6, and $CHCl_3$, dec. gradually > 150°, thermographic curve; IR spectrum. The complex was prepd. from Ph_2AsO_2H + $W(CO)_6$ in THF under reflux with UV irradiation, 15-32% yield (877).

Phenyl(3-phenylpropyl)arsinic Acid, Hydroxyphenyl(3-phenylpropyl)arsine Oxide, $AsC_{15}H_{17}O_2$, $Ph(PhCH_2CH_2CH_2)AsO_2H$

Synth.: From $PhAsCl_2$ by adding dropwise to a warm soln. of NaOH in H_2O, followed by $Ph(CH_2)_3Br$ and heating under reflux, 36% yield (1513).

Prop.: m. 141-142° (1513).

Rxn. with: H_2SO_4 by heating at 100°, cooling, neutralizing with 30% NaOH, extg. the oily arsinoline oxide deriv. with $CHCl_3$ and reducing with SO_2 in the presence of aq. HCl contg. KI → 1-phenyl-1,2,3,4-tetrahydroarsinoline (1513).

Poly(arsinico-p-phenylene), $(AsC_6H_5O_2)_2$, (M-375)

$$\{p\text{-}C_6H_4\overset{\overset{O}{\|}}{\underset{\underset{OH}{|}}{As}}\}_n,$$

Synth.: From $4\text{-}H_2NC_6H_4AsO$ by diazotization, dec. of the diazonium salt with NaOH and $CuSO_4$ at 1°, and acidification of the rxn. mixt. (739).

Prop.: Black powder, insol in org. solvents and aq. acids, sol. in aq. alkali, gives a H_2O-sol. NH_4 salt (739). Thermally stable and infusible up to 300° (524); during heating gives off some volatile matl.; may be useful as an ion-exchange resin (739). Semiconducting props. (524).

Poly(arsinico-p-phenylene-arsinicoethylene), $(As_2C_8H_{10}O_4)_n$,

$$\{\overset{\overset{O}{\|}}{\underset{\underset{OH}{|}}{As}}\text{-}C_6H_4\text{-}\overset{\overset{O}{\|}}{\underset{\underset{OH}{|}}{As}}\text{-}CH_2CH_2\}_n$$

Synth.: From $p\text{-}C_6H_4(AsO)_2 + BrCH_2CH_2Br$ in the presence of aq. alkali (739).

Prop.: Insol. in org. solvents, sol. in alkali, does not melt up to 380°, above this temp. turns dark (739).

Poly(arsinico-p-phenylenearsinico-p-xylylene), $(As_2C_{14}H_{14}O_4)_n$,

$$\{\overset{\overset{O}{\|}}{\underset{\underset{OH}{|}}{As}}\text{-}C_6H_4\text{-}\overset{\overset{O}{\|}}{\underset{\underset{OH}{|}}{As}}\text{-}CH_2C_6H_4CH_2\}_n$$

Synth.: From $p\text{-}C_6H_4(AsO)_2\cdot$hydrate + $p\text{-}C_6H_4(CH_2Br)_2$ + NaOH in aq. MeOH under reflux (739).

Prop.: Solid, stable up to 320°, shrinks and does not melt < 400°, insol. in org. solvents but sol. in aq. alkali (739).

Table 29. Arsinic Acids

R_2AsO_2H or $RR'AsO_2H$	Prepd. from (Yield, %)	Prop. and Rxn.	Ref.
C$_3$			
$EtMeAsO_2H$	(a)		1149
C$_4$			
Et_2AsO_2H (M-374, 378)	Et_3As in air or by the action of H_2O_2 (4-5)	Crystals, m. 137.5-138°, (a, b, c)	1057
C$_7$			
$MePhAsO_2H$ (M-375)	$MePhAsO_2Bu$ by hydrolysis $Ph(HO_2CCH_2)AsO_2H > 150°$	m. 176°, (d)	958 958
C$_9$			
$2\text{-}MeC_6H_4(HO_2CCH_2)AsO_2H$		IR spectrum interpretation	1223
$4\text{-}MeC_6H_4(HO_2CCH_2)AsO_2H$		IR spectrum interpretation	1223
C$_{10}$			
$(n\text{-}C_5H_{11})_2AsO_2H$	$n\text{-}C_5H_{11}MgBr + Me_2NAsCl_2$ in Et_2O, (e) (50)	Crystals, m. 132°	779
	$(n\text{-}C_5H_{11})_3As$ by exposure to air or to H_2O_2 (3)	Crystals, m. 131°, IR spectrum	1057
$Bu(C_6H_{13})AsO_2H$		Rxn. with $PhC{\equiv}CH \rightarrow Bu(C_6H_{13})AsC{\equiv}CPh$	911
C$_{12}$			
$Ph(2,4\text{-}Cl_2C_6H_3)AsO_2H$	$PhN_2 \cdot AsF_6$ by thermal dec. followed by hydrolysis with NaOH and pptn. with HCl (86)	m. 154-156°, rxn. with SO_2 in HCl $\rightarrow$ $Ph(2,4\text{-}Cl_2C_6H_3)AsCl$	1184
$(4\text{-}FC_6H_4)_2AsO_2H$ (M-380)	$PhF + AsF_5$ under reflux, followed by hydrolysis with NaOH and pptn. with HCl	White solid, m. 138°, NMR spectrum	1384
$(4\text{-}O_2NC_6H_4)_2AsO_2H$ (M-375)		A catalyst for esterification of H_3PO_4	190

Ph(4-O$_2$NC$_6$H$_4$)AsO$_2$H (M-381)	Ph(4-O$_2$NC$_6$H$_4$)AsSPh + H$_2$O$_2$	m. 176-177°	820
(4-H$_2$NC$_6$H$_4$)$_2$AsO$_2$H (M-382)		Addn. of > 0.01% to basal diet of farmyard animals enhances their growth and feed utilization	1609
(n-C$_6$H$_{13}$)$_2$AsO$_2$H	n-C$_6$H$_{13}$MgBr + Me$_2$NAsCl$_2$ in Et$_2$O, (e) (50)	Crystals, m. 131-132°, (f)	779
C$_{13}$			
Ph(2-MeC$_6$H$_4$)AsO$_2$H (M-383)		m. 158-160°, (g)	1184
Ph(4-MeC$_6$H$_4$)AsO$_2$H (M-383)		m. 149-151°, (g)	1184
Ph(4-MeOC$_6$H$_4$)AsO$_2$H		m. 167-169°, (g)	1184
C$_{14}$			
[4,2-F(Me)C$_6$H$_3$]$_2$AsO$_2$H	3-MeC$_6$H$_4$N$_2$·AsF$_6$ by thermal dec., followed by hydrolysis with NaOH and pptn. with HCl (24)	m. 195°, NMR spectrum	1384
[-CH$_2$(Ph)AsO$_2$H]$_2$ (M-385)		IR spectrum interpretation	1223
(n-C$_7$H$_{15}$)$_2$AsO$_2$H	n-C$_7$H$_{15}$MgBr + Me$_2$NAsCl$_2$ (2:1) in Et$_2$O, (e) (69)	Crystals, m. 126-127°	779
C$_{16}$			
(n-C$_8$H$_{17}$)$_2$AsO$_2$H	n-C$_8$H$_{17}$MgBr + Me$_2$NAsCl$_2$ (2:1) in Et$_2$O, (e) (80); n-C$_8$H$_{17}$MgBr + AsCl$_3$ (2:1) in Et$_2$O, (e) (9)	White cryst. flakes, m. 129°	779
C$_{18}$			
(n-C$_9$H$_{19}$)$_2$AsO$_2$H	n-C$_9$H$_{19}$MgBr + Me$_2$NAsCl$_2$ (2:1) in Et$_2$O, (e) (60)	Crystals, m. 126°	779

(a) Defoliation of living cotton and soybean plants (1149). (b) Rxn. with PhC≡CH (1:2) in C$_6$H$_6$ or MePh → Et$_2$AsC≡CPh (911). (c) Rxn. with PCl$_3$ at 25-30° → Et$_2$AsCl (939). (d) Herbicidal effect on rape and millet (100). (e) The Grignard rxn. is followed by hydrolysis with aq. HCl and oxidn. with H$_2$O$_2$ (779). (f) Zn[OAs(O)(C$_6$H$_{13}$)$_2$]$_2$, prepd. from (C$_6$H$_{13}$)$_2$AsO$_2$Na + ZnSO$_4$ (2:1) in H$_2$O, forms a polymer in C$_6$H$_6$ soln., viscosity data (348). (g) Rxn. with SO$_2$ in HCl → RR'AsCl (1184).

Table 29 Continued

R_2AsO_2H or $RR'AsO_2H$	Prepd. from	Prop. and Rxn.	Ref.
C_{20} $(n-C_{10}H_{21})_2AsO_2H$	$n-C_{10}H_{21}MgBr + Me_2NAsCl_2$ (2:1) in Et_2O, (e) (65)	Crystals, m. 127°	779
C_{22} $(n-C_{11}H_{23})_2AsO_2H$	$n-C_{11}H_{23}MgBr + Me_2NAsCl_2$ (2:1) in Et_2O, (e) (70)	Crystals, m. 123-124°	779
C_{24} $(n-C_{12}H_{25})_2AsO_2H$	$n-C_{12}H_{25}MgBr + Me_2NAsCl_2$ (2:1) in Et_2O, (e) (70)	Crystals, m. 125°	779
C_{26} $(n-C_{13}H_{27})_2AsO_2H$	$n-C_{13}H_{27}MgBr + Me_2NAsCl_2$ (2:1) in Et_2O, (e) (60)	Crystals, m. 123-124°	779
C_{28} $(n-C_{14}H_{29})_2AsO_2H$	$n-C_{14}H_{29}MgBr + Me_2NAsCl_2$ (2:1) in Et_2O, (e) (70)	Crystals, m. 124°	779
C_{30} $(n-C_{15}H_{31})_2AsO_2H$	$n-C_{15}H_{29}MgBr + Me_2NAsCl_2$ (2:1) in Et_2O, (e) (60)	m. 123-124°	779
C_{32} $(n-C_{16}H_{33})_2AsO_2H$	$n-C_{16}H_{33}MgBr + Me_2NAsCl_2$ (2:1) in Et_2O, (e) (60)	m. 122-123°	779
C_{34} $(n-C_{17}H_{35})_2AsO_2H$	$n-C_{17}H_{35}MgBr + Me_2NAsCl_2$ (2:1) in Et_2O, (e) (69)	Crystals, m. 120-121°	779
C_{36} $(n-C_{18}H_{37})_2AsO_2H$	$n-C_{18}H_{35}MgBr + Me_2NAsCl_2$ (2:1) in Et_2O, (e) (72)	Crystals, m. 121-122°	779
C_{38} $(n-C_{19}H_{39})_2AsO_2H$	$n-C_{19}H_{39}MgBr + Me_2NAsCl_2$ (2:1) in Et_2O, (e) (61)	Crystals, m. 120°	779
C_{40} $(n-C_{20}H_{41})_2AsO_2H$	$n-C_{20}H_{41}MgBr + Me_2NAsCl_2$ (2:1) in Et_2O, (e) (60)	Fine cryst. solid, m. 119-120°	779

2.4.6.1.2 ARSINIC ACID ESTERS

A few compounds of this class and their methods of synthesis were described in the main volume on page 388. Three more new esters are now reported which were prepared from the acids by the reaction with alcohols in benzene, with azeotropic distillation of water.

Butoxydimethylarsine Oxide, $AsC_6H_{15}O_2$, $Me_2As(O)OBu$
Synth.: From Me_2AsO_2H + BuOH (1:2) in C_6H_6 under reflux, with azeotropic removal of H_2O, theoret. yield (969).
Prop.: b_1 132° (959).

Butoxymethylphenylarsine Oxide, $AsC_{11}H_{17}O_2$, $MePhAs(O)OBu$
Synth.: From $Ph(HO_2CCH_2)AsOH$ + BuOH (1:2) in C_6H_6 under reflux, with azeotropic distn. of H_2O, 85% yield (958).
Prop.: b_2 164°, d_4^{20} 1.1726, n_D^{20} 1.5045 (958).
Rxn. with: $H_2O \rightarrow MePhAsO_2H$ (958).

Butoxydiphenylarsine Oxide, $AsC_{16}H_{19}O_2$, $Ph_2As(O)OBu$
Synth.: From Ph_2AsO_2H + BuOH (1:2) in C_6H_6 under reflux with azeotropic removal of H_2O, ~ 100% yield (959).
Prop.: m. 67° (959).

2.4.7 COMPOUNDS OF ARSENIC BONDED TO GROUP VA ELEMENTS

2.4.7.1 COMPOUNDS WITH ARSENIC-NITROGEN BONDS

Diaminodiphenylarsonium Chloride, $AsC_{12}H_{14}ClN_2$, $[Ph_2(H_2N)_2As]Cl$
Synth.: From Ph_3As + H_2NCl in C_6H_6, a mixt. of the title compd. with $[Ph_3(H_2N)As]Cl$ was obtd. (1421).
Prop.: White solid, IR spectrum (1421).
Rxn.: When the mixt. of the title compd. with $[Ph_3(H_2N)As]Cl$ was heated in DMF at 150°, upon cooling, a white ppt., probably $[Ph_3As=N-As(NH_2)Ph_2]Cl$, was obtd. (1421).

As-Amino-N-(aminophenylarsino)-As,As-diphenylarsine Imide Phenochloride,
$As_2C_{24}H_{24}ClN_3$, $[Ph_2As(NH_2) \cdots N \cdots As(NH_2)Ph_2]Cl$
Synth.: From Ph_2AsCl by treating with $H_2NCl-NH_3$ mixt. and recrystg. from $CHCl_3$ (1421).
Prop.: Crystd. from $CHCl_3$ with 2/3 $CHCl_3$, m. 130-160°, dec. in storage to $(Ph_2AsN)_3$; on heating in DMF $\rightarrow [Ph_2AsN]_3 \cdot DMF$. IR spectrum (1421).
Rxn. with: HCl $\rightarrow Ph_2AsCl_3$ (1421).

2.5 PRIMARY DERIVATIVES

2.5.5.2 COMPOUNDS WITH ARSENIC-HALOGEN BONDS

A few derivatives of this class were described in the main volume on page 389.

Tetrachlorophenylarsenic, $AsC_6H_5Cl_4$, $PhAsCl_4$, (M-389)
Synth.: From $PhAs(O)Cl_2$ + $SOCl_2$ at 60° (870).

Dichlorophenylarsine Oxide, $AsC_6H_5Cl_2O$, $PhAs(O)Cl_2$
Rxn. with: $SOCl_2$ at 60° → $PhAsCl_4$ (870).

2.5.6 COMPOUNDS WITH ARSENIC BONDED TO GROUP VIA ELEMENTS

2.5.6.1 COMPOUNDS WITH ARSENIC-OXYGEN BONDS

2.5.6.1.1 ALIPHATIC ARSONIC ACIDS

The aliphatic and arylaliphatic acids and their methods of synthesis were compiled in the main volume on pages 390-97.

The synthesis of these acids from sodium arsenite, Na_3AsO_3, in aqueous solution by the reaction with alkyl bromides is improved by keeping the reaction temperature 20° below the boiling point of the alkyl bromide used, but never above 100°, very thoroughly stirring the reaction mixture and, in the case of long-chain alkyl bromides, adding up to 40% methanol. In some cases, induction periods of more than 100 hours were observed, but the reaction can be accelerated by raising the temperature within the limits given above.

According to a new method, the acids are prepared from chloroarsylenebis-(diethylamine), $(Et_2N)_2AsCl$, by a condensation with an alkylmagnesium bromide in dry ether under nitrogen, followed by heating under reflux, hydrolysis with 3.2-3.6 N hydrochloric acid, and oxidation with 30% hydrogen peroxide solution.

Methanearsonic Acid, $AsCH_5O_3$, $MeAsO_3H_2$, (M-390)
Synth.: From Na_3AsO_3 + MeI in alk. soln. by heating (339).
From aq. Na_3AsO_3 soln. in a mixt. of MeCOEt and Varnolene (1:9) by injecting MeCl continuously at 60-65°/80 psig. $MeAsO_3HNa$ was obtd. in ~98% yield (1034).
Prop.: m. 154-155°, pK_1 4.58, pK_2 7.82 (339). Raman and IR spectra of the $MeAsO_3^{--}$ ion (1026). Arsenic detn. - analyt. method (652). $MeAs(O)(ONa)_2$ forms complexes with $Hg(OAc)_2$ which are insol. in H_2O and sol. in AcOH. The complex formation is useful in quant. detn. of the arsonate (250). Complexometric detn. of the acid after pptn. with Cd acetate (575). The movement, adsorption, and dec. of $MeAs(O)(ONa)_2$ in various soils was studied (475, 476). The stability of $H_3{}^{14}CAs(O)(OH)ONa$ in nonsteril soil was detd.; in 60 days the degradation reached 1.7 to 10%. In steril soil only 0.7% of the arsonate was

degraded. In pure cultures of soil organisms 3-20% of the arsonate was degraded. The degraded prod. was identified as the arsenate ion (1563).

Rxn. with: ROH under reflux → MeAsO(OR)$_2$ (335, 338).

Ph$_2$PbCl$_2$, used MeAsO(ONa)$_2$ → a compd. of uncertain structure, contg. the repeating {o-Pb(Ph)$_2$-O-As(O)Me}$_n$ units (726).

Biol. Prop. and Uses: The following entries include MeAsO$_3$H$_2$ as well as its salts.

Mixtures of the acid with N-(3,4-dichlorophenyl)methylacrylamide in a ratio of 1:5 to 5:1 produce a synergistic herbicidal effect on Bermuda grass and on weeds in cotton fields (1509).

The di-Na salt in postemergent applications for selected cotton varieties has a herbicidal effect, without any detrimental effect on the fiber quality (1326).

The di-Na salt added to Bosket silt loam in herbicidal concentrations inhibited the growth of cotton, rice, soybean, oats, and corn plants. The phytotoxic effect disappeared in 16 to 32 weeks (1377).

The di-Na and mono-Na salts have a selective herbicidal effect on Johnson grass in the greenhouse and field tests (667, 1050 1380), while very little harm is done to Bermuda grass (1380). The phytotoxic effect of both salts on young sorghum plants, purple nut sedge, and Johnson grass depends on the soil moisture, temp., and pH value (1399).

The mono-Na salt in postemergence herbicidal applications is toxic to bluegrass (1418).

The mono-Na and mono-NH$_4$ salts are effective in secondary postemergent applications in cotton (587).

The di-Na salt has a phytotoxic effect on Paspalum conjugation (1278).

The di-Na salt alone as well as in mixtures with diuron and with morea is effective in secondary postemergent applications in cotton (587). The di-Na salt is phytotoxic to broad leaf and grassy weeds (682).

The di-Na salt alone or with a toxaphene-DDT mixture as well as the mono-Na salt in the tests in cotton fields at a rate of 4.5 lb./acre showed a herbicidal effect and reduced the infestation of suckling insects for a number of days after the application. An increased fruit setting was also observed (1456).

The di-Na salt has a herbicidal effect on Axonopus compressus and Paspalum conjugatum between rows of rubber trees; the activity is influenced by the light intensity (926).

The mono-Na salt is effective in aquatic weed control but permits Bermuda grass to grow (933).

The acid is phytotoxic to Chenoprodium, Senecio, and tomato plants (1155).

The herbicidal activity of the mono-Na and di-Na salts is enhanced by Me_2SO (932).

Mixtures of the mono-Na salt with Cotaran, CP-4397, H-210, H-232, diuron, linuron, and prometryne, resp., are effective postemergent herbicides in the control of young weeds in cotton (588).

Mixtures of the di-Na salt with chloroxuron and diuron, resp., were tested as postemergent herbicides in soybean cultures (589).

Mixtures of the di-Na salt with $p\text{-}ClC_6H_4SCH_2CH_2NEt_2$ have a synergistic and selective herbicidal effect on Johnson grass, wild oats, mustard, crabgrass, pigweed, and watergrass (429).

The di-Na, di-NH$_4$, and mono-Ca salts {$Ca[OAs(OH)Me]_2$} are effective postemergence herbicides for crabgrass (271, 589, 599). A toxicity test of the acid to barley (1122).

The di-Na salt is a weak, oral sterilization agent for adult houseflies (597).

Mixtures of the mono-Na and di-Na salts with 1-(tetrahydrodicyclopentadienyl)-3,3-dimethylurea are effective herbicides for broadleaf and grassy weeds (1267).

The acid has a good prophylactic and curative effect against Xanthomonas oryzae (1493).

Fe(III) methanearsonate is one of the best fungicides for Pellicularia sasakii (1493).

Compositions contg. $MeAsO_3H_2$, $MeAs(O)(OH)ONa$, or $MeAs(O)(OH)ONH_4$, the mono-phosphate esters of ethoxylated alcohols and wetting agents in aqueous solution are used as herbicides for crabgrass and other undesirable grasses (1448).

Herbicidal concentrates contg. $MeAs(O)(OH)ONa$ and surfactants in aqueous solutions are claimed (1624).

Postemergent applications of $MeAs(O)(OH)ONa$ in combinations with various hydrotropes and surfactants as synergistic herbicides for crabgrass (1545).

$MeAsO_3H_2$ is a constituent of strychnoarsine preparations (1154).

Propanearsonic Acid, $AsC_3H_9O_3$, $PrAsO_3H_2$,   (M-392)
Rxn. with: γ-rays $\rightarrow$ $MeCHCH_2AsO_3H_2$ and Et· radicals, ESR spectra (284).
Biol. Prop. and Uses: The Ca salt is toxic to bluegrass in postemergence herbicidal applications (1418).

Compn. contg. $PrAs(O)O_2Ca$, the monophosphate esters of ethoxylated alcohols, and wetting agents are applied as aqueous solutions in pre- and post-emergence herbicidal treatments (1448).

Table 30. Aliphatic Arsonic Acids

$RAsO_3H_2$	Prepd. from (Yield, %)	Prop. and Rxn.	Ref.
C_2			
$EtAsO_3H_2$ (M-392)		m. 94-95°, pK_1 4.72, pK_2 8.00	339
		IR spectrum interpretation, (a)	1223
C_3			
$CH_2=CHCH_2AsO_3H_2$ (M-394)	Na_3AsO_3 + RBr	m. 126-127°, pK_1 4.48, pK_2 7.51	339
		IR spectrum interpretation, (a)	1223
i-$PrAsO_3H_2$ (M-394)	Na_3AsO_3 + RBr	m. 119°, pK_1 4.81, pK_2 8.36, (a)	339
C_4			
$BuAsO_3H_2$ (M-393)	Na_3AsO_3 + RBr	m. 152-153°, pK_1 4.76, (a)	339
	RMgBr + $(Et_2N)_2AsCl$, (b) (50.5)	m. 160-162°, (c, d)	1035
i-$BuAsO_3H_2$ (M-395)	Na_3AsO_3 + RBr	m. 169-170°, pK_1 4.79, pK_2 8.18, (a)	339
C_5			
$C_5H_{11}AsO_3H_2$ (M-395)	RMgBr + $(Et_2N)_2AsCl$, (b) (55)	m. 163-165°	1035
$EtMeCHCH_2AsO_3H_2$ (M-395)	RMgBr + $(Et_2N)_2AsCl$, (b) (90)	m. 190-191°	
C_6			
$C_6H_{13}AsO_3H_2$ (M-396)	RMgBr + $(Et_2N)_2AsCl$, (b) (62)	m. 163-165°	1035

(a). Rxn. with $R'OH \rightarrow RAs(O)(OR')_2$ (335, 338). (b) The rxn. is carried out in dry Et_2O at 0°, completed under reflux and is followed by hydrolysis with 3.6 N HCl and oxidn. of the Et_2O soln. with 30% H_2O_2 (1035). (c) IR spectrum interpretation (1223). (d) Rxn. with Zn-Hg + HCl in $Et_2O \rightarrow RAsH_2$ (1536). (e) The acid lowers the surface potential and surface tension of aq. KCl soln. (829). (f) A mixt. of $C_8H_{17}As(O)(ONH_4)_2$ with $C_{12}H_{25}As(O)$-$(ONH_4)_2$ has a herbicidal effect on crabgrass in postemergence application (271). (g) The condensation is carried out in dry Et_2O at 0° under N and completed under reflux; after cooling to 0°, the rxn. mixt. is hydrolyzed with 3.2 N HCl and then oxidized with 30% H_2O_2.

Table 30 Continued

$RAsO_3H_2$	Prepd. from (Yield, %)	Prop. and Rxn.	Ref.
C_7			
$PhCH_2AsO_3H_2$ (M-396)	Na_3AsO_3 + RCl	m. 180-182°, pK_1 4.43, pK_2 7.51, (a, e)	339
$n-C_5H_{11}CCl=CHAsO_3H_2$		Potential carcinogen in a prolongedq	769
		medicinal application	
$C_7H_{15}AsO_3H_2$ (M-396)	$RMgBr$ + $(Et_2N)_2AsCl$, (b) (55)	m. 157-159°	1035
C_8			
$C_8H_{17}AsO_3H_2$	$RMgBr$ + $(Et_2N)_2AsCl$, (b) (77)	m. 162-163°, (f)	1035
C_9			
$C_9H_{19}AsO_3H_2$	$RMgBr$ + $(Et_2N)_2AsCl$, (b) (80)	m. 157-158°	1035
C_{10}			
$C_{10}H_{21}AsO_3H_2$	$RMgBr$ + $(Et_2N)_2AsCl$, (b) (75)	m. 155-157°	1035
C_{11}			
$C_{11}H_{23}AsO_3H_2$	$RMgBr$ + $(Et_2N)_2AsCl$, (b) (60)	m. 152-153°	1035
C_{12}			
$C_{12}H_{25}AsO_3H_2$	$RMgBr$ + $(Et_2N)_2AsCl$, (b) (82)	m. 151-152°	1035
C_{13}			
$C_{13}H_{27}AsO_3H_2$	$RMgBr$ + $(Et_2N)_2AsCl$, (b) (72)	m. 146-147°, (f)	1035
C_{14}			
$C_{14}H_{29}AsO_3H_2$	$RMgBr$ + $(Et_2N)_2AsCl$, (g) (60)	m. 146-147°	1035
C_{15}			
$C_{15}H_{31}AsO_3H_2$	$RMgBr$ + $(Et_2N)_2AsCl$, (g) (31)	m. 143-145°	1035
C_{16}			
$C_{16}H_{33}AsO_3H_2$	$RMgBr$ + $(Et_2N)_2AsCl$, (g) (68)	m. 142-146°	1035

	Formula	Prep.	m.p.	Ref.
C_{17}	$C_{17}H_{35}AsO_3H_2$	RMgBr + $(Et_2N)_2AsCl$, (g) (52)	m. 138–140°	1035
C_{18}	$C_{18}H_{37}AsO_3H_2$	RMgBr + $(Et_2N)_2AsCl$, (g) (74)	m. 143–144°	1035
C_{19}	$C_{19}H_{39}AsO_3H_2$	RMgBr + $(Et_2N)_2AsCl$, (g) (65)	m. 135–137°	1035
C_{20}	$C_{20}H_{41}AsO_3H_2$	RMgBr + $(Et_2N)_2AsCl$, (g) (51)	m. 137–139°	1035

2.5.6.1.2 AROMATIC ARSONIC ACIDS

The methods of synthesis of aromatic arsonic acids were summarized in the main volume on pages 397-8.

An innovation in the synthesis of the arsonic acids is an electrolytic diazotization of aniline derivatives in the presence of sodium nitrite, a copper salt, and sodium arsenite.

2.5.6.1.2.1 BENZENEARSONIC ACID AND DERIVATIVES WITH ALIPHATIC AND ARYLALIPHATIC SUBSTITUENTS

Benzenearsonic acid and substituted benzene-arsonic acid containing not more than one substituent linked to the benzene ring by an aliphatic carbon atom were described in the main volume on pages 399-408.

Benzenearsonic Acid, $AsC_6H_7O_3$, $PhAsO_3H_2$, (M-399)

Prop.: The acid crystallizes in the form of minute square platelets (1235). Sepn. from mixts. by paper chromatography and paper electrophoresis (1422). Identification of the [74]As-labeled compd. by paper chromatography and electrophoresis (1185). Dissocn. consts. in H_2O (816) and in H_2O-MeOH mixts. were detd. (664, 814, 816). Thermodynamic data of dissocn. in H_2O-MeOH mixt. (816). IR spectrum, qual. interpretation (1223). The acid has a lowering effect on the surface potential and surface tension of aq. KCl soln. (829).

Rxn. with: ROH → $PhAs(O)(OR)_2$ (335, 959). Neutron bombardment → enrichment in [76]As by means of the hot atom effect of n capture (1467, see also 350).

Cu, Zn, Cd, Co, Mn, and Cr, resp. → complex salts extractable with org. solvents from aq. mineral acid solns. (47).

Zr in 2 N HNO_3 contg. 0.075 N $PhAsO_3H_2$ → [95]Zr is present in a radiocolloidal state as shown by the adsorption, centrifuging, dialysis and electrolysis tests (925).

Ta in a soln. contg. a tartrate and complexon at pH 0.7 - 2.5 and 80° → a ppt. (1195a).

Biol. Prop.: Evaluation of the effect on hatching of Heterodera schachtii (366). A reversible and time-dependent inhibition of chymotrypsin, trypsin, and Novo and Carlsberg subtilisins (622). Induction of the abscision of sweet oranges (724). Heterogeneity of immune globuline to $PhAsO_3^{--}$ hapten (763). Genetics of light chains in antibodies to $PhAsO_3H_2$ (764). Induction of antibodies to the acid (659). A preservative effect on spermatozoa in boar semen (1072). Deactivation tests on Adenovirus type 5 gave a negative result (1639).

Uses: Sepn. of Zr from Cs, Sr, Y, and Ce in HNO_3 soln. on activated carbon modified by $PhAsO_3H_2$ (923). Adsorption of Zr from 2 N HNO_3 on polymeric particles treated with $PhAsO_3H_2$ (922). Pptn. of Zr in 2 N HNO_3 and its detn. by a method of distribution dependent on concn. (924). Sepn. of Pa from many elements by adsorption on activated charcoal satd. with $PhAsO_3H_2$ (1117). Adsorption of Zr, Nb, Th, U^{6+}, and Po on activated charcoal satd. with $PhAsO_3H_2$ (1117). Quant. adsorption of Pa, Po, and (to large extent) Zr, Th, Ti, Nb, and U from acid soln. to charcoal satd. with $PhAsO_3H_2$; Pa and Po can be sepd. by desorbing the other elements (1190). Extn. of Pa as benzenearsonate with isoamyl alc. from 0.1 to 9 N acid solns. and back-extn. from the org. phase with oxalic, tartaric, and hydrofluoric acids and H_2SO_4-H_2O_2 soln.- a procedure for the sepn. of Pa from other elements (1116, 1117). Extn. of Zr and Nb as benzenearsonates with org. solvents from acid solns. (1116, 1117). Sepn. of Ta from other metals by pptn. from a soln. contg. tartrate and complexon in the pH range of 0.7 to 2.5 at 80° (1195a). Pptn. of Nb in spectrophotometric detn. of the metal in mild and low alloy steels (1606). Activation of tungstenite for flotation with oxidized petroleum (1032).

Salts: The compn. of Zr benzenearsonate formed from Zr sulfate solns. depends on the acidity, metal concn., and percent pptn. (1556).

Zr benzenearsonate tagged with ^{95}Zr, detn. of the soly. prod. (1011).

Neutron bombardment of Zr benzenearsonate → enrichment in ^{76}As by a factor of 1.2 x 10^4; the radiochem. purity of ^{76}As was 99.88-99.98% (1468).

o-Toluenearsonic Acid, $AsC_7H_9O_3$, $2\text{-}MeC_6H_4AsO_3H_2$, (M-401)
Prop.: Irregularly shaped crystallites (1232). IR spectrum, qual. interpretation (1223).
Salts: The Ca^{++} and Pb^{++} salts are crystalline, while the Sr, Ba, Cu^{++}, Zn, Cd, Co^{++}, Mn^{++}, and Cr^{3+} salts form powdery or flocculent ppts. (1232).

m-Toluenearsonic Acid, $AsC_7H_9O_3$, $3\text{-}MeC_6H_4AsO_3H_2$, (M-401)
Prop.: Crystallizes in long needles (1232). IR spectrum, qual. interpretation (1223).
Salts: The Ca, Pb^{++}, and Cu^{++} salts form cryst. ppts., while the salts of Sr, Ba, Zn, Cd, Co^{++}, Mn^{++}, and Cr^{3+} form flocculent or powdery ppts. (1232).
Biol. Prop.: The Co salt is effective against Ascaridia galli and Heterakis gallinae in poultry (688).

p-Toluenearsonic Acid, $AsC_7H_9O_3$, $4-MeC_6H_4AsO_3H_2$, (M-401)

Prop.: The acid forms needle-shaped crystallites (1232). IR spectrum, qual. interpretation (1223). The acid lowers the surface potential and surface tension of aq. KCl solns. (829).

Rxn. with: $ROH \rightarrow 4-MeC_6H_4As(O)(OR)_2$ (335).

Biol. Prop.: A Co salt of $4-MeC_6H_4AsO_3H_2$ is effective against Ascaridia galli and Heterakis gallinae in poultry (688).

Use:: Flotation of cassiterite, effect on the induction time during the attachment of the mineral particles to air bubbles (184).

Salts: The Ca and Pb^{++} salts form cryst. ppts., while the salts of Sr, Ba, Cu^{++}, Zn, Cd, Co^{++}, Mn^{++}, and Cr^{3+} form flocculent or powdery ppts. (1232).

2.5.6.1.2.2 HALOBENZENEARSONIC ACIDS

The halogen-substituted acids were described in the main volume on pages 408-9.

o-Bromobenzenearsonic Acid, $AsC_6H_6BrO_3$, $2-BrC_6H_4AsO_3H_2$, (M-408)

Synth.: From $2-H_2NC_6H_4AsO_3H_2$ in aq. HBr by diazotization and adding the soln. to aq. HBr contg. Cu^{++} (1659).

Prop.: Yellow crystals (1659).

Rxn. with: SO_2 in concd. HCl contg. KI $\rightarrow 2-BrC_6H_4AsCl_2$ (1659).

2-Chlorobenzenearsonic Acid, $AsC_6H_6ClO_3$, $2-ClC_6H_4AsO_3H_2$, (M-408)

Prop.: Rectangular plates (1232).

Biol. Prop.: The Co salt is effective against Ascaridia galli and Heterakis gallinae in poultry (688).

Salts: The Pb and Cd salts form cryst. ppts., while the salts of Cu, Zn, Co, Mn, and Cr^{3+} form flocculent ppts. (1232).

m-Chlorobenzenearsonic Acid, $AsC_6H_6ClO_3$, $3-ClC_6H_4AsO_3H_2$, (M-409)

Prop.: Needle-shaped crystals (1232).

Salts: The Ca, Pb^{++}, and Co^{++} salts form cryst. ppts., while the Cu^{++}, Mn^{++}, and Cr^{3+} salts are flocculent or powdery ppts. (1232).

p-Chlorobenzenearsonic Acid, $AsC_6H_6ClO_3$, $4-ClC_6H_4AsO_3H_2$, (M-409)

Prop.: Needle-shaped crystals (1232). Dissocn. consts. in H_2O (816) and in H_2O-MeOH mixt. (664, 816). Thermodynamic data of dissocn. in H_2O-MeOH mixt. (816). Soly. and standard chem. potentials of the acid and its monoanion in H_2O-MeOH mixt. (815).

Salts: The Ca and Pb^{++} salts form cryst. ppts., while the salts of Cu^{++}, Zn, Cd, Co^{++}, Mn^{++}, and Cr^{3+} form powdery or flocculent ppts. (1232).

p-Iodobenzenearsonic Acid, $AsC_6H_6IO_3$, $4-IC_6H_4AsO_3H_2$, (M-409)

Biol. Prop.: Binding consts. of anti-p-azobenzenearsonate antibody fractions (856). A Co salt of the acid is effective against Ascaridia galli and Heterakis gallinae in poultry (688).

2.5.6.1.2.3 HYDROXYBENZENEARSONIC ACIDS AND O-SUBSTITUTED DERIVATIVES

The arsenic derivatives of this type were described in the main volume on pages 410-14.

p-Hydroxybenzenearsonic Acid, $AsC_6H_7O_4$, $4-HOC_6H_4AsO_3H_2$, (M-410)
Prop.: Dissocn. consts. in H_2O-MeOH mixts. (664, 816). Thermodynamic data of dissocn. in H_2O-MeOH (816). Identification of the acid tagged with ^{74}As by paper chromatography and electrophoretic methods (1185).
Rxn. with: ROH in C_6H_6 under reflux → $4-HOC_6H_4As(O)(OR)_2$ (959).
Biol. Prop.: Combinations with $MeSO_3Et$ have a synergistic effect on the chromosome aberrations of Hordeum sativum (1092, 1094).

2.5.6.1.2.5 NITROGEN-SUBSTITUTED BENZENEARSONIC ACIDS

2.5.6.1.2.5.1 NITROSO- AND NITRO-BENZENEARSONIC ACIDS

o-Nitrobenzenearsonic Acid, $AsC_6H_6NO_5$, $2-O_2NC_6H_4AsO_3H_2$, (M-419)
Synth.: From $2-O_2NC_6H_4NH_2$ by electrolysis at a Pt or C electrode in the presence of $NaNO_2$, a Cu salt, and $NaAsO_2$, ~ 73% and 69% yield, resp., was obtd. The prod. was purer than from a chem. rxn. (1633).
From $2-O_2NC_6H_4NH_2$ in 5 N HCl by diazotization, adding the diazonium salt soln. to a soln. of As_2O_3 in aq. Na_2CO_3 contg. $CuSO_4$ at 5°, slowly warming up and acidifying with HCl, 93-97% yield (1659).
Prop.: Pale yellow crystals, m. 228-232° (1659). Dissocn. consts. in H_2O (816) and in H_2O-MeOH mixt. (664, 816). Thermodynamic data of dissocn. in H_2O-MeOH mixt. (816). The acid lowers the elec. surface potential of H_2O (830).
Rxn. with: ROH in C_6H_6 under reflux → $2-O_2NC_6H_4As(O)(OR)_2$ (959).
Electrolytic redn. in alk. soln. at a Cu cathode and Cu powder as catalyst → $2-H_2NC_6H_4AsO_3H_2$ (1631, 1632).
Fe in aq. NaCl soln. under reflux → $2-H_2NC_6H_4AsO_3H_2$ (1659).
Biol. Prop.: Immunochem. tests and antibodies formation (659). Deactivation of Adenovirus type 5 gave a negative result (1639).

m-Nitrobenzenearsonic Acid, $AsC_6H_6NO_5$, $3-O_2NC_6H_4AsO_3H_2$, (M-419)
Biol. Prop.: Immunochem. tests; antibodies formation (659).

p-Nitrobenzenearsonic Acid, $AsC_6H_6NO_5$, $4-O_2NC_6H_4AsO_3H_2$, (M-420)
Prop.: Dissocn. consts. in H_2O (816) and in H_2O-MeOH mixts. (664, 816). Thermodynamic data of dissocn. in H_2O-MeOH soln. (816). Explosibility data (492). Identification in feeds by extn. with MeOH, followed by TLC (1088). Identification of the acid tagged with ^{74}As by paper chromatographic and electrophoretic methods (1185). The acid increases elec. surface potential of H_2O (830).
Rxn. with: ROH in C_6H_6 under reflux → $4-O_2NC_6H_4As(O)(OR)_2$ (959).
Biol. Prop.: Prevention of histomoniasis in turkey poults (98). A reversible

and time-dependent inhibition of chymotrypsin, trypsin, and Novo and Carlsberg subtilisins (622). A Co salt of the acid is effective against Ascaridia galli and Heterakis gallinae for poultry (688). A growth-promoting agent effective against histomoniasis in hens. After intramuscular injections, the acid was rapidly excreted mainly unchanged, but partly reduced to $H_2NC_6H_4$-AsO_3H_2; after oral administration, the absorption was poor, excretion was slow, and the redn. was more extensive (1078). Immunochemical tests, formation of antibodies (659). A hapten activity for antibodies producing agglutination of human erythrocytes (1409). A study of antibody solutions for the affinity labeling rxn. with p-arsonobenzenediazonium fluoroborate (886).
Use: In diazo-type printing-plate emulsions for an improved adhesion (818).

2.5.6.1.2.5.2 o-ARSANILIC ACID AND N-SUBSTITUTED DERIVATIVES

o-Arsanilic Acid, $AsC_6H_8NO_3$, $2-H_2NC_6H_4AsO_3H_2$,   (M-421)
Synth.: From $2-O_2NC_6H_4AsO_3H_2$ by electrolytic redn. in 3 N NaOH soln., at a Cu cathode with Cu powder a catalyst at r.t., ~ 57% yield (1631); with a Pb electrode at 60-65°, in N NaCl soln. and $FeSO_4 \cdot 7 H_2O$-Fe powder catalyst, 96% yield; with Fe electrode, 90% yield (1632).
From $2-O_2NC_6H_4AsO_3H_2$ by redn. with Fe powder in aq. NaCl soln. under reflux, 67-73% yield (1659).
Prop.: Pale brown needles, m. 149-151° (1659). Identification of the acid tagged with ^{74}As by paper-chromatographic, electrophoretic, and ion exchange methods (1185).
Rxn. with: $NaNO_2$ and coupling with imidazolidine derivs. → the corr. 2-imidazolidinylazobenzenearsonic acids (654, 655).
$NaNO_2$ in acidic media and coupling with 3-bromochromotropic acid → 3-[(o-arsonophenyl)azo]-6-bromo-4,5-dihydroxy-2,7-naphthalenedisulfonic acid (916).
$NaNO_2$ in aq. HBr and dec. in 60% HBr soln. contg. Cu powder under reflux → $2-BrC_6H_4AsO_3H_2$ (1659).
Cyanurocellulose → [N-(o-arsonophenyl)amino]triazinylcellulose (1518).
HCHO → 5,5'-methylenebis(2-aminobenzenearsonic acid) (916).
HCHO + resorcinol → an ion-exchange polymer with a high selectivity for Ca^{++}, Zn^{++}, and Ce^{3+} (937).
Aq. H_2SO_4 → cleavage of the C-As bond (1073).
Cellulose or poly(vinyl alcohol) in the form of fibers or fabrics → hydrophilic complexes (1296).

o-Arsonoanilinomethanesulfonic Acid Monosodium Salt, $AsNaC_7H_9NO_6S$,
     $2-(NaO_3SCH_2NH)C_6H_4AsO_3H_2$
Synth.: From $2-H_2NC_6H_4AsO_3H_2$ in EtOH + HCHO (35%) at 18-20°, followed by 36.5% $NaHSO_3$, 80% yield (990).
Rxn. with: $4-O_2NC_6H_4N_2BF_4$ → $2,5-HO_3SCH_2NH(4-O_2NC_6H_4N=N)C_6H_3AsO_3H_2$ (990).

o-Arsonophenyliminodiacetic Acid, $AsC_{10}H_{12}NO_7$,

$2-H_2O_3AsC_6H_4N(CH_2CO_2H)_2$, (M-422)

Prop.: Half-wave potential of the complexes formed from the acid and metal ions at pH 2.5-2.7 and 9.3-9.4 (936).

N-(2,4,6-Trinitrophenyl)-o-arsanilic Acid, $AsC_{12}H_9N_4O_9$,

$2-[2,4,6-(O_2N)_3C_6H_2NH]C_6H_4AsO_3H_2$

Biol. Prop.: Uncoupling of oxidative phosphorylation in rat liver mitochondria and inhibition of the rxn. of 2,4,6-trinitrobenzaldehyde with albumine (1607).

N-(2,4-Dinitrophenyl)-o-arsanilic Acid, $AsC_{12}H_{10}N_3O_7$,

$2-[2,4-(O_2N)_2C_6H_3NH]C_6H_4AsO_3H_2$

Prop.: m. 260° (dec.) (1607).

Biol. Prop.: Uncoupling of oxidative phosphorylation in rat liver mitochondria and inhibition of the 2,4,6-trinitrobenzaldehyde rxn. with albumin (1607).

4-[N-(o-Arsonophenyl)amino]-6-hydroxy-2-triazinyloxy-Cellulose Derivatives,

Synth.: From partly chlorinated cyanurocellulose by a condensation with o-arsanilic acid Na salt in aq. soln. (1518).

Prop.: Potentiometric titration gave 2 waves, at pH 6.2 and 7.9. The prod. has ion-exchanging props.; it absorbs divalent Cu, Fe, Co, Ni, Zn, Cd, Mn, and Ca ions. Absorption spectra of the metal-resin derivs. at 400-700 mμ indicated the formation of complex salts (1518).

2.5.6.1.2.5.3 m-ARSANILIC ACID AND N-SUBSTITUTED DERIVATIVES

m-Arsanilic Acid, $AsC_6H_8NO_3$, $3-H_2NC_6H_4AsO_3H_2$, (M-424)

Prop.: Identification of the acid tagged with ^{74}As by paper chromatographic, electrophoretic, and ion-exchange methods (1185).

Rxn. with: Cellulose tosylate, cellulose nitrate, and with chlorodeoxycellulose, resp. (320).

2.5.6.1.2.5.4 p-ARSANILIC ACID AND N-SUBSTITUTED DERIVATIVES

p-Arsanilic Acid, p-Aminobenzenearsonic Acid, $AsC_6H_8NO_3$,

$4-H_2NC_6H_4AsO_3H_2$, (M-425)

Synth.: From aniline at 140-155° by slowly adding arsenic acid (2.5:1), removing excess aniline under a reduced pressure, dilg. the residue with H_2O, and isolating the prod. with or without tars, $\sim 30\%$ yield (173).

The ^{74}As-tagged acid is prepd. by the Bart rxn. from $4-O_2NC_6H_4N_2BF_4$ +

$Na_3{}^{74}AsO_3$, followed by a redn. of the nitroarsonic acid with Fe and HCl (590).
Arsanilic acid-1-^{14}C is prepd. from aniline-4-^{14}C + Na_3AsO_3 by fusion (590).
<u>Prop.</u>: NMR spectrum - analysis in terms of the additivity theory of Diehl and
a correlation of the substituent effects with the Hammett parameter and with
the group electric dipole moment (1429). K-edge x-ray absorption spectrum
(962). Detn. of As - anal. method (652). Detn. of 50 ppm arsanilic acid in
animal feeds - anal. procedure (1253). Identification in feeds by extn. with
MeOH and TLC (1088). Identification of the acid tagged with ^{74}As by chromatographic, electrophoretic, and ion-exchange methods (1185).
<u>Rxn. with</u>: Pentoses on paper chromatograms → pink color (25).
Hexoses on paper chromatograms → yellow color (25).
Aq. H_2SO_4 → cleavage of the As-C bond (1073).
BzCl → 4-(BzNH)$C_6H_4AsO_3H_2$ (1375).
$\overline{CH_2CH_2O}$ in H_2O contg. AcOH → $(HOCH_2CH_2)_2NC_6H_4AsO_3H_2$ (441).
AcCHXCONHR (X = Cl, Br, or I) → 4-$H_2O_3AsC_6H_4NHCH(Ac)CONHR$ (1524).
NCCHBrCONHR → 4-$H_2O_3AsC_6H_4NHCH(CN)CONHR$ (1524).
4,6-Dichloro-5-nitropyrimidine → N,N'-(5-nitropyrimidine-4,6-diyl)diarsanilic
acid (1638).
2,4,6-Trichloro-5-nitropyrimidine → N,N'-(2-chloro-5-nitropyrimidine-4,6-
diyl)diarsanilic acid (1638).
5-Amino-4,6-dichloropyrimidine → N-(5-amino-6-chloropyrimidin-4-yl)arsanilic
acid (1638).
HCONMe$_2$ in the presence of p-MeC$_6H_4SO_2Cl$ → $Me_2NCH=NC_6H_4AsO_3H_2$ (1449, 1450).
Riboflavin (1:1) → a charge-transfer complex (1168a).
6-Chloropurine + HCl in H_2O → N-purin-6-ylarsanilic acid (1638).
$NaNO_2$ in an acid medium, followed by coupling with trisubstd. pyrimidines →
the corr. trisubstd. p-arsonophenylazopyrimidines (681).
$NaNO_2$, followed by albumines, sheep red cell stroma or keyhole limpet hemo-
cyanine → a proteino-azobenzenearsonic acid, an antigenic agent (64).
$NaNO_2$, followed by various amino-acid polymers → p-arsonophenylazo conjugates
having antigenic prop. in inbred guinea-pigs (151, 152).
$NaNO_2$, followed by poly-L-lysine → conjugates of poly-L-lysine with 0.8-10.7
p-azobenzenearsonate moieties/mol. (205).
$NaNO_2$, followed by human γ-globulin → p-arsonobenzeneazo substd. γ-globulin
contg. a certain amt. of arsanilic acid groups linked to the protein by other
than -N=N- groups (477, 478).
$NaNO_2$, followed by coupling with antihapten antibodies having modified amino
groups in the lysyl residues (591).
Bovin serum albumine in the presence of [Me$_3$(4-$H_2NC_6H_4$)N]Cl → a deriv. which,
upon injection into rats, elicited 2 hapten-specific antibodies (626).
$NaNO_2$ in acid soln., followed by coupling with CM-cellulose-tyramine conjugates → an arsonobenzenediazo-cellulose deriv. active as an adsorbent for
antibenzenearsonic acid antibodies from antiserum (1552).
<u>Biol. Prop.</u>: The effect on the fertility, hatchability, growth, and blood
spot incidence in poultry (70). The effect on the egg production and body
weight of poultry (69, 437). As a feed additive for hens has no effect on
the transaminase and leucine amino peptidase activities and on direct and

total bilirubin but increases the erythrocyte count (1427, 1428) and improves
the egg production (1428). Metabolism in chickens and affinity for body
tissues (1186). Metabolism of the [14]C- or [14]C- and [74]As-labeled arsanilic
acid in chickens (1187). In growth-promoting administration to hens with
food is poorly absorbed and, after intramuscular injections, is excreted
practically unchanged (1076). Inhibitory effect on the growth of Lactobacillus leichmanii, L. arabinosus, Streptococcus faecalis, and Saccharomyces
carlsbergensis (1071). Ca arsanilate is effective against Ascaridia galli
and Heterakis gallinae in poultry and stimulates poultry growth (688). Inhibition of plasma esterase activity in rats (835). Na arsanilate inhibits
esterases in Tetrahymena pyriformis cultures (868). Mono-Na arsanilate is
highly effective in vitro against bovine ruminal protozoa for control of
bloat (1614). Arsanilic acid is a potential carcinogen in prolonged medicinal
applications (769). Combinations of mono-Na arsanilate with $MeSO_3Et$ produce
a synergistic effect on the chromosome aberrations of Hordeum sativum (1092,
(1094). Metabolic effect of mono-Na arsanilate on retinal cultures (678).
Induction of antibody production in rabbits and the effect on the amino acid
compn. and sequence in the antibodies (878, 879). Injection of arsanilic
acid combined with bovin serum albumine and $[Me_3(4-H_2NC_6H_4)N]Cl$ into rats
produced 2 hapten-specific antibodies (626). Excretion of [74]As by humans
after consumption of chickens fed arsanilic acid-[74]As (285). Reversible and
time-dependent inhibition of chymotrypsin, trypsin, and Novo and Carlsberg
subtilisins (622).

Uses: Food additive for poultry, FDA's dosage regulation (4, 5, 7, 12, 14,
15, 16), revised FDA's regulation (19). Prevention of coccidiosis (15) and
growth enhancement in chickens (69, 70). Food additive for swine, FDA regulation (7). Prevention of chronic respiratory disease in poultry (14). Differentiation of γ-amts. of pentoses and hexoses on paper chromatograms (25).
$H_2NC_6H_4-^{74}AsO_3H_2$ is useful in tracing the metabolic pathways of As in animal
feed as a growth promoter and feed efficiency enhancer for poultry and swine.
The ^{14}C-1-tagged arsanilic acid is useful in tracing the metabolism of the
aromatic ring in the chicken feed (590).

p-Carbamidobenzenearsonic Acid, Carbarsone, N-Carbamoylarsanilic Acid,
      $AsC_7H_9N_2O_4$, $4-(H_2NCONH)C_6H_4AsO_3H_2$,   (M-428)

Prop.: Detn. by x-ray diffractometry (908). Anal. detn. by UV spectrophotometry (890, 891). Identification in feeds by extn. with MeOH and TLC
(1088).

Biol. Prop.: Potential carcinogen in prolonged medicinal applications (769).
Prevention of histomoniasis in turkey poults (98). Antihistomonal effect in
turkey poults infected by Heterakis gallinae (1042). Antiprotozoal action
in bovine ruminal fluid in control of bloat (1614). Excretion of As after
oral administration to man as amoebicide (1617). Partial inhibition of the
oxygen consumption by Entamoeba histolytica (1494). Lethal and growth-inhi-
biting effect on Acanthamoeba and Entamoeba hystolytica (1225). Carcinogenic
activity in rats (656). In rat diet produced no carcinogenic effect, but at
a concn. of 2000 ppm depressed the growth and caused intestinal inflammation

resulting in death (1183).

Uses: Food additive for poultry, FDA's dosage regulation (18), and FDA's revocation of exemption from certification (17). Prevention of blackhead in poultry (17).

N-(Chloroacetyl)arsanilic Acid, $AsC_8H_9ClNO_4$, $4-(ClCH_2CONH)C_6H_4AsO_3H_2$, (M-435)

Rxn. with: 6% aq. $Na_2CO_3 \rightarrow HOCH_2CONHC_6H_4AsO_3H_2$ (1258).

N-(Iodoacetyl)arsanilic Acid, $AsC_8H_9INO_4$, $4-(ICH_2CONH)C_6H_4AsO_3H_2$, (M-435)

Synth.: From arsanilic acid dissolved in 10% NaOH by treating with ICH_2COCl, theoret. yield (441).

Prop.: White crystals, m. 204-205°, IR spectrum (441).

N-Acetylarsanilic Acid, $AsC_8H_{10}NO_4$, $4-(AcNH)C_6H_4AsO_3H_2$, (M-429)

Biol. Prop.: Poor absorption after oral administration, but considerable deacetylation after intramuscular injection to hens (1076).

N-Glycolylarsanilic Acid, $AsC_8H_{10}NO_5$, $4-(HOCH_2CONH)C_6H_4AsO_3H_2$, (M-429)

Synth.: From $ClCH_2CONHC_6H_4AsO_3H_2$ by hydrolysis with 6% aq. Na_2CO_3 at r.t., 27% yield (1258).

Rxn. with: Aq. $NaHCO_3$, followed by $Bi(NO_3)_3$ in glycerol-$H_2O \rightarrow HOCH_2CONHC_6H_4-As(O)(OH)OBiO$ (1258).

Biol. Prop.: The mono-Na salt is highly effective, in vitro, against bovine ruminal protozoa for control of bloat (1614). $HOCH_2CONHC_6H_4As(O)(OH)OBiO$ is highly effective, in vitro, against bovine ruminal protozoa for control of bloat (1614). Amoebicidal activity (836). Excretion of As after amoebicidal administration of the Bi salt to man and rat (1617). Combinations of the Bi salt with chloroquine phosphate have an additive effect on trophozoites of Entamoeba histolytica (1554). The Bi salt is effective against Trichuris vulpis in dogs (518).

Uses: The mono-Na salt is a growth promoter for poultry and swine administered with feed (465). The Bi salt is approved by FDA as anthelmintic for dogs (8).

N-(Carbamoylmethyl)arsanilic Acid, Tryparsamide, $AsC_8H_{11}N_2O_4$, $4-(H_2NCOCH_2NH)C_6H_4AsO_3H_2$, (M-435)

Biol. Prop.: The mono-Na salt is a potential carcinogen in prolonged medicinal applications (769). The salt has a selective action against Trypanosoma rhodensiense and T. brucei cultures (949). The concn. of arsenate-[74]As in the brain and in tumor and muscle tissues after injection of the mono-Na salt to C-57 mice was detd. (1048).

N-(4,6-Diamino-2-triazin-2-yl)arsanilic Acid, Malarsen, $AsC_9H_{11}N_6O_3$,

$$H_2N-\underset{\underset{NH_2}{|}}{\text{triazine}}-NH-C_6H_4-AsO_3H_2, \qquad (M-431)$$

Biol. Prop.: The di-Na salt has a trypanocidal effect against cultures of Trypanosoma rhodensiense and T. brucei (949). The salt inhibits phosphopyruvic carboxylase (290).

N-(2-Chloroethyl)-N-methylarsanilic Acid, $AsC_9H_{13}NO_3$,
  $Me(ClCH_2CH_2)NC_6H_4AsO_3H_2$

Synth.: From $PhN(Me)CH_2CH_2OH + POCl_3$, followed by nitrosation of the PhN(Me)-CH_2CH_2Cl intermediate, reduction of the nitroso compd., and conversion of the aniline deriv. by the Bart method, 12.5% yield (122, 441).
Prop.: Light brown needles, m. 141-142°, IR spectrum (441).

N-(Dimethylaminomethylene)arsanilic Acid, $AsC_9H_{13}N_2O_3$,
  $4-(Me_2NCH=N)C_6H_4AsO_3H_2$,   (M-442)

Synth.: From arsanilic acid + DMF in the presence of $p-MeC_6H_4SO_2Cl$ by stirring and heating gradually from 60-105° (1449, 1450).
Prop.: Crystals, m. 221-222° (1449, 1450).
Biol. Prop.: Effective in combatting bacteria, protozoa, viruses, and helminthic pathogens (1450).

N-(2-Amino-4-chloro-6-pyrimidinyl)arsanilic Acid, $AsC_{10}H_{10}ClN_4O_3$,

$$H_2N-\underset{N}{\text{pyrimidine}}(Cl)-NH-C_6H_4-AsO_3H_2, \qquad (M-445)$$

Synth.: From 2-amino-4,6-dichloropyrimidine + arsanilic acid in the presence of HCl (1638).
Rxn. with: BuONa in BuOH at 180° → the 4-butoxypyrimidinyl deriv. (1638). $HOCH_2CH_2OH$ in the presence of NaOH at 170° → the 4-(2-hydroxyethoxy)pyrimidinyl deriv. (1638).

N-(5-Amino-4-chloro-6-pyrimidinyl)arsanilic Acid, $AsC_{10}H_{10}ClN_4O_3$,

$$\text{pyrimidine}(Cl)\underset{NH_2}{\overset{NH}{<}}C_6H_4-AsO_3H_2$$

Synth.: From 5-amino-4,6-dichloropyrimidine and arsanilic acid (1:1) in the presence of HCl under reflux (1638).
Prop.: m. 253-255° (dec.) (1638).
Rxn. with: $HC(OEt)_3 + Ac_2O$ under reflux → p-(6-chloro-9H-purin-9-yl)benzenearsonic acid (1638).

N,N-Bis(2-chloroethyl)arsanilic Acid, $AsC_{10}H_{14}Cl_2NO_3$,
    $(ClCH_2CH_2)_2NC_6H_4AsO_3H_2$
Synth.: From $p-[(ClCH_2CH_2)_2N]C_6H_4NH_2$ by the Bart method, 37% yield (122, 441).
From $p-[(ClCH_2CH_2)_2N]C_6H_4AsO$ in alk. soln. + H_2O_2 soln. (441).
From $\{p-[(ClCH_2CH_2)_2N]C_6H_4As\lessdot\}_n$ in alk. soln. + H_2O_2 soln. (441).
Prop.: Yellow crystals, m. 158-161°, IR spectrum (122, 441).
Rxn. with: SO_2 in EtOH contg. aq. HI → $p-[(ClCH_2CH_2)_2N]C_6H_4AsO$ (122, 441).
$SnCl_2$ in concd. HCl in MeOH contg. KI → $\{p-[(ClCH_2CH_2)_2N]C_6H_4As\lessdot\}_n$ (122, 441).
$PhNHNH_2$ → unidentified prod. (441).
PCl_3 → unidentified prod. (441).

N,N-Diethylarsanilic Acid, $AsC_{10}H_{16}NO_3$, $Et_2NC_6H_4AsO_3H_2$
Synth.: From $PhNEt_2$ + $AsCl_3$ on a steam bath, followed by hydrolysis and
oxidn. of the $Et_2NC_6H_4AsO$ with H_2O_2 in aq. NaOH, 45% yield (122), 64% yield
based on $Et_2NC_6H_4AsO$ (441).
Prop.: Needles, m. 167-169°, IR and NMR spectra (122, 441).

N,N-Bis(2-hydroxyethyl)arsanilic Acid, $AsC_{10}H_{16}NO_5$,
    $p-[(HOCH_2CH_2)_2N]C_6H_4AsO_3H_2$
Synth.: From arsanilic acid + ethylene oxide in H_2O contg. AcOH, 88% yield
(122, 441).
Prop.: White crystals, m. 163-164°, IR spectrum (122, 441).
Rxn. with: $SOCl_2$ in $CHCl_3$ at 0° → $p-H_2NC_6H_4AsO$, $p-[(HOCH_2CH_2)_2N]C_6H_4Cl$, and
$p-ClC_6H_4N(CH_2CH_2Cl)_2$ (441).
$POCl_3$ or PCl_5 → tarry prods. contg. inorg. As (441).
$PhNHNH_2$ in MeOH under reflux, followed by 2 N NaOH, extn. with Et_2O, and a
treatment of the aq. phase with $[NH_4]Cl$ → $p-[(HOCH_2CH_2)_2N]C_6H_4AsO$ (441).

N-Purin-6-ylarsanilic Acid, $AsC_{11}H_{10}N_5O_3$,
Synth.: From arsanilic acid + 6-chloro-
purine + HCl in H_2O under reflux (1638).
Prop.: m. > 250° (1638).

N-(2,4-Dinitrophenyl)arsanilic Acid, $AsC_{12}H_{10}N_3O_7$,
    $4-[2,4-(O_2N)_2C_6H_3NH]C_6H_4AsO_3H_2$,   (M-442)
Prop.: m. 298° (1607).
Biol. Prop.: Uncoupling of oxidative phosphorylation in rat liver mitochon-
dria and inhibition of the 2,4,6-trinitrobenzaldehyde rxn. with albumin (1607).

N-[2-Amino-4-(2-hydroxyethoxy)-6-pyrimidinyl]arsanilic Acid,
    $AsC_{12}H_{15}N_4O_5\cdot H_2O$,
Synth.: From N-(2-amino-4-chloro-6-pyrimi-
dinyl)arsanilic acid + $HOCH_2CH_2OH$ contg.
NaOH by heating at 170° (1638).
Prop.: Crystd. from H_2O with 1 mol. H_2O, m. > 250° (1638).

N-Benzoylarsanilic Acid, $AsC_{13}H_{12}NO_4$, 4-(BzNH)C$_6$H$_4$AsO$_3$H$_2$

Synth.: From arsanilic acid by a condensation with BzCl (1375).
Biol. Prop.: Herbicidal and fungicidal props. (1375).

N-(2-Amino-4-butoxy-6-pyrimidinyl)arsanilic Acid, $AsC_{14}H_{19}N_4O_4 \cdot 2\ H_2O$

Synth.: From N-(2-amino-4-chloro-6-pyrimi-
dinyl)arsanilic acid by a rxn. with BuONa in
BuOH in a sealed tube at 180° (1638).
Prop.: The dihydrate m. 175° (dec.) (1638).

N-[2-Oxo-1-(phenylcarbamoyl)propyl]arsanilic Acid, $AsC_{16}H_{17}N_2O_5$,
4-[PhNHCOCH(Ac)NH]C$_6$H$_4$AsO$_3$H$_2$, (M-439)

Synth.: From arsanilic acid + PhNHCOCOClCOMe by heating in alc. (1524).

N-[2-Oxo-1-(p-tolylcarbamoyl)propyl]arsanilic Acid, $AsC_{17}H_{19}N_2O_5$,
4-[4-MeC$_6$H$_4$NHCOCH(Ac)NH]C$_6$H$_4$AsO$_3$H$_2$, (M-440)

Synth.: From arsanilic acid + 4-MeC$_6$H$_4$NHCOCHClCOMe by heating in alc. (1524).

N-[(6-Chloro-2-methoxy-9-acridinyl)arsanilic Acid, $AsC_{20}H_{16}ClN_2O_4$,
Rxn. with: Albumins and antibodies, fluorescence
studies of the Na salts (174).

2.5.6.1.2.5.6 AZOBENZENEARSONIC ACID DERIVATIVES

As described in the main volume on pages 454-457, the compounds of this
class have been prepared from o-, m-, and p-arsanilic acids by diazotization
and coupling the diazonium salts with various reactive molecules. Some of
the compounds prepared by the reaction with chromotropic acid and its derivatives constitute an important group of reagents which form complex salts with
metals. These salts are readily determined by spectrophotometric and colorimetric methods. Some of the compounds are described individually hereafter,
while the others are compiled in Table 31.

Arsono group-containing, biologically active substances have been prepared
from arsanilic acid by diazotization and coupling the diazonium intermediates
with peptides, albumins, and proteins. These products are entered following
Table 31.

5-Chloro-3-[(2,4-dihydroxyphenyl)azo]-2-hydroxybenzenearsonic Acid, Rezarson,
$AsC_{12}H_{10}ClN_2O_6$,
Synth.: From 5,3,2-Cl(H$_2$N)(HO)C$_6$H$_2$AsO$_3$H$_2$ by
diazotization and coupling with resorcinol
(988a).

Prop.: Orange-red to brownish yellow powder, insol. in H_2O, sol. in alkali, strong acids, and 50% EtOH; electronic absorption spectrum. The acid is extractable with i-$C_5H_{11}OH$ from aq. soln. (992).

Use: A reagent for spectrophotometric detn. (anal. procedure) of Mo^{6+} in steel (989), of Nb in the presence of Fe, Sb, Sn, Ce, Al, Th, W, V, Ge, Ta, Zr, and Ti (991), and of Ge (988a). A reagent for the detn. of Ge by a luminescence method - anal. procedure (992).

Complex salt with Ge (1:1) involves the arsono as well as the 2- and 2'-hydroxy groups. The salt is useful in a spectrophotometric detn. of Ge (988). Its soln. in H_2SO_4-H_3PO_4 gives a strong fluorescence (992).

4-[(o-Arsonophenyl)azo]-3-hydroxy-2,7-naphthalenedisulfonic Acid, Thorin, Thoron, Thoronol, $AsC_{16}H_{13}N_2O_{10}S_2$, (M-455)

Prop.: Visible spectrum of a 10^{-3} molar soln. at pH 1-12, isosbestic point at 580 mμ (49). Dissocn. consts. (48). Identification of the ionic species (pH, color, pK; T = the pentavalent anion): H_5T, strongly acid, raspberry red, 2.37; H_4T^-, 0-3, orange, 2.37; H_3T^{2-}, 4-7, red, 4.35; H_2T^{3-}, 8-10, yellow, 8.26; T^{5-}, 12, raspberry red, 11.16 (52).

Complex Salts: The following lanthanides (474) and actinides form salts with Thoron, with the metal/Thoron ratio of 1:2 (metal, color, stability range, $\lambda_{max.}$, ref.): Ce^{3+}, pink red, pH 3.5-7.5, 500 mμ, (1319); Eu^{3+}, --, pH 3.5-7.5, 500 mμ, (1322); La^{3+}, --, pH 4.0-8.0, 515 mμ, (1320); Lu^{3+}, --, pH 3.5-7.5, 500 mμ, (1322); Tb^{3+}, --, pH 3.5-7.5, 500 mμ, (1322); Th^{4+}, --, pH 1.0-6.5, 515 mμ, (1320); Yb^{3+}, --, pH 3.5-8.0, 500 mμ, (1322). The stability consts. of the preceding complex salts were calcd. (1319, 1320, 1322).

Pd^{++} forms a 1:1 salt, stable at pH 1.0-12.5, λ_{max} 525 mμ, stability const. calcd. (1320).

Hf^{4+} forms a pink-colored salt with 2 moles Thoron, which is stable at pH 0.75-2.5, λ_{max} 500 mμ, stability const. calcd. (1321).

Sc^{3+} forms a 1:2 salt, stable at pH 3.5-7.5, absorption spectrum and stability const. were detd. (1323).

UO_2^{++} forms a 1:1 salt, stable at pH 2.5-5.0, λ_{max} 510 mμ, stability const. calcd. (1320).

UO_2-Thoron-8-hydroxy-7-iodoquinolinesulfonic acid system, spectroscopic study (792).

Y^{3+} forms a pink-colored salt with 2 mols Thoron, which is stable at pH 3.0-6.5, λ_{max} 500 mμ, stability const. calcd. (1318).

Uses: A color indicator for a spectrophotometric titration of SO_4^{--} with $Ba(ClO_4)_2$ (261, 1082). Qual. and quant. colorimetric detn. of Th (0.02γ) and U in acid soln. (914). Spectrophotometric detn. of μg amts. of Li in the presence of other ions, anal. procedure (942). Mixt. with methylene blue is an indicator for the titration of Bi^{3+} with EDTA (1082). Detn. of trace amts. of B in $SiCl_4$ by measuring the fluorescence of the B-Thoron complex (1279). Spectrophotometric detn. of: Y (1318), Ce (1319), Eu, Tb, Yb, and Lu (1322), and Sc (1323). In the latter detn., Y as well as lanthanum, all other lanthanides, Ti^{4+}, Zr^{4+}, Hf^{4+}, Th^{4+}, Fe^{3+}, Pd^{++}, U^{6+}, PO_4^{3-}, AcO^-, $C_2O_4^{--}$, tartrate^{--}, and borate ions interfere when present at any concn. (1323).

3-[(o-Arsonophenyl)azo]-4,5-dihydroxy-2,7-naphthalenedisulfonic Acid, Arsenazo I, Neothoron, Uranon, $AsC_{16}H_{13}N_2O_{11}S_2$, (M-455)

Synth.: From chromotropic acid by a condensation with $2-H_2O_3AsC_6H_4N_2Cl$ in alk. soln. contg. Py at -5 to +5° (260).

Prop.: IR spectra in H_2O and in $n-C_5H_{11}OH$ (1517). Electronic spectrum (910). Spectrophotometric detn. in the presence of Arsenazo III (1134). Absorption spectra of the ionic forms in concd. H_2SO_4 and in weakly alk. media calcd. by the MOLCAO and perturbation methods (1334). The dissocn. consts. assignments suggest 5 successive steps: 1st both SO_3H groups, 2nd $AsO_3H_2 \rightarrow AsO_3H^-$, 3rd OH in position 4, 4th $AsO_3H^- \rightarrow AsO_3^{--}$, 5th the other OH group (1582). Titration with 0.0529 N KOH: 3 equiv. KOH neutralize simultaneously both SO_3H and 1 H of AsO_3H_2 at pH 6.4, the 4th equiv. KOH neutralizes AsO_3H^- to AsO_3^{--} at pH 9.45 (1336). A math. formula for the calcn. of the medium acidity at which a metal reacts with the reagent as a function of the metal hydroxide soly. was derived (1111).

Rxn. with: Diazonium salts derived from partly aminated polystyrene $\rightarrow$ a polymer contg. on the average three 3-[-o-arsonophenyl)azo]-4,5-dihydroxy-2,7-disulfonaphthyl-6-azo groups per 10 benzene rings of the polystyrene (1337). Optimum pH values for the rxn. with numerous metal ions were detd. (257).

Complex Salts: With Al, absorption spectra in H_2O and in $n-C_5H_{11}OH$ (1517). γ-Sensitivity index of the rxn. of Arsenazo I with Al^{3+} (977). The relation of the complex salt stability with $\Delta E/E_{20°}$ and the temp. effect on the light absorption of the salt in aq. soln. were detd. (920).

Be-Arsenazo I salt - a structural study by the LCAOMO method; electronic spectrum (910).

Mg, Ca, Cr, and Ba form 1:1 salts, the compns. of which were confirmed by spectrophotometry and potentiometric titration (1582).

In-Arsenazo I salt, spectrophotometric detn. - effect of the time of standing on the change in absorption max. of the absorption spectra (1014).

La and Sm form complex salts with Arsenazo I at pH 6.3-7 with the metal/acid ratio of 2:3 (1496). La^{3+} forms the LaH_2-Arsenazo I salt; spectrophotometric data (254).

NpO_2^+ forms a salt at pH 8.7-10.1; electronic absorption spectrum (344).

Nitrosylruthenium nitrate and nitrite react with Arsenazo I to form complex salts extractable from HNO_3 soln. with 30% Bu_3PO_4 in n-dodecane (813).

Ta-Arsenazo I salt in 0.1 and 1.0·N HCl soln., a redn. of its absorbance by H_2O_2, NaF, oxalic, tartaric, and citric acids and Tiron was detd. (103).

ScO_2^+-Arsenazo I salt - a structural study by the LCAOMO and spectrophotometric methods (910).

The Yb-Arsenazo I salt formation is favored by the pH range of 6-8; spectrophotometric data for $\Delta\lambda$, optimum length wave for Yb detn.: 570 mμ, molar absorptivity at pH 8 = 2.5 x 10^4; the complex is negatively charged (51).

V^{4+} forms a 1:1 salt in aq. soln. at pH $\leq$ 4 (985).

Zr^{4+} forms a 1:1 salt at pH 1.3, 1.7, and 3.5; molar absorptivity data (1176). The Zr^{4+} salt is unstable in alk. soln. (1015).

Arsenazo I salts of numerous metals and diphenylguanidine form systems extractable with alcs. from aq. solns. Thus, the Al and Be salts in the systems are extractable at pH 6-8 with $n-C_5H_{11}OH$, $i-C_5H_{11}OH$, and BuOH. Electronic spectra of the exts. were obtd. (1517). A Ti-Arsenazo I-diphenylguanidine system is extd. at pH 4-8 from aq. soln. with $n-C_5H_{11}OH$, $i-C_5H_{11}OH$, and BuOH. Electronic spectra of the exts. were detd. (1517). The systems of diphenylguanidine with the Arsenazo I salts of Fe(III), Zr, Hf, Th, Nb, and Ta are extd. with $n-C_5H_{11}OH$ from aq. acid media, while those of Sc, Yt, Ce, La, Fe(III), and Cu are extd. from neutral aq. media. Electronic spectra of the ternary systems with Ce, La, Be, Sc, Al, Th, Ti, Zr, and Hf were obtd. (1517).

Uses of Arsenazo I: Rapid colorimetric detn. of Be in Al-Mg-Be alloys; Be forms the 1:2 and 1:1 chelates at pH 12.2 and 5, resp. (781).

Spectrophotometric detn. of Ca and Mg in blood serum (929).

Copptn. of Cm with Arsenazo I + crystal violet at pH 4-7.5 (915).

Sepn. of Cm from Mg with Arsenazo I + urotropine at pH 4.5, followed by crystal violet (915).

Photometric detn. of Ga (7.56 γ Ga in 25 ml soln.) by itself and in the presence of Fe, Al, and In (50).

Photometric detn. of La and Sm in the presence of urotropine at pH 6.3-7 (1496).

Spectrophotometric detn. of La^{3+} as $LaH_2\cdot$Arsenazo I (254).

Spectrophotometric detn. of 10-100 γ Mg in Al, Zn, Co, and Ni alloys by the absorbance at 570 mμ at pH 10.0 (700).

Pptn. of Pa with Arsenazo I + crystal violet at pH 1.5-5.5 (917).

Spectrophotometric detn. of trace concns. of Pu (666, 1212).

Selective extn. of Pu(IV) from 5-7 N HNO_3 contg. Zr(IV) and U(VI) with 30% $(BuO)_3PO$ soln. (168).

Qualitative and quantitative colorimetric detn. of Th and U in acid soln. (914).

Colorimetric detn. (anal. procedure) of rare earth metals in the prods. from processing V-rare earth ores (896).

Indicator for chelometric microdetn. of rare earths (771).

Detn. of the total content of rare earth elements by bilateral differential spectrophotometry in electrolytes and Al alloys (598).

Indicator for the titration of Th^{4+} and U^{4+} with di-Na EDTA (716).

A colorimetric reagent for Ti detn. (794).

Photometric detn. of Sc (1311).

Sepn. of U from Pu and Zr by extn. with $(BuO)_3PO$ in C_6H_6 from aq. HNO_3 (921).

Spectrophotometric detn. of U including samples contg. interfering ions - anal. procedure (1446). Spectrophotometric detn. of U after extn. with $(BuO)_3PO$ in i-BuCOMe (462).

Polarographic detn. of UO_2^{2+} (326).

Spectrophotometric detn. of V^{4+} in the presence of V^{5+} after extn. with $CHCl_3$ from aq. quinine·HCl soln. (984, see also 985).

Spectrophotometric detn. of Yb at pH 7.0 (102).

Spot test for the identification of Zr and Hf and differentiation between Zr and Hf (805).

Arsenazo I-urotropine soln. - a developer for paper chromatograms in the sepn. of rare earth elements and Zr-Hf; effective in the sepn. of the Ce group metals (938).

Spectrophotometric detn. of F by fading of the blue color of the Arsenazo I-Th complex at pH 2 - anal. procedure (1396).

3-[(p-Arsonophenyl)azo]-4,5-dihydroxy-2,7-naphthalenedisulfonic Acid,
$AsC_{16}H_{13}N_2O_{11}S_2$,   (M-460)

<u>Synth.</u>: From arsanilic acid by diazotization and coupling with chromotropic acid di-Na salt in NaOAc soln. (1439).

<u>Use</u>: A color test reagent for: (a) detn. of Cu^{++} at pH 5.1 in a NaOAc buffer, (b) detn. of Be^{++} at pH 6.4 with a few drops of 1 N hexamethylenetetramine buffer, and (c) detn. of lanthanides at pH 6.4 with a drop of 1 N hexamethylenetetramine buffer. The test sensitivity is reduced by the F^- and oxalate^{--} anions (1439).

3-[(o-Arsonophenyl)azo]-4,5-dihydroxy-6-phenylazo-2,7-naphthalenedisulfonic
    Acid, Monoarsenazo III, $AsC_{22}H_{17}N_4O_{11}S_2$,   (M-464)

<u>Complex salts</u> with Zr^{4+}, Cd^{++}, La^{3+}, Th^{4+}, U^{4+}, and UO_2^{2+}, their compn. and optimum pH values for their formation were detd. (257).
[$LaH_2(Monoarsenazo III)_2$], spectrophotometric data (254).

Molar extinction coeffs. of the complex salts with: Sc^{3+}, Zr^{4+}, Hf^{4+}, Th^{4+}, U^{4+}, UO_2^{2+}, La^{3+}, and Y^{3+} were detd. (260).

2-{[6-[(o-Arsonophenyl)azo]-4,5-dihydroxy-2,7-disulfo-3-naphthyl]azo}benzoic
    Acid, $AsC_{23}H_{17}N_4O_{13}S_2$,   (M-465)

<u>Prop.</u>: Titration with 0.0529 N KOH: 4 equivs. KOH neutralize simultaneously 2 SO_3H, CO_2H, and AsO_3H_2 (to AsO_3H^-) at pH 7 (1336).

<u>Uses</u>: A color indicator for a pptn.-titration of SO_4^{--} with $Ba(ClO_4)_2$ (261).

Complex Salts: Cyclic salts with Al, Sc, Th, La, Nd, Sm, Ho, Er, Y, and Yb were formed; their relative stabilities and the temp. effect on the light absorption of the aq. solns. were detd. (920, see also 919).

6-[(o-Arsonophenyl)azo]-4,5-dihydroxy-3-[(3,8-disulfo-1-naphthyl)azo]naphtha-
    lene-2,7-disulfonic Acid, Arsenazo AE, $AsC_{26}H_{19}N_4O_{17}S_4$,

Synth.: From 1-amino-3,8-naphthalene-
disulfonic acid by diazotization and coupling with chromotropic acid, followed by
coupling with $[2-H_2O_3AsC_6H_4N_2]^+$ salts (47a).
Rxn. with: La, Nd, Pr, Sm, Gd, Yb, Er, and
Dy chlorides at pH 3.7-4.5 and at pH 6.0 →
the corr. 1:1 and 3:2 complex salts (47a).
Uses: Spectrophotometric detn. of Sc (1311) and of La, Nd, Pr, Sm, Gd, Yb, Er, and Dy (47a).

3-[(o-Arsonophenyl)azo]-4,5-dihydroxy-6-[(4-hydroxy-2,5-disulfo-3-naphthyl)-
    azo]-2,7-naphthalenedisulfonic Acid, Arsenazoamino-ε-Acid,
    $AsC_{26}H_{19}N_4O_{18}S_4$,   (M-466)

Prop.: Violet-red soln. in HNO_3
(1335). Electronic absorption
spectra in HNO_3 and in HCl (1335).
Use: Extn.-spectrophotometric
detn. of 1-10 γ/ml Pu(IV) - the crimson colored Pu(IV) complex salt is extd.
from 0.1 N HNO_3 with BuOH contg. $(PhCH_2)_3N$ and is detd. spectrophotometrically
(1069, see also 1335).
Complex salt with Pu(IV) is formed in strongly acid media. Spectroscopic
detn. of the complex gave 2 maxima at 500 and 620 mμ in 0.1-1.0 $NHNO_3$ and at
630 mμ in 0.1-1.0 N HCl. Th, U, Zr, and rare earth metals interfere (1335).

3-[(o-Arsonophenyl)azo]-6-{[5-(diethylsulfamoyl)-2-methoxyphenyl]azo}-4,5-di-
    hydroxy-2,7-naphthalenedisulfonic Acid, Arsenazo-2-methoxy-5-diethyl-
    sulfamic Acid, $AsC_{27}H_{28}N_5O_{14}S_3$,

Prop.: Violet red soln. in acids
(1335).
Use: Extn. of 1-10 γ/ml Pu(IV)
from 0.1 N HNO_3 with BuOH contg.
$(PhCH_2)_3N$ and spectrophotometric
detn. (1069, see also 1335).
Complex salt with Pu(IV) in acid media is blue-green when excess Pu is present and violet when excess of the reagent is present; λ_{max} at 635, 420, and
700 mμ (1335).

Table 31. Azobenzenearsonic Acid Derivatives

x-(RN=N)$C_6H_4AsO_3H_2$ in which x and R are:	Prepd. from (Yield %)	Prop. and Rxn.	Ref.
C_8			
2-[NCCH$_2$-	(A) *-N=NCH(CN)CO$_2$Et by boiling in 15% H$_2$O-EtOH KOH and acidifying with concd. HCl at r.t.	Pink plates, dec. 198°, sparingly sol. in EtOH, H$_2$O, and Me$_2$CO	532
C_9			
2-[OCNHC(S)SCH-	(A)*-N$_2$Cl + rhodanine in aq. NH$_4$OH, followed by acidification (76)	Light yellow crystals, m. 215-217° gives red-colored soln. in aq. NaOH and in H$_2$SO$_4$; absorption spectrum	654
2-[HO$_2$CCH(CN)-	(A)* -N=NCH(CN)CO$_2$Et by boiling in 15% H$_2$O-EtOH KOH soln. and acidifying with concd. HCl at 0° (74)	m. 209-210°, UV spectrum, (a)	532
2-[SCNHC(S)NHCH-	(A)*-N$_2$Cl + 2,4-dithioxoimidazolidine in dioxan (42)	Yellow, m. 209-210°	655
C_{10}			
4-[H$_2$N—(ring: NH$_2$, N, N, OH)—OH]	(A)*-N$_2$Cl + 2,4-diamino-6-hydroxy-pyrimidine (80)	m. 200° (dec.), biol. tests for anticancer activity	681
C_{11}			
4-[MeS—(ring: NH$_2$, N, N, NH$_2$)—NH$_2$]	(A)*-N$_2$Cl + 4,6-diamino-2-methyl-thiopyrimidine (36)	m. 317°, biol. tests for anti-cancer activity	681
2-[EtO$_2$CCH(CN)-	(A)*-N$_2$Cl + NCCH$_2$CO$_2$Et at pH 7 (95)	m. 200-202° (from EtOH), UV spectrum, (b)	532

348

C12		
4-[4-H2NC6H4- (M-458)	621	Reversible deactivation of Nevo and Carlsberg subtilisins
C14		
4-[4-Me2NC6H4- (M-458)	1429	NMR spectrum, (c, d)
C18		
4-[HO3S] (OH)	918	Tri-Na salt promotes the extn. of [Ag(Py)2]+ and [Zn(Py)4]++ with i-BuOH-CHCl3 (1:9) mixt. from H2O soln.
4-[2-H2NC10H6-1- (M-461)	918	Na salt promotes the extn. of [Cd(Py)4]++ and [Zn(Py)4]++ with i-BuOH-CHCl3 (1:9) mixt. from H2O soln.
2-[(NH2, SO3H)	256	Dissocn. consts., forms a 1:1 complex salt with Cu++

(A)*-N2Cl + 4-H2NC10H6-1-SO3Na in aq. soln. contg. NaOAc at -5 to +5° (64)

*(A) - designates the arsonophenyl moiety, -C6H4AsO3H2, linked to the diazonium group at the position indicated in the first column.
(a) Rxn. with RN=NCl in alk. soln. → RN=NC(CN)=N-NHC6H4AsO3H2 (532). (b) Rxn. with alc. KOH under reflux → 2-(H2O3As)C6H4N=NCH(CN)CO2H, when acidified at 0°, and 2-(H2O3As)C6H4N=NCH2CN, when acidified at r.t. (532). (c) Analysis of the NMR spectrum in terms of the additivity theory of Diehl and a correlation of the substituents effects with the Hammett parameter and with the group elec. dipole moment (1429). (d) Its Na salt promotes the extn. of [Cd(Py)4]++ and [Zn(Py)4]++ with i-BuOH-CHCl3 (1:9) mixt. from aq. soln. (918).

Table 31 Continued

x-(RN=N)C$_6$H$_4$AsO$_3$H$_2$
in which x and R are:

x and R	Prepd. from (Yield %)	Prop. and Rxn.	Ref.
C$_{18}$			
2-[Br, HO$_3$S, OH OH, SO$_3$H]	(A)*-N$_2$Cl + 3-bromochromotropic acid	Absorption spectrum; forms a cyclic salt with Al, (e)	916
4-{4-[4,3-HO(HO$_3$S)C$_6$H$_3$N=N]C$_6$H$_4$-		The effect of the H bridge structure and electrostatic reciprocal action on the rate of protolytic rxns.	519
4-[4-(4-H$_2$NC$_6$H$_4$SO$_2$NH(C$_6$H$_4$-	4-[(A)*-N=N]C$_6$H$_4$NH$_2$ + 4-H$_2$NC$_6$H$_4$-SO$_2$Cl	m. 198°	1490
4-[1-(Et$_2$NCH$_2$CH$_2$NH)C$_{10}$H$_6$-2-	(A)*-N$_2$Cl + 1-(Et$_2$NCH$_2$CH$_2$NH)C$_{10}$H$_7$	Dark orange solid, m. 158° (dec.), useful as a parasiticide, insecticide, and fungicide	530
C$_{22}$			
2-[HO$_3$S N=N, HO$_3$S, OH OH, SO$_3$H, Cl]		Ionization consts., (f)	1336
2-[PO$_3$H$_2$, N=N, HO$_3$S, OH OH, SO$_3$H, Br]	Chromotropic acid Ca salt + [5,2-Br(H$_2$O$_3$P)C$_6$H$_3$N$_2$]BF$_4$ followed by (A)*-N$_2$BF$_4$ (70-76)		994

Structure	Preparation	Use	Ref.
2-[Cl... (chromotropic acid azo structure with PO_3H_2, OH, OH, SO_3H, HO_3S, N=N, Cl)	Chromotropic acid + 4,2-Cl(H_2O_3P)-$C_6H_3N_2Cl$ (1:1:1) in aq. soln. satd. with Ca^{++}, Mg^{++}, or Sr^{++} at -5 to +5° (57)	An indicator for complexometric titration of U and rare earth elements	259
	Chromotropic acid + (A)*-N_2Cl, followed by 4,2-Cl(H_2O_3P)C_6H_3-N_2Cl (72)	Complex salts with: Sc, Zr, Hf, Th, U, UO_2, La, Lu, and Y and their molar extinction coeffs.	260
	Chromotropic acid Ca salt + 4,2-Cl(H_2O_3P)$C_6H_3N_2BF_4$, followed by (A)*-N_2BF_4 (78-84)		994
2-[(structure with PO_3H_2, OH, OH, SO_3H, HO_3S, N=N, Cl)	Chromotropic acid Ca salt + 5,2-Cl(H_2O_3P)$C_6H_3NBF_4$, followed by (A)*-N_2BF_4 (70-75)		994
2-[(M-464) (structure with SO_3H, OH, OH, SO_3H, HO_3S, N=N, SO_3H)		Photometric detn. of Sc and of rare earth elements	1331
C_{24} 2-[(structure with OMe, OH, OH, SO_3H, HO_3S, N=N, Me_2NSO_2) Arsenazo T		Photometric detn. of Sc	1311
		Photometric detn. of Ca in a flotation slurry of Li and Be pegmatite, (g)	1557

(e) The relation of the salt stability to $\Delta E/E_{20}°$ and the temp. effect on the light absorption of aq. soln. were detd. (920). (f) Titration with 0.0529 N KOH: 4 equivs. neutralized simultaneously 3 SO_3H and AsO_3H_2 (to AsO_3H^-) groups at pH 5.45, the 5th equiv. neutralized one OH at pH 7.78, and the 6th neutralized AsO_3H^- at pH 9.9 (1336). (g) Analytical procedure.

Table 31 Continued

x-(RN=N)C₆H₄AsO₃H₂

x-(RN=N)C$_6$H$_4$AsO$_3$H$_2$
in which x and R are:

	Prepd. from (Yield %)	Prop. and Rxn.	Ref.

C$_{26}$

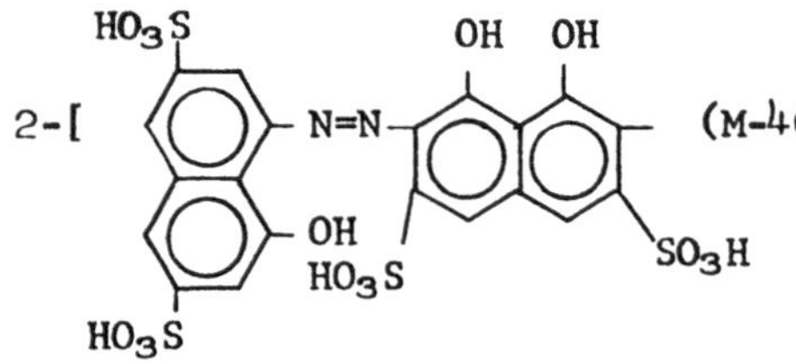

2-[...] (M-466)

Photometric detn. of Sc

1311

C$_{27}$

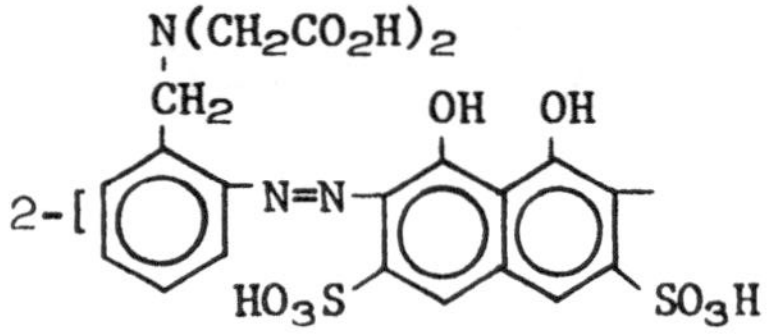

2-[...]

Chromotropic acid + (A)*-N₂Cl
followed by 2-[(HO₂CCH₂)₂-
NCH₂]C₆H₄N₂Cl (72)

Forms complex salts with: Sc^{3+},
Zr^{4+}, Th^{4+}, U^{4+}, UO$_2^{2+}$, La^{3+},
Lu^{3+}, and Y^{3+}; molar extinction
coeffs.

260

352

Conjugates of Arsonobenzeneazo Moieties with Peptides, Proteins, and Other Polymeric Substrates

The conjugates are formed from p-arsanilic acid by diazotization and coupling the arsonobenzenediazonium salts with various amino acids, peptides, albumins, and proteins. These compounds have been used in immunochemical studies.

Immunofiltration of antigens by means of p-arsonobenzeneazo-γ-globulins conjugates (1528).

Immune response to p-arsonobenzeneazo-human γ-globulin conjugates by mice immune to human γ-globulin (477, 478).

Hyperimmunization of rabbits with p-arsonobenzeneazo-γ-globulin conjugates produced antibodies from which two fractions were isolated: one of them appeared to be directed mainly against the singly ionized (pH 5) and the other against the doubly ionized (pH 9.5) arsonate groups (856). Determination of multiple antibody proteins and of the active sites and subunits of the antibodies isolated from the serum of rabbits hyperimmunized with p-arsonobenzeneazo-bovine γ-globulin conjugates (1177-1179). Structural differences of the antibodies produced by individual rabbits were established (1297, see also 879). The conjugates of p-arsonobenzeneazo-human γ-globulin failed to produce detectable antigenic fragments either in the circulation or in the spleens of rabbits (935).

The conjugates of p-arsonobenzeneazo-bovine γ-globulin induce different 7 S anti-arsonate antibodies in different lymph tissues in guinea pigs and the antibodies are different from those induced by bovine γ-globulin (887). The antibodies and delayed hypersensitivity of guinea pigs sensitized to the arsonate conjugates are highly specific for the homologous group (952).

The conjugates of thyroglobulin with diazotized arsanilic and sulfanilic acids in a soluble form or incorporated into incomplete Freund's adjuvant injected into rabbits produced antibody to native thyroglobulin and thyroid lesions (1592, 1593). A transient appearance of 19 S antibody was observed (1593).

The conjugates of bovine serum albumin with diazotized arsanilic and sulfanilic acids terminate an acquired tolerance to serum protein antigens in rabbits. Simultaneous injections of bovine serum albumin and the conjugates interfere with the inhibitory action of the latter (1591).

The conjugates of bovine serum albumin with equivalent amounts of diazotized arsanilic acid and p-aminophenyltrimethylammonium ions produced two hapten-specific rabbit antibodies in different ratios (862, 863) and a nonspecific γ-globulin antibody (863).

A study of the variability among anti-p-azobenzenearsonate antibodies by means of affinity labeling with p-arsonobenzenediazonium fluoroborate (886).

Peptide analysis of antibodies isolated from the serum of rabbits injected with the conjugates of diazotized arsanilic acid with bovine serum albumin (660).

Electrophoretic studies of antibodies induced in guinea pigs immunized with the p-arsonobenzeneazo-bovine serum albumin conjugates (1160).

The conjugates of diazotized arsanilic acid with casein, porcine insulin, and bovine gelatin produce antisera in rabbits containing various anti-azo-benzenearsonate antibodies, of which only certain parts can be removed by the immunizing antigen (1121). ^{131}I-labeled conjugates of insulin with p-arsono-benzenediazonium moieties are useful as the test antigens for anti-R_p antibody activity of various components of rabbit antiserum (1178, 1179).

A determination of the active sites in the polypeptide chains of anti-p-azo-benzenearsonate antibodies (1060).

The conjugates of diazotized arsanilic acid with the 1-24 peptide of ACTH injected into guinea pigs induce a specificity oriented toward the carrier of the haptenic groups. The 17-24 sequence of 8 amino acids is the active part of the molecule with respect to the induction and detection of delayed hyper-sensitivity in the test animals (1426).

A chromatographic resolution of the conjugates of diazotized arsanilic acid and bovine chymotrypsinogen A formed at pH 5.5; 85% of the azo linkages occur at the tyrosyl residues at two different positions in the protein molecule (1589).

The conjugates of diazotized arsanilic acid with fibroin are specific adsor-bents for anti-azobenzenearsonate antibodies from the serum of rabbits immu-nized with bovine/serum albumin, human γ-globulin, and azobenzenearsonate con-jugate antigens (1597).

The conjugates of casein with diazotized arsanilic acid and poly(aminosty-rene) are useful as insoluble immuno-adsorbents for the separation of anti-bodies from rabbit antiserum in the evaluation of the relative capacities of 7 S and 19 S antibodies for agglutination of human erythrocytes conjugated with arsanilic acid. p-Nitrobenzenearsonic acid has an inhibitory effect on the agglutination of the hapten-coupled red cells (1409).

The conjugates of poly[ε-(N-trifluoroacetyl)-L-lysyl]ribonuclease with di-azotized arsanilic acid, involving the tyrosine and histidine residues, after removal of the CF_3CO groups, were used for immunization of rabbits. The anti-bodies thus formed reflect the overall net charge of the molecule, rather than the charge within the limited area of the azobenzenearsonate groups (1306). The conjugates of diazotized arsanilic acid with tri-L-tyrosine, with N-acetyl-D-tyrosine amide, and with poly-D-tyrosines, resp., induce antibodies in guinea pigs (206, 208, see also 208a), and in rabbits (207). However, other studies indicate that the conjugates with poly-D-amino acids are neither immunogenic nor capable of eliciting delayed reactions (208a, 953).

The conjugates of diazotized arsanilic acid with polytyrosine, poly-Lys[51]-

Tyr[49], poly-Glu[52]Tyr[48], and poly-Ala[34]Tyr[30] injected in guinea pigs produce delayed cross reactions with the conjugates of the arsonate with bovine serum albumin (204a).

The conjugates of diazotized arsanilic acid with the polymers of L-tyrosine, glutamic acid (151, 152, 953, 954), alanine (953, 954), and L-lysine (151, 152, 205), resp., with or without adjuvants, are immunogenic and can elicit hapten-specific, delayed hypersensitivity in sensitized guinea pigs.

Determination of the As concentration in the tumor tissue, brain, muscles, liver, and kidney of C-57 mice after injection of the conjugates of diazotized arsanilic acid with polylysine (1048). The effect of the As/lysine ratio on the toxicity and uptake by blood, tumor tissue, and organ cells was detd. (1053).

The conjugates of diazotized arsanilic acid with: polytyrosine, polyhystidine, poly-L-lysine, poly-D-lysine, glutamic-lysine, glutamic-lysine-tyrosine, L-glutamic-alanine-tyrosine, D-glutamic-alanine-tyrosine, gelatin, elastoidin, insulin, bovine serum albumin (BSA), oxidized BSA, bovine ribonuclease (RNase), oxidized RNase, H-haemoglobin, pepsin, guinea pig serum albumin, and guinea pig Y-globulin were tested in the immunization of guinea pigs for the production of hapten-specific delayed sensitivity. The tests showed the role of the carriers in the development of delayed sensitivity to the azobenzenearsonate groups (812).

The conjugates of diazotized arsanilic acid with proteins have anticoagulant and antilymphoma activities in mice and rabbits. The anticoagulant effect may be due to the complex formation with fibrinogen which is resistant to the action of thrombin (242).

The conjugates of diazotized arsanilic acid with CM-cellulose-tyramine derivatives are effective adsorbents for anti-p-azobenzenearsonate antibodies from anti-sera (1552).

The conjugates of diazotized p-(p-aminophenylazo)benzenearsonic acid and ε-dinitrophenyllysine afford total removal of antibodies from the serum of rabbits immunized with bovine serum albumin, human Y-globulin, or p-azobenzenearsonate antigens (1597).

2.5.6.1.2.5.8 FORMAZANOBENZENEARSONIC ACID DERIVATIVES

A new type of benzenearsonic acid derivatives containing a cyanoformazano moiety has been prepared by a condensation of arylazocyanoacetic acid with arenediazonium salts, where one of the reagents contains an arsonophenyl group.

$$RN=NC(CN)CO_2H + R'N_2^+ \rightarrow RN=NC(CN)=NNHR'$$

$$R \text{ or } R' = C_6H_4AsO_2H$$

o-(3-Cyano-5-phenyl-1-formazano)benzenearsonic Acid, $AsC_{14}H_{12}N_5O_3$,
   2-[PhNHN=C(CN)N=N]$C_6H_4AsO_3H_2$

<u>Synth.</u>: From PhN=NCH(CN)CO$_2$H by a rxn. with [2-H$_2$O$_3$AsC$_6$H$_4$N$_2$]Cl in alk. soln.
at 0°, isolated as the di-Na salt, 79% yield (532).
<u>Prop.</u>: The di-Na salt, m. > 250°, UV spectrum (532).

o-[3-Cyano-5-(o-hydroxyphenyl)-1-formazano]benzenearsonic Acid,
   $AsC_{14}H_{12}N_5O_4$,   2-[2-HOC$_6$H$_4$NHN=C(CN)N=N]$C_6H_4AsO_3H_2$

<u>Synth.</u>: From 2-HOC$_6$H$_4$N=NCH(CN)CO$_2$H by a rxn. with [2-H$_2$O$_3$AsC$_6$H$_4$N$_2$]Cl in alk.
soln., isolated as a di-Na salt, 48% yield (532).
<u>Prop.</u>: The di-Na salt, m. > 250°, UV spectrum (532).

m-[5-(o-Arsonophenyl)-3-cyano-1-formazano]-o-hydroxybenzenesulfonic Acid,
   $AsC_{14}H_{12}N_5O_7S$,

<u>Synth.</u>: From 2-H$_2$O$_3$AsC$_6$H$_4$N=NCH(CN)CO$_2$H by a rxn. with [2,3-HO(HO$_3$S)C$_6$H$_3$N$_2$]Cl
in alk. soln., followed by acidification and pptn. with KCl, the di-K salt
was obtd. in 33% yield (532).
<u>Prop.</u>: The di-K salt, m. > 300°, UV spectrum (532).

o-[5-(o-Arsonophenyl)-3-cyano-1-formazano]benzoic Acid, $AsC_{15}H_{12}N_5O_5$,
   2-[2-HO$_2$CC$_6$H$_4$NHN=C(CN)N=N]$C_6H_4AsO_3H_2$

<u>Synth.</u>: From 2-H$_2$O$_3$AsC$_6$H$_4$N=NCH(CN)CO$_2$H by a rxn. with [2-HO$_2$CC$_6$H$_4$N$_2$]Cl in
alk. soln., isolated as a di-Na salt, 64% yield (532).
<u>Prop.</u>: The di-Na salt, m. > 250°, UV spectrum (532).

o-[3-Cyano-5-(2-hydroxy-3,5-xylyl)-1-formazano]benzenearsonic Acid,
   $AsC_{16}H_{16}N_5O_4$,   2-[2,5,3-HO(Me)$_2$C$_6$H$_2$NHN=C(CN)N=N]$C_6H_4AsO_3H_2$

<u>Synth.</u>: From 2-H$_2$O$_3$AsC$_6$H$_4$N=NC(CN)CO$_2$H by a rxn. with [2,3,5-HO(Me)$_2$C$_6$H$_2$N$_2$]Cl
in alk. soln., followed by acidification with AcOH, 70% yield (532).
<u>Prop.</u>: m. 206-208°, UV spectrum (532).

o-(1,3-Diphenyl-5-formazano)benzenearsonic Acid, $AsC_{19}H_{16}N_4O_3$,
   2-[PhN=NC(Ph)=NNH]$C_6H_4AsO_3H_2$

<u>Use</u>: Analytical color reagent for metal ions (840).

o-[1-(2-Benzothiazolyl)-3-phenyl-5-formazano]benzenearsonic Acid,
   $AsC_{20}H_{16}N_5O_3S$,

<u>Use</u>: Analytical color reagent for metal ions (840).

2.5.6.1.2.6 BENZENEARSONIC ACID DERIVATIVES CONTAINING HETEROCYCLIC SUBSTITUENTS

Benzenearsonic acids containing heterocyclic substituents linked to the benzene ring by a carbon or heteroatom of the heterocyclic group were compiled in the main volume on pages 468-73.

Several new purin-9-ylbenzenearsonic acid derivatives have recently been prepared. p-(6-Chloro-9H-purin-9-yl)benzenearsonic acid is formed from N-(5-amino-6-chloro-4-pyrimidinyl)arsanilic acid by a reaction with triethyl orthocarbonate in acetic anhydride under reflux. This 6-chloro derivative has been used as an intermediate in the preparation of several substituted purinylbenzenearsonic acids which are compiled in Table 32 on page 359.

2.5.6.1.2.7 DI- TO PENTA-SUBSTITUTED BENZENEARSONIC ACIDS

The methods of synthesis of di-, tri-, tetra-, and pentasubstituted benzenearsonic acids were briefly discussed in the main volume on page 474. Some disubstituted derivatives were described individually on pages 474-478, 504, while others were compiled in Table 61 on pages 479-500. The trisubstituted derivatives were arranged in Table 62 on pages 501-503; tetra- and pentasubstituted benzenearsonic acids were described on page 504.

Since 1964 several disubstituted benzenearsonic acids have been prepared. Physical constants, reactions, and/or biological properties and/or uses of the newly prepared and of the previously known compounds have been studied. Three disubstituted and three trisubstituted derivatives are described individually hereafter and many more disubstituted derivatives are compiled in Table 33 on page 362.

4-Hydroxy-3-nitrobenzenearsonic Acid, $AsC_6H_6NO_6$,
    $4,3-HO(O_2N)C_6H_3AsO_3H_2$,   (M-474)
Prop.: Dissocn. const. in aq. MeOH (664). Explosibility data (492). Identification in feeds by extn. with MeOH, followed by TLC (1088). Detn. of As - anal. method (652).
Biol. Prop.: A Co salt is effective against Ascaridia galli and Heterakis gallinae in poultry (688). A growth-promoting agent and coccidiostat for hens is poorly absorbed, after oral administration, but is very slowly excreted. After intramuscular injection, the excretion is faster. A metabolic redn. of the NO_2 group was detected (1077).
Uses: Photometric detn. of small amts. of Sn in Pb-Sb alloys (1550). A food additive for poultry, FDA specification and dosage (1, 2, 3, 6, 9, 10, 11, 13, 15), revised FDA regulation (19), for prevention of coccidiosis (10, 11, 13, 15).

3-Acetamido-4-hydroxybenzenearsonic Acid, Acetarsone, Stovarsol, Spirocid,
    Osarsol, $AsC_8H_{10}NO_5$, $4,3-HO(AcNH)C_6H_3AsO_3H_2$,   (M-477)
Prop.: Detn. of As - anal. method (652). Refractometric detn. in drugs (1263).

Biol. Prop.: Highly effective against Trichomonas intestinalis (233). Negative results in antitrichomona tests (997). Combinations with $MeSO_3Et$ have a synergistic effect on the chromosome aberration of barley (1092). Prevention of coccidiosis in poultry by administration of furazolidone, followed by the arsonic acid (1328). Excretion of As after oral administration to man as an amoebicide (1617). A salt with 1,3-bis(2-bromo-4-amidinophenoxy)propane is effective as a spirillicide and bactericide in a treatment of gingivitis (459).

5-Nitro-2-{3-[p-[(p-sulfophenyl)azo]phenyl]-1-triazeno}benzenearsonic Acid,
    Sulfarsazen, $AsC_{18}H_{15}N_6O_8S$,   (M-499)

Prop.: Thermodynamic consts. of dissocn.
and molar absorptivity (27).

Rxn. with: Hg^{++} at pH 9.5-10.5 in NH_4 citrate → a 1:1 complex (993).
Hg^+ → a colored complex (993).

Uses: Spectrophotometric detn. of Np at pH 8.75 (344). An indicator for the titration of Hg^{++}, Zn^{++}, Cd^{++}, Ni^{++}, Co^{++}, and Ca^{++} (993) and Pb in pigments (1010) with Trilon B. Colorimetric and photometric detn. of Hg^{++} in the presence of various metal ions (993). A complexometric indicator for potentiometric titration of Zn, Cd, Ni, and Pb - anal. procedure; dissocn. and stability consts. of the complex salts were detd. (1195).

Trisubstituted Derivatives

2,4,5-Trichlorobenzenearsonic Acid, $AsC_6H_4Cl_3O_3$, $2,4,5-Cl_3C_6H_2AsO_3H_2$,
    (M-501)

Salts: The acid forms a cryst. salt with Pb^{++} and flocculent or powdery salts with Cu^{++}, Mn^{++}, and Cr^{3+} (1232).

2,4,6-Trichlorobenzenearsonic Acid, $AsC_6H_4Cl_3O_3$, $2,4,6-Cl_3C_6H_2AsO_3H_2$,
    (M-501)

Salts: The acid forms a cryst. salt with Pb^{++} and flocculent or powdery salts with Cu^{++}, Cd, Co^{++}, Mn^{++}, and Cr^{3+} (1232).

3,5-Dibromo-o-arsanilic Acid, $AsC_6H_6Br_2NO_3$, $2,3,5-H_2N(Br)_2C_6H_2AsO_3H_2$

Biol. Prop.: Uncoupling of oxidative phosphorylation in rat liver mitochondria (1607).

Table 32. Heterocycle-Substituted Benzenearsonic Acids

x and R in $C_6H_4AsO_3H_2$	Prepd. from	Prop. and Rxn.	Ref.
C_{11}			
4-[purine with Cl, Cl, NH₂]	$NHC_6H_4AsO_3H_2 + HC(OEt)_2 + Ac_2O$ under reflux	m. > 250°, (a–d)	1638
4-[purine with OH]	From the corr. 6-chloro deriv. + 90% HCO_2H under reflux, followed by NaOH	m. > 250°	1638
4-[purine with NH₂]	From the corr. 6-chloro deriv. + NH_3 in EtOH at 140°	m. > 250°	1638
C_{12}			
4-[purine with MeO]	From the corr. 6-chloro deriv. + NaOH in MeOH under reflux	m. > 250°, active against Newcastle virus	1638
4-[purine with MeNH]	From the corr. 6-chloro deriv. + $MeNH_2$ in EtOH at 140°	m. > 250°	1638

(a) Rxn. with NH_3 or amines (RR′NH) → the 6-amino deriv. (1638). (b) Rxn. with HCO_2H, followed by NaOH → the 6-hydroxy deriv. (1638). (c) Rxn. with NaOH in MeOH → the 6-methoxy deriv. (1638). (d) Rxn. with NaOH + PhOH → the 6-phenoxy deriv. (1638).

Table 32 Continued

x and R in $C_6H_4AsO_3H_2$	Prepd. from	Prop. and Rxn.	Ref.
C_{15} $CH_2(CH_2)_3N$ 4-[	From the corr. 6-chloro deriv. + excess pyrrolidine under reflux	m. > 250°	1638
$CH_2CH_2NCH_2CH_2O$ 4-[	From the corr. 6-chloro deriv. + morpholine in methyl cellosolve under reflux	m. > 250°	1638
C_{16} $CH_2N(CH_2)_3CH_2$ 4-[	From the corr. 6-chloro deriv. + excess piperidine under reflux	m. > 250°	1638
$HNCH_2CH=CHCH=CHO$ 4-[	From the corr. 6-chloro deriv. + furfurylamine in methyl cellosolve under reflux	m. > 250°	1638
C_{17} p-ClC_6H_4NH $p-ClC_6H_4NH$ 4-[	From the corr. 6-chloro deriv. + p-$ClC_6H_4NH_2$ in H_2O contg. HCl under reflux	m. > 250°	1638

PhO 4-[...]	From the corr. 6-chloro deriv. + PhOH + NaOH in H_2O under reflux	m. > 250°	1638
PhNH 4-[...]	From the corr. 6-chloro deriv. + aniline in H_2O contg. HCl under reflux	m. > 250°	1638
$C_6H_{11}NH$ 4-[...]	From the corr. 6-chloro deriv. + cyclohexyl-amine in methyl cellosolve under reflux	m. > 250°	1638
C_{18} PhCH$_2$NH 4-[...]	From the corr. 6-chloro deriv. + PhCH$_2$NH$_2$ in methyl cellosolve under reflux	m. > 250°	1638
MeOC$_6$H$_4$NH 4-[...]	From the corr. 6-dichloro deriv. + p-anisidine in H_2O contg. HCl under reflux	m. > 250°	1638

Table 33. Disubstituted Benzenearsonic Acids, RR'$C_6H_3AsO_3H_2$

R and R' in RR'$C_6H_3AsO_3H_2$	Prepd. from (Yield %)	Prop. and Rxn.	Ref.
C_6			
3,5-Br_2-		Co salt, (a)	688
4,3-Cl(O_2N)-	(M-474)	Dissocn. const. in H_2O-MeOH mixt.	664,816
		Thermodynamic data of dissocn. in H_2O-MeOH mixt.	816
2,5-Cl_2-		Interaction with auxins	653
3,4-Cl(H_2N)-		Co salt, (a)	
2,5-$H_2N(O_2N)$-	(M-480)	Rxn. with $NaNO_2$ and coupling with 3-bromo-chromotropic acid, (b)	916
C_7			
5,2-Me(O_2N)-		IR spectrum, qual. interpretation	1223
4,2-Me(H_2N)-		Rxn. with $NaNO_2$, followed by MeSNa → 2,4-MeS-(Me)$C_6H_3AsO_3H_2$	129
C_8			
4,2-Me(MeS)-	2,4-H_2N(Me)$C_6H_3AsO_3H_2$, (c)	Almost colorless crystals, m. 164-166°, soly. data, IR spectrum	129
	2,4-MeS(Me)$C_6H_3NH_2$ by the Bart rxn. (5.5)	A floatation agent for Kassiterite	129
2,3-Me_2-	(M-483)	Forms cryst. salts with Ca, Pb^{++}, Zn, and Co^{++} and powdery or flocculent ppts. with Sr, Ba, Cd, Mn^{++} and Cr^{3+}	1232
2,4-Me_2-	(M-483)	Forms cryst. salts with Ca and Zn and powdery or flocculent ppts. with Sr, Ba, Cd, Mn^{++}, and Cr^{3+}, (a)	1232
2,5-Me_2-		Forms cryst. salts with Ca, Co^{++} and Mn^{++} and powdery or flocculent ppts. with Sr, Ba, Cu^{++}, Cd, and Cr^{3+}	1232
2,6-Me_2-	(M-483)	Forms cryst. salts with Pb^{++}, Zn, Cd, and Mn^{++} and flocculent or powdery ppts. with Ca, Sr, Ba, Co^{++}, and Cr^{3+}	1232

	Preparation	Reactions	Ref.
3,4-Me$_2$- (M-483)		Forms cryst. salts with Ca and Pb^{++} and flocculent or powdery ppts. with Sr, Ba, Cu^{++}, Zn, Cd, Co^{++}, Mn^{++}, and Cr^{3+}	1232
C$_{12}$			
2,5-H$_2$N(4-O$_2$NC$_6$H$_4$N=N)-	2,5-HO$_3$SCH$_2$NH(4-O$_2$NC$_6$H$_4$-N=N)C$_6$H$_3$AsO$_3$H$_2$ by hydrolysis with 3% NaOH (52)	Rxn. with NaNO$_3$, followed by 4-H$_2$NC$_6$H$_4$SO$_3$H → 5,2-(4-O$_2$NC$_6$H$_4$N=N)(4-HO$_3$SC$_6$H$_4$N=N)C$_6$H$_3$AsO$_3$H$_2$	990
C$_{13}$			
2,5-(HO$_3$SCH$_2$NH)(4-O$_2$N-C$_6$H$_4$N=N)-	2-(NaO$_3$SCH$_2$NH)C$_6$H$_4$AsO$_3$H$_2$ + [4-O$_2$NC$_6$H$_4$N$_2$]BF$_4$ in 40% NaOAc at 10° (20)	Rxn. with 3% NaOH at 50-60° and acidification → 2,5-H$_2$N(4-O$_2$NC$_6$H$_4$N=N)C$_6$H$_3$AsO$_3$H$_2$	990
C$_{16}$			
5,2-O$_2$N(Br- [structure: OH OH, N=N, SO$_3$H, HO$_3$S])	2,5-H$_2$N(O$_2$N)C$_6$H$_3$AsO$_3$H$_2$ by diazotization and coupling with 2-bromochromotropic acid, (d)	Rxn. with Sc^{3+} → rose-pale blue, 1:1 salt, useful in spectrophotometric detn. of Sc at pH 4.5-5.5	277, 916
C$_{18}$			
2,5-(4-NaO$_3$SC$_6$H$_4$N=NNH)-(4-O$_2$NC$_6$H$_4$N=N)-	2,5-H$_2$N(4-O$_2$NC$_6$H$_4$N=N)C$_6$H$_3$-AsO$_3$H$_2$ by diazotization and coupling with 4-H$_2$N-C$_6$H$_4$SO$_3$H	Rxn. with Pb ions → colored salt, spectrophotometric detn. of Ph	990
2,5-O$_2$N[4-(4-HO$_3$SC$_6$H$_4$-N=N)C$_6$H$_4$NHN=N]- (M-499)		Rxn. with Hg^{++} → weak, nonspecific color rxn.	993
C$_{19}$			
5,2-O$_2$N[4-(4-HO$_3$SC$_6$H$_4$-N=N)C$_6$H$_4$N(Me)N=N]- (M-500)		Rxn. with Hg^{++} → weak, nonspecific color rxn.	993

(a) The Co salt is effective against Ascaridia galli and Heterakis gallinae in poultry (688). (b) The rxn. gave 3-[(2-arsono-4-nitrophenyl)azo]-6-bromo-4,5-dihydroxy-2,7-naphthalenedisulfonic acid (916). (c) By diazotization and a rxn. with MeSNa in NaOH soln. at 70-80°, 53% yield (129). (d) Absorption spectrum (916).

2.5.6.1.4 DI-, TRI-, AND TETRAARSONIC ACIDS

The methods of synthesis of diarsonic acids were summarized in the main volume on page 526; the compounds were described on pages 527-530 and in Table 65.

This supplement contains only diarsonic acids. No tri- and tetraarsonic acids were reported in the period covered here. As mentioned in the main volume, Arsenazo III and related compounds constitute an important group of reagents for colorimetric and spectrophotometric determination of trace amounts of various metals, particularly rare earth and transition metals. Moreover, the diarsonic acids are useful in extracting certain metals from aqueous solutions into organic solvents.

2,2'-(4,5-Dihydroxy-2,7-disulfo-3,6-naphthylenebisazo)dibenzenearsonic Acid, 3,6-Bis[(o-arsonophenyl)azo]-4,5-dihydroxy-2,7-naphthalenedisulfonic Acid, Arsenazo III, $As_2C_{22}H_{18}N_4O_{14}S_2$, (M-525)

Synth.: From chromotropic acid Ca salt in a $Ca(OH)_2$ paste by treating with 2-H_2O_3As-$C_6H_4N_2^+$ at 0° (1148, see also 1630).

Prop.: The compd. forms a red-crimson aq. soln. with an absorption max. at 540 mμ and an isosbestic point at 585 mμ. No protonation could be observed in $\leq$ N HCl soln. Potentiometric titration with NaOH gave 2 equiv. points, the 1st at pH 6.4, corr. to the neutralization of 4 more acidic groups, and the 2nd point corr. to the 1st phenolic group. Aq. soln. has ~4.5 acid groups dissocd. (1215). Spectrophotometric detn. of the stepwise dissocn. consts. of all 8 H^+ ions (1188). Titration with 0.0529 N KOH: 2 SO_3H and 2 AsO_3H_2 (to AsO_3H^-) groups are simultaneously neutralized at pH 6.29; one of the AsO_3H^- is titrated at pH 8.7 (1336). A study of the states of Arsenazo III in strongly acidic, neutral, and alk. media by means of MOLCAO. Electronic absorption spectra in 95% H_2SO_4 and at pH 9.0 were obtd. In the range from 1 N H_2SO_4 to pH 8 the reagent exists in the azo form. A single stage ionization of the acidic or basic groups leads to a disruption of the symmetry of the system and to the appearance of a quinone-hydrazone form, together with the azoide (1333). Electronic absorption spectrum in 0.6 N HCl (1527, see also 987). Spectrophotometric detn. in mixts. with Arsenazo I (1134). The pH effect on the protonation in concd. $HClO_4$ and H_2SO_4 solns. and changes in electronic absorption (1110). Principal spectrophotometric data (333). IR spectrum (1213). Polarographic data (1527). Paper chromatographic sepn. from the monoazo analog - anal. procedure (1410). Soly. data in $HClO_4$ (1237). Elemental anal. for As (1216) and for N (1217). A comparison with Phosphonazo III with respect to its sensitivity in the detn. of Th, U, Zr, La, Sc, and Ca (986, 987).

Rxn. with: U(VI) in $HClO_4$ → a 1:1 complex, spectrophotometric detn.; Th, Zr, Pu(IV) and U(IV) interfere in the detn. of U(VI) (1137).
Zr traces in 2 N acidic soln. in the presence of sorbents such as glass or teflon powder → sorption of the complex salt (1237).
Zr(^{95}Zr) in 2 N HNO_3 → a ppt. of variable compns., as detd. by radiometric

measurements; its soly. depends on the colloidal and semicolloidal props. of the chelate (1259).

Uses: Spectrophotometric detn. of Al (60), of Yb at pH 3 (101), of Zr(IV) in 6.5 N HClO$_4$ as [Zr(Arsenazo III)H$_9$]$_2$ (253), of NpO$_2^+$ at pH 2-5 (344), of Hf in iron and steel (831), of Li (uγ amts.) in the presence of other cations - anal. procedure (942), of Th^{4+} in 4-10 N HCl, using H$_2$C$_2$O$_4$ as a masking agent for Zr and Ti or in 3.5-6.0 N HNO$_3$ in the presence of urea (1630), of UO$_2^{++}$ at pH 3 in a buffer soln., after sepn. from Th, Zr, V, Al, Cr, and rare earth metals (1630), of U^{6+} by first reducing it or directly in 6 N HNO$_3$ (with urea); in 6 N acids Al, Fe^{3+}, Mg, Zn, Ca, Ba, Cr^{3+}, Mn^{++}, Ni, Co, and other elements do not interfere (1630), of U^{6+} in 5.6 M HClO$_4$, anal. procedure (1137), of U in urine, anal. procedure (1159), and of Pu(III) and Pu(VI) as a Pu(IV)-Arsenazo III complex (1135, 1136).

Extn.-spectrophotometric detn. of 0.1 γ Pu(IV) in the presence of Ag, Cu, Ni, Pb, Al, and La ions; the Pu(IV)-Arsenazo III complex is extd. with BuOH contg. (PhCH$_2$)$_3$N from 0.1 to 1.0 N solns. of HNO$_3$, HCl, or H$_2$SO$_4$ (1068).

Extn.-spectrophotometric detn. of trace amts. of rare earth elements from rocks - anal. procedure (1074).

Extn. and detn. of Sc and Th - anal. procedure (1546).

Extn. of Pa from U ores and sepn. from mixts. (449).

Copptn. of Cm with Arsenazo III + crystal violet at pH 1-6 (915).

Copptn. of Pa with Arsenazo III + crystal violet at pH 2-4 and with Arsenazo III + Rhodamine B in 0.1-3.0 N HCl (917).

An indicator for a complexometric titration of 1-250 mg Mn^{++}/100 ml at pH 8.5-10.0)1255).

Quant. adsorption of Pa from acid soln. with charcoal treated with Arsenazo III (1190).

Adsorption of trace amts. Zr from acid soln. with teflon or glass powder in the presence of Arsenazo III (1237).

Complex Salts: The formation of the Al salt, its stoichiometry, molar adsorptivity, dissocn. consts., and stability were detd. (60).

Lu^{3+} forms a 1:1 complex salt; the nuclearity of the complex was detd. by a competitive effect of H$^+$ (255).

Arsenazo III functions as a quadridentate and forms octahedral hexacoordinate 1:1 complexes and cubic octacoordinate 1:2 complexes with alk. earth metals, several lanthanides, Y, Zr, Hf, Th, and U (258).

La forms a cyclic salt; light absorption and color of its solns. at various pH values were detd. (919).

The Sc complex salt changes its color at pH 2.2 from pink to blue (1311). For the salts of Sc^{3+}, Y^{3+}, and La^{3+}, dissocn. consts., stoichiometric coeffs., molar absorptivities, the values of the electron shifts, and sensitivity of the rxns. were detd. (333).

Gd^{3+} forms a 1:1 complex salt at pH 1-4, with the highest yield at pH 2.4; electronic absorption max. at 650 mμ (539).

Pd(II) forms the 2:1 and 1:1 Pd-Arsenazo III chelates at pH 2-4; electronic absorption spectra of both chelates were obtd. Numerous cations interfere in the formation of the Pd chelates (1215).

U(VI) forms the 1:1 and 2:1 U/Arsenazo III complex salts in acid solns. Polarographic redn. gave two waves, the 1st corr. to 2 electron redn. of the -N=N- groups to -NHNH- and the 2nd consumes 7 electrons, while the $-AsO_3H_2$ groups are reduced to AsH_3. The instability consts. of both complexes were detd. Both $-AsO_3H_2$ groups participate in the formation of the 1:1 complex. The 1:1 complex is sol., while the 2:1 is a fine cryst. ppt.; electronic absorption spectra were obtd. (1527).

Zr in $HClO_4$ forms two complex salts, one with intermolec. rings having a mol. absorptivity of 1.48 x 10^5 at 665 mμ (1109).

A math. relation is developed for the function of the acidity of a medium at which a metal reacts with Arsenazo III as a function of the soly. prod. of the metal hydroxide (1111).

3,6-Bis[(p-arsonophenyl)azo]-4,5-dihydroxy-2,7-naphthalenedisulfonic Acid,
    Palladiazo, $As_2C_{22}H_{18}N_4O_{14}S_2$,

Synth.: From $[4-H_2O_3AsC_6H_4N_2]Cl$ by coupling with chromotropic acid, 44% yield (1214).

Prop.: A dark red powder with greenish irridescence, readily sol. in H_2O with a violet-purple color, sparingly sol. in EtOH and in HCl, insol. in most org. solvents. Aq. soln. is very stable. Dec. rapidly > 300°. In concd. HCl the acid is protonized, while the color changes reversibly to deep blue and deep emerald green (7-13 N HCl). Soln. in concd. NaOH is deep blue. Potentiometric titrn. gives 2 equiv. points. Electronic absorption spectrum (1214). IR spectrum (1213). Elemental anal. for As (1216) and for N (1217) - anal. procedures.

Rxn. with: Pd(II) → pale blue or greyish blue color, depending on the pH of the medium (1214).

Pb(II) in HCl soln. → blue color with a slow ppt. formation; NO_3^- ions inhibit the rxn. (1214).

Bi^{3+} in HCl soln. → blue color and a slow pptn. (1214).

Complex Salt with Pd(II): a 2:3 (Pd/Palladiazo) ratio was confirmed spectrophotometrically. The complex salt has an anionic nature; it can be extd. with diphenylguanidine chloride or quaternary ammonium salts dissolved in BuOH or higher alcs. (1214).

4,5,5'-Trihydroxy-3,4'-azobis{[6-(o-arsonophenyl)azo]-2,7-naphthalenedisulfonic Acid}, Arsenazo IV, $As_2C_{32}H_{24}N_6O_{21}S_4$,

Prop.: Max. absorbance at 520-530 mµ (1553).

Use: Spectrophotometric detn. of Zr, Th, and U^{4+} - anal. procedure (1641).

Complex Salt of Th(IV) contains 3 mol. Arsenazo III/Th, is stable in the pH range of 1-10 and thus pro-

vides a greater sensitivity for the detn. of Th^{4+} ion than does Arsenazo III. The salt has the max. absorbance at 660 mµ (1553).

4-{[4,6-Bis(p-arsonoanilino)-s-triazin-2-ylamino]-4'-[(4,6-diamino-s-triazin-2-yl)amino]-2,2'-stilbenedisulfonic Acid,

$As_2C_{32}H_{30}N_{12}O_{12}S_2$,

Rxn. with: CH_2O in alk. soln. → a polymeric prod. useful as a tanning agent (433).

Table 34. Diarsonic Acids

Diarsonic Acid	Prepd. from	Prop. and Rxn.	Ref.
C_{12}			
4-$H_2O_3AsC_6H_4N=NC_6H_4AsO_3H_2$-4 (M-532)		Inhibition of antibodies specific for the diarsonic acid	659
C_{13}			
	2-$H_2NC_6H_4AsO_3H_2$ in EtOH + EtOH + 33% HCHO	Light yellow solid, dec. at elevated temp. without melting, (a, b)	916
C_{14}			
	2-$H_2O_3AsC_6H_4N_2Cl$ + 2-H_2O_3As-$C_6H_4N=NCH(CN)CO_2H$ in alk. soln.	m. 230-233°, UV spectrum	532
2,2'-(-$CH_2OC_6H_4AsO_3H_2)_2$	(2-$H_2NC_6H_4OCH_2$-)$_2$ in AcOH-H_2SO_4 + $AsCl_3$ + $NaNO_2$ at 1-3°, followed by dec. with CuBr on a steam bath and oxidn. with H_2O_2 in alk. soln.	White ppt., rxn. with SO_2 in HCl contg. KI → (2-$Cl_2AsC_6H_4$-OCH_2-)$_2$	292
4,4'-(-$CH_2NHC_6H_4AsO_3H_2)_2$ (M-534)		Toxic to Dicrocoelium lanceolatum flukes, (b)	153
C_{16}			
	2,4,6-Cl_3-5-O_2N-pyrimidine + 4-$H_2NC_6H_4AsO_3H_2$, (c)	Yellow ppt., m. > 250°	1638

 C₁₇	4,6-Cl$_2$-5-O$_2$N-pyrimidine + 4-H$_2$N$_2$C$_6$H$_4$AsO$_3$H$_2$, (c)	Yellow ppt., m. > 250°	1638
	4-(6-Cl-9H-purin-9-yl)C$_6$H$_4$-AsO$_3$H$_2$ + 4-H$_2$NC$_6$H$_4$AsO$_3$H$_2$ in H$_2$O contg. HCl under reflux	m. > 250°	1638
C$_{22}$	4,2-Cl(H$_2$O$_3$As)C$_6$H$_3$N$_2^+$ + chromotropic acid Ca salt in Ca(OH)$_2$ paste at 0°		1148
	Chromotropic acid + [2,4-H$_2$O$_3$-As(O$_2$N)C$_6$H$_3$N$_2$]Cl in alk. soln. satd. with Ca^{++}, Mg^{++}, or Sr^{++} at -5 to 0°	An indicator for complexo-metric titration of rare earth metals	259

(a) Diazotization and coupling with chromotropic acid deriv. → the corr. 5,5′-methylenebis(naphthylazobenzene-arsonic acid) deriv. (916). (b) A salt with 1,3-bis(2-bromo-4-amidinophenoxy)propane is an effective spirilli-cide and bactericide for gingivitis treatment (459). (c) The condensation is carried out in aq. Me$_2$CO contg. HCl under reflux (1638).

Table 34 Continued

Diarsonic Acid	Prepd. from	Prop. and Rxn.	Ref.
C_{22} (Cont.)			
	Chromotropic acid + [5,2-Cl-$(H_2O_3As)C_6H_3N_2$]Cl, followed by [2-$H_2O_3AsC_6H_4N_2$]Cl (1:1:1)	Complex salts, (d)	260
	Chromotropic diamide + [2-H_2O_3-$AsC_6H_4N_2$]Cl in aq. soln. contg. Py and CaO at -3 to 0°		312
C_{23}			
	Chromotropic acid + [5,2-HO_2C-$(H_2O_3As)C_6H_3N_2$]Cl, followed by [2-$H_2O_3AsC_6H_4N_2$]Cl (1:1:1)	Complex salts, (d)	260
	5,5'-Methylenebis(2-aminoben-zenearsonic acid) by diazoti-zation and coupling with chro-motropic acid	Electronic absorption spectrum	916

Structure	Preparation	Notes	Ref.
(structure: Me-phenyl, AsO_3H_2, H_2O_3As, OH OH, N=N, N=N, SO_3H, HO_3S)	Chromotropic acid + [5,2-Me-$(H_2O_3As)C_6H_3N_2$]Cl, followed by [2-$H_2O_3AsC_6H_4N_2$]Cl (1:1:1)	Complex salts, (d, e)	260
C24 (structure: CO_2H, HO_2C, AsO_3H_2, H_2O_3As, OH OH, N=N)	Chromotropic acid + repeated [5,2-$HO_2C(H_2O_3As)C_6H_3N_2$]Cl reagent (1:1:1)	Complex salts, (d)	260
(structure: Me, AsO_3H_2, H_2O_3As, OH OH, N=N, SO_3H, HO_3S)	Chromotropic acid + [5,2-HO_2C-$(H_2O_3As)C_6H_3N_2$]Cl (1:2) in one step in alk. soln. satd. with Ca^{++}, Mg^{++}, or Sr^{++} at -5 to 0°	Useful as an indicator in complexometric titrn. of rare earth metals	259
(structure: Me, AsO_3H_2, H_2O_3As, OH OH, N=N, SO_3H, HO_3S)	Chromotropic acid + [5,2-Me-$(H_2O_3As)C_6H_3N_2$]Cl (1:2)	Complex salts, (d)	260
C25 (structure: Pr, AsO_3H_2, H_2O_3As, OH OH, N=N, SO_3H, HO_3S)	Chromotropic acid + [5,2-Pr-$(H_2O_3As)C_6H_3N_2$]Cl, followed by [2-$H_2O_3AsC_6H_4N_2$]Cl (1:1:1)	Complex salts, (d)	260

(d) Complex salts of the acid with Sc, Zr, Hf, Th, U^{4+}, UO_2^{++}, La, Lu, and Y were prepd. and their molar extinction coeffs. were obtd. (260). (e) The nuclearity of the 1:1 La salt was detd. by a competitive effect of H^+ (255).

Table 34 Continued

Diarsonic Acid	Prepd. from	Prop. and Rxn.	Ref.
C_{28}			
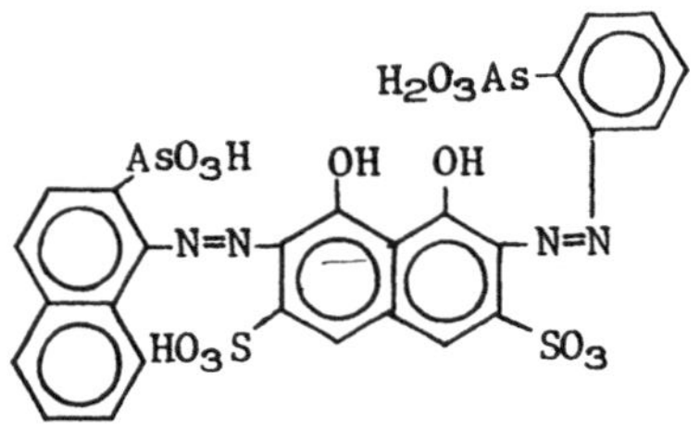	Chromotropic acid + [2-H_2O_3As-$C_{10}H_6$-1-N_2]Cl, followed by [2-$H_2O_3AsC_6H_4N_2$]Cl (1:1:1)	Complex salts, (d)	260
C_{32}			
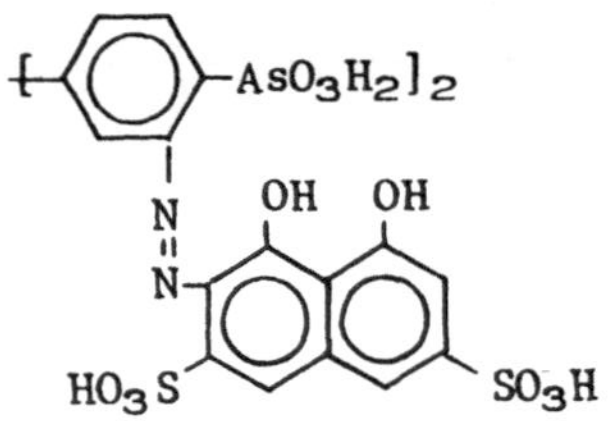		Copptn. of Cm with Arsenazo II + crystal violet	915
		Spectrophotometric detn. of Sc	1311
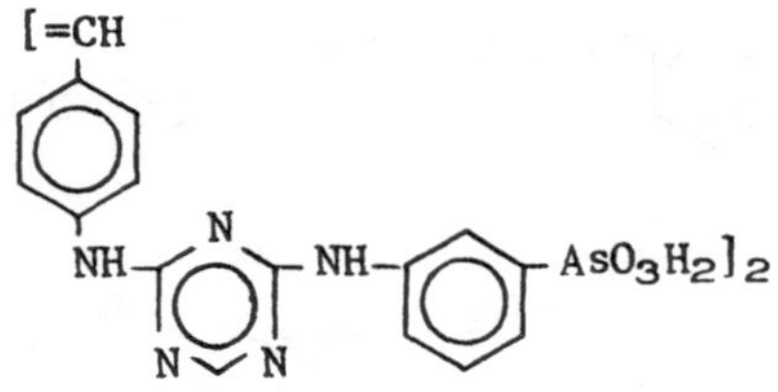		Rxn. with CH_2O → a fluorescent polymer	433

Arsenazo II (M-529)

C₃₃		916

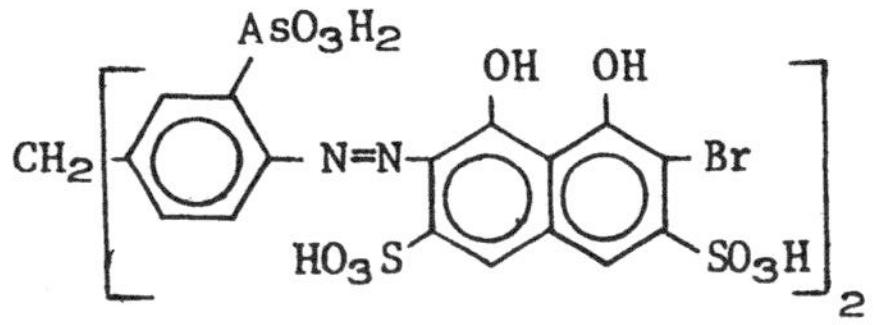

C_{33}

5,5'-Methylenebis(2-aminoben-zenearsonic acid) by diazoti-zation and coupling with 3-bromochromotropic acid

Absorption spectrum

916

C_{34}

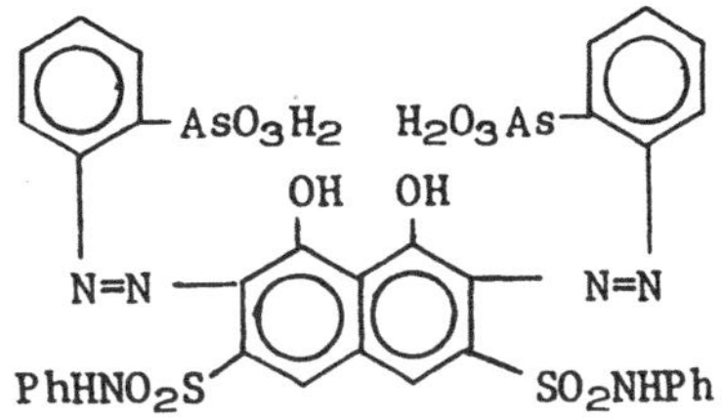

Chromotropic dianilide + [2-H₂O₃AsC₆H₄N₂]Cl in aq. soln. contg. Py and CaO at -3 to 0°

(f)

312

C_{58}

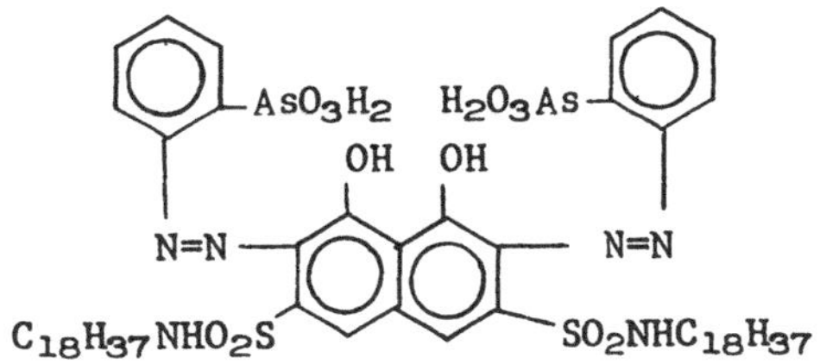

Chromotropic dioctadecyl-amide + [2-H₂O₃AsC₆H₄N₂]Cl in H₂O contg. Py and CaO at -3 to 0°

312

(f) A selective reagent for extractive photometric detn. of trace amts. of Th (262).

2.5.6.1.5 POLYMERS CONTAINING ARSONO GROUPS

Several compounds of this type were described in the main volume on page 542.

Poly(arsenazo-azostyrene)

Synth.: From polystyrene by nitration with fuming HNO_3 at -30° to obtn. a degree of nitrn. of 0.6, reducing the nitro groups and coupling with Arsenazo I in the presence of CaO and Py (1332, 1337).

Prop.: Dark blue or dark red powder, almost insol. in H_2O, org. solvents, and HCl, but partly sol. in alkalis. The polymer swells in concd. H_2SO_4 and turns green. It has a high sorption capacity for Th, Zr, Hf, UO_2, and Pu from strongly acid solns. and for Cu, Mo, UO_2, and rare earth elements from weakly acid solns. The polymer is useful for concn. of these elements from highly dild. solns. (1332, 1337).

2.5.6.1.6 ARSONIC ACID ESTERS

The methods of synthesis of the esters were summarized in the main volume on page 543, and the compounds were compiled in Table 66 on pages 544-547.

The esters reported in this supplement (Table 35) were prepared from arsonic acids and alcohol by heating and continuously removing water being set free.

2.5.6.2.5 TRITHIOARSONIC ACIDS

The main volume contains methanetrithioarsonic acid, the only representative of this type (M-550).

p-Toluenetrithioarsonic Acid, $AsC_7H_9S_3$, $4\text{-}MeC_6H_4AsS_3H_2$

Synth.: From $4\text{-}MeC_6H_4AsO(ONH_4)_2$ + H_2S in aq. soln., followed by a treatment of the resulting sulfide with NaSH; the prod. was isolated as the di-Na salt (128).

Prop.: The di-Na salt forms white needles contg. H_2O of crystn., m. $> 45°$, dec. $\sim 280°$, very sol. in H_2O; IR spectrum (128).

Table 35. Arsonic Acid Esters

RAsO(OR')₂	Prepd. from	Prop. and Rxn.	Ref.
MeAsO(OR)₂			
MeAsO(OEt)₂ (M-544)	MeAsO₃H₂ + EtOH in C₆H₈ with azeotropic removal of H₂O	At 187° isomerizes to MeOAs(OEt)₂, (a)	338
MeAsO(O-i-Pr)₂ (M-544)	MeAsO₃H₂ + i-PrOH, (b)	b_{22} 121°, n_D^{20} 1.4420, at 227° isomerizes to MeOAs(Oi-Pr)₂	338
MeAsO(OC₅H₁₁)₂	MeAsO₃H₂ + n-C₅H₁₁OH, (c)	b_2 144–146°, d_{20} 1.1245, n_D^{20} 1.4551 At 200° isomerizes to MeOAs(OC₅H₁₁)₂, (d)	335, 336, 338
EtAsO(OR)₂			
EtAsO(OEt)₂ (M-545)	EtAsO₃H₂ + EtOH in C₆H₈ with azeotropic removal of H₂O	> 217°, isomerizes to (EtO)₃As, (a, d)	338
EtAsO(O-i-Pr)₂	EtAsO₃H₂ + i-PrOH, (b)	b_{12} 113°, d_4^{20} 1.784, n_D^{20} 1.4420, at 265° isomerizes to EtOAs(Oi-Pr)₂, (a)	338
EtAsO(OC₅H₁₁)₂	EtAsO₃H₂ + n-C₅H₁₁OH under reflux, with removal of H₂O	b_5 152–154°, d_{20} 1.1036, n_D^{20} 1.4535 At 200° isomerizes to EtOAs(OC₅H₁₁)₂, (d, e)	336
CH₂=CHCH₂AsO(OR)₂			
CH₂=CHCH₂AsO(OEt)₂	RAsO₃H₂ + EtOH, (f)	b_2 88°, d_{20} 1.2677, n_D^{20} 1.4690 At 155° isomerizes to CH₂=CHCH₂OAs(OEt)₂	335, 338
CH₂=CHCH₂AsO(O-i-Pr)₂	RAsO₃H₂ + i-PrOH, (b)	b_6 104.5°, d_4^{20} 1.292, n_D^{20} 1.4468	338

(a) IR spectra in the solid state and in various solvents (1387b). (b) By heating in a reactor with a H₂O trap contg. CuSO₄ (338). (c) By heating in a flask equipped with a H₂O trap (335). (d) The isomerization occurs at much lower temp. in the presence of electrophilic reagents (336). (e) A noncatalytic isomerization starts at 245° (338). (f) The rxn. is carried out in a flask with a trap for H₂O contg. CuSO₄ or CaC₂ (335). (g) At 240° isomerizes to BuOAs(OEt)₂ (338, see also 335), but in the presence of EtI the isomerization occurs at 91° (336) and proceeds via the [BuAs(OEt)₃]I intermediate (337).

Table 35 Continued

RAsO(OR')$_2$	Prepd. from	Prop. and Rxn.	Ref.
i-PrAsO(OR)$_2$			
i-PrAsO(OEt)$_2$	i-PrAsO$_3$H$_2$ + EtOH, (f)	b$_4$ 89-91°, d$_{20}$ 1.2176, n$_D^{20}$ 1.4450	335
		At 115° isomerizes to i-PrOAs(OEt)$_2$	338
BuAsO(OR)$_2$			
BuAsO(OEt)$_2$	BuAsO$_3$H$_2$ + EtOH, (f)	b$_{13}$ 132-133°, d$_{20}$ 1.2009, n$_D^{20}$ 1.4532, (g)	335
		Rxn. with EtI at 91° → As(OEt)$_3$	337
BuAsO(OCH$_2$CH=CH$_2$)$_2$	BuAsO$_3$H$_2$ + CH$_2$=CHCH$_2$OH, (f)	b$_5$ 144°, d$_{20}$ 1.1935, n$_D^{20}$ 1.4789	335
BuAsO(O-i-Pr)$_2$	BuAsO$_3$H$_2$ + i-PrOH, (f)	b$_{2.5}$ 100°, d$_{20}$ 1.1282, n$_D^{20}$ 1.4463	335
BuAsO(OBu)$_2$ (M-546)		At 250° → (BuO)$_3$As, (h)	338
BuAsO(OC$_5$H$_{11}$)$_2$	BuAsO$_3$H$_2$ + C$_5$H$_{11}$OH, (c)	b$_4$ 156-158°, d$_{20}$ 1.0781, n$_D^{20}$ 1.4560, (h,i)	335
i-BuAsO(OR)$_2$			
i-BuAsO(OEt)$_2$	i-BuAsO$_3$H$_2$ + EtOH, (f)	b$_4$ 98-99°, d$_{20}$ 1.1990, n$_D^{20}$ 1.4525	335
		At 257° isomerizes to i-BuOAs(OEt)$_2$	338
i-BuAsO(O-i-Pr)$_2$	i-BuAsO$_3$H$_2$ + i-PrOH, (b)	b$_4$ 102°, d$_4^{20}$ 1.1292, n$_D^{20}$ 1.4468, at 282°	338
		isomerizes to i-BuOAs(O-i-Pr)$_2$	
i-BuAsO(OC$_5$H$_{11}$)$_2$	i-BuAsO$_3$H$_2$ + C$_5$H$_{11}$OH, (c)	b$_{2.5}$ 143°, d$_{20}$ 1.0758, n$_D^{20}$ 1.4553, (j)	335
O$_2$NC$_6$H$_4$AsO(OR)$_2$			
o-O$_2$NC$_6$H$_4$AsO(OBu)$_2$	RAsO$_3$H$_2$ + BuOH, (k)	d$_4^{20}$ 1.3311, n$_D^{20}$ 1.5300	959
p-O$_2$NC$_6$H$_4$AsO(OBu)$_2$	RAsO$_3$H$_2$ + BuOH, (k)	d$_4^{20}$ 1.3300, n$_D^{20}$ 1.5312	959
PhAsO(OR)$_2$			
PhAsO(OCH$_2$CH$_2$Cl)$_2$	PhAsO$_3$H$_2$ + ClCH$_2$CH$_2$OH, (k)	d$_4^{20}$ 1.5438, n$_D^{20}$ 1.5560	959
PhAsO(OCH$_2$CH=CH$_2$)$_2$	PhAsO$_3$H$_2$ + CH$_2$=CHCH$_2$OH, (f)	b$_3$ 173-175°, d$_{20}$ 1.1919, n$_D^{20}$ 1.5038	335
PhAsO(OCH$_2$CH$_2$OMe)$_2$	PhAsO$_3$H$_2$ + MeOCH$_2$CH$_2$OH, (f)	d$_4^{20}$ 1.3804, n$_D^{20}$ 1.5280	959
PhAsO(OBu)$_2$	PhAsO$_3$H$_2$ + BuOH, (f)	d$_4^{20}$ 1.2054, n$_D^{20}$ 1.5060	959
PhAsO(OCH$_2$CH$_2$OPh)$_2$	PhAsO$_3$H$_2$ + PhOCH$_2$CH$_2$OH, (f)	n$_D^{20}$ 1.5787	959

p-HOC$_6$H$_4$AsO(OBu)$_2$	RAsO$_3$H$_2$ + BuOH, (f)	n$_D^{20}$ 1.5570		959
p-MeC$_6$H$_4$AsO(OR)$_2$				
p-MeC$_6$H$_4$AsO(OEt)$_2$	RAsO$_3$H$_2$ + EtOH, (f)	b$_3$ 154–156°, d$_{20}$ 1.2790, n$_D^{20}$ 1.5223		335
p-MeC$_6$H$_4$AsO(OCH$_2$CH=CH$_2$)$_2$	RAsO$_3$H$_2$ + CH$_2$=CHCH$_2$OH, (f)	b$_3$ 164–166°, d$_{20}$ 1.2685, n$_D^{20}$ 1.5360		335
p-MeC$_6$H$_4$AsO(OBu)$_2$	RAsO$_3$H$_2$ + BuOH, (f)	b$_3$ 168–170°, d$_{20}$ 1.1674, n$_D^{20}$ 1.5080		335
PhCH$_2$AsO(OC$_5$H$_{11}$)$_2$	RAsO$_3$H$_2$ + C$_5$H$_{11}$OH, (c)	d$_{20}$ 1.1284, n$_D^{20}$ 1.4982		335
		At 177° isomerizes to PhCH$_2$OAs(OC$_5$H$_{11}$)$_2$		338

(h) IR spectra in the solid state and in C$_6$H$_6$ soln. (1387b). (i) At 257° isomerizes to BuOAs(OC$_5$H$_{11}$)$_2$ (338, see also 335); in the presence of EtI isomerizes at 107° (336). (j) At 272° isomerizes to i-BuOAs(OC$_5$H$_{11}$)$_2$ (338, see also 335); in the presence of C$_5$H$_{11}$I the isomerization occurs at 121° (336). (k) The rxn. is carried out in C$_6$H$_6$ under reflux with azeotropic removal of H$_2$O (959).

2.5.6.2.7 ALKYL- AND ARYLARSENIC DISULFIDES

A few compounds of this type were reported in the main volume (M-550).

Methyldithioxoarsenic, $AsCH_3S_2$, $MeAsS_2$, (M-550)
Biol. Prop.: Highly protective and curative effects on the control of rice
sheath blight caused by Pellicularia sasakii (1493).

3. HETEROCYCLIC ARSENIC COMPOUNDS

3.2 FOUR-MEMBERED RINGS

3.2.1 RING SYSTEMS WITH ARSENIC HETEROATOMS

3.2.1.4 TETRARSETANE DERIVATIVES

Tetrasubstituted tetrarsetanes are formed from primary dihaloarsines by a
reduction with mercury. Arsenobenzene, when heated with hexacarbonylmolybde-
num affords tetraphenyltetrarsetane complexes with tetracarbonylmolybdenum.

Tetrakis(trifluoromethyl)tetrarsetane, $As_4C_4F_{12}$, $\quad$ CF$_3$As——As-CF$_3$
Synth.: From CF_3AsI_2 by shaking with Hg in a sealed $\qquad$ | $\quad$ |
tube at 25°, sepg. the solid fraction by a magnetic $\quad$ CF$_3$As——As-CF$_3$
separatory funnel and crystg. from hexane at -78° (408).
Prop.: m. 98.2°, sensitive to basic impurities, $\log P = 12.4514-4003/T$, mol.
wt. detn.; IR, UV, and mass spectra (408).
Rxn. with: Oxygen → a nonvolatile white solid (408).
Alkaline soln. → HCF_3 (408).

Tetrakis(pentafluorophenyl)tetrarsetane, $As_4C_{24}F_2$, $\quad$ C$_6$F$_5$As—— AsC$_6$F$_5$
Synth.: From $C_6F_5AsCl_2$ by shaking with Hg, 74% yield $\quad$ | $\quad$ |
(650, 651). $\qquad$ C$_6$F$_5$As——AsC$_6$F$_5$
Prop.: m. 141°, NMR spectrum (650, 651).

Tetracarbonyl(tetraphenyltetrarsetane)molybdenum, $\qquad$ $As_4MoC_{28}H_{20}O_4$,
$\quad$ (PhAs)$_4$Mo(CO)$_4$
Synth.: From arsenobenzene + $Mo(CO)_6$ in C_6H_6 under reflux for 5 days (583).
Prop.: IR spectrum (583).

Octacarbonyl(tetraphenyltetrarsetane)dimolybdenum, $\qquad$ $As_4Mo_2C_{32}H_{20}O_8$,
$\quad$ (PhAs)$_4$[Mo(CO)$_4$]$_2$
Synth.: From arsenobenzene + $Mo(CO)_6$ in MePh under reflux (583).
Prop.: IR spectrum (583).

3.2.6 RING SYSTEMS WITH ARSENIC AND SULFUR HETEROATOMS

4-Cyanoimino-2-phenyl-1,3,2-dithiarsetane, $AsC_8H_5N_2S_2$,
Synth.: From $PhAsI_2$ + $NCN=C(SK)_2$ in DMF (1089).
Prop.: m. 93-95.5° (1089).

PhAs⟨S⟩=NCN (with S atoms shown at top and bottom of ring)

3.2.9 RING SYSTEMS WITH ARSENIC, OXYGEN, AND NITROGEN HETEROATOMS

2,2-Dihydro-2,2,2-triisobutyl-1-phenyl-1,3,2-azoxarsetan-4-one,
$AsC_{19}H_{32}NO_2$,   i-Bu$_3$AsN(Ph)C(O)O
Synth.: From i-Bu$_3$AsO + PhNCO in C_6H_6 at r.t. (223).
Prop.: White crystals, m. 87° (223).

3.3   FIVE-MEMBERED RING SYSTEMS

3.3.1   RING SYSTEMS WITH ARSENIC HETEROATOMS

3.3.1.1 ARSOLANE AND ARSOLE DERIVATIVES

The methods of synthesis of arsolane and arsole derivatives were summarized in the main volume on page 552. A new method for the synthesis of arsole derivatives consists of a condensation of primary arsines with 1,3-butadiyne derivatives in the presence of butyllithium. Moreover, arsole derivatives are formed from primary dihaloarsines by a condensation with acetylene derivatives complexed with iron carbonyl in methylene dichloride under ultraviolet irradiation. The arsolane and arsole derivatives are compiled in Table 36.

Table 36. Arsolane and Arsole Derivatives

Compound	Prepd. from (Yield %)	Prop. and Rxn.	Ref.
C_{10} (As=O, Ph structure)	(M-553)	A catalyst for the trimerization of isocyanates	728
C_{12} (cyclohexyl arsolane)	$[CH_2(CH_2)_3As(C_6H_{11})_2]Cl$ by distn. in vacuum (86)	b_8 116°	1533
(Me, As Ph, Me arsole)	$PhAsH_2$ + $MeC\equiv CC\equiv CMe$ in the presence of BuLi (33)	$b_{0.5}$ 81°, solns. show a blue or green fluorescence, UV and NMR spectra (a, b)	1002
C_{16} (Ph, As K, Ph arsole)	1,2,3-Triphenylarsole + K in $MeOCH_2CH_2OMe$ under reflux	Rxn. with RX(alkyl halides) → $RAsC(Ph)=CHCH=CPh$	1002
$[\overline{\quad}As(C_6H_{11})_2]Cl$	$Cl(CH_2)_4As(C_6H_{11})_2$ at 170°	Crystals, m. 200°, sol. in EtOH, distn. in vacuum → $\overline{CH_2(CH_2)_3AsC_6H_{11}}$	1533
C_{17} (Ph, As Me, Ph arsole)	$PhC=CHCH=C(Ph)AsK$ + Me halide (52)	m. 100-101.5°, UV and NMR spectra	1002
C_{18} (Ph, As Et, Ph arsole)	$PhC=CHCH=C(Ph)AsK$ + Et halide (43)	m. 77.5-79°, UV and NMR spectra	1002

 PhC=CHCH=C(Ph)AsK + MeOCH$_2$ halide (33)	m. 82.5-84.5°, UV and NMR spectra	1002
C$_{22}$ PhAsH$_2$ + p-ClC$_6$H$_4$C≡CC≡CC$_6$H$_4$Cl-p in the presence of BuLi (83)	Lemon-yellow crystals, m. 160-161°, solns. give a green or blue fluorescence; UV and NMR spectra	1002
 PhAsH$_2$ + PhC≡CC≡CPh in the presence of BuLi (83)	Lemon-yellow crystals, m. 186.5-187.5°, solns. give a strong green or blue fluorescence; UV and NMR spectra, (c, d)	1002
C$_{24}$ PhAsH$_2$ + p-MeC$_6$H$_4$C≡CC≡CC$_6$H$_4$Me-p in the presence of BuLi (58)	Lemon-yellow crystals, m. 179.5-180.5°, solns. show blue or green fluorescence; UV and NMR spectra	1002
C$_{28}$ Pentaphenylarsole + Na by heating in MePh under reflux	Grayish brown suspension, (e)	221

(a) Rxn. with MeO$_2$CC≡CCO$_2$Me → 3,6-dimethylphthalic acid di-Me ester (1002). (b) Rxn. with (NC)$_2$C=C(CN)$_2$ → 1,4-dimethyl-7-phenyl-7-arsabicyclo[2.2.1]hept-5-ene-2,2,3,3-tetracarbonitrile (1002). (c) Rxn. with MeO$_2$CC≡CCO$_2$Me → 3,6-diphenylphthalic acid di-Me ester (1002). (d) Rxn. with K in MeOCH$_2$CH$_2$OMe under reflux → PhC=CHCH=C(Ph)AsK (1002). (e) Rxn. with MeI → 1-methyl-2,3,4,5-tetraphenylarsole (221).

Table 36 Continued

Compound	Prepd. from (Yield %)	Prop. and Rxn.	Ref.
C_{29}	$PhC=C(Ph)C(Ph)=C(Ph)AsNa$ + MeI in MePh (68)	m. 212–212.5°	221
C_{30}	$PhAsH_2$ + 1,4-di-α-naphthyl-1,3-butadiyne in the presence of BuLi (86)	Lemon yellow crystals, m. 203.5–205.0°, solns. show blue or green fluorescence; UV and NMR spectra	1002
C_{34} (M-554)	$PhAsCl_2$ + $[Fe_2(CO)_6(PhC{\equiv}CPh)_2]$ or $[Fe_2(CO)_5(PPh_3)(PhC{\equiv}CPh)_2]$ in CH_2Cl_2 under UV irradiation at 30°	Yellow fluorescent, m. 213–314.5°, (f)	767

(f) Rxn. with Na in MePh under reflux → $[PhC=C(Ph)C(Ph)=C(Ph)As]^-Na^+$ (221).

3.3.1.5 DIARSOLANE AND BENZODIARSOLE DERIVATIVES

This group of compounds described in the main volume (M-560) included benzodiarsolium salts only. The following 1,2-diarsolane derivative is a new compound.

1,2-Diphenyl-1,2-diarsolane, $As_2C_{15}H_{16}$,

Synth.: From PhAsHK + Cl(CH$_2$)$_3$Cl in THF by heating, dec. with H$_2$O, sepg. the solid phase and distg. the liq., a mixt. of the title compd. with PhAsH(CH$_2$)$_3$AsHPh was obtd. The mixt. was dissolved in Et$_2$O, treated with BuLi, followed by dioxan, 73% yield (1535). From PhAs(Li)(CH$_2$)$_3$As(Li)Ph·2 Dioxan in THF-C$_6$H$_6$ mixt. by adding BrCH$_2$CH$_2$Br in C$_6$H$_6$, 56% yield (1535).

By-prod. from the rxn. of PhAsHK with Br(CH$_2$)$_3$Br in C$_6$H$_6$ under reflux (1535).

Prop.: b$_3$ 193-195°, sol. in Et$_2$O, C$_6$H$_6$, dioxan, and THF, sparingly sol. in petr. ether (1535).

Rxn. with: PhLi in Et$_2$O under reflux, followed by dioxan → Ph$_2$As(CH$_2$)$_3$As-(Li)Ph·Dioxan (1535).

BuLi in Et$_2$O under reflux, followed by dioxan → BuPhAs(CH$_2$)$_3$As(Li)Ph·Dioxan (1535).

MeI in EtOH-Et$_2$O → [PhAs(I)(CH$_2$)$_3$AsMe$_2$Ph]I (1535).

LiAlH$_4$ in dioxan under reflux → PhAsH$_2$ + PrPhAsH (790).

2,3-Dihydro-1,1,3,3-tetramethyl-1H-1,3-benzodiarsolium Salts,
  $[As_2C_{11}H_{18}]^{++}2\ X^-$,   (M-560)

Synth.: From o-C$_6$H$_4$(AsMe$_2$)$_2$ + CH$_2$Br$_2$ and CH$_2$I$_2$, resp., in a sealed tube, the dibromide and diiodide were obtd. Metathesis of the dibromide with KClO$_4$ gave the diperchlorate (372).

Prop.: For m. of the salts see (M-560). The salts are readily sol. in H$_2$O and the halide ions are freely titratable with AgNO$_3$. Mol. cond. data correspond closely to values expected for the 2:1 electrolytes (372).

X = Br, I, ClO$_4$

3.3.1.6 DIBENZARSOLE DERIVATIVES

The methods of synthesis of dibenzarsole derivatives were summarized in the main volume on page 560. In the four-year period covered by this supplement a few new methods of synthesis of this ring system were developed. Thus, dibenzarsole derivatives have been prepared by the condensation of arsenic trichloride with 2,2'-biphenylylenedilithium and 2,2'-biphenylylenecadmium, respectively, from primary dihaloarsines by the condensation with 2,2'-biphenylylenedilithium, and from 5,5'-spirobi(5H-dibenzarsole) derivatives by pyrolysis. 5-Alkyl- and 5-aryldibenzarsoles have been converted to quaternary dibenzarsolium salts by alkyl halides. The quaternary salts are readily reverted to 5-alkyl- and 5-aryldibenzarsoles by pyrolysis. Moreover, quaternary dibenzarsolium salts have been prepared from 5-alkyl-5,5'-spirobi(5H-dibenzarsole) by cleaving the spiro structure with a hydrohalic acid.

Table 37. Dibenzarsole Derivatives

Structure with positions 8 9 1 2 3 4, As, 5, 6, 7, where R is:

Deriv. of	Prepd. from (Yield %)	Prop. and Rxn.	Ref.
C_{12}			
5-Chloro- (M-562)	By-prod. from the rxn. of $AsCl_3$ + $(2-LiC_6H_4-)_2$ in Et_2O	m. 159-160°, (a)	723
	2,2′-Biphenylylene-Cd + $AsCl_3$ in Et_2O (82)	Rxn. with BuLi → 5-Eu-dibenzarsole	723
5-Iodo-	2,2′-Biphenylylene-Cd + $AsCl_3$ (1:1) in the presence of I^- in Et_2O followed by hydrolysis (72)	m. 168-169°	723
C_{13}			
5-Methyl- (M-363)	5-Cl-dibenzarsole + MeLi in Et_2O (68)	Rxn. with chloramine T, (b)	728
	5,5-EtMe-dibenzarsolium iodide by thermal dec., (c)		57
C_{14}			
5-Ethyl-	$EtAsI_2$ in C_6H_6 by adding to a rxn. mixt. of $(2-BrC_6H_4-)_2$ with Li in Et_2O and heating under reflux	bo.oz 120°, (d)	57
	5,5-MeEt-dibenzarsolium iodide by thermal dec., (c)		57
	5,5,8,8-Et_4-5,6,7,8-tetrahydrodibenzo-[e,g][1,4]diarsocinium dibromide by thermal dec.		57
	9,9,16,16-Et_4-9,10,15,16-tetrahydrotri-benzo[b,d,h][1,6]diarsecinium dibro-mide by thermal dec., (e)		57
5-[(MeO)$_2$P(S)S]- (M-564)		m. 82-83°, fungicidal prop.	68

Compound	m.p. / conditions	Ref.
C$_{15}$		
5-[EtOC(S)S]-	m. 94–95°	68
5-[Me$_2$NC(S)S]- (M-564)	m. 139–140°	68
	5-Et-dibenzarsole + MeI under reflux	
	m. 156.5–157° (dec.), (c)	57
	The picrate forms yellow crystals, m. 182–183°	
C$_{16}$		
5-[ProC(S)S]-	m. 81–82°	68
5-[i-ProC(S)S]-	m. 69–70°	68
5-Butyl-	5-Cl-dibenzarsole + BuLi in Et$_2$O (78)	
	m. 51–52°, (f)	723
5-[(EtO)$_2$P(S)S]- (M-564)	m. 85–86°, fungicidal prop.	68
C$_{17}$		
5-[O(CH$_2$CH$_2$)$_2$NC(S)S]-	m. 178–180°	68
5-[BuOC(S)S]-	m. 73–75°	68
5-[Et$_2$NC(S)S]-	m. 180–181°	68
C$_{18}$		
5-[CH$_2$(CH$_2$)$_4$NC(S)S]-	m. 163–164°	68
5-[(PrO)$_2$P(S)S]- (M-565)	n_D^{20} 1.6527, fungicidal prop.	68
5-[(i-PrO)$_2$P(S)S]- (M-565)	m. 73–74°, fungicidal prop.	68

(a) m. 157–158°; fungicidal activity (68).   (b) Rxn. with chloramine T, followed by (2-LiC$_6$H$_4$-)$_2$ in Et$_2$O and hydrolysis → 5-Me-5,5′-spirobi(5H-dibenzarsole) (723).   (c) A thermal dec. gave a 1:2 mixt. of 5-Me- and 5-Et-dibenzarsole (57).   (d) Rxn. with MeI → 5,5-MeEt-dibenzarsolium iodide (57).   (e) The thermal dec. gave a mixt. of 5-Et-dibenzarsole and [2-(Et$_2$As)C$_6$H$_4$-]$_2$ (57).   (f) Rxn. with chloroamine T in THF, followed by (2-LiC$_6$H$_4$-)$_2$ in Et$_2$O and hydrolysis → 5-Bu-5,5′-spirobi(5H-dibenzarsole) (723).

Table 37 Continued

Deriv. of where R is:	Prepd. from (Yield %)	Prop. and Rxn.	Ref.
C$_{20}$			
2,3,7,8-Bis(methylenedioxy)-5-phenyl-	PhAsCl$_2$ + 2,2'-dibromo-4,5,4',5'-bis(methylenedioxy)biphenyl + BuLi in Et$_2$O-C$_6$H$_6$ (51)	Colorless conglomerate of needles, m. 231-232°	434
5-[(BuO)$_2$P(S)S]-		n$_D^{20}$ 1.6335, fungicidal prop.	68
5-[(i-BuO)$_2$P(S)S]-		m. 64-65°, fungicidal prop.	68
C$_{24}$			
5-(2-Biphenylyl)- (M-567)	5,5-Alkyl(2-biphenylyl)dibenzarsolium iodides by thermal dec. at 150-180° (75-92)	m. 130-131°, rxn. with MeI → 5,5-Me(2-PhC$_6$H$_4$)-dibenzarsolium iodide	723
C$_{25}$	5-Me-5,5'-spirobi(5H-dibenzarsole) + aq. HCl under reflux and heating with KI (80)	m. 135-141° (dec.), thermal dec., (g)	723
	5-(2-PhC$_6$H$_4$)-dibenzarsole + MeI under reflux (77)		
C$_{28}$	5-Bu-5,5″-spirobi(5H-dibenzarsole) + aq. HCl under reflux and heating with KI (72)	Crystal, m. 107-108° (dec.), dec. at 150° → 5-(2-PhC$_6$H$_4$)-dibenzarsole + BuI	723

	Preparation	Properties	Ref.
C30 	5-Ph-5,5'-spirobi(5H-dibenzarsole) + Br in CH_2Cl_2 (93)	Crystals, m. 219–221°, (h)	723
	5-(p-ClC_6H_4)-5,5'-spirobi(5H-dibenzarsole) + aq. HCl under reflux and heating with KI (63)	Crystals, m. 207–208° (dec.)	723
	(2-LiC_6H_4)$_2$ + $AsCl_3$ (1:1 or 2:3) in Et_2O at −70° (76 and 86, resp.)	m. 289–290°	723
	5-Cl-dibenzarsole + (2-LiC_6H_4-)$_2$ in Et_2O, followed by hydrolysis (68)		723
C31 	5-p-tolyl-5,5'-spirobi(5H-dibenzarsole) + aq. HCl under reflux and heating with KI (64)	Crystals, monohydrate, m. 210–212°	723

(g) Dec. at 180° → 5-(2-PhC_6H_4)-dibenzarsole + MeI; at 250° → 17-Me-17H-tetrabenz[b,d,f,h]arsonin and 5-(2-PhC_6H_4)-dibenzarsole (723). (h) Rxn. with BuLi in Et_2O → 5-Bu-5,5'-spirobi[5H-dibenzarsole] (723).

3.3.1.7 SPIROBI(DIBENZARSOLE) DERIVATIVES AND TRIS(2,2'-BIPHENYLYLENE) ARSENATES

Several compounds of this class were described in the main volume on pages 571-73. Now this class includes alkali metal and alkaloid tris(2,2'-biphenylylene)arsenates.

5,5'-Spirobi(5H-dibenzarsolium) Iodide, $AsC_{24}H_{16}I$,   (M-571)
<u>Synth.</u>: From $(2-PhC_6H_4)_2AsCl_3$ + aq. KI, 88% yield (722).
From 5,5'-spirobi(5H-dibenzarsolium) tris(2,2'-biphenyly-
lene)arsenate + NaI in Me_2CO, 72% yield (722).
<u>Prop.</u>: Yellow ppt., dec. 310-313° (722).
<u>Rxn. with</u>: Bu_4Te (1:1) → 5-Bu-5,5'-spirobi(5H-dibenzar-
sole) (720).
Bu_4Te (1:2) → 5-Bu-5,5'-spirobi(5H-dibenzarsole) + $[Bu_3Te]I$ (720).
$(2-LiC_6H_4-)_2$ in Et_2O, followed by hydrolysis → 5,5'-spirobi(5H-dibenzarsolium)
tris(2,2'-biphenylylene)arsenate(1-) (722).
$(2-LiC_6H_4-)_2$ in Et_2O-THF-petr. ether at 20° → Li tri(2,2'-biphenylylene)arse-
nate(1⁻) (722).
$Na\{[(2-C_6H_4-)_2]_3As\}$ → $\{[(2-C_6H_4-)_2]_2As\}\{[(2-C_6H_4-)_2]_3As\}$ (722).
$2-PhC_6H_4Li$ in Et_2O → 5-(2-biphenylyl)-5,5'-spirobi(5H-dibenzarsole) (722).
RMgX or RLi → 5-R-5,5'-spirobi(5H-dibenzarsole) (723).

5-Methyl-5,5'-spirobi(5H-dibenzarsole), $AsC_{25}H_{19}$,   Formula I
<u>Synth.</u>: From 5,5'-spirobi(5H-dibenzarsolium)iodide +
MeLi or MeMgI in Et_2O, 89 and 84% yield, resp. (723).
From 5-Me-dibenzarsole + chloramine T in THF, followed
by $(2-LiC_6H_4-)_2$ in Et_2O and hydrolysis, 19% yield (723).
From 5-R-5,5'-spirobi(5H-dibenzarsole), where R = Bu-,
$p-MeC_6H_4-$, $p-ClC_6H_4-$, $p-Me_2NC_6H_4-$, or Ph-, by the rxn.
with excess MeLi in Et_2O. (723).   R = Me
From $\{[(2-C_6H_4-)_2]_2As\}\{[(2-C_6H_4-)_2]_3As\}$ + MeLi in Et_2O, 86% yield (722).
<u>Prop.</u>: Colorless crystals, m. 215-215.5° (723), m. 213-215° (722).
<u>Rxn. with</u>: RLi (R = Bu-, $p-MeC_6H_4-$, $p-Me_2NC_6H_4-$) → exchange of the Me group
for R (721, 723).
Aq. HCl under reflux, followed by heating with KI → 5,5-Me(2-PhC_6H_4)-dibenzar-
solium iodide (723).

5-Ethyl-5,5'-spirobi(5H-dibenzarsole), $AsC_{26}H_{21}$,   for the structural for-
<u>Synth.</u>: From 5,5'-spirobi(5H-dibenzarsolium)iodide   mula see Formula I
+ EtLi or EtMgBr in Et_2O, 91 and 78% yield, resp.   above, where R = Et
(723).
<u>Prop.</u>: m. 173-174°, dec. at 200° under N → 5-(2-PhC_6H_4)-dibenzarsole (723).

5-Butyl-5,5'-spirobi(5H-dibenzarsole), $AsC_{28}H_{25}$,   for the structural for-
<u>Synth.</u>: From 5,5'-spirobi(5H-dibenzarsolium)iodide   mula see Formula I
+ BuLi or BuMgBr in Et_2O, 94 and 85% yield, resp.   above, where R = Bu
(723).

From 5-Bu-dibenzarsole + chloramine T in THF, followed by $(2\text{-LiC}_6\text{H}_4\text{-})_2$ in Et$_2$O and hydrolysis, 57% yield (723).

From 5-R-5,5'-spirobi(5H-dibenzarsole), where R = Me-, p-MeC$_6$H$_4$-, p-ClC$_6$H$_4$-, p-Me$_2$NC$_6$H$_4$-, or Ph-, with excess BuLi in Et$_2$O (723).

From 5,5'-spirobi(5H-dibenzarsolium) iodide + Bu$_4$Te (1:1) in Et$_2$O, 65% yield (720).

Prop.: m. 166° (dec.) (720), m. 166-166.5° (dec.) (723).

Rxn. with: Aq. HCl under reflux, followed by heating with KI → 5-(2-PhC$_6$H$_4$-)-5-Bu-dibenzarsolium iodide (723).

RLi in Et$_2$O → exchange of Bu for R (723, see also 721).

5-(p-Chlorophenyl)-5,5'-spiro(5H-dibenzarsole), $\quad$ AsC$_{30}$H$_{20}$Cl, $\quad$ for the structural formula see Formula I (page 388), where R = p-ClC$_6$H$_4$-.

Synth.: From 5,5'-spirobi(5H-dibenzarsolium) iodide + p-ClC$_6$H$_4$Li in Et$_2$O, 79% yield (723).

From 5-R-5,5'-spirobi(5H-dibenzarsole), where R = Me-, Bu-, p-MeC$_6$H$_4$-, p-Me$_2$NC$_6$H$_4$-, or Ph-, with excess of p-ClC$_6$H$_4$Li in Et$_2$O (723).

Prop.: m. 211-212° (723).

Rxn. with: Aq. HCl, followed by KI → 5-(2-PhC$_6$H$_4$-)-5-(p-ClC$_6$H$_4$-)-dibenzarsolium iodide (723).

RLi in Et$_2$O → exchange of the ClC$_6$H$_4$ group for R (721, 723).

5-Phenyl-5,5'-spirobi(5H-dibenzarsole), $\quad$ AsC$_{30}$H$_{21}$, $\quad$ (M-572), $\quad$ for the structural formula see Formula I (page 388), where R = Ph-.

Synth.: From 5,5'-spirobi(5H-dibenzarsolium) iodide with PhMgBr in Et$_2$O, 83% yield (723) or with the rxn. prod. formed from Ph$_5$As + PhLi (1:1) in THF at -70°, 68% yield (722).

From 5,5'-spirobi(5H-dibenzarsolium) tris(2,2'-biphenylylene)arsenate(1⁻) + PhLi or MeLi in Et$_2$O, 79% yield (722).

From 5-R-5,5'-spirobi(5H-dibenzarsole), where R = Me-, Bu-, p-MeC$_6$H$_4$-, or p-ClC$_6$H$_4$-, with excess PhLi in Et$_2$O (723).

Prop.: m. 233-235° (722, 723).

Rxn. with: RLi (R = Me-, Bu-, p-MeC$_6$H$_4$-, p-ClC$_6$H$_4$-, or p-Me$_2$NC$_6$H$_4$-) → exchange of the Ph group for R (721, 723).

5-p-Tolyl-5,5'-spirobi(5H-dibenzarsole), $\quad$ AsC$_{31}$H$_{23}$, $\quad$ for the structural formula see Formula I (page 388), where R = p-MeC$_6$H$_4$-.

Synth.: From 5,5'-spirobi(5H-dibenzarsolium) iodide + p-MeC$_6$H$_4$Li in Et$_2$O, 83% yield (723).

From 5-R-5,5'-spirobi(5H-dibenzarsole), where R = Me-, Bu-, or p-Me$_2$NC$_6$H$_4$-, with excess p-MeC$_6$H$_4$Li (723).

Prop.: m. 211-212° (723).

Rxn. with: Aq. HCl, followed by KI → 5-(2-PhC$_6$H$_4$-)-5-(p-MeC$_6$H$_4$)-dibenzarsolium iodide (723).

RLi in Et$_2$O → exchange of the p-MeC$_6$H$_4$- group for R (721, 723).

5-[p-(Dimethylamino)phenyl]-5,5'-spirobi(5H-dibenzarsole), $AsC_{32}H_{30}N$,
    (M-573),   for the structural formula see Formula I (page 388), where
    R = p-Me$_2$NC$_6$H$_4$-.

Synth.: From 5,5'-spirobi(5H-dibenzarsolium) iodide + p-Me$_2$NC$_6$H$_4$Li in Et$_2$O, 82% yield (723).

From 5-R-5,5'-spirobi(5H-dibenzarsole), where R = Me-, Bu-, p-ClC$_6$H$_4$-, or p-MeC$_6$H$_4$-, by reacting with excess p-Me$_2$NC$_6$H$_4$Li in Et$_2$O (723).

Prop.: m. 234-236° (from THF-MeOH) (723).

Rxn. with: RLi (R = Me-, Bu-, p-MeC$_6$H$_4$-) in Et$_2$O → exchange of the Me$_2$NC$_6$H$_4$-group for R (721, 723).

5-(2-Biphenylyl)-5,5'-spirobi(5H-dibenzarsole), $AsC_{36}H_{25}$,   for the
    structural formula see Formula I (page 388), where R = 2-PhC$_6$H$_4$-.

Synth.: From Li tris(2,2'-biphenylylene)arsenate(1⁻) by dec. with MeOH, 90% yield (722).

From 5,5'-spirobi(5H-dibenzarsolium) tris(2,2'-biphenylylene)arsenate(1⁻) + PhLi in Et$_2$O, 58% yield (722).

From 5,5'-spirobi(5H-dibenzarsolium) iodide + 2-PhC$_6$H$_4$Li in Et$_2$O, followed by hydrolysis, 77% yield (722).

From methylbrucinium tris(2,2'-biphenylylene)arsenate(1⁻) diastereomers by dec. with MeOH or HCl in Me$_2$CO, 55-84% yield (722).

Prop.: Colorless crystals, m. 230-233° (722).

Lithium Tris(2,2'-biphenylylene)arsenate(1⁻), $AsLiC_{36}H_{24}$

Synth.: From 5,5'-spirobi(5H-dibenzarsolium) iodide + (2-LiC$_6$H$_4$-)$_2$ (1:1) in Et$_2$O-THF-petr. ether at 20°, 75% yield (722).

From 5,5'-spirobi(5H-dibenzarsolium) tris(2,2'-biphenylylene)arsenate(1⁻) + LiI in Et$_2$O-THF-petr. ether, 75% yield (722).

Prop.: Colorless cryst. ppt. contg. 1 mol. THF, dec. 215-219° (722).

Rxn. with: MeOH → 5-(2-PhC$_6$H$_4$)-5,5'-spirobi(5H-dibenzarsole) (722).

Methylbrucinium iodide in MeOH → methylbrucinium tris(2,2'-biphenylylene)arsenate(1⁻) diastereomers (722).

Sodium Tris(2,2'-biphenylylene)arsenate(1⁻), $AsNaC_{36}H_{24}$,   for the
    structural formula see the corr. formula of the preceding lithium salt.

Synth.: From 5,5'-spirobi(5H-dibenzarsolium) tris(2,2'-biphenylylene)arsenate-(1⁻) + NaI in Me$_2$CO, 52% yield (722).

Prop.: Colorless crystals, dec. 218-220° (722).

Rxn. with: 5,5'-spirobi(5H-dibenzarsolium) iodide → 5,5'spiro(5H-dibenzarsolium tris(2,2'-biphenylylene)arsenate (722).

Methylbrucinium Tris(2,2'-biphenylylene)arsenate(1⁻) Diastereomers,

    $AsC_{60}H_{53}N_2O_4$

<u>Synth.</u>: From the corr.
Li salt by the rxn. with
brucine methiodide in
MeOH, a quant. pptn. of
both stereomers, which upon
crystn. from Me_2CO gave the
levorotary isomer, and from

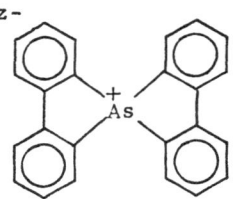 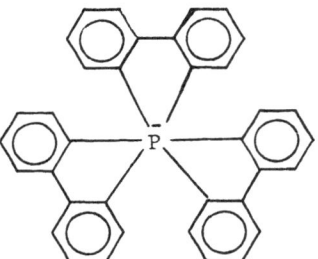

the mother liquor the noncrystg. dextrorotary isomer was isolated (722).
<u>Prop.</u>: The levorotary isomer dec. 207-209°, $[\alpha]_{578}^{24.5}$ -1220 ± 20°; the dextro-
rotary isomer $[\alpha]_{578}^{24.5}$ +850 ± 40° (722).
<u>Rxn. with</u>: 5,5'-spirobi(5H-dibenzarsolium)iodide → the l- and d-5,5'-spirobi-
(5H-dibenzarsolium) tri(2,2'-biphenylylene)arsenate (722).
MeOH in Me_2CO, using the l-isomer → 5-(2-PhC_6H_4)-5,5'-spirobi(5H-dibenzarsole),
optically inactive (722).
1 N HCl in Me_2CO, both stereomers → optically inactive 5-(2-PhC_6H_4)-5,5'-
spirobi(5H-dibenzarsole) (722).

5,5'-Spirobi(5H-dibenzarsolium) Tris(2,2'-biphenylylene)phosphate(1⁻),

    $AsC_{60}H_{40}P$,

<u>Synth.</u>: From 5,5'-spirobi(5H-dibenz-
arsolium) iodide + Na tris(2,2'-bi-
phenylylene)phosphate (1⁻), 90%
yield (718).
<u>Prop.</u>: Yellow crystals, m. 243-
245° (dec.) (718).

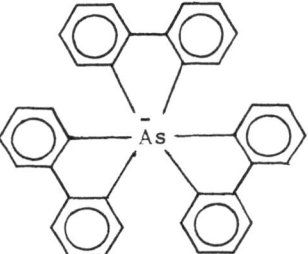

5,5'-Spirobi(5H-dibenzarsolium) Tris(2,2'-biphenylylene)arsenate(1⁻),

    $As_2C_{60}H_{40}$,

<u>Synth.</u>: From 5,5'-spirobi(5H-dibenz-
arsolium) iodide + (2-LiC_6H_4-)$_2$ in
Et_2O, 81% yield (722).
From 5,5'-spirobi(5H-dibenz-
arsolium) iodide + (-)- or (+)-
methylbrucinium tris(2,2'-bi-
phenylylene)arsenate, the l-
and d-diastereomers were obtd. (722)
<u>Prop.</u>: Yellow solid, dec. 209-210° (722), the (-)-salt m. 207-209°, $[\alpha]_{578}^{24.5}$
= -1180 ± 10°; the (+)-salt dec. 209-210°, $[\alpha]_{578}^{24.5}$ = +850°, $[M]_{578}^{24.5}$ = +7742°
(722).
<u>Rxn. with</u>: NaI in Me_2CO → 5,5'-spirobi(5H-dibenzarsolium) iodide + Na tris-
(2,2'-biphenylylene)arsenate(1⁻) (722).
PhLi in Et_2O → 5-(2-PhC_6H_4)-5,5'-spirobi(5H-dibenzarsole) and 5-Ph-5,5'-spiro-
bi(5H-dibenzarsole) (722).

MeLi in Et$_2$O, followed by hydrolysis → 5-(2-PhC$_6$H$_4$)-5,5'-spirobi(5H-dibenzar-sole) and 5-Me-5,5'-spirobi(5H-dibenzarsole) (722).

3.3.1.8 SPIROBI(BENZODIARSOLE) DERIVATIVES

A few compounds of this type were described in the main volume on page 574.

1,1',3,3'-Tetrahydro-1,1,1',1',3,3,3',3'-octamethyl-2,2'-spirobi(2H-1,3-benzo-
diarsolium) Salts,   [As$_4$C$_{21}$H$_{32}$]$^{4+}$4 X^{-4},   (M-574),

Synth.: From o-C$_6$H$_4$(AsMe$_2$)$_2$ + CBr$_4$ or CL$_4$
in a sealed tube, the tetrabromide and tetra-
iodide, resp., was obtd. Metathesis of the
tetrabromide with KClO$_4$ gave the tetraper-
chlorate (372, see also 1163).
Prop.: For m. see M-574. The salts are very
easily sol. in H$_2$O. The bromide and iodide ions are readily titratable.
Molar cond. data indicate extensive ion assocn. (372).

X = Br, I, ClO$_4$

3.3.1.9 MISCELLANEOUS HETEROCYCLIC ARSENIC DERIVATIVES CONTAINING FIVE-MEMBERED RINGS

3,4-Dihydrospiro[arsinoline-1(2H),2'-isoarsindolinium Salts,   [AsC$_{17}$H$_{18}$]$^{+}$X^{-},
Synth.: From 1-methyl-1,2,3,4-tetrahydroarsinoline
by mixing with o-C$_6$H$_4$(CH$_2$Br)$_2$, pulverizing the resulting
glassy prod., heating slowly to 200°/13 mm till effervescence is over, dissolving the prod. in EtOH and
treating with alc. NaI to form the iodide (1513).
Prop.: The iodide, m. 241-243°; the picrate, m.
163-164° (1513).

X^{-} = I^{-}, picrate^{-}

1,4-Dimethyl-7-phenyl-7-arsabicyclo[2.2.1]hept-5-ene-2,2,3,3-tetracarbonitrile,
AsC$_{18}$H$_{13}$N$_4$,
Synth.: From 2,5-dimethyl-1-phenylarsole by Diels-
Alder condensation with (NC)$_2$C=C(CN)$_2$ (1002).
Prop.: Colorless needles, m. 157°, NMR spectrum (1002).

3.3.5 RING SYSTEMS WITH ARSENIC AND OXYGEN HETEROATOMS

3.3.5.1 OXARSOLANE, OXADIARSOLANE, AND BENZOXADIARSOLE DERIVATIVES

Several oxadiarsolane and benzoxadiarsole derivatives were described in the
main volume on pages 575-78.

2,2,2-Triphenyl-1,2-oxarsolane,   AsC$_{21}$H$_{21}$O,
Synth.: From {Ph$_3$[HO(CH$_2$)$_3$]As}I + NaOH in THF under reflux,
71% yield (676).

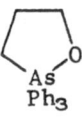

<u>Prop.</u>: m. 154°, IR spectrum; dec. at 160-180° → Ph$_3$As + allyl alc. (676).
<u>Rxn. with</u>: HI in MeOH → [Ph$_3$(HOCH$_2$CH$_2$CH$_2$)As]I (676).

Poly(1,3-p-phenylylene-1,3-dihydro-2,1,3-benzoxadiarsolylene),
    (As$_2$C$_{12}$H$_8$O)$_n$,   (M-576),

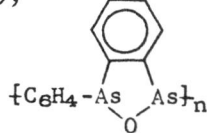

<u>Synth.</u>: From p-C$_6$H$_4$Li$_2$ + o-C$_6$H$_4$[As(OH)Cl]$_2$ in Et$_2$O at -25° and completing the rxn. under reflux in absence of O (739).

<u>Prop.</u>: A polymeric prod., the highest mol. wt. fraction of which is insol. in common org. solvents and softens at 280° (739).

3.3.5.2 DIOXARSOLANE AND DIOXARSINDAN DERIVATIVES

1,3,2-Dioxarsolane derivatives and their methods of preparation were described in the main volume on pages 576-78. Now this class also includes 1,3,2-dioxarsindan and naphth[2.3]-1,3,2-dioxarsole derivatives. The latter two ring systems are formed from arsenoso derivatives by a condensation with pyrocatechol and 2,3-naphthalenediol, respectively.

Table 38. Dioxarsolane and Dioxarsindan Derivatives

Structure	Prepd. from (Yield %)	Prop. and Rxn.	Ref.
C$_3$ MeAsOCH$_2$CH$_2$O	MeAsI$_2$ in C$_6$H$_6$ + HOCH$_2$-CH$_2$OH in C$_6$H$_6$ contg. Et$_3$N at 80° (48)	b$_{12}$ 61°, m. 28-30°, NMR spectrum	1608
C$_5$ MeAsOCH(Me)CH(Me)O	MeAsI$_2$ in C$_6$H$_6$ + [MeCH-(OH)-]$_2$ in C$_6$H$_6$ contg. Et$_3$N at 80° (71)	b$_{12}$ 54°, m. 37°, NMR spectrum	1608
C$_7$ (AsMe ring)	MeAsI$_2$ in C$_6$H$_6$ + o-C$_6$H$_4$-(OH)$_2$ in C$_6$H$_6$ contg. Et$_3$N at 80° (71)	b$_{0.1}$ 87°, m. 87°, NMR spectrum	1608
MeAsOC(Me)$_2$C(Me)$_2$O	MeAsI$_2$ in C$_6$H$_6$ + [Me$_2$C-(OH)-]$_2$ in C$_6$H$_6$ contg. Et$_3$N at 80° (89)	b$_{12}$ 75°, m. 59°, NMR spectrum	1608
C$_8$ PhAsOCH$_2$CH$_2$O (M-577)		IR and Raman spectra	1389
C$_9$ PhAsOCH$_2$CH(Me)O (M-577)		IR and Raman spectra	1389

Table 38 Continued

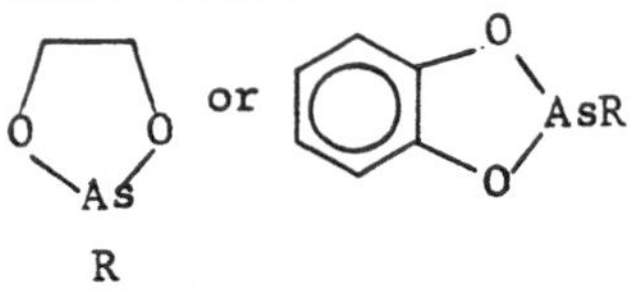 (structure)	Prepd. from (Yield %)	Ref.

C₁₂

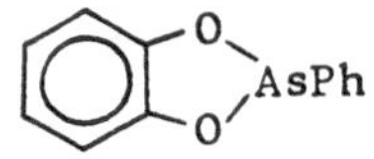

| (structure) AsPh | PhAsO + o-C₆H₄(OH)₂ in C₆H₆, (a) | 741 |

C₁₃

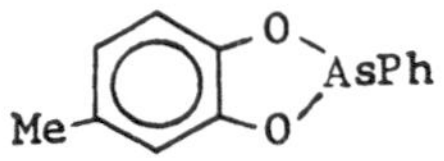

| Me (structure) AsPh | PhAsO + 4,1,2-Me(HO)₂C₆H₃ in C₆H₆, (a) | 741 |

| (structure) AsC₆H₄OMe-p | p-MeOC₆H₄AsO + o-C₆H₄(OH)₂ in C₆H₆, (a) | 741 |

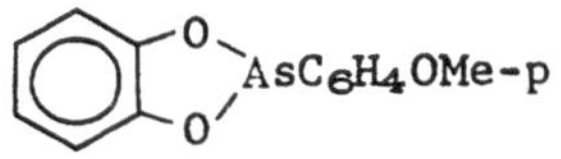

C₁₄

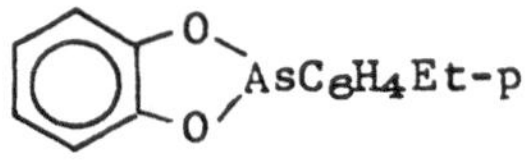

| (structure) AsC₆H₄Et-p | p-EtC₆H₄AsO + o-C₆H₄(OH)₂ in C₆H₆, (a) | 741 |

| (structure) AsC₆H₄NMe₂-p | p-Me₂NC₆H₄AsO + o-C₆H₄(OH)₂ in C₆H₆, (a) | 741 |

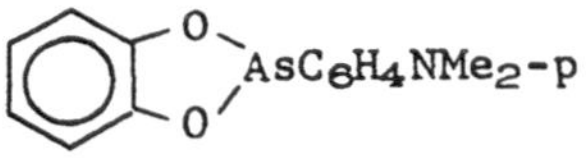

C₁₆

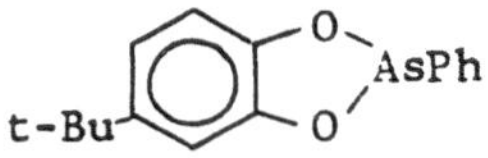

| t-Bu (structure) AsPh | PhAsO + 4,1,2-t-Bu(HO)₂C₆H₃ in C₆H₆, (a) | 741 |

| (structure) AsPh | PhAsO + 2,3-C₁₀H₆(OH)₂ in C₆H₆, (a) | 741 |

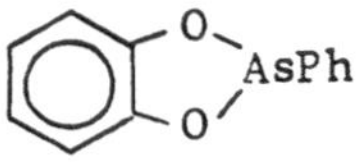

(a) By heating in the presence of catalytic amts. of p-MeC₆H₄SO₃H, Py, 2,4,6-trimethylpyridine, or Bu₃N, with azeotropic removal of H₂O (741).

3.3.6   RING SYSTEMS WITH ARSENIC AND SULFUR HETEROATOMS

3.3.6.1   DITHIARSOLANE AND DITHIARSINDAN DERIVATIVES

The methods of synthesis of dithiarsolane and dithiarsindan derivatives were summarized in the main volume on page 582, and the compounds were compiled in Table 72, pages 583-88. Besides the methods described there that have been used for the preparation of the cyclic dithio esters which appear in Table 39, 2,2-dihydro-1,3,2-dithiarsole derivatives have been prepared from tertiary dihaloarsenic compounds, R₃AsX₂, by a condensation with vinylenedithiole disodium salt derivatives.

Table 39. 1,3,2-Dithiarsolane and 1,3,2-Dithiarsindan Derivatives

Compound	Prepd. from (Yield %)	Prop. and Rxn.	Ref.
C_3 MeAsSCH$_2$CH$_2$S (M-583)	MeAsI$_2$ + HSCH$_2$CH$_2$SH + Et$_3$N (1:1:1) in C$_6$H$_6$ at 80° (81)	b$_{12}$ 120°, NMR spectrum	1608
	MeAsI$_2$ + [SCH$_2$CH$_2$SAsSCH$_2$-]$_2$		1438
	SCH$_2$CH$_2$SAsCl in THF at -30° + MeMgI in Et$_2$O and warming to r.t.	b$_{0.4}$ 60°	1438
C_4 AcAsSCH$_2$CH$_2$S	SCH$_2$CH$_2$SAsCl + anhyd. NaOAc in Ac$_2$O at 80° (58)	Colorless liquid, b$_{0.6}$ 104°	1438
C_7 (NC, CN, S, S, As, Me$_3$)	Me$_3$AsBr$_2$ + NaSC(CN)=C(CN)SNa in MeCN under reflux		1090
C_8 (Me, S, S, AsMe)	MeAsI$_2$ + o-C$_6$H$_4$(SH)$_2$ + Et$_3$N (1:1:1) in C$_6$H$_6$ at 80° (89)	b$_{0.01}$ 106°, NMR spectrum	1608
PhAsSCH$_2$CH$_2$S (M-583)	PhAsCl$_2$ + [SCH$_2$CH$_2$SAsSCH$_2$-]$_2$	b$_{0.4}$ 128°, m. 47°	1438
p-H$_2$NC$_6$H$_4$AsSCH$_2$CH$_2$S (M-584)	p-H$_2$NC$_6$H$_4$AsO + HSCH$_2$CH$_2$SH in EtOH under reflux and chilling in CO$_2$-Me$_2$CO (89)	m. 69-70.5°, IR spectrum, (a)	122, 441

(a) Rxn. with CH$_2$CH$_2$O in EtOH contg. p-MeC$_6$H$_4$SO$_3$H at r.t. → (HOCH$_2$CH$_2$)$_2$NC$_6$H$_4$AsSCH$_2$CH$_2$S (122, 441).

Table 39 Continued

Compound	Prepd. from (Yield %)	Prop. and Rxn.	Ref.
C$_{10}$ NC, CN dithiole ring $\overset{S}{\underset{S}{\diagup}}$As–Ph	PhAsI$_2$ + NaSC(CN)=C(CN)SNa in CH$_2$(OMe)$_2$	m. 131-132°	1090
2,4-H$_2$N(HO)C$_6$H$_3$As dithiolane(CO$_2$H, CO$_2$H) (M-584)	2,4-H$_2$N(HO)C$_6$H$_3$AsO + HO$_2$CCH(SH)CH(SH)CO$_2$H in Na$_2$CO$_3$ in H$_2$O at r.t. (86)	(b)	593
C$_{12}$ 3,4-AcNH(HO)C$_6$H$_3$As dithiolane(CO$_2$H, CO$_2$H)	3,4-AcNH(HO)C$_6$H$_3$AsO + HO$_2$CCH(SH)CH(SH)CO$_2$H + Na$_2$CO$_3$ in H$_2$O at 30° (80)	(b)	593
4-AcNHC$_6$H$_4$As dithiolane(CO$_2$H, CO$_2$H)	4-AcNHC$_6$H$_4$AsO + HO$_2$CCH(SH)CH(SH)CO$_2$H + Na$_2$CO$_3$ in H$_2$O at 30° (82)	(b)	593
2,4-HO(AcNH)C$_6$H$_3$As dithiolane(CO$_2$H, CO$_2$H)		(b)	593
triazine-substituted arsenic dithiolane (CH$_2$OH)		Trypanosoma resistance to the cyclic ester	702
4-[(ClCH$_2$CH$_2$)$_2$N]C$_6$H$_4$AsSCH$_2$CH$_2$S	(ClCH$_2$CH$_2$)$_2$NC$_6$H$_4$AsO + HSCH$_2$CH$_2$SH in EtOH under reflux	Oil, IR spectrum	122, 441

4-[(HOCH$_2$CH$_2$)$_2$N]C$_6$H$_4$AsSCH$_2$CH$_2$S	p-[(HOCH$_2$CH$_2$)$_2$N]C$_6$H$_4$AsO + HSCH$_2$CH$_2$SH in EtOH under reflux	Needles, m. 79.5-80°, IR and NMR spectra, (c)	122, 441
	p-H$_2$NC$_6$H$_4$AsSCH$_2$CH$_2$S + CH$_2$CH$_2$O in EtOH in the presence of p-MeC$_6$H$_4$SO$_3$H at CO$_2$-Me$_2$CO and warming to r.t.		122, 441
p-(Et$_2$N)C$_6$H$_4$AsSCH$_2$CH$_2$S	p-(Et$_2$N)C$_6$H$_4$AsO + HSCH$_2$CH$_2$SH in EtOH under reflux (65)	White needles, m. 74-75.5°, (d)	122, 441

C$_{13}$

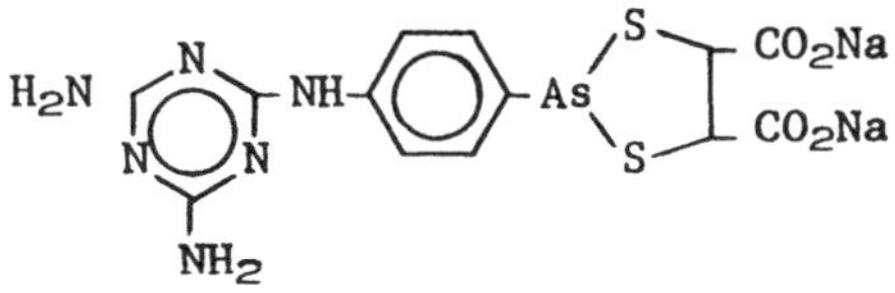

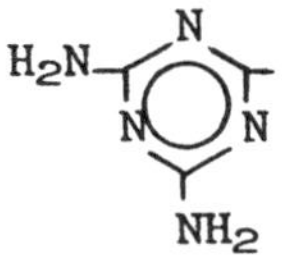

H$_2$N—[ring]—NH—[ring]—NH$_2$ ... As with CO$_2$Na / CO$_2$Na dithiolane

H$_2$N...NHC$_6$H$_4$AsO + HO$_2$CCH(SH)CH(SH)CO$_2$H + NaCO$_3$ (81) ... NH$_2$

The di-Na and di-K salts are effective against African sleeping sick-ness parasite | 593 |

C$_{24}$

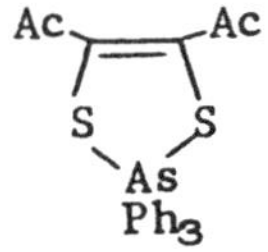

Ac, Ac — dithiole ring — As Ph$_3$

Ph$_3$AsCl$_2$ + AcC(SNa)=C(SNa)Ac in MeCN under reflux

1090

(b) The diacid and its mono- or di-Na salt are effective against African sleeping sickness parasite (593).
(c) The crude prod. on standing in EtOH soln. gave shiny needles which were identified as SCH$_2$CH$_2$SAsSCH$_2$CH$_2$S-AsSCH$_2$CH$_2$S (441). (d) IR and NMR spectra (441).

3.3.7 RING SYSTEMS WITH ARSENIC AND NITROGEN HETEROATOMS

1,3-Azarsolidine derivatives, a new type of heterocyclic arsenic compounds, are formed from 2-arsinoethylamine by a condensation with ketones.

2,2-Dimethyl-3-phenyl-1,3-azarsolidine, $AsC_{11}H_{16}N$,
Synth.: From $PhAsHCH_2CH_2NH_2$ by a condensation with acetone, 74% yield (1532).
Prop.: b_{1-2} 98-100° (1532).

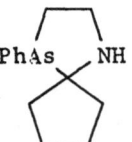

2,3-Diphenyl-1,3-azarsolididine, $AsC_{15}H_{16}N$,
Synth.: From $PhAsHCH_2CH_2NH_2$ by a condensation with BzH, 88% yield (1532).
Prop.: m. 66°, pK_a 6.20; HCl salt m. 116° (1532).

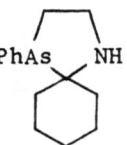

2,2-Tetramethylene-3-phenyl-1,3-azarsolidine, $AsC_{13}H_{18}N$,
Synth.: From $PhAsHCH_2CH_2NH_2$ by a condensation with cyclopenta-none, 81% yield (1532).
Prop.: b_{1-2} 128-132°; HCl salt m. 105° (dec.) (1532).

2,2-Pentamethylene-3-phenyl-1,3-azarsolidine, $AsC_{14}H_{20}N$,
Synth.: From $PhAsHCH_2CH_2NH_2$ by a condensation with cyclo-hexanone, 72% yield (1532).
Prop.: b_{1-2} 135-138° (1532)

3.3.9   RING SYSTEMS WITH THREE DIFFERENT HETEROATOMS

3.3.9.1 COMPOUNDS WITH ARSENIC, OXYGEN, AND SULFUR HETEROATOMS

2-Methyl-1,3,2-oxathiarsolane, AsC_3H_7OS, $Me\overline{AsOCH_2CH_2S}$
Synth.: From $\overline{SCH_2CH_2OAsCl}$ in THF at -30° by adding dropwise MeMgI in Et_2O and warming the mixt. to r.t., 40% yield (1438).
Prop.: $b_{0.8}$ 39-40° (1438).

3.3.9.3 COMPOUNDS WITH ARSENIC, SULFUR, AND NITROGEN HETEROATOMS

2-Methyl-1,3,2-azothiarsolane, AsC_3H_8NS, $Me\overline{AsNHCH_2CH_2S}$
Synth.: From $MeAsCl_2$ + $HSCH_2CH_2NH_2$ in the presence of Et_3N (1:1:2) in $ClCH_2CH_2Cl$, followed by Et_2O, 62% yield, along with N,N'-(methylarsylene)bis-(2-methyl-1,3,2-thiazarsolane) (1438).
Prop.: Colorless oil, $b_{1.5}$ 64° (1438).

3.4   SIX-MEMBERED RING SYSTEMS

3.4.1   RING SYSTEMS WITH ONE HETEROATOM

3.4.1.2   ARSINOLINE DERIVATIVES

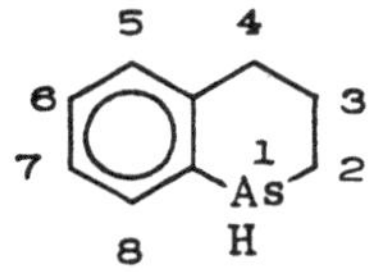

1,2,3,4-Tetrahydroarsinoline

The arsinoline derivatives and their homologs were described in the main
volume on pages 591-4. 1,2,3,4-Tetrahydro-1-phenylarsinoline was prepared
from phenyl(3-phenylpropyl)arsinic acid by the ring closure in sulfuric acid,
followed by the reduction of the arsinoline oxide with SO_2.

1-Chloro-1,2,3,4-tetrahydroarsinoline, $AsC_9H_{10}Cl$*
Rxn. with: PhMgBr → 1-phenyl-1,2,3,4-tetrahydroarsinoline (1513).

1,2,3,4-Tetrahydro-1-methylarsinoline, $AsC_{10}H_{13}$,   (M-592)*
Prop.:   Stable in soln., does not undergo intramol. dissocn. (1513).
Rxn. with: MeBr at 5° → 1,2,3,4-tetrahydro-1,1-dimethylarsinolinium bromide
(1513).
Picric acid → 1,2,3,4-tetrahydro-1-hydroxy-1-methylarsinolinium picrate (1513).
o-$C_6H_4(CH_2Br)_2$, followed by heating at 200°/13 mm and treating with ethanolic
NaI → 3,4-dihydrospiro[arsinoline-2[2H],2'-isoarsindolinium iodide (1513).
Quaternary Salts: The methobromide m. 250-251° (with effervescence); at 260-
270°/23 mm → 1,2,3,4-tetrahydro-1-methylarsinoline. A rxn. with satd. aq. KI
soln. → exchange of the Br⁻ for I⁻ (1513).

The methopicrate, m. 119-120°, and methochloroplatinate, m. 218-219°, were
prepd. from the methobromide in an analogous manner (1513).

1,2,3,4-Tetrahydro-1-hydroxy-1-methylarsinolinium picrate, prepd. as described
above, m. 151-152° (1513).

1,2,3,4-Tetrahydro-1-phenylarsinoline, $AsC_{15}H_{15}$*
Synth.: From 1-chloro-1,2,3,4-tetrahydroarsinoline + PhMgBr in Et_2O under
reflux, 75% yield (1513).
From Ph(PhCH_2CH_2CH_2)AsO_2H by heating with concd. H_2SO_4 at 100°, cooling, neutralizing with 30% NaOH, extg. the oily arsinoline oxide with $CHCl_3$ and reducing it with SO_2 in the presence of HCl and KI, 48% yield (1513).
Prop.: $b_{0.4}$ 140-144°, b_1 156-162°, slowly crystd. (1513).
Rxn. with: Br in $CHCl_3$, followed by H_2S → 1,2,3,4-tetrahydro-1-phenylarsino-
line sulfide (1513).

_____

* A structural formula of the ring system is shown above.

Li in THF under reflux, followed by o-BrCH$_2$C$_6$H$_4$CH$_2$CH$_2$Br in THF, removal of
the THF, and a treatment with NaI in Me$_2$CO → 3,3´,4,4´-tetrahydrospiro[arsino-
line-1(2H),2´(1´)-isoarsinolinium] iodide (1514).
MeI → 1,2,3,4-tetrahydro-1-methyl-1-phenylarsinolinium iodide (1513).
p-ClC$_6$H$_4$COCH$_2$Br in C$_6$H$_6$ under reflux → 1-(p-chlorophenacyl)-1,2,3,4-tetra-
hydro-1-phenylarsinolinium bromide (1513).
Br in CHCl$_3$, followed by shaking with dild. NH$_4$OH, sepn. of the CHCl$_3$ layer,
concn. to dryness, and treating the residue with picric acid in hot H$_2$O →
1,2,3,4-tetrahydro-1-hydroxy-1-phenylarsinolinium picrate (1513).
<u>Quaternary Salts</u>: The methiodide, m. 165-166° (1513).
The methopicrate, m. 105.5-107° (1513).
The chlorophenacyl bromide, m. 210-211° (1513).
The chlorophenacyl iodide, m. 160-162°, prepd. from the bromide in CHCl$_3$ by
treating with satd. aq. Na picrate, followed by KI in aq. Me$_2$CO (1513).
The hydroxy picrate, m. 113.5-115° (1513).

1,2,3,4-Tetrahydro-1-phenylarsinoline Sulfide,   AsC$_{15}$H$_{15}$S*
<u>Synth.</u>: From 1,2,3,4-tetrahydro-1-phenylarsinoline in CHCl$_3$ by treating with
Br and then bubbling H$_2$S through the soln. (1513).
<u>Prop.</u>: m. 103.5-105° (1513).

3.4.1.3 ISOARSINOLINE DERIVATIVES

   Several isoarsinoline derivatives and their methods of synthesis were
described in the main volume on pages 594-5. A new method for the preparation of isoarsinoline derivatives consists of a condensation of arsylenedi-
lithium derivatives with 2-[o-(bromomethyl)phenyl]ethyl bromide.

1,2,3,4-Tetrahydro-2-phenylisoarsinoline,   AsC$_{15}$H$_{15}$,   (M-595)
<u>Synth.</u>: From PhAsLi$_2$ in THF by adding a soln. of o-BrCH$_2$-
C$_6$H$_4$CH$_2$CH$_2$Br in THF at r.t. and completing the rxn. under
reflux, 90% yield (1514).
<u>Prop.</u>: b$_{0.3}$ 126-130° (1514).
<u>Quaternary Salt</u>: 1,2,3,4-Tetrahydro-2-hydroxy-2-phenylisoarsolinium nitrate,
m 154-154.5°, prepd. from the isoarsoline deriv. by adding H$_2$O and concd.
HNO$_3$ (1514).

3.4.1.8 SPIROARSINOLINEISOARSINOLINE DERIVATIVES

   Two spiro-arsinoline and -isoarsinoline derivatives were described in the
main volume on pages 599-600.

---

* A structural formula of the ring system is shown on the preceding page.

3,3′,4,4′-Tetrahydrospiro[arsinoline-1(2H),2′(1H)-isoarsinolinium]Iodide,
    AsC$_{18}$H$_{20}$I,

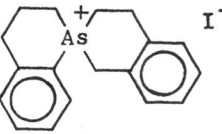

Synth.: From 1,2,3,4-tetrahydro-1-phenylarsinoline in
THF by treating with Li under reflux, followed by
o-(BrCH$_2$)C$_6$H$_4$CH$_2$CH$_2$Br in THF, replacing the solvent
with CHCl$_3$, removing LiBr and CHCl$_3$, and treating the
residue with NaI in Me$_2$CO (1514).
Prop.: m. 137-139° (1514).

3.4.2   RING SYSTEMS WITH TWO ARSENIC HETEROATOMS

3.4.2.1   DIARSENANE AND DIARSENIN DERIVATIVES

This group of compounds described in the main volume comprised 1,4-diarse-
nane, 1,4-diarsenin, and 1,4-benzodiarsenin derivatives (M-601 to 603). 1,2-
Diarsenane and 1,2-diarsenin derivatives described hereafter are new compounds.

1,2,3,6-Tetrahydro-1,2,4-trimethyl-1,2-diarsenin,   As$_2$C$_7$H$_{14}$,
Synth.: From (MeAs<)$_5$ + CH$_2$=C(Me)CH=CH$_2$ by irradiation with
a Hg lamp, 65% yield (1357).
Prop.: b$_{11}$ 108° (1357).

1,2,3,6-Tetrahydro-1,2,4,5-tetramethyl-1,2-diarsenin,   As$_2$C$_8$H$_{16}$,
Synth.: From (MeAs<)$_5$ + CH$_2$=C(Me)C(Me)=CH$_2$ by irradiation
with a Hg lamp (1357).
Prop.: b$_{0.25}$ 70-72°, NMR spectrum (1357).

1,2-Diphenyl-1,2-diarsenane,   As$_2$C$_{16}$H$_{18}$,   PhAs(CH$_2$)$_4$AsPh
Synth.: From PhAsHK·2 Dioxan in dioxan soln. by adding Cl(CH$_2$)$_4$Cl in C$_6$H$_6$, a
mixt. of the title compd. with PhAsH(CH$_2$)$_4$AsHPh was obtd. which upon a treatment with BuLi in dioxan gave the title compd. with 73% yield (1535).
From PhAs(Li)(CH$_2$)$_4$As(Li)Ph·2 Dioxan in C$_6$H$_6$-THF by adding BrCH$_2$CH$_2$Br in C$_6$H$_6$,
68% yield (1535).
Prop.: b$_3$ 196-198°, b$_4$ 203-206°, m. 49°, relatively insensitive to air, sol.
in Et$_2$O, C$_6$H$_6$, dioxan, and THF, sparingly sol. in EtOH and H$_2$O (1535).
Rxn. with: RLi in Et$_2$O under reflux, followed by dioxan → PhRAs(CH$_2$)$_4$As(Li)Ph
(1535).
MeI in EtOH-Et$_2$O → [PhAs(I)(CH$_2$)$_4$AsMe$_2$Ph]I (1535).

3.4.2.2   ARSANTHRENE DERIVATIVES

Arsanthrene, 5,10-alkanoarsanthrene, and 5,10-o-benzoarsanthrene derivatives
and their methods of synthesis were described in the main volume on pages 603-8.

5,10-Dihydro-5,10-dimethylarsanthrene, $As_2C_{14}H_{14}$, (M-604),
Prop.: X-ray crystallographic data; a "butterfly"
conformation with cis-methyl groups is postulated (843).

Poly(p-phenylene-5,10-arsanthrylene), $(As_2C_{18}H_{12})_n$, (M-606)
Synth.: From $p-C_6H_4Li_2$ + 5,10-dichloro-5,10-dihydro-
arsanthrene in THF at -35 to -45° and completing the
rxn. under reflux (739).
Prop.: A fraction with the highest mol. wt. is insol.
in the usual org. solvents, softens at 280°; another
fraction sol. in $CHCl_3$ but insol. in C_6H_6, softens at
260°, mol. wt. 13,300 (739).

3.4.5   RING SYSTEMS WITH ARSENIC AND OXYGEN HETEROATOMS

3.4.5.1 OXARSINANE AND PHENOXARSINE DERIVATIVES

The arsenic derivatives of this class and their methods of synthesis were
described in the main volume on pages 609-617. 10-Chlorophenoxarsine was
recently prepared from diphenyl ether by a reaction with arsenic trichloride
at 240-285° in the presence of aluminum chloride or bromide and from diphenyl
ether by heating with arsenic trioxide in the presence of aluminum powder
while introducing chlorine gas into the reaction mixture.

The 10-chloro derivative and 10,10'-oxybisphenoxarsine have been extensively
used as intermediates in the preparation of numerous 10-substituted phenoxarsine derivatives. Many of the phenoxarsine derivatives are potent biocides.
Their fungicidal and bactericidal activities are 10-20 times higher than those
of the corresponding phenarsazine analogs (1473).

10-Chlorophenoxarsine, $AsC_{12}H_8ClO$, (M-610),
Synth.: From $AsCl_3$ + Ph_2O in the presence of $AlCl_3$ or $AlBr_3$
at 240-285°, 78% yield (1041).
From As_2O_3 + Ph_2O + Al powder by heating and introducing Cl
gas (212).
Prop.: m. 124° (1041), m. 124-125° (68), m. 125° (712).
Rxn. with: RSNa → 10-RS-phenoxarsine (316, 1166).
$NaSC(O)NR_2$ → 10-[$R_2NC(O)S$]-phenoxarsine (1166).
RSO_2SK → 10-RSO_2S-phenoxarsine (1166).
$(RO)_3P$ → 10-[$(RO)_2P(O)$]-phenoxarsine (1166).
Alkali metal xanthates → S-(10-phenoxarsinyl) xanthates (1166).
NaF in Me_2CO → 10-fluorophenoxarsine (710).
$Me_2NC(S)SNa$ in Me_2CO → 10-($Me_2NC(S)S$)-phenoxarsine (708).
$RNHC(S)NHR$ → 10-[$RNHC(=NR)S$]-phenoxarsine (503).
RSO_3M or RSO_2SM (M = alkali metal) → 10-(RSO_3)-phenoxarsine and 10-(RSO_2S)-

phenoxarsine, resp. (504).

<u>Biol. Prop.</u>: Antifungal and antibacterial effect against Fusarium oxysporum, F. lycopersicum, Rhizoctania solani, Aspergillus terreus, Pullularia pullulans, and Erwinia carotovora (1041). A broad antifungal (Trichophyton asteroides and Candida albicans) and antibacterial activity spectrum in vitro (1473, see also 712).

<u>Uses:</u> A biocidal agent for polymeric latexes useful as film-forming coatings (1040). A fungicide and bactericide for vinyl resins (1634).

10,10′-Thiobisphenoxarsine, $As_2C_{24}H_{16}O_2S$,

<u>Use:</u> A fungicide and bactericide for vinyl resins (1634).

$$[O \quad As-]_2S$$

10,10′-Oxybisphenoxarsine, $As_2C_{24}H_{16}O_3$, (M-610)

<u>Synth.</u>: From 10-chlorophenoxarsine by treating with K_2CO_3 in DMF at 50-60° for 1-2 hrs. (1622).

<u>Prop.</u>: White cryst. solid, m. 180° (1622).

<u>Rxn. with:</u> RSH → 10-RS-phenoxarsine (505).

KI in H_2O + H_2SO_4 under reflux → 10-iodophenoxarsine (707).

HF in H_2O under reflux → 10-fluorophenoxarsine (709).

$Me_2NC(S)SNa$ + H_2SO_4 → 10-[$Me_2NC(S)S$]-phenoxarsine (711).

RCO_2H → 10-(RCO_2)-phenoxarsine (1166).

$$[O \quad As]_2O$$

<u>Biol. Prop.</u>: A fungicide (Trichophyton asteroides and Candida albicans) and bactericide with a broad activity spectrum in vitro (1473). Highly effective against the germination of Cochliobolus miyabeanus spores and against growth of Xanthomonas oryzae (1493). Fungicidal activity in the preservation of wood (246).

<u>Uses:</u> A biocidal constituent of polymer latexes for film-forming coatings (1040). A fungistat for latex paints (1621). A fungistat and bactericide for exterior latex and asphalt coatings (1622). A fungicide and bactericide for vinyl resins (1634).

10,10′-(Oxalyldithio)bisphenoxarsine, $As_2C_{26}H_{16}O_4S_2$, (M-611)

<u>Synth.</u>: From 10-Cl-phenoxarsine in Me_2CO by adding dropwise aq. NaSCOCOSNa soln. at 15-20°, dilg. the rxn. mixt. with ice-water and recovering the prod. by filtration, 49% yield (1166).

<u>Prop.</u>: m. 181-182° (dec.), IR spectrum (1166).

$$[O \quad AsSC(O)]_2$$

Table 40. Phenoxarsine Derivatives

Deriv. of where R is:	Prepd. from 10-Chlorophenoxarsine (A) or 10,10'-Oxybisphenoxarsine (B) (Yield %)	Prop. and Rxn.	Ref.
C$_{12}$			
10-Br-		A biocidal agent for polymeric latexes useful as film-forming coatings	1040
10-F-	(B) in H$_2$O + 25% HF under reflux	m. 116-117°	709
	(A) + NaF in Me$_2$CO	Needles, m. 155.5-156.5°, (a, b)	710
10-I-	(B) + KI in H$_2$O + 10% H$_2$SO$_4$ under reflux	m. 142-143°, (b, c)	707
C$_{13}$			
10-NCS-		A fungicide and bactericide for vinyl resins, (d, e)	1634
10-[H$_2$NC(=NH)S]-	10-Cl- or 10-Br-phenoxarsine + C(S)(NH$_2$)$_2$ in boiling EtOH	The HCl salt, m. 164-164.5°, the HBr salt, m. 160-162°, the picrate, m. 195.5-196.5° (dec.), (f)	503
10-Me- (M-612)		(g)	1040
10-MeSO$_2$S-	(A) + MeSO$_2$SK in aq. Me$_2$CO under reflux (66)	m. 122.5-125°, IR spectrum	1166
C$_{14}$			
10-CCl$_3$CO$_2$-	(B) + CCl$_3$CO$_2$H, (h) (97)	m. 116.5-117.5°, IR spectrum, (g)	1166
10-CF$_3$CO$_2$-	(B) + CF$_3$CO$_2$H, (h) (74)	m. 128-131°, IR spectrum	1166
10-ClCH$_2$CO$_2$-	(B) + ClCH$_2$CO$_2$H, (h) (74)	m. 139-141.5°, IR spectrum	1166
10-MeCOS-	(B) + AcSH, (h) (74)	m. 91.5-92.5°, IR spectrum	1166
10-MeSO$_3$-	(B) + MeSO$_3$H, (h) (63)	m. 159-162°, IR spectrum	1166
10-[MeOC(S)S]- (M-612)	(A) in Me$_2$CO + MeOC(S)SK, (i) (58)	m. 110-112°, IR spectrum, (g)	1166
		m. 107-108°	68

Compound	Preparation	Properties	Ref.
10-AcO-	(B) + AcOH under reflux or (h), (42)	m. 126-127°, IR spectrum	1166
	(A) + AcOH + NaOAc under reflux (42)		1166
10-[MeNHC(=NH)S]-	(A) + MeNHCSNH$_2$ in EtOH under reflux	Picrate, m. 181.5-182° (dec.), (f)	503
10-Et-		(b, g)	1140
10-EtO- (M-613)		(b)	1473
10-HOCH$_2$CH$_2$S-		(b)	1473
10-EtSO$_2$S-	(A) + EtSO$_2$SK in aq. Me$_2$CO under reflux (82)	m. 116-118°, IR spectrum	1166
10-[(MeO)$_2$P(S)S]- (M-613)	10-NCS-phenoxarsine + (MeO)$_3$P in C$_6$H$_6$ at 30° (100)	m. 90-93°, fungicidal activity	68
10-[(MeO)$_2$P(O)]-		m. 84-85°	871
C$_{15}$			
10-NCSCH$_2$CH$_2$S-	10-NCS-phenoxarsine + $\overline{\text{CH}_2\text{CH}_2\text{S}}$ in C$_6$H$_6$ at 30-40° (99)	m. 114°, fungicidal activity	873
10-[$\overline{\text{HNCH}_2\text{CH}_2\text{N=CS}}$]-	(A) + 2-imidazolethione in EtOH	Picrate, yellow crystals, m. 136-140.5° (dec.), (f)	503
10-EtOC(S)S- (M-613)	(A) + EtOC(S)SK, (i) (69)	m. 79-81°, IR spectrum, (b, g)	1166
		m. 79-80°, fungicidal activity	68
10-EtCOS-	(B) + EtCOSH, (h) (52)	m. 72-73°, IR spectrum	1166

(a) Bactericidal, fungicidal, and nematocidal activity (709). (b) A broad antifungal (Trichophyton asteroides and Candida albicans) and antibacterial spectrum in vitro (1473). (c) Bactericidal and fungicidal activity (707). (d) Rxn. with (RO)$_3$P → 10-(RO)$_2$P(O)-phenoxide (871). (e) Rxn. with $\overline{\text{CH}_2\text{CH}_2\text{S}}$ at 30-40° in C$_6$H$_6$ → 10-NCSCH$_2$CH$_2$S-phenoxarsine (873). (f) Antiparasitic and plant growth inhibiting activity (503). (g) A biocidal agent for polymeric latexes useful as film-forming coatings (1040). (h) In C$_6$H$_6$ under reflux, with azeotropic removal of H$_2$O (1166). (i) The reagent in aq. soln. is added dropwise at 15-20°, the rxn. mixt. is dild. with ice-H$_2$O and the solid prod. is recovered by filtration, while the liquid is extd. with CH$_2$Cl$_2$ (1166).

Table 40 Continued

Deriv. of where R is:

R–As (10) phenoxarsine ring system, positions 8,7,6 (left ring), O (5), 2,3,4.

Deriv. of where R is:	Prepd. from 10-Chlorophenoxarsine (A) or 10,10'-Oxybisphenoxarsine (B) (Yield %)	Prop. and Rxn.	Ref.
C$_{15}$ (Cont.)			
10-[Me$_2$NC(S)S]- (M-613)	(A) + Me$_2$NC(S)SNa in Me$_2$CO under reflux	m. 158.5-160°, a fungicide (b,j)	708
	(B) + Me$_2$NC(S)SNa + H$_2$SO$_4$ at pH 1-2 and 50-60°		711
10-Pr-		(g)	1040
10-[MeNHC(=NMe)S]-	(A) + (MeNH)$_2$CS in EtOH under reflux and pptg. with picric acid	(f)	503
C$_{16}$			
10-[CH$_2$=CHCH$_2$OC(S)S]- (M-613)	(A) in Me$_2$CO + CH$_2$=CHCH$_2$OC(S)SK, (i) (45)	m. 99-100°, IR spectrum	1166
10-EtOCOCOS-	(B) + EtOCOCOSH, (h)	m. 67.5-68°, IR spectrum	1166
10-[CH$_2$=CHCH$_2$NHC(=NH)S]-	(A) + CH$_2$=CHCH$_2$NHC(S)NH$_2$ in EtOH under reflux, (k)	The HCl salt, m. 187.5-188.5°	503
10-PrCOS-	(B) + PrCOSH, (h) (61)	The picrate, m. 201.5-202.5° (dec.), (f)	503
10-[i-PrCOS]-	(B) + i-PrCOSH, (h) (83)	m. 72-72.5°	1166
10-[ProC(S)S]- (M-614)	(A) in Me$_2$CO + ProC(S)SK, (i) (70)	m. 62.5-63.5°	1166
		m. 78-81°, IR spectrum	1166
10-[i-ProC(S)S]-	(A) in Me$_2$CO + i-ProC(S)SK, (i) (49)	m. 72-73.5°, fungicidal activity	68
10-[MeOCH$_2$CH$_2$OC(S)S]-	(A) + MeOCH$_2$CH$_2$OC(S)SK in C$_6$H$_6$ at 20-35°	m. 80.5-82°, IR spectrum	1166
10-Bu-		(l)	1460
		(g)	1040
10-BuSO$_2$S-	(A) in Me$_2$CO + BuSO$_2$SK in aq. Me$_2$CO under reflux (64)	m. 89.5-90.5°, IR spectrum	1166

Compound	Preparation	Properties	Ref.
10-[(EtO)$_2$P(S)S]- (M-614)	(B) + (EtO)$_2$P(S)SH, (h) (80)	m. 63.5-64°, IR spectrum, (b, m, n, o)	1166
10-[(EtO)$_2$P(O)]-	10-NCS-phenoxarsine + Pt(OEt)$_3$ in C$_6$H$_6$ at 30° (98)	m. 63-65°, fungicidal prop.	68
		m. 35-36°, d$_4^{20}$ 1.7115,, n$_D^{20}$ 1.6005	871
	(A) + P(OEt)$_3$ by heating (98)	n$_D^{25}$ 1.6044, IR spectrum	1166
C$_{17}$			
10-[OCH=CHCH=CCOSH]-	(B) + OCH=CHCH=CCOSH, (h) (84)	m. 116-116.5°, IR spectrum	1166
10-(4-Methyl-2-pyrimidyl-thio)-	(A) + 4-Me-2-pyrimidinethiol·HCl by heating with NaOEt in EtOH (37)	Yellow crystals, m. 108-110°	316
10-[O(CH$_2$CH$_2$)$_2$NC(S)S]-(M-614)		m. 148-149°	68
10-[i-BuCOS]-	(B) + i-BuCOSH, (h) (87)	m. < 20°, IR spectrum	1166
10-[BuOC(S)S]-	(A) in Me$_2$CO + BuOC(S)SK, (i) (78)	m. 69-70°, IR spectrum	1166
10-[i-BuOC(S)S]-	(A) in Me$_2$CO + i-BuOC(S)SK, (i) (94)	m. 40.0-41.5°, IR spectrum	1166
10-[s-BuOC(S)S]-	(A) in Me$_2$CO + s-BuOC(S)SK, (i) (79)	m. 82.5-83.5°, IR spectrum	1166
10-[EtOCH$_2$CH$_2$OC(S)S]-	(A) in Me$_2$CO + EtOCH$_2$CH$_2$OC(S)SK, (i) (78)	m. 67.0-68.5°, IR spectrum (g, p)	1166
10-[Et$_2$NC(S)S]- (M-614)		m. 121-122°, fungicidal prop.	68
10-[EtNHC(=NEt)S]-	(A) + (EtNH)$_2$CS in boiling EtOH, (q)	The picrate, yellow cryst. solid, m. 151-153° (dec.), (f)	503
10-[BuNHC(=NH)S]-	(A) + BuNHCSNH$_2$ in boiling EtOH, (q)	The picrate, yellow cryst. solid, m. 151-153° (dec.), (f)	503
10-C$_5$H$_{11}$O-		(g)	1040
10-[Et$_2$NC(O)S]-	(A) in Me$_2$CO + Et$_2$NC(O)SNa, (i) (62)	m. 90-93°, IR spectrum	1166

(j) Antifungal and antibacterial therapeutic effect in guinea pigs (1473). (k) Isolated as hydrochloride and picrate, resp. (503). (1) Useful as a pesticide for the control of various insects and plant species (1460). (m) Toxicity data: harmful in agricultural applications as insecticide due to irritation and a cumulative effect (1558). (n) Useful as a seed disinfectant in the control of cotton diseases (889). (o) Microdetn. of S (126). (p) Aq. soln. contg. 500 ppm of the phenoxarsine deriv. gives a complete control of tomato late blight, coontail, green algae, nematodes, and of the soil dwelling organisms producing root rot and decay (1460). (q) Isolated as a picrate by pouring the rxn. mixt. into alc. picric acid soln. (503).

Table 40 Continued

Deriv. of [phenoxarsine structure: positions 8,7 / 6,5 on one ring, As at 10, O, R; positions 2,3 / 4 on other ring] where R is:	Prepd. from 10-Chlorophenoxarsine (A) or 10,10'-Oxybisphenoxarsine (B) (Yield %)	Prop. and Rxn.	Ref.
C₁₈			
10-Ph-		(g)	1040
10-PhSO₂S-	(A) in Me₂CO + PhSO₂SK in aq. Me₂CO under reflux (65)	m. 117.5-119.5°, IR spectrum	1166
10-PhSO₃-	(B) + PhSO₃H, (h) (80)	m. 160-161°, IR spectrum	1166
10-[(4,6-Me₂-pyrimidin-2-yl) S-]	4,6-Me₂-2-pyrimidinethiol·HCl + NaOEt in EtOH at 70° and boiling with (A) (64)	Yellow crystals, m. 135-138°	316
10-[$\overline{OCH_2CH_2CH_2CHCH_2OC(S)}$S]- (M-615)		(g)	1040
10-[$\overline{CH_2(CH_2)_4NC(S)}$S]- (M-615)		m. 78-79°, fungicidal prop.	68
10-[$\overline{CH_2(CH_2)_4NC(O)}$S]-	(A) in Me₂CO + $\overline{CH_2(CH_2)_4NC(O)}$SNa, (l) (69)	m. 98-100°, IR spectrum	1166
10-[C₅H₁₁OC(S)S]-	(A) in Me₂CO + C₅H₁₁OC(S)SK, (l) (84)	n_D^{25} 1.6693, IR spectrum	1166
10-[1-C₅H₁₁OC(S)S]-	(A) in Me₂CO + 1-C₅H₁₁OC(S)SK, (l) (86)	m. 68.0-68.5°, IR spectrum	1166
10-[s-C₅H₁₁OC(S)S]-	(A) in Me₂CO + s-C₅H₁₁OC(S)SK, (l) (86)	m. 68.0-68.5°, IR spectrum, (g)	1166
10-[1-PrOCH₂CH₂OC(S)S]-	(A) + Me₂CHOCH₂CH₂OC(S)SK in C₆H₆ at 20-35°	(l)	1460
10-[(PrO)₂P(S)S]- (M-615)		High boiling liquid, fungicidal activity	68
10-[(1-PrO)₂P(S)S]- (M-615)	(B) + (1-PrO)₂P(S)SH, (h) (77)	m. 61-61.5°, IR spectrum	1166
		n_D^{20} 1.6213, fungicidal activity	68
10-[(1-PrO)₂P(O)-]	(A) + (1-PrO)₃P at 125° (94)	White waxy solid, (g)	1040, 1166
		IR spectrum	1166

C$_{18}$

10-[5-Br-2-benzothiazol S]-	(B) + 5-Br-2-benzothiazolethiol, (r)	(s)	505
10-[5-Cl-2-benzimidazol S]-	(B) + 5-Cl-2-benzimidazolethiol, (r)	(s)	505
10-[6-Cl-2-benzoxazol S]-	(B) + 6-Cl-2-benzoxazolethiol, (r)	(s)	505
10-[2,4,5-Cl$_3$C$_6$H$_2$NHC(=NH)S]-	(A) + 2,4,5-Cl$_3$C$_6$H$_2$NHC(S)NH$_2$ in EtOH, isolated as a picrate	(f)	503
10-[2-benzothiazol S]-	(B) + 2-benzothiazolethiol, (r)	Cryst. solid, m. 169-176°, (s)	505
10-[2-benzoxazol S]-	(B) + 2-benzoxazolethiol, (r)	Crystals, m. 114-115°, (s)	505
10-[2-benzimidazol S]-	(B) + 2-benzimidazolethiol, (r)	Cryst. solid, m. 211-212°, (s)	505
10-[PhCOS]-	(B) + PhCOSH, (h) (80)	m. 109-110°, IR spectrum, (g)	1166
10-[2-(HCONH)C$_6$H$_4$SO$_2$S]-	(A) + 2-(HCONH)C$_6$H$_4$SO$_2$SK	Antiparasitic and herbicidal activity	504
10-[PhNHC(=NH)S]-	(A) + PhNHC(S)NH$_2$ in EtOH under reflux, isolated as a picrate	Picrate, yellow cryst. solid, m. 168-172° (dec.), (f)	503
10-[PhOCH$_2$COS]-		(g)	1040
10-[cyclo-C$_6$H$_{11}$C(O)S]-	(B) + cyclo-C$_6$H$_{11}$COSH, (h) (65)	m. 78-79°, IR spectrum	1166

(r) In C$_6$H$_6$ under reflux (505).   (s) Antiparasitic and slimicidal activity (505).

Table 40 Continued

Deriv. of $\begin{smallmatrix}R\\As\end{smallmatrix}$ where R is: (positions 8,7,10,6,5,4,3,2,O)	Prepd. from 10-Chlorophenoxarsine (A) or 10,10'-Oxybisphenoxarsine (B) (Yield %)	Prop. and Rxn.	Ref.
C_{19} (Cont.)			
10-[cyclo-$C_6H_{11}OC(S)S$]-	(A) in Me_2CO + cyclo-$C_6H_{11}OC(S)SK$, (i) (59)	m. 91.5-92.5°, IR spectrum	1166
10-[$BuOCH_2CH_2OC(S)S$]-	(A) + $BuOCH_2CH_2OC(S)SK$ in C_6H_6 at 20-35°	Crystals, m. 51-52°, (1)	1460
C_{20}			
10-[2,4,5-$Cl_3C_6H_2OCH_2C(O)S$]-	(B) + 2,4,5-$Cl_3C_6H_2OCH_2COSH$, (h) (75)	m. 159-159.5°, IR spectrum	1166
10-[(7-Me-2-benzothiazolyl) S]-	(B) + 7-Me-2-benzothiazolethiol, (r)	(s)	505
10-[$PhCH_2C(O)S$]-	(B) + $PhCH_2COSH$, (h) (93)	m. 123-124°, IR spectrum	1166
10-[$PhCH_2OC(S)S$]-	(A) + $PhCH_2OC(S)SK$, (i) (71)	m. 124.0-125.5°, IR spectrum	1166
10-[$PhOCH_2C(O)S$]-	(B) + $PhOCH_2COSH$, (h) (71)	m. 108-108.5°, IR spectrum	1166
10-[4-$MeOC_6H_4C(O)S$]-	(B) + 4-$MeOC_6H_4COSH$, (h) (85)	m. 118-118.5°, IR spectrum, (g)	1166
10-[2-$AcNHC_6H_4SO_2S$]-	(A) + 2-$AcNHC_6H_4SO_2SK$	Antiparasitic and herbicidal activity	504
10-[4-$AcNHC_6H_4SO_2S$]-	(A) + 4-$AcNHC_6H_4SO_2SK$ in DMF at 80-100°	Cryst. solid, m. 177.5-178.5°, antiparasitic and herbicidal prop., (g)	504
10-[4-$AcNHC_6H_4SO_3$]-	(B) + 4-$AcNHC_6H_4SO_3H$	Antiparasitic and herbicidal activity	504
10-[n-$C_5H_{11}OCH_2CH_2OC(S)S$]-	(A) + n-$C_5H_{11}OCH_2CH_2OC(S)SK$ in C_6H_6 at 20-35°	(1)	1160
10-[$C_7H_{15}NHC(=NH)S$]-	(A) + $C_7H_{15}NHC(S)NH_2$ in EtOH, isolated as a picrate	The picrate, m. 140-142° (dec.), (f)	503

Compound	Preparation	Properties	Ref.
10-[(BuO)₂P(S)S]- (M-617)		n_D^{20} 1.6218, fungicidal activity	68
10-[(i-BuO)₂P(S)S]-		Fungicidal activity	68
C₂₁			
10-[2,4,5-Cl₃C₆H₂OCH₂CH₂C(O)S]-	(B) + 2,4,5-Cl₃C₆H₂OCH₂CH₂C(O)SH, (h) (82)	m. 136.5-137.5°, IR spectrum	1166
10-[S]- (EtO-benzoxazole)	(B) + 5-EtO-2-benzoxazolethiol, (r)	(s)	505
10-[PhCH₂CH₂OC(S)S]-	(A) + PhCH₂CH₂OC(S)SK, (i) (60)	m. 86.5-89.5°, IR spectrum	1166
10-[PhOCH₂CH₂OC(S)S]-	(A) + PhOCH₂CH₂OC(S)SK in Me₂CO, C₆H₆, H₂O, or xylene	Cryst. solid, m. 94.5-96.5°, (t)	1461
10-[m-(EtCONH)C₆H₄SO₃H]-	(B) + m-(EtCONH)C₆H₄SO₃H	Antiparasitic and herbicidal activity	504
10-[cyclo-C₆H₁₁CH₂CH₂C(O)S]-	(B) + cyclo-C₆H₁₁CH₂CH₂COSH, (h) (85)	m. 84-84.5°, IR spectrum	1166
10-[C₈H₁₇C(O)S]- (M-617)	(B) + C₈H₁₇COSH, (h) (85)	m. 52-53°, IR spectrum	1166
10-[C₈H₁₇OC(S)S]- (M-617)		(g)	1040
10-[BuNHC(=NBu)S]-	(A) + (BuNH)₂CS by heating in EtOH in the presence of picric acid, isolated as a picrate	The picrate, (f)	503
C₂₂			
10-[S]- (i-Pr-benzoxazole)	(B) + 5-i-Pr-2-benzoxazolethiol, (r)	(s)	505
2,10-Me[EtO S]- (benzothiazole)	10,10'-Oxybis(2-Me-phenoxarsine) + 6-EtO-2-benzothiazolethiol, (r)	(s)	505
10-[3,4,5-(MeO)₃C₆H₂C(O)S]-		(g)	1040
10-[p-(PrCONH)C₆H₄SO₃H]-	(B) + p-(PrCONH)C₆H₄SO₃H	Antiparasitic and herbicidal activity	504

(t) At concn. of 500 ppm the compd. is effective in controlling tomato late blight, Salvinia, green algae, southern army worms and root rot organisms (1461).

Table 40 Continued

R Deriv. of 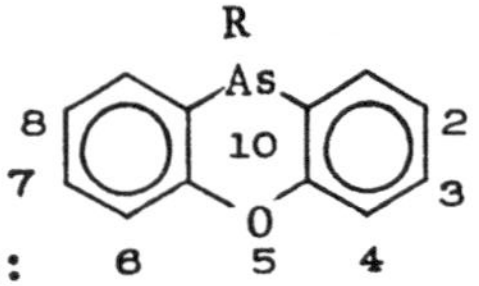 where R is:	Prepd. from 10-Chlorophenoxarsine (A) or 10,10'-Oxybisphenoxarsine (B) (Yield %)	Prop. and Rxn.	Ref.
C_{23} 10-[$C_{10}H_7$NHC(=NH)S]-	(A) + $C_{10}H_7$NHC(S)NH_2 in boiling EtOH in the presence of picric acid	The picrate, m. 171–173.5°, (f)	503
2,8,10-Me$_2$[(structure, EtO)] 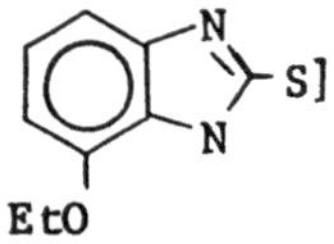	10,10'-oxybis(2,8-Me$_2$-phenoxarsine) + 7-EtO-2-benzimidazolethiol, (r)	(s)	505
10-[p-t-BuC$_6$H$_4$C(O)S]-	(B) + p-t-BuC$_6$H$_4$COSH, (h) (72)	m. 117–118°, IR spectrum	1166
10-[p-BuOC$_6$H$_4$C(O)S]-	(B) + p-BuOC$_6$H$_4$COSH, (h) (75)	m. 89.5–90.5°, IR spectrum	1166
10-[m-(BuCONH)C$_6$H$_4$SO$_2$S]-	10-Br-phenoxarsine + m-(BuCONH)C$_6$H$_4$-SO$_2$SK	Antiparasitic and herbicidal activity	504
C_{24} 10-[p-(n-C$_5$H$_{11}$O)C$_6$H$_4$C(O)S]-	(B) + p-(n-C$_5$H$_{11}$O)C$_6$H$_4$COSH, (h) (82)	m. 102–103°, IR spectrum	1166
10-[o-(C$_5$H$_{11}$CONH)C$_6$H$_4$SO$_3$]-	(B) + o-(C$_5$H$_{11}$CONH)C$_6$H$_4$SO$_3$H in C$_6$H$_6$ under reflux	Cryst. solid, antiparasitic and herbicidal activity	504
C_{25} 10-[(structure, p-BrC$_6$H$_4$)]	(B) + 7-(p-BrC$_6$H$_4$)-2-benzimidazole-thiol, (r)	(s)	505
10-[(structure, p-ClC$_6$H$_4$)]	(B) + 7-(p-ClC$_6$H$_4$)-2-benzothiazole-thiol, (r)	(s)	505

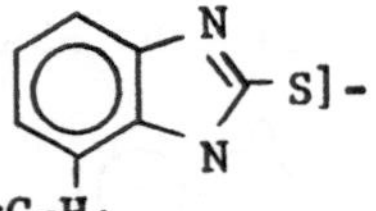

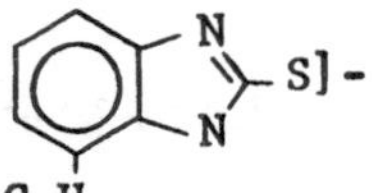

412

10-[
 (benzothiazole structure) S]-
 Ph | (B) + 7-Ph-2-benzothiazolethiol, (r) | (s) | 505

10-[p-ClC$_6$H$_4$NHC(=NC$_6$H$_4$Cl-p)S]- | (A) + (p-ClC$_6$H$_4$NH)$_2$CS in EtOH under reflux, isolated as a picrate | The picrate, yellow cryst. solid, m. 149-150.5° (dec.), (g) | 503

10-[3,4,5-(EtO)$_3$C$_6$H$_2$C(O)S]- | (B) + 3,4,5-(EtO)$_3$C$_6$H$_2$COSH, (h) (83) | m. 130.5-131.5°, IR spectrum | 1166

10-[C$_{12}$H$_{25}$NHC(=NH)S]- | (A) + C$_{12}$H$_{25}$NHC(S)NH$_2$ by heating in alc., isolated as a picrate | The picrate, m. 105.5-106.5° (g) | 503

C$_{27}$
 p-EtC$_6$H$_4$
 10-[(benzimidazole structure) S]- | (B) + 4-(p-EtC$_6$H$_4$)-2-benzimidazole-thiol, (r) | (s) | 505

10-[PhCH$_2$NHC(=NCH$_2$Ph)S]- | (A) + (PhNH)$_2$CS by heating in alc., isolated as a picrate | The picrate, yellow cryst. solid, m. 156-158.5° (dec.), (g) | 503

C$_{31}$
 10-[p-(C$_{12}$H$_{25}$O)C$_6$H$_4$C(O)S]- | | (g) | 1040

10-[C$_{18}$H$_{37}$NHC(=NH)S]- | (A) + C$_{18}$H$_{37}$NHC(S)NH$_2$ by heating in alc., isolated as a picrate | The picrate, m. 103-104°, (f) | 503

413

3.4.5.2 TETRAOXADIARSASPIRO[5,5]HENDECANE DERIVATIVES

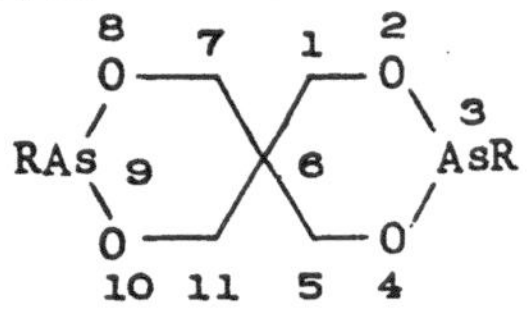

Three derivatives of this ring system were described in the main volume on page 618.

3,9-Diethyl-2,4,8,10-tetroxa-3,9-diarsaspiro[5,5]hendecane, $As_2C_9H_{18}O_4$,
    (M-618)
Prop.: IR spectrum (1387).

3,9-Diphenyl-2,4,8,10-tetroxa-3,9-diarsaspiro[5,5]hendecane, $As_2C_{17}H_{18}O_4$,
    (M-618)
Prop.: IR spectrum (1387).

3.4.7 RING SYSTEMS WITH ARSENIC AND NITROGEN HETEROATOMS

3.4.7.2 PHENARSAZINE DERIVATIVES

Numerous phenarsazine derivatives were described in the main volume on pages 620-25. 10-Chloro-5,10-dihydrophenarsazine has been used in the synthesis of 10-alkylthio- and 10-arylthio-dihydrophenarsazines as well as of 5,10-dihydrophenarsazinyl xanthates, thiocarbamates, and dithiophosphates derivatives. Some of these compounds have fungicidal and bactericidal properties. Three of them are described hereafter individually, while others are compiled in Table 41 on page 416.

10-Chloro-5,10-dihydrophenarsazine, Adamsite, $AsC_{12}H_9ClN$,   (M-620) 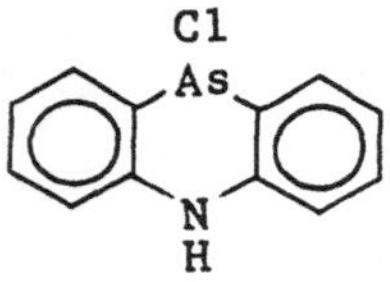
Prop.: The compd. crystd. from xylene with 1/2 mol. of the
solvent in a monoclinic form. Upon losing the solvent of
crystn., the crystals become an orthorhombic, bright yellow
powder or yellowish green single crystals (286, 287). The
crystal and mol. structures were studied by x-ray diffrac-
tion analysis (286, 287, 595). Mass spectrum (279).
Rxn. with: RSNa → 10-RS-5,10-dihydrophenarsazine (316).
NaOMe in MeOH, followed by RSH → 10-RS-5,10-dihydrophenarsazine (314).
NaOMe in MeOH, followed by HSP(S)(OR)₂ → the o-[(RO)₂P(S)S]- deriv. (315).
Py by heating → 10-chloro-5(5H),10'(5'H),5',10"(5"H)-terphenarsazine (1324).
Phenarsazine oxide in Py by treating with HCl or by heating in $PhNO_2$ → 10-
chloro-5(5H),10'(5'H),5',10"(5"H)-terphenarsazine (1324).
Biol. Prop.: Inhibition of cholinesterase from human plasma (307). Inhibition
of xanthine oxidase (894). A broad spectrum antifungal (Trichophyton asteroi-
des and Candida albicans) and antibacterial activity in vitro (1473).
Uses: Antifouling agent for coatings (1325). Inhibition of xanthine oxidase -
an indirect chromopotentiometric detn. of the As compd. (894).

10,10'-Oxybis(5,10-dihydrophenarsazine), $As_2C_{24}H_{18}N_2O$,   (M-621)

Synth.: 10-Butylthio-5,10-dihydrophenarsazine by boiling
with H_2O (314). The compd. is formed from 10-chloro- or
10-fluoro-5,10-dihydrophenarsazine in antifouling paints
by hydrolysis (1325).

Prop.: Colorless crystals, m. 300° (314).

Rxn. with: 10-Chloro-5,5-dihydrophenarsazine by heating in
$PhNO_2$ or by passing HCl through its soln. in Py → 10-chloro-
5(5H),10'(5'H),5',10"(5"H)-terphenarsazine (1324).

Biol. Prop.: A broad-spectrum antifungal (Trichophyton asteroides and Candida
albicans) and antibacterial activity in vitro (1473).

Use: Antifouling agents for coatings (1325).

10-Chloro-5(5H),10'(5'H),5',10"(5"H)-terphenarsazine, $As_3C_{32}H_{25}ClN_3$

Synth.: From 10-Cl-5,10-dihydrophenarsazine by heating
in Py, by adding H_2O to a soln. of 10-Cl-5,10-dihydro-
phenarsazine in Me_2CO or alc., and by passing HCl through
a hot soln. of phenarsazine oxide in Py (1324).
From a 1:1 mixt. of 5-Cl-5,10-dihydrophenarsazine with
phenarsazine oxide in $PhNO_2$ by heating (1324).

Prop.: m. 260-263° (1324).

Use: Antifouling agent for paints applicable to Al alloys,
concrete, and wood (1324).

3.4.7.3 BENZOPHENARSAZINE DERIVATIVES

7,12-Dihydrobenzo[a]-
phenarsazine

7,12-Dihydrobenzo[c]-
phenarsazine

The methods of synthesis of 7,12-dihydrobenzo[a]- and 7,12-dihydrobenzo[c]-
phenarsazine derivatives were described in the main volume on page 626, and
these compounds were compiled on pages 627-34. The benzophenarsazine derivatives which appear in Table 42 on page 421 of this volume have been prepared
by the previously reported methods.

Table 41. 5,10-Diphenarsazine Derivatives

Deriv. of	Prepd. from* (Yield %)	Prop. and Rxn.	Ref.
C_{12}			
10-Br-		Crystal and mol. structures by x-ray analysis	595
10-F-		Crystal and mol. structures by x-ray analysis	595
		Antifungal activity, (a, b)	796
10-I- (M-622)		Crystal and mol. structures by x-ray analysis, (a)	595
10-HO-	10-BuS-5,10-dihydrophenarsazine + H$_2$O$_2$ in Me$_2$CO at 50°	Colorless, cryst. powder, m. 350° (dec.)	314
	10-[N=C(Me)CH=C(Me)N=CS]-5,10-di-hydrophenarsazine + H$_2$O$_2$ in Me$_2$CO at 50°		314
C_{13}			
10-NC- (M-622)		(b)	1120
10-NCS- (M-622)		(b)	1120
10-Me- (M-622)		Mass spectrum	279
10-MeO- (M-622)		Rxn. with RSH → the 10-RS- deriv., (b)	317
C_{14}			
10-Ac-	10-Ac-5,10-dihydrophenarsazine + HF	Rxn. with HF → 10-F-5,10-dihydro-phenarsazine acetate	817
10-F-, acetate		Brown-yellow solid, m. 273°, (c)	817
10-Et- (M-623)		(a)	1473
10-EtO-		(a)	1473
10-HOCH$_2$CH$_2$S- (M-623)		(a)	1473

10-EtS-	(A) + NaOMe in MeOH, followed by EtSH (75)	Green crystals, m. 99-100°	314
5-Acetyl-10-chloro- (M-623)	10-(4,6-Me$_2$-2-pyrimidylthio)-5,10-dihydrophenarsazine + AcCl in C$_6$H$_6$	m. 228-229°, (d, e)	316
10-[(MeO)$_2$P(S)S]-	(A) + NaOMe in MeOH, followed by (MeO)$_2$P(S)SH (71)	m. 145-146°	315
C$_{15}$			
10-[EtOC(S)S]- (M-624)		(a, b, f)	1120
10-[Me$_2$NC(S)S]-		(a, b)	1473
10-PrS-	(A) + NaOPr in MeOH (66)	Green crystals, m. 82-83°, (g)	314
10-(i-PrS)-	(A) + NaOMe in MeOH under reflux, followed by i-PrSH (91)	Green crystals, m. 71-72°	314
C$_{16}$			
10-BuS-	(A) + NaOMe in MeOH, followed by BuSH (81)	Green crystals, m. 88°, (g,h,i,j)	314
10-(t-BuS)-	(A) + NaOMe in MeOH, followed by t-BuSH (67)	Green crystals, m. 126-127°	314
10-[(EtO)$_2$P(S)S]-	(A) + NaOMe in MeOH, followed by (EtO)$_2$P(S)SH in MeOH (80)	Yellowish-green crystals, m. 105-106°, (a, g)	315

* (A) in this column stands for 10-chloro-5,10-dihydrophenarsazine.
(a) A broad-spectrum antifungal and antibacterial activity in vitro (1473). (b) Useful as an antifouling agent for coatings (1325). Antifouling agent for paints used on ships' bottoms made of Al and Mg alloys (1120). (c) Antifouling agent for ship paints; toxicity data (817). (d) Rxn. with KSP(S)(OR)$_2$ in C$_6$H$_6$ → the 5-Ac-10-[(RO)$_2$-P(S)S]- deriv. (315). (e) Rxn. with RSNa → the 5-Ac-10-RS-deriv. (316). (f) Fungicidal activity against the spores of Alternaria solani and Botrytis cinerea (645). (g) Fungistatic and fungicidal activity against Alternaria radicina, Botrytis cinerea, Fascicladium dendriticum, Trichophyton gypseum, and Epidermiphyton; toxicity to mice and lethal doses (1133). (h) Rxn. with H$_2$O under reflux → 10,10'-oxybis(5,10'-dihydrophenarsazine) (314). (i) Rxn. with H$_2$O$_2$ in Me$_2$CO → 10-HO-5,10-dihydrophenarsazine 10-oxide (314, 316). (j) Rxn. with BzCl → BzSBu and 10-Bu-5,10-dihydrophenarsazine (314).

Table 41 Continued

Deriv. of 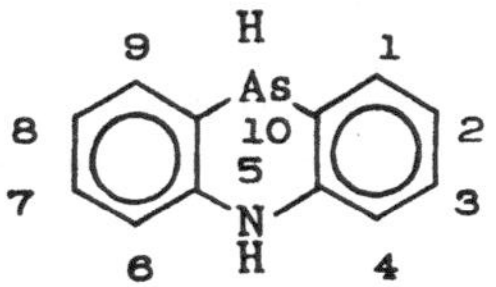	Prepd. from* (Yield %)	Prop. and Rxn.	Ref.
C_{17}			
10-[$Et_2NC(S)S$]-		(f)	645
10-(1-$PrSCH_2CH_2S$)-	5-10-dihydro-10-MeO-phenarsazine + 1-$PrSCH_2CH_2SH$ by heating (74)	m. 89-91°	317
10-[4-Me-2-pyrimidinyl S]- 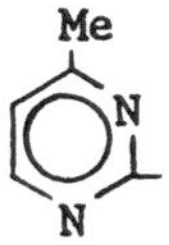	4-Me-2-pyrimidinethiol + NaOMe in MeOH at 50° + (A) (50)	Yellow crystals, m. 187-189°	316
10-[BuOC(S)S]-		(f)	645
C_{18}			
10-PhS-	(A) + NaOMe in MeOH under reflux, followed by PhSH (65)	Green crystals, m. 104-105°, (g)	314
10-(o-$H_2NC_6H_4S$)-	(A) + NaOMe in MeOH under reflux, followed by o-$H_2NC_6H_4SH$ (74)	Green crystals, m. 170°	314
	(A) + o-$H_2NC_6H_4SH$ in dry C_6H_6 under reflux (93)	(k)	314
10-[4,6-Me_2-2-pyrimidinyl S]- 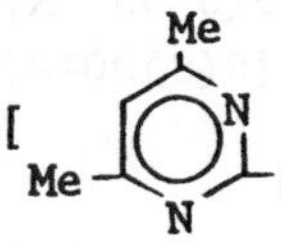	4,6-Me_2-2-pyrimidinethiol·HCl + NaOMe in MeOH at 50°, followed by (A) in MeOH (52)	Yellow crystals, m. 233-234°,	316
10-[1-$C_5H_{11}OC(S)S$]-		Effective against the spores and micelia of Alternaria solani and Botrylis cinerea	645

418

10-[(PrO)$_2$P(S)S]-	(A) + NaOMe in MeOH under reflux, followed by HSP(S)(OPr)$_2$ (89)	m. 72-74°	315
10-(Et$_2$NCH$_2$CH$_2$S)-	(A) + NaOMe in MeOH under reflux, followed by Et$_2$NCH$_2$CH$_2$SH (98)	Oily prod.	314
	(A) + Et$_2$NCH$_2$CH$_2$SH in C$_6$H$_6$ under reflux, isolated as a HCl salt (92)	HCl salt, m. 114-117°, (g)	314
5,10-Ac[(EtO)$_2$P(S)S]-	5,10-Ac(Cl)-5,10-dihydrophenarsazine + KSP(S)(OEt)$_2$ in Me$_2$CO (90)	m. 123-124°	315

C$_{19}$

5-Benzoyl-10-chloro-		Rxn. with KSP(S)(OR)$_2$ in Me$_2$CO → the 5,10-Bz[(RO)$_2$P(S)S]- deriv., (m)	315
10-PhCH$_2$S-	(A) + NaOMe in MeOH under reflux, followed by PhCH$_2$SH (72)	Green crystals, m. 131-132°, (g)	314

C$_{20}$

5,10-Ac[ ] 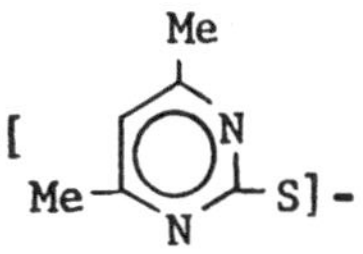 -	5,10-Ac(Cl)-5,10-dihydrophenarsazine + Na 4,6-Me$_2$-2-pyrimidinethiolate in DMF at 74° (77)	m. 194.5-196.5°	316,
10-[(i-BuO)$_2$P(S)S]-	(A) + NaOMe in MeOH under reflux, followed by (i-BuO)$_2$P(S)SH (87).	m. 97°	315

C$_{23}$

5,10-Bz[(EtO)$_2$P(S)S]-	5,10-Bz(Cl)-5,10-dihydrophenarsazine + KSP(S)(OEt)$_2$ in Me$_2$CO (91).	m. 135-136°	315

* (A) in this column stands for 10-chloro-5,10-dihydrophenarsazine.

(k) The HCl salt has a fungistatic and fungicidal activity against the species listed under (g) above (1133).

(l) Rxn. with AcCl → 5-Ac-10-Cl-5,10-dihydrophenarsazine (316). (m) Rxn. with RSNa → the 5,10-Bz(RS)- deriv. (316).

Table 41 Continued

Deriv. of 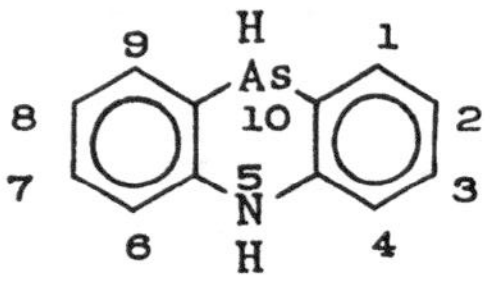 8	Prepd. from* (Yield %)	Prop. and Rxn.	Ref.
C$_{24}$ 10-[(PhO)$_2$P(S)S]-	(A) + NaOMe in MeOH under reflux, followed by HSP(S)(OPh)$_2$ (91)	m. 150-152°	315
C$_{25}$ 5,10-Bz[]-	5,10-Bz(Cl)-5,10-dihydrophenarsazine + Na 4,6-Me$_2$-2-pyrimidinethiolate in DMF at 75° (80)	m. 192-194°	316

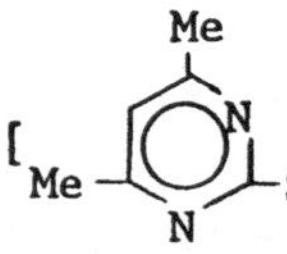

* (A) in this column stands for 10-chloro-5,10-dihydrophenarsazine.

420

Table 42. 7,12-Dihydrobenzophenarsazine Derivatives

7,12-Dihydrobenzo[a]-phenarsazine Deriv.	Prepd. from 2-$(RNH)C_{10}H_{7-x}R'_x$*		Prop. and Rxn.	Ref.
	R	R'		
C_{16}				
12-Chloro-	PhNH-	H	Mass spectrum	280
C_{20}				
12-Chloro-8,10-diethyl-	$2,4\text{-}Et_2C_6H_3$-	H	Yellow needles, m. 221°	281
12-Chloro-2,3,9,10-tetramethyl-	$3,4\text{-}Me_2C_6H_3$-	$6,7\text{-}Me_2$-	Inhibition of paralysis in rats induced by zoxazolamine	277
7,12-Dihydrobenzo[c]-phenarsazine Deriv.	Prepd. from 1-$(RNH)C_{10}H_{7-x}R'_x$*			
	R	R'		
C_{16}				
7-Chloro-	PhNH-	H	Yellow needles, m. 219°, (a, b)	275
			Mass spectrum	280
C_{17}				
7-Methyl-	(c)	(c)	Mass spectrum	280
7-Methoxy-			Inhibition of the spore germination of Cochliobolus miyabeanus	1493
C_{18}				
7-Acetoxy-			Inhibition of the spore germination of Cochliobolus miyabeanus	1493
C_{20}				
7-$MeCH=CHCO_2$-			Inhibition of the spore germination of Cochliobolus miyabeanus	1493
7-Chloro-9,11-diethyl-	$2,4\text{-}Et_2C_6H_3$-	H	Golden yellow leaflets, m. 185°	281
C_{32}				
7,7'-Oxybis-			Inhibition of the spore germination of Cochliobolus miyabeanus	1493

* By the condensation with $AsCl_3$ in o-$C_6H_4Cl_2$ under reflux.

(a) Bacteriostatic prop. (278). (b) Inhibition of paralysis in rats induced by zoxazolamine (277). (c) Prepd. from 7-chloro-7,12-dihydrobenzo[c]phenarsazine + MeMgI (280)

3.4.7.5 DIBENZO-, TRIBENZO-, BENZONAPHTHENO-, DINAPHTHO-, AND BENZOPHENALENO-PHENARSAZINE DERIVATIVES

Several dibenzophenarsazine derivatives and isomers were described in the main volume on pages 636-38. The following compounds were prepared by the same method as that described in the main volume, by a condensation of appropriate secondary aromatic amines with arsenic trichloride.

14-Chloro-1,2,3,4,9,14-hexahydro-12-methyldibenzo[a,c]phenarsazine, AsC$_{21}$H$_{19}$ClN,
Synth.: From 9-(p-toluylamino)-1,2,3,4-tetrahydro-
phenanthrene + AsCl$_3$ in o-C$_6$H$_4$Cl$_2$ under reflux (282).
Prop.: Yellow needles, m. 274° (282).

14-Chloro-7,14-dihydrobenzo[a]naphtho[1,2-j]phenarsazine,   AsC$_{24}$H$_{15}$ClN,
Synth.: From 2-(β-naphthylamino)phenanthrene and
AsCl$_3$ in o-C$_6$H$_4$Cl$_2$ by heating under reflux (1504).
Prop.: Yellow needles, m. 343°, brown-red soln.
in H$_2$SO$_4$ (1504).

16-Chloro-7,16-dihydrobenzo[c]naphtho[1,2-j]phenarsazine,   AsC$_{24}$H$_{15}$ClN,
Synth.: From 2-(α-naphthylamino)phenanthrene soln.
in o-C$_6$H$_4$Cl$_2$ and AsCl$_3$ under reflux (1504).
Prop.: Golden-yellow needles, m. 325°, brown-red
soln. in H$_2$SO$_4$ (1504).

16-Chloro-5,6,7,8,9,16-hexahydrotribenzo[a,c,f]phenarsazine,   AsC$_{24}$H$_{19}$ClN,
Synth.: From 9-(α-naphthylamino)-5,6,7,8-tetrahydro-
phenanthrene + AsCl$_3$ in o-C$_6$H$_4$Cl$_2$ under reflux (282).
Prop.: Yellow green needles, m. 232° (282).

9-Chloro-5,6,7,8,9,16-hexahydrotribenzo[a,c,h]phenarsazine,   AsC$_{24}$H$_{19}$ClN,
Synth.: From 9-(β-naphthylamino)-5,6,7,8-tetrahydro-
phenanthrene + AsCl$_3$ in o-C$_6$H$_4$Cl$_2$ under reflux (282).
Prop.: Mustard yellow needles, m. 261° (282).

7-Chloro-7,14-dihydrobenzo[c]phenaleno[1,9-f,g]phenarsazine, $C_{26}H_{15}AsClN$,
Synth.: From 1-(α-naphthylamino)pyrene + $AsCl_3$
in o-$C_6H_4Cl_2$ under reflux (283).
Prop.: Orange prisms (283).

7-Chloro-7,16-dihydrobenzo[a]phenaleno[1,9-h,i]phenarsazine, $C_{26}H_{15}AsClN$,
Synth.: From 1-(β-naphthylamino)pyrene + $AsCl_3$
in o-$C_6H_4Cl_2$ under reflux (283).
Prop.: Orange needles, m. 321-322° (283).
Biol. Prop.: Inactive as inducer of zoxazolamine
hydroxylase synthesis in rats (276).

17-Chloro-8,17-dihydrodinaphtho[2,3-a;2′,3′-j]phenarsazine, $AsC_{28}H_{17}ClN$,
Biol. Prop.: Inhibition of paralysis in young rats
caused by zoxazolamine (277).

3.4.7.7 BENZARSAZINOTHIANAPHTHEN DERIVATIVES

This is a new type of heterocyclic arsenic derivatives.

6-Chloro-6,11-dihydro[1,4]benzarsazino[3,2-b]thianaphthen, $AsC_{14}H_9ClNS$,
Synth.: From 3-anilinothianaphthen + $AsCl_3$ in
o-$C_6H_4Cl_2$ under reflux (273).
Prop.: Shiny yellow prisms, m. 248° (dec. > 225°), soln.
in H_2SO_4 shows an orange-red halochromy (273).
Biol. Prop.: Bacteriostatic activity (278).

6-Chloro-6,11-dihydro-8-methyl[1,4]benzarsazino[3,2-b]thianaphthen,
    $AsC_{15}H_{11}ClNS$,
Synth.: From 3-(p-tolylamino)thianaphthen by condensation with $AsCl_3$ in o-$C_6H_4Cl_2$ under reflux, 80% yield
(274).
Prop.: Deep yellow needles, m. 269° (dec. > 250°)
(274). Mass spectrum (280).
Biol. Prop.: Inhibition of paralysis in young rats caused by zoxazolamine
(277).

6-Chloro-6,11-dihydro-10-methyl[1,4]benzarsazino[3,2-b]thianaphthen,
    AsC$_{15}$H$_{11}$ClNS,
<u>Synth.</u>: From 3-(o-tolylamino)thianaphthen + AsCl$_3$
in o-C$_6$H$_4$Cl$_2$ under reflux, 80% yield (274).
<u>Prop.</u>: Deep yellow needles, m. 232° (dec. > 210°) (274).

6-Chloro-6,11-dihydro-9-methyl[1,4]benzarsazino[3,2-b]thianaphthen,
    AsC$_{15}$H$_{11}$ClNS,
<u>Synth.</u>: From 3-(m-tolylamino)thianaphthen + AsCl$_3$
in o-C$_6$H$_4$Cl$_2$ under reflux, 80% yield (274).
<u>Prop.</u>: Deep yellow needles, m. 286° (dec. > 260°) (274).

3.4.7.8 TRIAZATRIARSENINE DERIVATIVES

2,2,4,4,6,6-Hexahydro-2,2,4,4,6,6-hexaphenyl-s-triazatriarsenine,
    As$_3$C$_{36}$H$_{30}$N$_3$,
<u>Synth.</u>: From Ph$_2$AsCl$_3$ + NH$_3$ and from [Ph$_2$As(NH$_2$)∴N∴As-
(NH$_2$)Ph$_2$]Cl·CHCl$_3$ on standing in CHCl$_3$. When the latter complex salt is heated at 150° in DMF, (Ph$_2$AsN)$_3$·DMF is formed
(1421).
<u>Prop.</u>: m. 155-157°; the complex with DMF m. 155-157°; IR spectrum (1421).

3.5 SEVEN-MEMBERED RING SYSTEMS

3.5.1 RING SYSTEMS WITH ONE HETEROATOM

3.5.1.1 ARSEPIN DERIVATIVES

Through 1964 two dibenzarsepin isomers and their substitution derivatives
and quaternary salts were reported (M-641-42).

9-Methyl-9H-tribenz-[b,d,f]arsepin,   AsC$_{19}$H$_{15}$,
<u>Synth.</u>: From 5,6,7,8-tetrahydro-5,5,8,8-tetramethyl-
tribenzo[e,g,i][1,4]diarsecinium dibromide (the low- or
high-melting isomer) by heating at 200-205°/14 mm in a
stream of N or at 250-270°/14 mm in a sealed tube (1016).
From 10,11,12,13-tetrahydro-9,9,13,13-tetramethyl-9H-
tribenzo[f,h,j][1,5]-diarsacycloundecinium dibromide
and from 13,14,19,20-tetrahydro-13,13,20,20-tetra-

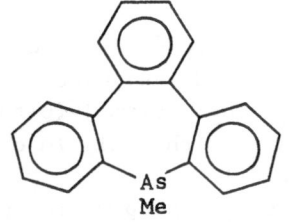

methyltetrabenzo[c,g,i,k][1,6]diarsacyclododecinium dibromide by heating at
220°/0.1 mm and at 245-250°/14 mm, resp. (1016).
<u>Prop.</u>: Crystals, m. 159-160° (1016).
<u>Quaternary Salts</u>: The methiodide monohydrate, m. 218-221°, prepd. from the
arsepin by boiling with MeI in MeOH (1016).
The methotoluenesulfonate, m. 174-182°, prepd. from the arsepin and methyl p-

toluenesulfonate by heating at 125° (1016).
The methopicrate, m. 190-191°, prepd. from the preceding methotoluenesulfonate
(1016).

3.6   EIGHT-MEMBERED RING SYSTEMS

3.6.2   RING SYSTEMS WITH TWO ARSENIC HETEROATOMS

3.6.2.1   DIBENZODIARSOCIN DERIVATIVES

Several arsenic derivatives of this class were described in the main volume
on pages 645-7.

5,5,8,8-Tetraethyl-5,6,7,8-tetrahydrodibenzo[e,g][1,4]diarsocinium,
  $[As_2C_{22}H_{32}]^{++}2X^-$,   (M-646)

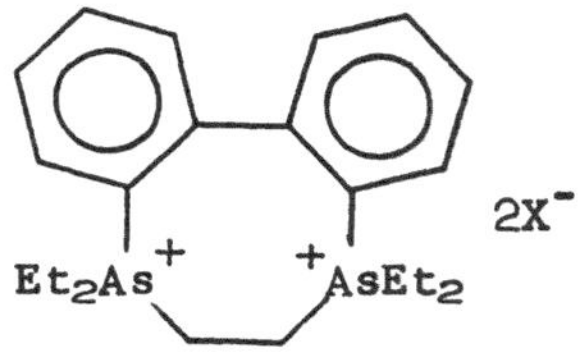

Synth.: From $[2-(Et_2As)C_6H_4-]_2$ by heating with
$BrCH_2CH_2Br$ in the presence of trace EtOH at 100°
in a sealed tube for 300 hrs., the dibromide was
obtd. which was converted to the dipicrate (57).
Prop.: The dibromide is a highly hydroscopic solid
which dec. at elevated temp. to 5-ethyldibenzarsole
(57). The dipicrate, yellow crystals, m. 153-155° (57).

$2X^-$

X = Br, picrate

3.6.6   RING SYSTEMS WITH ARSENIC AND SULFUR HETEROATOMS

3.6.6.4   TETRATHIATETRARSOCANE DERIVATIVES

The following eight-membered ring with alternating arsenic and sulfur atoms
is a new compound.

2,4,6,8-Tetrakis(pentafluorophenyl)-1,3,5,7,2,4,6,8-tetrathiatetrarsocane,
  $As_4C_{24}F_{20}S_4$,

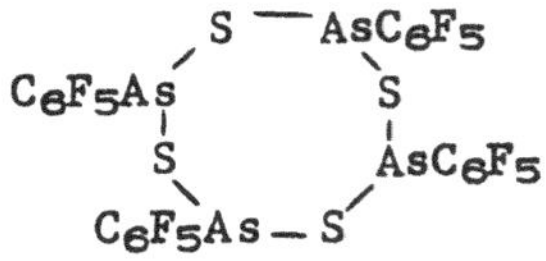

Synth.: From $C_6F_5AsCl_2$ + Ag_2S suspension in C_6H_6
under reflux, 32% yield (651).
Prop.: m. 169-169.5°, IR and NMR spectra (651).
Rxn. with: $HgCl_2$ → $C_6F_5AsCl_2$ (651).

3.6.7   RING SYSTEMS WITH ARSENIC AND NITROGEN HETEROATOMS

3.6.7.4   TETRARSATETRAZOCINE DERIVATIVES

The following eight-membered ring with alternating arsenic and nitrogen atoms
is a new compound.

1,1,3,3,5,5,7,7-Octaphenyl-1,3,5,7,2,4,6,8-tetrarsatetrazocine Diphenylarsenic Nitride Tetramer, $As_4C_{48}H_{40}N_4$,

Synth.: From Ph_2AsCl_3 + NH_4Cl + NH_3 at $-35°$ (679).
Prop.: Diamagnetic, nonionizable substance, m. 320°; a planar structure is postulated (679). The cryst. and mol. structures were detd. by three-dimensional single crystal x-ray analysis (622a).

3.7   NINE-MEMBERED RING SYSTEMS

3.7.1   RING SYSTEMS WITH ONE HETEROATOM

3.7.1.1   TETRABENZ[b,d,f,h]ARSONIN DERIVATIVES

Tetrabenz[b,d,f,h]arsonin derivatives described in the main volume (M-648-49) were prepared by a thermal rearrangement of 5,5'-spirobi(dibenzarsole) derivatives. Now the ring system can also be obtained from 5-(2-biphenylyl)-dibenzarsolium salts by a thermal rearrangement.

17-Methyl-17H-tetrabenz[b,d,f,h]arsonin, $AsC_{25}H_{19}$,

Synth.: From 5-(2-biphenylyl)-5-methyldibenzarsolium iodide by thermal dec. at 250° under N, 94% yield, along with 1% 5-(2-biphenylyl)dibenzarsole (723).
Prop.: Yellowish needles, m. 145-146°, NMR spectrum (723).

Quaternary Salt: The methiodide prepd. from the arsonin by treating with MeI (86% yield), forms colorless crystals, m. 260-264° (dec.), NMR spectrum (723).

3.7.2   RING SYSTEMS WITH TWO ARSENIC HETEROATOMS

3.7.2.2   TRIBENZO[d,f,h][1,3]DIARSONIN DERIVATIVES

The following tribenzodiarsonin derivatives are new compounds.

5,6-Dihydro-5,7-dimethyl-7H-tribenzo[d,f,h][1,3]diarsonin, $As_2C_{21}H_{20}$,

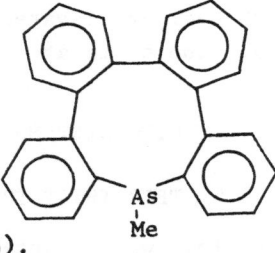

Synth.: From 5,6-dihydro-5,5,7,7-tetramethyl-7H-tribenzo[d,f,h][1,3]diarsoninium dibromide by heating at 245°/0.1 mm (1016).
Prop.: Crystals, m. 137-140° (1016).
Coordination Deriv. with $PdBr_2$, 1:1, bright yellow flocculent ppt., m. 300°, by heating from r.t., and at 304°, when immersed in a bath preheated to 290°. The complex was prepd. from the diarsine by heating with K_2PdBr_4 in alc. (1016).

5,6-Dihydro-5,5,7,7-tetramethyl-7H-tribenzo[d,f,h][1,3]diarsoninium Salts,
   [As$_2$C$_{23}$H$_{26}$]$^{++}$2X$^-$,   X = Br, picrate   (For the molecular structure
   of the diarsonin see the preceding entry.)
<u>Synth.</u>: From o-terpheny1-2,2"-ylenebis(dimethylarsine) + MeBr by heating in
a sealed tube at 100°, the dibromide was obtd. The dipicrate was prepd. from
the dibromide dissolved in hot aq. Me$_2$CO by mixing with aq. Na picrate (1016).
<u>Prop.</u>: The dibromide, m. 258-264°; pyrolysis at 245°/0.1 mm gave the preceding
diarsonin deriv. (1016). The dipicrate, m. 266° (dec.) (1016).

3.8   TEN-MEMBERED RING SYSTEMS

3.8.2   RING SYSTEMS WITH TWO ARSENIC HETEROATOMS

3.8.2.1 TRIBENZODIARSECIN DERIVATIVES

   The diarsecin derivatives reported in the main volume (M-651-2) included di-
benzo[g,i][1,6]diarsecin and tribenzo[b,d,h][1,6]diarsecin derivatives.

9,10,11,12-Tetrahydro-9,12-dimethyltribenzo[e,g,i][1,4]diarsecin,   As$_2$C$_{22}$H$_{22}$
<u>Synth.</u>: From the high melting 9,10,11,12-
tetrahydro-9,9,12,12-tetramethyltribenzo-
[e,g,i][1,4]diarsecinium dibromide by heating
at 200° and slowly raising the temp. to 220°/
0.1 mm; the compd. was identified as the dimethiodide dihydrate (1016).

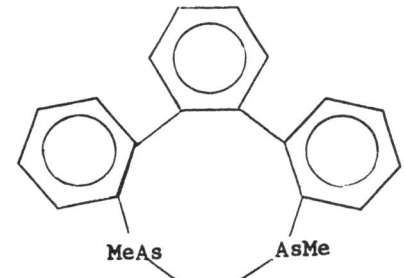

9,10,11,12-Tetrahydro-9,9,12,12-tetramethyltribenzo[e,g,i][1,4]diarsecinium
   Salts,   [As$_2$C$_{24}$H$_{28}$]$^{++}$2X$^-$,   X = Br, I, picrate   (For the molecular structure of the diarsecin see the preceding entry.)
<u>Synth.</u>: From o-terpheny1-2,2"-ylenebis(dimethylarsine) + BrCH$_2$CH$_2$Br in MeOH
by heating in a sealed tube at 100° a low melting dibromide was obtd., whereas
the quaternization under reflux at 150-155° in N gave a high melting dibromide.
The dibromides were converted to the diiodides by a treatment with LiI and to
the dipicrates by a treatment with Na picrate (1016).
<u>Prop.</u>: The low melting dibromide form crystd. as a dihydrate, m. 201-202°
(dec.), after recrystn. from MeOH, m. 208° (dec.). On heating at 200-205°/14
mm → 9-Me-1H-tribenz[b,d,f]arsepin (1016). The high melting dibromide, purified by trituration with Me$_2$CO, m. 266-277° (dec.). When the dibromide, suspended in hot Me$_2$CO, was treated with MeOH and pptd. with Et$_2$O, a dihydrate
was obtd. which, after drying over P$_2$O$_5$, m. 264-266° and after heating at
90°/0.1 mm, m. 272-275° (dec.). Pyrolysis at 250-260°/14 mm → 9-Me-9H-tribenz-
[b,d,f]arsepin, whereas at 200-220°/0.1 mm → 9,10,11,12-tetrahydro-9,12-di-
methyltribenzo[e,g,i][1,4]diarsecin (1016). The low melting diiodide forms

hydroscopic crystals, m. 256-257° (dec.). The high melting diiodide crystd.
with 2 H₂O, m. 286-289° (dec.), after heating at 80°/0.1 mm, m. 289-292°,
m. 292-294° (1016). The low melting dipicrate m. 256-257° (dec.) (1016).

9,9,16,16-Tetraethyl-9,10,15,16-tetrahydrotribenzo[b,d,h][1,6]diarsecinium

Dibromide, $As_2C_{28}H_{36}Br_2$,

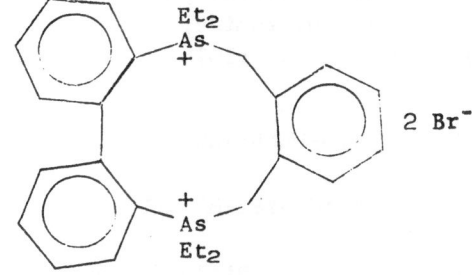

Synth.: From [2-(Et₂As)C₆H₄-]₂ by
heating with o-xylylene dibromide in
MeOH at 100° in a sealed tube (57).
Prop.: m. 267-270° (dec.); thermal
dec. → 5-Et-dibenzarsole + [2-(Et₂As)-
C₆H₄-]₂ (57).

3.9   LARGE RING SYSTEMS

3.9.1 ELEVEN-MEMBERED RING SYSTEMS

The following two eleven-membered rings are new compounds.

10,11,12,13-Tetrahydro-9,13-dimethyl-9H-tribenzo[f,h,j][1,5]diarsacycloundecin, $As_2C_{23}H_{24}$,

Synth.: From 10,11,12,13-tetrahydro-9,9,13,13-
tetramethyl-9H-tribenzo[f,h,j][1,5]diarsacyclo-
undecinium dibromide by heating at 220°/0.002 mm,
isolated as the dimethiodide by a treatment with
hot ethanolic MeI (1016).

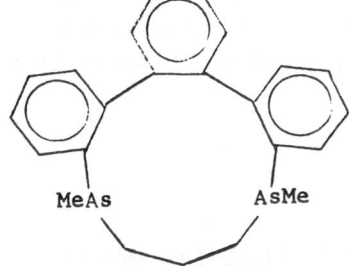

10,11,12,13-Tetrahydro-9,9,13,13-tetramethyl-9H-tribenzo[f,h,j][1,5]diarsacy-

cloundecinium Salts, $[As_2C_{25}H_{30}]^{++}2X-$, X = Br, I (For the
molecular structure of the base see the preceding entry.)

Synth.: From o-terphenyl-2,2"-ylenebis(dimethylarsine) + Br(CH₂)₃Br by heating under N at 120-125°, the dibromide dihydrate was obtd., which was converted to the diiodide monohydrate by treating the dibromide in aq. suspension
with ethanolic LiI or with Me₂CO-NaI (1016).

Prop.: The dibromide dihydrate m. 225-226° (dec.), at 90°/0.1 mm loses H₂O
and forms a hydroscopic, anhydrous dibromide, with unchanged m. temp. At
220°/0.002 mm two MeBr molecules are eliminated, leaving 10,11,12,13-tetra-
hydro-9,13-dimethyl-9H-tribenzo[f,h,j][1,5]diarsacycloundecin. A similar
heating of the dibromide at 220°/0.1 mm gives 1-methyltribenzo[b,d,f]arsepin.
At 220-225°/14 mm the dibromide is cleaved, giving a golden distillate consisting of o-terphenyl-2,2'-ylenebis(dimethylarsine), which can be identified
by quaternization with MeI in Me₂CO (1016).

The diiodide monohydrate m. 240-246° (dec.) remains unchanged when dried at 80°/0.1 mm. The diiodide prepd. directly from the diarsacycloundecin base by quaternization with MeI m. 270° (1016).

3.9.2 TWELVE-MEMBERED RING SYSTEM

The following compound has not been described before.

13,14,19,20-Tetrahydro-13,13,20,20-tetramethyltetrabenzo[c,g,i,k][1,6]diarsa-
cyclododecinium Salts, $[As_2C_{30}H_{32}]^{++}2X^-$,   X = Br, I

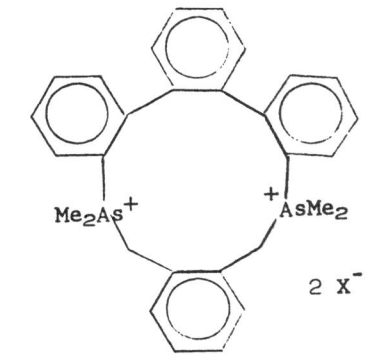

Synth.: From o-terphenyl-2,2″-ylenebis-
(dimethylarsine) + o-xylylene dibromide in
xylene under reflux at 135-140°, the dibromide was obtd. A metathetic rxn. of the
dibromide in aq. EtOH with ethanolic LiI
soln. gave the corr. diiodide (1016).
Prop.: The dibromide m. 195-197° (dec.).
At 185-190°/14 mm two MeBr molecules are
cleaved, leaving 13,14,19,20-tetrahydro-
13,20-dimethyltetrabenzo[c,g,i,k][1,6]-
diarsacyclododecin (1016).

The diiodide m. 181-182° (dec.); thermal dec. at 245-250°/14 mm → 1-methyltri-
benzo[b,d,f]arsepin (1016).

ANTIMONY

1. COMPOUNDS OF TRIVALENT ANTIMONY

1.1 TERTIARY STIBINES

1.1.1 SYMMETRIC DERIVATIVES

The methods of synthesis of tertiary symmetric stibines reported through 1964 were summarized in the main volume on page 653, and the compounds were described in pages 653 to 668 and in Table 80 on pages 669-76.

Since then, several new methods for the synthesis of the stibines have been developed. Trimethylstibine was prepared from antimony trichloride by a reaction with methyl chloride in a mixture formed from $NaAlCl_4$ by electrolysis at 220° and by a reaction with ammonium pentafluoromethylsilicate in aqueous solution. Triethylstibine is formed from antimony powder by a treatment with ethyl bromide and sodium, according to the equation: $Sb_n + 3nNa + 3nEtBr \rightarrow nEt_3Sb$. Stibylidynetriacetic acid esters are formed from trialkyl antimonides by the insertion of ketene and from antimony trichloride by a reaction with trialkylstannylacetic acid esters. A condensation of stibine with allyl chloride and mercuric oxide in acetic acid leads to the addition of antimony and mercury to each carbon atom of the olefinic bond to form 2,2′,2″-stibylidynetris[2-(chloromethyl)-ethylmercury acetate]. Triphenylstibine was prepared from antimony powder by heating with triphenylbismuthine at 300°, from pentaphenylantimony by heating with methyl iodide in a sealed tube, and from tris(triethylgermyl)stibine by metathesis with phenyl bromide. The latter mehhod was used in the preparation of tricyclopentyl- and tribenzylstibine from trialkylgermyl- and trialkylstannylstibine derivatives.

Trimethylstibine, SbC_3H_9, Me_3Sb,   (M-654)
<u>Synth.</u>: From $SbCl_3$ + MeMgI in Et_2O (484, 888).
From $SbCl_3$ + $(NH_4)_2(MeSiF_5)$ in aq. soln., isolated as Me_3SbBr_2 by adding methanolic Br soln. (1106).
From $SbCl_3$ by adding at 200° to MeCl suspended in a mixt. formed from $NaAlCl_4$ by electrolysis at 220°, 48% yield (1464).
From Sb_2O_3 in hexane + excess Me_3Al at 60°, distg. the prod. from the equilibrium mixt. under a high vacuum, converting the prod. to Me_3SbBr_2 by a treatment with Br in CCl_4, and reducing the dibromide to Me_3Sb (1045).
<u>Prop.</u>: b_{763} 80-81° (1045), spontaneously flammable in air (1106). Mass spectrum (881, 1619). Ionization and appearance potential data for the principal positive ions in the mass spectrum (1619). NMR spectrum (594) and its correlation with those of Me_3N, Me_3P, Me_3As, and Me_3Bi (880). IR spectrum, Sb-C and C-H vibrations (209) and the effect of the adjacent groups on the rocking and symmetric deformation frequencies of the Me groups and of CD_3 in $(CD_3)_3Sb$ (1491).
<u>Rxn. with</u>: Halogens (X) $\rightarrow Me_3SbX_2$ (484, 888, 1106, 1343).

CHI$_3$ in Me$_2$CO → Me$_3$Sb(CH$_2$I)(CHI$_2$) (1030).

(BzS-)$_2$ in C$_6$H$_6$ under reflux → Me$_3$Sb(SBz)$_2$ (1031).

NH$_2$Cl in Et$_2$O → [Me$_3$SbCl]$_2$NH + Me$_3$SbCl$_2$ (1045).

SbCl$_3$ in DMF at 72° and at 100° → an exchange of ligands giving Me$_2$SbCl and MeSbCl$_2$, resp.; kinetic studies of the rxn. by NMR spectroscopy gave 2nd order rate consts. (1595, 1596).

Chloromethylated styrene-divinylbenzene copolymers swollen in ClCH$_2$CH$_2$Cl under reflux, followed by a treatment with alc. NaOH at 60° → a polymeric sti- bonium base (888).

Coordination Deriv.: Hg[Co(SbMe$_3$)(CO)$_3$]$_2$, crystals, m. 145-147°, prepd. from Hg[Co(CO)$_4$]$_2$ + Me$_3$Sb in C$_6$H$_6$. IR spectrum - vibration assignment (1020).

{Ph$_3$SnCo[ON=C(Me)C(Me)=NOH]$_2$SbMe$_3$}, orange-red crystals, dec. 164-168°, sol. in org. solvents and alkali, dec. in strong acids, prepd. from Ph$_3$SnX + {HCo[ON=C(Me)C(Me)=NOH]$_2$SbMe$_3$} (1369).

Mn(SbMe$_3$)$_2$(CO)$_3$Br, prepd. from Mn(CO)$_5$Br + Me$_3$Sb; sol. in hydrocarbons and useful as an antiknock additive (1612).

cis-[Pt(SbMe$_3$)Cl$_2$]$_2$ (M-655), NMR spectrum (594).

cis-[Me$_2$Pt(SbMe$_3$)$_2$], NMR spectrum (594).

Use: A chain-transfer agent in a free-radical polymerization of vinyl mono- mers to low mol.-wt. polymers (1218). Mixts. with transition metal carbonyls and N-halogenated org. compds. catalyze the polymerization of vinyl monomers containing electron-withdrawing groups (773).

Triethylnylstibine,   SbC$_6$H$_3$,   (HC≡C)$_3$Sb
Synth.: From SbCl$_3$ in Et$_2$O by adding HC≡CMgBr in THF at -30°, warming up to r.t., hydrolyzing immediately the rxn. mixt., and drying the soln., 63% yield (1564).
Prop.: Colorless crystals, m. 71-72°, subl$_{13}$ 60°, IR and NMR spectra (1564). IR and Raman spectra, frequencies assignment, and mol. symmetry (950, 1064).

Triethylstibine,   SbC$_6$H$_{15}$,   Et$_3$Sb,   (M-655)
Synth.: From SbCl$_3$ + EtMgBr in Et$_2$O at r.t. (888).
From Sb by treating with Na, followed by EtBr, according to equation Sb$_n$ + 3nNa → nNa$_3$Sb $\xrightarrow{3nEtBr}$ Et$_3$Sb (1407).
From Sb$_2$O$_3$ + Et$_3$Al (1:2) in hexane under reflux, 89% yield; the rxn. proceeds via the (Et$_2$Sb)$_2$O intermediate (1444).
By electrolysis of NaBEt$_4$ in H$_2$O soln. at a sacrificial Sb anode (with a Hg cathode), 82% yield (1647).
By electrolysis of molten KAlEt$_4$ at a sacrificial Sb anode (with a Hg cathode) and 100-110°, ~ 100% yield. The rxn. leads to complete utilization of KAlEt$_4$, with the formation of K-Hg and Al particles in Hg (1648).
By-prod. from a rxn. of SbCl$_3$ with excess Et$_3$Al in n-C$_6$H$_{14}$ (1488).
Prop.: b$_{10}$ 43° (1647). Gas chromatographic detn. (980). IR spectrum (1045). NMR spectrum (343). Elec. cond. data for mixts. of Et$_3$Sb with Et$_3$Al, Et$_2$AlCl, and EtAlCl$_2$, resp. (1488).

<u>Rxn. with:</u> Br in CCl_4, followed by pyrolysis $\rightarrow Et_3SbBr_2 \rightarrow Et_2SbBr$ (175).

$Et_3Al \rightarrow [Et_3SbAlEt_3]$ (487).

$Et_3In \rightarrow [Et_3Sb \cdot InEt_3]$ and $[Et_3Sb(InEt_3)_2]$ (487).

$SbCl_3$ (3:2) $\rightarrow Et_3SbCl_2 + Sb$ (1488).

$SbBr_3$ (2:1) in $AcNEt_2$ at $100° \rightarrow Et_2SbBr + EtSbBr_2$ (1596).

$TiCl_4$ (0.2-4.0:1) $\rightarrow [TiCl_3(SbEt_3Cl_2)_n(Et_3Sb)_m]$, n = 0.2-0.44, m = 0.09-0.62 (1477).

Et_3GeH (1:3) at $200° \rightarrow (Et_3Ge)_3Sb$ (1576, 1577).

Et_3SnH (1:3) at $170° \rightarrow (Et_3Sn)_3Sb$ (1576, 1577).

R_3SiH (1:3) at $230° \rightarrow (R_3Si)_3Sb$ (1271, 1575, 1576, 1577).

$[Et_3AsTiCl_3]$ at $60° \rightarrow$ a black solid and a gas contg. C_2H_6, C_2H_4, H_2, and C_4 hydrocarbons. The black solid promotes the polymerization of α-olefins (1475).

NH_2Cl in $Et_2O \rightarrow (Et_3SbCl)_2NH$ (1045).

Chloromethylated styrene-divinylbenzene copolymers swollen in MePh at 110°, followed by aq. NaOH $\rightarrow$ a polymeric stibonium base (888).

<u>Coordination Deriv.:</u> $[Et_3Sb \cdot AlCl_3]$ (M-656), yellowish green liquid, prepd. from $Et_3Sb + AlCl_3$ (1:1) at 60° (1488, 1489). Elec. cond. data and NMR spectrum (1474). A mixt. of the complex with $Et_3Al + TiCl_3$ (0.005:1:1-6) catalyzes the polymerization of C_3H_6 to a stereoregular, cryst. prod. (1489).

$[Et_3Sb \cdot Et_2SbCl \cdot AlCl_3]$, yellowish brown crystals, sol. in $n-C_6H_{14}$ (1488). Light greenish brown crystals, m. 160° (1489). The complex was prepd. from a large excess of $Et_3Sb + AlCl_3$ (1488). A mixt. of the complex with $Et_3Al + TiCl_3$ in $n-C_6H_{14}$ catalyzes the polymerization of C_3H_6 to a stereoregular, highly cryst. polymer (1489).

$[Et_3Sb \cdot AlEtCl_2]$, orange liquid, sparingly sol. in $n-C_6H_{14}$, slightly sol. in C_6H_6, prepd. from $Et_3Sb + EtAlCl_2$ (1489).

$[Mo(SbEt_3)(CO)_5]$ (M-656), IR spectrum (435).

cis-$[Mo(SbEt_3)_2(CO)_4]$ (M-656), IR spectrum - the intensity of the C-O extension bands and its relation to the interaction force const. (154).

cis-$[Mo(SbEt_3)_3(CO)_3]$ (M-656), IR spectrum - the intensity of the C-O extension bands and its relation to the interaction force const. (154).

$[Ni(SbEt_3)(CO)_3]$ (M-656), IR and Raman spectra (213).

$[Ni(SbEt_3)_2(CO)_2]$ (M-657), IR and Raman spectra (213).

$[Ni(SbEt_3)_3(CO)]$, IR and Raman spectra (213).

cis-$[Pt(SbEt_3)_2Br_2]$ (M-657), IR spectrum (803).

cis-$[Pt(SbEt_3)_2Cl_2]$ (M-657), IR spectrum (803).

$[Ru(SbEt_3)_2(NO)Cl_3]$ (M-657), orange-red prisms, m. 109-110°, prepd. from $RuCl_3(NO) \cdot 5 H_2O + Et_3Sb$ by heating under reflux in $MeOCH_2CH_2OH$. IR spectrum (331).

$[Et_3Sb \cdot ZnCl_2]$, a mixt. with $TiCl_3$ and Et_3As (1.25:2.5:5) catalyzes the polymerization of C_3H_6 to a colorless, powdery prod. (1480).

Uses: A catalyst for the polymerization of ε-caprolactam (1168) and the formation of polyurethan foams from polyesters and diisocyanates (757). A chain-transfer agent for a free-radical polymerization of vinyl monomers to low mol. wt. polymers (1218). A black solid formed from Et_3Sb + $TiCl_3$ + Et_3Al catalyzes the polymerization of α-olefins to isotactic products (1475). Mixts. with alkali or alkaline earth metal hydrides and halides of Ti, V, Zr, Cr, or Mo catalyze the polymerization of α-olefins (386). Mixts. with $TiCl_3$ or $MnCl_2$ catalyze the polymerization of Me methacrylate (909). Mixts. with transition metal carbonyls and N-halogenated organic compds. catalyze the polymerization of vinyl monomers containing electron-withdrawing groups (773).

Tripropylstibine, SbC_9H_{21}, Pr_3Sb, (M-658)
Synth.: From $SbCl_3$ + PrMgI in Et_2O (888).
From Sb_2O_3 in hexane + excess Pr_3Al at 60° and distg. the prod. from an equilibrium mixt. in a high vacuum, 41% yield (1045).
Prop.: $b_{15.5}$ 93-95°, IR spectrum (1045).
Rxn. with: Br in CCl_4 → Pr_3SbBr_2 (175).
Chloromethylated styrene-divinylbenzene copolymers, swollen in $ClCH_2CH_2Cl$ at 90°, followed by alc. NaOH → polymeric stibonium bases (888).
NH_2Cl in Et_2O → $[Pr_3SbCl]_2NH$ (1045).
Coordination Deriv.: $[Pr_3Sb·Pr_2SbCl·AlCl_3]$, light greenish brown crystals, prepd. from excess Pr_3Sb + $AlCl_3$ (1489).

trans-$[Pt(SbPr_3)PyCl_2]$, yellow crystals, m. 97-98.5°, prepd. from $[Pt(SbPr_3)-Cl_2]_2$ by treating with Py in CH_2Cl_2; NMR spectrum (332).

Biol. Prop.: Bactericidal and fungicidal activity (537).
Uses: Complexes with $M[AlR_4]$ (M = alkali metals) and Ti, V, Zr, Cr, or Mo halides catalyze the polymerization of α-olefins (385). Mixts. with Ni^o and nonaromatic hydroxy compds. catalyze the oligomerization of linear 1,3-dienes (1382a).

Stibylidynetriacetic Acid Triethyl Ester, $SbC_{12}H_{21}O_6$, $Sb(CH_2CO_2Et)_3$
Synth.: From $Sb(OEt)_3$ + $CH_2=C=O$ in C_6H_6 soln., 76% yield (175, 581).
From $SbCl_3$ + $Et_3SnCH_2CO_2Et$ in C_6H_6 at 90-100°, 74% yield (176).
Prop.: $b_{0.05}$ 133-135°, d_4^{20} 1.4406, n_D^{20} 1.5150, atomic refraction data (175); $b_{0.05}$ 140°, d_4^{20} 1.4446, n_D^{20} 1.5149 (581). Spectra: IR (175, 581), NMR (175).
Rxn. with: H_2O → $Sb(OH)_3$ + EtOAc (176).
EtOH at 60-80° → $Sb(OEt)_3$ + EtOAc (176).
AcCl in CH_2Cl_2 + Et_3N in petr. ether → $AcCH_2CO_2Et$ + $SbCl_3$ (176).

Tributylstibine, $SbC_{12}H_{27}$, Bu_3Sb, (M-659)
Synth.: From Sb_2O_3 + Bu_3Al (1:2) in hexane under reflux, 87% yield, along with $(BuAlO)_n$; the rxn. proceeds via $(Bu_2Sb)_2O$ intermediate (1444).
Prop.: IR spectrum (209, 1045). NMR spectrum (1045).
Rxn. with: $SbCl_3$ in C_6H_6 at 50° → Bu_3SbCl_2 (1045).
NH_2Cl in Et_2O → $[Bu_3SbCl]_2NH$ (1045).
Biol. Prop.: Bactericidal and fungicidal activities (537). Mixts. with

peroxides catalyze the polymerization of epoxy compds. (902).

Uses: A catalyst for the polyurethan foam formation from diisocyanates and polyesters (757). A catalyst for the polymerization of ethylene sulfide (801). A chain transfer agent in a free-radical polymerization of styrene, Me methacrylate, acrylonitrile, and vinyl chloride (1218). A rxn. prod. with $HSCH_2$-CO_2H stabilizes the mixts. of poly(vinyl chloride) with acrylonitrile-butadiene-styrene copolymers (834). Catalysts for the polymerization olefins, particularly propylene and $\sim$-olefins, consist of mixts. and/or complexes of Bu_3Sb with: (a) alkali or alkaline earth hydrides and Ti, V, Zr, Cr, or Mo halides (386), (b) Na, K, Li, Mg, or Zn and Ti or V halides (387), (c) $RAlCl_2$ and $(RO)_4V$ (383), (d) organoaluminum halides and Ti, V, Zr, Cr, or Mo compds. (388), (e) $AlTi_3Cl_{12}$ and $EtAlCl_2$ (1228), and (f) $Ti(BH_4)_n \cdot (THF)_m$ (1070).

Triphenylstibine, $SbC_{18}H_{15}$, Ph_3Sb,   (M-660)

Synth.: From $SbCl_3$ + PhBr (1:4) by adding to excess Na in a $PhMe-C_6H_6$ mixt. and heating at 220°, 72% yield (1175).
From Ph_5Sb by heating with MeI at 100° in a sealed tube, 95% yield (1269).
From Ph_2SbF + $PhSiF_3$ (1:6) under reflux, 52% yield (1108).
From Ph_3Bi + Sb powder (1:1) by heating at 300° (379).
From $(Et_3Sn)_3Sb$ + PhBr (1:2) at 30°, 49% yield (1577).
Prop: m. 51-52° (1577), m. 48–49°, b_{10} 220° (1175), m. 50° (1658), m. 52°, $b_{0.3}$ 185° (1108). Spectra: IR (209, 803, 1001), far-IR (244). Half-wave potential (467, 749). Force consts. (832). Phase diagrams and eutectic points of the binary systems of Ph_3Sb with BzOH, Ph_2, and naphthalene, resp. (238). Zone refining of the binary systems with naphthalene (239) and with biphenyl (240). Thermionic response of K, Rb, and Cs in gas-chromatographic detn. of Ph_3Sb (795). Microanalysis by gas-chromatographic detn. of C_6H_{12} after hydrogenolysis over Pd/Al_2O_3 at 200° (1508).
Rxn. with: Halogens (X) → Ph_3SbX_2 and Ph_3SbX_4 (178, 482, 1108).
ClI → impure Ph_3SbClI (481).
I in $CHCl_3$ → Ph_3SbI_2, a charge transfer complex - UV and visible spectra (179). I soln. with Ph_3Sb soln. on filter paper → a brown or yellow spot of $Ph_3SbI_2 \cdot$ - nI_2, a spot test for Ph_3Sb (555).
ROCl → $Ph_3Sb(OR)Cl$ (432).
R_3SiOCl → $Ph_3Sb(OSiR_3)Cl$ (431, 432).
RI + HgI_2 → $[RPh_3Sb]HgI_3$ (616).
$SbCl_3$ (1:2) at 35-60° in CH_2Cl_2 → $PhSbCl_2$ (1098, 1505).
$SbCl_3$ (2:1) at 35-60° in CH_2Cl_2 → Ph_2SbCl (1098).
$PhSbCl_2$ (1:1) in CH_2Cl_2 under reflux → Ph_2SbCl (1098).
$TiCl_3$ or $ZrCl_3$ + Na in MePh at 80° under H → a complex which catalyzes the polymerization of propylene (1000).
NH_2Cl in Et_2O → $[Ph_3SbCl]_2NH$ (1045).
MeI in MeOH at 25°, the rate of rxn. and relative nucleophilic reactivity parameter were detd. by means of the change in elec. cond. (1203).
trans-$[Pt(Py)_2Cl_2]$ in MeOH at 25°, the rate of rxn. and relative nucleophilic reactivity parameter were detd. by UV spectroscopy (1203).
Na in liq. NH_3, followed by NH_4Br and $RnSnCl_{4-n}$ → $(Ph_2Sb)_{4-n}SnR_n$ (1372).

Na in liq. NH_3, followed by NH_4Br and Ph_3MCl (M = Ge, Pb) → Ph_2SbMPh_3 (1374).
H_2 over Pd/Al_2O_3 at 200° → hydrogenolysis and redn.; microanalysis by gas-chromatographic detn. of C_6H_{12} (1508).
$KNH_2 + NH_3$ → $K[Sb(NH)_2] + 3$ C_6H_6 (1362, 1363, 1364).
$MeSO_2NSO$ in C_6H_6 → $MeSO_2N=SbPh_3$ (?), a crude prod. which was converted to $Ph_3SbO \cdot H_2O$ by alk. hydrolysis (1385).
Hexamethylbenzene → solid 1:1 complex (1393).
Naphthalene, no complex formation (1393).
HCl-MeOH under reflux, followed by 20% KOH → $(Ph_2As)_2O$ (1505).
$Ph_2P(O)N_3$ in $Me(OCH_2CH_2)_3OMe$ → $Ph_2P(O)N=SbPh_3$ (1583).
30% H_2O_2 → $Ph_3SbO \cdot H_2O_2$ (1305).
$PhSn(N_3)_3$ (3:1) → $PhSn(N=SbPh_3)_3$ (1584).
p-MeC_6H_4Li in Et_2O in a sealed tube at r.t., followed by pouring into a CO_2 suspension in Et_2O → a mixt. of BzOH and p-$MeC_6H_4CO_2H$; the rxn. may proceed through equilibria: $Ph_3Sb + MeC_6H_4Li \rightleftharpoons [Ph_3SbC_6H_4Me]Li \rightleftharpoons Ph_2SbC_6H_4Me + PhLi$ (1620).
Perfluoropiperidine → Ph_3SbF_2 (118).
$$\left[M \left\{ \begin{matrix} S-CCF_3 \\ S-CCF_3 \end{matrix} \right\}_2 \right]_2 \rightarrow Ph_3SbM \left\{ \begin{matrix} S-CCF_3 \\ S-CCF_3 \end{matrix} \right\}_2 \quad \text{(M = Fe, Co) (110).}$$
^{60}Co γ-radiation → radio-sensitized dec. and formation of Ph_2, phenylcyclohexadiene, and tetrahydroquaterphenyl (1219).
Neutron bombardment → Sb isotope hot-atom rxns. (1059, see also 350).
<u>Coordination Deriv.:</u> $[Ph_3Sb \cdot C_6Me_6]$, a 1:1 complex, m. 137°, prepd. from Ph_3Sb + hexamethylbenzene by the phase-diagram technique (1393).

$[Cr(SbPh_3)(CO)_5]$ (M-662), prepd. from $Cr(CO)_6 + Ph_3Sb$ (1:1) in n-decane/cyclohexane at 122.5 ± 0.3° - the rate of the complex formation was followed by IR spectroscopy; a S_N1-dissocn. mechanism is proposed (1599).

$[Cr(SbPh_3)_2(CO)_4]$, yellow crystals, prepd. from $[MeSCH_2CH_2SMeCr(CO)_4]$ + excess Ph_3Sb in $ClCH_2CH_2Cl$ under reflux. IR spectrum - a cis configuration and octahedral structure are postulated (486).

$[Co_3(SbPh_3)(CO)_{11}]$, crystals, prepd. from $[Co(CO)_3]_4 + Ph_3Sb$ in petr. ether (14% yield). IR spectrum indicated the replacement of one CO group by Ph_3Sb (313).

$[Co(SbPh_3)(NO)_2Cl]$, stability in various solvents was detd. (136).

$[Co(SbPh_3)(NO)_2SEt]$, black-brown solid, m. 51-52°, prepd. from $[Co(NO)_2SEt]_2$ + Ph_3Sb in n-C_5H_{12} by heating; in soln. dissoc. to $[Co(NO)_2SEt]_2 + Ph_3Sb$ (729).

$[Co(SbPh_3)(CO)_2NO]$ (M-662), IR and NMR spectra; kinetics of the complex formation from $[Co(CO)_3NO] + Ph_3Sb$ in C_6H_6 was detd. (297).

$$\left[Ph_3SbCo \left(\begin{matrix} -S-C-CF_3 \\ -S-C-CF_3 \end{matrix} \right)_2 \right]$$, brown, paramagnetic crystals, m. 211-212°, prepd.
from $\{Co[S_2C_2(CF_3)_2]_2\}_2 + Ph_3Sb$ in pentane. Polarographic data, ESR and electronic spectra; a square-pyramidal structure of 5-coordinate Co is postulated (110).

$$\left[Ph_3SbCo \begin{pmatrix} S-CCN \\ \| \\ S-CCN \end{pmatrix}_2 \right], \text{ polarography (110)}.$$

$Hg[Co(SbPh_3)(CO)_3]_2$ (M-662), crystals, m. 185-187° (dec.), prepd. from $Hg[Co-(CO)_4]_2 + Ph_3Sb$ in C_6H_6 at 50°. IR spectrum - interpretation (1020).

$Cl_2Sn[Co(SbPh_3)(CO)_3]_2$, brick red crystals, m. 188°, prepd. from $Cl_2Sn[Co-(CO)_4]_2 + Ph_3Sb$ in C_6H_6 under reflux. IR spectrum (203).

$[Cu(SbPh_3)_3(B_3H_8)]$, white crystals, prepd. from $[Cu(SbPh_3)_3Cl] + Cs[B_3H_8]$ in Me_2CO at r.t., followed by pptn. with H_2O. The hydroborate, a nonelectrolyte in $CHCl_3$, is sol. in $CHCl_3$, Me_2CO, DMF, and MeCN, insol. in EtOH and H_2O. A soln. in a 1:1 $CHCl_3$-Me_2CO mixt. dec. at 24°, giving a black, metallic solid. IR spectrum (972).

$[Cu(SbPh_3)_3Cl]$, prepd. from Ph_3Sb in CHCl by mixing with Cu_2Cl_2 (792).

$[Au(SbPh_3)Cl]$, an unstable compd., prepd. from $AuCl + Ph_3Sb$ in EtOH (832).

$[Au(SbPh_3)Mn(CO)_5]$, prepd. from $[Au(SbPh_3)Cl] + Na[Mn(CO)_5]$ in THF and pptd. with C_5H_{12} (832). IR spectrum (226, 832).

$[Ir(SbPh_3)_3(CO)Cl]$, red crystals, dec. 136°, prepd. from $Ph_3Sb + [Ir(CO)_2-(p-H_2NC_6H_4Me)Cl]$ (3:1) in C_6H_6. IR spectrum (730).

$[HIr(SbPh_3)(PPh_3)_2Br_2]$, yellow crystals, m. 182°, prepd. from $Ph_3Sb + [HIr-(PPh_3)_2Br_2]$ in C_6H_6 under reflux. IR spectrum (83).

$[HIr(SbPh_3)_3Br_2]$, orange yellow crystals, m. 226°, prepd. from $[Ir(SbPh_3)_3-Br_3]$ in $C_6H_6 + NaBH_4$ in i-PrOH. The complex was also obtd. as by-prod. from the rxn. of $K_3[IrBr_3] + Ph_3Sb$ (84).

$[HIr(SbPh_3)_3Cl_2]$, yellow crystals, m. 197°, a by-prod. from the rxn. of aq. $Na_3[IrCl_6]$ soln. with Ph_3Sb in EtOH, isolated by chromatography. The complex was also prepd. from cis- or trans-$[Ir(SbPh_3)_3Cl_3]\cdot2$ C_6H_6 in $CH_2Cl_2 + NaBH_4$ in i-PrOH (86). Useful as a homogeneous hydrogenation catalyst for α-olefins (1628) and for unsatd. nonaromatic org. compds. (582).

$[HIr(SbPh_3)_2(CO)Cl_2]$, light yellow crystals, dec. 163°, prepd. from $[MeOCOIr-(SbPh_3)_3(CO)]$ and from $[Ir(SbPh_3)_3(CO)Cl]$ in C_6H_6 by treating with dry HCl. IR spectrum (731).

$[DIr(SbPh_3)_2(CO)Cl_2]$, crystals, prepd. from $[Ir(SbPh_3)_3(CO)Cl]$ in C_6H_6 by heating with 10% DCl in D_2O. IR spectrum (731).

$[Ir(SbPh_3)_2Br_3]$, maroon rhombohedral crystals, m. 198°, prepd. from $[Ir(SbPh_3)_2Br_3](MeOH)_x$ by heating in MePh at 100° (84).

$[Ir(SbPh_3)_2Br_3]\cdot(MeOH)_x$, yellow crystals, prepd. from $K[Ir(SbPh_3)_2Br_4]$ by dissolving in boiling MeOH and pptn. by cooling. The complex dec. at elevated temps., without melting (84).

$[Ir(SbPh_3)_2Cl_3]\cdot(MeOH)_{1.5}$, brown crystals, sinter at 120-130°, m. 170-171°, prepd. from $K[Ir(SbPh_3)_3Cl_4]$ soln. in $MeOH-Me_2CO$ (8:1) by boiling for 10 min. and cooling slowly; IR spectrum (86).

[Ir(SbPh$_3$)$_2$(CO)Br$_3$], yellow crystals, m. 220°, prepd. from [Ir(SbPh$_3$)$_2$Br$_3$] + CO in C$_6$H$_6$ under reflux; IR spectrum (84).

[Ir(SbPh$_3$)$_2$(CO)Br$_3$]·2 C$_6$H$_6$, yellow crystals, m. 205°, prepd. from trans-[Ir-(SbPh$_3$)$_3$Br$_3$]·2 C$_6$H$_6$ dissolved in warm C$_6$H$_6$ by treating with CO and pptg. with EtOH; IR spectrum and dipole moment data (86).

[Ir(SbPh$_3$)$_2$(CO)Cl$_3$]·2 C$_6$H$_6$, dark yellow crystals, m. 200°, prepd. from cis-or trans-[Ir(SbPh$_3$)$_3$Cl$_3$]·2 C$_6$H$_6$ in warm C$_6$H$_6$ by treating with CO and from [Ir(SbPh$_3$)$_2$Cl$_3$]$_2$ in C$_6$H$_6$ soln. by treating with CO, while heating, and purifying by chromatography. IR spectrum and dipole moment data (86).

[Ir(SbPh$_3$)$_2$(NH$_3$)Br$_3$], red crystals, m. 218°, prepd. from [Ir(SbPh$_3$)$_2$Br$_3$] + NH$_3$ in C$_6$H$_6$ under reflux; IR spectrum (84).

[Ir(SbPh$_3$)$_2$(Py)Br$_3$], red crystals, m. 238°, prepd. from [Ir(SbPh$_3$)$_2$Br$_3$] in C$_6$H$_6$ + Py under reflux, followed by pptn. with EtOH (84).

[Ir(SbPh$_3$)$_3$Br$_3$], red crystals, m. 233-235°, prepd. from [Ir(SbPh$_3$)$_2$Br$_3$] + Ph$_3$Sb in C$_6$H$_6$ under reflux and pptd. with EtOH; from [HIr(SbPh$_3$)$_3$Br$_2$] in CH$_2$Cl$_2$ + Br in CCl$_4$ under reflux and pptd. with MeOH; and a by-prod. in the rxn. of [Ir(SbPh$_3$)$_3$Br$_3$] in C$_6$H$_6$ soln. by reducing with NaBH$_4$, isolated by chromatography on Al$_2$O$_3$ (84).

[Ir(SbPh$_3$)$_3$Br$_3$]·CHCl$_3$, m. 235°, prepd. from trans-[Ir(SbPh$_3$)$_3$Br$_3$]·2 C$_6$H$_6$ by recrystn. from CHCl$_3$ (86).

trans-[Ir(SbPh$_3$)$_3$Br$_3$]·2 C$_6$H$_6$, salmon-pink crystals, m. 253°, a by-prod. from the rxn. of [Ir(SbPh$_3$)$_2$Br$_3$] + Ph$_3$Sb in C$_6$H$_6$, followed by chromatographic sepn., concn. in vacuo, and pptn. with EtOH; dipole moment data (86).

cis-[Ir(SbPh$_3$)$_3$Cl$_3$], orange crystals, m. 204-205°, prepd. from [Ir(SbPh$_3$)$_2$Cl$_3$]$_2$ as well as from [Ir(SbPh$_3$)$_2$Cl$_3$]·(MeOH)$_{1.5}$ solns. in C$_6$H$_6$ by treating with Ph$_3$Sb and pptg. with EtOH. The complex was also obtd. as by-prod. from the rxn. of Na$_3$[IrCl$_6$] + Ph$_3$Sb, followed by extn. with CHCl$_3$ and chromatography; dipole moment data (86).

[Ir(SbPh$_3$)$_3$Cl$_3$]·CHCl$_3$, m. 243°, prepd. from trans-[Ir(SbPh$_3$)$_3$Cl$_3$]·2 C$_6$H$_6$ by recrystn. from CHCl$_3$-EtOH (86).

trans-[Ir(SbPh$_3$)$_3$Cl$_3$]·2 C$_6$H$_6$, dark salmon-colored crystals, m. 243°, isolated from the rxn. prod. of Na$_3$[IrCl$_6$] with Ph$_3$Sb by extn. with CH$_2$Cl$_2$, pptn. with EtOH, dissolution of the ppt. in C$_6$H$_6$, and purification by chromatography; dipole moment data (86).

[Ir(SbPh$_3$)$_2$Cl$_3$]$_2$, red-maroon ppt., m. 170-171°, prepd. from [Ir(SbPh$_3$)$_2$Cl$_3$]·(MeOH)$_{1.5}$ by heating in vacuum at 100°; a nonelectrolyte in PhNO$_2$, IR spectrum (86).

[MeO$_2$CIr(SbPh$_3$)$_3$(CO)], yellow crystals, dec. 156°, prepd. from [Ir(SbPh$_3$)$_3$-(CO)$_2$]AlCl$_4$ in MeOH by treating with MeOH-KOH. IR spectrum (731).

[EtO$_2$CIr(SbPh$_3$)$_3$(CO)], clusters of yellow crystals, dec. 148°, prepd. from

[Ir(SbPh$_3$)$_3$(CO)$_2$]AlCl$_4$ in EtOH by treating with EtOH-KOH. IR spectrum (731).

[Ir(SbPh$_3$)$_2$(1,5-Cyclooctadiene)Cl], m. 172-174°, a nonelectrolyte in ClCH$_2$-CH$_2$Cl, prepd. from [Ir(SbPh$_3$)(1,5-Cyclooctadiene)Cl] + Ph$_3$Sb in EtOH (1573).

[Ir(SbPh$_3$)$_3$(CO)$_2$]HCl$_2$, the formation of the complex from [MeO$_2$ClIr(SbPh$_3$)$_3$(CO)] in Et$_2$O by treating with dry HCl at -60° was detected by IR spectrum; the complex dec. at -40°. A trigonal bipyramidal structure of the cation, with the Ph$_3$Sb ligands in the equatorial positions, is postulated (731).

[Ir(SbPh$_3$)$_3$(CO)$_2$]AlCl$_4$, light yellow platelets, prepd. from [Ir(SbPh$_3$)$_3$(CO)Cl] + excess AlCl$_3$ in C$_6$H$_6$ by treating with CO. IR spectrum - a trigonal bipyramidal structure of the cation, with the Ph$_3$Sb ligands in the equatorial positions, is postulated (731).

[Ir(SbPh$_3$)$_3$(CO)$_2$]FeCl$_4$, yellow crystals, prepd. in the same manner as the preceding complex, using FeCl$_3$ instead of AlCl$_3$ (731).

[Ir(SbPh$_3$)$_3$(CO)$_2$]PF$_6$, light yellow crystals, dec. 174°, prepd. from [Ir(SbPh$_3$)$_3$(CO)$_2$]AlCl$_4$ or the tetrachloroferrate in MeOH by treating with NaPF$_6$ at 0°; molar cond. data (731).

H[Ir(SbPh$_3$)$_2$Br$_4$], brown crystals, m. 154°, prepd. from K[Ir(SbPh$_3$)$_2$Br$_4$] + HBr in Me$_2$CO (84).

Me$_4$N[Ir(SbPh$_3$)$_2$Br$_4$], brown ppt., dec. without melting, prepd. from K[Ir(SbPh$_3$)$_2$-Br$_4$] by treating with NH$_4$Cl in MeOH (84).

K[Ir(SbPh$_3$)$_2$Br$_4$], brown crystals, m. 300° (dec.), prepd. from K$_3$[IrBr$_6$] in aq. KBr + Ph$_3$Sb in EtOH (84).

Py·H[Ir(SbPh$_3$)$_2$Br$_4$], brown-black flakes, insol. in org. solvents, m. 216-219°, prepd. from H[Ir(SbPh$_3$)$_2$Br$_4$] + Py and from K[Ir(SbPh$_3$)$_2$Br$_4$] + Py·HBr in MeOH. IR spectrum (84).

H[Ir(SbPh$_3$)$_2$Cl$_4$], brown crystals, m. 156-157°, prepd. from K[Ir(SbPh$_3$)$_2$Cl$_4$] in Me$_2$CO + concd. HCl under reflux. Mol. cond. in EtOCH$_2$CH$_2$OH was detd. (86).

K[Ir(SbPh$_3$)$_2$Cl$_4$], brown crystals, m. 261°, prepd. from Py·H[Ir(SbPh$_3$)$_2$Cl$_4$] in Me$_2$CO + KOH; mol. cond. in Me$_2$CO was detd. (86).

Py·H[Ir(SbPh$_3$)$_2$Cl$_4$], pink irridescent laminar crystals, m. 205°, prepd. from aq. Na$_3$[IrCl$_6$] by a rxn. with Ph$_3$Sb and Py·HCl in EtOH under reflux; from Na[Ir(SbPh$_3$)$_2$Cl$_4$] in Me$_2$CO + warm methanolic Py·HCl soln.; and from H[Ir(SbPh$_3$)$_2$Cl$_4$] in MeOCH$_2$CH$_2$OH + Py. IR spectrum (86).

Na[Ir(SbPh$_3$)$_2$Cl$_4$], pink crystals, m. 261°, prepd. from aq. Na$_3$[IrCl$_6$] soln. by adding to boiling Ph$_3$Sb soln. in EtOH and isolating with CH$_2$Cl$_2$ as insol. residue. The salt was also prepd. from Py·H[Ir(SbPh$_3$)$_2$Cl$_4$] in Me$_2$CO by treating with methanolic NaOH (86).

Ph$_4$As[Ir(SbPh$_3$)$_2$Cl$_4$], brown crystals, m. 197°, prepd. from Na[Ir(SbPh$_3$)$_2$Cl$_4$] in MeOH by treating with [Ph$_4$As]Cl. The salt is sol. in Me$_2$CO, warm EtOH, and CHCl$_3$, insol. in C$_6$H$_6$ (86).

[Fe(SbPh$_3$)(CO)$_4$] (M-662), yellowish brown crystals, m. 136°, prepd. from Ph$_3$Sb + Fe$_3$(CO)$_{12}$ in THF under reflux or in dioxan at 90-95° and isolated from the more sol. fraction (20% yield). IR spectrum (367); low frequency IR spectrum (1419). A trigonal bipyramidal mol. structure with the stibine ligand in one of the apical positions is postulated (1419).

[Fe(SbPh$_3$)$_2$(CO)$_3$] (M-662), yellowish brown, m. 196°, prepd. from Ph$_3$Sb + Fe$_3$(CO)$_{12}$ in THF under reflux or in dioxan at 90-95° and isolated from the less sol. fraction. IR spectrum (367). Rxn. with HgX$_2$ → Fe(SbPh$_3$)$_2$(CO)$_3$·HgX$_2$ (31).

[Fe(SbPh$_3$)$_2$(CO)$_3$·HgBr$_2$], dec. 118-125°, prepd. from [Fe(SbPh$_3$)$_2$(CO)$_3$] + HgBr$_2$ in EtCOMe-C$_6$H$_6$. IR spectrum (31).

[Fe(SbPh$_3$)$_2$(CO)$_3$·HgCl$_2$], yellow ppt., darkens at 87°, dec. 118-120°, prepd. from [Fe(SbPh$_3$)$_2$(CO)$_3$] + HgCl$_2$ in EtCOMe-C$_6$H$_6$. IR spectrum (31).

[Fe(SbPh$_3$)(CO)(NO)$_2$] (M-662), IR spectra, NO valence vibration (135), CO and NO valence vibrations and bond angles (137), solvent effect on the CO and NO vibrations (136).

Na$_3$[Fe(SbPh$_3$)(CN)$_5$], light yellow, fine crystals, prepd. from Na$_3$[Fe(CN)$_5$-NH$_3$]·2 H$_2$O + Ph$_3$Sb in MeOH. Elec. cond. data and IR spectrum (1124). Mössbauer spectrum (573, 1124).

$$\left[Ph_3SbFe \begin{pmatrix} S-C-CF_3 \\ S-C-CF_3 \end{pmatrix}_2 \right]$$, black, diamagnetic prisms, m. 171-173°, prepd. from Ph$_3$Sb + {Fe[S$_2$C$_2$(CF$_3$)$_2$]$_2$}$_2$ in pen ane. Polarographic data and electronic and ESR spectra. A square-pyramidal structure for the 5-coordinate Fe complex is postulated (110).

$$\left[Ph_3SbFe \begin{pmatrix} S-CCN \\ S-CCN \end{pmatrix}_2 \right]$$, polarographic data (110).

[Mn(SbPh$_3$)(CO)$_4$]$_2$ (M-663), prepd. from H$_3$Mn$_3$(CO)$_{12}$ + Ph$_3$Sb in cyclohexane under reflux (807).

[Mn(SbPh$_3$)$_2$(CO)$_3$F], prepd. from [Mn(CO)$_4$F]$_2$ by heating with Ph$_3$Sb. The complex is sol. in hydrocarbons and useful as an antiknock additive (1612).

[Mn(SbPh$_3$)(NO)$_3$] (M-663), prepd. from [Mn(NO)$_3$(CO)] + Ph$_3$Sb in C$_6$H$_{12}$. IR spectrum, NO valence vibration (135).

trans-[Mn(SbPh$_3$)$_2$(CO)$_3$NCS], yellow solid, prepd. from Mn(CO)$_5$SCN + Ph$_3$Sb in CHCl$_3$ at 80° or in MeCN at 25°. IR spectrum (553).

cis-[Mn(SbPh$_3$)$_2$(CO)$_3$SCN] (M-663), yellow solid, prepd. from Mn(CO)$_5$SCN + excess Ph$_3$Sb in CHCl$_3$ or MeCN at 25°. IR spectrum (553). Rxn. with BiPy in CHCl$_3$ at 50° → displacement of the Ph$_3$Sb ligands with BiPy (553).

[Mn(SbPh$_3$)(CO)$_4$SCN], yellow solid, prepd. from Mn(CO)$_5$SCN + Ph$_3$Sb in CHCl$_3$ at 25°. IR spectrum (553).

cis-[MeMn(SbPh$_3$)(CO)$_4$], orange crystals, m. 114.5-116°, prepd. from MeMn(CO)$_5$ + Ph$_3$Sb in C$_6$H$_6$ under reflux. IR and NMR spectra (893).

cis-[AcMn(SbPh$_3$)(CO)$_4$], m. 101.5-102°, prepd. from MeMn(CO)$_5$ + Ph$_3$Sb at r.t. (119). IR spectrum (119, 893). NMR spectrum (893).

[PhMn(SbPh$_3$)(CO)$_4$], m. 105-106°, prepd. from PhMn(CO)$_5$ + Ph$_3$Sb in CH$_2$Cl$_2$ at r.t. IR spectrum (119).

[π-C$_5$H$_5$Mn(SbPh$_3$)(CO)$_2$], pale yellow ppt., prepd. from [π-C$_5$H$_5$Mn(CO)$_3$] + Ph$_3$Sb in THF under UV light at 75°. Stable in air; IR and NMR spectra and magnetic moment data (120).

[π-C$_5$H$_5$Mn(SbPh$_3$)$_2$(CO)], red-brown crystals, prepd. from [π-C$_5$H$_5$Mn(CO)$_3$] + Ph$_3$Sb in EtOH under UV light. The complex dec. on exposure to oxygen. IR and NMR spectra; magnetic moment data (120).

[Ph$_3$GeMn(SbPh$_3$)(CO)$_4$], light yellow substance, m. 214.5-216.5°, prepd. from [Ph$_3$GeMn(CO)$_5$] by heating with Ph$_3$Sb at 170°. IR spectrum (1138).

[Mo(SbPh$_3$)(CO)$_5$] (M-664), the rate of the complex formation from Mo(CO)$_6$ + Ph$_3$Sb (1:1) in n-decane-cyclohexane mixt. at 97.8 ± 0.2° and in dioxan-THF mixt. at 82.3 ± 0.2° was followed by IR spectroscopy. A S$_N$1-dissocn. mechanism is postulated (1599). The rate of the CO substitution by Ph$_3$Sb in [Mo(SbPh$_3$)(CO)$_5$] was too slow to be measured (1653). IR spectrum (435, 439). Low-frequency IR spectrum (1419). A tetragonal bipyramidal structure with the Ph$_3$Sb ligand in one of the apical positions is postulated (1419).

cis-[Mo(SbPh$_3$)$_2$(CO)$_4$] (M-664), prepd. from [Mo(CO)$_4$(Cyclooctadiene)] + Ph$_3$Sb in n-C$_7$H$_{16}$. IR spectrum (646, 1651). The rate of the cyclooctadiene displacement with Ph$_3$Sb was followed by the spectrum changes in the UV and visible ranges. The rxn. rate const. and the activation parameters were calcd. (1651). The rate of the substitution of the Ph$_3$Sb ligands with α, α'-BiPy was studied (646).

[Mo(SbPh$_3$)$_2$(CO)$_3$]Br$_2$, orange crystals, prepd. from Mo(CO)$_4$Br$_2$ + Ph$_3$Sb in Me$_2$CO. IR spectrum (376).

[Mo(SbPh$_3$)$_2$(CO)$_3$]Cl$_2$, prepd. from Mo(CO)$_4$Cl$_2$ + Ph$_3$Sb in Me$_2$CO. IR spectrum (376).

[Mo(SbPh$_3$)$_2$(NO)$_2$Br$_2$], green crystals, prepd. from [Mo(SbPh$_3$)$_2$(CO)$_3$]Br$_2$ + NO in CH$_2$Cl$_2$. Two cis isomers were established by IR spectroscopy (77).

[Mo(SbPh$_3$)$_2$(NO)$_2$Cl$_2$], green crystals, prepd. from [Mo(SbPh$_3$)$_2$(CO)$_3$]Cl$_2$ + NO in CH$_2$Cl$_2$. Two cis isomers were established by IR spectroscopy (77).

[Mo(SbPh$_3$)(Ph$_2$PCH$_2$-)$_2$(CO)$_3$], yellow ppt., prepd. from [Mo(CO)$_4$(Ph$_2$PCH$_2$-)$_2$] + excess Ph$_3$Sb in xylene under reflux. IR spectrum; a cis configuration and octahedral structure are postulated (486).

[Mo(SbPh$_3$)(CO)$_3$(o-Phenanthroline)], purple-black crystals, prepd. from [Mo(CO)$_4$(o-Phenanthroline)] + Ph$_3$Sb in xylene (758). IR spectrum (436, 758).

[π-C$_5$H$_5$MoH(SbPh$_3$)(CO)$_2$], very sensitive to air, prepd. from [π-C$_5$H$_5$Mo(SbPh$_3$)-(CO)$_2$I] by a redn. with 1% Na-Hg in THF, followed by a treatment with p-MeC$_6$H$_4$-SO$_3$H, was not isolated from the THF soln. IR and NMR spectra (1019).

[π-C₅H₅Mo(SbPh₃)(CO)₂Br], crystals, m. 170°, prepd. from [π-C₅H₅Mo(CO)₃Br] + Ph₃Sb by heating in C₆H₆ and purifying by chromatography. IR spectrum (1522).

[π-C₅H₅Mo(SbPh₃)(CO)₂Cl], orange crystals, m. 155°, prepd. from [π-C₅H₅Mo-(CO)₃Cl] by heating with Ph₃Sb in C₆H₆ under reflux and purifying by chromatography (1522) and in C₆H₆-cyclohexane soln. under UV irradiation (670). A nonelectrolyte; soly. data (670). Spectra: IR (670, 1522), visible and NMR (670).

[π-C₅H₅Mo(SbPh₃)(CO)₂I], red crystals, dec. 163° (670), m. 165° (1522), m. 185° (dec.) (1018), prepd. from [π-C₅H₅Mo(CO)₃I] + Ph₃Sb in C₆H₆ by heating under reflux (1018, 1522) and purifying by chromatography (1522) and in C₆H₆-cyclohexane mixt. by UV irradiation (670). A nonelectrolyte; soly. data (670). Spectra: IR (670, 1018, 1019, 1522), visible (670), NMR (670, 1018, 1019). The substance consists of a mixt. of the cis and trans isomers distinguishable by their NMR spectra (1018). A redn. of the iodo complex with 1% Na-Hg in THF → [π-C₅H₅Mo(SbPh₃)(CO)₂]⁻, identified by its NMR spectrum (1019).

[π-C₅H₅Mo(SbPh₃)₂(CO)Cl], orange crystals, dec. 170°, a nonelectrolyte, prepd. from [π-C₅H₅Mo(CO)₃Cl] + excess Ph₃Sb in C₆H₆-cyclohexane mixt. under UV irradiation. IR, visible, and NMR spectra (670).

[π-C₅H₅Mo(SbPh₃)₂(CO)I], red crystals, dec. 145°, a nonelectrolyte, prepd. from [π-C₅H₅Mo(CO)₃I] + excess Ph₃Sb in C₆H₆-cyclohexane under UV irradiation. Soly. data; IR, visible, and NMR spectra (670).

[π-C₅H₅Mo(SbPh₃)(CO)₂SnPh₃], white crystals, stable in air, m. 242-243° (dec.), prepd. from [π-C₅H₅Mo(SbPh₃)(CO)₂I] by a redn. with Na-Hg in THF, followed by a rxn. with Ph₃SnCl. IR and NMR spectra (1019). A trans arrangement of the CO ligands was identified by IR spectroscopy (1023).

[Ni(SbPh₃)₂(CO)₂] (M-664), a catalyst for the dimerization of butadiene to 1,3,6-octatriene, vinylcyclohexene, cyclooctadiene, and other isomers (556).

[Ni(SbPh₃)₄] (M-664), green yellow, diamagnetic crystals, prepd. from K₄[Ni(CN)₄] + Ph₃Sb (1:6) at 80° in liq. NH₃ (147) and from Ni acetylacetonate dissolved in C₆H₆ by a redn. with Et₂AlOEt in the presence of Ph₃Sb (1611). A nonelectrolyte (147), useful as a catalyst for the oligomerization of 1,3-dienes (1611).

[Ni(SbPh₃)₂(CH₂=CHCHO)₂], prepd. from [Ni(CH₂=CHCHO)₂] + Ph₃Sb in Et₂O under reflux. The complex catalyzes a trimerization of butadiene to trans,trans,trans-1,5,9-cyclododecatriene (557).

[Ni(SbPh₃)(CH₂=CHCN)₂], a catalyst for a dimerization of butadiene to vinyl-cyclohexene; addition of m-cresol enhances the catalytic activity about three-fold (556).

[Ni(SbPh₃)₂(PhCH=CHCN)₂], prepd. from [Ni(PhCH=CHCN)₂] + Ph₃Sb in Et₂O under reflux (557). The complex promotes the dimerization and trimerization of butadiene to vinylcyclohexene, 2,6-octadiene, and cyclododecatriene, resp. (556, 557).

[π-C₅H₅Ni(SbPh₃)I], dark red-violet crystals, m. 101-102° (dec.), prepd. from Ni(CO)₄ by mixing with Ph₃Sb (1:2) in THF at r.t., completing the rxn., after the CO evolution subsides, under reflux, then cooling with ice, adding dropwise (π-C₅H₅)₂Ni followed by I in THF at r.t., and heating under reflux (52% yield). Rxn. with RMgX → [π-C₅H₅Ni(SbPh₃)R] (1629).

[π-C₅H₅Ni(SbPh₃)Ph], dark green, diamagnetic crystals, m. 96-98° (dec.), prepd. from [π-C₅H₅Ni(SbPh₃)I] in C₆H₆ + PhMgX in Et₂O (32% yield) (1629).

[π-C₅H₅Ni(SbPh₃)C₆H₂Me₃-2,4,6], yellow-green, diamagnetic crystals, m. 134-135°, prepd. from [π-C₅H₅Ni(SbPh₃)I] in C₆H₆ + 2,4,6-Me₃C₆H₂MgX in THF (42% yield) (1629).

[Nb(SbPh₃)Cl₅], colored ppt., prepd. from NbCl₅ + Ph₃Sb in CCl₄, n-C₆H₁₄, or cyclohexane (466).

[Nb(SbPh₃)₂Cl₅], colored ppt., prepd. from NbCl₅ + Ph₃Sb in C₆H₆ (466).

[π-C₅H₅Nb(SbPh₃)(CO)₃], red crystals, dec. 157° without melting, prepd. from [π-C₅H₅Nb(CO)₄] + Ph₃Sb in n-C₅H₁₂ under UV light (58% yield). IR and NMR spectra (869a).

[Os(SbPh₃)₃Cl₃], green, paramagnetic crystals, prepd. from Na₂OsCl₆ + Ph₃Sb in warm EtOH. A nonelectrolyte; soly. data; IR spectrum. An octahedral structure is postulated. The complex was converted to [Os(SbPh₃)₄Cl₂], [Os(SbPh₃)₃(CO)Cl₂], [Os(SbPh₃)₂(CO)₂Cl₂], and [Os(SbPh₃)₂(NO)Cl₃] as described below (82).

[Os(SbPh₃)₄Cl₂], red-orange crystals, m. 227°, prepd. from [Os(SbPh₃)₃Cl₃] by a redn. with NaBH₄ in CH₂Cl₂-EtOCH₂CH₂OH in the presence of Ph₃Sb. IR spectrum (82).

[Os(SbPh₃)₂(CO)₂Br₂], cryst. substance, prepd. from Os(CO)₃Br₂ + Ph₃Sb in C₆H₆ by heating under reflux. IR spectrum (674). Solvent effect on IR spectrum (673).

[Os(SbPh₃)₂(CO)₂Cl₂], cryst. substance, prepd. from Os(CO)₃Cl₂ + Ph₃Sb in C₆H₆ under reflux (674) and from [Os(SbPh₃)₃Cl₃] + CO in CHCl₃-EtOAc or in EtOCH₂CH₂OH. Chromatographic purification of the prod. gave two isomers: a cis isomer, white cryst. contg. 0.5 C₆H₆ of crystn., m. ~ 280°, the crystals free of the solvent m. ~ 280°; and a trans isomer, yellow crystals, m. 268°. IR spectra and dipole moments of both isomers were obtd. (82, see also 674). Solvent effects on IR spectra (673). The ligands arrangement in the isomers are considered (82).

[Os(SbPh₃)₂(CO)₂I₂], cryst. substance, prepd. from Os(CO)₃I₂ + Ph₃Sb in C₆H₆ under reflux. IR spectrum (674). The effects of solvents on IR spectrum (673).

[Os(SbPh₃)₃(CO)Cl₂], pale green crystals, m. 280° (dec.), obtd. as a by-prod. from [Os(SbPh₃)₃Cl₃] + CO in CHCl₃-EtOAc soln. IR spectrum and dipole moment. The ligands arrangement is considered (82).

[Os(SbPh$_3$)$_2$(NO(Cl$_3$], red crystals, m. 243° (dec.), prepd. from [Os(SbPh$_3$)$_3$Cl$_3$] + NO in CHCl$_3$ under reflux (81, 82). A nonelectrolyte in PhNO, soly. data (81). IR spectrum and dipole moment data; the ligands arrangement is considered (81, 82) and a trans-stibine structure is suggested (81).

[Pd(SbPh$_3$)$_2$Cl$_2$] (M-664), catalyzes the polymerization of ketene to a resinous polyester (1394). A mixt. of the complex with SnCl$_2$·2 H$_2$O is ineffective in promoting hydrogenation of methyl linoleate, but induces its stereoisomerization (793).

[Pd(SbPh$_3$)$_2$(NCO)$_2$], crystals, nonconducting in MeNO$_2$, prepd. from [Pd(SbPh$_3$)$_2$-Cl$_2$] + AgNCO in Me$_2$CO. IR spectrum - a coordination of the NCO groups through the N atoms is postulated (1156, 1157).

[Pd(SbPh$_3$)$_2$(SCN)$_2$] (M-664), far-IR spectrum (841).

[Pd(SbPh$_3$)$_2$(PPh$_3$)$_2$], prepd. from [Pd(PPh$_3$)$_4$] by a rxn. with Ph$_3$Sb (1472).

[Pd(SbPh$_3$)$_2$(NO$_2$)$_2$], solid substance, prepd. from K$_2$[Pd(NO$_2$)$_4$] in H$_2$O + Ph$_3$Sb (1:2) in EtOH at 70°. IR spectrum indicates a N-bonding of the NO$_2$ groups (268).

[π-CH$_2$=CHCH$_2$Pd(SbPh$_3$)Cl], pale yellow prisms, dec. 114-116°, prepd. from [π-CH$_2$=CHCH$_2$PdCl]$_2$ + Ph$_3$Sb in warm C$_6$H$_6$ (1248).

[π-CH$_2$=C(Me)CH$_2$Pd(SbPh$_3$)Cl], pale yellow prisms, relatively unstable, dec. 150-158°, prepd. from [π-CH$_2$=C(Me)CH$_2$PdCl]$_2$ + Ph$_3$Sb in warm C$_6$H$_6$ (1248). The formation of the complex from [π-CH$_2$=C(Me)CH$_2$PdCl]$_2$ + Ph$_3$Sb (1:2) in CDCl$_3$ at -73° was studied by NMR spectroscopy. The proton shifts and coupling consts. were detd. (1569, see also 1248). At higher temp. and with excess Ph$_3$Sb, the changes of the spectrum were explained in terms of a rotation of the π-allyl group in its own plane. For the systems where only a part of the dimer has been converted into the monomer, NMR spectra were obtd. which indicated the occurrence of both monomer-monomer and monomer-dimer reactions; the spectra changed with increasing temp. (1569, see also 1568). The kinetics and mechanism of the rxns. are discussed (1568). The reversible interconversions may occur through a short-lived π-methallyl to σ-methallyl conversions caused either by the rxn. of the monomeric complex with excess free Ph$_3$Sb ligand or by a rxn. of the complex with [π-CH$_2$=C(Me)CH$_2$PdCl]$_2$ (1568).

[Pt(SbPh$_3$)(PPh$_3$)$_3$], prepd. from [Pd(PPh$_3$)$_4$] + Ph$_3$Sb (1472).

[Pt(SbPh$_3$)$_2$Cl$_2$] (M-664), a complex with SnCl$_2$·2 H$_2$O catalyzes the reduction of Me linoleate to a mono-ene and a stereoisomerization of the former (109).

[Pt(SbPh$_3$)$_2$(NCO)$_2$], deep yellow crystals, prepd. from [Pt(SbPh$_3$)$_2$Cl$_2$] by a rxn. with AgNCO in Me$_2$CO. IR and electronic spectra - a coordination of the NCO groups through the N atoms is postulated (1157).

[Pt(SbPh$_3$)$_2$(NO$_2$)$_2$], solid substance, prepd. from K$_2$[Pt(NO$_2$)$_4$] in H$_2$O + Ph$_3$Sb (1:2) in EtOH at 70°. IR spectrum indicated a N-bonding of the NO$_2$ groups (268).

[Pt(SbPh$_3$)$_4$](NO$_3$)$_2$, cream-colored crystals, m. 147-151° (dec.), prepd. from [Pt(SbPh$_3$)$_2$Cl$_2$] + Ph$_3$Sb (1:2) in Me$_2$CO by shaking with finely powdered AgNO$_3$. Molar cond. in MeOH and in DMF is lower than that of a 2:1 electrolyte (1602).

[Re(SbPh$_3$)$_2$Cl$_4$], royal blue crystals, sparingly sol. in CH$_2$Cl$_2$, prepd. from [Re(MeCN)$_2$Cl$_4$] + Ph$_3$Sb by fusion at 100° in vacuum (1302, see also 1301).

cis-[Re(SbPh$_3$)(CO)$_4$Cl], m. 183°, prepd. from [Re(CO)$_4$Cl]$_2$ + Ph$_3$Sb in CCl$_4$ at r.t. IR spectrum (1652).

cis-[Re(SbPh$_3$)(PPh$_3$)(CO)$_3$Cl], prepd. from cis-[Re(SbPh$_3$)(CO)$_4$Cl] + Ph$_3$P in MePh at 70°; the rxn. kinetics was detd. by means of changes observed in its IR spectrum (1652).

[Ph$_3$GeRe(SbPh$_3$)(CO)$_4$], white solid, m. 216-218°, prepd. from Ph$_3$GeRe(CO)$_5$ by heating with Ph$_3$Sb at 220°. IR spectrum (1143).

[Ph$_3$SnRe(SbPh$_3$)(CO)$_4$], white solid, m. 210-212°, prepd. from Ph$_3$SnRe(CO)$_5$ by heating with Ph$_3$Sb at 215-220°. IR spectrum (1143).

[Rh(SbPh$_3$)$_3$Cl], purple powder, dec. 170° (1006), prepd. from [Rh(CH$_2$=CHCH$_2$OH)$_2$-Cl]$_2$ + Ph$_3$Sb in EtOH under reflux (940) and from [Rh(CH$_2$=CH$_2$)$_2$Cl]$_2$ + Ph$_3$Sb in EtOH (1559) or in MeOH under reflux in N$_2$ atm. (1006).

[Rh(SbPh$_3$)$_2$(CO)Cl] (M-665), red crystals, prepd. from [Rh(SbPh$_3$)$_3$Cl] + an aldehyde in C$_6$H$_6$ under reflux. IR spectrum (1006).

[Rh(SbPh$_3$)$_2$(CO)(NCS)], bright orange solid, m. 181-183° (dec.), prepd. from [Rh(SbPh$_3$)$_2$(CO)Cl] + KSCN in Me$_2$CO. IR spectrum (802).

[Rh(SbPh$_3$)$_2$(F$_2$C=CF$_2$)Cl], yellow-orange crystals, m. 154-156°, prepd. from [Rh(SbPh$_3$)$_3$Cl] + F$_2$C=CF$_2$ in CH$_2$Cl$_2$ in a sealed tube at r.t. IR and NMR spectra. A square-planar Rh(I) structure is postulated. Rxn. with CO → [Rh-(SbPh$_3$)$_4$(CO)Cl] (1007).

[Rh(SbPh$_3$)(Norbornadiene)Cl], far-IR spectrum (160).

[Rh(SbPh$_3$)(1,5-Cyclooctadiene)Cl], a nonelectrolyte, dec. when heated, without melting, monomeric in ClCH$_2$CH$_2$Cl at 37°, prepd. from [Rh(1,5-Cyclooctadiene)-Cl]$_2$ + Ph$_3$Sb (1:2). IR spectrum indicated the absence of free olefinic bonds. NMR spectrum in CDCl$_3$ showed a nonequivalence of the "olefinic" protons. With rising temp. the proton signals broadened and finally coalesced, due to the Ph$_3$Sb ligand exchange between [Rh(1,5-Cyclooctadiene)Cl]$_2$ and]Rh(SbPh$_3$)(1,5-Cyclooctadiene)Cl]. The rate of the rxn. was detd. The exchange probably proceeds via a 5-coordinate Rh intermediate (1573).

{Rh(SbPh$_3$)(CO)[SP(S)F$_2$]}$_n$, brownish red solid, stable in air, prepd. from {Rh(CO)$_2$[SP(S)F$_2$]}$_2$ + Ph$_3$Sb at -196°, then adding C$_6$H$_6$ and warming to r.t. IR spectrum; mol. wt. detn. indicated a polymeric structure (693).

[Rh(SbPh$_3$)$_2$(CF$_3$C≡CCF$_3$)$_2$Cl], yellow-orange crystals, m. 208-210°, stable in air, prepd. from [Rh(SbPh$_3$)$_3$Cl] + CF$_3$C≡CCF$_3$ in C$_6$H$_6$ in a sealed tube at 75-80°. IR, NMR, and mass spectra. A 5-coordinate Rh(III) structure is postulated in which the two CF$_3$C≡CCF$_3$ groups and Rh form a rhodiacyclopentadiene ring (1007).

[Rh(SbPh$_3$)$_3$(CF$_3$C≡CCF$_3$)Cl], dark red-orange crystals, m. 173-175°, prepd. from [Rh(SbPh$_3$)$_3$Cl] in CH$_2$Cl$_2$ by shaking with excess CF$_3$C≡CCF$_3$ in a sealed tube at r.t. IR and NMR spectra. The suggestion was made that the CF$_3$C≡CCF$_3$ ligand may occupy two coordination positions about the Rh atom. Rxn. with CO → [Rh(SbPh$_3$)$_4$(CO)Cl] (1007).

[Cl$_3$SnRh(SbPh$_3$)$_2$(Norbornadiene)], orange crystals, dec. ∼ 200°, a nonelectrolyte in MeNO$_2$, prepd. from [Cl$_3$SnRh(Norbornadiene)$_2$] in CH$_2$Cl$_2$ by adding Ph$_3$Sb (1637).

[Rh(SbPh$_3$)$_3$(CO)$_2$]PF$_6$, yellow needles, stable up to 159°, prepd. from [Rh-(SbPh$_3$)$_3$(CO)Cl]AlCl$_3$ by treating it with CO, followed by NaPF$_6$. Molar cond. data. A trigonal bipyramidal structure with the Ph$_3$Sb ligands arranged in a plane is proposed (731).

[Ph$_2$(PhCH$_2$)$_2$P][Rh(SbPh$_3$)(CF$_3$C≡CCF$_3$)$_2$(CN)Cl], fine yellow crystals, prepd. from [Rh(SbPh$_3$)$_2$(CF$_3$C≡CCF$_3$)$_2$Cl] in CH$_2$Cl$_2$ by treating with the CN$^-$ ion and pptg. with [Ph$_2$(PhCH$_2$)$_2$P]$^+$. IR spectrum (1007).

[MeO$_2$CRh(SbPh$_3$)$_3$(CO)], yellow-brown crystals, dec. 144°, prepd. from [Rh-(SbPh$_3$)$_3$(CO)$_2$]AlCl$_4$ + MeOH-KOH. IR spectrum (731).

[π-CH$_2$=CHCH$_2$Rh(SbPh$_3$)$_2$Cl], orange, paramagnetic ppt., m. 190-205°, a non-electrolyte in ClCH$_2$CH$_2$Cl and in MeNO$_2$. IR spectrum indicated a π-bonded allyl ligand (1559).

[RhH$_2$(SbPh$_3$)$_2$Cl], unstable in the solid state, prepd. from [Rh(SbPh$_3$)$_3$Cl] in CHCl$_3$ by treating with H$_2$ at 25°/1 atm. pressure. NMR spectra in CHCl$_3$ and in C$_6$H$_6$ (1006).

[RhH(SbPh$_3$)$_2$Cl$_2$], cryst. solid, unstable in the solid state, prepd. from [Rh-(SbPh$_3$)$_3$Cl] in CHCl$_3$ or in CH$_2$Cl$_2$ soln. by treating with HCl. NMR spectrum (1006).

[Rh(SbPh$_3$)$_3$Cl$_3$], m. 155-156°, a catalyst for the oxidn. of EtPh to AcPh (195).

[Rh(SbPh$_3$)$_4$(CO)Br], prepd. from [Rh(CO)$_2$Cl]$_2$ by treating with Ph$_3$Sb (1541).

[Rh(SbPh$_3$)$_4$(CO)Cl], prepd. from [Rh(SbPh$_3$)$_2$(F$_2$C=CF$_2$)Cl] and from [Rh(SbPh$_3$)$_3$-(CF$_3$C≡CCF$_3$)Cl] by treating with CO (1007) and from [Rh(CO)$_2$Cl]$_2$ by treating with Ph$_3$Sb (1541).

[HRh(SbPh$_3$)$_2$(CO)Cl$_2$], unstable compd., its formation from [MeO$_2$CRh(SbPh$_3$)$_3$-(CO)] or from [Rh(SbPh$_3$)$_3$(CO)Cl] in C$_6$H$_6$ and dry HCl was detected by IR spectrum. The complex dec. during recrystn. from C$_6$H$_6$-petr. ether satd. with HCl to [Rh(SbPh$_3$)$_3$(CO)Cl]. An octahedral structure with the Ph$_3$Sb ligands in trans- and Cl ligands in cis-positions is postulated (731).

[DRh(SbPh$_3$)$_2$(CO)Cl$_2$], prepd. from [Rh(SbPh$_3$)$_3$(CO)Cl] in C$_6$H$_6$ + DCl in D$_2$O (731).

[Rh(SbPh$_3$)(CF$_3$C≡CCF$_3$)$_2$(Py)$_2$Cl], lemon yellow needles, prepd. from [Rh(SbPh$_3$)$_2$-(CF$_3$C≡CCF$_3$)$_2$Cl] in CH$_2$Cl$_2$ + Py. IR spectrum; a 6-coordinate Rh(III) structure with a rhodiacyclopentadiene ring is postulated (1007).

[Rh(SbPh$_3$)$_2$(CF$_3$C≡CCF$_3$)$_2$(CO)Cl], pale yellow needles, m. 189-192° (dec.), prepd. from [Rh(SbPh$_3$)$_2$(CF$_3$C≡CCF$_3$)$_2$Cl] in CH$_2$Cl$_2$ by treating with CO. IR spectrum; a 6-coordinate Rh(III) structure is postulated in which the two CF$_3$C≡CCF$_3$ groups and Rh form a rhodiacyclopentadiene ring (1007).

[Rh(SbPh$_3$)$_2$(CF$_3$C≡CCF$_3$)$_2$(PF$_3$)Cl], lemon yellow crystals, prepd. from [Rh-(SbPh$_3$)$_2$(CF$_3$C≡CCF$_3$)$_2$Cl] in CH$_2$Cl$_2$ + PF$_3$ at 25°. The prod. contd. orange-yellow and lemon-yellow crystals - a mixt. of the starting matl. and the P contg. complex. IR spectrum indicated the presence of both complexes. A 6-coordinate Rh(III) structure is postulated in which the two CF$_3$C≡CCF$_3$ groups and Rh form a rhodiacyclopentadiene ring (1007).

[π-CH$_2$=C(Me)CH$_2$Rh(SbPh$_3$)$_2$Cl$_2$], orange, paramagnetic ppt., which dec., without melting, at elevated temp.; a nonelectrolyte in ClCH$_2$CH$_2$Cl and in MeNO$_2$. IR and NMR spectra and dipole moment data indicated π-bonding of the allylic ligand; a configuration is postulated in which the Ph$_3$Sb ligands are trans to the allylic group (1559). NMR spectrum in CDCl$_3$ showed no broadening of the proton signals at C-1 and C-3 with rising temp. (1571). A rxn. with CO in CHCl$_3$ consumed 3 moles CO and gave a black prod. of uncertain structure (1560).

[Ru(SbPh$_3$)$_3$Cl$_2$], dark red microcrystals, m. 162°, prepd. from RuCl$_3$·3 H$_2$O in MeOH + excess Ph$_3$Sb and from RuCl$_3$ + excess Ph$_3$Sb by refluxing in MeOH or MeOCH$_2$CH$_2$OH. The complex is sparingly sol. in Me$_2$CO, EtOH, and C$_6$H$_6$ and more sol. in hot Py and quinoline (1453).

[Ru(SbPh$_3$)$_3$(CO)Cl$_2$], orange brown, diamagnetic crystals, m. 238-240° (85), m. 234-236° (1459), prepd. from [Ru(SbPh$_3$)$_4$Cl$_2$] + CO in CHCl$_3$ (85) and from RuCl$_3$·3 H$_2$O dissolved in boiling EtOH by treating with CO and adding 3-fold excess Ph$_3$Sb in Me$_2$CO. The complex is insol. in MeOH, Me$_2$CO, C$_6$H$_6$, EtOAc, MeNO$_2$, and H$_2$O; sparingly sol. in CH$_2$Cl$_2$ (1453); a nonelectrolyte in PhNO$_2$ (85). IR spectrum (85, 1453).

[Ru(SbPh$_3$)$_4$Cl$_2$], diamagnetic crystals, m. 240-242°, a nonelectrolyte in PhNO$_2$, prepd. from aq. K$_2$[RuCl$_6$] soln. + alc. Ph$_3$Sb soln. Rxn. with CO → [Ru-(SbPh$_3$)$_3$(CO)Cl$_2$] and [Ru(SbPh$_3$)$_2$(CO)$_2$Cl$_2$] (85).

[Ru(SbPh$_3$)$_2$(CO)$_2$Cl$_2$], diamagnetic crystals, m. 248-250°, a nonelectrolyte in PhNO$_2$ (85), white crystals, m. 260°, for which a cis configuration is postulated (1453), prepd. from [Ru(SbPh$_3$)$_4$Cl$_2$] + CO in C$_6$H$_6$ under reflux (85) and from RuCl$_3$·3 H$_2$O soln. in boiling EtOH by treating with CO, then refluxing with excess SnCl$_2$ under N to give a pale yellow soln., adding a 6-fold excess of Ph$_3$Sb in Et$_2$O, refluxing gently, and recrystg. the yellow crystals from CHCl$_2$-MeOH mixt. (1453). Dipole moment data (85). IR spectrum (85, 1453).

[Ru(SbPh$_3$)$_2$(NO)Cl$_3$] (M-665), this molecular formula has been attributed to the following two substances: (a) diamagnetic crystals, m. 195-197°, a non-electrolyte in PhNO$_2$, prepd. from [Ru(SbPh$_3$)$_4$Cl$_2$] + NO in CHCl$_3$ under reflux and from aq. K$_2$[Ru(NO)Cl$_5$] soln. + Ph$_3$Sb in EtOH under reflux; IR spectrum (85), and (b) yellow, paramagnetic solid, turned black at 220°, prepd. from Ru(NO)Cl$_3$·3 H$_2$O + Ph$_3$Sb in EtOH; IR spectrum (543).

[Tc(SbPh₃)₂(CO)Cl], colorless crystals, a nonelectrolyte in Me₂CO, prepd. from Tc(CO)₅Cl + Ph₃Sb (1:2) in EtOH at 60-70° (737).

[Ti(SbPh₃)₂Cl₄], dark red substance, prepd. from TiCl₄ + Ph₃Sb in C₆H₆ or petr. ether (65).

[W(SbPh₃)₂(CO)₃Br₂], orange, diamagnetic crystals, prepd. from W(CO)₆ + Br in CH₂Cl₂, followed by Ph₃Sb. IR spectrum; a pentagonal bipyramidal structure with the Br atoms in the apical positions is suggested (76). Rxn. with NO → [W(SbPh₃)₂(NO)₂Br₂] (77).

[W(SbPh₃)₂(CO)₃Cl₂], golden yellow, diamagnetic ppt., prepd. from W(CO)₆ + Cl, followed by exposure to air, extn. with EtOH, and adding Ph₃Sb to the EtOH soln. IR spectrum; a pentagonal bipyramidal structure for the 7-coordinate W complex is suggested (76). Rxn. with NO in CH₂Cl₂ → [W(SbPh₃)₂(NO)₂Cl₂] (77).

[W(SbPh₃)₂(NO)₂Br₂], green crystals, prepd. from [W(SbPh₃)₂(CO)₃Br₂] + NO in CH₂Cl₂. IR spectrum (77).

[W(SbPh₃)₂(NO)₂Cl₂], green crystals, prepd. from [W(SbPh₃)₂(CO)₃Cl₂] + NO in CH₂Cl₂. IR spectrum (77).

[π-C₅H₅W(SbPh₃)(CO)₂Cl], prepd. in trace amts. from [π-C₅H₅W(CO)₃Cl] + excess Ph₃Sb in C₆H₆ under reflux. IR spectrum (1522).

[π-C₅H₅W(SbPh₃)(CO)₃]PF₆, yellow powder, prepd. from [π-C₅H₅W(CO)₃Cl] by a treatment with Ph₃Sb and AlCl₃ (1:2:2) in C₆H₆ under reflux and pptn., after cooling, with NH₄PF₆. IR spectrum (1522).

[V(SbPh₃)₂(CO)₄], yellow-green, paramagnetic, pyrophorous ppt., prepd. from V(CO)₆ + Ph₃Sb in n-C₆H₁₄ at r.t., recrystd. from CH₂Cl₂ by pouring the soln. into a CO₂-MeOH mixt. Useful in the vapor phase plating of metals with V at 500-550° under reduced pressure (1600). Rxn. with NO in petr. ether → [V(SbPh₃)(CO)₄(NO)] (1601).

[V(SbPh₃)(CO)₅], rxn. with Na-Hg in EtOH → [V(SbPh₃)(CO)₅]⁻ (1601).

Biol. Prop.: Bactericidal and fungicidal activity (537).

Uses: A flame inhibitor or retardant for polyesters (319), for polyurethan foams (740), and for epoxy resins at a concn. of 3 Sb/1000 C in the presence or absence of halogens (1025). An initiator for anionic polymerization of HCHO (903). A photo-initiator for the polymerization of ethylenically unsatd. compds. by UV irradiation (1024). A catalyst for the ring-opening polymeriza- tion of cyclic bis(ethylene isophthalate) (1636).
An inhibitor of a thermal polymerization of triethylene glycol dimethacrylate by radiation (1578).
A fog inhibitor for photothermographic sheets (743, 744). An activator for photographic direct-positive emulsion (745). A constituent of the photosen- sitive layers for lithographic plates (746).
A chain-transfer agent in a free-radical polymerization of PhCH=CH₂, CH₂=C(Me)- CO₂Me, CH₂=CHCN, and CH₂=CHCl (1218).

Admixture of 2-5% of Ph_3Sb to polyimide compns. makes them resistant to high-temp. corona discharge (1046).

A synergistic enhancer for the stabilizing effect of secondary aromatic amines on polyesters contg. hydroperoxide groups (550).

Mixts. with Ph_3P catalyze the formation of β-hydroxy esters from epoxides and carboxylic acids (497) such as epoxy-novolaks and acrylic acid to form poly-merizable thermosetting polyhydroxy polyacrylate esters (1281).

Compns. contg. Ph_3Sb, aryl dialkylphosphinates, and polymeric viscosity en-hancers are useful as noncorrosive hydraulic fluids (1039).

Mixts. with GeO_2 catalyze the polycondensation of dicarboxylic acid esters with $p-HOCH_2CH_2OC_6H_4CO_2Me$ at 270-280° in vacuo to form polymers suitable for making fibers (1365).

Mixts. with Fe acetylacetonate and R_3Al catalyze the condensation of 1,3-con-jugated C_{4-6} dienes with C_2H_2 to form cis-1,4-dienes (639).

Complexes with $Ni^°$ and nonaromatic OH compds. catalyze the oligomerization of linear 1,3-dienes (1382a).

Complexes with $TiCl_3$ or $ZrCl_3$ and Na formed in MePh at 80° (1000) and with $RAlX_2$ and $(RO)_4Zr$ (383) catalyze the polymerization of propylene.

Complexes with $EtAlCl_2$ and $PdCl_2$ (1:25:5.5) or $RhCl_3 \cdot 3 H_2O$ catalyze the poly-merization and oligomerization of propylene (771).

Mixts. with $R_{3-n}AlX_n$ and Ti, V, Zr, Cr, or Mo compds. (388), with $M[AlR_4]$ (M = alkali metal) and Ti, V, Zr, Cr, or Mo halides (385), with alkali or alkaline earth metal hydrides and Ti, V, Zr, Cr, or Mo halides (386), and with Na, Li, Al, Mg, Zn, or Cu chloride and Ti, V, Zr, Cr, or Mo compds. (1036) catalyze the polymerization of α-olefins.

Mixts. with transition metal carbonyls and N-halogenated organic compds. cata-lyze the polymerization of vinyl monomers contg. electron-withdrawing groups (773).

A complex with $CuCl_2$ (1:1) is useful as a semiconductor for thermistors based on organophosphorus compds. (1433).

Tri-n-hexylstibine, $SbC_{18}H_{39}$, $(n-C_6H_{13})_3Sb$, (M-666)
Synth.: From $SbCl_3$ + $n-C_6H_{13}MgBr$ in Et_2O under reflux, 67% yield (1497).
Prop.: Colorless, oily liquid, b_{10} 191-192°, d_{10} 1.0860, n_D^{20} 1.4920, its combustion leaves a dull deposit (1497).
Rxn. with: Br in $CHCl_3$ → $(n-C_6H_{13})_3SbBr_2$ (1497).
Coordination Deriv.: $\{Mn[Sb(C_6H_{13})_3]_2(CO)_3Br\}$, prepd. from $[Mn(CO)_4Br]_2$ by heating with $(C_6H_{13})_3Sb$; sol. in hydrocarbons, useful as an antiknock agent (1612).
Use: Complexes with $M[AlR_4]$ (M = alkali metal) and Ti, V, Zr, Cr, or Mo halides catalyze the polymerization of α-olefins (385).

Tri-p-tolylstibine, $SbC_{21}H_{21}$, $(4-MeC_6H_4)_3Sb$, (M-666)
Prop.: m. 124° (1398).
Rxn. with: Br in $CHCl_3$ → $(4-MeC_6H_4)_3SbBr_2$ (1398).
Coordination Deriv.: $\{Rh[Sb(C_6H_4Me-4)_3](1,5-cyclooctadiene)Br\}$, orange ppt., prepd. from $[Rh(1,5-cyclooctadiene)Br]_2$ + $(4-MeC_6H_4)_3Sb$ in CH_2Cl_2. Rxn. with

Py or its derivs. → displacement of the stibine by the base; the rxn. kinetics were detd. (309).

{Rh[Sb(C$_6$H$_4$Me-4)$_3$](1,5-cyclooctadiene)Cl}, yellow ppt., prepd. from [Rh(1,5-cyclooctadiene)Cl]$_2$ + (4-MeC$_6$H$_4$)$_3$Sb in CH$_2$Cl$_2$. Rxn. with Py or its deriv. → displacement of the stibine ligand by the base; kinetics of the rxn. were detd. (309).

Tris(p-vinylphenyl)stibine,   SbC$_{24}$H$_{21}$,   (4-CH$_2$=CHC$_6$H$_4$)$_3$Sb
Synth.: From SbCl$_3$ + 4-CH$_2$=CHC$_6$H$_4$MgCl in THF at -20° in the presence of hydroquinone, followed by warming to r.t., 36% yield (572).
Prop.: Colorless needles, m. 72-73° (572).
Use: A stabilizer for vinyl copolymers (572).
Polymers and Copolymers: The stibine polymerizes at elevated temp. to a transparent yellow mass, insol. in org. solvents and infusible, due to its crosslinked structure. On heating the polymer to 300° under Ar it loses ~ 9.4% of its wt. (884). A mixture of the stibine with styrene polymerizes, when heated at 40-100° in a sealed tube, to a transparent, colored, insol., and infusible mass. A polymerization of the stibine-styrene mixt. in the presence of (t-BuO-)$_2$ at 120-125° gives an insol. and infusible, colorless mass. A copolymerization with Me methacrylate in the presence of a free-radical initiator gives a prod. which even after extn. with MeOH is insol. and infusible colorless powder (884).

Tris(p-allylphenyl)stibine,   SbC$_{27}$H$_{27}$,   (4-CH$_2$=CHCH$_2$C$_6$H$_4$)$_3$Sb
Synth.: From SbCl$_3$ + 4-CH$_2$=CHCH$_2$C$_6$H$_4$MgBr in Et$_2$O under reflux, followed by dec. with H$_2$O, 85% yield (1640).
Rxn. with: H$_2$O$_2$ in N KOH → (HOCH$_2$CHOHCH$_2$C$_6$H$_4$)$_3$Sb(OH)$_2$ (1640).
Br in CHCl$_3$ → (BrCH$_2$CHBrCH$_2$C$_6$H$_4$)$_3$SbBr$_2$ (1640).

Tridodecylstibine,   SbC$_{36}$H$_{75}$,   (C$_{12}$H$_{25}$)$_3$Sb,   (M-668)
Coordination Deriv.: {Mn[Sb(C$_{12}$H$_{25}$)$_3$]$_2$(CO)$_3$Cl} (M-668), prepd. from Mn(CO)$_5$Cl by heating with the stibine. The complex is sol. in hydrocarbons and useful as an antiknock agent (1612).
Use: Mixts. with Na, K, Li, Mg, or Zn and Ti or V halides catalyze the polymerization of α-olefins (387).

Tris(2-phenyl-1,2-dicarbododecaboran(12)-1-yl)stibine,   SbB$_{30}$C$_{24}$H$_{45}$
Synth.: From SbCl$_3$ in Et$_2$O + 2-phenyl-1,2-dicarbododecaboran(12)-1-yllithium (1:3) in C$_6$H$_6$ under reflux, 57% yield (1642).
Prop.: m. 313-316° (dec.) (1642). IR spectrum (222).

Table 43. Symmetric Tertiary Stibines

R_3Sb	Prepd. from (Yield %)	Prop. and Rxn.	Ref.
C_3			
$(CF_3)_3Sb$ (M-653)		Rxn. with $(CF_3)_2NCl \rightarrow SbF_3$, $CF_3N=CF_2$ + CF_3Cl	71,72
C_6			
$(CH_2=CH)_3Sb$ (M-655)		IR spectrum, (a)	209
		Bactericidal and fungicidal activity	537
C_9			
$(CH_2=CHCH_2)_3Sb$ (M-657)		IR spectrum, (a)	209,210
$(MeO_2CCH_2)_3Sb$	$SbCl_3 + Et_3SnCH_2CO_2Me$ in C_6H_6 at 90-100° (58)	$b_{0.04}$ 134-135°, d_4^{20} 1.6009, n_D^{20} 1.5400, IR and NMR spectra	176
C_{12}			
$i-Bu_3Sb$ (M-659)	$Sb_2O_3 + i-Bu_3Al$ (1:2) in hexane under reflux, (b) (71)		1444
C_{15}			
$[AcOHgCH_2CH(CH_2Cl)]_3Sb$ $(\overline{CH_2CH_2CH_2CH_2CH})_3Sb$	$SbH_3 + CH_2=CHCH_2Cl + HgO$ in AcOH	Fungicidal and bactericidal prop.	1447
	$(Et_3Ge)_3Sb + RBr$ (1:3) at 160° (66-80)	$b_{1.5}$ 109-114°, d_4^{20} 1.293	1576,1577
	$(Et_3Sn)_3Sb + RBr$ (1:3) at 150° (75-82)		1576,1577
$(PrO_2CCH_2)_3Sb$	$Sb(OPr)_3 + CH_2=C=O$ in C_6H_6 (80)	$b_{0.05}$ 153-155°, d_4^{20} 1.3338, n_D^{20} 1.5059, atomic refraction data, IR and NMR spectra	175
		$b_{0.05}$ 155-156°, d_4^{20} 1.3340, n_D^{20} 1.5085, IR spectrum, (c, d)	581
$i-PrO_2CCH_2)_3Sb$	$Sb(Oi-Pr)_3 + CH_2=C=O$ in C_6H_6 (93)	$b_{0.03}$ 131-132°, d_4^{20} 1.3132, n_D^{20} 1.4968, (c, e)	175
	$SbCl_3 + Et_3SnCH_2CO_2-i-Pr$ in C_6H_6 at 90-100° (60)	(f)	176

C_{18}	Preparation	Properties	Ref.
$(C_6F_5)_3Sb$	$SbCl_3 + C_6F_5MgBr$ in Et_2O at -10 to $-20°$ (32)	Colorless needles, m. 74°, IR spectrum	566
	$SbCl_3$ in Et_2O by adding to C_6F_5Li in Et_2O at $-78°$ and warming to r.t. (75)	m. 73-75°, IR spectrum	842
$(p\text{-}BrC_6H_4)_3Sb$ (M-659)	$SbCl_3 + RMgBr$ in Et_2O under reflux and dec. with H_2O (92)	m. 136-137° (dec.) from Et_2O, m. > 260° (recrystd. from Me_2CO-AcOEt), d^{90} 1.8160	1640
$(p\text{-}ClC_6H_4)_3Sb$ (M-660)		Mössbauer spectrum, (g)	665
$(cyclo\text{-}C_6H_{11})_3Sb$ (M-666)	$SbCl_3 + RMgCl$ in Et_2O (80)	White crystals, m. 64-65°, (h, i)	782
$(BuO_2CCH_2)_3Sb$	$Sb(OBu)_3 + CH_2{=}C{=}O$ in C_6H_6 (51)	$b_{0.03}$ 175-176°, d_4^{20} 1.2650, n_D^{20} 1.4988, (e)	175
$(i\text{-}BuO_2CCH_2)_3Sb$	$Sb(Oi\text{-}Bu)_3 + CH_2{=}C{=}O$ in C_6H_6 (75)	$b_{0.05}$ 200-201°, d_4^{20} 1.2846, n_D^{20} 1.4985, IR spectrum (e)	581
		$b_{0.28}$ 145-147°, d_4^{20} 1.2658, n_D^{20} 1.4982, (e)	175
$(i\text{-}C_6H_{13})_3Sb$	$Sb_2O_3 + (i\text{-}C_6H_{13})_3Al$ in C_6H_{14} under reflux, (j) (62)		1444

(a) See also (345). (b) $(i\text{-}BuAlO)_n$ is formed as a by-prod.; the rxn. proceeds via $(i\text{-}Bu_2Sb)_2O$ (1444). (c) Elemental analysis (1502). (d) Rxn. with AcCl → $AcCH_2CO_2Pr + SbCl_3$ (176). (e) Atomic refraction data and IR and NMR spectra (175, 176). (f) Rxn. with Ag_2O in Et_2O → $(i\text{-}PrO_2CCH_2)_3SbO$ (176). (g) $\{Rh[(p\text{-}ClC_6H_4)_3Sb]_2(CO)(NCS)\}$, prepd. from $p\text{-}ClC_6H_4)_3Sb + Bu_4N[Rh(CO)_2(NCS)_2]$ in MeOH at $-78°$; IR spectrum (802). (h) Rxn. with Cl in Et_2O → $(C_6H_{11})_3SbCl_2$ (782). (i) $Mn[Sb(C_6H_{11})_3]_2(CO)_3F$, prepd. by heating $(C_6H_{11})_3Sb + [Mn(CO)_4F]_2$, sol. in hydrocarbons, useful as an antiknock agent (1612). (j) $(i\text{-}C_6H_{13}AlO)_n$ is formed as by-prod.; the rxn. proceeds via $[(i\text{-}C_6H_{13})_2Sb]_2O$ intermediate (1444).

Table 43 Continued

R_3Sb	Prepd. from (Yield %)	Prop. and Rxn.	Ref.
C_{21}			
$(PhCH_2)_3Sb$ $(M-666)$	$(Et_3Ge)_3Sb + PhCH_2Br$ (1:3) at $100°$ (57)	Identified as $(PhCH_2)_3SbBr_2$	1576, 1577
	$(Et_3Sn)_3Sb + PhCH_2Br$ (1:3) at $100°$ (62)	(k)	1576, 1577
$(4-MeOC_6H_4)_3Sb$ $(M-670)$		Mössbauer spectrum	665
$(n-C_7H_{15})_3Sb$ $(M-670)$	$SbCl_3 + RMgBr$ in Et_2O under reflux (58)	b_{20} $220°$, d_4^{19} 1.0610, n_D^{19} 1.4880, burns with a characteristic flame and leaves an Sb deposit, (l, m)	1497
C_{24}			
$(PhCH_2CH_2)_3Sb$ $(M-667)$		Complexes with $M[AlR_4]$ (M = alkali metal) and Ti, V, Zr, Cr, or Mo halides catalyze the polymerization of α-olefins	385
$(n-C_8H_{17})_3Sb$ $(M-667)$	$SbCl_3 + RMgBr$ in Et_2O under reflux (30)	Colorless liquid, b_6 192–193°, d_4^{19} 1.0140, n_D^{19} 1.4740, burns with a characteristic flame and leaves an Sb deposit (l, n, o, p)	1497
C_{27}			
$(p-MeCH=CHC_6H_4)_3Sb$	$SbCl_3 + RMgBr$ in THF under reflux (74)	Yellowish solid, m. 127°, sol. in org. solvents, (q)	1498
$[p-CH_2=C(Me)C_6H_4]_3Sb$	$SbCl_3 + RMgBr$ in THF (75)	Cream-colored substance, m. 120°, (q)	1498
$(n-C_9H_{19})_3Sb$	$SbCl_3 + RMgBr$ in Et_2O (33)	Oily liquid, b_6 199–200°, d_4^{19} 0.9975, n_D^{19} 1.4768, burns with a characteristic frame, (l, q)	1497

C_{30}		
$(1\text{-}C_{10}H_7)_3Sb$ (M-668)		Complexes with $M[AlR_4]$ (M = alkali metal) and Ti, V, Zr, Cr, or Mo halides catalyze the polymerization of α-olefins	385
$[p\text{-}(EtCH{=}CH)C_6H_4]_3Sb$ (M-671)	$SbCl_3$ + RMgBr in THF (73)	Yellow solid, m. 136°, (q)	1498
$(n\text{-}C_{10}H_{21})_3Sb$	$SbCl_3$ + RMgR in Et_2O under reflux in H_2 atm. (38)	Colorless oil, b_6 203–204°, d_4^{19} 0.9522, n_D^{19} 1.4615, (1, q)	1497
C_{38}		
$(p\text{-}PhOC_6H_4)_3Sb$ (M-671)	$SbCl_3$ + RMgBr in Et_2O under reflux and dec. with H_2O (44)	m. 133–140°, d^{90} 1.3396	1640
$(p\text{-}cyclo\text{-}C_6H_{11}C_6H_4)_3Sb$	$SbCl_3$ + RMgBr (2:10) in Et_2O under reflux and dec. with H_2O (91)		1640

(k) Mixts. with alkali or alkaline earth metal hydrides and Ti, V, Zr, Cr, or Mo halides catalyze the polymerization of α-olefins (386). (1) Mol. and atm. refraction, surface tension, and parachor data (1497). (m) Rxn. with Br in $CHCl_3 \rightarrow (n\text{-}C_7H_{15})_3SbBr_2$ (1497). (n) Rxn. with Br in $CHCl_3 \rightarrow (n\text{-}C_8H_{17})_3SbBr_2$ (1497). (o) The stilbine catalyzes the formation of polyurethan foams from diisocyanates and polyesters (757). (p) Mixts. with Li, Na, K, Mg, or Zn with Ti or V halides catalyze the polymerization of α-olefins (387). (q) Rxn. with Br in $CHCl_3 \rightarrow$ the corr. dibromotriarylantimony, R_3SbBr_2 (1497, 1498).

1.1.2 UNSYMMETRIC $R_2R'Sb$ DERIVATIVES

The unsymmetric tertiary stibines and their methods of synthesis were reported in the main volume on pages 672-6. Besides the synthesis methods previously described, the stibines can be prepared as follows: (a) from distibines by a cleavage with organic halides and organolithium derivatives respectively; (b) from secondary stibinolithium and stibinomagnesium derivatives by metathesis with alkyl halides; (c) from secondary stibines by a condensation with acetylene derivatives; (d) from stibinous and stibonous esters by the insertion of ketene to form stibinoacetic acid esters and stibylenediacetic acid esters, respectively; and (e) from secondary halostibines by metathesis with trialkylstannylacetic acid esters to form stibinoacetic acid esters. The stibines containing at least one ethynyl group are capable of condensing with tertiary stannanes to form the corresponding stannylvinylstibine derivatives.

The unsymmetric stibines are compiled in Table 44 on page 455.

1.1.4 DITERTIARY STIBINES

Several compounds of this group were described in the main volume on pages 678-9. Many new ditertiary stibines, compiled in Table 45 on page 459, were prepared from secondarystibinolithium or sodium derivatives by metathesis with α,w-dihaloalkanes, from tertiary stibines containing an ethynyl group by a condensation with secondary stibines, and from secondary halostibines by the reaction with phenylenedilithium. The formation of trimethylphosphonium bis(dimethylstibino)methylide from trimethylphosphonium dimethylstibinomethylide by thermal decomposition at 120° was detected by NMR spectroscopy.

Table 44. Unsymmetric Tertiary Stibines, $R_2R'Sb$

$R_2R'Sb$	Prepd. from (Yield %)	Prop. and Rxn.	Ref.
C_5			
Et_2MeSb	$(Et_2Sb-)_2$ + MeI (1:1) in THF under reflux, a mixt. of Et_2MeSb + Et_2SbI is obtd.	Light orange oil, b_2 58-59°	783
C_6			
$Me_2(Me_3P=CH)Sb$	Me_2SbCl + $Me_2P(CH_2Li)=CH_2$ in Et_2O at 25° (40)	m. -26 to -25°, $b_{0.1}$ 40-42°, NMR and IR spectra, (a)	1355
C_7			
$Et_2(CH_2=CHCH_2)Sb$	$Et_2SbLi\cdot Dioxan$ + $CH_2=CHCH_2Cl$ in THF at -20° (78)	b_2 50-52°, soly. data, rxn. with MeI → $[Et_2Me(CH_2=CHCH_2)Sb]I$, (b)	783
C_8			
Me_2PhSb (M-673)		NMR spectrum, (c, d)	882
$Et_2SbCH_2CO_2Et$	Et_2SbOEt + $CH_2=C=O$ in C_6H_6 (78)	$b_{0.18}$ 60-61°, d_4^{20} 1.3646, n_D^{20} 1.5218, (e)	175
$Et_2[Cl(CH_2)_4]Sb$	$Et_2SbLi\cdot Dioxan$ + $Cl(CH_2)_4Cl$ (1:1) in THF at -50° (53)	Light yellow oil, b_2 76-77°, sensitive to air, (f)	783
C_9			
$Pr_2SbCH_2CO_2Me$	Pr_2SbOMe + $CH_2=C=O$ in C_6H_6 (79)	$b_{0.0085}$ 47-48°, d_4^{20} 1.3088, n_D^{20} 1.5160, (e)	175
C_{10}			
Et_2PhSb	$(Et_2Sb-)_2$ in Et_2O-dioxan + PhLi in Et_2O at -45°, (g) (38)	b_2 92-94°	783
$Bu_2(HC\equiv C)Sb$		Rxn. with R_2SbH → $(Bu_2SbCH=)_2$, (h)	1140, 1142

(a) Dec. at 120° → $Me_3P=CH_2$ + $Me_3P=C(SbMe_2)_2$ (1355).   (b) EtMgBr in C_6H_6 initiates the polymerization of the stibine to a cryst. polymer (56).   (c) Rxn. with $[4-(Me_2SiN_3)C_6H_4]_2O$ (2:1) → $\{4-[Me_2PhSb=NSiMe_2]C_6H_4\}_2O$ (1584-1586).   (d) Useful as a chain-transfer agent in a free-radical polymerization of vinyl monomers (1218).   (e) IR and NMR spectra and atm. refraction data (175).   (f) Rxn. with $(C_6H_{11})_2PLi$ in THF → $Et_2Sb(CH_2)_4P(C_6H_{11})_2$ (783). (g) A mixt. of Et_2PhSb (38%) and Et_2SbLi (97%) was obtd. (783).   (h) Rxn. with R_3MH (M = Sn, Ge) in a sealed tube → $R_3MCH=CHSbBu_2$ and some $R_3MCH=CHMR_3$ (1142).

Table 44 Continued

R₂R'Sb	Prepd. from (Yield %)	Prop. and Rxn.	Ref.
C_{10} (Cont.)			
$Pr_2SbCH_2CO_2Et$	$Pr_2SbOEt + CH_2=C=O$ in C_6H_6 (84)	$b_{0.05}$ 68-69°, d_4^{20} 1.2896, n_D^{20} 1.5097, sensitive to air and moisture, atm. refraction data, (i)	175
C_{11}			
$Bu_2(CH_2=CHCH_2)Sb$ (M-674)	$Bu_2SbLi + CH_2=CHCH_2Br$ in THF (73)		1444
	$(Bu_2Sb)_2Mg + CH_2=CHCH_2Br$ in THF (31)		1444
$Bu_2SbCH_2CO_2Me$	$Bu_2SbOMe + CH_2=C=O$ in C_6H_6 (76)	$b_{0.0085}$ 69-70°, d_4^{20} 1.2562, n_D^{20} 1.5108, (e)	175
C_{12}			
$Bu(EtO_2CCH_2)_2Sb$	$BuSb(OEt)_2 + CH_2=C=O$ in C_6H_6 (85)	$b_{0.018}$ 104-105°, d_4^{20} 1.3474, n_D^{20} 1.5142, (e)	175
$Bu_2SbCH_2CO_2Et$	$Bu_2SbOEt + CH_2=C=O$ in C_6H_6 (77)	$b_{0.006}$ 81-83°, d_4^{20} 1.2155, n_D^{20} 1.5039, (e, j)	175
$t-Bu_2[Cl(CH_2)_4]Sb$	$t-Bu_2SbLi \cdot Dioxan + Cl(CH_2)_4Cl$ (1:1) in THF at -40° (32)	b_2 93-100°	785
C_{13}			
Ph_2MeSb (M-674)	$(Ph_2Sb-)_2 + MeI$ in Et_2O-THF under reflux, (k)		784
$Bu_2(Me_3SnCH=CH)Sb$	$Bu_2SbC \equiv CH + Me_3SnH$ at 80° in a sealed tube, (l)	$b_{0.5}$ 102-104°, n_D^{20} 1.5400, unstable liquid, can be stored at low temp. in the dark under inert gas	1142
C_{14}			
$Ph_2(HC \equiv C)Sb$	$Ph_2SbCl + HC \equiv CMgBr$ in THF at -2° (48)	b_1 127-130°, n_D^{20} 1.6618, d_4^{20} 1.4786, (m, n)	1142
$t-Bu_2PhSb$	$(t-Bu_2Sb-)_2 + PhLi$ in Et_2O-dioxan, (o) (47)	b_1 96-100°, soly. data	785
	$t-Bu_2SbCl + PhMgBr$ in Et_2O (85)		785

C16			
$Ph_2SbCH_2CO_2Et$ (M-674)	$b_{0.025}$ 132-133°, d_4^{20} 1.4470, n_D^{20} 1.6293	$Ph_2SbCl + Et_3SnCH_2CO_2Et$ in C_6H_6 at 90-100° (73)	176
$Ph_2[Cl(CH_2)_4]Sb$	Yellowish oil, b_2 139-142°, soly. data	$Ph_2SbLi \cdot Dioxan + Cl(CH_2)_4Cl$ (1:1) in THF at 40° (59)	784
$(C_6H_{11})_2[Cl(CH_2)_4]Sb$	Colorless oil, b_{13} 151-154°	$(C_6H_{11})_2SbLi + Cl(CH_2)_4Cl$ (1:1) in THF-Et_2O at 30-40° (61)	782
C18			
$(C_6F_5)_2PhSb$	m. 89-91°, IR spectrum	C_6F_5Li in Et_2O at -78° + $PhSbCl_2$ in Et_2O and warming to r.t. (50)	842
$Ph_2(C_6F_5)Sb$	m. 30-33°, IR spectrum	C_6F_5Li in Et_2O at -78° + Ph_2SbCl in Et_2O and warming to r.t. (25)	842
$(4-ClC_6H_4)_2PhSb$	See the orig. paper	See the orig. paper	616
C20			
$Ph_2(PhCH=CH)Sb$	$b_{0.0035}$ 152-153°, n_D^{20} 1.6888, IR spectrum	$Ph_2SbH + PhC\equiv CH$ at 0° or r.t., then heating to ~ 50° (80)	1141
	b_1 198-199°, n_D^{20} 1.6890	$Ph_2SbCl + PhCH=CHMgBr$ in THF-Et_2O (24)	1141
$Ph_2[PhC(=CH_2)]Sb$	$b_{0.5}$ 170-173°, n_D^{20} 1.6802, IR spectrum	$Ph_2SbCl + PhC(=CH_2)MgBr$ in THF-Et_2O (34)	1141
$(4-MeC_6H_4)_2PhSb$		See the orig. paper	616
$(4-MeOC_6H_4)_2PhSb$		See the orig. paper	616

(i) Rxn. with PhOH at 50-100°/4-11 mm → $Pr_2SbOH + EtOAc$ (176). (j) Rxn. with Ag_2O in Et_2O → $Bu_2Sb(CH_2CO_2Et)O$ (176). (k) A mixt. of $Ph_2MeSb + Ph_2SbH$ was obtd. (784). (l) $Me_3SnSbBu_2$ is formed as by-prod. (1142). (m) Rxn. with Ph_3SnH → $Ph_3SnCH=CHSbPh_2$ (1142). (n) Rxn. with Ph_2SbH → $Ph_2SbCH=CHSbPh_2$ (1142). (o) t-$Bu_2SbLi \cdot Dioxan$ is formed as by-prod. (785).

Table 44 Continued

R_2R'Sb	Prepd. from (Yield %)	Prop. and Rxn.	Ref.
$Et_2[(C_6H_{11})_2P(CH_2)_4]$Sb	$Et_2Sb(CH_2)_4Cl + (C_6H_{11})_2PLi$ (1:1) in THF (97)	Light yellow oil, dec. during distn. in vacuum, (p)	783
C_{21}			
$Et_2Sb(CH_2)_4PMe(C_6H_{11})_2]I$	$Et_2Sb(CH_2)_4P(C_6H_{11})_2 + MeI$ (1:1) in Et_2O (90)	m. 71-76°, sensitive to air, soly. data	783
C_{28}			
$Bu_2(Ph_3SnCH=CH)$Sb	$Ph_3SnH + Bu_2SbC≡CH$ at 80° under Ar, (q)	b_1 169-172°, n_D^{20} 1.5910	1142
C_{32}			
$Ph_2(Ph_3SnCH=CH)$Sb	$Ph_3SnH + Ph_2SbC≡CH$ at 85-90° under Ar	$b_{0.001}$ 206-210°, n_D^{20} 1.6900	1142

(p) Rxn. with S in C_6H_6 under reflux → $Et_2Sb(S)(S)(CH_2)_4P(S)(C_6H_{11})_2$ (783). (o) $Ph_3SnCH=CHSnPh_3$ is formed as by-prod. (1142).

Table 45. Ditertiary Stibines

$R_2SbC_xH_ySbR'_2$	Prepd. from (Yield %)	Prop. and Rxn.	Ref.
C_8			
$Me_3P=C(SbMe_2)_2$	$Me_3P=CHSbMe_2$ by heating at 120°	The formation of the distibine was detected by NMR spectroscopy	1355
C_{10}			
$m-(Me_2Sb)C_6H_4SbMe_2$ (M-678)		Rxn. with $Ph_2Si(N_3)_2$ (1:1) → $[Ph_2SiN=SbMe_2C_6H_4SbMe_2=N]_n$	1584–1586
C_{12}			
$Et_2Sb(CH_2)_4SbEt_2$	$Et_2SbLi \cdot Dioxan + Cl(CH_2)_4Cl$ (2:1) in THF at -10° (82)	Colorless oil, sensitive to air, b_2 129-131°, soly. data, (a, b)	783
C_{13}			
$Et_2Sb(CH_2)_5SbEt_2$	$Et_2SbLi \cdot Dioxan + Cl(CH_2)_5Cl$ (2:1) in THF at -10° (74)	Colorless oil, sensitive to air, b_2 132-134°, soly. data, (a, c, d)	783
C_{14}			
$Et_2Sb(CH_2)_6SbEt_2$	$Et_2SbLi \cdot Dioxan + Cl(CH_2)_6Cl$ in THF under reflux (71)	Colorless oil, sensitive to air, b_2 144-146°, soly. data, (a, b)	783
C_{18}			
$Bu_2SbCH=CHSbBu_2$	$Bu_2SbC≡CH + Bu_2SbH$	$b_{0.0025}$ 108-109°, n_D^{20} 1.5490, IR spectrum	1140
	$Bu_2SbC≡CH + Bu_3GeH$ at 80° in Ar	$b_{0.5}$ 157-160°, n_D^{20} 1.5450	1142
C_{20}			
$t-Bu_2Sb(CH_2)_4Sb-t-Bu_2$	$t-Bu_2SbLi \cdot Dioxan + Cl(CH_2)_4Cl$ in THF (32)	Orange colored oil, sensitive to air, soly. data, dec. during distn. in vacuum, (b)	785
C_{22}			
$Bu_2SbCH=CHSbPh_2$	$Bu_2SbC≡CH + Ph_2SbH$	$b_{0.0035}$ 147-150°, n_D^{20} 1.5960	1140
C_{26}			
$Ph_2SbCH=CHSbPh_2$	$Ph_2SbC≡CH + Ph_2SbH$ at 80-83° under Ar	Viscous liquid, IR spectrum, (e)	1142

(a) Rxn. with S in C_6H_6 under reflux → $R_2Sb(S)C_xH_ySb(S)R_2$ (783, 784). (b) Rxn. with MeI in MeOH → $[MeR_2Sb(CH_2)_n-SbR_2MeI_2$ (783, 785). (c) Rxn. with air in Et_2O → $Et_2Sb(O)(CH_2)_5SbEt_2$ (783). (d) Rxn. with H_2O_2 in Me_2CO → $Et_2Sb(O)(CH_2)_5Sb(O)Et_2$ (783). (e) Rxn. with Cl in $CHCl_3$ at -78° → $Ph_2SbCl_2CH=CHSbCl_2Ph_2$ (1142).

Table 45 Continued

$R_2SbC_xH_ySbR'_2$	Prepd. from (Yield %)	Prop. and Rxn.	Ref.
C27			
$(C_6H_{11})_2Sb(CH_2)_3Sb(C_6H_{11})_2$	$(C_6H_{11})_2SbLi + Cl(CH_2)_3Cl$ (2:1) in THF-Et$_2$O at -45° (99)	Yellow oil, sensitive to air, soly. data, (f)	782
C28			
$Ph_2Sb(CH_2)_4SbPh_2$	$Ph_2SbLi\cdot Dioxan + Cl(CH_2)_4Cl$ (2:1) in THF at -20° (42)	Colorless needles, m. 108°, soly. data	784
	Ph_2SbNa in dioxan + $Cl(CH_2)_4Cl$ (2:1) (96)	(f)	784
$(C_6H_{11})_2Sb(CH_2)_4Sb(C_6H_{11})_2$	$(C_6H_{11})_2SbLi + Cl(CH_2)_4Cl$ (2:1) in THF-Et$_2$O at -45° (97)	Yellow oil, sensitive to air, soly. data, (f)	782
C29			
$Ph_2Sb(CH_2)_5SbPh_2$	$Ph_2SbLi\cdot Dioxan + Cl(CH_2)_5Cl$ (2:1) in THF at -20° (89)	Partly cryst. light yellow oil, dec. during distn., soly. data, (a)	784
$(C_6H_{11})_2Sb(CH_2)_5Sb(C_6H_{11})_2$	$(C_6H_{11})_2SbLi + Cl(CH_2)_5Cl$ (2:1) in THF-Et$_2$O at -45° (89)	Yellow oil, sensitive to air, soly. data, (f)	782
C30			
$m-(Ph_2Sb)C_6H_4SbPh_2$ (M-679)		Rxn. with $Ph_3GeN_3 \rightarrow Ph_3GeN=Sb(Ph)_2-C_6H_4Sb(Ph)_2=NGePh_3$	1584-1586
$p-(Ph_2Sb)C_6H_4SbPh_2$	Ph_2SbCl in Et$_2$O + $p-C_6H_4Li_2$ in petr. ether under reflux, (g) (28)	White crystals, m. 143-145°, soly. data, IR and UV spectra, dipole moment, (f, h)	1658
$Ph_2Sb(CH_2)_6SbPh_2$	$Ph_2SbLi\cdot Dioxan + Cl(CH_2)_6Cl$ (2:1) in THF at -20° (77)	Colorless crystals, m. 98-99°, soly. data, (a)	784
$(C_6H_{11})_2Sb(CH_2)_6Sb(C_6H_{11})_2$	$(C_6H_{11})_2SbLi + Cl(CH_2)_6Cl$ (2:1) in THF-Et$_2$O at -45° (99)	Yellow oil, sensitive to air, soly. data, (f)	782

(f) Rxn. with Br $\rightarrow$ $R_2SbBr_2C_xH_ySbBr_2R_2$ (782, 784). (g) 17% yield was obtd. with C_6H_6 instead of Et$_2$O. (h) Rxn. with H_2O_2 in Me$_2$CO $\rightarrow$ the Sb, Sb, Sb', Sb'-tetrahydroxide (1658).

1.2 SECONDARY STIBINES

Two secondary stibines were described in the main volume on pages 680-81. A few new secondary stibines were prepared, as previously, from secondary halostibines by a reduction with lithium tetrahydroaluminate. A secondary stibine containing acetoxymercuri substituents was prepared by a condensation of stibine with allyl chloride and mercury oxide in acetic acid.

Diethylstibine, SbC_4H_{11}, Et_2SbH
Synth.: From Et_2SbBr + $LiAlH_4$ in Et_2O at -78° in the dark, obtd. as a soln. in Et_2O (783).
Prop.: Unstable compd., readily dec. → $Et_2SbSbEt_2$ (783).
Rxn. with: PhLi in Et_2O-dioxan → $Et_2SbLi \cdot Dioxan$ + $Et_2SbSbEt_2$ (783).

Dibutylstibine, SbC_8H_{19}, Bu_2SbH
Prop.: IR spectrum (210).
Rxn. with: $Bu_2SbC \equiv CH$ → $Bu_2SbCH=CHSbBu_2$ (1140, 1142).
Use: Combined with tartaric acid is effective as a fire-retardant for epoxy resins (1615).

Di-tert-butylarsine, SbC_8H_{19}, $t-Bu_2SbH$
Synth.: From $t-Bu_2SbCl$ + $LiAlH_4$ in Et_2O at -78° and warming to 35°, 17.3% yield (785).
Prop.: Colorless oil, b_1 24-25° (with dec.), sensitive to air and moisture, sol. in org. solvents (785).
Rxn. with: PhLi in Et_2O-dioxan → $t-Bu_2SbLi \cdot Dioxan$ (785).

Diphenylstibine, $SbC_{12}H_{11}$, Ph_2SbH,   (M-680)
Synth.: From Ph_2SbCl + $LiAlH_4$ in Et_2O at -78°, 47% yield, along with $Ph_2SbSbPh_2$ (784).
Prop.: b_1 118-120° (784). IR spectrum (210).
Rxn. with: Na in dioxan → Ph_2SbNa (784).
K in dioxan → Ph_2SbK (784).
$R_2SbC \equiv CH$ → $Ph_2SbCH=CHSbR_2$ (1140, 1142).
$PhC \equiv CH$ → $Ph_2SbCH=CHPh$ (1141).
$Et_3SiC \equiv CH$ → $Ph_2SbSbPh_2$, $Et_3SiCH=CH_2$, and $Et_3SiC \equiv CH$ (1142).
CCl_4 at -10 to 0° → C_6H_6, $CHCl_3$, Ph_2SbCl, and probably antimonobenzene (1141).
$PhCCl_3$ in Et_2O → Ph_2SbCl + $PhCHCl_2$ (1141).
Dild. HCl in Me_2CO → Ph_2SbCl and evolution of H (1141).
AcOH in petr. ether + H_2SO_4 → Ph_2SbOAc + H (1141).
BzOH in Et_2O at 30-35° → Ph_2SbOBz (1141).
Use: With halides or oxyhalides of Ti, V, or Cr on SiO_2 or Al_2O_3 an effective polymerization catalyst (42).

Dicyclohexylstibine, $SbC_{12}H_{23}$, $(C_6H_{11})_2SbH$
Synth.: From $(C_6H_{11})_2SbCl$ + $LiAlH_4$ in Et_2O at -20° in the dark, 93% yield (782).

<u>Prop.</u>: Yellow-orange oil (782).
<u>Rxn. with</u>: PhLi in Et$_2$O → (C$_6$H$_{11}$)$_2$SbLi (782).

2,2'-Stibylenebis(3-chloropropylmercury Acetate), SbHg$_2$C$_{10}$H$_{17}$Cl$_2$O$_4$,
    HSb[CH(CH$_2$Cl)CH$_2$HgOAc]$_2$
<u>Synth.</u>: From SbH$_3$ by a condensation with allyl chloride and HgO in AcOH
(1447).
<u>Biol. Prop.</u>: A fungicide and bactericide (1447).

1.3 PRIMARY STIBINES

   Two primary stibines were described in the main volume on page 681.

Methylstibine,   SbCH$_5$,   MeSbH$_2$,   (M-681)
<u>Use</u>: Combinations with halides or oxyhalides of Ti, V, or Cr deposited on
SiO$_2$ or Al$_2$O$_3$ are effective polymerization catalysts (42).

1.4   COMPOUNDS WITH Sb-Sb BONDS

1.4.1 DISTIBINES

   Four distibines were described in the main volume on pages 681-2.

Tetramethyldistibine,   Sb$_2$C$_4$H$_{12}$,   Me$_2$SbSbMe$_2$,   (M-681)
<u>Uses</u>: Complexes with RAlX$_2$ and Ti, Zr, V, Cr, or Mo halides catalyze the
polymerization of α-olefins (384). Mixts. with transition metal carbonyls
and N-halogenated org. compds. catalyze the polymerization of vinyl monomers
containing electron-withdrawing groups (773). A chain-transfer agent in a
free-radical initiated polymerization of vinyl monomers (1218).

Tetraethyldistibine,   Sb$_2$C$_8$H$_{20}$,   Et$_2$SbSbEt$_2$
<u>Synth.</u>: From Et$_2$SbBr + LiAlH$_4$ in Et$_2$O and a slow dec. of the resulting
Et$_2$SbH in Et$_2$O at r.t. in the dark, 87% yield (783).
From Et$_2$SbLi·Dioxan + CH$_2$Br$_2$ in THF at -25°, 31% yield (783).
From elemental Sb, Na, and EtBr according to the formula: Sb$_n$ + 2nNa → n/2
Na$_2$SbSbNa$_2$ $\xrightarrow{\text{2nEtBr}}$ Et$_2$SbSbEt$_2$ (1407).
<u>Prop.</u>: Intensively yellow liquid, dec. during distn. in vacuum (783).
<u>Rxn. with</u>: MeI → Et$_2$SbMe and Et$_2$SbI (783).
PhLi → Et$_2$SbLi + Et$_2$SbPh (783).
<u>Use</u>: A chain-transfer agent in a free-radical catalyzed polymerization of
vinyl monomers (1218).

Tetra-tert-butyldistibine,   Sb$_2$C$_{16}$H$_{36}$,   t-Bu$_2$SbSb-t-Bu$_2$
<u>Synth.</u>: From t-Bu$_2$SbCl + t-Bu$_2$SbLi·Dioxan in Et$_2$O at r.t., 98% yield (785).
From t-Bu$_2$SbLi·Dioxan + Cl(CH$_2$)$_4$Cl in dioxan at 65°, 95% yield (785).
<u>Prop.</u>: Yellow-orange oil (785).

Rxn. with: PhLi in Et$_2$O-dioxan → t-Bu$_2$SbLi·Dioxan + t-Bu$_2$SbPh (785).

Tetraphenyldistibine, Sb$_2$C$_{24}$H$_{20}$, Ph$_2$SbSbPh$_2$, (M-682)
Synth.: From Ph$_2$SbOAc by electrolytic redn. and by interaction with Ph$_2$Sb$^-$
generated electrochemically from Ph$_2$SbSbPh$_2$ (467).
From Ph$_2$SbH by heating with Et$_3$SiC≡CH in a sealed tube at 80° (1142).
From Ph$_2$SbLi·Dioxan + BrCH$_2$CH$_2$Br or ClCH$_2$CH$_2$Cl in THF at -30°, 82 and 78%
yield, resp. (784).
By-prod. from the rxn. of Ph$_2$SbCl with LiAlH$_4$ in Et$_2$O at -78° (784).
Prop.: Yellow crystals, m. 121-124° (1142), m. 123° (784). Half-wave poten-
tial (467, 472).
Rxn. with: PhLi in Et$_2$O-dioxan → Ph$_2$SbLi·Dioxan + Ph$_3$Sb (784).
MeI → Ph$_2$SbI + Ph$_2$SbMe (784).
Polarographic redn. with 2e$^-$ → 2 Ph$_2$Sb$^-$ (467, 472).

Tetracyclohexyldistibine, Sb$_2$C$_{24}$H$_{44}$, (C$_6$H$_{11}$)$_2$SbSb(C$_6$H$_{11}$)$_2$
Synth.: From (C$_6$H$_{11}$)$_2$SbCl in Et$_2$O + Na in MePh, followed by pptn. with MeOH,
96% yield (782).
From (C$_6$H$_{11}$)$_2$SbLi + BrCH$_2$CH$_2$Br, CH$_2$Cl$_2$, or CH$_2$Br$_2$ in THF-Et$_2$O at -45°, 30%
yield (782).
Prop.: Yellow, voluminous crystals, m. 71-73° (782).
Rxn. with: Br → (C$_6$H$_{11}$)$_2$SbBr$_3$ (782).

1.5   COMPOUNDS WITH ANTIMONY-HALOGEN AND ANTIMONY-PSEUDOHALOGEN BONDS

1.5.1   SECONDARY HALOSTIBINES, R$_2$SbX AND RR'SbX

   Secondary halostibines and their methods of synthesis were described in the
main volume on pages 682-88. The halostibines have also been prepared from
distibines, R$_2$SbSbR$_2$, by a reaction with methyl iodide under reflux. The
latter reaction affords mixtures of secondary iodostibines, R$_2$SbI, and unsymmetric tertiary stibines, R$_2$SbMe. Moreover, the title compounds are formed
from oxybis(secondary stibines) by a treatment with concentrated hydrochloric
acid. Fluorodiphenylstibine is formed from antimony trifluoride or ammonium
pentafluoroantimonide by the reaction with trifluorophenylsilane, triethoxy-
phenylsilane, and ammonium pentafluorophenylsilicate, respectively.

Chlorodimethylstibine, SbC$_2$H$_6$Cl, Me$_2$SbCl, (M-686)
Synth.: From Me$_3$Sb by a rxn. with Cl to form Me$_3$SbCl$_2$, followed by a thermal
cleavage (1343).
From Me$_3$Sb + SbCl$_3$ in DMF at 72° and at 100°, a mixt. of the title compd. with
MeSbCl$_2$ was formed; the rate of the ligand exchange was detd. by NMR spectroscopy - second order rate consts. were obtd. (1595). When Me$_3$Sb + SbCl$_3$ were
used at a ratio of 2:1 at 100°, Me$_2$SbCl was obtd. with 88% yield, along with
6.5% MeSbCl$_2$. In the absence of DMF no equilibration takes place (1596)
Rxn. with: Br in CCl$_4$ → Me$_2$SbClBr$_2$ (481).

LiN(Me)SiMe$_2$ → Me$_2$SbN(Me)SiMe$_3$ (1343).
MeNCH$_2$CH$_2$N(Me)PN(Me)Li → MeNCH$_2$CH$_2$N(Me)PN(Me)SbMe$_2$ (1346).

Bromodiethylstibine,   SbC$_4$H$_{10}$Br,   Et$_2$SbBr,   (M-686)
Synth.: From Et$_3$Sb + Br in CCl$_4$ at -20 to -25°, then removing the solvent
and pyrolyzing the Et$_3$SbBr$_2$ at 195-200°, 84% yield (175).
From SbCl$_3$ + EtMgBr (1:3) in Et$_2$O, followed by Br and thermal dec. of the re-
sulting Et$_3$SbBr$_2$ at 190-210°/450 mm., 75% yield (783).
From Et$_3$Sb + SbBr$_3$ (2:1) in AcNEt$_2$ at 100°, 88% yield, along with EtSbBr$_2$
(1596).
Prop.: b$_1$ 44-46°, d$_4^{20}$ 1.9073 (175), b$_2$ 50-54° (783).
Rxn. with: ROH + Na → Et$_2$SbOR (175).
LiAlH$_4$ → Et$_2$SbH → Et$_2$SbSbEt$_2$ (783).

Diethyliodostibine,   SbC$_4$H$_{10}$I,   Et$_2$SbI
Synth.: From Et$_2$SbSbEt$_2$ + MeI (1:1) in THF under reflux, a mixt. of Et$_2$SbI
+ Et$_2$SbMe was obtd. (783).
Prop.: Red oil, b$_2$ 52-55° (783).

Bromodipropylstibine,   SbC$_6$H$_{14}$Br,   Pr$_2$SbBr
Synth.: From Pr$_3$Sb + Br in CCl$_4$ at -20 to -25°, then removing the solvent
and dec. the resulting Pr$_3$SbBr$_2$ at 195-200°, 69% yield (175).
Prop.: b$_{0.5-1.0}$ 63-64°, d$_4^{20}$ 1.7286 (175). Elemental analysis (1502, 1503).
Rxn. with: Na in ROH → Pr$_2$SbOR (175).

Bromodibutylstibine,   SbC$_8$H$_{18}$Br,   Bu$_2$SbBr,   (M-684)
Synth.: From Bu$_3$Sb + Br in CCl$_4$ at -20 to -25°, then removing the solvent
and dec. the Bu$_3$SbBr$_2$ thus formed at 195-200°, 90% yield (175).
Prop.: b$_{0.0085}$ 59-60°, d$_4^{20}$ 1.5926 (175). IR spectrum (209, 210).
Rxn. with: Na in ROH → Bu$_2$SbOR (175).

Dibutylchlorostibine,   SbC$_8$H$_{18}$Cl,   Bu$_2$SbCl
Synth.: From Bu$_3$SbCl$_2$ by heating > 200°/200 mm, 89% yield (1045, 1444).
Prop.: Orange liquid, b$_{10}$ 121-123°, IR spectrum (1045).
Rxn. with: Dithiocarbamates → Bu$_2$SbSC(S)NR$_2$ (1100).
NaSR in C$_6$H$_6$ → Bu$_2$SbSR (1102).
Li in THF → Bu$_2$SbLi (1444).
Mg in THF → (Bu$_2$Sb)$_2$Mg (1444).

Di-tert-butylchlorostibine,   SbC$_8$H$_{18}$Cl,   t-Bu$_2$SbCl,   (M-686)
Synth.: From SbCl$_3$ + t-BuMgBr in Et$_2$O at -25°, dec. the rxn. mixt., and iso-
lating the prod., 70% yield (785).
Prop.: b$_1$ 48-50° (785).
Rxn. with: LiAlH$_4$ in Et$_2$O → t-Bu$_2$SbH (785).
t-Bu$_2$SbLi·Dioxan → t-Bu$_2$SbSb-t-Bu$_2$ (785).
RMgX → t-Bu$_2$SbR (785).

Chlorobis(p-chlorophenyl)stibine, $SbC_{12}H_8Cl_3$, $(4-ClC_6H_4)_2SbCl$, (M-687)
Rxn. with: $NaSR \rightarrow (4-ClC_6H_4)_2SbSR$ (537).

Chlorodiphenylstibine, $SbC_{12}H_{10}Cl$, Ph_2SbCl, (M-684)
Synth.: From Ph_3SbCl_2 by dec. at 250-270°/150 mm (784).
From Ph_3Sb + HCl in MeOH under reflux, followed by a treatment with H_2O to
form $(Ph_2Sb)_2O$ and a conversion with concd. HCl, 50% yield (784).
From $(Ph_2Sb)_2O$ in EtOH by adding dropwise concd. HCl, 93% yield (1505).
From Ph_3Sb with $SbCl_3$ (2:1) or with $PhSbCl_2$ (1:1) in CH_2Cl_2 under reflux,
92 and 93% yield, resp. (1098).
Prop.: Brown oil, b_2 135-145°, m. 68° (784), m. 61-65° (1098), m. 65-69°
(1505), IR spectrum (209, 210).
Rxn. with: $LiAlH_4 \rightarrow Ph_2SbH$ (784).
$RLi \rightarrow Ph_2SbR$ (842).
$M(O_2CR)_n$ (M = alkali or alkaline earth metal) $\rightarrow Ph_2SbO_2CR$ (536, 948, 1101,
1102).
$NaSR \rightarrow Ph_2SbSR$ (537).
ROH in the presence of a base at 75-100° $\rightarrow Ph_2SbOR$ (537).
$NaOR \rightarrow Ph_2SbOR$ (1102).
Na or K in THF $\rightarrow Ph_3Sb$ + Sb (1658).
$NaSC(S)NR_2 \rightarrow Ph_2SbSC(S)NR_2$ (905, 1100).
NaN_3 at 25-116° in Py $\rightarrow Ph_2SbN_3$ (1272).
$p-C_6H_4Li_2 \rightarrow p-(Ph_2Sb)C_6H_4SbPh_2$ (1658).
$Li \cdot SnR_3$ in THF $\rightarrow Ph_2SbSnR_3$ (1372).
$NaSC(S)NHCH_2Ph$ (1:2) in $CHCl_3 \rightarrow (Ph_2Sb)_2S$, $(PhCH_2NH)_2CS$, and $PhSb[SC(S)NH-CH_2Ph]_2$ (1505).
$[NH_4]SC(S)NHPh$ (1:1) in $CHCl_3 \rightarrow PhSb[SC(S)NHPh]_2$, $(Ph_2Sb)_2S$, and PhNCS (1505).
$[Me_2NH_2]SC(S)NHR \rightarrow Ph_2SbSC(S)NHR$ (1100).
$RMgX \rightarrow Ph_2SbR$ (1141, 1142).
$(PrO)_3P$ at 50-80° $\rightarrow$ dec. (1174).
$Et_3SnCH_2CO_2Et \rightarrow Ph_2SbCH_2CO_2Et$ (176).
$SbCl_3$ (1:1) in $CH_2Cl_2 \rightarrow PhSbCl_2$ (1098).
Biol. Prop.: Bactericidal (537) and biocidal (1098) activity.

Fluorodiphenylstibine, $SbC_{12}H_{10}F$, Ph_2SbF
Synth.: From SbF_3 + $[NH_4]_2[PhSiF_5]$ (1:2) in aq. soln., 43% crude yield (1106).
From SbF_3 + $PhSiF_3$ (1:2) in aq. HF soln., 36% yield; addn. of EtOH to the rxn.
mixt. increased the yield to 48% (1106).
From SbF_3 to $PhSiF_3$ (2:3) by shaking in H_2O, 68% yield; in aq. EtOH the yield
was 78% (1106).
From SbF_3 + $PhSi(OEt)_3$ (1:2) in 40% HF cooled by ice, 50% yield (1106).
From SbF_3 + NH_4F in H_2O + $PhSiF_3$ a ppt. was isolated from which the prod. was
extd. with EtOH (1106).
From $[NH_4]_2SbF_5$ + $PhSiF_3$ (1:2) in aq. soln. contg. NH_4F, 34% yield (1106).
Prop.: m. 157-158° (recrystd. from EtOH), m. 160-161° (1106).
Rxn. with: NH_4F in AcOH + $EtSiF_3$, followed by KOH and 10% ethanolic I soln.
$\rightarrow (Ph_2EtSbI)_2O$ (1107).

PhSiF$_3$ under reflux → Ph$_3$Sb (1108).

Iododiphenylstibine,   SbC$_{12}$H$_{10}$I,   Ph$_2$SbI,   (M-687)
Synth.: From Ph$_2$SbSbPh$_2$ + MeI in Et$_2$O-THF under reflux, a mixt. of Ph$_2$SbI
and Ph$_2$SbMe was formed (784).
Prop.: Half-wave potential (467).
Rxn. with: (PrO)$_3$P at 50-80° → dec. (1174).

Chlorodicyclohexylstibine,   SbC$_{12}$H$_{22}$Cl,   (C$_6$H$_{11}$)$_2$SbCl,   (M-687)
Synth.: From (C$_6$H$_{11}$)$_3$SbCl$_2$ by heating at 230°/360 mm, 45% yield (782).
Prop.: Yellow oil, b$_1$ 136-142°, sensitive to light and air (782).
Rxn. with: LiAlH$_4$ in Et$_2$O → (C$_6$H$_{11}$)$_2$SbH (782).
Na in MePh → (C$_6$H$_{11}$)$_2$SbSb(C$_6$H$_{11}$)$_2$ (782).
NaSR → (C$_6$H$_{11}$)$_2$SbSR (537).

1.5.2 PRIMARY DIHALOSTIBINES

Primary dihalostibines were described in the main volume on pages 689-96.
The methods of synthesis discussed in the main volume (M-689) have been extended by using antimony tribromide, instead of the trichloride, in the
equilibration with tertiary stibines as well as with secondary halostibines.
Dichloroethoxystibine was converted to ethyl dichlorostibinoacetate by a
reaction with ketene.

Dichloromethylstibine,   SbCH$_3$Cl$_2$,   MeSbCl$_2$,   (M-693)
Synth.: From Me$_2$SbCl$_3$ by thermal dec. at 110-115° under exclusion of mois-
ture, 83% yield (1343).
From Me$_3$Sb + SbCl$_3$ in DMF at 72° and 100°, a mixt. of MeSbCl$_2$ and Me$_2$SbCl was
formed; the rate of the ligands exchange was followed by NMR spectroscopy
(1595).
Prop.: Water-clear liquid, with a strong light refraction, moderately sensi-
tive to air, crystd. at the temp. of liquid air (1343).
Rxn. with: H$_2$NCOC(SNa)=C(SNa)CONH$_2$ in MeCN → MeSbSC(CONH$_2$)=C(CONH$_2$)S (1090).
LiN(Me)SiMe$_3$ → MeSb[N(Me)SiMe$_3$]$_2$ (1343).

Dibromoethylstibine,   SbC$_2$H$_5$Br$_2$,   EtSbBr$_2$
Synth.: From Et$_3$Sb + SbBr$_3$ (1:2) in AcNEt$_2$ at 100° (1596).

Dichloroethylstibine,   SbC$_2$H$_5$Cl$_2$,   EtSbCl$_2$,   (M-693)
Use: Complexes with the halides of transition metals of Group IVB or VB and
with org. derivs. of metals of Group II or IIIA catalyze the polymerization
of ethylene to high density linear polymers (249a).

Dichlorostibinoacetic Acid Ethyl Ester,   SbC$_4$H$_7$Cl$_2$O$_2$,   Cl$_2$SbCH$_2$CO$_2$Et
Synth.: From Cl$_2$SbOEt + CH$_2$=C=O in C$_6$H$_6$, 2% yield (175).
From Sb(CH$_2$CO$_2$Et)$_3$ + SbCl$_3$ (1:2) in C$_6$H$_6$, 93% yield (176).
Prop.: White crystals, m. 88-90° (176).

Dibromobutylstibine, $SbC_4H_9Br_2$, $BuSbBr_2$
Synth.: From Bu_3Sb + Br in a sealed tube, 92% yield (175).
Prop.: m. 29-30°, $b_{0.018}$ 76-78° (175).

Butyldichlorostibine, $SbC_4H_9Cl_2$, $BuSbCl_2$
Rxn. with: $\overline{CH_2CH_2S}$ → $BuAs(SCH_2CH_2Cl)_2$ (961).
Biol. Prop.: Bactericidal and fungicidal activity (537).
Use: Antifouling agent for paints (537).

Dichloro(p-chlorophenyl)stibine, $SbC_6H_4Cl_3$, $4-ClC_6H_4SbCl_2$, (M-694)
Rxn. with: Me N-acetylcysteinate → interaction of 10% SH groups (1639).
Biol. Prop.: Negative results in the inactivation tests of Adenovirus type 5 (1639).

Dichlorophenylstibine, $SbC_6H_5Cl_2$, $PhSbCl_2$, (M-690)
Synth.: From Ph_3Sb + $SbCl_3$ (1:2) in CH_2Cl_2 under reflux, 95-99% yield (1098, 1505).
From Ph_2SbCl + $SbCl_3$ (1:1) in CH_2Cl_2 under reflux (1098).
Prop.: m. 53-57° (1098), m. 58-60° (1505).
Rxn. with: RLi → Ph_2SbR (842).
ROH in the presence of NH_3 → Ph_2SbOR (537).
$NaSC(S)NR_2$ or $NH_4SC(S)NR_2$ → $PhSb(Cl)SC(S)NR_2$ and $PhSb[SC(S)NR_2]_2$ (907, 1100, 1505).
2,2'-Dilithio-4,5,4',5'-bis(methylenedioxy)biphenyl → 2,3,7,8-bis(methylene-dioxy)-5-phenylstibole (434).
RCO_2Na (1:2) → $PhSb(O_2CR)_2$ (1101) and Ph_2SbO_2CR, $Sb(O_2CR)_3$ + 4 NaCl (1273).
Aq. NaOH → PhSbO (1505).
$NaSC(S)NR_2$ (1:1) in $CHCl_3$ → $PhSb(Cl)SC(S)NR_2$ (907, 1505).
$NH_4SC(S)NHR$ in $CHCl_3$ → $PhSb[SC(S)NHR]_2$ (1505).
$\overline{CH_2CH_2S}$ → $PhSb(SCH_2CH_2Cl)_2$ (961).
Biol. Prop.: Fungicidal and bactericidal activity (537, 1098).
Uses: A fungicide and bactericide for paints, plastics, and fibrous prods. (537). A catalyst for the formation of polyurethan foams from diisocyanates and polyesters (757). A flux activator in soldering by fusible solders (1224).

Dichloro-n-octylstibine, $SbC_8H_{17}Cl_2$, $n-C_8H_{17}SbCl_2$
Rxn. with: RCO_2Na (1:2) → $(n-C_8H_{17})_2SbO_2CR$ + NaCl + Sb_2O_3 (1273).

Dichloro-2-naphthylstibine, $SbC_{10}H_7Cl_2$, $2-C_{10}H_7SbCl_2$
Rxn. with: $NCN=C(SK)_2$ in DMF → $C_{10}H_7\overline{SbSC(=NCN)S}$ (1089).
$H_2NCOC(SNa)=C(SNa)CONH_2$ in MeCN → $C_{10}H_7\overline{SbSC(CONH_2)=C(CONH_2)S}$ (1090).

1.5.5 SECONDARY PSEUDOHALOSTIBINES

Azidodiphenylstibine, $SbC_{12}H_{10}N_3$, Ph_2SbN_3
Synth.: From Ph_2SbCl + NaN_3 in Py at 25-116°, 85-98% yield (1272).

<u>Prop.</u>: Light tan, viscous liquid; IR spectrum. Quant. anal. data deviated excessively from theoret. values. Pyrolysis at 220° → Ph_3Sb + $PhSb(N_3)_2$ (1272).

<u>Rxn. with</u>: Ph_3P at 60-80° → $Ph_2SbN=PPh_3$ + N_2 (1272).

1.5.6 PRIMARY PSEUDOHALOSTIBINES

Diisocyanatophenylstibine, $SbC_8H_5N_2O_2$, $PhSb(NCO)_2$
<u>Uses</u>: Mixts. with TiO_2 are effective stabilizers for urethan prepolymers (565, 1616).

1.5.9 MISCELLANEOUS HALOSTIBINE DERIVATIVES

Chloro[(dimethylthiocarbamoyl)thio]phenylstibine, $SbC_9H_{11}ClNS_2$,
    $PhSb(Cl)SC(S)NMe_2$
<u>Synth.</u>: From $PhSbCl_2$ + $NaSC(S)NMe_2$ (1:1) in $CHCl_3$ at 25°, 96% yield (907, 1505).
<u>Prop.</u>: m. 172-175°, UV and IR spectra (907, 1505).

Chloro[(diethylthiocarbamoyl)thio]phenylstibine, $SbC_{11}H_{15}ClNS_2$,
    $PhSb(Cl)SC(S)NEt_2$
<u>Synth.</u>: From $PhSbCl_2$ + $NaSC(S)NEt_2 \cdot 3\ H_2O$ (1:1) in $CHCl_3$ at 25°, 97% yield (907, 1505).
<u>Prop.</u>: m. 90-93°, UV and IR spectra (907, 1505).

Chloro[(diphenylthiocarbamoyl)thio]phenylstibine, $SbC_{19}H_{15}ClNS_2$,
    $PhSb(Cl)SC(S)NPh_2$
<u>Synth.</u>: From $PhSbCl_2$ + $NaSC(S)NPh_2$ (1:1) in $CHCl_3$ at 25°, 84% yield (1505).
<u>Prop.</u>: m. 176-178°, UV and IR spectra (907, 1505).

1.6   COMPOUNDS OF ANTIMONY BONDED TO GROUP VIA ELEMENTS

1.6.1   COMPOUNDS WITH ANTIMONY-OXYGEN BONDS

1.6.1.2 STIBINOUS ACID ESTERS

   Stibinous acid esters have recently been named as derivatives of stibine, e.g., Me_2SbOMe is called methoxydimethylstibine. The main volume contains three compounds of this group (M-697). The esters reported in the period 1965-1968 are compiled in Table 46.

Table 46. Stibinous Acid Esters, R$_2$SbOR'

R$_2$SbOR'	Prepd. from (Yield %)	Prop. and Rxn.	Ref.
C$_6$			
Et$_2$SbOEt	R$_2$SbBr + Na in EtOH at 60-70° (73)	b$_{4.5}$ 41-42°, d$_4^{20}$ 1.3730, (a)	175
C$_7$			
Pr$_2$SbOMe	R$_2$SbBr + Na in MeOH at 60° (73)	m. 31-32°, b$_1$ 48-49°, (a)	175
C$_8$			
Pr$_2$SbOEt	R$_2$SbBr + Na in EtOH at 60-70° (78)	b$_1$ 52-52.5°, d$_4^{20}$ 1.2679, (a,b)	175
C$_9$			
Bu$_2$SbOMe	R$_2$SbBr + Na in MeOH at 60° (88)	b$_{3.5}$ 83-84°, b$_{0.008}$ 48-49°, m. 38-39°, (a)	175
C$_{10}$			
Bu$_2$SbOEt	R$_2$SbBr + Na in EtOH at 60-70° (83)	b$_1$ 64-65°, d$_4^{20}$ 1.2437, (a)	175
C$_{16}$			
Ph$_2$SbOBu	R$_2$SbCl in C$_6$H$_6$ by bubbling NH$_3$ and treating the Ph$_2$Sb-Cl·NH$_3$ complex with BuOH under reflux (88)	b$_{0.5}$ 133-141°, biocidal activity	537
	R$_2$SbCl in C$_6$H$_6$ by bubbling NH$_3$, adding BuOH and distg. at 133-141°/0.5 mm (38)	Crystd. with 2 BuOH, b$_{0.3}$ 134-137°, fungicidal and bactericidal activity	1102

(a) Rxn. with CH$_2$=C=O in C$_6$H$_6$ → R$_2$SbCH$_2$CO$_2$R' (175). (b) Elemental analysis by a rxn. with S, followed by pyrolysis (1502).

1.6.1.3 STIBINOUS ACID ANHYDRIDES WITH CARBOXYLIC ACIDS

The anhydrides and their methods of synthesis were described in the main volume on pages 697-98. This type of antimony derivatives have been indexed by the Chemical Abstracts as stibine derivatives, e.g., Ph$_2$SbOAc has been named acetoxydiphenylstibine. Besides the methods of synthesis summarized in the main volume, the anhydrides can also be prepared from secondary halostibines by the reaction with alkali metal or ammonium carboxylates in methanol on a steam bath, from primary dihalostibines by heating with alkali metal or ammonium carboxylates in 95% ethanol for an extensive period of time, and from secondary stibines with carboxylic acid in ether at 30-35° or in petroleum ether treated with sulfuric acid.

Acetoxydiphenylstibine is described hereafter and all the other compounds of this group are compiled in Table 47 on page 471.

Acetoxydiphenylstibine, $SbC_{14}H_{13}O_2$, Ph_2SbOAc, (M-698)

Synth.: From Ph_2SbCl in MeOH or 95% EtOH + NaOAc·3 H_2O (1:1) on a steam bath, 99% yield (436, 948, 1101, 1102).

From $PhSbCl_2$ in EtOH by adding to a boiling soln. of NaOAc (2 moles) in 95% EtOH and heating the mixt. under reflux, 74% yield (1273).

From $(Ph_2As)_2O$ + AcOH by stirring at r.t., 97% yield (1505).

From Ph_2AsH in petr. ether + AcOH in petr. ether treated with H_2SO_4 (1141).

Prop.: White cryst. solid, m. 130-132° (536, 948, 1102, 1505), m. 128-131° (1101, 1273), m. 128-130° (1141). Half-wave potential (467, 471).

Rxn. with: $PrCO_2H$ → Ph_2SbO_2CPr (1099).

RSH → Ph_2SbSR (1099).

RNH_2 + CS_2 → $PhSb[SC(S)NHR]_2$, $PhCH_2NCS$, and $(Ph_2Sb)_2S$ (1505).

$NaSC(S)NR_2$ → $Ph_2SbSC(S)NR_2$ (905).

Electrolytic redn. → $Ph_2SbSbPh_2$ (467).

Polarographic redn. with $2e^-$ → Ph_2Sb^-, kinetics of the electron transfer to various acceptors was detd. (469).

Polarographic redn. ($2e^-$) in the presence of i-PrBr → Ph_2Sb^-, reactivity of the anion with i-PrBr was detd. (471).

Biol. Prop.: Bactericidal and fungicidal activity (536, 537, 948, 1101, 1102).

Use: A bactericide and fungicide for paints, plastics, and fibrous materials (537).

1.6.1.4 OXYBISSTIBINES, $(R_2Sb)_2O$

A few oxybisstibines were described in the main volume (M-699).

Oxybis(diphenylstibine), $Sb_2C_{24}H_{20}O$, $(Ph_2Sb)_2O$, (M-699)

Synth.: From Ph_3Sb by refluxing with MeOH contg. HCl and neutralizing with 20% KOH soln., 33% yield (1505).

Prop.: m. 78-80° (1505). Half-wave potential (467).

Rxn. with: RSH in C_6H_6 under reflux → Ph_2SbSR (537, 1097, 1102).

$[Ph_2CH_2NH_3]SC(S)NHCH_2Ph$ in $CHCl_3$ at 25° → $(Ph_2Sb)_2S$ (907, 1505).

RCO_2H → Ph_2SbO_2CR (1097, 1386, 1505).

Concd. HCl in EtOH → Ph_2AsCl (1505).

NH_3 in CS_2 → $(Ph_2Sb)_2S$ (905).

$PhNH_2$ + CS_2 in $CHCl_3$ → $(Ph_2Sb)_2S$ and $(PhNH)_2CO$ (1505).

$PhCH_2NH_2$ + CS_2 in $CHCl_3$ → $PhSb[SC(S)NHCH_2Ph]_2$, $(Ph_2Sb)_2S$, and $PhCH_2NCS$ (1505).

6-Mercaptopurine·H_2O in Me_2CO under reflux → 6-(diphenylstibino)mercapto-purine (906).

Table 47. Stibinous Acid Anhydrides with Carboxylic Acids

R_2SbO_2CR'	Prepd. from (Yield %)	Prop. and Rxn.	Ref.
C_{10}			
Bu_2SbOAc		Fungicidal and bactericidal activity, useful as an antifouling agent for paints	537
C_{14}			
$Ph_2SbO_2CCCl_3$	$(R_2Sb)_2O + Cl_3CCO_2H$	m. 126-127° (dec.)	1386
C_{16}			
$Ph_2SbO_2CC(=CH_2)Me$	$R_2SbCl + NaO_2CC(=CH_2)Me$ in MeOH		536
Ph_2SbO_2CPr	$R_2SbOAc + PrCO_2H$ (1:1) in MePH under reflux with a slow distn. (98)	Crystals, m. 68-71°, (a)	1099
	$R_2SbCl + NaO_2CPr$ in MeOH		536,948
C_{17}			
Ph_2SbO_2CBu	$R_2SbCl + NaO_2CBu$ in MeOH	(a)	536,948
C_{18}			
$Ph_2SbO_2C(CH_2)_4Me$	$R_2SbCl + NaO_2C(CH_2)_4Me$ in MeOH	(a)	536,948
C_{19}			
$Ph_2SbO_2CC_6H_4Cl-4$	$(R_2Sb)_2O + R'CO_2H$ in C_6H_6 under reflux with azeotropic distn.	m. 122-125°, (a)	1097
	$R_2SbCl + R'CO_2Na$ in aq. MeOH under reflux (~100)	m. 124-126°	536,948, 1101-2
	$RSbCl_2$ in i-PrOH + $R'CO_2Na$ (1:2) in i-PrOH under reflux		1273
Ph_2SbOBz (M-698)	$Ph_2SbH + BzOH$ in Et_2O at 30-35° (69)	m. 123-125°, soly. data	1141

(a) Fungicidal and bactericidal activity (536, 948, 1097, 1101, 1102).

Table 47 Continued

R_2SbO_2CR'	Prepd. from (Yield %)	Prop. and Rxm.	Ref.
C_{20} $Ph_2SbO_2CCH(Et)Bu$	$(R_2Sb)_2O + R'CO_2H$ in C_6H_6 under reflux with azeotropic distn.	m. 48-49°, (a)	1097
C_{23} $(n\text{-}C_8H_{17})_2SbO_2CC_6H_4Cl\text{-}4$	$R_2SbCl + R'CO_2Na$ in EtOH under reflux $RSbCl_2 + R'CO_2Na$ in C_6H_6 under reflux	(a)	1101 1273
C_{24} $(n\text{-}C_8H_{17})_2SbO_2CCH(Et)Bu$	$R_2SbCl + R'CO_2NH_4$ in C_6H_6 under reflux	(a)	536,948
C_{30} $Ph_2SbO_2C(CH_2)_7CH=CHCH_2\text{-}CH=CH(CH_2)_4Me$	$R_2SbCl + R'CO_2Na$ in appropriate solvent	(a)	536,948

1.6.1.6 STIBOSO COMPOUNDS

Numerous stiboso derivatives were described in the main volume on pages 699-703.

Oxophenylstibine, Stibosobenzene, SbC_6H_5O, PhSbO, (M-700)
Synth.: From $PhSbCl_2$ in EtOH by adding dropwise to aq. NaOH at ice temp., 93% yield (1505).
Prop.: m. 157-163° (1505).
Rxn. with: NH_3 in $CS_2 \rightarrow$ PhSbS (905).
$H_2NCSNH_2 \rightarrow$ PhSbS (1043).
$RNH_2 + CS_2 \rightarrow PhSb[SC(S)NHR]_2$ (1505).
$[PhCH_2NH_3]SC(S)NHCH_2Ph$ in $CHCl_3 \rightarrow$ PhSbS, $(PhCH_2NH)_2CS$, and $PhSb[SC(S)NHCH_2Ph]_2$ (1505).
Use: Complexes with $RAlX_2$ and Ti, Zr, V, Cr, or Mo halides catalyze the polymerization of α-olefins (384).

1.6.1.7 STIBONOUS ACID ESTERS

Stibonous acid esters have recently been indexed by the Chemical Abstracts as stibine derivatives, e.g., $BuSb(OEt)_2$ has been named butyldiethoxystibine. No compounds of this type were described in the main volume.

Butyldiethoxystibine, $SbC_8H_{19}O_2$, $BuSb(OEt)_2$
Synth.: From $BuSbBr_2$ + Na in EtOH at 60-70°, 80% yield (175).
Prop.: b_1 64-65°, d_4^{20} 1.4599 (175).
Rxn. with: $CH_2=C=O \rightarrow BuSb(CH_2CO_2Et)_2$ (175).

Dibutoxyphenylstibine, $SbC_{14}H_{23}O_2$, $PhSb(OBu)_2$
Synth.: From $PhSbCl_2$ in C_6H_6 by bubbling NH_3 and treating the $PhSbCl_2 \cdot NH_3$ adduct with BuOH under reflux (537).
Prop.: $b_{0.3}$ 134-147° (537).
Biol. Prop.: Biocidal activity (537).

1.6.2 COMPOUNDS WITH ANTIMONY-SULFUR BONDS

1.6.2.2 THIOSTIBINOUS ACID ESTERS WITH THIOLS

This new type of antimony compounds has been indexed by the Chemical Abstracts as stibine derivatives, e.g., Ph_2SbSBu has been named (butylthio)-diphenylstibine. These compounds are formed from secondary haloarsines by a reaction with thiol-sodium derivatives in benzene under reflux, from oxy-bis(secondary stibines) by a reaction with thiols in benzene, with azeotropic distillation of water, and from acetoxydiarylstibines and thiols in toluene under reflux with a slow distillation of acetic acid. These esters are compiled in Table 48.

Table 48. Thiostibinous Acid Esters with Thiols

R_2SbSR'	Prepd. from (Yield %)	Prop. and Rxn.	Ref.
C_{16}			
$(p\text{-}ClC_6H_4)_2SbSBu$	$R_2SbCl + R'SNa$ in C_6H_6 under reflux	Biocidal activity	537
$Ph_2SbSCH_2CO_2Et$	$(R_2Sb)_2O + R'SH$ in C_6H_6 under reflux with azeotropic distn. (98)	$b_{0.55}$ 210-222°, (a)	537, 1097, 1102
Ph_2SbSBu	$R_2SbCl + NaSR'$ in C_6H_6 under reflux	(a)	537
C_{17}			
Ph_2SbS [heterocyclic structure]	$(R_2Sb)_2O + R'SH \cdot H_2O$ in Me_2CO under reflux (95)	m. 214-216°, IR, far-IR, and UV spectra	906
$Ph_2SbSC_5H_{11}$	$(R_2Sb)_2O + R'SH$ in C_6H_6 under reflux with azeotropic distn. (94)	$b_{0.5}$ 218°, (a)	537, 1097, 1102
	$R_2SbOAc + R'SH$ in MePh under reflux with slow distn.	$b_{0.02}$ 164°	1099
C_{18}			
$Ph_2SbSC_6Cl_5$	$R_2SbOAc + R'SH$ in MePh under reflux with slow distn. (82)	m. 152-154°, (a)	1099
$(cyclo\text{-}C_6H_{11})_2SbSC_6H_{13}$	$R_2SbCl + n\text{-}R'SNa$ in C_6H_6 under reflux	(a)	537
C_{20}			
$Bu_2SbSC_{12}H_{25}$	$R_2SbCl + R'SNa$ in C_6H_6 under reflux	(a)	537, 1102
C_{21}			
$(PhCH_2)_2SbSCH_2Ph$	$R_2SbCl + R'SNa$ in C_6H_6 under reflux	(a)	537
C_{24}			
$Ph_2SbSC_{12}H_{25}$	$R_2SbCl + R'SNa$ in C_6H_6 under reflux	(a)	537

(a) Fungicidal and bactericidal activity (537, 1097, 1099, 1102).

1.6.2.3 THIOSTIBINOUS ACID ANHYDROSULFIDE WITH THIOACIDS

No compounds of this type were reported in the main volume.

Dibutyl[(diethylthiocarbamoyl)thio]stibine, $SbC_{13}H_{28}NS_2$, $Bu_2SbSC(S)NEt_2$
Synth.: From Bu_2SbCl + $[Et_2NH_2]SC(S)NEt_2$ (1:1) in THF under reflux (1100).

[(Dimethylthiocarbamoyl)thio]diphenylstibine, $SbC_{15}H_{16}NS_2$, $Ph_2SbSC(S)NMe_2$
Synth.: From Ph_2SbCl or Ph_2SbOAc + $NaSC(S)NMe_2$ in $CHCl_3$, 70 and 90% yield,
resp.(905).
From Ph_2SbCl + $[Me_2NH_2]SC(S)NMe_2$ in THF under reflux (1100).
Prop.: m. 115-117°, UV and IR spectra (905), m. 116-116.5° (1100).
Biol. Prop.: Bactericidal activity (1100).

[(Diethylthiocarbamoyl)thio]diphenylstibine, $SbC_{17}H_{20}NS_2$, $Ph_2SbSC(S)NEt_2$
Synth.: From Ph_2SbCl or Ph_2SbOAc + $NaSC(S)NEt_2$ in $CHCl_3$, 85 and 60% yield,
resp. (905).
Prop.: m. 68-69°, UV and IR spectra (905).

[(Diphenylthiocarbamoyl)thio]diphenylstibine, $SbC_{25}H_{20}NS_2$, $Ph_2SbSC(S)NPh_2$
Synth.: From Ph_2SbCl or Ph_2SbOAc + $NaSC(S)NPh_2$ in C_6H_6-MeCN, 70 and 87%
yield, resp. (905).
Prop.: m. 186-188°, UV and IR spectra (905).

Ethylenebis[imino(thiocarbonyl)thio]bis(diphenylstibine), $Sb_2C_{28}H_{26}N_2S_4$,
    $Ph_2SbSC(S)NHCH_2CH_2NHC(S)SSbPh_2$
Synth.: From Ph_2SbCl + $NaSC(S)NHCH_2CH_2NHC(S)SNa$ in THF under reflux (1100).
Biol. Prop.: Bactericidal activity (1100).

1.6.2.4 THIOBISSTIBINES

Thiobis(diphenylstibine), $Sb_2C_{24}H_{20}S$, $(Ph_2Sb)_2S$,   (M-704)
Synth.: From $(Ph_2Sb)_2O$ and from Ph_2SbCl dissolved in CS_2 by treating with
NH_3 gas, 61-62% yield (905).
From $(Ph_2Sb)_2O$ + $[PhCH_2NH_3]SC(S)NHCH_2Ph$ in $CHCl_3$ at 25°, 96% yield (907, 1505).
From Ph_2SbCl + $NH_4[SC(S)NHPh]$ (1:1) in $CHCl_3$, 62% yield along with $PhSb[SC-$
$(S)NHPh]_2$ and PhNCS (1505).
From Ph_2SbCl + $NaSC(S)NHCH_2Ph$ (1:2) in $CHCl_3$, 56% yield along with $(PhCH_2NH)_2CS$
$PhSb[SC(S)NHCH_2Ph]_2$, and $PhCH_2NCS$ (1505).
From $(Ph_2Sb)_2O$ with $PhNH_2$ (1:2) and CS_2 in $CHCl_3$, 68% yield along with
$(PhNH)_2CS$ (1505).
From $(Ph_2Sb)_2O$ with $PhCH_2NH_2$ (1:2) and excess CS_2 in $CHCl_3$, 44% yield along
with $PhSb[SC(S)NHCH_2Ph]_2$ and $PhCH_2NCS$ (1505).
From Ph_2SbOAc in CS_2 by adding dropwise $PhCH_2NH_2$ in $CHCl_3$, 80% yield along
with $PhSb[SC(S)NHCH_2Ph]_2$ and $PhCH_2NCS$ (1505).
Prop.: Crystals, m. 63-65° (1505).

1.6.2.6 THIOSTIBOSO COMPOUNDS

No compounds of this class were reported in the main volume.

Phenylthioxostibine, Thiostibosobenzene, SbC_6H_5S,   PhSbS
<u>Synth.</u>: From PhSbO in CS_2 by treating with NH_3, 51% yield (905).
From PhSbO by treating with H_2NCSNH_2 (1043).
From PhSbO by a rxn. with $[PhCH_2NH_3]SC(S)NHCH_2Ph$ in $CHCl_3$ at 25°, 36% yield
(907, 1505).
From $PhSb[SC(S)NHCH_2Ph]_2$ or $PhSb[SC(S)NHPh]_2$ in MeCN under reflux, 75 and
70% yield, resp. (907, 1505).
<u>Prop.</u>: Crystals, m. 58-60° (1505), m. 60-63° (1505).

1.6.2.7 DITHIOSTIBONOUS ACID ESTERS WITH THIOLS

Several dithioesters were reported in the main volume on pages 704-705.

Bis(dodecylthio)phenylstibine, $SbC_{30}H_{55}S_2$, $PhSb(SC_{12}H_{25})_2$
<u>Biol. Prop.</u>: Bactericidal and fungicidal activity (537).
<u>Use</u>: A bactericide for paints (537).

1.6.2.8 DITHIOSTIBONOUS ACID ANHYDROSULFIDES WITH THIOACIDS

A few anhydrosulfides of dithiostibonous acids were described in the main
volume on page 705.

Bis[(dimethylthiocarbamoyl)thio]phenylstibine, $SbC_{12}H_{17}N_2S_4$,
    $PhSb[SC(S)NMe_2]_2$
<u>Synth.</u>: From $PhSbCl_2$ + $NaSC(S)NMe_2$ (1:2) in $CHCl_3$ at 25°, 95% yield (907, 1505).
<u>Prop.</u>: m. 237-239°, UV and IR spectra (907, 1505).

Phenylbis[(propylthiocarbamoyl)thio]stibine, $SbC_{14}H_{21}N_2S_4$,
    $PhSb[SC(S)NHPr]_2$
<u>Synth.</u>: From $PhSbCl_2$ + $[NH_4]SC(S)NHPr$ (1:2) in THF under reflux (1100).
<u>Biol. Prop.</u>: Bactericidal activity (1100).

Bis[(diethylthiocarbamoyl)thio]phenylstibine, $SbC_{16}H_{25}N_2S_4$,
    $PhSb[SC(S)NEt_2]_2$
<u>Synth.</u>: $PhSbCl_2$ + $NaSC(S)NEt_2$ (1:2) in $CHCl_3$ at 25°, 97% yield (907, 1505).
<u>Prop.</u>: m. 148-150°, UV and IR spectra (907, 1505).

Phenylbis[(phenylthiocarbamoyl)thio]stibine, $SbC_{20}H_{17}N_2S_4$,
    $PhSb[SC(S)NHPh]_2$
<u>Synth.</u>: From $PhSbCl_2$ + $[NH_4]SC(S)NHPh$ (1:2) in $CHCl_3$ at 25°, 57% yield (907, 1505).
By-prod. from the rxn. of Ph_2SbCl with $[NH_4]SC(S)NHPh$ (1:1) in $CHCl_3$ (1505).

Prop.: m. 142-144° (907), m. 145-146° (recrystd. from C$_6$H$_6$-petr. ether) (1505). UV and IR spectra. Dec. during heating in MeCN under reflux → PhSbS and (PhNH)$_2$CS (907, 1505).

Bis[(benzylthiocarbamoyl)thio]phenylstibine,   SbC$_{22}$H$_{21}$N$_2$S$_4$,
    PhSb[SC(S)NHCH$_2$Ph]$_2$
Synth.: From PhSbCl$_2$ and PhCH$_2$NHC(S)SNa (1:2) in CHCl$_3$ at 25°, 83% yield (907, 1505).
From PhSbO with PhCH$_2$NH$_2$ (1:2) and excess CS$_2$ in CHCl$_3$, 86% yield (1505).
From (Ph$_2$Sb)$_2$O with PhCH$_2$NH$_2$ (1:2) and excess CS$_2$ in CHCl$_3$, 30% yield along with (Ph$_2$Sb)$_2$O and PhCH$_2$NCS (1505).
By-prod. from the rxn. of Ph$_2$SbOAc and CS$_2$ in CHCl$_3$ by dropwise addn. of PhCH$_2$NH$_2$ dissolved in CHCl$_3$, 11% yield along with (Ph$_2$Sb)$_2$S and PhCH$_2$NCS (1505).
By-prod. from the rxn. of Ph$_2$SbCl with NaSC(S)NHPh (1:2) in CHCl$_3$, 8% yield along with (Ph$_2$Sb)$_2$S and (PhCH$_2$NH)$_2$CO (1505).
Prop.: White solid, m. 135-136°, UV and IR spectra (907, 1505). Dec. when heated in MeCN under reflux → PhSbS and (PhCH$_2$NH)$_2$CS (907, 1505).

Bis[(diphenylthiocarbamoyl)thio]phenylstibine,   SbC$_{32}$H$_{25}$N$_2$S$_4$,
    PhSb[SC(S)NPh$_2$]$_2$
Synth.: From PhSbCl$_2$ + NaSC(S)NPh$_2$ (1:2) in CHCl$_3$ at 25°, 80% yield (907, 1505).
Prop.: m. 241-244°, UV and IR spectra (907, 1505).

1.7   COMPOUNDS OF ANTIMONY BONDED TO GROUP VA ELEMENTS

1.7.1   COMPOUNDS WITH ANTIMONY-NITROGEN BONDS

1.7.1.1   SECONDARY STIBINOAMINES, R$_2$SbNR$'_2$

No secondary stibinoamines were reported in the main volume.

(Dimethylstibino)methyl(trimethylsilyl)amine,   SbSiC$_6$H$_{18}$N,   Me$_2$SbN(Me)SiMe$_3$
Synth.: From Me$_2$SbCl by adding to LiN(Me)SiMe$_3$ in C$_6$H$_{14}$ cooled by ice and completing the rxn. at 20°, 72% yield (1343).
Prop.: Colorless liquid, b$_1$ 44-46°, NMR spectrum (1343).

2-[(Dimethylstibino)methylamino]-1,3-dimethyl-1,3,2-diazaphospholidine,
    SbC$_7$H$_{19}$N$_3$P,
Synth.: From 2-(methylamino-1,3-dimethyl-1,3,2-
diazaphospholidine in C$_6$H$_{14}$ by adding BuLi, followed
by Me$_2$SbCl (1:1:1), 60% yield (1346).
Prop.: b$_{0.01}$ 88-90°, NMR spectrum (1346).

[(Diphenylstibino)imino]triphenylphosphorane, $SbC_{30}H_{25}NP$, $Ph_3P=NSbPh_2$
Synth.: From $Ph_2SbN_3 + Ph_3P$ at 60-135° (1272).
Prop.: Colorless crystals, m. 123.5-124.0°, IR spectrum (1272).

1.7.1.6 STIBYLENEDIAMINES

No stibylenediamines, $RSb(NR'_2)_2$, were reported in the main volume.

Methylstibylenebis[methyl(trimethylsilyl)amine], $SbSi_2C_9H_{27}N_2$,
    $MeSb[N(Me)SiMe_3]_2$
Synth.: From $MeSbCl_2$ in C_6H_{14} by adding dropwise to $LiN(Me)SiMe_3$ in C_6H_{14}
at -20°, 54% yield (1343).
Prop.: Colorless liquid, $b_{0.1}$ 59-61°, NMR spectrum (1343).

p-Phenylenebis[stibylenebis(dimethylamine)], $Sb_2C_{14}H_{28}N_4$,
    $1,4-C_6H_4[Sb(NMe_2)_2]_2$
Rxn. with: $1,4-C_6H_4[P(Me)(O)N_3]_2$ (1:2) → $\{NP(Me)(O)C_6H_4P(Me)(O)N=Sb(NMe_2)_2-$
$C_6H_4Sb(NMe_2)_2\}_n$ (1583).

1.8   COMPOUNDS OF ANTIMONY BONDED TO METALS

1.8.1   COMPOUNDS WITH THE ELEMENTS OF GROUPS IA-IVA

1.8.1.1 COMPOUNDS WITH GROUP IA ELEMENTS

No secondary stibino-alkali metal derivatives were described in the main
volume.

Diethylstibinolithium, $SbLiC_4H_{10}$, Et_2SbLi
Synth.: From $Et_2SbSbEt$ in Et_2O-dioxan + PhLi in Et_2O at -45°, obtd. as a
complex with 1 mol. of dioxan, 97% yield (783).
From Et_2SbH in Et_2O-dioxan + PhLi in Et_2O, a mixt. of $Et_2SbLi\cdot$Dioxan with
$Et_2SbSbEt_2$ was obtd. (783).
Prop.: The complex with 1 mol. of dioxan forms a yellow amorphous, pyrophorous
ppt., sol. in THF and dioxan, sparingly sol. in Et_2O and petr. ether, dec.
in H_2O (783).
Rxn. with: $Cl(CH_2)_nCl$ (2:1) → $Et_2Sb(CH_2)_nSbEt_2$ (783).
$Cl(CH_2)_nCl$ (1:1) → $Et_2Sb(CH_2)_nCl$ (783).
RCl → Et_2SbR (783).
CH_2Br_2 → $Et_2SbSbEt_2$ (783).

Dibutylstibinolithium, $SbLiC_8H_{18}$, Bu_2SbLi
Synth.: From Bu_2SbCl by treating with Li (1:2) in THF (1444).
Rxn. with: RBr in THF → Bu_2SbR (1444).

Di-tert-butylstibinolithium, $SbLiC_8H_{18}$, $t-Bu_2SbLi$
Synth.: From $t-Bu_2SbH$ + PhLi in Et_2O-dioxan at -25°, 80% yield isolated as a

complex with 1 mol. dioxan (785).

From t-Bu$_2$SbSb-t-Bu$_2$ + PhLi in Et$_2$O-dioxan, isolated as a complex with 1 mol. dioxan, 80% yield along with t-Bu$_2$SbPh (785).

Prop.: The complex with 1 mol. dioxan forms orange-colored, amorphous, pyrophorous powder, sol. in THF and Et$_2$O, dec. in H$_2$O (785).

Rxn. with: Cl(CH$_2$)$_4$Cl in THF → t-Bu$_2$Sb(CH$_2$)$_4$Sb-t-Bu$_2$ and t-Bu$_2$Sb(CH$_2$)$_4$Cl (785).

t-Bu$_2$SbCl → t-Bu$_2$SbSb-t-Bu$_2$ (785).

I in C$_6$H$_6$ under reflux → (t-BuSb)$_4$ (785).

Diphenylstibinolithium, SbLiC$_{12}$H$_{10}$, Ph$_2$SbLi

Synth.: From Ph$_2$SbH + PhLi in Et$_2$O at -30°, followed by addn. of dioxan, 93% yield isolated as a complex with 1 mol. dioxan (784).

From Ph$_2$SbSbPh$_2$ + PhLi in Et$_2$O-dioxan at -35°, 98% yield isolated as a complex with 1 mol. dioxan (784).

Prop.: The complex with dioxan forms a yellow, pyrophorous ppt., sol. in THF, dioxan, and C$_6$H$_6$, sparingly sol. in petr. ether, dec. in H$_2$O and in EtOH (784).

Rxn. with: Cl(CH$_2$)$_n$Cl → Ph$_2$Sb(CH$_2$)$_n$SbPh$_2$ and Ph$_2$Sb(CH$_2$)$_n$Cl (784).

Dicyclohexylstibinolithium, SbLiC$_{12}$H$_{22}$, (C$_6$H$_{11}$)$_2$SbLi

Synth.: From (C$_6$H$_{11}$)$_2$SbH + PhLi in Et$_2$O at -35° (782).

Prop.: Yellow, pyrophorous crystals, sol. in THF, dioxan, and C$_6$H$_6$, dec. in H$_2$O (782).

Rxn. with: Cl(CH$_2$)$_n$Cl (2:1) → (C$_6$H$_{11}$)$_2$Sb(CH$_2$)$_n$Sb(C$_6$H$_{11}$)$_2$ (782).

Cl(CH$_2$)$_n$Cl (1:1) → (C$_6$H$_{11}$)$_2$Sb(CH$_2$)$_n$Cl (782).

BrCH$_2$CH$_2$Cl → [(C$_6$H$_{11}$)$_2$Sb-]$_2$ (782).

CH$_2$Cl$_2$ or CH$_2$Br$_2$ → [(C$_6$H$_{11}$)$_2$Sb-]$_2$ (782).

Coordination Deriv.: (C$_6$H$_{11}$)$_2$SbLi·1.5 Dioxan, voluminous yellow ppt., prepd. from (C$_6$H$_{11}$)$_2$SbLi in Et$_2$O by adding dioxan (782).

Diphenylstibinopotassium, SbKC$_{12}$H$_{10}$, Ph$_2$SbK

Synth.: From Ph$_2$SbH + K in dioxan at -22° (784).

Diphenylstibinosodium, SbNaC$_{12}$H$_{10}$, Ph$_2$SbNa

Synth.: From Ph$_2$SbH + Na in dioxan at -22°, 96% yield obtd. as an intensively red soln. (784).

Rxn. with: Cl(CH$_2$)$_n$Cl in Et$_2$O → Ph$_2$Sb(CH$_2$)$_n$SbPh$_2$ (784).

1.8.1.2 COMPOUNDS WITH GROUP IIA ELEMENTS

Bis(dibutylstibino)magnesium, Sb$_2$MgC$_{16}$H$_{36}$, (Bu$_2$Sb)$_2$Mg

Synth.: From Bu$_2$SbCl by treating with Mg (1:1); the compd. was not isolated, but the rxn. mixt. was used in further conversions (1444).

Rxn. with: RBr in THF → Bu$_2$SbR (1444).

R$_3$SiCl in THF → Bu$_2$SbSiR$_3$ (1444).

1.8.1.3 COMPOUNDS WITH GROUP IIIA ELEMENTS

Polymers containing phenylstibylene moieties linked to two decaborane(12) units were prepared by a condensation of oxophenylstibine, PhSbO, with decaborane, $(H_3O)_2B_{10}H_{10}$ (499).

1.8.1.4 COMPOUNDS WITH GROUP IVA ELEMENTS

Several compounds of this group were described in the main volume (M-706).

Dibutyl(trimethylsilyl)stibine, $SbSiC_{11}H_{27}$, $Bu_2SbSiMe_3$, (M-706)
Synth.: From $(Bu_2Sb)_2Mg$ + Me_3SiCl in THF at r.t., 26% yield (1444).
Prop.: $b_{0.03}$ 51° (1444).

Benzylbis(triethylgermyl)stibine, $SbGe_2C_{19}H_{37}$, $PhCH_2Sb(GeEt_3)_2$
Synth.: From $(Et_3Ge)_3Sb$ + $PhCH_2Br$ (1:1) at 100°, 59% yield (1576, 1577).
Prop.: $b_{1.5}$ 155-159°, n_D^{20} 1.5979, d_4^{20} 1.379; partial symmetrization of the compd. occurred during distn. (1576, 1577).

Diphenyl(triphenylgermyl)stibine, $SbGeC_{30}H_{25}$, $Ph_3GeSbPh_2$
Synth.: From Ph_3Sb + Na (1:1) in liq. NH_3, followed by NH_4Br and Ph_3GeCl, removing the solvent, dissolving the prod. in C_6H_6 and treating it, after distg. C_6H_6, with C_5H_{12}, 12% yield (1374).
Prop.: m. 120°, sensitive to oxygen (1374).

Dibutyl(trimethylstannyl)stibine, $SbSnC_{11}H_{27}$, $Me_3SnSbBu_2$
Synth.: By-prod. from the rxn. of $Bu_2SbC\equiv CH$ with Me_3SnH in a sealed tube at 80° (1142).
Prop.: $b_{0.5}$ 126-130°, n_D^{20} 1.5500, unstable compd., can be stored under an inert gas in the dark at a low temp. (1142).

Diphenyl(triphenylstannyl)stibine, $SbSnC_{30}H_{25}$, $Ph_3SnSbPh_2$
Synth.: From Ph_3Sb by treating with Na in liq. NH_3, followed by NH_4Br and Ph_3SnCl, removing the solvent, dissolving the residue in C_6H_6, concg. to dryness, and treating the residue with C_5H_{12}, 34% yield (1372).
From Ph_2SbCl + $LiSnPh_3$ in THF, removing the $Ph_3SnSnPh_3$ by-prod. and treating the residue as in the preceding method, 64% yield (1372).
Prop.: Colorless crystals, m. 116°, soly. data, sensitive to O, stable in H_2O free of O (1372).
Rxn. with: O or H_2O_2 → cleavage of the Sn-Sb and Sb-C bonds (1372).

Diphenylstannylenebis(diphenylstibine), $Sb_2SnC_{36}H_{30}$, $Ph_2Sn(SbPh_2)_2$
Synth.: From Ph_3Sb by treating with Na in liq. NH_3 followed by NH_4Br and Ph_2SnCl_2, removing the solvent, dissolving the residue in C_6H_6, concg. to dryness, and treating the oily residue with C_5H_{12}, 29% yield (1372).
Prop.: Colorless crystals, m. 150°, sol. in aromatic hydrocarbons and THF, sensitive to O, stable in H_2O free of O (1372).
Rxn. with: O or H_2O_2 → cleavage of the Sb-Sn and Sb-C bonds (1372).

Phenylbis(triphenylstannyl)stibine, $SbSn_2C_{42}H_{35}$, $PhSb(SnPh_3)_2$
Synth.: From $PhSbCl_2$ + $LiSnPh_3$ in THF, removing the solvent, dissolving the
prod. in C_6H_6, concg. to dryness, and treating the residue with C_5H_{12}, 40%
yield (1372).
Prop.: Colorless crystals, m. 120° (dec.), sol. in aromatic hydrocarbons and
THF, relatively stable to O (1372).

Phenylstannylidynetris(diphenylstibine), $Sb_3SnC_{42}H_{35}$, $PhSn(SbPh_2)_3$
Synth.: From Ph_3Sb by treating with Na in liq. NH_3, followed by NH_4Br and
$PhSnCl_3$, removing the solvent, dissolving the residue in C_6H_6, concg. to
dryness, and treating the residue with C_5H_{12}, 35% yield (1372).
Prop.: Colorless crystals, m. 90°, sol. in aromatic hydrocarbons and THF,
sensitive to O, stable in H_2O free of O (1372).
Rxn. with: O or H_2O_2 → cleavage of the Sb-Sn and Sb-C bonds (1372).

Stannatetrayltetrakis(diphenylstibine), $Sb_4SnC_{48}H_{40}$, $(Ph_2Sb)_2Sn$
Synth.: From Ph_3Sb + Na in liq. NH_3, followed by NH_4Br and $SnCl_4$, removing
the solvent, dissolving the prod. with C_6H_6, concg. to dryness, and treating
the residue with C_5H_{12}, 68% yield (1372).
Prop.: Colorless crystals, m. 75°, sol. in aromatic hydrocarbons and THF,
sensitive to O, stable in H_2O free of O (1372).
Rxn. with: O or H_2O_2 → cleavage of the Sb-Sn and Sb-C bonds (1372).

Diphenyl(triphenylplumbyl)stibine, $SbPbC_{30}H_{25}$, $Ph_3PbSbPh_2$
Synth.: From Ph_3Sb + Na in liq. NH_3, followed by NH_4Br and Ph_3PbCl, removing
the solvent, dissolving the prod. in C_6H_6, concg. to dryness, and treating
the residue with C_5H_{12}, 11% yield (1374).
Prop.: Dec. 115°, sensitive to O (1374).

1.8.2 COMPOUNDS WITH TRANSITION METALS

1.8.2.8 COMPOUNDS WITH GROUP VIII METALS

Bis(dimethylglyoximato)(diphenylstibino)(pyridine)cobalt, $SbCoC_{25}H_{29}N_5O_4$,
    $Ph_2SbCo[ON=C(Me)C(Me)=NOH]_2 \cdot Py$
Synth.: From Ph_2SbX by a rxn. with $HCo[ON=C(Me)C(Me)=NOH]_2 \cdot Py$ (1369).
Prop.: Brown crystals, dec. 185-195°, sol. in org. solvents, stable in O and
in alkali but unstable in strong acids (1369).

2. COMPOUNDS OF PENTAVALENT ANTIMONY

2.1 QUINQUENARY DERIVATIVES

Several quinquenary antimony compounds were described in the main volume, pp. 708-11.

(Diiodomethyl)(iodomethyl)trimethylantimony, $SbC_5H_{12}I_3$, $Me_3(ICH_2)(I_2CH)Sb$
Synth.: From Me_3Sb + CHI_3 in Me_2CO under reflux (1030).
Prop.: Yellow needle-shaped crystals, m. ~ 90° (dec.), IR and NMR spectra (1030).
Rxn. with: Aq. HBr under reflux → $(Me_3Sb)_2CH_2 \cdot Br_2 \cdot 2\ H_2O$ (1030).

Pentamethylantimony, SbC_5H_{15}, Me_5Sb, (M-708)
Synth.: From $Me_4SbOSiMe_3$ + $[Me_3SiOAlMe_2]_2$ (2:1) (1351).
Prop.: Liquid with a faint yellow tinge, m. -19°, vapor pressure at 25° ~ 8 mm, stable at r.t. in a clean glass container; UV, visible, IR, and Raman spectra. The structure of the mol. is considered (494).

Pentaethylantimony, $SbC_{10}H_{25}$, Et_5Sb
Synth.: From Et_3SbBr_2 by the rxn. with an ethylmetal deriv., 80% yield (96), with Et_2Mg in Et_2O at r.t. and completed under reflux, 80% yield (1476), or with Et_2Zn in $n-C_6H_{14}$ at 70°, 36% yield (1476).
Prop.: $b_{0.14}$ 50.5° (96), $b_{0.19}$ 55.8°, IR spectrum (1476).
Rxn. with: MeOH → Et_4SbOMe + C_2H_6 (1476).
Et_3Al (1:0.59) in $n-C_6H_{14}$ at 10-40° → white crystals: $[Et_4Sb][AlEt_4]$ or $Et_5Sb \cdot AlEt_3$ (1476, 1478).
Et_3Al (1:2.99) in $n-C_6H_{14}$ at 10-40° → colorless liquid: $[Et_4Sb][Al_2Et_7]$ or $Et_5Sb \cdot 2\ AlEt_3$ (1476, 1478).
Et_2AlCl (1:0.66) in $n-C_6H_{14}$ at -20 to +30° → pale yellow liquid: $[Et_4Sb][Et_3AlCl]$ (1476).
Et_2AlCl (1:3) in $n-C_6H_{14}$ at -20 to +30° → colorless liquid: $[Et_4Sb][Et_5Al_2Cl_2]$ and $[Et_4Sb][Et_4Al_2Cl_3]$ (1476).
$EtAlCl_2$ (1:0.59) in $n-C_6H_{14}$ at -20 to +30° → pale yellow liquid: $[Et_4Sb]-[Et_2AlCl_2]$ (1476).
$EtAlCl_2$ (1:2.99) in $n-C_6H_{14}$ at -20 to +30° → white crystals: $[Et_4Sb]AlCl_4$ (1476).
$AlCl_3$ (1:0.67) in C_6H_6 at 20-60° → pale yellow soft crystals: $[Et_4Sb][EtAlCl_3]$ (1476).
Complex Compd.: $[Et_5Sb \cdot AlEt_3]$, crystals insol. in C_6H_6, prepd. from Et_5Sb + Et_3Al in C_6H_{14} (96, see also 1476, 1478).
Use: Black crystals formed from Et_5Sb + $TiCl_3$ + Et_3Al catalyze a stereospecific polymerization of C_3H_6 (96). A complex formed from $[MeEt_6SbAlBr]$ + Et_3Al + $TiCl_3$ in $n-C_6H_{16}$ catalyzes a stereospecific (isotactic) polymerization of C_3H_6 (1486).

Pentaphenylantimony, SbC$_{30}$H$_{25}$, Ph$_5$Sb, (M-710)

Synth.: From Ph$_3$SbCl$_2$ in (MeOCH$_2$-)$_2$ at 10° by adding PhMgBr in Et$_2$O, 62% yield (1643).

A tritium-labeled Ph$_5$Sb is formed when Ph$_3$Sb is treated with [4-^{3}H]PhLi (1:6) in THF and the Li[SbPh$_6$] intermediate is hydrolyzed with H$_2$O (438).

Prop.: m. 157-159° (1643), IR spectrum (1001). Photolysis → Ph$_3$Sb, Ph$_2$, and quaterphenyl; pyrolysis → Ph$_3$Sb and Ph$_2$ only (1397).

Rxn. with: MeOH under reflux → Ph$_4$SbOMe (227, 228).

BuLi in THF → exchange of the Ph groups for Bu groups (438).

[4-^{3}H]PhLi in THF → Li[SbPh$_6$] $\xrightarrow{\text{H}_2\text{O}}$ tritium-labeled Ph$_5$Sb (438).

MeI at 100° in a sealed tube → Ph$_3$Sb, [Ph$_4$Sb]I, C$_6$H$_6$, MePh, PhI, and Ph$_2$ (1269).

Use: A chain-transfer agent in a free-radical polymerization of vinyl monomers (1218).

2.2 QUATERNARY DERIVATIVES

2.2.1 ANTIMONY YLIDS

The following antimony-ylid is a new compound.

Triphenylstibonium Tetraphenylcyclopentadienylide, SbC$_{47}$H$_{35}$,

Synth.: From Ph$_3$Sb and diazotetraphenylcyclopentadiene by heating at 140° under N, 87% yield (975).

Prop.: Ochre crystals, m. 196-198°, stable in air, UV spectrum is quite

similar to those of other ylids. Dec. upon heating in EtOH or MeNO$_2$ soln. → tetraphenylcyclopentadiene (975).

Rxn. with: HClO$_4$ → [Ph$_3$SbCHC(Ph)=C(Ph)C(Ph)=CPh]ClO$_4$ (975).

RCHO → no fulvene deriv. is formed (975).

2.2.2 QUATERNARY STIBONIUM BASES AND SALTS

The methods of synthesis of quaternary stibonium bases and salts were summarized in the main volume on page 711. The compounds were described individually on pages 712-713 and in Table 87, pages 714-21, respectively. In the period covered by this supplement, many new stibonium salts were prepared by a few methods different from those reported previously. Thus, stibonium nitrates were obtained from stibonium halides by metathesis with silver nitrate. Likewise, stibonium fluorides were prepared by metathesis of stibonium chlorides with hydrogen fluoride in the presence of alkali metal alkoxides. Quaternary stibonium iodides can be converted to the corresponding chlorides by a reaction with cuprous chloride, followed by a treatment with hydrogen sulfide. Moreover, stibonium hydroborates are formed from stibonium halides by a treatment with ammonium hydroborates. Pentaarylantimony derivatives react with alkyl iodides to form tetraarylstibonium iodides. A quaternization of tertiary stibines with alkyl iodides in the presence of mercury iodide affords the corresponding stibonium triiodomercurates.

Numerous quaternary stibonium aluminates and dialuminates have been prepared according to the following reaction equations: (a) $R_4SbX + R_{3-n}AlX_n \rightarrow [R_4Sb][R_{3-n}AlX_{n+1}]$; (b) $R_3SbX_2 + R_{3-n}AlX_n \rightarrow [R_4Sb][R_{3-n}AlX_{n+1}]$; (c) $R_3Sb + R_3Al + RX \rightarrow [R_4Sb]R_3AlX$; (d) $R_5Sb + R_{3-n}AlX_n \rightarrow [R_4Sb][R_{4-n}AlX_n]$; (e) $SbX_3 + R_{3-n}AlX_n$ excess $\rightarrow [R_4Sb]R_xAlX_y$; (f) $[R_4Sb][R'_{4-n}AlX_n] + R'_{3-n}AlX_n \rightarrow [R_4Sb]-[R_{7-n}Al_2X_n]$. Likewise, stibonium halides add titanium halides to form quaternary stibonium halotitanates and halodititanates, respectively.

Besides the following three stibonium salts described individually, numerous stibonium salts and bases are compiled in Table 49, page 486.

Tetraethylstibonium Chloride, $SbC_8H_{20}Cl$, $[Et_4Sb]Cl$
Synth.: From $[Et_4Sb]I$ by the rxn. with Cu_2Cl_2 in EtOH under reflux, followed by a treatment with H_2S (1329).
Prop.: Polarographic redn. showed 2 waves (1488). IR spectrum (1477).
Rxn. with: EtLi in $n-C_6H_{14} \rightarrow Et_5Sb$ (1476).
$TiCl_4 \rightarrow [Et_4Sb]_2TiCl_6$, $[Et_4Sb]TiCl_5$, and $[Et_4Sb]Ti_2Cl_9$ (1477).
$TiCl_3 \rightarrow [Et_4Sb]_2TiCl_5$ and $[Et_4Sb]TiCl_4$ (1477).
$TiCl_2$ at 80° $\rightarrow [Et_4Sb]_2TiCl_5$ and $[Et_4Sb]TiCl_4$ (1477).

Tetraethylstibonium Trichloroethylaluminate, $SbAlC_{10}H_{25}Cl_3$,
     $[Et_4Sb][EtAlCl_3]$
Synth.: From $[Et_4Sb]Cl + EtAlCl_2$ (1:1) in $n-C_6H_{14}$ at 60° (96, 1485, 1488).
From $Et_3SbCl_2 + Et_2AlCl$ (1:1 or 1:2) at r.t. (1488).
From $Et_5Sb + AlCl_3$ (1:0.67) in C_6H_6 at 20-60° (1476).
Prop.: Crystals, m. $\sim$ 35° (96, 1488, 1489), m. 34-36° (1485, 1487). Pale yellow, soft crystals (1476). Elec. cond. data (1488).
Rxn. with: $EtAlCl_2 \rightarrow [Et_4Sb]AlCl_4$ (1488).
$Et_3Al \rightarrow [Et_4Sb][(Et_2AlCl)_{1.33}(Et_3Al)_{0.56}Cl]$ (1488).
$Et_2AlCl \rightarrow$ no rxn. (1488).
Uses: A mixt. with $TiCl_4$ in xylene catalyzes the polymerization of ethylene to a cryst. prod. (1487). A mixt. with $Et_3Al + TiCl_3$ in $n-C_6H_{14}$ catalyzes a stereospecific polymerization of C_3H_6 (1489). A complex with $\delta-TiCl_3$ (2:1) in $n-C_6H_{14}$ catalyzes the polymerization of C_3H_6 to an isotactic polypropylene (1485).

Tetraethylstibonium Dichlorodiethylaluminate, $SbAlC_{12}H_{30}Cl_2$,
     $[Et_4Sb][Et_2AlCl_2]$
Synth.: From $Et_5Sb + EtAlCl_2$ (1:0.59) in $n-C_6H_{14}$ at -20 to +30° (1476).
From $[Et_4Sb]Cl + Et_2AlCl$ (1:3) in $n-C_6H_{14}$ at 60°, a mixt. of the title compd. with $[Et_4Sb][Et_4Al_2Cl_3]$ was obtd. (1488).
From $[Et_4Sb]Cl + Et_2AlCl$ (1:1) in $n-C_6H_{14}$ at 60° (1488, 1489).
From $Et_3SbCl_2 + Et_3Al$ (1:1) at r.t. (1488).
Prop.: Colorless liquid (1488, 1489). Pale yellow liquid (1476).
Rxn. with: $EtAlCl_2 \rightarrow [Et_4Sb]AlCl_4$ (1488).

$Et_3Al \rightarrow [Et_4Sb][Et_5Al_2Cl_2]$ (1488).

Uses: A mixt. with Et_3Al or Bu_3Al and $TiCl_3$ in $n-C_6H_{14}$ catalyzes a stereo-specific polymerization of C_3H_6, styrene, and other α-olefins (1483, 1489). A mixt. with $TiCl_4$ in xylene catalyzes the polymerization of C_2H_4 to a crystalline polyethylene (1487).

2.2.4 QUATERNARY ALKOXY ANTIMONY DERIVATIVES

Quaternary alkoxyantimony compounds are formed from quaternary stibonium salts by the reaction with alcohols in the presence of bases, from quaternary stibonium bases by the reaction with alcohols, and from quinquenary antimony compounds and alcohols under reflux.

Covalent bonding of the alkoxy groups to the metal atom has been postulated.

Two compounds are described individually hereafter, while other compounds appear in Table 50 on page 495.

Tetramethyl(trimethylsiloxy)antimony, $SbSiC_7H_{21}O$, $Me_4SbOSiMe_3$,   (M-712)
In the main volume (M-712) this compd. was erroneously formulated as a stibonium salt.
Rxn. with: $[Me_3SiOAlMe_2]_2$ (2:1) $\rightarrow$ Me_5Sb and $[(Me_3SiO)_2AlMe]_2$ (1351).
$[Me_3SiOInMe_2]_2$ (2:1) $\rightleftharpoons$ $[Me_4Sb]^+[Me_2InOSiMe_3]_2^-$ (1351).

Methoxytetra-p-tolylantimony, $SbC_{29}H_{31}O$, $(4-MeC_6H_4)_4SbOMe$
Synth.: From $(4-MeC_6H_4)_3Sb$ + Br in $CHCl_3$, followed by $4-MeC_6H_4MgBr$ in C_6H_6, and conversion of the resulting $[(4-MeC_6H_4)_4Sb]Br$ with NaOMe in MeOH (1397, 1398).
Prop.: White, hydroscopic solid, m. 64-65° (1397, 1398). NMR spectrum indicated that the compd. in soln. exists as an equilibrium of a trigonal bipyramide and square pyramide (1397, 1398).

2.3   TERTIARY DERIVATIVES

2.3.2 TRIALKYLSTIBONIUM SALTS

Trimethylstibonium Chloride, $SbC_3H_{10}Cl$, Me_3SbHCl
Use: The rxn. prods. formed with the halides or oxyhalides of Ti, V, or Cr deposited on SiO_2 or Al_2O_3 catalyze the polymerization of olefins (42).

Table 49. Quaternary Stibonium Bases and Salts

R_4Sb^+	X^-	Prepd. from (Yield %)	Prop. and Rxn.	Ref.
C₄				
Me_4Sb	Cl (M-714)		IR spectrum	417
	OH		An inhibitor of the ether formation in the esterification of terephthalic acid with ethylene glycol	1402
	I (M-712)		NMR spectrum, coupling consts. and solvent effect, (a)	1027
	HgI_3 (M-714)		IR spectrum	417
	NO_3	$[Me_4Sb]I + AgNO_3$ in moist Me_2CO by heating	Colorless crystals, m. > 260°, insol. in org. solvents; IR spectrum	1404
	Et_3AlI	$[Me_4Sb]I + Et_3Al$	Colorless liquid, sparingly sol. in $n-C_6H_{14}$ and C_6H_6, (b)	1489
C₇				
$CH_2(SbMe_3)_2$	$Br_2 \cdot 2 H_2O$	$Me_3Sb(CH_2)(CH_2I) + HBr$ in H_2O under reflux	Colorless crystals, m. 199° (dec.), NMR spectrum	1030
	$[NO_3]_2$	$[CH_2(SbMe_3)_2]Br_2 \cdot 2 H_2O$ + aq. $AgNO_3$	Colorless crystals, m. 223°, NMR spectrum	1030
C₈				
$Et_2Me(CH_2=CHCH_2)Sb$	I	$Et_2SbCH_2CH=CH_2 + MeI$ in MeOH under reflux (80)	Dec. 160°	783
Et_4Sb	I (M-715)		NMR spectrum and ion pair assocn. data, (c, d)	343
	$AlCl_4$	$SbCl_3 + Et_3Al$ 1:2 in $n-C_6H_{14}$ at 70°	Colorless or light yellow crystals, insol. in $n-C_6H_{14}$ m. 180-182°, (e)	1489
				1495

486

	$SbCl_3$ + excess Et_2AlCl in $n\text{-}C_6H_{14}$ under reflux	White crystals, m. 185°, (f)	1488
	$[Et_4Sb]Cl$ + $AlCl_3$ 1:1 in $n\text{-}C_6H_{14}$ at 60°	Rxn. with $Et_3Al \rightarrow [Et_4Sb]\text{-}[Et_5Al_2Cl_2]$	1488
	Et_3SbCl_2 + $EtAlCl_2$ 1:1 or 1:2 at r.t.	White crystals or pale yellow gelatinous solid when excess $EtAlCl_2$ is used	1488
	$[Et_4Sb]Cl$ + $EtAlCl_2$ 1:3 in $n\text{-}C_6H_{14}$ at 60°		1488
	$[Et_4Sb][Et_2AlCl_2]$ + $EtAlCl_2$ 1:2 in $n\text{-}C_6H_{14}$ at 60°	(g)	1488
	Et_5Sb + $EtAlCl_2$ 1:2.99 in $n\text{-}C_6H_{14}$ at -20 to +30°	White crystals	1476
Al_nCl_{3n+1}	$[Et_4Sb]Cl$ + $AlCl_3$ 1:3 in $n\text{-}C_6H_{14}$ at 60°	Orange brown liquid	1488
$AlCl_3I$	$[Et_4Sb]I$ + $EtAlCl_2$ 1:2	White crystals	1488
	$[Et_4Sb]I$ + $AlCl_3$ 1:1 in C_6H_6 at 50°	Brown crystals	1488
Al_2Cl_6I (?)	$[Et_4Sb]I$ + $AlCl_3$ 1:2 in C_6H_6 at 50°	Reddish orange liquid	1488
$EtAlCl_3$	$[Et_4Sb]Cl$ + $EtAlCl_2$ 1:1	m. 34-36°, (h)	1487
$EtAlCl_2I$	$[Et_4Sb]I$ + $EtAlCl_2$ 1:1 at 70°	White crystals	1488
Et_2AlClI	$[Et_4Sb]I$ + Et_2AlCl 1:1 at 70°	Colorless liquid	1488
Et_3AlBr	Et_3Sb + Et_3Al + $EtBr$ in hydrocarbons	Solid at -18°, molar cond. data	487

(a) IR spectrum (1404). (b) A mixt. with Et_3Al + $TiCl_3$ in $n\text{-}C_6H_{14}$ catalyzes a stereoregular polymerization of C_3H_6 (1489). (c) NMR spectrum, coupling consts. and solvent effect (1027). (d) Rxn. with $CuCl$ in $EtOH \rightarrow [Et_4Sb]Cl$ (1329). (e) A rxn. with Et_3Al + $TiCl_3$ in $n\text{-}C_6H_{14}$ gave a complex which catalyzed a stereospecific polymerization of C_3H_6 (1489). (f) m. 182-185° (1489). (g) A mixt. with Et_2AlCl + $AlCl_3 \cdot 3$ $TiCl_3$ in $n\text{-}C_6H_{14}$ catalyzed a stereospecific polymerization of C_3H_6 (1489). (h) A mixt. with $TiCl_4$ catalyzes polymerization of C_2H_4 to a cryst. polyethylene (1487).

Table 49 Continued

R_4Sb^+	X^-	Prepd. from (Yield %)	Prop. and Rxn.	Ref.
C_8 (Cont.)				
Et_4Sb	Et_3AlCl	Et_3Sb + Et_3Al + EtCl at 50°	Colorless liquid, molar cond. data	487
		$[Et_4Sb]Cl$ + Et_3Al 1:1 in n-C_6H_{14} at 50°	Colorless liquid, molar cond. data, (i, j)	1488
		Et_5Sb + Et_2AlCl 1:0.66 in n-C_6H_{14} at -20 to +30°	Pale yellow liquid	1476
	Et_3AlI	Et_3Sb + Et_3Al + EtI at 100°	Colorless liquid, solid at 0°, molar cond. data	487
		$[Et_4Sb]I$ + Et_3Al 1:1 at 70°	Colorless liquid, (g, k, l)	1487-1489
	Et_4Al, (m)	Et_5Sb + Et_3Al 1:0.59 in n-C_6H_{14} at 10-40°	White crystals, (n, o)	1476, 1478
	$Et_4Al_2Cl_3$	$[Et_4Sb]Cl$ + Et_2AlCl 1:3 in n-C_6H_{14} at r.t.	Colorless liquid contg. $[Et_4Sb]$-$[Et_2AlCl_2]$	1488
		$[Et_4Sb][Et_2AlCl_2]$ + Et_2AlCl 1:1 in n-C_6H_{14} at 60°		1488
		Et_5Sb + Et_2AlCl 1:3 in n-C_6H_{14} at -20 to +30°	Colorless liquid contg. $[Et_4Sb]$-$[Et_5Al_2Cl_2]$	1476
	$Et_4Al_2Cl_2I$	$[Et_4Sb]I$ + Et_2AlCl 1:2 at 70°	Colorless liquid	1488
	$Et_5Al_2Cl_2$	Et_5Sb + Et_2AlCl 1:3 in n-C_6H_{14} at -20 to +30°	Colorless liquid contg. $[Et_4Sb]$-$[Et_4Al_2Cl_3]$	1476
		$SbCl_3$ + excess Et_3Al in n-C_6H_{14}	Colorless liquid contg. Et_3Sb, metallic Sb, and Et_nAlCl_{3-n}	1488, 1489
		Et_3SbCl_2 + Et_3Al 1:2.4 at r.t.	Colorless liquid	1488
		$[Et_4Sb][Et_3AlCl]$ + Et_2AlCl 1:1 in n-C_6H_{14} at 60°		1488
		$[Et_4Sb][Et_2AlCl_2]$ + Et_3Al 1:1 in n-C_6H_{14} at 60°		1488

Compound	Preparation	Properties	Ref.
	$[Et_4Sb][EtAlCl_3]$ + excess Et_3Al in $n\text{-}C_6H_{14}$ at 60°		1488
	$[Et_4Sb][AlCl_4]$ + excess Et_3Al in $n\text{-}C_6H_{14}$ at 70°	A mixt. with Et_3Al + $TiCl_3$ in $n\text{-}C_7H_{16}$ catalyzes a stereospecific polymerization of C_3H_6	1488, 1489
Et_6Al_2Cl	$SbCl_3$ + Et_3Al 1:4.5 in $n\text{-}C_6H_{14}$ under reflux	Colorless fuming liquid, insol. in $n\text{-}C_6H_{14}$, d^{35} 1.112, n_D^{35} 1.4995, (p)	1487
	$[Et_4Sb]Cl$ + Et_3Al 1:2	d^{35} 1.056, n_D^{35} 1.4990, (p)	1487
	$[Et_4Sb]Cl$ + Et_3Al 1:3 in $n\text{-}C_6H_{14}$ at r.t.	Colorless liquid, molar cond. data	1488, 1489
	$[Et_4Sb][Et_3AlCl]$ + Et_3Al 1:1 in $n\text{-}C_6H_{14}$ at 60°		1488
Et_6AlI	$[Et_4Sb]I$ + Et_3Al 1:2 at 70°	Colorless liquid contaminated with $[Et_4Sb][Et_3AlI]$	1488
Et_7Al_2, (q)	Et_5Sb + Et_3Al 1:3 in $n\text{-}C_6H_{14}$ at 10–40°	Colorless liquid, (r, s)	1476, 1478
$Et_3(i\text{-}C_6H_{13})_3Al_2Cl$	$[Et_4Sb]Cl$ + $(i\text{-}C_6H_{13})Al$ + Et_3Al in $n\text{-}C_6H_{14}$ at 50°	Colorless liquid, slightly sol. in C_6H_6 and $n\text{-}C_6H_{14}$, (t)	1489
$TiCl_4$	$[Et_4Sb]Cl$ + $TiCl_3$ 1:1 in C_6H_6 at r.t.	Brown crystals, dec. 200–204°, (u, v, w)	1477, 1488

(i) Rxn. with Et_2AlCl → $[Et_4Sb][Et_5Al_2Cl_2]$ (1488). (j) Rxn. with Et_3Al → $[Et_4Sb][Et_6Al_2Cl]$ (1488). (k) Molar cond. data (1488). (l) A mixt. with Et_3Al + $TiCl_3$ in $n\text{-}C_6H_{14}$ catalyzes a stereospecific polymerization of C_3H_6 (1489). (m) It is not certain whether the compd. is a stibonium salt or a $[Et_5Sb \cdot AlEt_3]$ complex (1476, 1478). (n) Rxn. with $TiCl_3$ (1:10) in C_6H_6 at 30°, followed by Et_3Al at 60° → black powder contg. Ti 34.28, Al 5.6, Sb 0.68, and Cl 51.93, an effective catalyst for olefin polymerization (1478). (o) Rxn. prod. with $TiCl_4$ is effective catalyst for polymerization of C_2H_4 (1487). (p) A mixt. with $TiCl_4$ in xylene at 60° catalyzes polymerization of olefins (1479). (q) It is uncertain whether the compd. is a stibonium salt or an $[Et_5Sb \cdot 2\ AlEt_3]$ complex. (r) Rxn. with $TiCl_3$ followed by Et_3Al → a compd. contg. Ti:Al:Sb = 1:0.05–0.5:0.001–0.1, catalyzes polymerization of olefins (1478). (s) Rxn. prod. with $TiCl_3$ catalyzes a stereospecific polymerization of olefins (1479). (t) A mixt. with Pr_3Al + $TiCl_3$ in xylene at 90° catalyzes a stereospecific polymerization of C_3H_6 (1489). (u) IR spectrum and x-ray refraction data (1477). (v) A complex with Et_3Al and $TiCl_3$ (0.7:2.5:5) prepd. at -80° catalyzes polymerization of α-olefins (1484). (w) A complex with Et_3Al and $TiCl_3$ (1482).

Table 49 Continued

R_4Sb^+	X^-	Prepd. from (Yield %)	Prop. and Rxn.	Ref.
C_8 (Cont.)				
Et_4Sb	$TiCl_5$	$[Et_4Sb]Cl + TiCl_4$ 1:1 in C_6H_6 at r.t.	Yellow powder, dec. 147-152°, (u, x)	1477
	Ti_2Cl_9	$[Et_4Sb]Cl + TiCl_4$ 0.45:1 in C_6H_6 at r.t.	Yellow powder, dec. 91-102°, (u, v, x)	1477
$[Et_4Sb]_2$	$TiCl_5$	$[Et_4Sb]Cl + TiCl_3$ 2.5:1 in C_6H_6 at r.t.	Yellow crystals, dec. 165-200°, (u)	1477
		$[Et_4Sb]Cl + TiCl_2$ in C_6H_6 at 80° under pressure	Yellowish brown powder contg. $[Et_4Sb]TiCl_4$	1477
	$TiCl_6$	$[Et_4Sb]Cl + TiCl_4$ 2.2:1 in C_6H_6 at r.t.	Yellow powder, dec. 220-225°, (u, y)	1477
C_9				
$Me_3(o-O_2NC_6H_4)Sb$	ClO_4	By-prod. from a rxn. of $[Me_3PhSb]NO_3$ + fuming HNO_3 + $NaClO_4$	UV and NMR spectra	600
$Me_3(m-O_2NC_6H_4)Sb$	ClO_4	$[Me_3PhSb]NO_3$ + fuming HNO_3, followed by $NaClO_4$ (87)	UV and NMR spectra	600
$Me_3(p-O_2NC_6H_4)Sb$	ClO_4	By-prod. from the preceding rxn.	UV and NMR spectra	600
Me_3PhSb	NO_3		Rxn. with HNO_3 followed by $NaClO_4 \rightarrow$ the preceding 3 compds.	600
$Et_3(MeO_2CCH_2)Sb$	BPh_4	$Et_3Sb + BrCH_2CO_2Me$ in Et_2O followed by BPh_4^-	Crystals, m. 145-146°, pK_a 8.5	788
Et_3PrSb	Cl	$[Et_3PrSb]I + CuCl$ in Et_2O under reflux, followed by H_2S		1329
	Et_3InBr	$Et_3Sb + Et_3In + PrI$ at 90-110°	Colorless liquid, stable < 125°, solid at -67°, molar cond. data	487
C_{12}				
Pr_4Sb	$Et_2AlBrCl$	$[Pr_4Sb]Br + Et_2AlCl$	Colorless liquid	1489

C₁₄				
Bu₃EtSb	Pr₃AlCl	[Bu₃EtSb]Cl + Pr₃Al in n-C₆H₁₄	Colorless liquid, (b)	1489
	Cl	Bu₃Sb + EtCl at 100°		1489
[Et₂MeSbCH₂CH₂-]₂	I₂	(Et₂SbCH₂CH₂-)₂ + MeI in MeOH under reflux (71)	Microcrystals, m. 215-217° (dec.), soly. data	783
C₁₆				
Bu₄Sb	Cl	[Bu₄Sb]I + CuCl in EtOH under reflux, followed by H₂S	(z)	1329
[Et₂MeSbCH₂CH₂CH₂-]₂	Pr₃AlBr	[Bu₄Sb]Br + Pr₃Al	(h)	1487
	I₂	Et₂Sb(CH₂)₆SbEt₂ + MeI in MeOH under reflux (90)	Microcrystals, m. 219-221° (dec.), soly. data	783
C₁₉				
MePh₃Sb	HgI₃	Ph₃Sb + MeI + HgI₂		616
	Q (aa)	[MePh₃Sb] + Li⁺Q⁻ 1:1 in EtOH (56)	m. 154-156°	725
	Q·Q (aa)	[MePh₃Sb]I + Li⁺Q⁻ 1:2 in MeCN (13)	m. 207-210°	725
C₂₀				
EtPh₃Sb	HgI₃	Ph₃Sb + EtI + HgI₂		616
C₂₂				
Me(p-MeC₆H₄)₃Sb	BF₄ (M-716)		m. 126-127°	725
	Q (aa)	[Me(p-MeC₆H₄)₃Sb]BF₄ + Li⁺Q⁻ 1:1 in EtOH (43)	m. 188-192°	725
MeEt₂Sb(CH₂)₄P(C₆H₁₁)₂Me	I₂	Et₂Sb(CH₂)₄P(C₆H₁₁)₂ + MeI 1:2 in MeOH under reflux (84)	Colorless crystals, m. 135-137°, soly. data	783
[t-Bu₂MeSbCH₂CH₂CH₂-]₂	I₂	t-Bu₂Sb(CH₂)₄Sb-t-Bu₂ + MeI 1:2 in MeOH under reflux (48)	Light yellow crystals, m. 124-126°	785

(x) A mixt. with $TiCl_3$ and R_2AlX or R_3Al catalyzes a stereospecific polymerization of **α**-olefins (1481). (y) A mixt. with Et_3Al and $TiCl_3$ (0.1:1.5) catalyzes a stereospecific polymerization of C_3H_6 (1481). (z) A fungicide and bactericide for plastics, paints, and fibers (947). (aa) Q = $(NC)_2C=$ $=C(CN)_2$.

Table 49 Continued

R_4Sb^+	X^-	Prepd. from (Yield %)	Prop. and Rxn.	Ref.
C_{24}				
$Ph_3(p\text{-}BrC_6H_4)Sb$	BF_4	$Ph_3Sb + [p\text{-}BrC_6H_4N_2]BF_4$	Rxn. with excess NaOMe in MeOH → evolution of C_6H_6 and PhBr	616
$Ph_3(m\text{-}O_2NC_6H_4)Sb$	BF_4	$Ph_3Sb + [m\text{-}O_2NC_6H_4N_2]BF_4$	Rxn. with excess NaOMe in MeOH → evolution of C_6H_6 and $PhNO_2$	616
$Ph_3(p\text{-}O_2NC_6H_4)Sb$	BF_4	$Ph_3Sb + [p\text{-}O_2NC_6H_4N_2]BF_4$	Rxn. with NaOMe → $R_3R'SbOMe$; with excess NaOMe in MeOH → C_6H_6 and $PhNO_2$ (z)	616
Ph_4Sb	$CH_2{=}CHCO_2$			947
	B_3H_8	$[Ph_4Sb]Br + [Me_4N]B_3H_8$ in H_2O (63)	Colorless ppt., dec. during crystn. from MeOH	61
	BF_4 (M-718)	$Ph_3Sb + PhN_2BF_4$	Rxn. with NaOMe → Ph_4SbOMe, (ab-ad)	616
	Br (M-713)	$Ph_3Sb + PhBr$ under γ-radiation	m. 212-218°, (ac, ae-aj)	306
	(M-717)	$Ph_3Sb + PhBr$ in the presence of $AlCl_3$		214
	Cl (M-717)	$Ph_3SbCl_2 + PhBr$	IR spectrum	214
		$Ph_3Sb + PhCl + AlCl_3$ under reflux and salting out with NaCl	UV spectrum, (ah, ai)	483
	ClO_3		(ai)	197
	ClO_2		(ai)	197
	F (M-717)	$[Ph_4Sb]Cl + HF$ or its alc. or Et_2O soln., followed by heating with NaOR in vacuum	Useful in organometallic complexes for electrolytic processes	1547
		$[Ph_4Sb]Cl + NaOR +$ anhyd. HF or its alc. or Et_2O soln.	(ai)	1547
	^{18}F	$[Ph_4Sb]_2SO_4 + {}^{18}F^-$	Extractable with CCl_4 from aq. soln.	214

Compound / anion	Preparation	Properties	Ref.
OH (M-717)	[Ph$_4$Sb]Br + NH$_4$OH	At 70–80° in xylene → Ph$_3$SbO, (ak)	229, 1086
I (M-717)	[Ph$_4$Sb]BF$_4$ + NH$_4$OH in EtOH	Rxn. with ROH → Ph$_4$SbOR	616
I$_3$ (M-718)	Ph$_5$Sb + MeI	m. 198°, (ah, ai)	1269
NO$_3$ (M-718)		(ah)	1001
ClO$_4$ (M-718)	[Ph$_4$Sb]$_2$SO$_4$ + ClO$_4^-$ in amperometric titrn.	(ai)	197
		Ppt., (ai)	1086
Q·Q (aa) (M-718)	[Ph$_4$Sb]I + Q 1:2 in MeCN at 60°	Black rods 0.2 x 1.4 x 0.4 mm, m. 219–220° (dec.), elec. resistivity data (ai)	26, 507
[Ph$_4$Sb]$_2$			
Cr$_2$O$_7$			197
MoO$_4$	[Ph$_4$Sb]$_2$SO$_4$ + MoO$_4^{--}$ in acid soln.	Ppt. m. 235°, extractable with CH$_2$Cl$_2$ from aq. soln. at pH 1–7 m. 209°, useful in photometric detn. of Mo	1649
MoO$_2$(SCN)$_3$			1649
SO$_4$ (M-719)	[Ph$_4$Sb]OH + H$_2$SO$_4$	Amperometric titrn. (al)	1086
	[Ph$_4$Sb]Br in H$_2$O + NH$_4$OH, followed by H$_2$SO$_4$		214
ON(SO$_3$)$_2$ (M-719)	[Ph$_4$Sb]Br in H$_2$O soln. + solid K$_2$[ON(SO$_3$)$_2$]	Paramagnetic crystals, soly. data, ESR spectrum	567
C$_{25}$ **Ph$_3$(p-MeC$_6$H$_4$)Sb**			
OH	[Ph$_3$(p-MeC$_6$H$_4$Sb)]BF$_4$ + NH$_4$OH in EtOH	Unstable, rxn. with EtOH → Ph$_3$(p-MeC$_6$H$_4$)SbOEt	616

(ab) Rxn. with NH$_4$OH in EtOH → Ph$_4$SbOH (616). (ac) Mössbauer spectrum (665). (ad) Prolonged heating with excess NaOR in ROH → dec. with evolution of C$_6$H$_6$ (616). (ae) Rxn. with NaOR → Ph$_4$SbOR (483). (af) Rxn. with NaOMe in MeOH → Ph$_3$Sb(OMe)$_2$ (227, 228, 1397, 1398). (ag) Polarographic redn., 2e$^-$ → Ph$_3$Sb, Ph$_2$Hg, and C$_6$H$_6$ (1087). (ah) IR spectrum (1001). (ai) Extn. from aq. acid soln. with CHCl$_3$ (197). (aj) Inhibitory effect on the uptake of Et$_4$N$^+$ and N'-methylnicotinamide by slices of dog kidney and on the renal tubular excretion of N'-methylnicotinamide by the hen (1430). (ak) UV spectrum (483). (al) Extractive titrn. of F$^-$ into CHCl$_3$ (1180).

Table 49 Continued

R_4Sb^+	X^-	Prepd. from (Yield %)	Prop. and Rxn.	Ref.
C_{28} $(p\text{-MeC}_6H_4)_4Sb$	Br (M-719)	$R_3SbBr_2 + RMgBr$	m. 223-225°, NMR spectrum, (am)	1397, 1398
C_{47} Ph_3Sb (cyclopentadiene, Ph, Ph, Ph, Ph)	ClO_4	+ $HClO_4$, followed by Et_2O	Orange yellow plates, m. 195°	975
Polymers $\{CH_2CHCH_2CH\}_n$ — CH_2SbMe_3 $CH_2=CH$	OH	Chloromethylated styrene + divinylbenzene copolymer + Me_3Sb in $ClCH_2CH_2Cl$ under reflux, (an)	Anion-exchange resin	888
$\{CH_2CHCH_2CH\}_n$ — CH_2SbEt_3 $CH_2=CH$	OH	Chloromethylated styrene + divinylbenzene copolymer swollen in MePh at 110° + Et_3Sb, (an)	Anion-exchange resin	888
$\{CH_2CHCH_2CH\}_n$ — CH_2SbPh_3 $CH_2=CH$	OH	Chloromethylated styrene + divinylbenzene copolymer swollen in $ClCH_2CH_2Cl$ + Ph_3Sb under reflux, (an)	Anion-exchange resin	888

(am) Rxn. with NaOMe in MeOH → R_4SbOMe (1397, 1398). (an) Followed by alc. or aq. NaOH at 60° (888).

Table 50. Quaternary Alkoxyantimony Derivatives, R_4SbOR'

R_4SbOR'	Prepd. from (Yield %)	Prop. and Rxn.	Ref.
C_9			
Et_4SbOMe	$Et_5Sb + MeOH$ at r.t. (94)	Colorless liquid, sol. in MeOH and H_2O	1476
C_{25}			
Ph_4SbOMe	$[Ph_4Sb]OH + MeOH$		616
	$[Ph_4Sb]Br + NaOMe$ in MeOH	Colorless crystals, m. 132°, sensitive to moisture, (a)	1397, 1398
	$[Ph_4Sb]Br + NaOMe$ in MeOH (68)	Rxn. with MeOH under reflux → $Ph_3Sb(OMe)_2$	228
	$Ph_4SbBr + NaOMe$ in MeOH (90)	m. 218-220°, UV, IR, and NMR spectra	483
	$Ph_5Sb + MeOH$ under reflux (97)	m. 130-132°, NMR spectrum, a covalent bonding is postulated	227, 228
C_{26}			
$Ph_3(p-BrC_6H_4)SbOEt$	$[R_3R'Sb]BF_4 + NH_4OH$ in EtOH	Unstable compd.	616
$Ph_3(m-O_2NC_6H_4)SbOEt$	$[R_3R'Sb]BF_4 + NH_4OH$ in EtOH	Unstable compd.	616
$Ph_3(p-O_2NC_6H_4)SbOEt$	$[R_3R'Sb]BF_4 + NH_4OH$ in EtOH	Unstable compd.	616
Ph_4SbOEt	$[Ph_4Sb]Cl + EtOH$, (b) (51)	Crystals; UV, IR, and NMR spectra	483
	$[Ph_4Sb]OH$ by dissolving in EtOH		616
C_{27}			
$Ph_3(p-MeC_6H_4)SbOEt$	$[R_3R'Sb]OH$ by boiling in EtOH		616
Ph_4SbOPr	$[Ph_4Sb]Cl + PrOH$, (b) (45)	Crystals; UV, IR, and NMR spectra	483
$Ph_4SbO-i-Pr$	$[Ph_4Sb]Cl + i-PrOH$, (b) (29)	Crystals; UV, IR, and NMR spectra	483
$Ph_3(p-MeOC_6H_4)SbOEt$	$[R_3R'Sb]BF_4 + NH_4OH$ in EtOH	Unstable compd.	616
C_{28}			
Ph_4SbOBu	$[Ph_4As]Cl + BuOH$, (b)	Undistillable oil, IR and NMR spectra	483
$Ph_4SbO-s-Bu$	$[Ph_4As]Cl + s-BuOH$, (b)	Crude oil	483
$Ph_4SbO-t-Bu$	$[Ph_4As]Cl + t-BuOH$ under reflux (4)	Crude oil	483

(a) Single crystal x-ray diffraction analysis indicated a trigonal bipyramidal structure (1397, 1398).   (b) In the presence of NaH under anhyd. conditions under reflux.

2.3.5   COMPOUNDS WITH ANTIMONY-HALOGEN BONDS

2.3.5.2 DIHALO AND TETRAHALO DERIVATIVES

The methods of synthesis of tertiary dihaloantimony compounds were summarized in the main volume on pages 722-3. Since 1964 a few new reactions and reagents for the preparation of the antimony derivatives have been reported. Sulfuryl chloride is an effective chlorination agent for the conversion of tertiary stibines. Thionyl chloride and phosgene are useful in the conversion of tertiary stibine oxides to the dichloro derivatives. A conversion of secondary trihaloantimony derivatives with dialkylmercury gives mixed tertiary dihaloantimonies in moderate yields. Tertiary stibines on heating with antimony trichloride at a ratio of 3:2 in n-hexane afford the compounds of this group. The formation of tertiary dihaloantimonies having two different halogen atoms was detected by NMR spectroscopy in the equilibrium mixtures formed according to the reaction equation: $R_3SbX_2 + R_3SbY_2 \rightleftharpoons 2\ R_3SbXY$, where R's are identical and X and Y are two different halogen atoms. Moreover, the compounds with two different halogens are formed from tertiary stibines by a reaction with iodine bromide and chloride, respectively. The difluoroantimony derivatives, R_3SbF_2, have been prepared from the corresponding dibromo and dichloro compounds by a metathetic reaction with silver fluoride in aqueous hydrofluoric acid and with silver tetrafluoroborate in water, respectively. The difluoro compounds are also formed from tertiary stibines by a reaction with perfluoropiperidine. Tertiary bromoiodo- and diiodo-antimonies combine with iodine, and the corresponding bromotriiodo and tetraiodo derivatives are formed.

A few compounds of this group are described individually hereafter and on page 502, while the others are compiled in Table 51.

Dibromotrimethylantimony, $SbC_3H_9Br_2$, Me_3SbBr_2,   (M-723)
Synth.: From Me_3Sb + Br in Et_2O, 73% yield (484), 68% yield (888); in CCl_4, 76% yield (1045); in MeOH (1106).
From $SbCl_3$ + MeMgI, followed by a treatment of the Me_3Sb with Br in CCl_4 (352).
Prop.: White crystals (352), m. 185-186.5° (1045), m. 198° (dec.) (1404).
IR spectrum in $CHCl_3$ (1403). NMR spectra at different temps. (979, 1045, 1342). Sol. in 95% EtOH; dec. at elevated temp. → Me_2SbBr and MeBr (484).
Rxn. with: AgX (X = F, NO_3, ClO_4, 1/2 CO_3, 1/2 SO_4, 1/2 CrO_4, 1/2 C_2O_4) → Me_3SbX_2 (352, 1045).
NaOR → $Me_3Sb(OR)Br$ (430) and $Me_3Sb(OR)_2$ (1031).
ROH + NH_3 gas in $CHCl_3$ or C_6H_6 → $Me_3Sb(OR)_2$ (430).
Zn in H_2O → Me_3Sb (888).
Me_3SbX_2 (X = I, Cl, or F) in $CDCl_3$ ⇌ Me_3SbBrX (979).
Me_3SbS + $(Me_3SnS)_3$ 3:3:1 → Me_2SnBr_2·2 Me_3SbS (1403).

Dichlorotrimethylantimony, $SbC_3H_9Cl_2$, Me_3SbCl_2,   (M-724)
Synth.: From $SbCl_3$ + MeMgI in Et_2O and treating the resulting Me_3Sb with Cl gas, 64% yield (484).

From $(Me_3SbCl)_2O$ + concd. HCl (1045).

Prop.: Crystals, sol. in H_2O (484). IR spectrum (1403). NMR spectra at -32°, 30.5°, and 70° (979, 1045, 1342). Thermal dec. $\rightarrow$ Me_2SbCl + MeCl (484).

Rxn. with: NaOR $\rightarrow$ $Me_3Sb(OR)Cl$ (430).

ROH + NH_3 gas in $CHCl_3$ or C_6H_6 $\rightarrow$ $Me_3Sb(OR)_2$ (430).

Me_3SbX_2 (X = Br, F, or I) in $CDCl_3$ $\rightleftharpoons$ Me_3SbClX (979, 1083).

NaN_3 in $ClCH_2CH_2Cl$ $\rightarrow$ $Me_3Sb(N_3)_2$ (1356).

Me_3SbS + $(Me_2SnS)_3$ 3:3:1 $\rightarrow$ $[Me_2SnCl_2 \cdot 2\ Me_3SbS]$ (1403).

Use: A fungicide and bactericide for paints, plastics, and fibrous matl. (947).

Dichlorotris(2-chlorovinyl)antimony, $SbC_6H_6Cl_3$, $(ClCH=CH)_2SbCl_2$, (M-724)

Synth.: From $SbCl_5$ + C_2H_2 (1139).

Prop.: m. 61°, IR and NMR spectra as well as x-ray diffraction studies indicated the presence of only one cis-ClCH=CH- and two trans-ClCH=CH- groups. This isomer was previously (M-724) described as an all cis-compd. (1139).

Dichlorotriethylantimony, $SbC_6H_{15}Cl_2$, Et_3SbCl_2, (M-724)

Synth.: From Et_3Sb + $SbCl_3$ 3:2 in $n-C_6H_{14}$ at 70°, 80% yield (1488).

Prop.: b_3 123.5° (1488).

Rxn. with: NaNCO 1:2 in MeCN at 81° $\rightarrow$ $Et_3Sb(NCO)_2$ (1444).

$TiCl_4$ at 60° $\rightarrow$ $Et_3SbCl_2 \cdot (TiCl_4)_n$, n = 0.5-1.3 (1477).

$TiCl_3$ at 60°, no rxn. (1477).

$TiCl_2$ at 60° $\rightarrow$ $[TiCl_3(Et_3Sb)_{0.5}(Et_3SbCl_2)_{0.11}]$ (1477).

$AlCl_3$ $\rightarrow$ $[Et_3SbCl_2 \cdot AlCl_3]$ (1488).

$EtAlCl_2$ $\rightarrow$ $[Et_4Sb]AlCl_4$ (1488).

Et_3Al 1:2 $\rightarrow$ $[Et_4Sb][Et_5Al_2Cl_2]$ (1488).

Et_3Al $\rightarrow$ $[Et_4Sb][Et_2AlCl_2]$ (1488).

Et_2AlCl $\rightarrow$ $[Et_4Sb][EtAlCl_3]$ (1488).

Coordination Deriv.: $[Et_3SbCl_2 \cdot AlCl_3]$, white crystals, m. $\sim$ 50°, prepd. from Et_3SbCl_2 + $AlCl_3$ 1:1 at 20° (1488).

$[Et_3SbCl_2 \cdot nAlCl_3]$, dark green liquid, prepd. from Et_3SbCl_2 + $AlCl_3$ 1:2 at r.t. (1488).

$[Et_3SbCl_2(TiCl_4)_n]$, n = 0.5-1.3, yellowish dark brown liquid, probably consisting of a mixt. of $(Et_3SbCl_2)_2 \cdot TiCl_4$, $Et_3SbCl_2 \cdot TiCl_4$, and $Et_3SbCl_2 \cdot (TiCl_4)_2$, prepd. from Et_3SbCl_2 + $TiCl_4$ at various ratios at 60° in a sealed tube. IR spectrum. The complex showed remarkable elec. cond. (1477).

$[(Et_3SbCl_2)_n(Et_3Sb)_mTiCl_3]$, where n = 0.2-0.44 and m = 0.09-0.62, dark brown powder, prepd. from Et_3Sb + $TiCl_4$ in $n-C_6H_{14}$ at -20° to r.t. Hydrolysis of the complex gave nearly theoretical amts. of Et_3SbCl_2 and Et_3Sb (1477).

Use: A rxn. prod. formed from Et_3SbCl_2 + $TiCl_3$ + Et_3Al catalyzes polymerization of C_3H_6 (95).

(Three more compounds are entered on page 502.)

Table 51. Tertiary Dihalo- and Tetrahalo-Antimony Compounds

R_3SbX_2 or $_4$ or R_3SbXY_3	Prepd. from (Yield %)	Prop. and Rxn.	Ref.
C$_3$			
$Me_3SbBrCl$	$Me_3SbBr_2 + Me_3SbCl_2$ in $CDCl_3$	The existence of the compd. was detected by NMR spectroscopy, (b, c)	979
Me_3SbBrF	$Me_3SbBr_2 + Me_3SbF_2$ in $CDCl_3$, (a)	The compd. was detected by NMR spectroscopy, (b, c)	979
Me_3SbBrI	$Me_3SbBr_2 + Me_3SbI_2$ in $CDCl_3$, (a)	The compd. was detected by NMR spectroscopy, (b, c)	979
Me_3SbClF	$Me_3SbCl_2 + Me_3SbF_2$ in $CDBr_3$, (a)	The compd. was detected by NMR spectroscopy, (b, c)	979
Me_3SbClI	$Me_3SbCl_2 + Me_3SbI_2$ in $CDBr_3$, (a)	The compd. was detected by NMR spectroscopy, (b, c)	979
Me_3SbFI	$Me_3SbF_2 + Me_3SbI_2$ in $CDCl_3$, (a)	The compd. was detected by NMR spectroscopy, (b, c)	979
Me_3SbF_2 (M-729)	$Me_3SbBr_2 + AgF$ in aq. HF	Crystals sol. in H_2O, MeOH, and $CHCl_3$, (c, d, e)	352
Me_3SbI_2 (M-729)	$Me_3Sb + I$ in Et_2O (45-62)	Crystals, sol. in EtOH, thermal dec. → $Me_2SbI + MeI$, (c, d, f)	484
C$_6$			
$(CH_2=CH)_3SbBr_2$ (M-729)	R_3Sb, (g) + Br in $CHCl_3$ (15)	b_7 128°	1107
$(CH_2=CH)_3SbCl_2$ (M-729)		IR spectrum	209,210
$(CH_2=CH)_3SbI_2$ (M-729)	R_3Sb, (h) + I in Et_2O (15)	Yellow crystals, m. 35-36°	1107
Et_3SbBr_2 (M-729)	$SbCl_3 + EtMgBr$ 1:3 in Et_2O, followed by Br (95)	Thermal dec. at 190-210°/450 mm ↑ → Et_2SbBr, (i)	783
C$_9$			
$(CH_2=CHCH_2)_3SbCl_2$ (M-730)		IR spectrum	209,210
C$_{12}$			
Bu_3SbBr_2 (M-731)		IR spectrum	209,210

Bu$_3$SbCl$_2$	Bu$_3$Sb + SbCl$_3$ in C$_6$H$_6$ at 50° (83)	Colorless liquid, b$_{0.14}$ 120-125°, slowly hydrolyzes on exposure to the atmosphere, (j, k, l, m)	1045
i-Bu$_3$SbCl$_2$		Rxn. with NaNCO in MeCN under reflux → i-Bu$_3$Sb(NCO)$_2$, (n)	1443-4
C$_{14}$			
Ph$_2$(cis-ClCH=CH)SbCl$_2$	Ph$_2$SbCl$_3$ + (cis-ClCH=CH)$_2$Hg in C$_6$H$_6$ by heating (53)	m. 102°, IR and NMR spectra	1139
Ph$_2$(trans-ClCH=CH$_2$)SbCl$_2$	Ph$_2$SbCl$_3$ + (trans-ClCH=CH)$_2$Hg in C$_6$H$_6$ at r.t. (65)	m. 92°, IR and NMR spectra	1139
C$_{15}$			
(cyclo-C$_5$H$_9$)$_3$SbBr$_2$	R$_3$Sb + Br	m. 104-106°	1576
C$_{18}$			
Ph$_3$SbBrF	Ph$_3$SbBr$_2$ + Ph$_3$SbF$_2$, (a)	(b)	1083
Ph$_3$SbBrI	Ph$_3$Sb + IBr in MeCN	White crystals, m. 215°, a nonelectrolyte in MeCN, (o, p)	178
Ph$_3$SbClF	Ph$_3$SbCl$_2$ + Ph$_3$SbF$_2$ in CHCl$_3$, (a)	(b)	1083
Ph$_3$SbClI	Ph$_3$Sb + ICl in Et$_2$O at -70°	White, impure solid, soly. data	481
Ph$_3$SbFI	Ph$_3$SbF$_2$ + Ph$_3$SbI$_2$ in CHCl$_3$, (a)	(b)	1083

(a) By equilibration at -32, 30.5, and 70° (979). (b) Equilibrium consts. were detd. at different temps. (1083). (c) NMR spectrum (1342). (d) NMR spectra at -32, 30.5, and 70° (979). (e) Rxn. with Me$_3$SbX$_2$ (X = Cl, Br, or I) ⇌ Me$_3$SbFX (979, 1083). (f) Rxn. with Me$_3$SbX$_2$ (X = Cl, F, Br) in CDCl$_3$ ⇌ Me$_3$SbXI (979). (g) (CH$_2$=CH)$_3$Sb was prepd. from SbF$_3$ + [NH$_4$]$_2$[CH$_2$=CHSiF$_5$] 1:3 in aq. soln. (1107). (h) (CH$_2$=CH)$_3$Sb was prepd. from SbF$_3$ + CH$_2$=CHSi(OEt)$_3$ + 25% NH$_4$·HF$_2$ 3:9:15 (1107). (i) Rxn. with Na$_2$S in MeOH → Et$_3$SbS (1403). (j) IR spectrum (209, 210, 1045). (k) Pyrolysis → Bu$_2$SbCl + BuCl (1045, 1444). (l) Rxn. with NaNCO at 81° → Bu$_3$Sb(NCO)$_2$ (1444). (m) Useful as a fungicide and bactericide for plastics, paints, and fibrous matl. (947). (n) Rxn. with urea at 140° → i-Bu$_3$Sb(NCO)$_2$ (1445). (o) Rxn. with IBr or Br$_2$ → Ph$_3$SbBr$_2$ (178). (p) Rxn. with I$_2$ → [Ph$_3$SbBr]$^+$I$_3^-$ (178).

Table 51 Continued

R_3SbX_2 or 4 or R_3SbXY_3	Prepd. from (Yield %)	Prop. and Rxn.	Ref.
C_{18} (Cont.)			
Ph_3SbF_2 (M-732)	Ph_3SbCl_2 in aq. EtOH + AgF or AgBF$_4$ in H$_2$O (79)	m. 115-116°, IR spectrum	482
	$Ph_3Sb + \overline{CF_2(CF_2)_4NCl}$ in C_6H_6 at r.t. (69)	(r)	118
Ph_3SbBrI_3	$Ph_3SbBrI + I$ in MeCN, (s)	Molar cond. data	178
Ph_3SbI_4	Ph_3Sb (s) with I in MeCN	Molar cond. data, ionic $[Ph_3SbI]^+I_3^-$ is suggested, (t)	178
$(cyclo\text{-}C_6H_{11})_3SbBr_2$		Rxn. with Na$_2$S → $(cyclo\text{-}C_6H_{11})_3SbS$	1403
$(cyclo\text{-}C_6H_{11})_3SbCl_2$ (M-732)	$R_3Sb + Cl$ in Et$_2$O (85)	White crystals, m. 211-212°, hydrolytically unstable, dec. in EtOH, (u, v)	782
$(n\text{-}C_6H_{13})_3SbBr_2$	$R_3Sb + Br$ in CHCl$_3$ (63)	Colorless powder, m. 148°	1497
C_{21}			
$(p\text{-}MeC_6H_4)_3SbBr_2$ (M-733)	$R_3Sb + Br$ in CHCl$_3$, (w) (80)	m. 233°, NMR spectrum, (x)	1398
$(PhCH_2)_3SbClF$	$R_3SbCl_2 + R_3SbF_2$ by equilibration, (b)	NMR spectra in Me$_4$Si and in CFCl$_3$ at -60°	1083
$(PhCH_2)_3SbCl_2$		NMR spectrum at -60°, rxn. with AgF in MeCN → $(PhCH_2)_3SbF_2$	1083
$(PhCH_2)_3SbF_2$	R_3SbCl_2 in MeCN + excess AgF by heating (43)	Crystals, m. 113-115°, NMR spectra in Me$_4$Si and CFCl$_3$ at -60°	1083
$(o\text{-}MeOC_6H_4)_3SbCl_2$ (M-733)		(q)	665
$(n\text{-}C_7H_{15})_3SbBr_2$	$R_3Sb + Br$ in CHCl$_3$ (65)	White solid, m. 136°	1497
C_{24}			
$(n\text{-}C_8H_{17})_3SbBr_2$	$R_3Sb + Br$ in CHCl$_3$ (58)	White solid, m. 129°	1497
C_{26}			
$(Ph_2SbCl_2CH=)_2$	$(Ph_2SbCH=)_2 + Cl$ in CHCl$_3$ at -78°	White crystals, m. 168°	1142

C₂₇			
(p-MeCH=CHC₆H₄)₃SbBr₂	R₃Sb + Br in CHCl₃ (90)	Light yellow solid, m. 174°, sol. in org. solvents; IR spectrum	1498
{p-[CH₂=C(Me)]C₆H₄}₃SbBr₂	R₃Sb + Br in CHCl₃ (88)	Cream-colored solid, m. 168°, sol. in org. solvents	1498
[p-(BrCH₂CHBrCH₂)C₆H₄]₃SbBr₂	R₃Sb + Br in CHCl₃	m. 122-125°	1640
[(cyclo-C₆H₁₁)₂SbBr₂CH₂]₂CH₂	[(cyclo-C₆H₁₁)₂SbCH₂]₂CH₂ in Et₂O + Br in C₆H₆ at -45° (65)	Colorless crystals, m. 158-162°	782
(n-C₉H₁₉)₃SbBr₂	R₃Sb + Br in CHCl₃ (44)	White solid, m. 125°	1497
C₂₈			
Ph₂SbBr₂(CH₂)₄SbBr₂Ph₂	Ph₂Sb(CH₂)₄SbPh₂ + Br in Et₂O-C₆H₆ (90)	Colorless crystals, m. 164-165°, sol. in C₆H₆, insol. in Et₂O, dec. in H₂O	784
[(cyclo-C₆H₁₁)₂SbBr₂CH₂CH₂-]₂	[(cyclo-C₆H₁₁)₂SbCH₂CH₂-]₂ in Et₂O + Br in C₆H₆ at -45° (80)	Colorless crystals, m. 178°	782
C₂₉			
[(cyclo-C₆H₁₁)₂SbBr₂CH₂-CH₂]₂CH₂	(cyclo-C₆H₁₁)₂Sb(CH₂)₅Sb(cyclo-C₆H₁₁)₂ in Et₂O + Br in C₆H₆ at -45° (44)	Colorless crystals, m. 181-183°	783
C₃₀			
[(-(MeCH₂CH=CH)C₆H₄]₃SbBr₂	R₃Sb + Br in CHCl₃ (70)	Cream-colored solid, m. 185°	1498
[(cyclo-C₆H₁₁)₂SbBr₂CH₂CH₂CH₂CH₂-]₂	[(cyclo-C₆H₁₁)₂SbCH₂CH₂CH₂CH₂-]₂ in Et₂O + Br in C₆H₆ at -45° (74)	Colorless crystals, m. 161-162°	782
(n-C₁₀H₂₁)₃SbBr₂	R₃Sb + Br in CHCl₃ (50)	White crystals, m. 118°	1497
1,4-C₆H₄(SbBr₂Ph₂)₂	1,4-C₆H₄(SbPh₂)₂ + Br in C₆H₆	Dec. 265-270°, stable in H₂O, (y)	1658

(q) Mössbauer effect (665). (r) Rxn. with Ph₃SbX₂ (X = Br, Cl, or I) in CDCl₃ ⇄ Ph₃SbFX (1083). (s) By conductometric titrn. (178). (t) Unstable compd., dec. → [Ph₄Sb]I₃ (178). (u) Sparingly sol. in inert solvents (782). (v) Thermal dec. at 230°/360 mm → (C₆H₁₁)₂SbCl (782). (w) For the method of prepn. see also ref. 1397. (x) Rxn. with p-MeC₆H₄MgBr → (p-MeC₆H₄)₄SbBr (1397, 1398). (y) Hydrolysis → 1,4-C₆H₄[Sb(OH)₂Ph₂]₂ (1658).

Dibromotriphenylantimony, $SbC_{18}H_{15}Br_2$, Ph_3SbBr_2,   (M-726)

Synth.: From Ph_3Sb + Br in MeCN (178, 482).

Prop.: m. 212-214° (482), m. 214-215° (1577). Nonelectrolyte in MeCN (178).
Half-wave potential (467). IR spectrum (209, 210, 482, 1001). Far-IR spectrum (803).

Rxn. with: NaOMe in MeOH → $Ph_3Sb(OMe)_2$ (227, 228, 1397, 1398).
NaOR 1:1 → $Ph_3Sb(OR)Br$ (430).
ROH + NH_3 gas in $CHCl_3$ or C_6H_6 → $Ph_3Sb(OR)_2$ (430).
$Ph_3SbF_2 \rightleftarrows 2\ Ph_3SbBrF$ (1083).

Dichlorotriphenylantimony, $SbC_{18}H_{15}Cl_2$, Ph_3SbCl_2,   (M-726)

Synth.: From Ph_3SbO + HCl (229).
From $(Ph_3SbCl)_2O$ + HCl gas (1045).
From Ph_3SbO + $COCl_2$ or $SOCl_2$ in MeCN (79).
From Ph_3Sb + SO_2Cl_2 in MePh, 90% yield (116).
From Ph_3Sb + Cl_2 in MeCN (178) or in CCl_4 (1108).
From $SbCl_3$ + PhN_2Cl by dec. with Zn dust in org. solvents, a mixt. of $PhSbCl_2$
(the main prod.) along with Ph_2SbCl, Ph_3Sb, and Ph_3SbCl_2 was obtd. (1243).

Prop.: Colorless crystals (116), m. 142° (1108), m. 142-43° (1243). A nonelectrolyte in MeCN (178). Sol. in $n-C_6H_{14}$ (481). Half-wave potential (467).
Spectra: IR (209, 210, 1001), NMR (1045), Mössbauer (665). Crystal structure
by x-ray diffraction method: a trigonal bipyramid with the Ph group in the
equatorial plane. A dimeric "assocd. cation" structure suggested by other
authors is rejected (1243). Crystallographic data (1657). The energy of mol.
interaction in the gaseous state was calcd. (1657). Thermal dec. at 250-270°/
150 mm → Ph_2SbCl (784).

Rxn. with: Ph_3SbO → $(Ph_3SbCl)_2O$ (229).
NaOR → $Ph_3Sb(OR)Cl$ (430).
ROH + NH_3 gas in $CHCl_3$ or C_6H_6 → $Ph_3Sb(OR)_2$ (430).
$AgNO_3$ in EtOH → $[Ph_3Sb(ONO_2)]_2O$ (1045, 1521).
AgF or $AgBF_4$ → Ph_3SbF_2 (482).
$AgClO_4$ → $(Ph_3SbClO_4)_2O$ (482).
Electrolytic redn. → Ph_3Sb + 2 Cl^- (467).
$Ph_2NC(S)SNa$ in C_6H_6-MeCN → Ph_3Sb (905).
$Ph_3SbF_2 \rightleftarrows 2\ Ph_3SbCF$ (1083).
$NCN=C(SK)_2$ in DMF → $Ph_3\overline{SbSC(=NCN)S}$ (1089).
$NCC(SNa)=C(SNa)CN$ in $MeOCH_2CH_2OMe$ → $Ph_3\overline{SbSC(CN)=C(CN)S}$ (1090).
N_2O_4 → $Ph_3Sb(NO_3)_2$ (1521).

Use: A fungicide and bactericide for paints, plastics, and fibrous matls.
(947).

Diiodotriphenylantimony, $SbC_{18}H_{15}I_2$, Ph_3SbI_2,   (M-732)

Synth.: From Ph_3Sb + I in MeCN (178) or in $CHCl_3$ (179, 482).

Prop.: m. 153° (179), m. 163-164° (482). A nonelectrolyte in MeCN (178).
Sol. in $n-C_6H_{14}$ (481). IR spectrum (482). UV and visible spectra of the
prod. prepd. in $CHCl_3$ were attributed to an associative outer charge-transfer
complex (179). Kinetics and energy of the transformation of the outer charge

transfer complex to the inner $[Ph_3SbI]^+I^-$ complex were detd. (180).

Rxn. with: I $\rightarrow$ Ph_3SbI_4 (178).
Br $\rightarrow$ Ph_3SbBr_2 (178).
IBr $\rightarrow$ Ph_3SbBr_2 (178).
Ph_3SbF_2 $\rightleftarrows$ 2 Ph_3SbFI (1083).

2.3.5.3 HALOHYDROXY, HALOALKOXY, AND HALOSILOXY DERIVATIVES

A few tertiary halohydroxy-, haloalkoxy-, and halosiloxy-antimony derivatives, $R_3Sb(OR')X$, were described in the main volume on page 738. The haloalkoxy derivatives compiled in Table 52 were prepared by the previously known
method from tertiary dihaloantimony derivatives by a reaction with alkali
metal alkoxide at a molar ratio of 1:1 in chloroform or benzene. The halo-
siloxy and halo-tert-butoxy derivatives are formed from tertiary stibines by
coordination with siloxy halides and tert-butoxy halides, respectively.

Table 52. Tertiary Haloalkoxy- and Halosiloxy-Antimony Derivatives

$R_3Sb(OR')X$	Prepd. from	Prop. and Rxn.	Ref.
C_5			
$Me_3Sb(OEt)Br$	Me_3SbBr_2 + NaOEt 1:1, (a)	Rxn. with H_2O_2 $\rightarrow$ $(Me_3SbBrO-)_2$	430
$Me_3Sb(OEt)Cl$	Me_3SbCl_2 + NaOEt 1:1, (a)	Rxn. with H_2O_2 $\rightarrow$ $(Me_3SbClO-)_2$	430
C_{19}			
$Ph_3Sb(OMe)Br$	Ph_3SbBr_2 + NaOMe 1:1, (a)		430
$Ph_3Sb(OMe)Cl$ (M-738)	Ph_3SbCl_2 + NaOMe 1:1, (a)	Rxn. with H_2O_2 $\rightarrow$ $(Ph_3SbClO-)_2$	430
C_{21}			
$Ph_3Sb(OSiMe_3)Br$	Ph_3Sb + Me_3SiOBr	m. 104-106°	432
$Ph_3Sb(OSiMe_3)Cl$ (M-738)	Ph_3Sb + Me_3SiOCl	m. 89-90°	431, 432
C_{22}			
$Ph_3Sb(O-t-Bu)Cl$	Ph_3Sb + $t-BuOCl$	m. 153-156°	432
C_{36}			
$Ph_3Sb(OSiPh_3)Br$	Ph_3Sb + Ph_3SiOBr	m. 214-216°	432
$Ph_3Sb(OSiPh_3)Cl$	Ph_3Sb + Ph_3SiOCl	m. > 200° (dec.)	431, 432

(a) The reaction is carried out in $CHCl_3$ or C_6H_6.

2.3.5.6 DIPSEUDOHALO DERIVATIVES

The following tertiary diazido- and diisocyanato-antimony derivatives are new compounds.

Diazidotrimethylantimony, $SbC_3H_9N_6$, $Me_3Sb(N_3)_2$
Synth.: From Me_3SbCl_2 + NaN_3 1:4 in $ClCH_2CH_2Cl$ at r.t., 87% yield, probably formed through the Me_3SbClN_3 intermediate (1356).
Prop.: Colorless solid, m. 91°, almost insensitive to heat and shock; monomeric in C_6H_6 soln., quite stable to moist air. IR and NMR spectra - a trigonal bipyramidal structure is suggested (1356).

Triethyldiisocyanatoantimony, $SbC_8H_{15}N_2O_2$, $Et_3Sb(NCO)_2$
Synth.: From Et_3SbO by heating with urea at 30° or with HNCO at 80° (1444). From Et_3SbCl_2 by heating with NaNCO 1:2 in MeCNO at 81° (1444).

Tributyldiisocyanatoantimony, $SbC_{14}H_{27}N_2O_2$, $Bu_3Sb(NCO)_2$
Synth.: From Bu_3SbO by mixing with powdered urea, heating to 130°, and completing the rxn. at 140° in vacuum, while removing H_2O and NH_3 (1443-5). From Bu_3SbO by heating with NaNCO 1:2 at 80° (1444). From Bu_3SbCl_2 by heating with NaNCO 1:2 in MeCN at 81° (1444).
Prop.: Liquid, dec. ~ 190° (1443), $b_{0.2}$ 122° (1445). IR spectrum (1443).

Triisobutyldiisocyanatoantimony, $SbC_{14}H_{27}N_2O_2$, i-$Bu_3Sb(NCO)_2$
Synth.: From i-Bu_3SbCl_2 + NaNCO in MeCN under reflux, 75% yield (1443, 1444). From i-Bu_3SbCl_2 by heating with urea at 140° (1445). From i-Bu_3SbO by heating with NaNCO at 80° or with urea at 130° (1444).
Prop.: Colorless liquid, $b_{0.15}$ 122-124°, n_D^{20} 1.5128, IR spectrum (1443-5).

Diisocyanatotrioctylantimony, $SbC_{26}H_{51}N_2O_2$, $(C_8H_{17})_3Sb(NCO)_2$
Synth.: From $(C_8H_{17})_3SbO$ by heating with urea (1445).
Prop.: Viscous oil (1445).

2.3.6 COMPOUNDS WITH ANTIMONY BONDED TO GROUP VIA ELEMENTS

2.3.6.1 COMPOUNDS WITH ANTIMONY-OXYGEN BONDS

2.3.6.1.5 DIHYDROXY DERIVATIVES AND THE CORRESPONDING SALTS

A few tertiary dihydroxyantimony derivatives and their methods of synthesis were reported in the main volume on pages 739-40. Two new dihydroxy derivatives described hereafter were prepared by the known methods.

Several salt-like compounds, compiled in Table 53, may be considered as anhydrides of tertiary dihydroxyantimony derivatives with inorganic and carboxylic acids, respectively. These compounds have been prepared from tertiary dihydroxyantimonies by metathesis with silver salts of inorganic and organic acids and with alkali metal carboxylates, respectively, and from tertiary dihydroxyantimonies and stibine oxides by the reaction with carboxylic acids. Moreover, a reaction of tertiary stibines with fuming nitric acid and with dinitrogen tetroxide, respectively, leads to the corresponding dinitratoantimonies. With the exception of tertiary diperchloratoantimonies, all the salt-like compounds of this group are nonionic and some of them appear to be polymeric.

Dihydroxytriphenylantimony, $SbC_{18}H_{17}O_2$, $Ph_3Sb(OH)_2$, (M-740)
Rxn. with: $HCO_2H \rightarrow Ph_3Sb(O_2CH)_2$ (482).
$AcOH \rightarrow Ph_3Sb(OAc)_2$ (482).
ROOH in $C_6H_6 \rightarrow Ph_3Sb(OOR)_2$ (1276).
Phosphate, acetate, propionate, and benzoate anions $\rightarrow$ extn. from aq. media into $CHCl_3$ as $Ph_3Sb(OH)H_2PO_4$, $Ph_3Sb(OH)OAc$, $Ph_3Sb(OH)O_2CEt$, and $Ph_3Sb(OH)OBz$, resp. (1376).

Dihydroxytris[p-(2,3-dihydroxypropyl)phenyl]antimony, $SbC_{27}H_{35}O_8$,
    $[p-(HOCH_2CHOHCH_2)C_6H_4]_3Sb(OH)_2$
Synth.: From $(p-CH_2=CHCH_2C_6H_4)_3Sb$ in N KOH + excess 3% H_2O_2 (1640).
Prop.: m. > 260°, slowly dec. > 150° (1640).

1,4-Phenylenebis(dihydroxydiphenylantimony), $Sb_2C_{30}H_{28}O_4$,
    $Ph_2Sb(OH)_2C_6H_4Sb(OH)_2Ph_2$
Synth.: From $Ph_2SbBr_2C_6H_4SbBr_2Ph_2$ by hydrolysis (1658).
From $Ph_2SbC_6H_4SbPh_2$ by oxidn. with H_2O_2 in Me_2CO (1658).
Prop.: Dec. 240° (1658).

Table 53. Tertiary Dihydroxyantimony Anhydrides with Inorganic and Carboxylic Acids

R_3SbX_2	Prepd. from	Prop. and Rxn.	Ref.
C_3			
$Me_3Sb(BF_4)_2$	$Me_3SbBr_2 + AgBF_4$ in MeOH	White solid, subl vac. 50°, IR spectrum and x-ray diffraction analysis	352
$Me_3SbB_{12}F_{12}$	$Me_3SbBr_2 + Ag_2[B_{12}F_{12}]$ in MeOH	Crystd. with 1 mol. MeOH as a reddish solid, anal. impure, IR spectrum indicated the absence of MeOH	352
Me_3SbCO_3	$Me_3SbBr_2 + Ag_2CO_3$ in SO_2	Crystals sol. in H_2O and MeOH, IR spectrum, (a)	352
Me_3SbCrO_4	$Me_3SbBr_2 + Ag_2CrO_4$ in H_2O	Crystals sol. in H_2O, IR spectrum, (b)	352
$Me_3Sb(ONO_2)_2$ (M-741)	$Me_3SbBr_2 + AgONO_2$ in MeOH	Crystals sol. in H_2O, MeOH, and $CHCl_3$, UV and IR spectra, (c)	352
	$Me_3SbBr_2 + AgONO_2$ in H_2O	m. 149-150°, NMR spectrum	1045
	$Me_3SbO + 60\% HNO_3$ 1:2 in Me_2CO	Colorless crystals, m. 149°, soly. data, (d-f)	1404
$Me_3Sb(ClO_4)_2$	$Me_3SbBr_2 + AgClO_4$ in MeOH	White solid, explodes violently, (d)	352
Me_3SbSO_4 (M-741)	$Me_3SbBr_2 + Ag_2SO_4$ in H_2O	Crystals sol. in H_2O, IR spectrum, a structure analogous to that under (b)	352
$Me_3Sb(SbF_6)_2$	$Me_3SbBr_2 + AgSbF_6$ in MeOH or in SO_2	Hydroscopic white solid, IR spectrum	352
Me_3SbSiF_6	$Me_3SbBr_2 + Ag_2SiF_6$ in MeOH	Colorless, deliquescent crystals, subl. at 50°, IR spectrum and x-ray diffraction data	352
C_5			
$Me_3SbOC(O)C(O)O$	$Me_3SbBr_2 + Ag$ oxalate in H_2O	Crystals, sol. in H_2O, IR spectrum, a structure analogous to that under (a) is suggested	352
$Me_3Sb[OC(O)H]_2$	$Me_3SbO + $ excess HCO_2H, exo-thermic rxn.	Colorless scale-like crystals, m. 81°, monomeric in C_6H_6, soly. data, (g)	1404
C_7			
$Me_3Sb(OAc)_2$	$Me_3SbO + $ excess $AcOH$, exo-thermic rxn.	m. 80.5-81°, monomeric in C_6H_6, soly. data; IR spectrum, a structure analogous to that under (g) is postulated	1404
C_9			
$Me_3Sb(O_2CEt)_2$	$Me_3SbO + $ excess $EtCO_2H$	Colorless, slightly viscous liquid, b_6 110°, n_D^{25} 1.4795, soly. data; IR spectrum, a structure analogous to that under (g) is postulated	1404

C$_{11}$			
Me$_3$Sb(O$_2$CPr)	Me$_3$SbO + excess PrCO$_2$H	b$_{2.5-3}$ 128°, soly. data, IR spectrum, a structure analogous to that under (g) is postulated	1404
C$_{16}$			
Bu$_3$Sb(OAc)$_2$		A catalyst for the trimerization of isocyanates	728
i-Bu$_3$Sb(OAc)$_2$		A catalyst for the trimerization of isocyanates	728
C$_{17}$			
Me$_3$Sb(OBz)$_2$	Me$_3$SbO + BzOH 1:2 by heating in C$_6$H$_6$	Colorless crystals, m. 154°, soly. data, IR spectrum, a structure analogous to that under (g) is postulated	1404
C$_{18}$			
Ph$_3$Sb(ONO$_2$)$_2$	Ph$_3$Sb + fuming HNO$_3$ or excess N$_2$O$_4$	Crystals, m. 143-145°, m. 144-146°, a nonelectrolyte in MeCN, IR spectrum, (h)	1521
Ph$_3$SbSO$_4$		Crystals, m. > 300°, IR spectrum, a polymeric structure is suggested	482
C$_{20}$			
Ph$_3$Sb[OC(O)H]$_2$	Ph$_3$Sb(OH)$_2$ + hot HCO$_2$H	Crystals, m. 157-162° (dec.), IR spectrum, the Sb-formate bonds are discussed	482
C$_{22}$			
Ph$_3$Sb(OAc)$_2$ (M-741)	Ph$_3$Sb(OH)$_2$ + AcOH	m. 214-215°, IR spectrum, the Sb-OAc bonds are discussed	482
C$_{24}$			
Ph$_3$Sb(O$_2$CCH=CH$_2$)$_2$	Ph$_3$SbO + CH$_2$=CHCO$_2$H at 130° Ph$_3$SbCl$_2$ + NaO$_2$CCH=CH$_2$ 1:2	m. ~90° (i)	497 497
C$_{26}$			
Bu$_3$Sb(OBz)$_2$		A catalyst for trimerization of isocyanates	728

(a) A bridging function of the CO$_3$ group between 2 planar Me$_3$Sb groups in a polymeric structure is suggested (352). (b) A polymeric structure with the CrO$_4$ moiety covalently bonded to 2 Me$_3$Sb groups (352). (c) A covalent bonding of the ONO$_2$ groups and a planar arrangement of the Me$_3$Sb moiety in a trigonal bipyramid is suggested (352). (d) Raman spectrum in H$_2$O soln. indicated the existence of the Me$_3$Sb^{++} cation with a planar C$_3$Sb skeleton (495). (e) IR spectra in soln. and in the solid state indicated a trigonal bipyramid (1404). (f) Rxn. with KBr → Me$_3$SbBr$_2$ (1404). (g) IR spectra in the solid state and in soln. indicated a covalent bonding of the carboxy groups and a trigonal bipyramidal structure with the C$_3$Sb group in a plane (1404). (h) A covalent bonding of the ONO$_2$ groups is postulated (482). (i) Useful as a fungicide and bactericide for paints, plastics, and synth. fibers (947).

2.3.6.1.6 TERTIARY STIBINE OXIDES

A few compounds of this group were described in the main volume on pages 743-44.

Trimethylstibine Oxide, SbC_3H_9O, Me_3SbO,   (M-743)
Rxn. with: ROOH in C_6H_6 under reflux → $Me_3Sb(OOR)_2$ (1276).
H_2S in MeOH → Me_3SbS (1403).
RCO_2H excess → $Me_3Sb(O_2CR)_2$ (1404).

Triethylstibine Oxide, $SbC_6H_{15}O$, Et_3SbO
Rxn. with: $CO(NH_2)_2$ 1:2 at 130° → $Et_3Sb(NCO)_2$ (1444).
HNCO 1:2 at 80° → $Et_3Sb(NCO)_2$ (1444).
Use: A catalyst for the formation of polyurethan foams from polyesters and diisocyanates (757).

Allylethyl(o-nitrophenyl)stibine Oxide, $SbC_{11}H_{14}NO_3$,
    $Et(CH_2=CHCH_2)(2-O_2NC_6H_4)SbO$
Rxn.: Polymerization in the presence of BuMgCl in THF at 25° → amorphous solid polymer with a specific reduced viscosity of 0.055 (56).

Dibutylstibinoacetic Acid Ethyl Ester Oxide, $SbC_{12}H_{23}O_3$, $Bu_2Sb(O)CH_2CO_2Et$
Synth.: From $Bu_2SbCH_2CO_2Et + Ag_2O$ in Et_2O, 82% yield (176).
Prop.: $b_{0.003}$ 83-85°, d_4^{20} 1.1414, n_D^{20} 1.5016, IR and NMR spectra (176).

Tributylstibine Oxide, $SbC_{12}H_{27}O$, Bu_3SbO,   (M-743)
Rxn. with: CO_2 → $Bu_3\overline{SbOC(O)O}$ (223).
CS_2 → $Bu_3\overline{SbOC(S)S}$ (223).
COS → $Bu_3\overline{SbOC(O)S}$ or $Bu_3\overline{SbOC(S)O}$ (223).
RNCO → $Bu_3\overline{SbN(R)C(O)O}$ (223).
Urea by fusion → $Bu_3Sb(NCO)_2$ (1443, 1444).
HNCO 1:2 at 80° → $Bu_3Sb(NCO)_2$ (1444).
Uses: A catalyst for a trimerization of isocyanates (727, 728). A catalyst for the formation of polyurethan foams from polyesters and diisocyanates (757).

Triisobutylstibine Oxide, $SbC_{12}H_{27}O$, i-Bu_3SbO
Rxn. with: CO_2 → i-$Bu_3\overline{SbOC(O)O}$ (223).
CS_2 → i-$Bu_3\overline{SbOC(S)S}$ (223).
Urea 1:2 at 130° → i-$Bu_3Sb(NCO)_2$ (1444).
HNCO 1:2 at 80° → i-$Bu_3Sb(NCO)_2$ (1444).

[5-(Diethylstibino)pentyl]diethylstibine Oxide, $Sb_2C_{13}H_{30}O$,
    $Et_2Sb(CH_2)_5Sb(O)Et_2$
Synth.: From $Et_2Sb(CH_2)_5SbEt_2$ by oxidn. with air in Et_2O, 58% yield (783).
Prop.: Colorless amorphous solid, sol. in MeOH, insol. in Et_2O, C_6H_6, Me_2CO, and H_2O (783).

Pentamethylenebis(diethylstibine Oxide), $Sb_2C_{13}H_{30}O_2$,
     $Et_2Sb(O)(CH_2)_5Sb(O)Et_2$
Synth.: From $Et_2Sb(CH_2)_5SbEt_2$ + excess H_2O_2 in Me_2CO, 49% yield (783).
Prop.: Colorless amorphous solid, dec. at elevated temp., insol. in Me_2CO,
sparingly sol. in MeOH, sol. in H_2O (783).

Bis(p-chlorophenyl)vinylstibine Oxide, $SbC_{14}H_{11}Cl_2O$,
     $(4-ClC_6H_4)_2(CH_2=CH)SbO$
Rxn.:In the presence of BuMgBr in THF → cream-colored, amorphous polymer (56).

Stibylidynetriacetic Acid Triisopropyl Ester Oxide, $SbC_{15}H_{27}O_7$,
     $(i-PrO_2CCH_2)_3SbO$
Synth.: From $Sb(CH_2CO_2-i-Pr)_2$ + Ag_2O in Et_2O, 55% yield (176).
Prop.: $b_{0.0085}$ 137-138°, d_4^{20} 1.3056, n_D^{20} 1.4984, IR and NMR spectra (176).

Triphenylstibine Oxide, $SbC_{18}H_{15}O$, Ph_3SbO, (M-743)
Synth.: From $Ph_3Sb(OMe)_2$ + H_2O 1:1 in Me_2CO (227-29).
From $[Ph_4Sb]OH$ by heating at 70-80° in xylene, 90% yield (227, 229).
Prop.: m. 221.5-222° (227, 229). IR spectrum (227, 229, 803). NMR spectrum
(227, 229, see also 1045). The compd. was found to be different from that
formed by dehydration of $Ph_3Sb(OH)_2$ and by oxidn. of Ph_3Sb with H_2O_2 (229).
Rxn. with: O in p-xylene under reflux → $(Ph_2SbO)_2O$ (227).
HCl → Ph_3SbCl_2 (227, 229).
Ph_3SbCl_2 → $(Ph_3SbCl)_2O$ (229).
Uses: A catalyst for the polyurethan foam formation from polyesters and diisocyanates (757). A complex with H_2O_2: $(Ph_3SbO)_2 \cdot H_2O_2$ catalyzes a crosslinking of polyesters with styrene (1366). Complexes with picric or benzoic
acid are useful as semiconductors for the thermisistors based on organophosphorus compds. (1433).

Trioctylstibine Oxide, $SbC_{24}H_{51}O$, $(n-C_8H_{17})_3SbO$
Rxn. with: Urea → $(n-C_8H_{17})_3Sb(NCO)_2$ (1445).
Use: A catalyst for the polyurethan foam formation from polyesters and diiso-
cyanates (757).

2.3.6.1.7 DIALKOXY, DIARYLOXY, AND DIQUINOLINOLATO DERIVATIVES

This group of compounds reported in the main volume (M-744) included one
dialkoxy and two disiloxy derivatives.

Dimethoxytrimethylantimony, $SbC_5H_{15}O_2$, $Me_3Sb(OMe)_2$
Rxn. with: H_2O_2 → $Me_3Sb(OOH)_2$ (430).

Diethoxytrimethylantimony, $SbC_7H_{19}O_2$, $Me_3Sb(OEt)_2$
Synth.: From Me_3SbBr_2 + NaOEt in EtOH at r.t., 59% yield (1031).
Prop.: Colorless, hydroscopic liquid, b_5 66-67° (1031).
Rxn. with: RSH in C_6H_6 at r.t. → Me_3Sb + RSSR + 2 EtOH (1031).

RCOSH → $Me_3Sb(SCOR)_2$ (1031).

Tributyldiethoxyantimony, $SbC_{16}H_{37}O_2$, $Bu_3Sb(OEt)_2$
Use: A catalyst for trimerization of isocyanates (727, 728).

Dimethoxytriphenylantimony, $SbC_{20}H_{21}O_2$, $Ph_3Sb(OMe)_2$
Synth.: From Ph_4SbOMe + MeOH under reflux, 81% yield (227, 228).
From [Ph_4Sb]Br or Ph_3SbBr_2 + NaOMe in MeOH (227, 228, 1397, 1398).
Prop.: Colorless crystals, obtd. by subl., m. 100-102° (227, 228, 1397, 1398), sol. in nonpolar solvents (227, 228), sensitive to moisture (1397, 1398). NMR spectrum, a trigonal bipyramidal structure is postulated (227, 228). Single crystal x-ray diffraction data confirmed the trigonal bipyramidal structure with the MeO groups in the apical positions (1397, 1398).
Rxn. with: H_2O 1:1 in Me_2CO → Ph_3SbO (227-29).

Bis(pentachlorophenoxy)tripropylantimony, $SbC_{21}H_{21}Cl_2O_2$, $Pr_3Sb(OC_6Cl_5)_2$
Use: A fungicide and bactericide for paints, plastics, and synth. fibers (947).

Diethoxytriphenylantimony, $SbC_{22}H_{25}O_2$, $Ph_3Sb(OEt)_2$
Rxn. with: H_2O_2 → $Ph_3Sb(OEt)_2$ (430).

Triphenylbis(8-quinolinolato)antimony, $SbC_{36}H_{27}O_2$, $Ph_3Sb(OC_9H_6N)_2$
Synth.: From Ph_3SbO + 8-hydroxyquinoline in MePh under reflux, 60% yield (947).
Prop.: m. 197-199° (947).
Use: A fungicide and bactericide for paints, plastics, and synth. fibers (947).

2.3.6.1.8 TERTIARY STIBINE PEROXIDE DERIVATIVES

Five types of peroxyantimony derivatives were described in the main volume, pages 744-47. This supplement contains two new types of antimony derivatives with peroxy bonds: $R_3Sb(OOH)_2$ and $R_3Sb(X)OOSb(X)R_3$. The compounds are compiled in Table 54.

Table 54. Tertiary Stibine Peroxy Derivatives

Peroxy Compd.	Prepd. from (Yield %)	Prop.	Ref.
$Me_3Sb(OOH)_2$	$Me_3Sb(OMe)_2$ in C_6H_6 + H_2O_2 in Et_2O (92)	Colorless substance, m. 60-62° (dec.)	430
$Ph_3Sb(OOH)_2$	$Ph_3Sb(OEt)_2$ in C_6H_6 + H_2O_2 in Et_2O-C_6H_6 (85)	Colorless crystals contg. 1/3 C_6H_6 or 1/3 $CHCl_3$, becomes turbid at 120°, turns brown ~ 170°, m. 210-215° (dec.)	430

Table 54 Continued

Peroxy Compd.	Prepd. from (Yield %)	Prop.	Ref.
$Me_3Sb(OO-t-Bu)_2$ (M-746)	Me_3SbO + t-BuOOH in C_6H_6, (a) (68)	m. 82-84°	1276
$Ph_3Sb(OO-t-Bu)_2$ (M-746)	$Ph_3Sb(OH)_2$ + t-BuOOH in C_6H_6, (a) (87)	m. 101-103°	1276
$Ph_3Sb(OOCMe_2Ph)_2$ (M-746)	$Ph_3Sb(OH)_2$ + $PhCMe_2OOH$ in C_6H_6, (a) (42)	m. 87-88°	1276
$Me_3Sb(Br)OOSb(Br)Me_3$	$Me_3Sb(OEt)Br$ in C_6H_6 + H_2O_2 in Et_2O (74)	Colorless crystals, dec. 300°, unstable in storage and dec. during crystn. from C_6H_6	430
$Me_3Sb(Cl)OOSb(Cl)Me_3$	$Me_3Sb(OEt)Cl$ + H_2O_2 in Et_2O (67)	Colorless crystals, dec. 300°	430
$Ph_3Sb(Br)OOSb(Br)Ph_3$	$Ph_3Sb(OMe)Br$ + H_2O_2 in Et_2O (80)	Colorless crystals, softens > 190°, m. 205-210° (dec.), (b)	430
$Ph_3Sb(Cl)OOSb(Cl)Ph_3$	$Ph_3Sb(OMe)Cl$ + H_2O_2 in Et_2O (81)	Crystals, m. 170-172° (dec.)	430

(a) Under reflux with azeotropic removal of H_2O (1276). (b) Dec. during crystn. from $C_6H_6 \rightarrow (Ph_3SbBr)_2O$ (430).

2.3.6.1.9 OXYBIS(TERTIARY ANTIMONY) DERIVATIVES, $(R_3SbX)_2O$

Several compounds of this type were described in the main volume on page 748.

Oxybis[halotrialkyl(or triaryl)antimony] derivatives are formed from tertiary dihaloantimonies, R_3SbX_2, by hydrolysis in aqueous alcohol and from iminobis(halo-tertiary-antimonies), $(R_3SbX)_2NH$, by hydrolysis in atmospheric moisture. The oxybisnitrato analogs, $(R_3SbONO_2)_2O$, are formed from the oxybishalo derivatives by metathesis with silver nitrate in aqueous solution, from tertiary stibine oxides by a treatment with 60% nitric acid in moist acetone, from tertiary dihaloantimonies by a reaction with silver nitrate in alcoholic solution or in moist acetonitrile, and from tertiary dinitratoantimonies by hydrolysis with aqueous alcohol. Sulfatooxybis(triphenylantimony) was obtained from triphenylstibine sulfate, $Ph_3\overset{+}{Sb}OSO_3^-$, by hydrolysis in alcohol. The oxybisperchlorato derivatives can be obtained from tertiary dihaloantimonies by a reaction with silver perchlorate in hot alcoholic solution.

The derivatives of this group are compiled in Table 55, with the exception of one compound which appears on page 514 following the table.

Table 55. Oxybis(tertiary Antimony) Derivatives, $(R_3SbX)_2O$

$(R_3SbX)_2O$	Prepd. from (Yield %)	Prop. and Rxn.	Ref.
C_6			
$(Me_3SbCl)_2O$ (M-748)	$(Me_3SbCl)_2NH$ by hydrolysis in atm. moisture (~100)	m. > 200° (dec.), NMR spectrum, at 260° → Me_3Sb, Sb_2O_3, and MeCl, (a, b)	1045
$(Me_3SbONO_2)_2O$	$(Me_3SbCl)_2O$ + $AgNO_3$ in H_2O soln.	m. 262° (dec.), NMR and IR spectra	1045
	Me_3SbO + 60% HNO_3 1:1 in moist Me_2CO	White solid, dec. 267°, sol. in polar solvents; IR spectrum in the solid state, (c)	1404
C_{12}			
$[(CH_2=CH)_3SbI]_2O$	By-prod. from $(CH_2=CH)_3Sb$ + I in Et_2O recovered from the mother liquor at -78°	Dec. > 260°, soly. data	1107
$(Et_3SbCl)_2O$	$(Et_3SbCl)_2NH$ by hydrolysis in atm. moisture (~100)	m. 179-180° (from cyclohexane), IR and NMR spectra, (b)	1045
$(Et_3SbONO_2)_2O$	$(Et_3SbCl)_2O$ + $AgNO_3$ in H_2O	IR spectrum	1045
C_{18}			
$(Pr_3SbCl)_2O$	$(Pr_3SbCl)_2NH$ by hydrolysis in atm. moisture (~100)	m. 98-100° (from n-C_6H_{14}), IR and NMR spectra	1045
C_{24}			
$(Bu_3SbCl)_2O$	$(Bu_3SbCl)_2NH$ by hydrolysis in atm. moisture (~100)	m. 52-54°, IR spectrum	1045
C_{36}			
$(Ph_3SbBr)_2O$ (M-748)	Ph_3SbBr_2 by hydrolysis in aq. alc.	m. 252-253°, IR spectrum, (d)	482
	$Ph_3SbBrOOSbBrPh_3$ by crystn. from C_6H_6	m. 260-263°	430
	Ph_3Sb + Ph_3SbCl_2		229
$(Ph_3SbCl)_2O$ (M-748)	Ph_3SbCl_2 by hydrolysis in aq. alc.	m. 216-218°, IR spectrum	482
	$(Ph_3SbCl)_2NH$ by boiling in wet Me_2CO (~100)	IR and NMR spectra, (a)	1045

Compound	Preparation	Properties	Ref.
$(Ph_3SbClO_4)_2O$	Ph_3SbCl_2 + $AgClO_4$ in hot alc. soln. (explosion hazard)	White crystals, m. > 300°, IR and Raman spectra; ionic ClO_4 groups and tetrahedral geometry are postulated	482
$(Ph_3SbONO_2)_2O$ (M-748)	Ph_3SbCl_2 + $AgNO_3$ in alc. and crystg. from boiling H_2O	m. 226° (dec.), IR and NMR spectra	1045
	Ph_3SbCl_2 + $AgNO_3$ in moist MeCN and crystd. from MeCN	White needles, m. 224° (dec.), a non-electrolyte in MeCN	1521
	$Ph_3Sb(ONO_2)_2$ by hydrolysis with aq. alc.	m. 237-240°	482
$[(Ph_3Sb)_2SO_4]O$	Ph_3SbSO_4 by hydrolysis in alc.	Crystals, m. 255-259°	482
C_{48}			
$(Ph_3SbOPh)_2O$	See the original publication		227

(a) Rxn. with concd. HCl or HCl gas → R_3SbCl_2 (1045). (b) Rxn. with aq. $AgNO_3$ → $(R_3SbONO_2)_2O$ (1045). (c) An intermediate, between covalent and ionic, character of the $Sb-ONO_2$ bond is postulated (1404). (d) IR spectrum in the solid state (1404). (e) IR spectrum, a covalent bonding of the ONO_2 groups is postulated (482, 1521).

Oxybis(ethyliododiphenylantimony), $Sb_2C_{28}H_{30}I_2O$, $(EtPh_2SbI)_2O$
Synth.: From Ph_2SbF + NH_4F (1:3) in AcOH by treating with $EtSiF_3$, dilg. the
rxn. mixt. with H_2O, extn. of the prod. with Et_2O, concg. to remove Et_2O,
treating with KOH, dissolving in Et_2O and adding 10% I in Et_2O (1107).
Prop.: Almost colorless plates, m. 155-161° (1107).

2.3.6.2   COMPOUNDS WITH ANTIMONY-SULFUR BONDS

2.3.6.2.6 TERTIARY STIBINE SULFIDES

A few tertiary stibine sulfides and their methods of preparation were
described in the main volume on pages 748-49. Besides the methods reported
before, the sulfides are also formed from tertiary stibine oxides and from
tertiary dihaloantimonies by a reaction with hydrogen sulfide and with hydrated
sodium sulfide, respectively, in methanol. Several stibine sulfides are
listed in Table 56.

Trimethylstibine Sulfide, SbC_3H_9S, Me_3SbS
Synth.: From Me_3SbO + H_2S in MeOH (1403).
Prop.: m. 176° (dec.), monomeric in C_6H_6 and in $CHCl_3$ at 25°. IR spectra in
$CHCl_3$ and in the solid state. NMR spectrum in $CHCl_3$ (1403).
Rxn. with: Ph_3SnCl in MeOH, Me_2CO, or $CHCl_3$ → Me_3SbCl_2 + $(Ph_3Sn)_2S$ (1403).
$BuSnCl_3$ in MeOH, Me_2CO, or $CHCl_3$ → Me_3SbCl_2 + $(BuSnS)_2S$ (1403).
Ph_2SnCl_2 in MeOH, Me_2CO, or $CHCl_3$ → Me_3SnCl_2 + Ph_2SnS (1403).
$SnCl_4$ in MeOH, Me_2CO, or $CHCl_3$ → Me_3SnCl_2 + SnS_2 (1403).
Coordination Deriv.: $[Me_3SbS \cdot Et_2SnCl_2]$, prepd. from Me_3SbS + Et_2SnCl_2 in
Me_2CO, MeOH, or $CHCl_3$ (1403).

$[Me_3SbS \cdot Me_2SnCl_2]$, prepd. from Me_3SbS + Me_2SnCl_2 in MeOH, Me_2CO, or $CHCl_3$
(1403).

$[(Me_3SbS)_2Me_2SnCl_2]$, colorless crystals, m. 147-147.5°, prepd. from Me_3SbS +
Me_2SnCl_2 2:1 in MeOH on a steam bath (1405) and by mixing Me_3SbS with Me_3SbCl_2
and $(Me_2SnS)_3$ 3:3:1 in $CHCl_3$, followed by evapn. of the solvent (1403).
Monomeric in MePh but partly dissocd. in $CHCl_3$ and in DMF soln. IR spectra in
$CHCl_3$ and in the solid state; NMR spectrum in $CHCl_3$; x-ray powder diffraction
data (1403, 1405). A bonding through S to Sn is suggested (1405). In $CHCl_3$
soln. an equilibrium state: $[(Me_3SbS)_2SnMe_2Cl_2] \rightleftarrows Me_3SbS$ + Me_3SbCl_2 + 1/3
$(Me_2SnS)_3$ is postulated (1405).

$[(Me_3SbS)_2Me_2SnBr_2]$, colorless crystals, m. 123-124°, prepd. by mixing Me_3SbS
with Me_3SbCl_2 and $(Me_2SnS)_3$ 3:3:1 in $CHCl_3$, followed by evapn. Sol. in polar
solvents with dissocn. IR spectra in $CHCl_3$ and in the solid state (1403).

$[(Me_3SbS)_2Et_2SnCl_2]$, colorless crystals, m. 122-123°, prepd. by mixing Me_3SbS
with Me_3SbCl_2 and $(Et_2SnS)_3$ 3:3:1 in $CHCl_3$ and evapn. of the solvent. IR
spectrum in the solid state (1403).

$[(Me_3SbS)_2Et_2SnBr_2]$, colorless crystals, m. 121°, prepd. by mixing Me_3SbS with
Me_3SbBr_2 and $(Et_2SnS)_3$ 3:3:1 in $CHCl_3$ and evapn. Sol. in polar solvents with
dissocn. IR spectrum in the solid state (1403).

Triphenylstibine Sulfide, $SbC_{18}H_{15}S$, Ph_3SbS, (M-749)

<u>Prop.</u>: IR spectrum (803). Dipole moment and molar Kerr const.; the arrange-
ment of the Ph groups is discussed (94).

<u>Coordination Deriv.</u>: $[Cu(Ph_3SbS)_2I]$, pale yellow crystals, m. 125° (dec.),
prepd. from Ph_3SbS in Me_2CO + Cu_2I_2 in satd. KI soln. (851).

$[Hg(Ph_3SbS)I_2]$, yellow, unstable crystals, m. 162° (dec.), prepd. from
Ph_3SbS + HgI_2 in Me_2CO (851).

<u>Use</u>: Complexes with $RAlX_2$ and Ti, V, Zr, Cr, or Mo halides catalyze the
polymerization of α-olefins (384).

Tricyclohexylstibine Sulfide, $SbC_{18}H_{15}S$, $(C_6H_{11})_3SbS$, (M-749)

<u>Synth.</u>: From $(C_6H_{11})_3SbBr_2$ + hydrated Na_2S in MeOH (1403).

<u>Prop.</u>: m. 144°, monomeric in $CHCl_3$ soln. (1403).

<u>Rxn. with</u>: $Ph_3SnCl \rightarrow (C_6H_{11})_3SbCl_2$ + $(Ph_3Sn)_2S$ (1403).

$BuSnCl_3 \rightarrow (C_6H_{11})_3SbCl_2$ + $(BuSnS)_2S$ (1403).

$R_2SnCl_2 \rightarrow (C_6H_{11})_3SbCl_2$ + R_2SnS (1403).

$SnCl_4 \rightarrow (C_6H_{11})_3SbCl_2$ + SnS_2 (1403).

2.3.6.2.8 BIS(ACYLTHIO)ANTIMONY DERIVATIVES

No compounds of this type were reported in the main volume.

Trimethylbis(thioacetato)antimony, $SbC_7H_{15}O_2S_2$, $Me_3Sb(SAc)_2$

<u>Synth.</u>: From $Me_3Sb(OEt)_2$ + AcSH at a temp. < 5°, 71% yield (1031).

<u>Prop.</u>: Colorless needles, m. 51-52°, IR spectrum (1031).

Trimethylbis(thiobenzoato)antimony, $SbC_{17}H_{19}O_2S_2$, $Me_3Sb(SBz)_2$

<u>Synth.</u>: From $Me_3Sb(OEt)_2$ + BzSH at r.t., 92% yield (1031).

From Me_3Sb + BzSSBz in C_6H_6 under reflux (1031).

<u>Prop.</u>: Colorless scale-like crystals, m. 108°, IR spectrum, dec. when heated
in C_6H_6 under reflux $\rightarrow Me_3Sb$ + BzSSBz (1031).

Table 56. Tertiary Stibine Sulfides

R_3SbS	Prepd. from (Yield %)	Prop. and Rxn.	Ref.
C_8			
Et_3SbS (M-749)	Et_3SbBr_2 + hydrated Na_2S in MeOH	m. 119-120°, IR spectrum	1403
C_{12}			
Bu_3SbS (M-749)		Useful as a catalyst for tri-merization of isocyanates	728
C_{13}			
$Et_2Sb(S)(CH_2)_4Sb(S)Et_2$	$Et_2Sb(CH_2)_4SbEt_2$ + S in C_6H_6 under reflux (63)	Colorless needles, m. 137-139°, soly. data	783
C_{13}			
$Et_2Sb(S)(CH_2)_5Sb(S)Et_2$	$Et_2Sb(CH_2)_5SbEt_2$ + S in C_6H_6 under reflux (43)	Colorless needles, m. 136°, soly. data	783
C_{14}			
$Et_2Sb(S)(CH_2)_6Sb(S)Et_2$	$Et_2Sb(CH_2)_6SbEt_2$ + S in C_6H_6 under reflux (63)	Microcrystals, m. 109-110°, soly. data	783
C_{20}			
$(cyclo-C_6H_{11})_2P(S)(CH_2)_4Sb(S)Et_2$	$(cyclo-C_6H_{11})_2P(CH_2)_4SbEt_2$ + S in C_6H_6 under reflux (68)	Colorless needles, m. 125-126°, soly. data	783
C_{29}			
$Ph_2Sb(S)(CH_2)_5Sb(S)Ph_2$	$Ph_2Sb(CH_2)_5SbPh_2$ + S in C_6H_6 under reflux (27)	Colorless needles, m. 117-118°, soly. data	784
C_{30}			
$Ph_2Sb(S)(CH_2)_6Sb(S)Ph_2$	$Ph_2Sb(CH_2)_6SbPh_2$ + S in C_6H_6 under reflux (77)	Colorless crystals, m. 156-158°, soly. data	784

2.3.7 COMPOUNDS OF ANTIMONY BONDED TO GROUP VA ELEMENTS

2.3.7.1 COMPOUNDS WITH ANTIMONY-NITROGEN BONDS

2.3.7.1.4 IMINOBISANTIMONY DERIVATIVES

This group comprises tertiary haloantimony derivatives, $(R_3SbX)_2NH$, bridged by an imino group. No compounds of this type were described before. They are formed from tertiary stibines by a reaction with chloramine in ether.

Table 57. Iminobisantimony Derivatives, $(R_3SbX)_2NH$

$(R_3SbX)_2NH$	Prepd. from (Yield %)	Prop. and Rxn.	Ref.
C_6			
$(Me_3SbCl)_2NH$	$Me_3Sb + NH_2Cl$ in Et_2O (71)	m. > 200° (dec.), IR and NMR spectra, (a)	1045
C_{12}			
$(Et_3SbCl)_2NH$	$Et_3Sb + H_2NCl$ in Et_2O (87)	m. 128-129.5°, IR and NMR spectra, (a)	1045
C_{18}			
$(Pr_3SbCl)_2NH$	$Pr_3Sb + H_2NCl$ in Et_2O (76)	m. 83-85°, IR and NMR spectra, (a)	1045
C_{24}			
$(Bu_3SbCl)_2NH$	$Bu_3Sb + H_2NCl$ in Et_2O (51)	m. 49-50°, IR and NMR spectra, (a)	1045
C_{36}			
$(Ph_3SbCl)_2NH$	$Ph_3Sb + H_2NCl$ in Et_2O (55-71)	m. 219-220.5°, IR and NMR spectra, (b)	1045

(a) Rxn. with atm. moisture → $(R_3SbCl)_2O$ (1045). (b) Rxn. with wet Me_2CO → $(Ph_3SbCl)_2O$ (1045).

2.3.7.1.7 TERTIARY STIBINE IMIDES, $R_3Sb=NR'$

The imides, $R_3Sb=NR'$, described in the main volume, pages 750-52, include include compounds in which R' are acyl, sulfonyl, silyl, germyl, and stannyl moisties, respectively. This supplement contains a new stibine imide monomer and a polymer, in which R' are phosphinyl groups.

Table 58. Tertiary Stibine Imides

$R_3Sb=NR'$	Prepd. from	Prop.	Ref.
C_{19} $Ph_3Sb=NSO_2Me$	$Ph_3Sb + MeSO_2N=S=O$ in C_6H_6 under reflux	Crude prod., hydrolysis by 2 N $NaOH \rightarrow Ph_3SbO \cdot H_2O$	1385
C_{30} $Ph_3Sb=NP(O)Ph_2$	$Ph_3Sb + Ph_2P(O)N_3$ in $Me(OCH_2CH_2)_3OMe$ under reflux	Useful as insecticide, lub. oil additive, UV light stabilizer for polymers, and fire retardant	1583
C_{32} $\{4-[Me_2PhSb=NSiMe_2]C_6H_4\}_2O$ (M-752)	$Me_2PhSb + NaN_3$ in xylene, followed by $[4-(Me_2SiCl)C_6H_4]_2O$	(a)	1584-1586
C_{60} $(Ph_3Sb=N)_3SnPh$ (M-752)	$Ph_3Sb + PhSn(N_3)_3$ 3:1 in DMF under reflux	(a)	1584
C_{66} $m\text{-}C_6H_4(Ph_2Sb=NGePh_3)_2$ (M-752)	$m\text{-}(Ph_2Sb)_2C_6H_4 + Ph_3GeN_3$ 1:2	(a)	1584-1586
$(C_{22})_n$ (M-752)	$m\text{-}(Me_2Sb)_2C_6H_4 + Ph_2Si(N_3)_2$ 1:1 by heating in Py	Stable to oxidn., hydrolysis, and high temp., (b)	1584-1586
	$p\text{-}[(Me_2N)_2Sb]_2C_6H_4 + p\text{-}[MeP(O)(N_3)]_2\text{-}C_6H_4$ 1:2 in Py under reflux	Useful as insecticide, lub. oil additive, UV light stabilizer for polymers, and fire retardant	1583

(a) Useful as a UV light stabilizer for polymers (1584, 1585). (b) Useful as additives to lub. oils, hydraulic fluids, transformer oils, synth. fibers, sheets, and varnishes (1584-1586).

2.4   SECONDARY DERIVATIVES

2.4.5   COMPOUNDS WITH ANTIMONY-HALOGEN BONDS

2.4.5.1 TRIHALO DERIVATIVES

Numerous secondary trihaloantimony derivatives and their methods of preparation were described in the main volume, pages 753-59.

Dibromochlorodimethylantimony, $SbC_2H_6Br_2Cl$, Me_2SbBr_2Cl
Synth.: From Me_2SbCl in CCl_4 + Br (481).
Prop.: White, impure solid (481).

Trichlorodimethylantimony, $SbC_2H_6Cl_3$, Me_2SbCl_3
Synth.: From Me_2SbCl + Cl in CS_2 at -20°, 67% yield (1343).
Prop. Snow-white, odorless substance (1343). IR spectrum (481). Thermal
dec. at 110-115° → $MeSbCl_2$ (1343).
Rxn. with: AgF → Me_2SbF_3 (481).
Use: A fungicide and bactericide for paints, plastics, and synth. fibers
(947).

Trifluorodimethylantimony, $SbC_2H_6F_3$, Me_2SbF_3
Synth.: From Me_2SbCl_3 in EtOH + aq. AgF, 70% yield (481).
Prop.: White crystals, IR spectrum (481).

Trichlorobis(nitromethyl)antimony-Nitromethane Complex, $SbC_3H_7Cl_3N_3O_6$,
    $(O_2NCH_2)_2SbCl_3 \cdot MeNO_2$
Synth.: From $SbCl_5$ in $MeNO_2$ by heating under reflux (1199).
Prop.: Light brown solid, m. > 225° (1199).

Dibutyltrichloroantimony, $SbC_8H_{18}Cl_3$, Bu_2SbCl_3,   (M-756)
Use: Complexes with a transition metal halide of Group IVB or VB and with an
organic deriv. of Group IIA or IIIA metal catalyze polymerization of C_2H_4 to
high-density, linear polymers (249a).

Trichlorodiphenylantimony, $SbC_{12}H_{10}Cl_3$, Ph_2SbCl_3,   (M-754)
Synth.: From $PhSbCl_2$ + $[PhN_2][SbCl_4]$, followed by hydrolysis of the resulting
$Ph_2SbO_2H \cdot Sb_2O_3$ in 5 N HCl, isolated as a monohydrate (1244).
By-prod. from a dec. of $[PhN_2][SbCl_4]$ with Zn powder in org. solvents (1244).
Prop.: X-ray diffraction of the monohydrate confirmed the presence of H_2O; a
distorted octahedral structure for the monohydrate is postulated (1244).
Rxn. with: NH_3 + NH_4Cl at -35° → $(Ph_2Sb=N)_4$ (679).
$(trans-ClCH=CH)_2Hg$ → $Ph_2(trans-ClCH=CH)SbCl_2$ (1139).
$(cis-ClCH=CH)_2Hg$ → $Ph_2(cis-ClCH=CH)SbCl_2$ (1139).
Use: A fungicide and bactericide for paints, plastics, and synth. fibers
(947).

Tribromodicyclohexylantimony, $SbC_{12}H_{22}Br_3$, $(C_6H_{11})_2SbBr_3$
Synth.: From $(C_6H_{11})_2SbSb(C_6H_{11})_2$ in Et_2O + Br in C_6H_6 at $-45°$ to $-35°$, 82% yield (782).
Prop.: White platelets, dec. 63-65.5°; unstable, dec. slowly at r.t. → $C_6H_{11}SbBr_2$ and $C_6H_{11}Br$ (782).

2.4.6 COMPOUNDS OF ANTIMONY BONDED TO GROUP VIA ELEMENTS

2.4.6.1 COMPOUNDS WITH ANTIMONY-OXYGEN BONDS

2.4.6.1.2 TRIARYLOXY DERIVATIVES, $R_2Sb(OR')_3$

No compounds of this type were reported in the main volume.

Tris(pentachlorophenoxy)diphenylantimony, $SbC_{30}H_{10}Cl_{15}O_3$, $Ph_2Sb(OC_6Cl_5)_3$
Use: A fungicide and bactericide for paints, plastics, and synth. fibers (947).

Diphenyltris(8-quinolinolato)antimony, $SbC_{39}H_{28}N_3O_3$,
Use: A fungicide and bactericide for paints, plastics, and synth. fibers (947).

2.4.6.1.3 STIBINIC ACID ANHYDRIDES

Anhydrides of stibinic acids and orthostibinic acids with carboxylic acids were reported in the main volume on page 766.

Tris(acryloyloxy)dibutylantimony, $SbC_{17}H_{27}O_6$, $Bu_2Sb(O_2CCH=CH_2)_3$
Use: A fungicide and bactericide for plastics, paints, and synth. fibers (947).

2.4.6.1.9 OXYBIS(SECONDARY ANTIMONY) DERIVATIVES

Oxybis(oxodiphenylantimony), $Sb_2C_{24}H_{20}O_3$, $(Ph_2SbO)_2O$
Synth.: From Ph_3SbO by heating in p-xylene under reflux in the presence of air (227).

2.5 PRIMARY DERIVATIVES

2.5.5 COMPOUNDS WITH ANTIMONY-HALOGEN BONDS

2.5.5.2 TETRAHALO DERIVATIVES

Primary tetrahaloantimony derivatives and their methods of synthesis were reported in the main volume on pages 766-73.

The following three compounds: 2,5-Cl(O$_2$N)C$_6$H$_3$SbCl$_4$, 2-ClC$_6$H$_4$SbCl$_4$, and PhSbCl$_4$ (M-767) are useful as fungicides and bactericides for paints, plastics, and synth. fibers (947).

Tetrachloro(2-methoxy-2,2-diphenylethyl)antimony Ammonia Complex,
   SbC$_{15}$H$_{15}$Cl$_4$O·NH$_3$,   MeOCPh$_2$CH$_2$SbCl$_4$·NH$_3$
Synth.: By-prod. from a rxn. of Ph$_2$C=CH$_2$ + SbCl$_5$ in CH$_2$Cl$_2$ at -80°, followed by quenching the rxn. mixt. with MeOH-NH$_3$ (570).
Prop.: Colorless crystals, m. 158-160°, IR and NMR spectra (570).

2.5.5.3 TRIHALOENOLATO DERIVATIVES

No compounds of this type were reported before.

Acetylacetonatotrichlorophenylantimony,   SbC$_{11}$H$_{12}$Cl$_3$O$_2$,
Synth.: From PhSbO$_3$H$_2$ by treating with Ac$_2$CH$_2$ and HCl, followed by extn. with CH$_2$Cl$_2$ (838, 839).
Prop.: Transparent, needle-like crystals, m. 176-178° (dec.), IR and NMR spectra (838, 839).

2.5.6   COMPOUNDS OF ANTIMONY BONDED TO GROUP VIA ELEMENTS

2.5.6.1   COMPOUNDS WITH ANTIMONY-OXYGEN BONDS

2.5.6.1.2   AROMATIC STIBONIC ACIDS

The methods of synthesis of aromatic stibonic acids were described in the main volume on pages 774-75.

2.5.6.1.2.1 BENZENESTIBONIC ACID AND DERIVATIVES WITH ALIPHATIC AND ARYLALI-
             PHATIC SUBSTITUENTS

Benzenestibonic Acid,   SbC$_6$H$_7$O$_3$,   PhSbO$_3$H$_2$,   (M-776)
Rxn. with: Ac$_2$CH$_2$ + HCl → PhSb[OC(Me)=CHAc]Cl$_3$ (838, 839)
Pilocarpine and dibazole in HCl → cryst. ppts., of which the optical props. were detd. (1240).
Theophylline in HCl soln. → theophylline·H(PhSbCl$_5$), a cryst. ppt. (1241).
Phenothiazine derivs. (PSN) in HCl → PSN·PhSbCl$_5$ complexes, distinguishable by bright colors, sol. in EtOH, MeOH, Me$_2$CO, and Ac$_2$O (1238, 1240).
Salts of pilocarpine, ephedrine, quinine, theophylline, dibazole, quinosole, bigumal, corazol, choline chloride, and hexamethylenetetramine in concd. HCl → cryst. ppts. (1240).
Dibasole in H$_2$O soln. → cryst. ppt. (1240).
Uses: A color test reagent for the detn. of phenothiazine derivs. in HCl: aminazine - bright raspberry, dinezine - red, diprazine - raspberry rose, etapirazine - violet, and metarazine - raspberry (1238).
Detection of γ quant. of dibazole (1240).

o-Toluenestibonic Acid, $SbC_7H_9O_3$, $2\text{-MeC}_6H_4SbO_3H_2$, (M-778)

Rxn. with: Phenothiazine derivs. (PSN) in HCl → PSN·$MeC_6H_4SbCl_5$ complexes, distinguishable by bright colors, sol. in EtOH, MeOH, Me_2CO, and Ac_2O (1238). Caffeine and methylcaffeine, resp., in HCl soln. → cryst. ppts. with specific shapes and characteristic optical props. useful in identifying and differentiating these purine derivs. (1241).

Uses: A color test reagent for the detection and detn. of phenothiazine derivs. in HCl: aminazine - bright raspberry, dinezine - red, dispazine - raspberry rose, etapirazine - violet, and metarazine - raspberry (1238).

p-Toluenestibonic Acid, $SbC_7H_9O_3$, $4\text{-MeC}_6H_4SbO_3H_2$, (M-777)

Rxn. with: Phenothiazine derivs. (PSN) in HCl → PSN·$MeC_6H_4SbCl_5$ complexes, distinguishable by bright colors, sol. in EtOH, MeOH, Me_2CO, and Ac_2O (1238). Caffeine in HCl soln. → cryst. ppt. with specific shapes and optical props. useful in identifying this purine deriv. (1241).

Use: A color test reagent for the detection and detn. of phenothiazine derivs.: aminazine - bright raspberry, dinezine - red, diprazine - raspberry rose, etapirazine - violet, and metarazine - raspberry (1238).

2.5.6.1.2.3 HYDROXYBENZENESTIBONIC ACIDS AND O-SUBSTITUTED DERIVATIVES

A few compounds of this type were compiled in the main volume on page 781.

o-Hydroxybenzenearsonic Acid, $SbC_6H_7O_4$, $2\text{-HOC}_6H_4SbO_3H_7$

Rxn. with: Caffeine and methylcaffeine, resp., in HCl soln. → cryst. ppts. with characteristic crystal shapes and optical props. which are useful in identifying and distinguishing these purine derivs. (1241).

p-Hydroxybenzenestibonic Acid, $SbC_6H_7O_4$, $4\text{-HOC}_6H_4SbO_3H_2$, (M-781)

Rxn. with: Phenothiazine derivs. (PSN) in HCl → PSN·$HOC_6H_4SbCl_5$ complexes characterized by bright colors, sol. in EtOH, MeOH, Me_2CO, and Ac_2O (1238). Caffeine and methylcaffeine, resp., in HCl soln. → cryst. ppts. with characteristic crystal shapes and optical props. useful in identifying and distinguishing these purine derivs. (1241).

Use: A color test reagent for detection and detn. of phenothiazine derivs.: aminazine - bright raspberry, dinezine - red, diprazine - raspberry rose, etapirazine - violet, and metarazine - raspberry (1238).

2.5.6.1.2.5 NITROGEN-SUBSTITUTED BENZENESTIBONIC ACID DERIVATIVES

Numerous benzenestibonic acids carrying nitrogen substituents were described in the main volume on pages 786-90.

m-Nitrobenzenestibonic Acid, $SbC_6H_6NO_5$, $3\text{-}O_2NC_6H_4SbO_3H_2$, (M-788)

Rxn. with: Phenothiazine derivs. (PSN) in HCl → PSN·$m\text{-}O_2NC_6H_4SbCl_5$ complexes with characteristic colors, sol. in EtOH, MeOH, Me_2CO, and Ac_2O (1238). Papaverine and dibazole in H_2O soln. → ppts. (1240).

Use: A color test reagent for the detection and detn. of phenothiazine derivatives: aminazine - bright raspberry, dinezine - raspberry rose, etapirazine - violet, and metarazine - raspberry (1238).

p-Nitrobenzenestibonic Acid, $SbC_6H_6NO_5$, $4-O_2NC_6H_4SbO_3H_2$, (M-786)
Prop.: Dissolves in H_2O with an orange color (1239).
Rxn. with: Phenothiazine derivs. (PSN) in HCl → $PSN \cdot p-O_2NC_6H_4SbCl_5$ complexes with characteristic bright colors, sol. in EtOH, MeOH, Me_2CO, and Ac_2O (1238).
Mg^{++} in alk. soln. → a cornflower-blue ppt., sparingly sol. in H_2O (1239).
Cu, Hg, Ti, Pb, and Bi ions → colored ppts. (1239).
Dibazole in H_2O soln. → cryst. ppt. (1240).
Uses: A color test reagent for the detection and detn. of phenothiazine derivs.: aminazine - pale raspberry, dinezine - pale red, diprazine - pale raspberry, etapirazine - pale violet, and metarazine - pale raspberry (1238). Detn. of Mg^{++} in the presence of other cations (1239).

o-Aminobenzenestibonic Acid, $SbC_6H_8NO_3$, $2-H_2NC_6H_4SbO_3H_2$
Synth.: From $o-O_2NC_6H_4SbO_3H_2$ by catalytic redn. over Raney Ni (259a).
Rxn. with: $NaNO_2$ and coupling the diazonium compd. with chromotropic acid → 4,5-dihydroxy-3,6-bis[(2-stibonophenyl)azo]-2,7-naphthalenedisulfonic acid (259, 259a).

m-Aminobenzenestibonic Acid, $SbC_6H_8NO_3$, $3-H_2NC_6H_4SbO_3H_2$
Rxn. with: Caffeine, methylcaffeine, theobromine, and theophylline, resp., in HCl soln. → cryst. precipitates, with specific crystal shapes and characteristic optical props. useful in identifying and distinguishing these purine derivs. (1241).

p-Aminobenzenestibonic Acid, $SbC_6H_8NO_3$, $4-H_2NC_6H_4SbO_3H_2$, (M-786)
Rxn. with: Phenothiazine derivs. (PSN) in HCl → $PSN \cdot H_2NC_6H_4SbCl_5$ complexes with characteristic bright colors, sol. in EtOH, MeOH, Me_2CO, and Ac_2O (1238). Caffeine, methylcaffeine, and theophylline, resp., in HCl soln. → cryst. ppts., with specific crystal shapes and optical props., useful in identifying and distinguishing these purine derivs. The ppts. probably consist of the alkaloid $\cdot H[4-H_2NC_6H_4SbCl_5]$ salts (1241).
Use: A color test reagent for the detection and detn. of phenothiazine derivs.: aminazine - bright raspberry, dinezine - red, diprazine - raspberry rose, etapirazine - violet, and metarazine - raspberry (1238).

p-Acetamidobenzenestibonic Acid, $SbC_8H_{10}NO_4$, $4-AcNHC_6H_4SbO_3H_2$, (M-788)
Rxn. with: Caffeine and methylcaffeine, resp., in HCl soln. → cryst. ppts. with specific crystal shapes and optical props., useful in identifying and distinguishing these purine derivs. (1241).

2.5.6.1.4 DISTIBONIC ACIDS

This group of antimony derivatives described in the main volume (M-796-99) comprised numerous distibonic acids and one tristibonic acid.

In the period covered by this supplement only one new distibonic acid, an analog of Arsenazo III, was prepared.

4,5-Dihydroxy-3,6-bis[(2-stibonophenyl)azo]-2,7-naphthalenedisulfonic Acid, $SbC_{22}H_{18}N_4O_{14}S_2$,

Synth.: From $2\text{-}O_2NC_6H_4SbO_3H_2$ by catalytic hydrogenation over Raney Ni, followed by diazotization and coupling with chromotropic acid. 46% yield (259, 259a).

Complex Salts with Th^{4+}, UO_2^{++}, and Cd^{++}, UV spectra and the pH effect on their extinction coeffs. (259a).

2.5.6.1.9 TETRAKISARYLOXY DERIVATIVES

No compounds of this type were reported in the main volume.

Butyltetrakis(pentachlorophenoxy)antimony, $SbC_{28}H_9Cl_{20}O_4$, $BuSb(OC_6H_5)_4$
Use: A fungicide and bactericide for paints, plastics, and synth. fibers (947).

Phenyltetrakis(quinolinolato)antimony, $SbC_{42}H_{29}N_4O_4$,
Use: A fungicide and bactericide for paints, plastics, and synth. fibers (947). $PhSb(O\text{-quinolinolato})_4$

3. HETEROCYCLIC ANTIMONY COMPOUNDS

3.2.1 FOUR-MEMBERED RINGS WITH ANTIMONY HETEROATOMS

3.2.1.4 TETRASTIBETANE DERIVATIVES

The following stibetane derivative is a new compound.

Tetra-tert-butyltetrastibetane, $Sb_4C_{16}H_{36}$,

$$
\begin{array}{ccc}
\text{t-BuSb} & \!\!\!-\!\!\! & \text{Sb-t-Bu} \\
| & & | \\
\text{t-BuSb} & \!\!\!-\!\!\! & \text{Sb-t-Bu}
\end{array}
$$

<u>Synth.</u>: From t-Bu$_2$SbLi·Dioxan + iodine in C_6H_6 under reflux, 40% yield (785).
By-prod. in the rxn. of t-Bu$_2$SbCl + LiAlH$_4$ in Et$_2$O at -78°, followed by distn. (785).
<u>Prop.</u>: Red amorphous powder, sensitive to air, soly. data (785).

3.2.5 RING SYSTEMS WITH ANTIMONY AND OXYGEN HETEROATOMS

No four-membered rings with Sb and O heteroatoms were reported in the main volume.

2,2,2-Tributyl-2,2-dihydro-1,3,2-dioxastibetan-4-one, $SbC_{13}H_{27}O_3$,
$$Bu_3\overline{SbOC(O)O}$$
<u>Synth.</u>: From Bu$_3$SbO + CO$_2$ in Et$_2$O at r.t. (223).
<u>Prop.</u>: Highly viscous, clear, nondistillable oil, n_D^{20} 1.5141, dec. at 50-60° → Bu$_3$SbO + CO$_2$ (223).

2,2-Dihydro-2,2,2-triisobutyl-1,3,2-dioxastibetan-4-one, $SbC_{13}H_{27}O_3$,
$$i\text{-}Bu_3\overline{SbOC(O)O}$$
<u>Synth.</u>: From i-Bu$_3$SbO + CO$_2$ in Et$_2$O at r.t. (223).
<u>Prop.</u>: Nondistillable oil, n_D^{27} 1.4896 (223).

3.2.6 RING SYSTEMS WITH ANTIMONY AND SULFUR HETEROATOMS

The following dithiastibetane derivatives are new compounds.

4-Cyanoimino-2-(2-naphthyl)-1,3,2-dithiastibetane, $SbC_{12}H_7N_2S_2$,
<u>Synth.</u>: From 2-naphthyl-SbCl$_2$ + NCN=C(SK)$_2$ in DMF (1089).

$$C_{10}H_7Sb \overset{S}{\underset{S}{\diagup\!\!\!\diagdown}} {=}NCN$$

4-Cyanoimino-2,2-dihydro-2,2,2-triphenyl-1,3,2-dithiastibetane, $SbC_{20}H_{15}N_2S_2$,
<u>Synth.</u>: From Ph$_3$SbCl$_2$ + NCN=C(SK)$_2$ in DMF (1089).
<u>Prop.</u>: Ppt. (1089).

$$Ph_3Sb \overset{S}{\underset{S}{\diagup\!\!\!\diagdown}} {=}NCN$$

2,2,2-Tributyl-2,2-dihydro-1,3,2-dithiastibetane-4-thione, $SbC_{13}H_{27}S_3$,
    $Bu_3\overline{SbSC(S)S}$

<u>Synth.</u>: From $Bu_3Sb + CS_2 + S$ dissolved in CS_2 under reflux, 81% yield (223).
<u>Prop.</u>: Dark yellow, nondistillable oil, n_D^{23} 1.5630 (223).

3.2.9   RING SYSTEMS WITH THREE DIFFERENT HETEROATOMS

3.2.9.1   COMPOUNDS WITH ANTIMONY, OXYGEN, AND SULFUR HETEROATOMS

The following stibetane derivatives are new compounds.

2,2,2-Tributyl-2,2-dihydro-1,3,2-oxathiastibetan-4-one, $SbC_{13}H_{27}O_2S$,
    $Bu_3\overline{SbOC(O)S}$

<u>Synth.</u>: From $Bu_3SbO + COS$ in Et_2O, ~ 100% yield (223).
<u>Prop.</u>: Clear, amber, nondistillable oil, n_D^{24} 1.5317. The ring structure has
not been definitely established and the possibility of a 1,3,2-dioxastibetane-
4-thione ring should be taken into consideration (223).

2,2,2-Tributyl-2,2-dihydro-1,3,2-oxathiastibetane-4-thione, $SbC_{13}H_{27}OS_2$,
    $Bu_3\overline{SbSC(S)O}$

<u>Synth.</u>: From $Bu_3SbO + CS_2$ in Et_2O at r.t., ~ 100% yield (223).
<u>Prop.</u>: Clear, yellow, nondistillable oil, n_D^{26} 1.5430 (223).

2,2-Dihydro-2,2,2-triisobutyl-1,3,2-oxathiastibetane-4-thione, $SbC_{13}H_{27}OS_2$,
    $i-Bu_3\overline{SbOC(S)S}$

<u>Synth.</u>: From $i-Bu_3SbO + CS_2$ in Et_2O, ~ 100% yield (223).
<u>Prop.</u>: Clear, dark yellow oil, n_D^{27} 1.5318 (223).

3.2.9.2   COMPOUNDS WITH ANTIMONY, OXYGEN, AND NITROGEN HETEROATOMS

No compounds of this type were reported in the main volume.

2,2,2-Tributyl-3-(m-chlorophenyl)-2,2-dihydro-1,3,2-oxazastibetan-4-one,
    $SbC_{19}H_{31}ClNO_2$, $Bu_3\overline{SbOC(O)NC_6H_4Cl}$

<u>Synth.</u>: From $Bu_3SbO + 3-ClC_6H_4NCO$ in C_6H_6 (223).
<u>Prop.</u>: Yellow oil (223).

2,2,2-Tributyl-2,2-dihydro-3-phenyl-1,3,2-oxazastibetan-4-one,
    $SbC_{19}H_{32}NO_2$, $Bu_3\overline{SbOC(O)NPh}$

<u>Synth.</u>: From $Bu_3SbO + PhNCO$ in C_6H_6 (223).
<u>Prop.</u>: Yellow oil (223).

2,2,2-Tributyl-2,2-dihydro-3-n-octyl-1,3,2-oxazastibetan-4-one,
    $SbC_{21}H_{44}NO_2$, $Bu_3\overline{SbOC(O)NC_8H_{17}}$

<u>Synth.</u>: From $Bu_3SbO + n-C_8H_{17}NCO$ in C_6H_6 (223).
<u>Prop.</u>: Yellow oil (223).

2,2,2-Tributyl-3-ethyl-2,2-dihydro-1,3,2-oxazastibetan-4-one,
   SbC$_{15}$H$_{32}$NO$_2$,   Bu$_3$SbOC(O)NEt
<u>Synth.</u>: From Bu$_3$SbO + EtNCO in C$_6$H$_6$ at 30°, 98% yield (223).
<u>Prop.</u>: Thick, viscous oil (223).

3.3   FIVE-MEMBERED RING SYSTEMS

3.3.1   RING SYSTEMS WITH ONE HETEROATOM

3.3.1.6 DIBENZOSTIBOLE DERIVATIVES

   Numerous dibenzstibole derivatives were described in the main volume on
pages 800-805.

2,3,7,8-Bis(methylenedioxy)-5-phenyldibenzostibole,   SbC$_{20}$H$_{13}$O$_4$,
<u>Synth.</u>: From PhSbCl$_2$ + 2,2'-dibromo-4,5,4',5'-
bis(methylenedioxy)biphenyl + BuLi in Et$_2$O-C$_6$H$_6$,
46% yield (434).
<u>Prop.</u>: Colorless crystals. m. 242-244° (434).

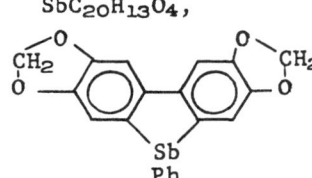

3.3.5   RING SYSTEMS WITH ANTIMONY AND OXYGEN HETEROATOMS

3.3.5.2 DIOXASTIBOLANE DERIVATIVES

   A few dioxastibolane derivatives were reported in the main volume on pages
806-7.

2,2-Dihydro-4,4,5,5-tetramethyl-2,2,2-tri-p-tolyl-1,3,2-dioxastibolane,
   SbC$_{27}$H$_{33}$O$_2$,
<u>Prop.</u>: NMR spectrum indicated that the compd.   (p-MeC$_6$H$_4$)$_3$Sb
in soln. exists as an equilibrium mixt. of tri-
gonal bipyramidal and square pyramidal configurations (1397).

3.3.6   RING SYSTEMS WITH ANTIMONY AND SULFUR HETEROATOMS

3.3.6.1 DITHIASTIBOLE DERIVATIVES

   Several compounds of this type were described in the main volume on pages
808-10.

4,5-Dicarbamoyl-2-methyl-1,3,2-dithiastibole,   SbC$_5$H$_7$N$_2$O$_2$S$_2$,
<u>Synth.</u>: From MeSbCl$_2$ + H$_2$NCOC(SNa)=C(SNa)CONH$_2$ in
MeCN under reflux (1090).   MeSb

2-(2-Naphthyl)-1,3,2-dithiastibole-4,5-dicarbonitrile, $SbC_{14}H_7N_2S_2$,
Synth.: From 2-naphthyl-$SbCl_2$ + $NaSC(CN)=C(CN)SNa$
in MeCN under reflux (1090).

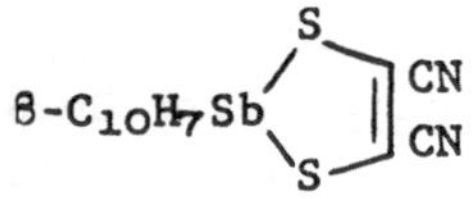

β-$C_{10}H_7Sb$

2,2-Dihydro-2,2,2-triphenyl-1,3,2-dithiastibole-4,5-dicarbonitrile,
     $SbC_{22}H_{15}N_2S_2$,
Synth.: From Ph_3SbCl_2 + $NaSC(CN)=C(CN)SNa$ in
$MeOCH_2CH_2OMe$ at 25-30° (1090).
Prop.: m. 171-172° (1090).

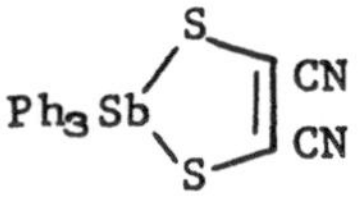

Ph_3Sb

3.6   EIGHT-MEMBERED RING SYSTEMS

3.6.7 RING SYSTEMS WITH ANTIMONY AND NITROGEN HETEROATOMS

The following eight-membered $(Sb-N)_4$ ring is a new compound.

1,1,3,3,5,5,7,7-Octaphenyl-2,4,6,8,1,3,5,7-tetrazatetrastibocine,
     $Sb_4C_{48}H_{20}N_4$,
Synth.: From Ph_2SbCl_3 + NH_4Cl + NH_3 at -35° (679).
Prop.: Diamagnetic, nonionizable solid, m. 282° (679).

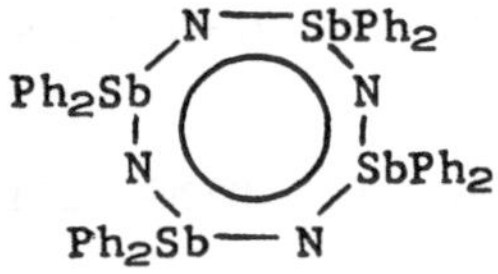

BISMUTH

1. COMPOUNDS OF TRIVALENT BISMUTH

1.1 TERTIARY BISMUTHINES

1.1.1 SYMMETRIC DERIVATIVES

Symmetric tertiary bismuthines and their methods of synthesis were reported in the main volume on pages 815-23. Besides the methods described in the main volume, the bismuthines are also formed from bismuth trichloride by the reactions with dialkylzinc and ferrocenylsilver derivatives, respectively, from bismuth oxide by the reaction with arenediazonium salts, from secondary halobismuthines by disproportionation in pyridine, and from bismuth hydroxide in aqueous hydrogen fluoride solution containing ammonium fluoride by a reaction with trifluorophenylsilane.

Some bismuthines have found application as biocides. For example, tripropyl-, tricyclohexyl-, and trioctyl-bismuthine are useful as bactericides and fungicides for surface application and for incorporation in various synthetic materials (944). Moreover, tri-p-chlorophenyl-, tris[m-(trifluoromethyl)-phenyl]- and tri-α-naphthyl-bismuthine have been useful as the active ingredients in marine antifouling compositions (661).

Trimethylbismuthine, BiC_3H_9, Me_3Bi,   (M-816)
Synth.: From $BiBr_3$ + Me_2Zn (excess) in Et_2O under N, 50-60% yield (1343).
Prop.: Colorless liquid, b_{760} 102-106° (1343). IR spectrum (209, 1491) - the effect of the adjacent groups on the rocking and symmetric deformation frequencies of the Me groups and of the deuteriated analog (1491). NMR spectrum and its correlation with the spectra of Me_3N, Me_3P, Me_3As, and Me_3Sb (880). Mass spectrum (881).
Rxn. with: Br in C_5H_{12} → Me_2BiBr and $MeBiBr_2$ (1343).

Triethylbismuthine, BiC_6H_{15}, Et_3Bi,   (M-816)
Prop.: IR and Raman spectra (209).
Rxn. with: R_3SiH 1:3 at 165-170° → $(R_3Si)_3Bi$ (1271, 1575, 1576, 1577).
Et_3GeH (excess) at 130-135° → $(Et_3Ge)_2BiEt$ (900).
Et_3GeH 1:1 at 145-150° → $Et_3GeBiEt_2$ (900).
Et_3GeH 1:3 at 140-145° → $(Et_3Ge)_3Bi$ (1576, 1577).
Et_3SnH 1:3 at 100° → $(Et_3Sn)_3Bi$ (1576, 1577).
Coordination Deriv.: $[Mo(BiEt_3)(CO)_5]$ (M-816), IR spectrum (435).
Uses: A chain-transfer agent in a free-radical polymerization of vinyl monomers to low mol.-wt. polymers (1218). A catalyst for polyurethan foam formation from diisocyanates and polyesters (757). Combinations with $TiCl_3$ or $MnCl_2$ catalyze a polymerization of Me methacrylate to highly cryst. polymers (909).

Tributylbismuthine, BiC$_{12}$H$_{27}$, Bu$_3$Bi, (M-821)

Use: A bactericide and fungicide for surface application and for incorporation with various substances (944). Complexes with transition metal hydroborates catalyze a polymerization of α-olefins (1070).

Tricyclopentadienylbismuthine, BiC$_{15}$H$_{15}$, (C$_5$H$_5$)$_3$Bi, (M-817)

Synth.: A procedure for obtg. the compd. labeled on the Bi atom includes the use of small-scale syntheses, isotope and nonisotopic exchange rxns. and β decay in the labeled product (1063).

Triphenylbismuthine, BiC$_{18}$H$_{15}$, Ph$_3$Bi, (M-818)

Synth.: From Ph$_2$BiBr dissolved in Py by adding to petr. ether, 87% yield, along with PhBiBr$_2\cdot$2 Py. Under these conditions Ph$_2$BiCl gave Ph$_3$Bi with a yield of 95% (1169).
From Bi(OH)$_3$ in a mixt. with 40% aq. HF and 32% aq. NH$_4$F by adding PhSiF$_3$ (0.5:2:6.1:1.5), 50% yield (1106).
From Ph$_3{}^{210}$PbI by β-disintegration, a mixt. of Ph$_3{}^{210}$Bi with Ph$_2{}^{210}$BiI, Ph^{210}PbI$_2$, and inorg. ^{210}Bi^{3+} is formed and the compds. are sepd. by paper chromatography (1627).
Prop.: White solid, m. 77-78° (1169), m. 78° (1106, 1658). Soly. in styrene at 20° and 60° was detd. (123). IR spectrum (803). Far-IR spectrum (244, 803). Raman spectrum (797). Half-wave potential data (467). The crystal and mol. structure by x-ray diffraction analysis (704, see also 706). A spot test for Ph$_3$Bi in soln. on a filter paper by adding iodine soln. produces a brown or yellow spot - Ph$_3$BiI$_2\cdot$nI$_2$ (555). Microanalysis by gas-liquid chromatography, after hydrogenolysis over Pd/Al$_2$O$_3$ at 200° (1508). Thermionic response of K, Rb, and Cs salts insensitive in gas-chromatographic detn. of Ph$_3$Bi (795).
Rxn. with: I$_2$ 1:3 → BiI$_3$ + 3 PhI (178).
IBr → Ph$_2$BiBr and PhBiBr$_2$ (178).
Sb at 300° → Ph$_3$Sb (379).
(C$_6$F$_5$)$_2$TlBr in Et$_2$O under reflux → Ph$_2$TlBr (456).
Diazotetraphenylcyclopentadiene → Ph$_3$Bi$\overline{\text{CCPh=CPhCPh=CPh}}$ (976).
Photodec. and thermodec. at 260-270° in ^{14}C$_6$H$_6$ → ^{14}C$_6$H$_5$Ph and Ph$_2$ (2.5%), but no exchange with Ph$_3$Bi (1270).
Photodec. and thermodec. at 260-270° in C$_6$D$_6$ → C$_6$H$_5$-Ph* ranging from C$_{12}$D$_5$H$_5$ to C$_{12}$DH$_9$, Ph$_2$, and C$_6$H$_5$*-C$_6$H$_5$* ranging from C$_{12}$D$_6$H$_4$ to C$_{12}$D$_{10}$, but no deuteriated Ph$_3$Bi (1270).
Hexamethylbenzene, no complex formation (1393).
BiCl$_3$ 2:1 in xylene, toluene, or mesitylene under reflux → Ph$_2$BiCl (1555).
RCO$_2$H in MePh under reflux → PhBi(O$_2$CR)$_2$ (1095, 1096).
NH$_3$ 1:3 → 3 C$_6$H$_6$ + Bi(NH$_2$)$_3$ → BiN + 2 NH$_3$ (1362, 1364).
H$_2$ over Pd/Al$_2$O$_3$ at 200° → hydrogenolysis useful in a microanalysis by gas-liquid chromatographic detn. of C$_6$H$_{12}$ (1508).
BiBr$_3$ 1:2 in Et$_2$O → PhBiBr$_2$ (1505).
BiCl$_3$ 3:2 in Et$_2$O → Ph$_2$BiCl (1505).
SO$_2$Cl$_2$ in petr. ether → Ph$_3$BiCl$_2$ (1505).

p-MeC$_6$H$_4$Li in Et$_2$O in a sealed tube at r.t., followed by CO$_2$ in Et$_2$O → a mixt.
of BzOH and p-MeC$_6$H$_4$CO$_2$H; the rxn. may proceed through the equilibria:
Ph$_3$Bi + p-MeC$_6$H$_4$Li ⇌ Li[Ph$_3$BiC$_6$H$_4$Me] ⇌ Ph$_2$BiC$_6$H$_4$Me + PhLi (1620).
^{60}Co γ-radiation → a radio-sensitized dec. to Ph$_2$, phenylcyclohexadienes, and
tetrahydroquaterphenyls (1219).
Szilard-Chalmers rxn. - η, γ rxn. → Ph$_3$Bi, Ph$_2$Bi$^+$, PhBi^{++}, and inorg. Bi
(1605).
Coordination Deriv.: [π-C$_5$H$_5$Mn(BiPh$_3$)(CO)$_2$], beige ppt., unstable in O,
prepd. from π-C$_5$H$_5$Mn(CO)$_3$ + Ph$_3$Bi in THF under UV light at 75°. IR and NMR
spectra. Dipole moment (120).

[Mo(BiPh$_3$)(CO)$_5$] (M-820), IR spectrum (435).

[Mo(BiPh$_3$)(CO)$_3$(Phenanthroline)], black crystals, prepd. from [Mo(CO)$_4$(Phenanthroline)] + Ph$_3$Bi in xylene. IR spectrum (758).

[Ph$_3$BiNbCl$_5$], colored ppt., prepd. from Ph$_3$Bi + NbCl$_5$ in CCl$_4$, n-C$_6$H$_{14}$, and
cyclo-C$_6$H$_{12}$ (466).

[(Ph$_3$Bi)$_2$NbCl$_5$], colored ppt., prepd. from Ph$_3$Bi + NbCl$_5$ in C$_6$H$_6$ (466).

Uses: A germicide for textile matls. (62). An active constituent of marine
antifouling compns. (661). In a combination with neomycin is a durable antibacterial finish for cellulosic textile fibers (662).
A photographic fog inhibitor (743). A background fog inhibitor for photothermographic sheets (744). An activator for photographic direct-positive
emulsion (745). A constituent of the photosensitive compns. for photographic
plates (746).
A chain-transfer agent in a free-radical polymerization of styrene, Me methacrylate, acrylonitrile, and vinyl chloride (1218).
The bismuthine catalyzes the formation of polyurethan foams from diisocyanates
and polyesters (757). It is effective as a photoinitiator of a polymerization of olefins by UV light (1024). A 1:1 complex with SnCl$_4$ or with a
chloride of a Group IVA or VA metal promotes a polymerization of polar monomers (800). Mixts. or complexes with halides, alkyl halides, or acetylace-
tonates of Ti, V, Zr, Cr, or Mo and with a chloride of Na, Li, K, Al, Zn, Mg,
or Cu catalyze a polymerization of ethylene or propylene to cryst., high-
mol. wt. polymers (1036). Mixts. with Cu$_2$Cl$_2$ and AlBr$_3$ catalyze a polymerization of olefins (554). Mixts. with ion-exchange resins and with halides of
Group IVB, VB, VIB, or VIII metals catalyze a polymerization of α-olefins
(549). A rxn. prod. formed from Ph$_3$Bi, EtAlCl$_2$, and RhCl$_3$·3 H$_2$O dimerizes
propylene (770).
Ph$_3$Bi catalyzes the ester formation of aromatic carboxylic acids with glycols
(1650).
The flammability of epoxy resins is moderately inhibited by the bismuthine
(1025). Incorporation of 2-5% Ph$_3$Bi in polyimides makes them resistant to a
high temp. corona discharge (1046).
Compns. contg. Ph$_3$Bi, aryl dialkylphosphinates, and polymeric thickeners are
effective corrosion inhibitors for hydraulic fluids (1039). Ph$_3$Bi is useful
as a stabilizer for poly(vinyl chloride) (220).

The bismuthine is useful as a flux for fusible solders such as Sn-Pb, Sn-In-Ag, Sn-Bi, and Sn-Cd (1224).

Complexes with $CuCl_2$ and with salicylic acid are useful as semiconductors for thermistors (1434). The bismuthine is a quencher for 9,10-diphenylanthracene scintillator in ethylnaphthalene (303). Scintillators for the detection of γ-radiation contain up to 40% Ph_3Bi in polystyrene along with 3% p-terphenyl and 0.08% 1,3,5-triphenyl-2-pyrazoline (123).

Tris(p-sulfamoylphenyl)bismuthine, $BiC_{18}H_{18}N_3O_6S_3$, $(4-H_2NO_2SC_6H_4)_3Bi$, (M-822)

Synth.: From $^{210}Bi_2O_3$ + $4-H_2NO_2SC_6H_4N_2^+$, a ^{210}Bi-tagged prod. was obtd. (642).

Prop.: β-decay of the ^{210}Bi-tagged prod. $\rightarrow$ $[(4-H_2NO_2SC_6H_4)_3Po]^+$ (642).

Tri-m-tolylbismuthine, $BiC_{21}H_{21}$, $(3-MeC_6H_4)_3Bi$

Synth.: From ^{210}Bi-labeled $BiCl_3$ + $3-MeC_6H_4MgX$, ^{210}Bi-labeled bismuthine was obtd. (1130).

Prop.: β-decay $\rightarrow$ $(3-MeC_6H_4)_3Po^+$ and $(3-MeC_6H_4)_2Po$ (1130).
"Slow beats" in the quadrupole spin echo for ^{209}Bi was detd. (46).

Tri-o-tolylbismuthine, $BiC_{21}H_{21}$, $(2-MeC_6H_4)_3Bi$, (M-822)

Synth.: From ^{210}Bi-labeled $BiCl_3$ + $2-MeC_6H_4MgX$, a ^{210}Bi-tagged bismuthine was obtd. (1130).

Prop.: β-decay $\rightarrow$ $(2-MeC_6H_4)_3Po^+$ + $(2-MeC_6H_4)_2Po$ (1130).

Tris(p-methoxyphenyl)bismuthine, $BiC_{21}H_{21}O_3$, $(4-MeOC_6H_4)_3Bi$, (M-823)

Synth.: From ^{210}Bi-tagged $BiCl_3$ + $4-MeOC_6H_4MgX$, a ^{210}Bi-tagged bismuthine was obtd. (1127).

Prop.: β-decay in Me_2CO, C_6H_6, and in the cryst. state $\rightarrow$ $(4-MeOC_6H_4)_3Po^+$ and $(4-MeOC_6H_4)_2P$ (1127).

Tris(p-vinylphenyl)bismuthine, $BiC_{24}H_{21}$, $(4-CH_2=CHC_6H_4)_3Bi$

Synth.: From $BiCl_3$ + $4-CH_2=CHC_6H_4MgCl$ in THF in the presence of hydroquinone, followed by stirring at r.t., 36% yield (572).

Prop.: Bright yellow crystals, m. 72-74° (572); at 100° polymerizes to a transparent yellow mass (884).

Use: A thermal stabilizer for vinyl copolymers (572).

Polymers and Copolymers: The bismuthine polymerizes when heated in a sealed tube at 100°, in the presence of an initiator, to a solid, transparent, yellow mass which is insol. in org. solvents and infusible due to crosslinking. The prod. loses 13.3% of its weight when heated at 300° (884). A copolymer with styrene, prepd. at 40-100° without any initiator, is a transparent, colored, insol. and infusible mass which loses 3-6% weight when heated at 250°. When the copolymerization is carried out at 120-125° in the presence of $(t-BuO)_2$, an insol. and infusible powdery prod. is obtd. A copolymer with Me methacrylate, prepd. at 120-125° in the presence of $(t-BuO)_2$ is an insol. and infusible colorless powder which at 250° loses 19% weight (884).

Tri-2,5-xylylbismuthine, $BiC_{24}H_{27}$, $(2,5-Me_2C_6H_3)_3Bi$
Synth.: From ^{210}Bi-tagged $BiCl_3$ + $2,5-Me_2C_6H_3MgX$, a ^{210}Bi-tagged bismuthine was obtd. (1129).
Prop.: β-decay → $(2,5-Me_2C_6H_3)_3Po^+$ and $(2,5-Me_2C_6H_3)_2Po$ (1129).

Tris(p-ethoxyphenyl)bismuthine, $BiC_{24}H_{27}O_3$, $(4-EtOC_6H_4)_3Bi$, (M-823)
Synth.: From ^{210}Bi-tagged $BiCl_3$ + $4-EtOC_6H_4MgBr$, a ^{210}Bi-tagged bismuthine was obtd. which was purified by thin layer and paper chromatography (1125).
Prop.: β-decay of the RaE species → $(4-EtOC_6H_4)_3Po^+$ and $(4-EtOC_6H_4)_2Po$ (1125).

Trimesitylbismuthine, $BiC_{27}H_{31}$, $(2,4,6-Me_3C_6H_2)_3Bi$, (M-823)
Synth.: From ^{210}Bi-tagged $BiCl_3$ + $2,4,6-MeC_6H_2MgBr$, a ^{210}Bi-tagged bismuthine was obtd. (1126).
Prop.: Cryst. substance; β-decay → $(2,4,6-Me_3C_6H_2)_3Po^+$, $(2,4,6-Me_3C_6H_2)_2Po^{++}$, and $(2,4,6-Me_3C_6H_2)_2Po$. The conjugation effect on the stability of the bismuthine is considered (1126).

Tris(2-chloroferrocenyl)bismuthine, $BiFe_3C_{30}H_{24}Cl_3$, $(C_5H_5FeC_5H_3Cl)_3Bi$
Synth.: From 2-chloroferrocenylsilver + $BiCl_3$ in C_6H_6 at r.t., 46% yield, purified by chromatography (1338).
Prop.: Dec. > 230°. IR spectrum (1338).
Rxn. with: HCl followed by H_2S → Bi_2S_3 + FcCl (1338).

Tris(1'-chloroferrocenyl)bismuthine, $BiFe_3C_{30}H_{24}Cl_3$, $(ClFcd)_3Bi$
Synth.: From ClFcdAg + $BiCl_3$ in C_6H_6 at r.t., 67% yield, purified by TLC (1338).
Prop.: m. 203.5-204.5°, IR spectrum (1338).
Rxn. with: Concd. HCl in C_6H_6, followed by H_2O and H_2S → Bi_2S_3 and FcCl (1338).

1.1.4 DITERTIARY BISMUTHINES

A few ditertiary bismuthines were described in the main volume on pages 826-27.

p-Phenylenebis(diphenylbismuthine), $Bi_2C_{30}H_{24}$, $Ph_2BiC_6H_4BiPh_2$
Synth.: From Ph_2BiCl in Et_2O + $p-C_6H_4Li_2$ in petr. ether under reflux, 14-20% yield (1658).
Prop.: White crystals, m. 160-162°, sol. in hydrocarbons, $CHCl_3$, Me_2CO, and Et_2O, sparingly sol. in EtOH and petr. ether. UV and IR spectra (1658).
Rxn. with: Br → the corr. tetrabromide (1658).

1.2 SECONDARY BISMUTHINES

Dimethylbismuthine, Me_2BiH (M-827), and diphenylbismuthine, Ph_2BiH, in mixtures or complexes with halides or oxyhalides of Ti, V, or Cr, deposited on SiO_2 or Al_2O_3, are effective polymerization catalysts (42).

1.4 COMPOUNDS WITH Bi-Bi BONDS

1.4.1 DIBISMUTHINES

No dibismuthines were reported in the main volume.

Tetraphenyldibismuthine, $Bi_2C_{24}H_{20}$, $Ph_2BiBiPh_2$
Synth.: From Ph_2BiCl by electrochem. redn. (467).
Prop.: Half-wave potential (467, 472).
Rxn. with: e^-, electrochem. redn. → a green soln. (467).
Polarographic redn., $2e^-$ → $2 Ph_2Bi^-$ (472).

1.5 COMPOUNDS WITH BISMUTH-HALOGEN AND BISMUTH-PSEUDOHALOGEN BONDS

1.5.1 SECONDARY HALOBISMUTHINES

The halobismuthines and their methods of synthesis were reported in the main volume on pages 828-30.

Numerous secondary halobismuthines are useful as fungicides and bactericides for surface applications and/or for incorporation in plastics, paints, and fibrous materials (945, 946). These compounds include: Me_2BiCl (M-829), Et_2BrBr, $Ph(CH_2=CH)BiCl$, $(p-ClC_6H_4)_2BiBr$ (M-829), $(p-ClC_6H_4)_2BiI$ (M-830), $(p-ClC_6H_4)_2BiCl$, Ph_2BiI (M-830), $(C_6H_{11})_2BiCl$, and $(p-MeC_6H_4)_2BiCl$ (M-830).

Bromodimethylbismuthine, BiC_2H_6Br, Me_2BiBr, (M-829)
Synth.: From $Me_3Bi + Br$ in C_5H_{12} at ice temp., a mixt. of Me_2BiBr and $MeBiBr_2$ was obtd. (1343).
Rxn. with: $LiN(Me)SiMe_3$ → $Me_2BiN(Me)SiMe_3$ (1343).
Use: A fungicide and bactericide for surface application and for incorporation in plastics and fibrous matls. (945).

Bromodiphenylbismuthine, $BiC_{12}H_{10}Br$, Ph_2BiBr, (M-829)
Rxn. with: Py in petr. ether → $PhBiBr_2 \cdot 2 Py + Ph_3Bi$ (1169).

Chlorodiphenylbismuthine, $BiC_{12}H_{10}Cl$, Ph_2BiCl, (M-828)
Synth.: From $Ph_3Bi + BiCl_3$ (3:2) in Et_2O, 78% yield (1505).
From $Ph_3Bi + BiCl_3$ (2:1) in MePh, xylene, or mesitylene under reflux, 88, 97, and 99% yield, resp. (1555).
Prop.: m. 180-184° (1505). Pale yellow solid, m. 185-187° (1555). Half-wave potential (467, 471).
Rxn. with: $NaSC(S)NMe_2$ 1:1 in $CHCl_3$ → $PhBi[SC(S)NMe_2]_2 + Ph_3Bi$ (1505).
$NaSC(S)NEt_2 \cdot 3 H_2O$ 1:1 in $CHCl_3$ → $PhBi[SC(S)NEt_2]_2 + Ph_3Bi + Bi[SC(S)NEt_2]_2$ (1505).
NaN_3 in Py → Ph_2BiN_3 (1272).
$CH_2=C(Me)CO_2NH_4$ in C_6H_6 → $Ph_2BiO_2CC(Me)=CH_2$ (1096).
Py in petr. ether → Ph_3Bi and $PhBiCl_2 \cdot 2 Py$ (1169).
Electrochem. redn., e^- → $Ph_2BiBiPh_2$ (467).

Polarographic redn., $2e^- \rightarrow Ph_2Bi^-$, followed by the electron transfer to i-PrBr (471).

Szilard-Chalmers rxn., n,γ rxn. $\rightarrow Ph_3Bi$, $PhBi^{++}$, and inorg. Bi (1605).

Use: A bactericide and fungicide for surface application and/or incorporation in plastics, fibrous matls. (945), and paints (946).

Iododiphenylbismuthine, $BiC_{12}H_{10}I$, Ph_2BiI, (M-830)

Synth.: $Ph_2{}^{210}BiI$ was isolated by paper chromatography from the prods. formed from $Ph_3{}^{210}PbI$ by β-disintegration (1627).

Use: A fungicide and bactericide for plastics, fibrous matls., and paints (946).

1.5.2 PRIMARY DIHALOBISMUTHINES

Several primary dihalobismuthines and their methods of synthesis were reported in the main volume on pages 830-31.

Many primary dihalobismuthines are useful as fungicides and bactericides for surface applications and for incorporation in fibrous materials (945) and in paints (946). These biocidal compounds include: $MeBiBr_2$, $MeBiCl_2$, $EtBiBr_2$, $EtBiCl_2$, $CH_2=CHBiCl_2$, $PrBiBr_2$, $PrBiCl_2$, $BuBiBr_2$, $BuBiCl_2$, $i\text{-}BuBiBr_2$, $i\text{-}BuBiCl_2$, $n\text{-}C_5H_{11}BiBr_2$, $n\text{-}C_5H_{11}BiCl_2$, $p\text{-}MeC_6H_4BiCl_2$, $CH_2=CHC_6H_4BiCl_2$, $BuCH(Et)CH_2BiCl_2$, $n\text{-}C_8H_{17}BiBr_2$, and $n\text{-}C_8H_{17}BiCl_2$ (945, 946).

Dibromomethylbismuthine, $BiCH_3Br_2$, $MeBiBr_2$, (M-831)

Synth.: By-prod. from the rxn. of Me_3Bi with Br in C_5H_{12} at ice temp. (1343).

Rxn. with: $LiN(Me)SiMe_3 \rightarrow MeBi[N(Me)SiMe_3]_2$ (1343).

Dibromophenylbismuthine, $BiC_6H_5Br_2$, $PhBiBr_2$, (M-831)

Synth.: From $Ph_3Bi + BiBr_3$ 1:2 in Et_2O at r.t., 30% yield (1505).

Prop.: m. 203-206° (1505).

Rxn. with: $NaSC(S)NMe_2 \cdot 2 H_2O$ in $CHCl_3 \rightarrow PhBi[SC(S)NMe_2]_2$ (907, 1505).

$NaSC(S)NEt_2 \cdot 3 H_2O$ in $CHCl_3 \rightarrow PhBi[SC(S)NEt_2]_2 + Bi[SC(S)NEt_2]_3$ (1505).

Coordination Deriv.: $PhBiBr_2 \cdot 2 Py$, pale yellow needles, m. 145-146°, prepd. from Ph_2BiBr dissolved in Py by adding petr. ether (1169).

Use: A fungicide and bactericide for surface applications and/or incorporation in plastics, fibrous matls. (945), and paints (946).

Dichlorophenylbismuthine, $BiC_6H_5Cl_2$, $PhBiCl_2$, (M-831)

Coordination Deriv.: $PhBiCl_2 \cdot 2 Py$, white crystals, m. 136-137°, prepd. from Ph_2BiCl dissolved in Py by adding to petr. ether (1169).

Use: A fungicide and bactericide for surface applications and/or for incorporation in plastics, fibrous matls. (945), and paints (946).

Diiodophenylbismuthine, $BiC_6H_5I_2$, $PhBiI_2$, (M-831)

Synth.: $Ph^{210}BiI_2$ was isolated by paper chromatography from a mixt. of prods. formed by β-disintegration of $Ph_3{}^{210}PbI$ (1627).

1.5.5 SECONDARY PSEUDOHALOBISMUTHINES

Cyanodiphenylbismuthine was the only compound of this type described in the main volume (M-832).

Dibutylcyanobismuthine, Bu_2BiCN, cyanodiphenylbismuthine, Ph_2BiCN, and diphenylthiocyanatobismuthine, Ph_2BiSCN, are useful as fungicides and bactericides for surface applications to, and/or incorporation in, plastics, fibrous materials, and paints (945, 946).

Azidodiphenylbismuthine, $BiC_{12}H_{10}N_3$, Ph_2BiN_3
Synth.: From Ph_2SbCl + NaN_3 in Py at r.t., removing the solvent and extg. the prod. with C_6H_6 (1272).
Prop.: Crude gummy solid, IR spectrum (1272).

1.5.6 PRIMARY DIPSEUDOHALOBISMUTHINES

No compounds of this type were reported in the main volume.

Butyl- and isobutyl-dicyanobismuthine, $BuBi(CN)_2$ and $i-BuBi(CN)_2$, are useful as fungicides and bactericides for plastics, fibrous materials, and paints (946).

1.6   COMPOUNDS WITH BISMUTH BONDED TO GROUP VIA ELEMENTS

1.6.1   COMPOUNDS WITH BISMUTH-OXYGEN BONDS

1.6.1.1 SECONDARY HYDROXYBISMUTHINES

No compounds of this type were described in the main volume.

Hydroxydimethylbismuthine, Me_2BiOH, and hydroxydiphenylbismuthine, Ph_2BiOH, are useful as bactericides and fungicides for surface applications and/or for incorporation in plastics, fibrous materials, and paints (945, 946).

1.6.1.2 SECONDARY ALKOXY- AND ARYLOXY-BISMUTHINES

An alkoxybismuthine was reported in the main volume (M-832).

Dibutylphenoxybismuthine, Bu_2BiOPh, is useful as a fungicide and bactericide for surface applications to, and/or for incorporation in, plastics, paints, and fibrous materials (945, 946).

1.6.1.3 SECONDARY ACYLOXYBISMUTHINES

Several compounds of this class are useful as fungicides and bactericides for surface applications and/or incorporation in plastics, fibrous materials, and paints (943, 945, 946). These biologically active compounds include: $(CH_2=CHCH_2)_2BiOAc$, Bu_2BiOAc, $Bu_2BiO_2CC(Me)=CH_2$, $Bu_2BiO_2CCH_2S-i-C_8H_{17}$, and Ph_2BiOAc.

Methacryloyloxydiphenylbismuthine, $BiC_{16}H_{17}O_2$, $Ph_2BiO_2CCMe=CH_2$
<u>Synth.</u>: From Ph_2BiCl + $CH_2=CMeCO_2NH_4$ in C_6H_6 under reflux (1096).
<u>Prop.</u>: m. 220° (dec.) (1096).
<u>Biol. Activity</u>: A bactericide effective against Staphylococcus aureus (1096).

1.6.1.6 PRIMARY OXOBISMUTHINES

Two compounds of this group were described in the main volume (M-832).

Alkyloxobismuthines are useful as fungicides and bactericides for surface applications to, and/or for incorporation in, plastics, fibrous materials, and paints (945, 946). These biocidal compounds include: MeBiO (945, 946), EtBiO, PrBiO, BuBiO, i-BuBiO, $n-C_5H_{11}BiO$, and $n-C_8H_{17}BiO$ (946).

1.6.1.7 PRIMARY DIALKOXY- AND DIARYLOXY-BISMUTHINES

No bismuthines of this type were reported in the main volume.

Several dialkoxy- and diaryloxy-bismuthines have been claimed to be effective fungicides and bactericides for surface applications to, and/or for incorporation in, plastics, fibrous materials, and paints (945, 946). These biocidal bismuthines include: $PrBi(OMe)_2$, $PrBi(OPh)_2$, $BuBi(OMe)_2$, $BuBi(OPh)_2$, $BuBi(OC_6H_4Ph-2)_2$, $i-BuBi(OMe)_2$, and $i-BuBi(OPh)_2$.

1.6.1.8 PRIMARY BIS(ACYLOXY)BISMUTHINES

Only two compounds of this group were reported in the main volume (M-832).

Primary bis(acyloxy)bismuthines are biologically active and have been claimed to be effective fungicides and bactericides useful for surface applications and for incorporation in plastics, fibrous materials, and paints (943, 945, 946). These compounds include: $MeBi(OAc)_2$, $EtBi(OAc)_2$, $PrBi(OAc)_2$, $BuBi(OAc)_2$, $i-BuBi(OAc)_2$, $BuBi(O_2CCH=CH_2)_2$, $BuBi(OBz)_2$, $cyclo-C_5H_9Bi(OAc)_2$, $n-C_5H_{11}Bi(OAc)_2$, $n-C_8H_{17}Bi(OAc)_2$, $BuCH(Et)CH_2Bi(OAc)_2$, and $PhBi(O_2CC_6H_4OH-o)_2$.

Bis(blycoloyloxy)phenylbismuthine, $BiC_{10}H_{11}O_6$, $PhBi(O_2CCH_2OH)_3$
<u>Rxn. with</u>: Diisocyanates → polyurethans resistant to microorganisms (1095).

Bis(methacryloyloxy)phenylbismuthine, $BiC_{14}H_{15}O_4$, $PhBi[O_2CC(Me)=CH_2]_2$
<u>Rxn.</u>: Homopolymers and copolymers with styrene are formed in the presence of Bz_2O_2. The prods. have bactericidal props. (1096).

Bis(p-hydroxybenzoyloxy)phenylbismuthine, $BiC_{20}H_{15}O_6$, $PhBi(O_2CC_6H_4OH-4)_2$
<u>Synth.</u>: From Ph_3Bi + $p-HOC_6H_4CO_2H$ in MePh under reflux (1095).
<u>Prop.</u>: White solid, m. > 260° (1095).
<u>Rxn. with</u>: Diisocyanates → polyurethans resistant to microorganisms (1095).
<u>Biol. Prop.</u>: A bactericide (1095).

Phenylbis(p-vinylbenzoyloxy)bismuthine, $BiC_{24}H_{19}O_4$, $PhBi(O_2CC_6H_4CH=CH_2)_2$
Synth.: From Ph_3Bi + p-CH_2=$CH_6H_4CO_2H$ in MePh under reflux (1096).
Prop.: m. 230° (dec.) (1096).
Rxn. with: Styrene in the presence of $Bz_2O_2 \rightarrow$ a copolymer with bactericidal
prop. (1096).
Biol. Prop.: Effective against Staphylococcus aureus (1096).

1.6.2   COMPOUNDS WITH BISMUTH-SULFUR BONDS

1.6.2.2   SECONDARY ALKYLTHIOBISMUTHINES

An arylthio homolog, Ph_2BiSPh, was described in the main volume on page 833.

Dibutyl(dodecylthio)bismuthine is the only compound of this type reported
in the period covered by this supplement. It is useful and effective as a
fungicide and bactericide for plastics, fibrous materials, and paints (945,
946).

1.6.2.6 PRIMARY THIOXOBISMUTHINES

No bismuthines of this type, RBiS, were reported in the main volume.

Several alkylthioxobismuthines are useful as fungicides and bactericides
for surface applications to, and/or incorporation in, plastics, paints, and
fibrous materials (945, 946). These biocidal compounds include: MeBiS,
EtBiS, PrBiS, BuBiS, i-BuBiS, n-$C_5H_{11}BiS$, and n-$C_8H_{17}BiS$.

1.6.2.7 PRIMARY BIS(THIO)BISMUTHINE DERIVATIVES, $RBi(SR')_2$

A few primary bis(arylthio)bismuthines were described in the main volume
(M-833).

Some compounds of this type have a microbiocidal activity.

Bis(2-hydroxyethylthio)phenylbismuthine, $BiC_{10}H_{15}O_2S_2$, $PhBi(SCH_2CH_2OH)_2$
Synth.: From Ph_3Bi + $HOCH_2CH_2SH$ in MePh under reflux (1095).
Rxn. with: 2,4-Tolylene diisocyanate $\rightarrow$ polyurethan resistant to microorganisms
(1095).
Biol. Prop.: A bactericide (1095).

Phenylbis(phenylthio)bismuthine, $BiC_{18}H_{15}S_2$, $PhBi(SPh)_2$, (M-833)
Use: An active constituent of marine antifouling compns. (661).

Isobutylbis(phenylthio)bismuthine, $BiC_{16}H_{19}S_2$, i-$BuBi(SPh)_2$
Use: A fungicide and bactericide for plastics, fibrous materials, and paints
(946).

Bis(dodecylthio)phenylbismuthine, $BiC_{30}H_{55}S_2$, $PhBi(SC_{12}H_{25})_2$

Use: A fungicide and bactericide for surface application to, and/or incorporation in plastics and fibrous materials (945, 946).

1.6.2.8 PRIMARY BIS(CARBAMOYL)BISMUTHINES

No compounds of this type were described in the main volume.

Bis[(dimethylthiocarbamoyl)thio]phenylbismuthine, $BiC_{12}H_{17}N_2S_4$,
    $PhBi[SC(S)NMe_2]_2$
Synth.: From $PhBiBr_2$ + $NaSC(S)NMe_2 \cdot 2 H_2O$ 1:2 in $CHCl_3$ at 25°, 98% yield
(907, 1505).
From Ph_2BiCl + $NaSC(S)NMe_2 \cdot 2 H_2O$ 1:1 in $CHCl_3$ at r.t., 84% yield along with
Ph_3Bi (1505).
Prop.: Yellow solid, m. 215-216° (1505), m. 210-211° (907). UV and IR spectra (907, 1505).

Bis[(diethylthiocarbamoyl)thio]phenylbismuthine, $BiC_{16}H_{25}N_2S_4$,
    $PhBi[SC(S)NEt_2]_2$
Synth.: From Ph_2BiCl + $NaSC(S)NEt_2 \cdot 3 H_2O$ 1:1 in $CHCl_3$ at r.t., 38% yield
along with Ph_3Bi and $Bi[SC(S)NEt_2]_3$ (1505).
From $PhBiBr_2$ + $NaSC(S)NEt_2 \cdot 3 H_2O$ 1:2 in $CHCl_3$ at r.t., 21% yield along with
60% $Bi[SC(S)NEt_2]_3$ (907, 1505).
Prop.: Yellow prisms, m. 131-133°, UV and IR spectra (907, 1505).

1.7   COMPOUNDS OF BISMUTH BONDED TO GROUP VA ELEMENTS

1.7.1 COMPOUNDS WITH BISMUTH-NITROGEN BONDS

No compounds with Bi-N bonds were reported in the main volume.

(Dimethylbismuthino)methyl(trimethylsilyl)amine, $BiSiC_6H_8$,
    $Me_2BiN(Me)SiMe_3$
Synth.: From Me_2BiBr in mixt. with $MeBiBr_2$ at 0° by adding $LiN(Me)SiMe_3$ in
C_6H_{14}, 37% yield along with $MeBi[N(Me)SiMe_3]_2$ (1343).
Prop.: Light yellow oil, $b_{0.1}$ 31-32°, NMR spectrum (1343).

Methylbismuthylenebis[methyl(trimethylsilyl)amine], $BiSi_2C_9H_{27}N_2$,
    $MeBi[N(Me)SiMe_3]_2$
Synth.: From $MeBiBr_2$ in a mixt. with Me_2BiBr at 0° by adding $LiN(Me)SiMe_3$
in C_6H_{14}, 43% yield along with 37% $Me_2BiN(Me)SiMe_3$ (1343).
Prop.: Light yellow oil, $b_{0.1}$ 70-71°, NMR spectrum (1343).

1.8   COMPOUNDS OF BISMUTH BONDED TO METALS

1.8.1   COMPOUNDS WITH ELEMENTS OF GROUPS IA TO IVA

1.8.1.3   COMPOUNDS WITH GROUP IIIA METALS

Polymers containing phenylbismuthylene groups linked to decaborane(12) units were prepared by a condensation of PhBiO with $(H_3O)_2B_{10}H_{10}$ (499).

1.8.1.4   COMPOUNDS WITH GROUP IVA METALS

Diethyl(triethylgermyl)bismuthine, $BiGeC_{10}H_{25}$, $Et_3GeBiEt_2$
Synth.: From Et_3Bi + Et_3GeH 1:1 at 145-150°, 8.5% yield (900).
Prop.: b_2 132-136°, stable < 200°, dec. at 270° (900).

Ethylbis(triethylgermyl)bismuthine, $BiGe_2C_{14}H_{35}$, $(Et_3Ge)_2BiEt$
Synth.: From Et_3Bi + Et_3GeH excess at 130-135°, 86% yield (900).
Prop.: Light yellow liquid, stable < 200°, dec. at 270° (900).

1.8.2   COMPOUNDS WITH TRANSITION METALS

1.8.2.7   COMPOUNDS WITH GROUP VIIB METALS

Bis(dimethylglyoximato)(diphenylbismuthino)(pyridine)cobalt,
    $BiCoC_{25}H_{29}N_5O_4$, $Ph_2BiCo[ON=C(Me)C(Me)=NOH]_2 \cdot Py$
Synth.: From Ph_2BiX by a rxn. with $HCo[ON=C(Me)C(Me)=NOH]_2 \cdot Py$ (1369).
Prop.: Brown crystals, dec. ~ 225-230°, sol. in org. solvents, stable in alkali but dec. in strong acids (1369).

2. COMPOUNDS OF PENTAVALENT BISMUTH

2.1 QUINQUENARY DERIVATIVES

Pentaphenylbismuth, $BiC_{30}H_{25}$, Ph_5Bi, (M-834)

Synth.: From ^{210}Bi-tagged Ph_3BiCl_2 + PhLi, a ^{210}Bi-tagged prod. was obtd. (1131).

A tritium-tagged prod. was obtd. from Ph_5Bi by a treatment with $[4-^3H]PhLi$ (1:6) in THF, followed by hydrolysis of the $Li[BiPh_6]$ with H_2O (438).

Prop.: The ^{210}Bi-tagged prod. undergoes β-decay → Ph_3Po^+, Ph_2Po, and Ph_4Po (1131). The normal Ph_5Bi forms dark purple crystals, m. 90-100°, the purple melt turns yellow (722).

Rxn. with: BuLi in THF → exchange of the Ph groups for Bu (438).

PhLi in Et_2O at -70° → yellow crystals which upon adding MeOH gave Ph_5Bi (722).

2.2 QUATERNARY DERIVATIVES

2.2.1 BISMUTH YLIDS

The following compound is the first bismuth ylid derivative.

Triphenylbismuthonium Tetraphenylcyclopentadienylide, $BiC_{47}H_{35}$,

Synth.: From Ph_3Bi + diazotetraphenylcyclo-pentadiene by heating at 140° (976).

Prop.: Deep blue solid, m. 195° (dec.), sol. in Et_2O from which it can be pptd. with petr. ether; unstable in soln.,
dissolves in Et_2O, C_6H_6, and Me_2CO, giving deep blue to red-purple soln.; electronic spectrum (976). It is rapidly dec. by NaOH in MeOH (976).

Rxn. with: $HClO_4$ or picric acid → no bismuthonium salts (976).

2.2.2 QUATERNARY BISMUTHONIUM SALTS

A few bismuthonium salts were described in the main volume, pages 834-35.

Tetramethylbismuthonium-^{210}Bi ion, $Me_4^{210}Bi^+$, is probably formed during β-decay of $Me_4^{210}Pb$, and the bismuthonium intermediate undergoes further degradation from which ^{210}Bi is deposited on the walls of the vessel (506).

Tetraphenylbismuthonium Octahydrotriborate, $BiBC_{24}H_{28}$, $[Ph_4Bi]B_3H_8$

Synth.: From $[Ph_4Bi]NO_3$ + $[NH_4]B_3H_8$ in H_2O, 92% yield (61).

Prop.: Colorless ppt., dec. 90°, unstable in soln., dec. during recrystn. from $EtOH-Et_2O$ or $MeOH-H_2O$ (61).

Tetraphenylbismuthonium Perchlorate, $BiC_{24}H_{20}ClO_4$, $[Ph_4Bi]ClO_4$, (M-835)

Prop.: Half-wave potential (467).

2.3 TERTIARY DERIVATIVES

2.3.5 COMPOUNDS WITH BISMUTH-HALOGEN BONDS

2.3.5.2 DIHALO DERIVATIVES

Several compounds of this type, R_3BiX_2, and their methods of synthesis were described in the main volume on pages 836-38.

Some tertiary dihalobismuth derivatives are useful as fungicides and bactericides for surface applications to, and/or incorporation in, plastics, fibrous materials, and paints (945, 946). These biocidal compounds include: $(BrC_6H_4)_3BiCl_2$, $(ClC_6H_4)_3BiCl_2$, $(O_2NC_6H_4)_3BiCl_2$, Ph_3BiCl_2, Ph_3BiF_2, Ph_3BiBr_2, $(MeC_6H_4)_3BiCl_2$, $(MeOC_6H_4)_3BiCl_2$, $(Me_2C_6H_3)_3BiCl_2$, $(\alpha-C_{10}H_7)_3BiCl_2$, $(PhC_6H_4)_3$-$BiCl_2$, and $Ph_2(p-MeC_6H_4)BiCl_2$.

Two compounds of this group are described individually hereafter, while the others are compiled in Table 59.

Dibromotriphenylbismuth, $BiC_{18}H_{15}Br_2$, Ph_3BiBr_2,   (M-837)
Synth.: From Ph_3Bi + Br in MeCN (178).
Prop.: A nonelectrolyte in MeCN (178). Half-wave potential (467). Far-IR spectrum (803).
Rxn.: Electrolytic redn. → Ph_3Bi + 2 Br⁻ (467).
Uses: A fungicide and bactericide for surface applications to, and/or incorporation in, plastics and fibrous materials (945). A catalyst for the esterification of aromatic dicarboxylic acids with glycols (1650).

Dichlorotriphenylbismuth, $BiC_{18}H_{15}Cl_2$, Ph_3BiCl_2,   (M-836)
Synth.: From Ph_3Bi + Cl in MeCN (178).
From Ph_3Bi in petr. ether by adding dropwise SO_2Cl_2, 96% yield (1505).
From ^{210}Bi-tagged $BiCl_3$ + PhMgX, a ^{210}Bi-tagged prod. was obtd. (1131).
Prop.: m. 134-138° (1505), a nonelectrolyte in MeCN (178). Crystal and mol. structure studies indicated 2 crystallographically independent molecules in an asymmetric unit but both adopted the same configuration. The molecule forms a trigonal bipyramid with the Cl atoms in the apical positions (705, see also 706).
Rxn. with: $NaSC(S)NR_2 \cdot nH_2O$ in $CHCl_3$ → Ph_3Bi + $[R_2NC(S)S-]_2$ (1505).
PhLi, using $Ph_3{}^{210}BiCl_2$ → ^{210}Bi-tagged Ph_3Bi (1131).
Uses: A germicide for textile matls. (62). An active constituent of marine antifouling compns. (661). A bactericide and fungicide in aerosol formulations for treatment of plastics and fibrous matls. (943, 945). A catalyst for the esterification of aromatic dicarboxylic acids with glycols (1650).

Table 59. Tertiary Dihalobismuth Derivatives

R_3BiX_2	Prepd. from	Prop. and Rxn.	Ref.
C_{18}			
$(3\text{-}ClC_6H_4)_3BiCl_2$		Effective as a marine antifoul-ing agent	661
Ph_3BiF_2 (M-837)		Effective as a marine antifoul-ing agent	661
C_{21}			
$(2\text{-}MeC_6H_4)_3{}^{210}BiBr_2$	$BiCl_3$ contg. ${}^{210}Bi$ + RMgX, (a)	β-decay $\rightarrow R_3PoBr + R_2PoBr_2$	1128
$(3\text{-}MeC_6H_4)_3{}^{210}BiBr_2$	$BiCl_3$ contg. ${}^{210}Bi$ + RMgX, (a)	β-decay $\rightarrow R_3PoBr + R_2PoBr_2$	1128
$(2\text{-}MeC_6H_4)_3{}^{210}BiCl_2$	$BiCl_3$ contg. ${}^{210}Bi$ + RMgX, (b)	β-decay $\rightarrow R_3PoCl + R_2PoCl_2$	1128
$(3\text{-}MeC_6H_4)_3{}^{210}BiCl_2$	$BiCl_3$ contg. ${}^{210}Bi$ + RMgX, (b)	β-decay $\rightarrow R_3PoCl + R_2PoCl_2$	1128
$(4\text{-}MeC_6H_4)_3{}^{210}BiCl_2$	$BiCl_3$ contg. ${}^{210}Bi$ + RMgX, (b)	β-decay $\rightarrow R_3PoCl + R_2PoCl_2$	1128
$(2\text{-}MeC_6H_4)_3{}^{210}BiF_2$	R_3BiCl_2 labeled with ${}^{210}Bi$ + AgF	β-decay $\rightarrow R_3PoF + R_2PoF_2$	1128
$(3\text{-}MeC_6H_4)_3{}^{210}BiF_2$	R_3BiCl_2 labeled with ${}^{210}Bi$ + AgF	β-decay $\rightarrow R_3PoF + R_2PoF_2$	1128
$(4\text{-}MeOC_6H_4)_3{}^{210}BiCl_2$		β-decay $\rightarrow R_3PoCl + R_2PoCl_2$	1127
C_{24}			
$(2,5\text{-}Me_2C_6H_3)_3{}^{210}BiBr_2$	$BiCl_3$ contg. ${}^{210}Bi$ + RMgX, (a)	β-decay $\rightarrow R_3PoBr + R_2PoBr_2$	1128
$(2,5\text{-}Me_2C_6H_3)_3{}^{210}BiCl_2$	$BiCl_3$ contg. ${}^{210}Bi$ + RMgX, (b)	β-decay $\rightarrow R_3PoCl + R_2PoCl_2$	1128, 1129
$(2,5\text{-}Me_2C_6H_3)_3{}^{210}BiF_2$	$R_3{}^{210}BiCl_2$ + AgF	β-decay $\rightarrow R_3PoCl + R_2PoX_2$	1128
C_{30}			
$(1\text{-}C_{10}H_7)_3BiCl_2$		Effective as a marine antifoul-ing agent	661
$1,4\text{-}C_6H_4(BiPh_2Br_2)_2$	$4\text{-}(Ph_2Bi)C_6H_4BiPh_2$ + Br in C_6H_6	Dec. 135-137°	1658

(a) Followed by bromination with 1% Br in $CHCl_3$ (1128).   (b) Followed by chlorination with SO_2Cl_2 in $CHCl_3$ (1128).

2.3.5.3 HALOHYDROXY DERIVATIVES

Chlorohydroxytriphenylbismuth, $Ph_3Bi(OH)Cl$, (M-839) is useful as a fungicide and bactericide for surface application to, and/or incorporation in, plastics, fibrous materials, and paints (945, 946).

2.3.5.6 DIPSEUDOHALO DERIVATIVES

No compounds of this type were reported in the main volume.

Dicyanotriphenylbismuth, $Ph_3Bi(CN)_2$, is useful as a fungicide and bactericide for surface applications to, and/or incorporation in, plastics, fibrous materials, and paints (945, 946).

2.3.5.9 MISCELLANEOUS TERTIARY BISMUTH DERIVATIVES

Acetoxychlorotriphenylbismuth, $Ph_3Bi(OAc)Cl$, is useful as a fungicide and bactericide for surface applications to, and/or incorporation in, plastics, fibrous materials, and paints (945, 946).

2.3.6 COMPOUNDS WITH BISMUTH BONDED TO GROUP VIA ELEMENTS

2.3.6.1 COMPOUNDS WITH BISMUTH-OXYGEN BONDS

2.3.6.1.3 DIHYDROXYBISMUTH DERIVATIVES AND THEIR ANHYDRIDES WITH INORGANIC AND CARBOXYLIC ACIDS

Several dinitrato, $R_3Bi(NO_3)_2$, and bis(acyloxy), $R_3Bi(O_2CR')_2$, derivatives were described in the main volume on pages 839-40.

Tertiary bis(acyloxyl)bismuth derivatives are effective as fungicides and bactericides for surface applications to, and/or incorporation in, plastics, fibrous materials, and paints (945, 946). The biocidal compounds of this group include: $Ph_3Bi(OH)_2$, $Ph_3Bi(O_2CCH_2Cl)_2$, $Ph_3Bi(OAc)_2$, $Ph_3Bi(O_2CCH_2CH_2SH)_2$, $Ph_3Bi[O_2CC(Me)=CH_2]_2$, $Ph_3Bi(O_2CCHOHCHOHCO_2H)_2$, $Ph_3Bi(OBz)_2$, $Ph_3Bi(O_2CC_6H_4OH-2)_2$, $Ph_3Bi(O_2CC_6H_4OH-4)_2$, $Ph_3Bi(O_2CC_6H_4NH_2-4)_2$, $Ph_3Bi(O_2CC_6H_4NH_4-4)_2 \cdot 2\ Me_2CO$, $(4-MeC_6H_4)_3Bi(OAc)_2$, $(4-MeC_6H_4)_3Bi(OBz)_2$, $(2-MeC_6H_4)_3Bi(O_2CC_6H_4OH-2)_2 \cdot C_6H_6$, and $(MeC_6H_4)_3Bi[O_2CCH_2S(CH_2)_5CHMe_2]_2$.

Carbonatotriphenylbismuth, $BiC_{19}H_{15}O_3$, Ph_3BiCO_3
Rxn. with: RCO_2H in MePh at 120° → $Ph_3Bi(O_2CR)_2$ (1096).

Bis(methacryloyloxy)triphenylbismuth, $BiC_{26}H_{25}O_4$, $Ph_3Bi[O_2CC(Me)=CH_2]_2$
Synth.: From Ph_3BiCO_3 + $CH_2=C(Me)CO_2H$ in MePh at 120°, 80% yield (1096).
Prop.: m. 170° (1096).
Biol. Prop.: Effective against Staphylococcus aureus (1096).
Use: A fungicide and bactericide for surface applications and/or incorporation in plastics and fibrous materials (945).

Bis[(p-aminobenzoyl)oxy]triphenylbismuth, $BiC_{32}H_{27}N_2O_4$,
　　$Ph_3Bi(O_2CC_6H_4NH_2-4)_2$,　　　(M-840)
<u>Rxn. with</u>: Diisocyanates → polyurethans resistant to microorganisms (1095).

Triphenylbis[(p-vinylbenzoyl)oxy]bismuth,　　$BiC_{36}H_{29}O_4$,
　　$Ph_3Bi(O_2CC_6H_4CH=CH_2-4)_2$
<u>Synth.</u>: From Ph_3BiCO_3 + $4-CH_2=CC_6H_4CO_2H$ 1:2 in MePh at 120° (1096).
<u>Prop.</u>: m. 170-180° (dec.) (1096).
<u>Biol. Prop.</u>: Effective against Staphylococcus aureus (1096).

2.3.6.1.6 TERTIARY BISMUTHINE OXIDES

No tertiary bismuthine oxides were reported in the main volume.

Dicyclohexylvinylbismuthine Oxide,　　$BiC_{14}H_{25}O$,　　$(C_6H_{11})_2(CH_2=CH)BiO$
<u>Rxn.</u>: In the presence of Grignard reagents in MePh → cryst. polymers, η_{spec}.
0.044 in $HCONMe_2$ at 30° (56).

3-Butenyldiphenylbismuthine Oxide,　　$BiC_{16}H_{17}O$,　　$Ph_2(CH_2=CHCH_2CH_2)BiO$
<u>Rxn.</u>: In $HCONMe_2$ in the presence of $BuMgCl$ → amorphous polymer (56).

Triphenylbismuthine Oxide,　　$BiC_{18}H_{15}O$,　　Ph_3BiO
<u>Use</u>: A fungicide and bactericide for surface application and/or incorporation in plastics, fibrous materials, and paints (945, 946).

2.3.6.1.7 DIALKOXY AND DIARYLOXY DERIVATIVES

No tertiary dialkoxy- and diaryloxy-bismuth derivatives were described in the main volume.

A few dialkoxy and diaryloxy derivatives are useful as fungicides and bactericides for surface application and/or incorporation in plastics, fibrous materials and paints. These include: $Ph_3Bi(OBu)_2$, $Ph_3Bi(OPh)_2$, $(MeC_6H_4)_3Bi(OEt)_2$, and $(MeC_6H_4)_3Bi(OC_6H_4Ph-2)_2$ (945, 946).

2.3.6.2 COMPOUNDS WITH BISMUTH-SULFUR BONDS

2.3.6.2.6 TERTIARY BISMUTHINE SULFIDES

No bismuthine sulfides were described in the main volume.

Triphenylbismuthine Sulfide,　　$BiC_{18}H_{15}S$,　　Ph_3BiS
<u>Use</u>: A fungicide and bactericide for surface application to, and/or incorporation in, plastics, fibrous materials, and paints (945, 946).

2.3.6.2.9 MISCELLANEOUS TERTIARY BISMUTH COMPOUNDS

Bis(dodecylthio)triphenylantimony, $Ph_3Bi(SC_{12}H_{25})_2$, is useful as a fungicide and bactericide for surface applications to, and/or incorporation in, plastics, fibrous materials, and paints (945, 946).

2.4 SECONDARY DERIVATIVES

2.4.6 COMPOUNDS WITH BISMUTH BONDED TO GROUP VIa ELEMENTS

2.4.6.1 COMPOUNDS WITH BISMUTH-OXYGEN BONDS

Dioctyltris(ricinoleoyloxy)bismuth, $BiC_{70}H_{133}O_9$,
   $(C_8H_{17})_2Bi[O_2C(CH_2)_7CH=CHCH_2CHOH(CH_2)_5Me]_3$
__Rxn. with:__ Diisocyanates → polyurethans resistant to microorganisms (1095).
__Biol. Prop.:__ A bactericide (1095).

Addition to page 32:

[Ru(Ars)(Py)Cl$_3$] (where Ars = Ph$_3$As), orange-brown crystals, m. 234-235°, prepd. from [Ru(Ars)$_2$(MeOH)Cl$_3$] by warming with Py and adding excess petroleum (1453).

Addition to page 42:

Trimethylphosphonium (Dimethylarsino)methylide,   AsC$_6$H$_{15}$P,   Me$_3$P=CHAsMe$_2$
<u>Synth.</u>: From Me$_3$SiOAsMe$_2$ + Me$_3$P=CHSiMe$_3$ in Et$_2$O at r.t., 57% yield (1355).
<u>Prop.</u>: m. -37 to -35°, b$_{12}$ 85-87°, NMR and IR spectra; dec. at 120° →
Me$_3$P=CH$_2$ and Me$_3$P=C(AsMe$_2$)$_2$ (1355).

Additions to page 43:

(o-Bromophenyl)dimethylarsine,   AsC$_8$H$_{10}$Br,   2-BrC$_6$H$_4$AsMe$_2$
<u>Synth.</u>: From 2-BrC$_6$H$_4$AsCl$_2$ + MeMgI (1:2) in Et$_2$O, 85% yield (1659).
<u>Prop.</u>: b$_{10}$ 132-135° (1659).
<u>Rxn. with</u>: BiLi in petrol → 2-LiC$_6$H$_4$AsMe$_2$ (1659).

(o-Lithiophenyl)dimethylarsine,   AsLiC$_8$H$_{10}$,   2-LiC$_6$H$_4$AsMe$_2$
<u>Synth.</u>: From 2-BrC$_6$H$_4$AsMe$_2$ + BuLi in petrol (1659).
<u>Prop.</u>: Pale yellow crystals (1659).
<u>Rxn. with</u>: MeAsCl$_2$ → [2-(Me$_2$As)C$_6$H$_4$]$_2$AsMe (1659).
AsCl$_3$ (3:1) → [2-(Me$_2$As)C$_6$H$_4$]$_3$As (1659).

Addition to page 134:

[Co(TTA)(CO)][Co(CO)$_4$] (where TTA = [2-(Ph$_2$As)C$_6$H$_4$]$_3$As, prepd. from [Co(CO)$_4$]$_2$ in THF by adding dropwise the tetraarsine and refluxing for 15 min. (691).

Addition to page 168:

Perchloratodiphenylarsine,   AsC$_{12}$H$_{10}$ClO$_4$,   Ph$_2$AsClO$_4$
<u>Prop.</u>: Half-wave potential (467).

Addition to page 253:

[Ph$_4$As][Cr(CO)$_5$SnCl$_3$], prepd. from [Ph$_4$As]SnCl$_3$ in CH$_2$Cl$_2$ + Cr(CO)$_6$ under irradiation. Yellow to orange solid, m. 124-125°, stable in air for a short time, sol. in polar org. solvents, a 1:1 electrolyte in MeNO$_2$ (1307).

Addition to page 272:

[Ph$_4$As]$_2$[Re$_3$(CNS)$_8$Cl$_3$], prepd. from [Ph$_4$As]Cl in EtOH + ReCl$_3$ + HSCN in EtOH. Yellow-green powder, molar cond. data, absorption spectrum. Rxn. with SCN$^-$ under reflux → replacement of the remaining Cl atoms with CNS (1282).

Addition to page 280:

2.3.2 TERTIARY ARSONIUM SALTS

Several ionic salts of the general formula [(R$_3$As)$_2$BH$_2$]X, which may be considered as trialkylborylarsonium salts containing a tertiary arsine ligand coordinated with the boron atom were reported.

[(R$_3$As)$_2$BH$_2$]X Salts

[(R$_3$As)$_2$BH$_2$]$^+$	X$^-$	Prepd. from	Prop.	Ref.
C$_6$				
(Me$_3$As)$_2$BH$_2$	Cl	[(Me$_3$As)$_2$BH$_2$][B$_{12}$H$_{11}$AsMe$_3$] dissolved in H$_2$O at 80° by passing through a column contg. a Cl anion-exchange resin		1067
	Br$_3$	[(Me$_3$As)$_2$BH$_2$]Cl by treating with Br		1067
	HgCl$_3$	[(Me$_3$As)$_2$BH$_2$]Cl + HgCl$_2$		1067
	PF$_6$	[(Me$_3$As)$_2$BH$_2$]Cl + [NH$_4$]PF$_6$	(a)	1067
	Reineckate	[(Me$_3$As)$_2$BH$_2$]Cl + [Cr(SCN)$_4$-(NH$_3$)$_2$]$^-$		1067
	B$_{12}$H$_{12}$	By-prod. from the rxn. of Me$_3$As with B$_2$H$_6$ at 175°, followed by extn. with diglyme and dec. of the solid residue with H$_2$O		1067
	B$_{12}$H$_{11}$AsMe$_3$	By-prod. from the rxn. of Me$_3$As with B$_2$H$_6$ at 175°, obtd. after sepn. of [B$_{12}$H$_{10}$(AsMe$_3$)$_2$] and dec. of the residue with H$_2$O	(b)	1067
C$_{12}$				
[(Et$_3$As)$_2$BH$_2$]	B$_{12}$H$_{12}$	Et$_3$AsBH$_3$ + B$_5$H$_9$ > 75°		1067
	B$_{12}$H$_{11}$AsEt$_3$	Et$_3$AsBH$_3$ + B$_5$H$_9$ > 75°		1067
[(i-BuMe$_2$As)$_2$BH$_2$]	B$_{12}$H$_{12}$	i-BuMe$_2$AsBH$_3$ + B$_5$H$_9$ > 75°		1067
	B$_{12}$H$_{11}$(AsMe$_2$i-Bu)	i-BuMe$_2$AsBH$_3$ + B$_5$H$_9$ > 75°		1067

(a) Colorless crystals. (b) Colorless, feathery crystals, m. 171-173°, IR spectra of the cation and anion were obtd. Useful in the prepn. of resistors from cellulosic matl.

Addition to page 439:

$Na_2[Fe(SbPh_3)(CN)_5]$, green, paramagnetic crystals, prepd. from $Na_3[Fe(SbPh_3)-(CN)_5]$ in MeOH by oxidn. with 0.1 N Br soln. in MeOH. Molar cond. in H_2O and magnetic prop. were detd. IR spectrum (1124). Mössbauer spectrum (573, 1124).

Addition to page 468:

Perchloratodiphenylstibine, $SbC_{12}H_{10}ClO_4$, Ph_2SbClO_4
Prop.: Half-wave potential (467).

Addition to page 532:

Tri-p-tolylbismuthine, $BiC_{21}H_{21}$, $(4\text{-}MeC_6H_4)_3Bi$,   (M-822)
Synth.: From ^{210}Bi-tagged $BiCl_3$ + $4\text{-}MeC_6H_4MgX$, a ^{210}Bi-tagged bismuthine was obtd. (1130).
Prop.: β-decay → $(4\text{-}MeC_6H_4)_3Po^+$ + $(4\text{-}MeC_6H_4)_2Po$ (1130). "Slow beats" in the quadrupole spin echo for ^{209}Bi was detd. (46).

BIBLIOGRAPHY

1. Anon., Fed. Regist. 29, 15814-16, Nov. 25, 1964; CA 62, 3317.
2. ---, ibid., 30, 3594, Mar. 18, 1965; CA 62, 13760.
3. ---, ibid., 6071-2, Apr. 29, 1965; CA 63, 1145.
4. ---, ibid., 6389-91, May 7, 1965; CA 63, 1145.
5. ---, ibid., 6732, May 18, 1965; CA 63, 3535.
6. ---, ibid., 12670-3, Oct. 5, 1965; CA 63, 18931.
7. ---, ibid., 31, 1069-70, Jan. 27, 1966; CA 64, 13285.
8. ---, ibid., 9471, July 12, 1966; CA 65, 8666.
9. ---, ibid., 11876-9, Sept. 9, 1966; CA 65, 14329.
10. ---, ibid., 12435, Sept. 20, 1966; CA 65, 19210.
11. ---, ibid., 14350-1, Nov. 8, 1966; CA 66, 27760.
12. ---, ibid., 32, 4170-1, Mar. 17, 1967; CA 67, 2122.
13. ---, ibid., 6970-1, May 6, 1967; CA 67, 20776.
14. ---, ibid., 9224-5, June 29, 1967; CA 67, 72489.
15. ---, ibid., 15012, Oct. 31, 1967; CA 68, 11754.
16. ---, ibid., 33, 312-3, Feb. 17, 1968; CA 68, 86170.
17. ---, ibid., 5871, Apr. 17, 1968; CA 68, 113383.
18. ---, ibid., 10207, July 17, 1968; CA 69, 66197.
19. ---, ibid., 11024-6, Aug. 2, 1968; CA 69, 66205.
20. Abel, E. W., and Armitage, D. A., J. Organometal. Chem. 5, 326-9 (1966); CA 64, 14213.
21. --- and Brady, D. B., ibid., 11, 145-9 (1968); CA 68, 44628.
22. ---, Crow, J. P., and Illingworth, S. M., Chem. Commun. 1968, 817; CA 69, 55910.
23. --- et al., J. Chem. Soc. A 1968, 1203-8; CA 69, 14333.
24. --- and Hutson, G. V., J. Inorg. Nucl. Chem. 30, 2339-44 (1968); CA 69, 113003.
25. Abuin, J., and Lobo, E. S., Arch. Bioquim., Quim. Farm., Tucuman 11, 109-13 (1963-64); CA 65, 10645.
26. Acker, D. S., and Blomstrom, D. C. (E. I. duPont de Nemours and Co.), U.S. 3,162,641, Dec. 22, 1964 (Cont. of U.S. 3,062,019 and 3,115,506); CA 63, 549.
27. Adamovich, L. P., and Oleinik, A. A., Zavodsk. Lab. 32, 387-91 (1966); CA 65, 4723.
28. Adams, D. M., J. Chem. Soc. 1964, 1771-6; CA 61, 1400.
29. --- and Chandler, P. J., ibid., A 1967, 1009-13; CA 67, 69025.
30. ---, Chandler, P. J., and Churchill, R. G., ibid., 1272-4; CA 67, 77581.
31. ---, Cook, D. J., and Kemmitt, R. D. W., ibid., A 1968, 1067-72; CA 68, 118925.
32. --- and Lock, P. J., ibid., A 1967, 620-3; CA 66, 109960.
33. --- and Morris, D. M., ibid., 1666-8; CA 67, 112516.
34. --- and Morris, D. M., ibid., 1669-70; CA 67, 112517.
35. --- and Morris, D. M., ibid., 2067-9; CA 68, 34406.
36. --- and Morris, D. M., ibid., A 1968, 694-5; CA 68, 82674.
37. Adams, R. W., Batley, G. E., and Bailar, J. C., Jr., J. Amer. Chem. Soc.

90, 6051-6 (1968); CA 69, 105783.

38. Addison, C. C., and Kilner, M., J. Chem. Soc. A 1968, 1539-44; CA 69, 48823.

39. Affsprung, H. E., and Archer, V. S., Anal. Chem. 36, 2512-13 (1964); CA 62, 3405.

40. --- and Robinson, J. L., Anal. Chim. Acta 37, 81-90 (1967); CA 66, 51937.

41. Aftandilian, V. D. (to Cabot Corp.), U.S. 3,222,296, Dec. 7, 1965; CA 64, 5225.

42. ---, U.S. 3,285,890, Nov. 15, 1966; CA 66, 11289.

43. Aguiar, A. M., and Archibald, T. G., J. Org. Chem. 32, 2627-8 (1967); CA 67, 63583.

44. ---, Archibald, T. G., and Kapicak, L. A., Tetrahedron Lett. 1967, 4447-50; CA 68, 22022.

45. --- et al., J. Org. Chem. 33, 1681-3 (1968); CA 68, 105342.

46. Ainbinder, N. E., et al., Fiz. Tverd. Tela 10, 2026-9 (1968); CA 69, 63422.

47. Akatsu, E., Nippon Genshiryoku Kenkyusho Kenkyu Hokoku No. 1099, 56 pp. (1965); CA 64, 18943.

47a. Akhmaeva, N. L., Dedkov, Yu. M., and Khramov, V. P., Zh. Anal. Khim. 22, 1482-6 (1967); CA 68, 56214.

48. Akhmedli, M. K., and Babaeva, T. R., Uch. Zap. Azerb. Gos. Univ., Ser. Khim. Nauk 1965 (3), 31-6; CA 65, 17780.

49. ---, Babaeva, T. R., and Imamverdieva, F. B., Azerb. Khim. Zh. 1965 (1), 104-13; CA 63, 10869.

50. ---, Bashirov, E. A., and Glushchenko, E. L., Uch. Zap. Azerb. Gos. Univ., Ser. Khim. Nauk 1965 (4), 9-15; CA 66, 61494.

51. --- and Granovskaya, P. B., Azerb. Khim. Zh. 1965 (5), 105-8; CA 64, 18403.

52. --- and Imamverdieva, ibid., 1966 (3), 122-6; CA 66, 6127.

53. Akhmedzyanov, M. A., et al., Zh. Nauch. Prikl. Fotogr. Kinematogr. 12, 462-3 (1967); CA 68, 55359.

54. Alexander, R., and Parker, A. J., J. Amer. Chem. Soc. 89, 5549-51 (1967); CA 68, 6859.

55. Allan, J. R., et al., J. Inorg. Nucl. Chem. 27, 1305-9 (1965); CA 63, 3780.

56. Allcock, H. R. (to Am. Cyanamid Co.), Fr. 1,367,820, July 24, 1964; CA 62, 6592.

57. Allen, D. W., Millar, I. T., and Mann, F. G., J. Chem. Soc. C 1967, 1869-75; CA 67, 100203.

58. Allen, E. A., Feenan, K., and Fowles, G. W. A., ibid., 1965, 1636-42; CA 62, 11402.

59. Allison, G. B., and Sheldon, J. C., Inorg. Chem. 6, 1493-7 (1967); CA 67, 78566.

60. Alykov, N. M., Kazokov, B. I., and Cherkesov, A. I., Fiz.-Khim. Issled. Prir. Sorbentov Ryada Anal. Sist. 1967, No. 2, 87-101; CA 69, 113200.

61. Amberger, E., and Gut, E., Chem. Ber. 101, 1200-4 (1968); CA 68, 104635.

62. American Cyanamid Co., Brit. 1,003,685, Sept. 8, 1965; CA 63, 18346.

63. Amici, D., Paparelli, M., and Tedeschi, G. G., Boll. Soc. Ital. Biol. Sper. 42, 1496-8 (1966); CA 67, 20308.

64. Amkraut, A. A., Garvey, J. S., and Campbell, D. H., J. Exptl. Med. 124, 293-306 (1966); CA 65, 19138.

65. Anagnostopoulos, A. K., Chim. Chronika (Athens, Greece) 31 (9), 141-6 (1966); CA 66, 82009.

66. Anderson, H. H., J. Chem. Eng. Data 9, 592-4 (1964); CA 62, 1551.

67. Andreev, S. N., Khaldin, V. G., and Andreeva, M. V., Dokl. Akad. Nauk SSSR 169, 95-7 (1966); CA 65, 17776.

68. Andreeva, E. I., et al., Khim. o Sel'sk. Khoz. 3 (4), 28-30 (1965); CA 63, 8979.

69. Andrews, D. K., Bird, H. R., and Sunde, M. L., Poultry Sci. 45, 838-47 (1966); CA 65, 7686.

70. ---, Bird, H. R., and Sunde, M. L., ibid., 1305-13 (1966); CA 66, 35795.

71. Ang, H. G., and Emeléus, H. J., Chem. Commun. 1966, 460; CA 65, 10618.

72. --- and Emeléus, H. J., J. Chem. Soc. A 1968, 1334-6; CA 69, 19271.

73. ---, Manoussakis, G., and El-Nigumi, Y. O., J. Inorg. Nucl. Chem. 30, 1715-17 (1968); CA 69, 87128.

74. --- and West, B. O., Aust. J. Chem. 20, 1133-42 (1967); CA 67, 32752.

75. Angelici, R. J., and Busetto, L., Inorg. Chem. 7, 1935-6 (1968); CA 69, 71172.

76. Anker, M. W., Colton, R., and Tomkins, I. B., Aust. J. Chem. 20, 9-12 (1967); CA 66, 43309.

77. ---, Colton, R., and Tomkins, I. B., ibid., 21, 1149-54 (1968); CA 69, 7928.

78. ---, Colton, R., and Tomkins, I. B., ibid., 1159-63 (1968); CA 69, 7929.

79. Appel, R., and Heinzelmann, W. (BASF), Ger. 1,192,205, May 6, 1965; CA 63, 8405.

80. Apsitis, A., Jansons, E., and Adiyane, E. E., Uch. Zap., Latv. Gos. Univ. 88, 83-8 (1967); CA 69, 64369.

81. Araneo, A., Gazz. Chim. Ital. 96, 1560-2 (1966); CA 66, 61380.

82. --- and Bianchi, C., ibid., 97, 885-97 (1967); CA 67, 104686.

83. --- and Mantinengo, S., ibid., 95, 61-6 (1965); CA 63, 6585.

84. --- and Mantinengo, S., ibid., 825-30 (1965); CA 63, 10981.

85. --- and Mantinengo, S., Rend. Ist. Lombardo Sci. Lettere, A 99, 829-35 (1965); CA 64, 1748.

86. --- et al., Gazz. Chim. Ital. 95, 1435-46 (1965); CA 67, 28855.

87. Archer, V. S., Dissert. Abstr. 25, 3231 (1964); CA 62, 12439.

88. Archibald, Th. G., Diss. Abstr. B 28, 4054 (1968); CA 69, 36216.

89. Armstrong, M., and Webb, M., Biochem. J. 103, 913-22 (1967); CA 67, 8916.

90. Armstrong, R., et al., Aust. J. Chem. 20, 2771-6 (1967); CA 68, 78390.

91. --- and Gibson, N. A., ibid., 21, 897-905 (1968); CA 69, 7893.

92. Arnett, E. M., et al., J. Amer. Chem. Soc. 87, 1541-53 (1965); CA 62, 13908.

93. --- and McKelvey, D. R., ibid., 1393-4 (1965); CA 63, 14148.

94. Aroney, M. J., LeFèvre, R. J. W., and Saxby, J. D., J. Chem. Soc., Suppl. 1964, 6180-5; CA 63, 9790.

95. Asahi Chemical Industry Co., Ltd., Fr. 1,386,468, Jan. 22, 1965; CA 62, 16405.

96. ---, Brit. 1,001,699, Aug. 18, 1965; CA 63, 15004.

97. Ashby, E. C. (Ethyl Corp.), U.S. 3,153,671, Oct. 20, 1964; CA 62, 258.

98. Atkinson, R. L., et al., Poultry Sci. 46, 1003-8 (1967); CA 67, 80089.

99. Augdahl, E., Grundnes, J., and Klaboe, P., Inorg. Chem. 4, 1475-80 (1965); CA 63, 14349.

99a. Baay, Y. L., and MacDiarmid, A. G., Inorg. Nucl. Chem. Lett. 3, 159-61 (1967); CA 67, 28835.

100. Babin, V. V., et al., Khim. Sel. Khoz. 4, 923-5 (1966); CA 66, 75186.

101. Babko, A. K., Akhmedli, M. K., and Granovskaya, P. B., Ukr. Khim. Zh. 32, 879-85 (1966); CA 66, 16182.

102. ---, Akhmedli, M. K., and Granovskaya, P. B., ibid., 1015-18 (1966); CA 66, 16206.

103. --- and Shtokalo, M. I., ibid., 30, 972-9 (1964); CA 62, 4685.

104. Baczuk, R. J., and Bolleter, W. T., Anal. Chem. 39, 93-5 (1967); CA 66, 43464.

105. Baddley, W. H., J. Amer. Chem. Soc. 90, 3705-10 (1968); CA 69, 87165.

106. Bagnall, K. W., Proc. Int. Conf. Coord. Chem., 8th, Vienna 1964, 424-6; CA 66, 111159.

107. --- et al., J. Chem. Soc. 1965, 350-3; CA 62, 6127.

108. --- and Laidler, J. B., ibid., A 1966, 516-20; CA 64, 18943.

109. Bailar, J. C., and Itatani, H., J. Amer. Chem. Soc. 89, 1592-9 (1967); CA 67, 2664.

110. Balch, A. L., Inorg. Chem. 6, 2158-62 (1967); CA 68, 35425.

111. Ball, W. J., Brit. 1,127,677, Sept. 18, 1968; CA 69, 105916.

112. Ballhausen, C. J., NASA Accession No. N65-18152; Rept. No. AD455103; CA 64, 4440.

113. Bally, R., C. R. Acad. Sci. Paris 261, 3617-18 (1965); CA 64, 5858; Acta Crystallogr. 23, 295-306 (1967); CA 67, 58061.

114. Bamford, W. R., Lovie, J. C., and Watt, J. A. C., J. Chem. Soc. C 1966, 1137-40; CA 65, 3898.

115. Bandoli, G., et al., Acta Crystallogr., Sect. B 24, 1129-30 (1968); CA 69, 55135.

116. Banister, A. J., and Moore, L. F., J. Chem. Soc. A 1968, 1137-8; CA 69, 10524.

117. Bankovskii, Yu. A., Gertner, M. D., and Yanson, E. Yu., U.S.S.R. 185,907, Sept. 12, 1966; CA 67, 2941; Uch. Zap., Latv. Gos. Univ. 1967, 57-8; CA 69, 92569.

118. Banks, R. E., Haszeldine, R. N., and Hatton, R., Tetrahedron Lett. 1967, 3993-6; CA 68, 78093.

119. Bannister, W. D., Green, M., and Haszeldine, R. N., Chem. Commun. (London) 1965 (4), 54-6; CA 62, 16291.

120. Barbeau, C., Can. J. Chem. 45, 161-6 (1967); CA 66, 46469.

121. Barclay, G. A., et al., Aust. J. Chem. 20, 1571-7 (1967); CA 67, 70182.

121a. ---, Harris, C. M., and Kingston, J. V., Chem. Ind. (London) 1965, 227-8.

122. Bardos, Th. J., Datta-Gupta, N., and Hebborn, P., J. Med. Chem. 9, 221-7 (1966); CA 64, 11246.

123. Baroni, E. E., et al., At. Energ. (USSR) 17, 497-500 (1964); CA 62, 10604.

124. Bartecki, A., Chem. Zvesti 19, 161-6 (1965); CA 63, 7883.

125. --- and Dembicka, D., Roczniki Chem. 39, 1783-98 (1965); CA 64, 18932.

126. Basargin, N. N., and Novikova, K. F., Zh. Analit. Khim. 21, 473-8 (1966); CA 65, 6297.

127. Basco, N., and Yee, K. K., Chem. Commun. 1968, 153-4; CA 68, 100616.

128. Baumgaertel, E., and Gruner, H., J. Prakt. Chem. 33, 108-12 (1966); CA 66, 46464.

129. --- and Gruner, H., ibid., 38, 34-9 (1968); CA 69, 87123.

130. Beachley, O. T., and Coates, G. E., J. Chem. Soc. 1965, 3241-7; CA 63, 4330.

131. Beck, W., and Fehlhammer, W. P., Angew. Chem., Int. Ed. Engl. 6, 169-70 (1967); CA 66, 91254.

132. --- et al., Chem. Ber. 100, 2335-61 (1967); CA 67, 78577.

133. --- et al., ibid., 3955-60 (1967); CA 68, 44339.

134. ---, Feldl, K., and Schuierer, E., Angew. Chem. 77, 458-9 (1965); CA 63, 9423.

135. --- and Lottes, K., Chem. Ber. 98, 2657-73 (1965); CA 63, 14229.

136. --- and Lottes, K., Z. Naturforsch. 19b, 987-94 (1964); CA 62, 3640.

137. ---, Melnikoff, A., and Stahl, R., Chem. Ber. 99, 3721-7 (1966); CA 66, 50499.

138. ---, Nitzschmann, R. E., and Smedal, H. S., J. Organometal. Chem. 8, 547-50 (1967); CA 67, 104697.

139. --- and Schuierer, E., Z. Anorg. Allg. Chem. 347, 304-9 (1966); CA 66, 51791.

140. ---, Schuierer, E., and Feldl, K., Angew. Chem., Int. Ed. Engl. 5, 249 (1966); CA 64, 18944.

141. --- et al., Z. Naturforsch. 21b, 811-12 (1966); CA 66, 25621.

142. --- and Smedal, H. S., Angew. Chem., Int. Ed. Engl. 5, 253 (1966); CA 64, 18945.

143. --- and Stetter, K., Inorg. Nucl. Chem. Lett. 2, 383-5 (1966); CA 66, 51770.

144. --- et al., Chem. Ber. 100, 3944-54 (1967); CA 68, 35427.

145. --- et al., ibid., 101, 2143-52 (1968); CA 69, 32565.

146. Behrends, K., and Kiel, G., Naturwissenschaften 54, 537 (1967); CA 67, 121866.

147. Behrens, H., and Mueller, A., Z. Anorg. Allgem. Chem. 341, 124-36 (1965); CA 64, 9214.

148. Belluco, U., et al., Gazz. Chim. Ital. 95, 576-82 (1965); CA 63, 12655.

149. Benaim, J., C. R., Acad. Sci. Paris, 261, 1996-9 (1965); CA 64, 17633.

150. ---, ibid., Ser. C 262, 937-9 (1966); CA 65, 16998.

151. Ben-Efraim, S., and Cunningham, M. J., Ann. Inst. Pasteur 112, 476-94 (1967); CA 67, 20033.

152. --- and Leskowitz, S., Nature 210, 1068-9 (1966); CA 65, 6060.

153. Benex, J., and Lamy, L., Compt. Rend. Soc. Biol. <u>161</u>, 592-3 (1967); CA <u>68</u>, 10686.

154. Benlian, D., and Bigorgne, M., Bull. Soc. Chim. Fr. <u>1967</u>, 4106-11; CA <u>68</u>, 91465.

155. Benner, G. S., Diss. Abstr. B <u>27</u>, 1074-5 (1966).

156. --- and Meek, D. W., Inorg. Chem. <u>6</u>, 1399-403 (1967); CA <u>67</u>, 49976.

157. Bennett, M. A., et al., J. Chem. Soc. A <u>1967</u>, 501-9; CA <u>66</u>, 115779.

158. --- and Clark, R. J. H., J. Chem. Soc., Suppl. No. 1, 5560-8 (1964); CA <u>63</u>, 2537.

159. ---, Clark, R. J. H., and Goodwin, A. D. J., Inorg. Chem. <u>6</u>, 1625-31 (1967); CA <u>67</u>, 86229.

160. ---, Clark, R. J. H., and Milner, D. L., ibid., 1647-52 (1967); CA <u>67</u>, 86233.

161. ---, Erskine, G. J., and Nyholm, R. S., J. Chem. Soc. A <u>1967</u>, 1260-3; CA <u>67</u>, 96401.

162. ---, Kneen, W. R., and Nyholm, R. S., Inorg. Chem. <u>7</u>, 552-6 (1968); CA <u>68</u>, 74727.

163. ---, Kneen, W. R., and Nyholm, R. S., ibid., 556-60 (1968); CA <u>68</u>, 74740.

164. --- et al., J. Chem. Soc. <u>1964</u>, 4570-7; CA <u>62</u>, 1308.

165. --- and Mason, R., Proc. Chem. Soc. <u>1964</u>, 395-6; CA <u>62</u>, 6125.

166. Benson, R. E. (E. I. duPont de Nemours and Co.), U.S. 3,255,195, June 7, 1966; CA <u>65</u>, 15555.

167. Bentz, M., and Macario, Ch., Bull. Soc. Pathol. Exotique <u>56</u>, 416-21 (1963); CA <u>60</u>, 11204.

168. Beran, M., Zdrazil, K., and Havelka, S., Collection Czech. Chem. Commun. <u>30</u>, 2850-4 (1965); CA <u>63</u>, 9039.

169. Berei, K., Kozlemen. <u>13</u>, 49-53 (1965); CA <u>65</u>, 10651; J. Chromatogr. <u>20</u>, 406-7 (1965); CA <u>64</u>, 5728.

170. ---, and Vasaros, L., Magy. Kem. Foly. <u>73</u>, 313-17 (1967); CA <u>67</u>, 96623.

171. Bergman, J. G., Jr., and Cotton, F. A., Inorg. Chem. <u>5</u>, 1208-13 (1966); CA <u>65</u>, 4758.

172. --- and Cotton, F. A., ibid., 1420-4 (1966); CA <u>65</u>, 8314.

173. Berndt, E. W., Hladky, W. F., and Fockler, J. D. (Salsbury Labs.), U.S. 3,296,290, Jan. 3, 1967; CA <u>66</u>, 95199.

174. Berns, D. S., and Singer, S. J., Immunochemistry <u>1</u>, 209-17 (1964); CA <u>62</u>, 3242.

175. Besolova, E. A., Foss, V. A., and Lutsenko, I. F., Zh. Obshch. Khim. <u>38</u>, 267-73 (1968); CA <u>69</u>, 52243.

176. ---, Foss, V. A., and Lutsenko, I. F., ibid., 1574-8 (1968); CA <u>69</u>, 106849.

177. Beveridge, A. D., and Harris, G. S., J. Chem. Soc., Suppl. <u>1964</u>, 6076-84; CA <u>63</u>, 9792.

178. ---, Harris, G. S., and Inglis, F., J. Chem. Soc. A <u>1966</u>, 520-8; CA <u>64</u>, 18510.

179. Bhaskar, K. R., et al., J. Inorg. Nucl. Chem. <u>28</u>, 1915-25 (1966); CA

65, 15193.

180. Bhat, S. N., and Rao, C. N. R., J. Amer. Chem. Soc. 88, 3216-19 (1966); CA 65, 8061.

181. Biagini, S., et al., Ric. Sci., Rend., Sez. A 6, 161-3 (1964); CA 62, 11180.

182. Bigorgne, M., and Benard, J., Rev. Chim. Miner. 3, 831-59 (1966); CA 66, 59211.

183. --- and Bouquet, G., C. R. Acad. Sci., Paris, Ser. C 264, 1485-7 (1967); CA 67, 59210.

184. Binder, G., Paitz, J., and Serf, E., Ger. 1,246,132, Aug. 3, 1967; CA 67, 104464.

185. Birchall, T., and Jolly, W. L., Inorg. Chem. 5, 2177-80 (1966); CA 66, 15296.

186. Bird, P. H., and Wallbridge, M. G. H., J. Chem. Soc. 1965, 3923-8; CA 63, 5219.

187. Bjerrum, J., et al., Proc. Symp. Coord. Chem., Tihany, Hung. 1964, 187-98 (Pub. 1965); CA 64, 18940.

188. Blair, A. J. F., and Michael, H., Ber. Kernforschungsanlage Juelich No. 434, 24 pp. (1966); CA 68, 6869.

189. Blandamer, M. J., Gough, T. E., and Symons, M. C. R., Trans. Faraday Soc. 62, 286-95 (1966); CA 64, 15215.

190. Bliznyuk, N. K., et al., USSR 178,819, Feb. 3, 1966; CA 65, 2126.

191. --- et al., USSR 188,971, Nov. 17, 1966; CA 67, 63768.

192. --- et al., USSR 210,150, Feb. 6, 1968; CA 69, 52302.

193. --- et al., USSR 216,718, Apr. 26, 1968; CA 69, 77508.

194. ---, Levskaya, G. S., and Matyukhina, E. N., Zh. Obshch. Khim. 35, 1247-50 (1965); CA 63, 11611.

195. Blum, J., Roseman, H., and Bergmann, E. D., Tetrahedron Lett. 1967, 3665-8; CA 68, 2644.

196. Blundell, T. L., and Powell, H. M., J. Chem. Soc. A 1967, 1650-7; CA 67, 112104.

197. Bock, R., and Grallath, E., Z. Anal. Chem. 222, 283-90 (1966); CA 67, 76678.

198. --- and Langrock, P., ibid., 219, 23-31 (1966); CA 65, 4975.

199. Boehland, H., and Malitzke, P., Z. Anorg. Allg. Chem. 350, 70-83 (1967); CA 66, 101231.

200. --- and Niemann, E., ibid., 336, 225-33 (1965); CA 63, 6592.

201. Bombieri, G., Volponi, L., and Sindellari, L., Ric. Sci. 36, 1226-7 (1966); CA 67, 101196.

202. --- et al., Chem. Commun. 1967, 977; CA 68, 7203.

203. Bonati, F., et al., J. Chem. Soc. A 1966, 1052-5; CA 65, 8315.

204. --- and Ugo, R., J. Organometal. Chem. 7, 167-80 (1967); CA 66, 38035.

204a. Borek, F., and Stupp, Y., Immunochemistry 2, 323-8 (1965); CA 64, 5602.

205. --- and Stupp, Y., ibid., 3, 339-40 (1966); CA 65, 11126.

206. ---, Stupp, Y., and Sela, M., Science 150, 1177-8 (1965); CA 64, 7196.

207. ---, Stupp, Y., and Sela, M., J. Immunol. 98, 739-44 (1967); CA 66, 93490.

208. Borek, F., Stupp, Y., and Sela, M., Biochim. Biophys. Acta 140, 360-2 (1967); CA 67, 52214.

208a. --- et al., Biochem. J. 96, 577-82 (1965); CA 63, 7500.

209. Borisov, A. E., et al., Dokl. Akad. Nauk SSSR 173, 855-8 (1967); CA 67, 37870.

210. --- et al., Ukr. Fiz. Zh. 13, 75-82 (1968); CA 69, 14351.

211. Bosnish, B., et al., J. Amer. Chem. Soc. 88, 3926-9 (1966); CA 65, 13757.

212. Bouquet, G., and Bigorgne, M., Spectrochim. Acta, Part A 23, 1231-4 (1967); CA 67, 38051.

213. ---, Loutellier, A., and Bigorgne, M., J. Mol. Struct. 1, 211-37 (1968); CA 68, 118176.

214. Bowen, L. H., and Rood, R. T., J. Inorg. Nucl. Chem. 28, 1985-90 (1966); CA 65, 13090.

215. Bower, L. M., and Stiddard, M. H. B., J. Chem. Soc. A 1968, 706-10; CA 68, 82673.

216. Bradford, C. W., Platinum Metals Rev. 11, 104-5 (1967); CA 67, 78597.

217. --- and Nyholm, R. S., Chem. Commun. 1967, 384-5; CA 67, 17462.

218. Brady, H. A., and Peevy, F. A., Proc. S. Weed Conf. 21, 218-21 (1968); CA 69, 1987.

219. Branden, C. I., Proc. Int. Conf. Coord. Chem., 8th, Vienna 1964, 114-16; CA 67, 26829.

220. Braun, W., et al., Ger. (East) 42,962, Jan. 5, 1966; CA 64, 16092.

221. Bray, E. H. (Union Carbide Corp.), U.S. 3,338,941, Aug. 29, 1967; CA 68, 39816.

222. Bregadze, V. I., Chumaevskii, N. A., and Shkirtil, E. B., Dokl. Akad. Nauk SSSR 181, 910-13 (1968); CA 69, 81954.

223. Breindel, A. W., and Herbstman, Sh. (Stauffer Chemical Co.), U.S. 3,317,575, May 2, 1967; CA 67, 32780.

224. Brewis, S., Dent, W. T., and Smith, R. D., J. Chem. Soc. 1965, 1539-41; CA 62, 13037.

225. Bridgland, B. E., Fowles, G. W. A., and Walton, R. A., J. Inorg. Nucl. Chem. 27, 383-9 (1965); CA 62, 6129.

226. Brier, P. N., et al., J. Chem. Soc. A 1967, 1889-94; CA 68, 17225.

227. Briles, G. H., Diss. Abstr. B 27, 1081 (1966).

228. --- and McEwen, W. E., Tetrahedron Lett. 1966, 5191-6; CA 66, 28155.

229. --- and McEwen, W. E., ibid., 5299-302; CA 66, 28859.

230. Brisdon, B. J., and Edwards, D. A., Chem. Commun. 1966, 278-9; CA 65, 3310.

231. --- and Edwardo, D. A., Inorg. Chem. 7, 1898-903 (1968); CA 69, 83062.

232. Brisden, B. J., et al., J. Chem. Soc. A 1967, 1825-31; CA 68, 26404.

233. Brisou, B., Bull. Soc. Pathol. Exotique 57, 1058-64 (1964); CA 63, 13907.

234. Brničević, N., and Djordjević, C., Inorg. Chem. 7, 1936-8 (1968); CA 69, 83056.

235. Brodie, A. M., et al., J. Chem. Soc. A 1968, 987-90; CA 68, 118922.

236. --- et al., Inorg. Chim. Acta 2, 195-8 (1968); CA 69, 56594.

237. Brodie, A. M., et al., J. Chem. Soc. A 1968, 2039-42; CA 69, 92473.

238. Brooks, M. S., AD 614612 (1965); CA 63, 14112.

239. ---, AD 630609 (1966); CA 66, 108896.

240. --- and Hancock, W. R., AD 615968 (1965); CA 63, 15623.

241. Brookes, P. R., and Shaw, B. L., J. Chem. Soc. A 1967, 1079-84; CA 67, 69276.

242. Broome, J. D., and Kidd, J. G., J. Exptl. Med. 120, 449-66, 467-90 (1964); CA 61, 16664.

243. Brown, D., Easey, J. F., and Jones, P. J., J. Chem. Soc. A 1967, 1698-1702; CA 67, 113273.

244. Brown, D. H., Mohammed, A., and Sharp, D. W. A., Spectrochim. Acta 21, 659-62 (1965); CA 62, 12611.

245. Brown, D. S., and Bushnell, G. W., Acta Crystallogr. 22, 296-9 (1967); CA 66, 59679.

246. Brown, F. L., and Gooch, R. M., Wood Sci. 1, 41-6 (1968); CA 69, 88079.

247. Brown, R. D., and Nunn, E. K., Australian J. Chem. 19, 1567-76 (1966); CA 65, 14640.

248. Brown, T. M., and Knox, G. F., J. Amer. Chem. Soc. 89, 5296-7 (1967); CA 67, 104667.

249. Bruce, M. I., and Stone, F. G. A., J. Chem. Soc. A 1967, 1238-41; CA 67, 87341.

249a. Bruce, J., MacM., Jr., and Robinson, I. M. (E. I. duPont de Nemours and Co.), U.S. 3,118,865, Jan. 21, 1964; CA 60, 10819.

250. Bruja, N. Z., Rev. Chim. (Bucharest) 17, 359-60 (1966); CA 65, 15163.

251. Bruker, A. B., Grinshtein, E. I., and Soborovskii, L. Z., Zh. Obshch. Khim. 36, 1133-8 (1966); CA 65, 12230.

252. Buckman, T. D., Griffith, O. H., and McConnell, H. M., J. Chem. Phys. 43, 2907-8 (1965); CA 64, 4477.

253. Budesinsky, B., Z. Anal. Chem. 206, 401-9 (1964); CA 62, 3401.

254. ---, ibid., 207, 105-10 (1965); CA 62, 4684.

255. ---, ibid., 178-81 (1965); CA 62, 7354.

256. ---, ibid., 241-7 (1965); CA 62, 7354.

257. ---, ibid., 247-56 (1965); CA 62, 7354.

258. ---, Talanta 15, 1063-4 (1968); CA 69, 110973.

259. --- and Haas, K., Czech. 122,379, March 15, 1967; CA 67, 73688; Fr. 1,474,471, March 24, 1967; CA 67, 73688.

259a. --- and Haas, K., Collection Czech. Chem. Commun. 29, 2758-66 (1964).

260. ---, Haas, K., and Vrzalova, D., ibid., 30, 2373-81 (1965); CA 63, 10657.

261. --- and Krumlova, L., Anal. Chim. Acta 39, 375-81 (1967); CA 67, 96572.

262. --- and Mendova, B., Talanta 14, 523-5 (1967); CA 66, 121814.

263. Bueding, E., and Fisher, J., Biochem. Pharmacol. 15, 1197-1211 (1966); CA 65, 10880.

264. Bullen, G. J., Acta Cryst. 18, 974 (1965); CA 63, 5042.

265. Bullock, J. I., J. Inorg. Nucl. Chem. 29, 2257-64 (1967); CA 67, 112499.

266. Burg, A. B., and Singh, J., J. Amer. Chem. Soc. 87, 1213-16 (1965); CA 62, 13177.

267. Burmeister, J. L., and Gysling, H. J., Inorg. Chim. Acta 1, 100-4 (1967); CA 67, 84954.

268. --- and Timmer, R. C., J. Inorg. Nucl. Chem. 28, 1973-8 (1966); CA 65, 13193.

269. Burns, D. Th., Fogg, A. G., and Higgens, C. T., J. Chromatogr. 30, 287-8 (1967); CA 67, 104954.

270. ---, Fogg, A. G., and Higgens, C. T., ibid., 32, 793-4 (1966); CA 68, 74894.

271. Butler, J. D., and Slife, F. W., Weeds 13, 370-1 (1965); CA 64, 4191.

272. Busev, A. I., Lunina, G. E., and Basargin, N. N., Zh. Anal. Khim. 21, 1414-19 (1966); CA 66, 61532.

273. Buu-Hoi, N. P., et al., J. Chem. Soc. 1965, 2646-8; CA 61, 16221.

274. --- et al., ibid., C 1966, 47-9; CA 64, 8117.

275. --- et al., ibid., 1792-4; CA 65, 20097.

276. --- and Hien, D.-P., Z. Naturforsch. B 22, 532-4 (1967); CA 67, 52624.

277. --- and Hien, D.-P., Biochem. Pharmacol. 17, 1227-36 (1968); CA 69, 34402.

278. --- and Lambelin, G., C. R. Acad. Sci., Paris, Ser. D 267, 247-9 (1968); CA 69, 57613.

279. ---, Mangane, M., and Jacquignon, P., J. Heterocyclic Chem. 3, 149-51 (1966); CA 65, 597.

280. ---, Mangane, M., and Jacquignon, P., ibid., 374-6 (1966); CA 65, 19981.

281. ---, Mangane, M., and Jacquignon, P., J. Chem. Soc. C 1967, 662-5; CA 67, 11423.

282. ---, Perche, J. C., and Saint-Ruf, G., Bull. Soc. Chim. Fr. 1968, 627-31; CA 69, 59079.

283. ---, Roussel, O., and Petit, L., J. Chem. Soc. 1963, 956-60; CA 59, 1637.

284. Cadena, D. G., Jr., and Rowlands, J. F., ibid., B 1967, 506-8; CA 67, 16542.

285. Calesnick, B., Wase, A., and Overby, L. R., Toxicol. Appl. Pharmacol. 9, 27-30 (1966); CA 65, 7886.

286. Camerman, A., Dissert. Abstr. 26, 4258 (1966).

287. --- and Trotter, J., J. Chem. Soc. 1965, 730-8; CA 62, 5960.

288. Camerman, N., Dissert. Abstr. 26, 4259 (1966).

289. Campion, R. J., et al., Inorg. Chem. 6, 672-81 (1967); CA 66, 88977.

290. Cannata, J. J. B., and Stoppani, A. O. M., Anales Asoc. Quim. Arg. 51, 265-71 (1963); CA 63, 13627.

291. Cannell, L. G., Slaugh, L. H., and Mullineaux, R. D. (to Shell Internat. Res.), Ger. 1,186,455, Feb. 4, 1965; CA 62, 16054.

292. Cannon, R. D., Chiswell, B., and Venanzi, L. M., J. Chem. Soc. A 1967, 1277-81; CA 67, 96388.

293. Canziani, F., Sartorelli, U., and Zingales, F., Rend. Ist. Lombardo Sci. Lett. A 99, 21-9 (1965); CA 65, 3297.

294. ---, Sartorelli, U., and Zingales, F., Chim. Ind. (Milan) 49, 469-73 (1967); CA 67, 49972.

295. ---, Sartorelli, U., and Zingales, F., Rend. Ist. Lombardo Sci. Lett. A 101, 227-32 (1967); CA 68, 83947.

296. Canziani, F., and Zingales, F., Rend. Ist. Lombardo Sci. Lett., A 96, 513-22 (1962); CA 62, 11401.

297. Cardaci, G., et al., Inorg. Chim. Acta 1, 340-6 (1967); CA 68, 43584.

298. Cariati, F., Ugo, R., and Bonati, F., Chem. Ind. (London) 1964, 1714-15; CA 62, 6127.

299. ---, Ugo, R., and Bonati, F., Inorg. Chem. 5, 1128-32 (1966); CA 65, 4993.

300. Carmichael, W. M., and Edwards, D. A., J. Inorg. Nucl. Chem. 30, 2641-6 (1968); CA 69, 113034.

301. --- et al., Inorg. Chim. Acta 1, 93-6 (1967); CA 67, 78590.

302. Carpenter, W. R., and Palenik, G. J., J. Org. Chem. 32, 1219-20 (1967); CA 67, 81636.

303. Carter, J. G., and Christophorou, L. G., J. Chem. Phys. 46, 1883-90 (1967); CA 66, 81684.

304. Carty, A. J., and Tuck, D. G., J. Chem. Soc., Suppl. 1964, 6012-17; CA 63, 9422.

305. Casey, A. T., and Clark, R. J. H., Inorg. Chem. 7, 1598-602 (1968); CA 69, 55707.

306. Caspari, G., and Drawe, H., Z. Naturforsch. B 22, 574-9 (1967); CA 67, 103904.

307. Castro, J. A., Biochem. Pharmacol. 17, 295-303 (1968); CA 68, 94430.

308. Cattalini, L., et al., Gazz. Chim. Ital. 95, 567-75 (1965); CA 63, 12655.

309. ---, Ugo, R., and Orio, A., J. Amer. Chem. Soc. 90, 4800-3 (1968); CA 69, 62049.

310. Cavell, R. G., and Dobbie, R. C., J. Chem. Soc. A 1967, 1308-10; CA 67, 82247.

311. --- and Dobbie, R. C., ibid., A 1968, 1406-10; CA 69, 15499.

312. Ceskoslovenska Akademie Ved., Neth. Appl. 6,609,641, Jan. 9, 1967; CA 67, 32519.

313. Cetini, G., et al., Inorg. Chem. 7, 609-10 (1968); CA 68, 92570.

314. Chadaeva, N. A., Kamai, G. Kh., and Mamakov, K. A., Zh. Obshch. Khim. 36, 916-20 (1966); CA 65, 13707.

315. ---, Kamai, G. Kh., and Mamakov, K. A., ibid., 1994-9 (1966); CA 66, 95136.

316. ---, Kamai, G. Kh., and Mamakov, K. A., ibid., 37, 1402-7 (1967); CA 68, 69098.

317. ---, Kamai, G. Kh., and Mamakov, K. A., Izv. Akad. Nauk SSSR, Ser. Khim. 1967, 2554-6; CA 69, 106838.

318. ---, Kamai, G. Kh., and Usacheva, G. M., Zh. Obshch. Khim. 36, 704-8 (1966); CA 65, 10618.

319. Chae, Y. C., et al., Proc. Ann. Tech. Conf., SPI, Reinf. Plast. Div. 1967, 6E/8 pp.; CA 67, 54673.

320. Chaikina, E. A., Gal'braikh, L. S., and Rogovin, Z. A., Cellul. Chem. Technol. (Jassy) 1, 625-39 (1967); CA 68, 88321.

321. Chalchat, B. P., et al., Bull. Soc. Pathol. Exotique 58, 73-80 (1965); CA 64, 10283.

322. Chalmers, A. A., Lewis, J., and Wild, S. B., J. Chem. Soc. A 1968, 1013-18; CA 68, 118906.

322a. Chambers, R. D., and Cunningham, J., Tetrahedron Letters 1965, 2389-91.

323. Chan, S. S., and Willis, C. J., Can. J. Chem. 46, 1237-48 (1968); CA 68, 114717.

324. Chandrasegaran, L., and Rodley, G. A., Inorg. Chem. 4, 1360-1 (1965); CA 63, 15824.

325. Chansler, J. F., and Pierce, D. A., J. Econ. Entomol. 59, 1357-9 (1966); CA 66, 10169.

326. Chapron, Y., Graziani, E., and Francois, H., C. R., Acad. Sci. Paris, C 262, 1247-9 (1966); CA 65, 2106.

327. Chatt, J., and Heaton, B. T., Spectrochim. Acta, Part A 23, 2220-1 (1967); CA 67, 103711.

328. ---, Johnson, N. P., and Shaw, B. L., J. Chem. Soc. A 1967, 604-7; CA 66, 115781.

329. --- et al., Chem. Commun. 1968, 419-20; CA 68, 118389.

330. --- and Shaw, B. L., J. Chem. Soc. A 1966, 1437-42; CA 65, 18142.

331. --- and Shaw, B. L., ibid., 1811-12; CA 66, 18747.

332. --- and Westland, A. D., ibid., A 1968, 88-90; CA 68, 43343.

333. Cherkesov, A. I., and Alykov, N. M., Zh. Analit. Khim. 20, 1312-20 (1965); CA 64, 11833.

334. Chernokal'skii, B. D., Bairamov, R. B., and Kamai, G., Dokl. Akad. Nauk SSSR 180, 1406-7 (1968); CA 69, 85878.

335. ---, Gamoyurova, V. S., and Kamai, G. Kh., Izv. Vysshikh. Uchebn. Zavedenii, Khim. i Khim. Tekhnol. 8, 959-62 (1965); CA 64, 17630.

336. ---, Gamoyurova, V. S., and Kamai, G. Kh., Dokl. Akad. Nauk SSSR 166, 144-7 (1966); CA 64, 12718.

337. ---, Gamayurova, V. S., and Kamai, G., ibid., 384-7 (1966); CA 64, 12491.

338. ---, Gamayurova, V. S., and Kamai, G. Kh., Z. Obshch. Khim. 36, 1673-6 (1966); CA 66, 75498.

339. ---, Gamayurova, V. S., and Kamai, G. Kh., ibid., 1677-9 (1966); CA 66, 49726.

340. ---, Gel'fond, A. S., and Kamai, G., ibid., 37, 1396-9 (1967); CA 68, 13105.

341. --- et al., Izv. Vyssh. Ucheb. Zaved., Khim. Khim. Tekhnol. 9, 768-74 (1966); CA 66, 76112.

342. --- et al., ibid., 918-23 (1966); CA 67, 11554.

343. Chuck, R. J., et al., Nucl. Magnetic Resonance Chem., Proc. Symp., Cagliari, Italy 1964, 189-98; CA 66, 6920.

344. Chudinov, E. G., Zh. Analit. Khim. 20, 805-11 (1965); CA 64, 1324.

345. Chumaevskii, N. A., Tr. Komis. po Spektroskopii, Akad. Nauk SSSR 3, 84-91 (1964); CA 64, 10592.

346. --- and Borisov, A. E., Dokl. Akad. Nauk SSSR 161, 366-9 (1965); CA 63, 462.

347. Ciampolini, M., and Bertini, I., J. Chem. Soc. A 1968, 2241-4; CA 69, 81901.

348. Ciana, A., Ric. Sci. 37, 835-9 (1967); CA 68, 115042.

349. Claeys, E. G., J. Organometal. Chem. 5, 446-53 (1966); CA 65, 602.

349a. --- and Kelen, G. P. van der, Spectrochim. Acta 22, 2095-2101 (1966); CA 66, 23734.

349b. --- and Kelen, G. P. van der, ibid., 2103-9 (1966); CA 66, 33735.

350. Claridge, R. F. C., Merz, E., and Riedel, H. J., Nukleonik 7, 53-8 (1965); CA 62, 16293.

351. Clark, H. C., and Dixon, K. R., Chem. Commun. 1967, 717; CA 68, 2971.

352. --- and Goel, R. G., Inorg. Chem. 5, 998-1003 (1966); CA 65, 1599.

353. --- and Pickard, A. L., J. Organometal. Chem. 13, 61-71 (1968); CA 69, 77312.

354. Clark, J. P., Langford, V. M., and Wilkins, C. J., J. Chem. Soc. A 1967, 792-4; CA 67, 21985.

355. Clark, R. J. H., Spectrochim. Acta 21, 955-63 (1965).

356. ---, J. Chem. Soc. 1965, 5699-704; CA 63, 15840.

357. ---, Record Chem. Progr. (Kresge-Hooker Sci. Lib.) 26, 269-82 (1965); CA 64, 7537.

358. --- et al., J. Chem. Soc. A 1966, 989-90; CA 65, 11743.

359. ---, Greenfield, M. L., and Nyholm, R. S., J. Chem. Soc. A 1966, 1254-9; CA 65, 14819.

360. --- et al., J. Chem. Soc. 1965, 2865-71; CA 62, 14160.

361. ---, Kepert, D. L., and Nyholm, R. S., ibid., 2877-83; CA 62, 14160.

362. --- et al., Spectrochim. Acta 22, 1697-700 (1966); CA 65, 17911.

363. --- and Negrotti, R. H. U., Chem. Ind. (London) 1968, 154; CA 68, 65286.

364. ---, Nyholm, R. S., and Scaife, D. E., J. Chem. Soc. A 1966, 1296-302; CA 65, 18143.

365. ---, Nyholm, R. S., and Taylor, F. B., ibid., A 1967, 1802-4; CA 68, 26131.

366. Clarke, A. J., and Shepherd, A. M., Nematologica 10, 431-53 (1964); CA 62, 8171.

367. Clifford, A. F., and Mukherjee, A. K., Inorg. Syn. 8, 185-91 (1966); CA 65, 18138.

368. Coates, G. E., and Lauder, A., J. Chem. Soc. 1965, 1857-64; CA 62, 13169.

368a. --- and Mukherjee, R. N., ibid., 1964, 1295-1303; CA 60, 14530.

369. --- and Ridley, D., ibid., 166-73; CA 60, 6347.

370. Coffey, R. S. (Imperial Chem. Ind. Ltd.), Brit. 1,121,643, July 31, 1968; CA 69, 60537.

371. Cole, J. J., and Pflaum, R. T., Proc. Iowa Acad. Sci. 71, 145-50 (1964); CA 63, 14349.

372. Collings, R. N., Nyholm, R. S., and Tobe, M. L., NASA Accession No. N64-28665, Rept. No. AD 441241; CA 63, 4330.

373. Collman, J. P., et al., Inorg. Chem. 7, 1298-303 (1968); CA 69, 36245.

374. --- and Roper, W. R., J. Amer. Chem. Soc. 87, 4008-9 (1965); CA 63, 14357.

375. Colton, R., and Tomkins, I. B., Australian J. Chem. 19, 1143-6 (1966); CA 65, 6698.

376. --- and Tomkins, I. B., ibid., 1519-21 (1966); CA 65, 19664.

377. Conix, A. J. (Gevaert Photo-Producten N.V.), Ger. 1,159,646, Dec. 19, 1963; CA 64, 12836.

378. --- and Dohmen, L. M. (Gevaert Photo-Producten N.V.), Ger. 1,199,500, Mar. 26, 1962; CA 63, 16502.

379. Considine, W. J., and Ventura, J. J., J. Organometal. Chem. 3, 420 (1965); CA 63, 1814.

380. Cook, C. D., Nyholm, R. S., and Tobe, M. L., J. Chem. Soc. 1965, 4194-9; CA 63, 9427.

381. Cooke, M., Green, M., and Kirkpatrick, D., J. Chem. Soc. A 1968, 1507-10; CA 69, 56616.

382. Cooper, D., and Plane, R. A., Inorg. Chem. 5, 2209-12 (1966); CA 66, 16097.

383. Coover, H. W., Jr. (Eastman Kodak Co.), U.S. 3,222,337, Dec. 7, 1965; CA 64, 5227.

384. --- and Joyner, F. B. (Eastman Kodak Co.), Brit. 1,000,348, Aug. 4, 1965; CA 64, 2191.

385. --- and Shearer, N. W., Jr. (Eastman Kodak Co.), U.S. 3,196,140, July 20, 1965; CA 63, 11724.

386. --- and Shearer, N. W., Jr. (Eastman Kodak Co.), U.S. 3,216,988, Nov. 9, 1965; CA 64, 3714.

387. --- and Shearer, N. W., Jr. (Eastman Kodak Co.), U.S. 3,220,997, Nov. 30, 1965.

388. --- and Shearer, N. W., Jr. (Eastman Kodak Co.), U.S. 3,284,427, Nov. 8, 1966; CA 66, 11292.

389. Copley, D. B., Fairbrother, F., and Thompson, A., J. Less-Common Metals 8, 256-61 (1965); CA 62, 15746.

390. Corr, J. W., Diss. Abstr. B 27, 2638 (1967).

391. Corvaja, C., and Nordio, P. L., Ric. Sci. 38, 44-5 (1968); CA 69, 14760.

392. Coskran, K. J., Huttermann, T. J., and Verkade, J. G., Advan. Chem. Ser. No. 62, 590-603 (1967); CA 66, 101184.

393. Cotton, F. A., et al., Inorg. Chem. 4, 326-30 (1965); CA 62, 8646.

394. --- and Harris, C. B., ibid., 7, 2140-4 (1968); CA 69, 100757.

395. ---, Johnson, B. F. G., and Wing, R. M., ibid., 4, 502-7 (1965); CA 62, 11404.

396. ---, Legzdins, P., and Lippard, S. J., J. Chem. Phys. 45, 3461-2 (1966); CA 66, 33368.

397. --- and Lippard, S. J., Inorg. Chem. 4, 1621-9 (1965); CA 63, 17459.

398. --- and Lippard, S. J., Chem. Commun. 1965, 245-6; CA 63, 7881.

399. --- and Lippard, S. J., Inorg. Chem. 5, 9-16 (1966); CA 64, 4567.

400. --- and Lippard, S. J., ibid., 416-23 (1966); CA 64, 10494.

401. --- and McCleverty, J. A., ibid., 6, 229-32 (1967); CA 66, 51771.

402. ---, Oldham, C., and Walton, R. A., ibid., 214-23 (1967); CA 66, 43304.

403. ---, Robinson, W. R., and Walton, R. A., ibid., 223-8 (1967); CA 66, 43321.

404. --- et al., ibid., 929-35 (1967); CA 66, 121637.

405. ---, Wing, R. M., and Zimmerman, R. A., ibid., 11-15 (1967); CA 66, 33216.

406. Coucouvanis, D., and Fackler, J. P., Jr., J. Amer. Chem. Soc. 89, 1346-51 (1967); CA 66, 81990.

407. Cousins, D. R., and Hart, F. A., J. Inorg. Nucl. Chem. 29, 2965-74 (1967); CA 68, 35462.

408. Cowley, A. H., Burg, A. B., and Cullen, W. R., J. Amer. Chem. Soc. 88, 3178-9 (1966); CA 65, 13752.

409. Craft, L. N., Diss. Abstr. B 27, 1109-10 (1966).

410. Creemers, H. M. J. C., Verbeek, F., and Noltes, J. G., J. Organometal. Chem. 8, 469-77 (1967); CA 67, 53186.

411. Crutchfield, D. M., Proc. S. Weed Conf. 20, 183-9 (1967); CA 67, 10610.

412. Csaszar, J., Magy. Kem. Folyoirat 71, 114-20 (1965); CA 62, 15570.

413. Cullen, W. R., Can. J. Chem. 41, 317-21 (1963); CA 60, 5547.

414. ---, ibid., 322-8 (1963); CA 60, 5547.

415. ---, Dawson, D. S., and Styan, G. E., J. Organometal. Chem. 3, 406-13 (1965); CA 63, 629.

416. ---, Dawson, D. S., and Styan, G. E., J. Chem. Soc. 43, 3392-9 (1965); CA 64, 3591.

417. ---, Deacon, G. B., and Green, J. H. S., Can. J. Chem. 43, 3193-200 (1965); CA 64, 2870.

418. ---, Deacon, G. B., and Green, J. H. S., ibid., 44, 717-24 (1966); CA 64, 12049.

419. --- and Dhaliwal, P. S., ibid., 45, 379-82 (1967); CA 66, 55562.

420. --- and Dhaliwal, P. S., ibid., 719-24 (1967); CA 66, 94465.

421. ---, Dhaliwal, P. S., and Stewart, C. J., J. Organometal. Chem. 6, 364-72 (1966); CA 65, 18611.

422. ---, Dhaliwal, P. S., and Stewart, C. J., Inorg. Chem. 6, 2256-9 (1967); CA 68, 35442.

423. ---, Green, B. R., and Hochstrasser, R. M., J. Inorg. Nucl. Chem. 27, 641-51 (1965); CA 62, 8517.

424. --- et al., J. Amer. Chem. Soc. 90, 3293-5 (1968); CA 69, 96858.

425. --- and Leeder, W. R., Inorg. Chem. 5, 1004-8 (1966); CA 65, 2293.

426. --- and Styan, G. E., J. Organometal. Chem. 4, 151-6 (1965); CA 63, 13313.

427. --- and Styan, G. E., Can. J. Chem. 44, 1225-7 (1966); CA 65, 2292.

428. Cunninghame, R. G., Nyholm, R. S., and Tobe, M. L., J. Chem. Soc., Suppl. No. 1, 5800-7 (1964); CA 63, 232.

429. Curtis, R., and Snell, Th. L. (Stauffer Chem. Co.), U.S. 3,287,101, Nov. 22, 1966; CA 66, 27941.

430. Dahlmann, J., and Rieche, A., Chem. Ber. 100, 1544-9 (1967); CA 68, 2969.

431. ---, Rieche, A., and Austenat, L., Angew. Chem. Intern. Ed. Engl. 5, 727 (1966); CA 65, 16993.

432. ---, Rieche, A., and Austenat, L., Monatsber. Deut. Akad. Wiss. Berlin 9, 105-8 (1967); CA 68, 49680.

433. D'Alelio, G. F. (Dal Mon Research Co.), U.S. 3,327,018, June 20, 1967; CA 67, 44934.

434. Dallacker, F., and Adolphen, G., Ann. Chem. 694, 110-16 (1966); CA 65,

12186.

435. Dalton, J., et al., J. Chem. Soc. A 1968, 1195-9; CA 69, 14331.

436. --- et al., ibid., 1208-11; CA 69, 14334.

437. Damron, B. L., Waldroup, P. W., and Harms, R. H., Poultry Sci. 45, 151-6 (1966); CA 64, 11606.

438. Daniel, H., and Paetsch, J., Chem. Ber. 101, 1451-6 (1968); CA 68, 104326.

439. Darensbourg, D. J., and Brown, Th. L., Inorg. Chem. 7, 959-66 (1968); CA 68, 118157.

440. Das, V. G. K., and Kitching, W., J. Organometal. Chem. 13, 523-8 (1968); CA 69, 77381.

441. Datta-Gupta, N., Diss. Abstr. B 27, 1435 (1966).

442. Davies, N., Bird, P. H., and Wallbridge, M. G. H., J. Chem. Soc. A 1968, 2269-72; CA 69, 80942.

443. Davison, A., et al., Inorg. Chem. 2, 1227-32 (1963); CA 60, 1314.

444. --- and Holm, R. H., Inorg. Syn. 10, 8-26 (1967); CA 68, 56104.

445. --- and Howe, D. V., Chem. Commun. 1965, 290-1; CA 63, 9424.

446. --- et al., Inorg. Chem. 3, 814-23 (1964).

447. ---, Howe, D. V., and Shawl, E. T., ibid., 6, 458-63 (1967); CA 66, 91244.

448. --- and Shawl, E. T., Chem. Commun. 1967, 670; CA 67, 90902.

449. Davydov, A. V., et al., Tr. Komis. po Analit. Khim., Akad. Nauk SSSR, Inst. Geokhim. i Analit. Khim. 15, 64-79 (1965); CA 63, 14331.

450. Dawes, J. L., and Kemmitt, R. D. W., J. Chem. Soc. A 1968, 2093-5; CA 69, 92510.

451. Day, M. W., Mich. State Univ., Agr. Expt. Sta., Quart. Bull. 47, 383-6 (1965); CA 63, 6253.

452. Deacon, G. B., J. Organometal. Chem. 9, P1-P2 (1967); CA 67, 64498.

453. --- and Green, J. H. S., Spectrochim. Acta, 24 A, 959-64 (1968); CA 69, 55860.

454. --- and Green, J. H. S., ibid., 1125-33 (1968); CA 69, 55894.

455. ---, Green, J. H. S., and Kynaston, W., Aust. J. Chem. 19, 1603-7 (1966); CA 65, 14673.

456. ---, Green, J. H. S., and Nyholm, R. S., J. Chem. Soc. 1965, 3411-25; CA 63, 4319.

457. ---, Green, J. H. S., and Taylor, F. B., Aust. J. Chem. 20, 2069-76 (1967); CA 68, 7782.

458. --- and Nyholm, R. S., J. Chem. Soc. 1965, 6107-16; CA 64, 747.

459. Debarre, F., and Cometti, A. (Soc. des Usines Chimiques Rhone-Poulenc), Fr. M 3623, Nov. 22, 1965.

460. Deeming, A. J., and Shaw, B. L., J. Chem. Soc. A 1968, 1887-9; CA 69, 64211.

461. Deitrich, R. A., Arch. Biochem. Biophys. 119, 253-63 (1967); CA 66, 91839.

462. Delgado, F., Palomino, J. V., and Petrement, J., Anales Real Soc. Espan. Fis. Quim. (Madrid) Ser. B 62, 275-92 (1966); CA 65, 11720.

463. Dellepiane, G., and Zerbi, G., J. Mol. Spectrosc. 24, 62-86 (1967); CA

$\underline{67}$, 112612.

464. Dening, R. G., Hartley, F. R., and Venanzi, L. M., J. Chem. Soc. A $\underline{1967}$, 324-8; CA $\underline{66}$, 59550.

465. Dennis, E. W., and Arnold, A. (Sterling Drug Inc.), U.S. 3,271,159, Sept. 2, 1966; CA $\underline{65}$, 19234.

466. Desnoyers, J., and Rivest, R., Can. J. Chem. $\underline{43}$, 1879-80 (1965); CA $\underline{63}$, 7880.

467. Dessy, R. E., Chivers, T., and Kitching, W., J. Amer. Chem. Soc. $\underline{88}$, 467-70 (1966); CA $\underline{64}$, 9571.

468. --- et al., ibid., $\underline{90}$, 2001-4 (1968); CA $\underline{68}$, 101150.

469. --- and Pohl, R. L., ibid., 2005-8 (1968); CA $\underline{68}$, 101180.

470. --- and Pohl, R. L., ibid., 1995-2001 (1968); CA $\underline{68}$, 101187.

471. ---, Pohl, R. L., and King, R. B., ibid., $\underline{88}$, 5121-4 (1966); CA $\underline{66}$, 54850.

472. ---, Weissman, P. M., and Pohl, R. L., ibid., 5117-21 (1966); CA $\underline{66}$, 43138.

473. Dewhirst, K. C., Inorg. Chem. $\underline{5}$, 319-21 (1966); CA $\underline{64}$, 14215.

474. Dey, A. K., et al., U. S. Dept. Com., Clearinghouse Sci. Tech. Inform. AD 627221, 151-62 (1965); CA $\underline{65}$, 7981.

475. Dickens, R., Diss. Abstr. B $\underline{27}$, 2565 (1967).

476. --- and Hiltbold, A. E., Weeds $\underline{15}$, 299-304 (1967); CA $\underline{68}$, 11888.

477. Dietrich, F. M., J. Immunol. $\underline{97}$, 216-23 (1966); CA $\underline{65}$, 9516.

478. ---, Nature $\underline{213}$, 496-7 (1967); CA $\underline{66}$, 63842.

479. DiSipio, L., et al., Coord. Chem. Rev. $\underline{2}$, 129-35 (1965); CA $\underline{67}$, 87329.

480. Djordjevic, C., et al., J. Chem. Soc. A $\underline{1966}$, 16-17; CA $\underline{64}$, 9227.

481. Doak, G. O., and Long, G. G., Trans. N. Y. Acad. Sci. $\underline{28}$, 403-11 (1966); CA $\underline{65}$, 12232.

482. ---, Long, G. G., and Freedman, L. D., J. Organometal. Chem. $\underline{4}$, 82-91 (1965); CA $\underline{63}$, 2534.

483. ---, Long, G. G., and Freedman, L. D., ibid., $\underline{12}$, 443-50 (1968); CA $\underline{69}$, 52248.

484. ---, Long, G. G., and Key, M. E., Inorg. Synth. $\underline{9}$, 92-7 (1967); CA $\underline{67}$, 21987.

485. Dobbie, R. C., and Cavell, R. G., Inorg. Chem. $\underline{6}$, 1450-3 (1967); CA $\underline{67}$, 77275.

486. Dobson, G. R., and Houk, L. W., Inorg. Chim. Acta $\underline{1}$, 287-93 (1967); CA $\underline{68}$, 8849.

487. Doetzer, R. (Siemens-Schuckertwerke A.G.), Ger. 1,200,817, Sept. 16, 1965; CA $\underline{63}$, 15896.

488. Dolcetti, G., and Peloso, A., Gazz. Chim. Ital. $\underline{97}$, 230-42 (1967); CA $\underline{67}$, 70161.

489. --- and Peloso, A., ibid., 1540-50 (1967); CA $\underline{69}$, 13288.

490. ---, Peloso, A., and Sindellari, L., ibid., $\underline{96}$, 1648-60 (1966); CA $\underline{67}$, 15503.

491. ---, Peloso, A., and Tobe, M. L., J. Chem. Soc. $\underline{1965}$, 5196-203; CA $\underline{63}$, 15837.

492. Dorsett, H. G., Jr., and Nagy, J., U. S. Bur. Mines, Rep. Invest. $\underline{1968}$,

No. 7132, 23 pp.; CA 69, 44997.

493. Dove, M. F. A., Chem. Commun. 1965, 23-4; CA 64, 6685.

494. Downs, A. J., Schmutzler, R., and Steer, I. A., ibid., 1966, 221-2; CA 65, 1594.

495. --- and Steer, I. A., J. Organometal. Chem. 8, P21-P24 (1967); CA 67, 38048.

496. Dowsing, R. D., et al., Nature 1968, 219; CA 69, 91701.

497. Doyle, Th. E., et al. (H. H. Robertson Co.), U.S. 3,317,465, May 2, 1967; CA 67, 12096.

498. Drawe, H., and Caspari, G., Angew. Chem., Intern. Ed. Engl. 5, 317-18 (1966); CA 64, 19668.

499. Drinkard, W. C., Jr. (E. I. duPont de Nemours and Co.), U.S. 3,373,001, Mar. 12, 1968; CA 68, 115214.

500. DuBois, Th. D., and Meek, D. W., Inorg. Chem. 6, 1395-8 (1967); CA 67, 49975.

501. Duckworth, V. F., Harris, C. M., and Stephenson, N. C., Inorg. Nucl. Chem. Lett. 4, 419-25 (1968); CA 69, 100685.

502. Duffy, N. V., et al., Nature 212, 177-8 (1966); CA 66, 28868.

503. Dunbar, J. E. (Dow Chemical Co.), U.S. 3,166,579, Jan. 19, 1965; CA 62, 7800.

504. ---, U.S. 3,162,661, Dec. 22, 1964; CA 63, 7800.

505. ---, U.S. 3,287,462, Nov. 22, 1966; CA 66, 28894.

506. Duncan, J. F., and Thomas, F. G., J. Inorg. Nucl. Chem. 29, 869-90 (1967); CA 67, 7118.

507. DuPont de Nemours and Co., E. I., Brit. 937,571, Sept. 25, 1963; CA 60, 1223.

508. Durand, M., and Laurent, J.-P., C. R., Acad. Sci. Paris 261, 3793-5 (1965); CA 64, 5908.

509. Dyer, G., et al., J. Chem. Soc. A 1966, 1110-23; CA 65, 8317.

510. --- and Meek, D. W., J. Amer. Chem. Soc. 89, 3983-7 (1967); CA 67, 78580.

510a. --- et al., Abstr. Papers 151th ACS Meeting, Pittsburgh, Pa., Mar. 28-31, 1966, H64.

511. ---, Hartley, J. G., and Venanzi, L. M., J. Chem. Soc. 1965, 1293-7; CA 62, 8645.

512. --- and Meek, D. W., Inorg. Chem. 4, 1398-402 (1965); CA 63, 12654.

513. --- and Meek, D. W., ibid., 6, 149-53 (1967); CA 66, 43297.

514. --- and Venanzi, L. M., J. Chem. Soc. 1965, 2771-8; CA 62, 14056.

515. ---, Workman, M. O., and Meek, D. W., Inorg. Chem. 6, 1404-7 (1967); CA 67, 49979.

516. Eberle, S. H., and Robel, W., Inorg. Nucl. Chem. Lett. 4, 113-17 (1968); CA 68, 83988.

517. Edwards, P. R., and Johnson, C. E., J. Chem. Phys. 49, 211-16 (1968); CA 69, 72236.

518. Edwin, J. P., Southwestern Vet. 17, 218-20 (1964); CA 64, 18273.

519. Eigen, M., and Kruse, W., Z. Naturforsch. 18b, 857-65 (1963); CA 60, 4851.

520. Einstein, F. W. B., Cullen, W. R., and Trotter, J., J. Amer. Chem. Soc.

88, 5670-1 (1966); CA 66, 69712.

521. Einstein, F. W. B., and Rodley, G. A., J. Inorg. Nucl. Chem. 29, 347-51 (1967); CA 66, 59705.

522. --- and Trotter, J., J. Chem. Soc. A 1967, 824-8; CA 67, 6489.

523. Eisenberg, R., et al., Inorg. Chem. 7, 741-8 (1968); CA 68, 90699.

524. Eley, D. D., and Pacini, B. M., Polymer 9, 159-72 (1968); CA 68, 105621.

525. Ellermann, J., and Dorn, K., Chem. Ber. 100, 1230-4 (1967); CA 66, 115772.

526. --- and Dorn, K., ibid., 101, 643-56 (1968); CA 68, 69097.

527. --- and Schirmacher, D., Angew. Chem., Int. Ed. Engl. 7, 738-9 (1968); CA 69, 106833.

528. Elmes, P. S., and West, B. O., Coord. Chem. Rev. 3, 279-84 (1968); CA 69, 48797.

529. Elslager, E. F. (Parke, Davis & Co.), U.S. 3,342,844, Sept. 19, 1967; CA 68, 68711.

530. ---, Worth, D. F., and Short, F. W. (Parke, Davis & Co.), U.S. 3,218,309, Nov. 16, 1965; CA 64, 14148.

531. Ercoli, N., Nature 216, 398-9 (1967); CA 68, 58467.

532. Ermakova, M. I., Vorontsova, L. N., and Latosh, N. I., Zh. Obshch. Khim. 37, 649-52 (1967); CA 67, 43893.

533. Ertel, H., and Horner, L., Ann. Univ. Ferrara, Sez. 5, Suppl. 4, 71-9, 80 (1966); CA 67, 26301.

534. Ettore, R., Dolcetti, G., and Peloso, A., Gazz. Chim. Ital. 97, 1681-8 (1967); CA 68, 99231.

535. ---, Peloso, A., and Dolcetti, G., ibid., 968-79 (1967); CA 68, 43791.

536. Evans, D. C. (M. & T. Chemicals, Inc.), Brit. 1,079,659, Aug. 16, 1967; CA 68, 49771.

537. ---, Brit. 1,106,035, March 13, 1968; CA 68, 105369.

538. Evans, D., Osborn, J. A., and Wilkinson, G., Inorg. Synth. 11, 99-101 (1968); CA 69, 40858.

539. Evseeva, T. I., and Cherstvenkova, E. P., Zh. Anal. Khim. 22, 34-9 (1967); CA 66, 119262.

540. Fackler, J. P., Jr., and Coucouvanis, D., Chem. Commun. 1965, 556-7; CA 64, 456.

541. --- and Coucouvanis, D., J. Amer. Chem. Soc. 88, 3913-20 (1966); CA 65, 15422.

542. ---, Dolbear, G. E., and Coucouvanis, D., J. Inorg. Nucl. Chem. 26, 2035-7 (1964); CA 62, 2491.

543. Fairy, M. B., and Irving, R. J., J. Chem. Soc. A 1966, 475-9; CA 64, 18946.

544. Faithful, B. D., et al., J. Chem. Soc. A 1966, 1185-8; CA 65, 14820.

545. --- and Tuck, D. G., Chem. Ind. (London) 1966, 992-3; CA 65, 5482.

546. --- and Wallwork, S. C., Chem. Commun. 1967, 1211; CA 68, 54312.

547. Faraglia, G., et al., J. Organometal. Chem. 10, 363-8 (1967); CA 68, 21971.

548. Faraone, G., et al., Gazz. Chim. Ital. 98, 480-7 (1968); CA 69, 30533.

549. Farbenfabriken Bayer A.-G., Brit. 959,499, June 3, 1964; CA 61, 8437.

550. ---, Neth. Appl. 6,506,043, Nov. 15, 1965; CA 64, 17808.

551. ---, Neth. Appl. 6,600,826, July 22, 1966; CA 66, 11401.

552. Farnand, J. R., Meadus, F. W., and Puddington, I. E. (Canad. Patents and Develop. Ltd.), U.S. 3,401,794, Sept. 17, 1968; CA 69, 98649.

553. Farona, M. F., and Wojcicki, A., Inorg. Chem. 4, 1402-9 (1965); CA 63, 12653.

554. Feay, D. C., Mackey, J. C., and Byrne, J. B. (Dow Chemical Co.), U.S. 3,179,649, Apr. 20, 1965; CA 63, 5771.

555. Feigl, F., and Goldstein, D., Mikrochim. Acta 1966, 1-3; CA 65, 6281.

556. Feldman, J., et al., Am. Chem. Soc., Div. Petrol. Chem., Preprints 9, A55-A64 (1964); CA 64, 14076.

557. ---, Saffer, B. A., and Thomas, M. (National Distillers and Chemical Corp.), U.S. 3,251,893, May 17, 1966; CA 65, 2147.

558. Feltham, R. D., Kasenally, A., and Nyholm, R. S., J. Organometal. Chem. 7, 285-8 (1967); CA 66, 65598.

559. ---, Metzger, H. G., and Silverthorn, W., Inorg. Chem. 7, 2003-6 (1968); CA 69, 92491.

560. --- and Silverthorn, W., Inorg. Syn. 10, 159-64 (1967); CA 68, 78374.

561. --- and Silverthorn, W., Inorg. Chem. 7, 1154-8 (1968); CA 69, 14303.

562. Fergusson, J. E., and Hickford, J. H., J. Inorg. Nucl. Chem. 28, 2293-6 (1966); CA 65, 19667.

563. ---, Robinson, B. H., and Wilkins, C. J., J. Chem. Soc. A 1967, 486-90; CA 66, 81988; Proc. Int. Conf. Coord. Chem., 8th, Vienna, 1964, 146-7; CA 66, 101148.

564. Field, R. A., and Kepert, D. L., J. Less-Common Metals 13, 378-84 (1967); CA 67, 87333.

565. Fielding, H. C., and Williamson, W. I. (Imperial Chemical Industries Ltd.), Brit. 1,001,774, Aug. 18, 1965, Addn. to Brit. 979,272; CA 63, 18477.

565a. Figgis, B. N., et al., J. Chem. Soc. A 1966, 1411-21; CA 65, 17875.

566. Fild, M., Glemser, O., and Christoph, G., Angew. Chem. 76, 953 (1964); CA 62, 7794.

567. Fillmore, D. L., and Wilson, B. J., Inorg. Chem. 7, 1592-4 (1968); CA 69, 55687.

568. Fischer, E. O., and Schneider, R. J. J., J. Organometal. Chem. 12, P27-P30 (1968).

569. --- and Strametz, H., ibid., 10, 323-30 (1967); CA 68, 39779.

570. Fleischfresser, B. E., et al., J. Amer. Chem. Soc. 90, 2172-4 (1968); CA 69, 51312.

571. Fleming, P. B., Dougherty, Th. A., and McCarley, R. E., J. Amer. Chem. Soc. 89, 159-60 (1967); CA 66, 51794.

572. Florinskii, F. S., and Koton, M. M., USSR 165,716, Oct. 26, 1964; CA 62, 6516.

573. Fluck, E., and Kuhn, P., Z. Anorg. Allgem. Chem. 350, 263-70 (1967); CA 66, 120557.

574. Fok, J. Sh.-K., Diss. Abstr. 25, 3815-16 (1965).

575. Formanek, I., Fulop, L., and Szantho, C., Rev. Med. 10, 312-4 (1964); CA 62, 14423.

576. Forster, D., and Goodgame, D. M. L., J. Chem. Soc. 1965, 262-7; CA 62, 6018.

577. --- and Goodgame, D. M. L., ibid., 454-8; CA 62, 6125.

578. --- and Goodgame, D. M. L., Inorg. Chem. 4, 715-18 (1965); CA 62, 14053.

579. --- and Goodgame, D. M. L., ibid., 823-9 (1965); CA 63, 1461.

580. --- and Horrocks, W. DeW., ibid., 5, 1510-14 (1966); CA 65, 11737.

581. Foss, V. L., Besolova, E. A., and Lutsenko, I. F., Zh. Obshch. Khim. 35, 759-60 (1965); CA 63, 4330.

582. Fotis, P., Jr., and McCollum, J. D. (Standard Oil Co. Indiana), U.S. 3,324,018, June 6, 1967; CA 67, 53616.

583. Fowles, G. W. A., and Jenkins, D. K., Chem. Commun. (London) 1965, 61-2; CA 62, 16293.

584. ---, Lewis, D. F., and Walton, R. A., J. Chem. Soc. A 1968, 1468-73; CA 69, 48846.

585. --- and Walton, R. A., ibid., 1964, 4330-4; CA 62, 1309.

586. Frankel, E. N., et al., J. Org. Chem. 32, 1447-50 (1967); CA 67, 2552.

587. Frans, R. E., and Holifield, E. L., Arkansas Univ. (Fayetteville) Agr. Expt. Sta. Mimeo., Ser. No. 142, 24 pp. (1965); CA 63, 3553.

588. --- and Smith, H. R., ibid., No. 156, 29 pp. (1967); CA 69, 1963.

589. ---, Smith, H. R., and Fullerton, T. M., ibid., No. 152, 28 pp. (1966); CA 65, 11259.

590. Fredrickson, R. L., et al., J. Assoc. Offic. Agr. Chemists 48, 10-17 (1965); CA 62, 16293.

591. Freedman, M. H., Grossberg, A. L., and Pressman, D., Immunochemistry 5, 367-81 (1968); CA 69, 65781.

592. ---, Grossberg, A. L., and Pressman, D., Biochemistry 7, 1941-50 (1968); CA 69, 1390.

593. Friedheim, E. A. H., Ger. 1,229,543, Dec. 1, 1966; CA 66, 28895.

594. Fritz, H. P., and Schwarzhans, K. E., J. Organometal. Chem. 5, 103-5 (1966); CA 64, 7552.

595. Fukuyo, M., Nakatsu, K., and Shimada, A., Bull. Chem. Soc. Japan 39, 1614-15 (1966); CA 65, 14552.

596. Furlani, C., Zinato, E., and Furlan, F., Atti Accad. Nazl. Lincei, Rend., Classe Sci. Fis., Mat. Nat. 38, 517-24 (1965); CA 63, 15817.

596a. Fusco, R., Peri, C. A., and Corradini, V. (Montecatini), Ger. 1,189,079, Mar. 18, 1965.

597. Fye, R. L., LaBrecque, G. C., and Gouck, H. K., J. Econ. Entomol. 59, 485-7 (1966); CA 64, 14895.

598. Ganopol'skii, V. I., Barkovskii, V. F., and Ganopol'skaya, T. A., Zavodsk. Lab. 30, 267-70 (1964); CA 60, 18875.

599. Gaskin, T. A., Agron. J. 56, 340-2 (1964); CA 61, 6293.

600. Gastaminza, A., et al., J. Chem. Soc. B 1968, 534-9; CA 69, 2229.

601. Gatilov, Yu. F., and Ionov, L. B., Zh. Obshch. Khim. 38, 2561-5 (1968); CA 70, 68477.

602. ---, Ionov, L. B., and Kamai, G. Kh., ibid., 372-4 (1968); CA 69, 76078.

603. Gatilov, Yu. F., and Kamai, G. Kh., Zh. Obshch. Khim. 36, 55-7 (1966);
     CA 64, 14212.

604. ---, Kamai, G. Kh., and Ionov, L. B., ibid., 37, 1904-6 (1967); CA 68,
     29825.

605. ---, Kamai, G. Kh., and Ionov, L. B., Nekot. Aspekty Stereokhim. Org.
     Proizvod. Mysh'yaka 1966, 50-61; CA 69, 87121.

606. ---, Kamai, G. Kh., and Ionov, L. B., Zh. Obshch. Khim. 38, 370-1
     (1968); CA 69, 77377.

607. ---, Kamai, G. Kh., and Shagidullin, R. R., ibid., 36, 1670-2 (1966);
     CA 66, 46462.

608. --- and Sidorova, L. P., Nekot. Aspekty Stereokhim. Org. Proizvod.
     Mysh'yaka 1966, No. 1, 71-5; CA 69, 67789.

609. --- and Kralichkina, M. G., Zh. Obshch. Khim. 38, 1798-800 (1968); CA
     70, 78097.

610. Gaudiano, G., et al., Chim. Ind. (Milan) 49, 1343-6 (1967); CA 68, 2941.

611. Gehrke, H., Jr., Diss. Abstr. 25, 4952 (1965).

612. Gel'man, N. E., Sheveleva, N. S., and Shakhova, N. I., Zh. Anal. Khim.
     23, 1067-70 (1968); CA 69, 83195.

613. General Electric Co., Neth. Appl. 287,353, Feb. 25, 1965; CA 64, 6848.

614. Gevaert Photo-Producten N.V., Neth. Appl. 6,415,250, Mar. 25, 1965;
     CA 63, 5864.

615. Gibson, N. A., and Hosking, J. W., Australian J. Chem. 18, 123-5 (1965);
     CA 62, 15479.

616. Giddings, B. E., Diss. Abstr. 28 B, 4061-2 (1968).

617. Giesen, M., Federation Assoc. Techniciens Ind. Peintures, Vernis,
     Emaux Encres Imprimerie Europe Continental, Congr. 8, 185-96 (1966);
     CA 65, 17625.

618. No entry.

619. Ginsberg, A. P., Chem. Commun. 1968, 857-8; CA 69, 83064.

620. Gladshtein, B. M., Kulyulin, I. P., and Soborovskii, L. J., Zh. Obshch.
     Khim. 36, 488-92 (1966); CA 65, 743.

621. Glazer, A. N., Proc. Nat. Acad. Sci. U.S. 59, 996-1002 (1968); CA 68,
     111805.

622. ---, J. Biol. Chem. 243, 3693-701 (1968); CA 69, 24754.

622a. Glick, M. D., Diss. Abstr. 25, 5546 (1965).

623. Glover, D. J., and Rosen, J. M., Anal. Chem. 37, 306-7 (1965); CA 62,
     8366.

624. Goetz, H., Schulze, H., and Werner, K., Phys. Status Solidi 24, K95-K96
     (1967); CA 68, 25308.

625. Going, J. E., and Pflaum, R. T., Proc. Iowa Acad. Sci. 72, 117-22 (1965);
     CA 69, 15949.

626. Gold, E. F., Knight, K. L., and Haurowitz, K. L., Biochem. Biophys.
     Res. Commun. 18, 76-80 (1965); CA 62, 12311.

627. Goldwhite, H., Rowsell, D. G., and Valdez, C., J. Organometal. Chem. 12,
     133-41 (1968); CA 69, 19285.

628. Goodall, D. C., J. Chem. Soc. A 1966, 1562-4; CA 66, 7966.

629. Goodfellow, R. J., et al., ibid., A 1968, 1604-9; CA 69, 47734.

630. Goodfellow, R. J., Goggin, P. L., and Duddell, D. A., J. Chem. Soc. A 1968, 504-6; CA 68, 82676.

631. ---, Goggin, P. L., and Venanzi, L. M., ibid., A 1967, 1897-900; CA 68, 7827.

632. --- and Venanzi, L. M., ibid., 1965, 7533-4; CA 64, 6077.

633. Goodgame, D. M. L., and Goodgame, M., Inorg. Chem. 4, 139-43 (1965); CA 62, 6018.

634. ---, Goodgame, M., and Weeks, M. J., J. Chem. Soc. A 1967, 1676-9; CA 67, 113304.

635. --- and Hitchman, M. A., Inorg. Chem. 4, 721-5 (1965); CA 62, 14157.

636. --- and Hitchman, M. A., ibid., 7, 1404-7 (1968); CA 69, 40838.

637. --- and Malerbi, B. W., Spectrochim. Acta, Part A 24, 1254-5 (1968); CA 69, 63138.

638. --- and Marsham, D. F., J. Chem. Soc. A 1966, 1167-9; CA 65, 14814.

639. Goodrich Co., B. F., Neth. Appl. 6,609,478, Jan. 9, 1967; CA 67, 32323.

640. Govindappa, M. H., and Grewal, J. S., Indian J. Agr. Sci. 35, 210-15 (1965); CA 64, 20547.

641. --- and Grewal, J. S., Indian J. Hort. 22, 80-6 (1965); CA 68, 28707.

642. Gracheva, L. M., Nefedov, V. D., and Grachev, S. A., Radiokhimiya 9, 738-41 (1967); CA 68, 83305.

643. Graham, J. R., and Angelici, R. J., Inorg. Chem. 6, 2082-5 (1967); CA 68, 6779.

644. Graham, W. A. G., ibid., 7, 315-21 (1968); CA 68, 53453.

645. Granin, E. I., et al., Khim. v Sel'sk. Khoz. 3, 26-30 (1965); CA 63, 4887.

646. Graziani, M., Zingales, F., and Belluco, U., Inorg. Chem. 6, 1582-6 (1967); CA 67, 87328.

647. Graziani, R., et al., Chem. Commun. 1967, 1284-5; CA 68, 54350.

648. ---, Zarli, B., and Bandoli, G., Ric. Sci. 37, 984-5 (1967); CA 69, 71298.

649. Green, J. H. S., Kynaston, W., and Rodley, G. A., Spectrochim. Acta, Part A 24, 853-62 (1968); CA 69, 31675.

650. Green, M., and Kirkpatrick, D., Chem. Commun. 1967, 57-8; CA 66, 76109.

651. --- and Kirkpatrick, D., J. Chem. Soc. A 1968, 483-6; CA 68, 59677.

652. Griepink, B., and Krijgsmann, W., Mikrochim. Acta 1968, 574-81; CA 69, 8300.

653. Griffiths, J., et al., Ann. Appl. Biol. 58, 183-91 (1966); CA 66, 27889w.

654. Grishchuk, A. P., and Baranov, S. N., Zh. Organ. Khim. 1, 762-4 (1965); CA 63, 8523.

655. --- and Perova, T. V., USSR 196,866, May 31, 1967; CA 68, 95815.

656. Griswold, D. P., Jr., et al., Cancer Res. 26A, 619-25 (1966); CA 64, 18288.

657. Grobe, J., Z. Anorg. Allgem. Chem. 361, 32-46 (1968); CA 69, 102631.

658. --- and Sheppard, N., Z. Naturforsch. B 23, 901-5 (1968); CA 69, 63194.

659. Groff, J. L., Ferber, J. M., and Shulman, S., Immunology 12, 219-24 (1967); CA 66, 84166.

660. Groff, J. L., and Haurowitz, F., Immunochemistry 1, 31-6 (1964); CA 60, 14990.

661. Gross, F. J. (American Cyanamid Co.), U.S. 3,197,314, July 27, 1965; CA 63, 8610.

662. ---, U.S. 3,395,212, July 30, 1968; CA 69, 78425.

663. Grundnes, J., Klaeboe, P., and Plahte, E., Selec. Top. Struct. Chem. 1967, 265-75; CA 69, 63210.

664. Gueguen, N., and Juillard, J., C. R. Acad. Sci., Paris, Ser. C 263, 616-17 (1966); CA 66, 14531.

665. Gukasyan, S. E., and Shpinel, V. S., Phys. Status Solidi 29, 49-52 (1968); CA 69, 82081.

666. Guyon, J. C., Stokely, J. R., and Shults, W. D., U.S. At. Energy Comm. 1968 (ORNL-TM-2108), 13 pp.; CA 69, 7686.

667. Hadden, W. A., et al., Louisiana State Univ. Eng. Res. Bull. No. 87, 80 pp. (1965); CA 65, 7923.

668. Hadzi, D., Klofatur, C., and Oblak, S., J. Chem. Soc. A 1968, 905-8; CA 68, 99211.

669. --- and Kobilarov, N., ibid., A 1966, 439-45; CA 64, 15705.

670. Haines, R. J., Nyholm, R. S., and Stiddard, M. H. B., ibid., A 1967, 94-8; CA 66, 38030.

671. ---, Nyholm, R. S., and Stiddard, M. H. B., ibid., A 1968, 46-7; CA 68, 45772.

672. Hala, J., Navratil, O., and Nechuta, V., J. Inorg. Nucl. Chem. 28, 553-61 (1966); CA 64, 13452.

673. Hales, L. A. W., and Irving, R. J., Spectrochim. Acta A 23, 2981-7 (1967); CA 68, 25236.

674. --- and Irving, R. J., J. Chem. Soc. A 1967, 1932-5; CA 68, 44336.

675. Hamada, Sh., and Kobayashi, K., Eiyo to Shokuryo 21, 64-6 (1968); CA 69, 57771.

676. Hands, A. R., and Mercer, A. J. H., J. Chem. Soc. C 1967, 1099-100; CA 67, 32748.

677. --- and Mercer, A. J. H., ibid., C 1968, 1331-7; CA 69, 10520.

678. Hansson, H. A., Exp. Eye Res. 5, 335-54 (1966); CA 66, 45169.

679. Haque, R. U., and Din, B. U., Pakistan J. Sci. Ind. Res. 9, 121-4 (1966); CA 66, 28851.

680. Harmon, K. M., and Lake, R. R., Inorg. Chem. 7, 1921-3 (1968); CA 69, 81913.

681. Harmon, R. E., Dutton, F. E., and Warren, H. D., J. Med. Chem. 11, 627-9 (1968); CA 69, 96636.

682. Harnden, R. C., and Moore, J. O. (Ansul Co.), U.S. 3,342,584, Sept. 19, 1967; CA 67, 116069.

683. Harris, G. S., Proc. Chem. Soc. 1961, 65.

684. --- and Ali, M. F., Inorg. Nucl. Chem. Lett. 4, 5-8 (1968); CA 68, 95905.

685. --- and Inglis, F., J. Chem. Soc. A 1967, 497-500; CA 66, 85837.

686. --- et al., Chem. Commun. 1967, 442-4; CA 67, 103432.

687. --- et al., Chem. Ind. (London) 1968, 1483.

688. Harshaw Chemical Co. and Dr. Mayfield Lab., Brit. 1,081,260, Aug. 31, 1967; CA 68, 6166.
689. Hart, F. A., and Newbery, J. E., J. Inorg. Nucl. Chem. 28, 1334-6 (1966); CA 65, 3305.
690. --- and Newbery, J. E., ibid., 30, 318-19 (1968); CA 68, 82693.
691. Hartley, J. G., Kerfoot, D. G. E., and Venanzi, L. M., Inorg. Chim. Acta 1, 145-8 (1967); CA 67, 87311.
692. Hartman, F. A., Kilner, M., and Wojcicki, A., Inorg. Chem. 6, 34-40 (1967); CA 66, 34393.
693. --- and Lustig, M., Inorg. Chem. 7, 2669-70 (1968); CA 70, 25316.
694. Hartmann, H., Ann. Chem. 714, 1-7 (1968); CA 69, 52229.
695. ---, Karbstein, B., and Reiss, W., Naturwissenschaften 52, 59 (1965); CA 62, 11852.
696. Harvey, A. B., J. Phys. Chem. 70, 3370-1 (1966); CA 66, 5852.
697. ---, Spectrosc. Lett. 1, 197-203 (1968); CA 69, 101311.
698. --- and Wilson, M. K., J. Chem. Phys. 44, 3535-46 (1966); CA 65, 186.
699. Hassell, W. F., Jarvie, A., and Walker, S., J. Chem. Phys. 46, 3159-62 (1967); CA 67, 6582.
700. Hattori, T., Tsukahara, I., and Yamamoto, T., Bunseki Kagaku 15, 35-41 (1966); CA 64, 14952.
701. Hawking, F., Ann. Trop. Med. Parasitol. 57, 255-61 (1963); CA 60, 16365.
702. ---, ibid., 262-82 (1963); CA 61, 16495.
703. --- and Walker, P. J., Exptl. Parasitol. 18, 63-86 (1966); CA 64, 16348.
704. Hawley, D. M., and Ferguson, G., J. Chem. Soc. A 1968, 2059-63; CA 69, 81460.
705. --- and Ferguson, G., ibid., 2539-43; CA 69, 111060.
706. ---, Ferguson, G., and Harris, G. S., Chem. Commun. 1966, 111-12; CA 64, 18555.
707. Hayashi, E., and Kado, M. (Ihara Agricult. Chem. Co., Ltd.), Japan. 16,552, July 29, 1965; CA 63, 18122.
708. --- and Kado, M. (Ihara Agricult. Chem. Co., Ltd.), Japan. 17,027, Aug. 3, 1965; CA 63, 18121.
709. --- and Kado, M. (Ihara Agricult. Chem. Co., Ltd.), Japan. 17,028, Aug. 3, 1965; CA 63, 18122.
710. --- and Kado, M. (Ihara Agricult. Chem. Co., Ltd.), Japan. 23,793, Oct. 19, 1965; CA 64, 3602.
711. --- and Kado, M. (Ihara Agricult. Chem. Co., Ltd.), Japan. 23,794, Oct. 19, 1965; CA 64, 3602.
712. --- and Kado, M. (Ihara Agricult. Chem. Co., Ltd.), Japan. 3,454, Feb. 28, 1966; CA 64, 19682.
713. Hayter, R. G., and Williams, L. F., J. Inorg. Nucl. Chem. 26, 1977-83 (1964); CA 62, 215.
714. Hazeldean, G. S. F., Nyholm, R. S., and Parish, R. V., J. Chem. Soc. A 1966, 162-5; CA 64, 10734.
715. Heaney, H., and Millar, I. T., ibid., 1965, 5132-4; CA 63, 16380.
716. Heer, B. H. J. de, Plas, Th. van der, and Hermans, M. E. A., Anal. Chim. Acta 32, 292-3 (1965); CA 62, 11148.

717. Hegarty, B. F., and Kitching, W., J. Organometal. Chem. 6, 578-82 (1966); CA 66, 15212.

718. Hellwinkel, D., Chem. Ber. 98, 576-87 (1965); CA 62, 14719.

719. ---, Angew. Chem., Int. Ed. Engl. 5, 725 (1966); CA 65, 16276.

720. --- and Fahrbach, G., Chem. Ber. 101, 574-84 (1968); CA 68, 68939.

721. --- and Kilthau, G., Angew. Chem., Int. Ed. Engl. 5, 969-70 (1966); CA 66, 18750.

722. --- and Kilthau, G., Ann. Chem. 705, 66-75 (1967); CA 67, 73035.

723. --- and Kilthau, G., Chem. Ber. 101, 121-37 (1968); CA 68, 49700.

724. Hendershott, C. H., Proc. Am. Soc. Hort. Sci. 85, 201-9 (1964); CA 63, 3552.

725. Hennig, D., et al., Z. Chem. 7, 763 (1967); CA 68, 49696.

726. Henry, M. C., et al., U. S. Dept. Com., Office Tech. Serv., AD 286,653, 8 pp. (1962); CA 60, 6859.

727. Herbstman, Sh., J. Org. Chem. 30, 1259-60 (1965); CA 63, 4297.

728. ---, (Stauffer Chem. Co.), U.S. 3,278,492, Oct. 11, 1966; CA 66, 66142.

729. Hieber, W., Bauer, I., and Neumair, G., Z. Anorg. Allgem. Chem. 335, 250-7 (1965); CA 62, 12732.

730. --- and Frey, V., Chem. Ber. 99, 2607-13 (1966); CA 65, 16466.

731. --- and Frey, V., ibid., 2614-19 (1966); CA 65, 16466.

732. --- and Kaiser, K., Z. Anorg. Allg. Chem. 358, 271-81 (1968); CA 69, 24095.

733. ---, Klingshirn, W., and Beck, W., Chem. Ber. 98, 307-10 (1965); CA 62, 7794.

734. --- and Kummer, R., Z. Naturforsch. 20b, 271 (1965); CA 63, 12659.

735. --- and Kummer, R., Z. Anorg. Allg. Chem. 344, 292-305 (1966); CA 65, 13187.

736. --- and Kummer, R., Chem. Ber. 100, 148-59 (1967); CA 66, 51756.

737. ---, Lux, F., and Herget, C., Z. Naturforsch. 20b, 1159-65 (1965); CA 64, 13735.

738. --- and Opavsky, W., Chem. Ber. 101, 2966-8 (1968); CA 69, 59353.

739. Hobin, T. P. (Minister of Aviation, London), Brit. 1,015,501, Jan. 5, 1966; CA 64, 16013.

740. Holmquist, H. E. (E. I. duPont de Nemours Co.), Fr. 1,379,500, Jan. 20, 1964; CA 64, 766.

741. Holter, S. N., Diss. Abstr. B 28, 4497-8 (1968).

742. Hopkins, T. E., et al., Inorg. Chem. 5, 1427-31 (1966); CA 65, 8120.

743. Horizons Inc., Neth. Appl. 6,409,367, Feb. 15, 1965; CA 63, 181.

744. ---, Neth. Appl. 6,602,979, Sept. 12, 1966; CA 66, 33529.

745. ---, Neth. Appl. 6,603,912, Oct. 3, 1966; CA 66, 42327.

746. ---, Neth. Appl. 6,607,628, Jan. 20, 1967; CA 67, 7048.

747. Horner, L., Beck, P., and Roettger, F. (Farbwerke Hoechst A.-G.), Ger. 1,183,338, Dec. 10, 1964; CA 63, 7980.

748. --- and Haufe, J., Chem. Ber. 101, 2903-20 (1968); CA 69, 59352.

749. --- and Haufe, J., ibid., 2921-4 (1968); CA 69, 66660.

750. --- and Hofer, W., Tetrahedron Letters 1965, 3281-5; CA 63, 16185.

751. --- and Hofer, W., ibid., 4091-6; CA 64, 9563.

752. Horner, L., and Hofer, W., Tetrahedron Letters 1966, 3321-2; CA 65, 13486.

753. --- and Hofer, W., ibid., 3323-8; CA 65, 13486.

754. --- and Winkler, H., ibid., 1964, 3271-4; CA 62, 1546.

755. ---, Winkler, H., and Meyer, E., ibid., 1965, 789-92; CA 62, 16024.

756. Horner, S. M., et al., Inorg. Chem. 7, 1859-63 (1968); CA 69, 83054.

757. Hostettler, F., and Cox, E. F. (Union Carbide Corp.), U.S. 3,235,518, Feb. 15, 1966; CA 64, 12902.

758. Houk, L. W., and Dobson, G. R., Inorg. Chem. 5, 2119-23 (1966); CA 66, 25560.

759. Howell, G. V., and Williams, R. L., J. Chem. Soc. A 1968, 117-18; CA 68, 39757.

760. Howell, I. V., and Venanzi, L. M., ibid., A 1967, 1007-9; CA 67, 60527.

761. Hoyer, E., et al., Z. Chem. 7, 354-5 (1967); CA 67, 104687.

762. --- et al., Inorg. Nucl. Chem. Lett. 3, 457-61 (1967); CA 68, 8876.

763. Hoyer, L. W., et al., J. Expt. Med. 127, 589-603 (1968); CA 68, 85750.

764. --- and Mage, R., J. Immunol. 99, 25-30 (1967); CA 67, 42041.

765. Hsieh, H. L. (Phillips Petroleum Co.), U.S. 3,354,133, Nov. 21, 1967; CA 68, 13906.

766. Huang, Y.-T., Ting, W.-Y., and Cheng, H.-Sh., Hua, Hsueh Hsueh Pao 31, 38-41 (1965); CA 63, 629.

767. Hubel, K. W., and Braye, E. H. (Union Carbide Corp.), U.S. 3,280,017, Oct. 18, 1966; CA 66, 2462.

768. Hudson, M. J., Nyholm, R. S., and Stiddard, M. H. B., J. Chem. Soc. A 1968, 40-3; CA 68, 45800.

769. Hueper, W. C., Potential Carcinogenic Hazards Drugs, Proc. Symp. 1965, 79-104; CA 69, 75175.

770. Hung, Ph. Nhu, and Lefebvre, G., C. R. Acad. Sci., Paris, Ser. C 265, 519-21 (1967); CA 68, 12350.

771. Hung, Sh.-Ch., and Jen, H.-T., Hua Hsueh Hsueh Pao 31, 91-5 (1965); CA 62, 15421.

772. Hunt, R. L., Roundhill, D. M., and Wilkinson, G., J. Chem. Soc. A 1967, 982-4; CA 67, 43903.

773. Imperial Chemical Industries, Ltd., Neth. Appl. 6,514,643, May 12, 1966; CA 65, 13841.

774. ---, Neth. Appl. 6,602,062, Aug. 19, 1966; CA 66, 10556.

775. ---, Neth. Appl. 6,603,612, Sept. 19, 1966; CA 66, 28511.

776. ---, Neth. Appl. 6,605,627, Oct. 31, 1966; CA 66, 55581.

777. ---, Neth. Appl. 6,608,122, Dec. 12, 1966; CA 67, 26248.

778. Irgolic, K., Zingaro, R. A., and Linder, D. E., J. Inorg. Nucl. Chem. 30, 1941-55 (1968); CA 69, 70552.

779. ---, Zingaro, R. A., and Smith, M. R., J. Organometal. Chem. 6, 17-24 (1966); CA 65, 5482.

780. Irving, R. J., and Laye, P. G., J. Chem. Soc. A 1966, 161; CA 64, 10729.

781. Ishiguro, Yo., Niimi, Yo., and Shibata, Sh., Nagoya Kogyo Gijutsu Shikensho Hokoku 15, 8-13 (1966); CA 64, 10388.

782. Issleib, K., and Hamann, B., Z. Anorg. Allg. Chem. 332, 179-88 (1964);

CA <u>62</u>, 9170.

783. Issleib, K., and Hamann, B., Z. Anorg. Allg. Chem. <u>339</u>, 289-97 (1965);
CA <u>64</u>, 5129.

784. --- and Hamann, B., ibid., <u>343</u>, 196-203 (1966); CA <u>64</u>, 17631.

785. ---, Hamann, B., and Schmidt, L., ibid., <u>339</u>, 298-303 (1965); CA <u>64</u>,
5130.

786. --- and Krech, F., ibid., <u>328</u>, 21-33 (1964); CA <u>62</u>, 9169.

787. --- and Kuemmel, R., J. Organometal. Chem. <u>3</u>, 84-91 (1965); CA <u>62</u>, 8981.

788. --- and Lindner, R., Ann. Chem. <u>707</u>, 120-9 (1967); CA <u>68</u>, 12316.

789. ---, Lux, G., and Stolz, R., Z. Anorg. Allg. Chem. <u>350</u>, 44-50 (1967);
CA <u>66</u>, 99079.

790. ---, Tzschach, A., and Schwarzer, R., ibid., <u>338</u>, 141-6 (1965); CA <u>63</u>,
11611.

791. ---, Tzschach, A., and Block, H., Chem. Ber. <u>101</u>, 2931-7 (1968); CA <u>69</u>,
67506.

792. Isvoranu-Panait, C., An. Univ. Bucuresti, Ser. Stiint. Natur. <u>14</u>, 95-
104 (1965); CA <u>67</u>, 68147.

793. Itatani, H., and Bailar, J. C., Jr., J. Amer. Oil Chem. Soc. <u>44</u>, 147-51
(1967); CA <u>67</u>, 2663.

794. Ivanova, A. N., and Mustafin, I. S., Peredovye Metody Khim. Tekhnol. i
Kontrolya Proizv. (Rostov-on-Don:Rostovsk. Univ.) Sb. <u>1964</u>, 196-9;
CA <u>62</u>, 13813.

795. Ives, N. F., and Giuffrida, L., J. Ass. Offic. Anal. Chem. <u>50</u>, 1-4
(1967); CA <u>66</u>, 61612.

796. Iwamoto, H., and Kikuchi, M., Hakko Kyokaishi <u>22</u>, 218-22 (1964); CA <u>63</u>,
18706.

797. Jackson, J. A., and Nielsen, J. R., J. Mol. Spectry. <u>14</u>, 320-41 (1964);
CA <u>62</u>, 1209.

798. Jain, S. R., and Sisler, H. H., Inorg. Chem. <u>7</u>, 2204-7 (1968); CA <u>69</u>,
113063.

799. Jenkins, J. M., and Shaw, B. L., J. Chem. Soc. <u>1965</u>, 6789-96; CA <u>64</u>,
6077.

800. Jenkins, L. T. (Monsanto Co.), U.S. 3,210,329, Oct. 5, 1965; CA <u>64</u>,
2187.

801. Jennings, B. E. (Imperial Chemical Industries, Ltd.), Brit. 1,077,958,
Aug. 2, 1967; CA <u>67</u>, 82542.

802. Jennings, M. A., and Wojcicki, A., Inorg. Chem. <u>6</u>, 1854-9 (1967); CA
<u>67</u>, 104725.

803. Jensen, K. A., and Nielsen, H., Acta Chem. Scand. <u>17</u>, 1875-85 (1963).

804. Jex, V. B. (Union Carbide Corp.), U.S. 3,257,440, June 21, 1966; CA
<u>65</u>, 8960.

805. Johnson, A. R., Jr., J. Chem. Educ. <u>42</u>, 439 (1965); CA <u>63</u>, 12298.

806. Johnson, A. W., and Martin, J. O., Chem. Ind. (London) <u>1965</u>, 1726-7;
CA <u>64</u>, 750.

807. Johnson, B. F. G., J. Chem. Soc. A <u>1967</u>, 475-8; CA <u>66</u>, 81969.

808. --- et al., J. Organometal. Chem. <u>10</u>, 105-9 (1967); CA <u>67</u>, 120948.

809. Johnson, B. F. G., Lewis, J., and Subramanian, M. S., J. Chem. Soc. A 1968, 1993-2001; CA 69, 64219.

810. Johnson, M. P., Shriver, D. F., and Shriver, S. A., J. Amer. Chem. Soc. 88, 1588-9 (1966); CA 64, 17022.

811. Jones, K., and Lappert, M. F., J. Organometal. Chem. 3, 295-307 (1965); CA 62, 14715.

812. Jones, V. E., and Leskowitz, S., Nature 207, 596-7 (1965); CA 63, 12139.

813. Joon, K., Kjeller Rep. No. 119, 6 pp.; CA 68, 72857.

814. Juillard, J., Bull. Soc. Chim. France 1964, 3069-71; CA 62, 9859.

815. ---, ibid., 1968, 1894-9; CA 69, 54851.

816. --- and Simonet, N., ibid., 1883-94; CA 69, 54850.

817. Kageyama, I., Miyamura, N., and Okamura, K., Bochu Kagaku 25, 134-7 (1960); CA 61, 4591.

818. Kalle, A.-G., Brit. 1,125,150, Aug. 28, 1968; CA 69, 112213.

819. Kamai, G., Azerbaev, I. N., and Zaripov, P. K., Izv. Akad. Nauk Kaz. SSR, Ser. Khim. 16, 85-9 (1966); CA 65, 13752.

820. ---, Chadaeva, N. A., and Osipova, M. P., Zh. Obshch. Khim. 37, 2754-7 (1967); CA 69, 36229.

821. --- and Gatilov, Yu. F., ibid., 35, 987-8 (1965); CA 63, 9982.

822. --- and Gatilov, Yu. F., ibid., 1239-40 (1965); CA 63, 11611.

823. --- and Gatilov, Yu. F., Nekot. Aspecty Stereokhim. Org. Proizvod. Mysh'yaka 1966, 67-70; CA 69, 77517.

824. --- and Miftakhova, R. G., Zh. Obshch. Khim. 35, 546-8 (1965); CA 63, 629.

825. --- and Miftakhova, R. G., USSR 169,528, Mar. 17, 1965; CA 63, 2899.

826. --- and Miftakhova, R. G., Zh. Obshch. Khim. 35, 2001-3 (1965); CA 64, 6685.

827. ---, Miftakhova, R. G., and Karunnaya, L. A., ibid., 38, 1565-8 (1968); CA 69, 96843.

828. --- and Usacheva, G. M., Zh. Obshch. Khim. 36, 2000-2 (1966); CA 66, 76118.

829. Kamienski, B., and Krauss, E., Electrochim. Acta 10, 879-81 (1965); CA 63, 9449.

830. --- et al., Abh. Deut. Akad. Wiss. Berlin, Kl. Chem., Geol. Biol. 1966, 705-9 (Pub. 1967); CA 68, 53593.

831. Kammori, O., Taguchi, I., and Komiya, R., Bunseki Kagaku 14, 249-52 (1965); CA 63, 10669.

832. Kasenally, A. S., et al., J. Chem. Soc. 1965, 3407-11; CA 63, 1457.

832a. ---, Nyholm, R. S., and Stiddard, M. H. B., ibid., 5343-6; CA 63, 15834.

833. --- et al., Nature 204, 871-2 (1964); CA 62, 6129.

834. Kato, F., and Yatsu, M. (Katsuta Chem. Ind. Co., Ltd.), Japan. 2903, Feb. 7, 1967; CA 67, 33319.

835. Katz, S., and Kaplun, B., Medicina (Buenos Aires) 24, 224-6 (1964); CA 62, 2149.

836. Kaushiva, B. S., Intern. Congr. Chemotherapy, Proc., 3rd, Stuttgart 1963, 1539-40 (Pub. 1964); CA 65, 9395.

837. Kawai, M., Rev. Int. Serv. Sante Armees Terre Mer Air 39, 861-74 (1966);

CA <u>66</u>, 93545.

838. Kawasaki, Yo., and Okawara, R., Bull. Chem. Soc. Jap. <u>40</u>, 428 (1967); CA <u>67</u>, 21986.

839. ---, Tanaka, T., and Okawara, R., ibid., 1562-5 (1967); CA <u>68</u>, 7938.

840. Kawase, A., Bunseki Kagaku <u>16</u>, 1364-9 (1967); CA <u>69</u>, 64285.

841. Keller, R. N., Johnson, N. B., and Westmoreland, L. L., J. Amer. Chem. Soc. <u>90</u>, 2729-30 (1968).

842. Kemmitt, R. D. W., Nichols, D. I., and Peacock, R. D., J. Chem. Soc. A <u>1968</u>, 2149-52; CA <u>69</u>, 83002.

843. Kennard, O., et al., Chem. Commun. <u>1968</u>, 269-71; CA <u>68</u>, 99559.

844. Kenski, Z., Tr. Komis. po Spektroskopii, Akad. Nauk SSSR <u>1964</u>, 213-25; CA <u>63</u>, 9253.

845. Kepert, D. L., and Mandyczewsky, R., J. Chem. Soc. A <u>1968</u>, 530-3; CA <u>68</u>, 83961.

846. Kettle, S. F. A., Theoret. Chim. Acta <u>4</u>, 150-4 (1966); CA <u>64</u>, 7386.

847. Khattak, M. A., and Magee, R. J., Chem. Commun. <u>1965</u>, 400; CA <u>63</u>, 12658.

848. Kidby, D. K., Plant Physiol. <u>41</u>, 1139-44 (1966); CA <u>65</u>, 18927.

849. Kilbourn, B. T., Blundell, T. L., and Powell, H. M., Chem. Commun. <u>1965</u>, 444-5; CA <u>64</u>, 95.

850. Kim, Y. J., et al., Kisul Yon'guso. Pogo <u>3</u>, 129-32 (1964); CA <u>65</u>, 11323.

851. King, M. G., and McQuillan, G. P., J. Chem. Soc. A <u>1967</u>, 898-901; CA <u>67</u>, 851.

852. King, R. B., and Eggers, C. A., Inorg. Chem. <u>7</u>, 340-5 (1968); CA <u>68</u>, 59704.

853. Kingston, J. F., Inorg. Nucl. Chem. Lett. <u>4</u>, 65-6 (1968); CA <u>68</u>, 74743.

854. Kingston, J. V., and Wilkinson, G., J. Inorg. Nucl. Chem. <u>28</u>, 2709-13 (1966); CA <u>66</u>, 16129.

855. Kirkham, W. J., et al., J. Chem. Soc. <u>1965</u>, 550-3; CA <u>62</u>, 6127.

856. Kitagawa, M., et al., Immunochemistry <u>4</u>, 197-202 (1967); CA <u>67</u>, 72011.

857. Klamann, K., et al., Chem.-Ing.-Tech. <u>39</u>, 1024-30 (1967); CA <u>68</u>, 39153.

858. Klassen, D. M., and Crosby, G. A., J. Mol. Spectrosc. <u>25</u>, 348-405 (1968); CA <u>68</u>, 91512.

859. Kligerman, M. M., Lofgren, S., and Dean, C., Radiology <u>88</u>, 993-4 (1967); CA <u>67</u>, 18402.

860. Klingman, G. C., and Spain, J. M., Proc. Southern Weed Conf. <u>18</u>, 145-52 (1965); CA <u>63</u>, 1167.

861. Klofutar, C., Krasovec, F., and Kusar, M., Croat. Chem. Acta <u>40</u>, 23-8 (1968); CA <u>69</u>, 86163.

862. Knight, K. L., Lopez, M. A., and Haurowitz, F., J. Biol. Chem. <u>241</u>, 2286-92 (1966); CA <u>64</u>, 18204.

863. ---, Roelofs, M. J., and Haurowitz, F., Biochim. Biophys. Acta <u>133</u>, 333-7 (1967); CA <u>66</u>, 63807.

864. Knock, F. E., J. Am. Geriat. Soc. <u>13</u>, 313-24; CA <u>62</u>, 15323.

865. ---, Galt, R. M., and Oester, Y. T., ibid., <u>15</u>, 882-99 (1967); CA <u>67</u>, 106717.

866. Koehler, H., and Seifert, B., Z. Anorg. Allgem. Chem. <u>344</u>, 63-71 (1966); CA <u>65</u>, 4982.

867. Koehler, H., and Seifert, B., Z. Naturforsch. B 22, 238-41 (1967); CA 67, 78583.

868. Koehler, L. D., and Fennell, R. A., J. Morphol. 114, 209-23 (1964); CA 64, 1064.

869. Kolar, F. L., Zingaro, R. A., and Irgolic, K., J. Inorg. Nucl. Chem. 28, 2981-5 (1966); CA 66, 32410.

869a. Kolobova, N. E., and Pysynskii, A. A., Izv. Akad. Nauk SSSR, Ser. Khim. 1966, 2231-3; CA 66, 85840.

870. Kolomiets, A. F., and Levskaya, G. S., Zh. Obshch. Khim. 36, 2024-5 (1966); CA 66, 76115.

871. ---, Levskaya, G. S., and Bliznyuk, N. K., USSR 192,203, Feb. 6, 1967; CA 69, 3008.

872. ---, Levskaya, G. S., and Bliznyuk, N. K., USSR 210,154, Feb. 6, 1968; CA 69, 52300.

873. --- et al., USSR 192,809, Mar. 2, 1967; CA 68, 105186.

874. --- et al., USSR 206,569, Dec. 8, 1967; CA 69, 51637.

875. Kono, I., Mem. Fac. Agr. Kagoshima Univ. 5, 9-17 (1965); CA 63, 15481.

876. Koromogawa, J., Tamura, B., and Suzuki, Sh. (Takeda Chem. Industries, Ltd.), Japan. 12,925, July 8, 1964; CA 62, 11854.

877. Korshak, V. V., et al., Dokl. Akad. Nauk SSSR 177, 1348-51 (1967); CA 68, 59921.

878. Koshland, M. E., Biol. Conf. "Oholo" 9, 29-31 (1964); CA 65, 1215.

879. ---, Englberger, F. M., and Shapanka, R., Biochemistry 5, 641-51 (1966); CA 64, 10229.

880. Kostyanovskii, R. G., et al., Izv. Akad. Nauk SSSR, Ser. Khim. 1967, 1629; CA 68, 73888.

881. --- and Yakshin, V. V., ibid., 2363; CA 68, 77526.

882. --- et al., ibid., 2128; CA 68, 25353.

883. --- et al., ibid., 1968, 677-89; CA 69, 87118.

884. Koton, M. M., and Florinskii, F. S., Dokl. Akad. Nauk SSSR 163, 598-9 (1966); CA 65, 18692.

885. Kovacs, T., Zentr. Bakteriol., Parasitenk. Abt. I, Orig. 187, 382-90 (1962); CA 62, 5781.

886. Koyama, J., Grossberg, A. L., and Pressman, D., Biochemistry 7, 1935-41 (1968); CA 69, 1389.

887. --- and Miyajima, N., J. Biochem. (Tokyo) 61, 283-9 (1967); CA 66, 144120.

888. Kozhevnikova, N. E., et al., USSR 173,924, Aug. 6, 1965; CA 64, 2248.

889. Kozlova, L. N., Byull. Nauch.-Tekh. Inform. Tadzh. Nauch.-Issled. Inst. Sel.-Khoz. No. 4, 60-3 (1966); CA 67, 90027.

890. Kracmar, J., and Kracmarova, J., Cesk. Farm. 15, 121-9 (1966); CA 65, 3668.

891. ---, Kracmarova, J., and Sladkova, J., Pharmazie 21, 460-3 (1966); CA 66, 22229.

892. Kraihanzel, C. S., and Maples, P. K., J. Amer. Chem. Soc. 87, 5267-8 (1965); CA 64, 6686.

893. --- and Maples, P. K., Inorg. Chem. 7, 1806-15 (1968); CA 69, 66600.

894. Kramer, D. N., Guilbault, G. G., and Cannon, P. L., Jr. (U.S. Dept. of the Army), U.S. 3,366,555, Jan. 30, 1968; CA 68, 84188.

895. Kranz, J., Phytopathol. Z. 52, 59-72 (1965); CA 63, 3559; ibid., 335-48 (1965); CA 63, 936.

896. Krasil'nikova, L. N., and Vostroknutova, M. Ya., Sb. Nauchn. Tr. Vses. Nauchn.-Issled. Gorno-Met. Inst. Tsvetn. Metal. No. 9, 41-8 (1965); CA 63, 15533.

897. Krasovec, F., and Klofutar, C., Solvent Extr. Chem., Proc. Int. Conf., Goteborg 1966, 509-16; CA 69, 70572.

898. Kreisman, P., et al., J. Amer. Chem. Soc. 90, 1067-8 (1968); CA 68, 73216.

899. Krokhv, V. V., USSR 166,335, Nov. 19, 1964; CA 62, 14728.

900. Kruglaya, O. A., Vyazankin, N. S., and Razuvaev, G. A., Zh. Obshch. Khim. 35, 394 (1965); CA 62, 14722.

901. Kudinova, V. V., Foss, V. L., and Lutsenko, I. F., Zh. Obshch. Khim. 36, 1863-4 (1966); CA 66, 65596.

902. Kudisch, N., and Leebrick, J. R. (M & T Chemicals, Inc.), Fr. 1,362,703, June 5, 1964; CA 62, 5405.

903. Kuenzel, E., Giefer, A., and Kern, W., Makromol. Chem. 96, 17-29 (1966); CA 65, 12290.

904. Kuno, S., Nichidai Igaku Zasshi 18, 1236-46 (1959); CA 61, 11231.

905. Kupchik, E. J., and Calabretta, P. J., Inorg. Chem. 4, 973-8 (1965); CA 63, 4318.

906. --- and McInerney, E. F., J. Organometal. Chem. 11, 291-8 (1968); CA 68, 59675.

907. --- and Theisen, C. T., ibid., 627-30 (1968); CA 68, 87363.

908. Kuroda, K., Hashizume, G., and Fukuda, K., Yakugaku Zasshi 87, 1175-83 (1967); CA 68, 24588.

909. Kutner, A. (Hercules Powder Co.), U.S. 3,193,540, July 6, 1965; CA 63, 16501.

910. Kuzin, E. L., Savvin, S. B., and Gribov, L. A., Zh. Anal. Khim. 23, 490-9 (1968); CA 69, 22050.

911. Kuz'min, K. I., and Panfilovich, Z. U., Zh. Obshch. Khim. 38, 1344-5 (1968); CA 69, 86485.

912. --- and Pavlova, L. A., ibid., 36, 1478-80 (1966); CA 66, 2626.

913. --- and Pavlova, L. A., ibid., 37, 1399-401 (1967); CA 68, 13100.

914. Kuznetsov, V. I., USSR 61,967, Jan. 4, 1965; CA 62, 9802.

915. ---, Akimova, T. G., and Eliseeva, O. P., Radiochim. Metody Opred. Mikroelementov, Akad. Nauk SSSR, Sb. Statei 1965, 44-9; CA 63, 12318.

916. ---, Bazargin, N. N., Zh. Obshch. Khim. 35, 879-83 (1965); CA 63, 6927.

917. ---, Fisenko, L. P., and Akimova, T. G., Radiokhim. Metody Opred. Mikroelementov, Akad. Nauk SSSR, Sb. Statei 1965, 49-54; CA 63, 12318.

918. --- and Moseev, L. I., Radiokhimiya 6, 433-9 (1964); CA 62, 11151.

919. --- and Petrova, T. V., Latvijas PSR Zinatnu Akad. Vestis, Khim. Ser. 1964, 263-70; CA 62, 9855.

920. --- and Petrova, T. V., ibid., 1966, 62-9; CA 66, 30011.

921. Kyrs, M., et al., Czech. 111,913, Aug. 15, 1964; CA 62, 7431.

922. ---, Pistek, P., and Selucky, P., Sb. Ref. Celostatni Radiochem. Konf., 3, Liblice, Czech. 1964, 45-73; CA 64, 18461.

923. ---, Pistek, P., and Selucky, P., Collection Czech. Chem. Commun. 31, 2689-94 (1966); CA 65, 17739.

924. ---, Pistek, P., and Selucky, P., ibid., 32, 747-56 (1967); CA 66, 61553.

925. ---, Selucky, P., and Pistek, P., Radio Chim. Acta 6, 72-5 (1966); CA 67, 16896.

926. Kzn, P. R., J. Rubber Res. Inst. Malaya 19, 81-4 (1965); CA 65, 11260.

927. LaMar, G. N., Fischer, R. N., and Horrocks, W. DeW., Jr., Inorg. Chem. 6, 1798-803 (1967); CA 67, 95413.

928. Lambert, J. B., and Jackson, G. F., III, J. Amer. Chem. Soc. 90, 1350-1 (1968); CA 69, 35236.

929. Lamkin, E. G., and Williams, M. B., Anal. Chem. 37, 1029-31 (1965); CA 63, 7337.

930. Lane, A. P., and Burg, A. B., J. Amer. Chem. Soc. 89, 1040-1 (1967); CA 67, 11558.

931. Langford, C. H., and Aplington, J. P., J. Organometal. Chem. 4, 271-7 (1965); CA 63, 11289.

932. Lapham, V. T., Proc. Southern Weed Conf. 19, 438-42 (1966); CA 65, 6216.

933. ---, Louisiana Conservationist 18 (11/12), 2 (1966); CA 66, 18229.

934. Larkworthy, L. F., and Phillips, D. J., J. Inorg. Nucl. Chem. 29, 2101-2 (1967); CA 68, 16596.

935. Larose, A., Rose, B., and Richter, M., Can. J. Biochem. 42, 833-9 (1964); CA 61, 3557.

936. Lastovskii, R. P., et al., Tr. Soveshch. po Fiz. Metodam Issled. Organ. Soedin. i Khim. Protsessov, Akad. Nauk Kirg. SSR, Inst. Organ. Khim., Frunze 1962, 39-46 (Pub. 1964); CA 62, 6121.

937. --- et al., Issled. Svoistv Ionoobmen. Materialov, Akad. Nauk SSSR, Inst. Fiz. Khim. 1964, 104-7; CA 62, 4162.

938. Lauer, R. S., Kononenko, L. I., and Poluektov, N. S., Molekul. Khromatogr., Akad. Nauk SSSR, Inst. Fiz. Khim. 1964, 99-105; CA 62, 9749.

939. Laughlin, R. G. (Procter & Gamble Co.), U.S. 3,222,287, Dec. 7, 1965; CA 64, 3859.

940. Lawson, D. N., Osborn, J. A., and Wilkinson, G., J. Chem. Soc. A 1966, 1733-6; CA 66, 18760.

941. Layton, A. J., et al., Nature 214, 1109-10 (1967); CA 67, 49968.

942. Lazarev, A. I., and Lazareva, V. I., Zh. Anal. Khim. 23, 36-8 (1968); CA 68, 101551.

943. Leebrick, J. R. (M & T Chemicals, Inc.), Fr. 1,397,139, Apr. 30, 1965; CA 63, 5459.

944. --- (M & T Chemicals, Inc.), Brit. 1,000,755, Aug. 11, 1965; CA 63, 13973.

945. --- (M & T Chemicals, Inc.), U.S. 3,239,411, Mar. 8, 1965; CA 64, 20554.

946. --- (M & T Chemicals, Inc.), U.S. 3,247,050, Apr. 19, 1966; CA 65, 1328.

947. --- (M & T Chemicals, Inc.), U.S. 3,287,210, Nov. 22, 1966; CA 66, 85070.

948. Leebrick, J. R., and Remes, N. L. (M & T Chemicals, Inc.), U.S. 3,367,954, Feb. 6, 1968; CA 68, 105367.

949. Lehmann, D. L., Ann. Trop. Med. Parasitol. 58, 189-91 (1964); CA 62, 948.

950. Lemmon, D. H., Diss. Abstr. B 28, 632 (1967).

951. Leroy, M. J. F., et al., C. R. Akad. Sci. Paris, Ser. C 263, 601-4 (1966); CA 66, 24038.

952. Leskowitz, S., Immunochemistry 4, 91-4 (1967); CA 67, 114117.

953. ---, Science 155, 350-2 (1967); CA 66, 53884.

954. --- and Zak, S. J., Nature 211, 246-8 (1966); CA 65, 11195.

955. Levinson, H. S., and Garber, E. B., Nature 207, 751-2 (1965); CA 63, 11296; see also U. S. Dept. Comm., Office Tech. Serv. AD 611869, Vol. 2, 169-83 (1965); CA 64, 2398.

956. Levinson, J. J., and Robinson, S. D., Inorg. Nucl. Chem. Lett. 4, 407-9 (1968); CA 69, 56629.

957. Levskaya, G. S., et al., USSR 196,816, May 31, 1967; CA 68, 95976.

958. --- and Kolomiets, A. F., Zh. Obshch. Khim. 36, 2024 (1966); CA 66, 65604.

959. --- and Kolomiets, A. F., ibid., 37, 905-7 (1967); CA 67, 108722.

960. --- et al., USSR 196,831, May 31, 1967; CA 68, 77987.

961. --- et al., USSR 196,823, May 31, 1967; CA 68, 68466.

962. Levy, R. M., Van Wazer, J. R., and Simpson, J., Inorg. Chem. 5, 332-3 (1966); CA 64, 16818.

963. Lewis, J., Nyholm, R. S., and Rodley, G. A., J. Chem. Soc. 1965, 1483-8; CA 62, 8528.

964. ---, Nyholm, R. S., and Rodley, G. A., Nature 207, 72-3 (1965); CA 63, 7874.

965. Linder, D. E., Diss. Abstr. B 28, 520 (1967).

966. ---, Zingaro, R. A., and Irgolic, K., J. Inorg. Nucl. Chem. 29, 1999-2006 (1967); CA 67, 94454.

967. Lindner, E., and Kranz, H., Chem. Ber. 101, 3438-44 (1968); CA 69, 106837.

968. --- and Schless, H., ibid., 99, 3331-6 (1966); CA 66, 11002.

969. Lindoy, L. F., Livingstone, S. E., and Lockyer, T. N., Nature 211, 519 (1966); CA 65, 13188.

970. ---, Livingstone, S. E., and Lockyer, T. N., Australian J. Chem. 19, 1391-1400 (1966); CA 65, 13192.

971. ---, Livingstone, S. E., and Lockyer, T. N., Inorg. Chem. 6, 652-6 (1967); CA 66, 91253.

972. Lippard, S. J., and Ucko, D. A., Inorg. Chem. 7, 1051-6 (1968); CA 69, 15461.

973. Lippincott, E. R., Nagarajan, G., and Stutman, J. M., J. Phys. Chem. 70, 78-84 (1966); CA 64, 5769.

974. Lloyd, D., and Singer, M. I. C., Chem. Ind. (London) 1967, 510-11; CA 67, 32757.

975. --- and Singer, M. I. C., ibid., 787; CA 67, 54238.

976. --- and Singer, M. I. C., Chem. Commun. 1967, 1042-3; CA 68, 13124.

977. Lobanova, E. F., Kleshchel'skaya, E. I., and Korchenenkova, L. I., Sb. Tr. Vses. Zaoch. Politekh. Inst. 1967, 46-54; CA 69, 15632.

978. Lock, C. J. L., and Wilkinson, G., J. Chem. Soc. 1964, 2281-5; CA 61, 7924.

979. Long, G. G., et al., Inorg. Chem. 5, 1358-61 (1966); CA 65, 8218.

980. Longi, P., and Mazzocchi, R., Chim. Ind. (Milan) 48, 718-20 (1966); CA 65, 14431.

981. Lorenz, J. W., and Fischer, H., Ann. Univ. Ferrara, Sez. 5, Chim. Pura Appl., Suppl. 4, 81-90, Discussion 91-92 (1966); CA 66, 121473.

982. --- and Fischer, H., Ber. Bunsenges. Physik. Chem. 69, 689-99 (1965); CA 64, 289.

983. Lovie, J. C., and Watt, J. A. (Imperial Chemical Industries, Ltd.), Brit. 1,023,797, March 23, 1966; CA 64, 19824.

984. Lozanovskaya, I. N., and Petrashen, V. I., Zh. Anal. Khim. 22, 1196-200 (1967); CA 68, 18449.

985. --- and Popova, D. S., Akust. Magn. Obrab. Veshchestva, Novocherkassk 1966, 47-50; CA 66, 119463.

986. Lukin, A. M., et al., Dokl. Akad. Nauk SSSR 173, 361-3 (1967); CA 67, 55143.

987. --- et al., Tr. Vses. Nauch.-Issled. Inst. Khim. Reaktivov Osobo Chist. Khim. Veshchestv 30, 42-9 (1967); CA 69, 15596.

988. ---, Efremenko, O. A., and Petrova, G. S., Zh. Anal. Khim. 22, 1234-8 (1967); CA 68, 84040.

988a. ---, Efremenko, O. A., and Podolskaya, B. L., ibid., 21, 970-5 (1966).

989. ---, Kaslina, N. A., and Zavarikhina, G. B., USSR 200,294, July 13, 1967; CA 68, 65454.

990. --- and Petrova, G. S., Tr. Vses. Nauch.-Issled. Inst. Khim. Reaktivov Osobo Chist. Khim. Veshchestv 28, 227-32 (1966); CA 67, 64496.

991. ---, Podolskaya, B. L., and Zavorikhina, G. B., USSR 171,657, May 27, 1965; CA 65, 2744.

992. --- et al., Tr. Vses. Nauch.-Issled. Inst. Khim. Reaktivov Osobo Chist. Khim. Veshchestv 30, 161-6 (1967); CA 69, 8127.

993. ---, Smirnova, K. A., and Petrova, G. S., ibid., 29, 290-8 (1966); CA 68, 18242.

994. ---, Zavarikhina, G. B., and Bolotina, N. A., Zh. Obshch. Khim. 37, 478-83 (1967); CA 67, 74441.

995. Lupin, M. S., and Shaw, B. L., J. Chem. Soc. A 1968, 741-9; CA 68, 92593.

996. Lustig, M., and Ruff, J. K., Inorg. Chem. 6, 2115-17 (1967); CA 67, 121900.

997. Lyubimova, L. K., Antibiotiki 8, 741-44 (1963); CA 63, 18880.

998. Maccioni, A., and Secci, M., Rend. Seminario Fac. Sci. Univ. Cagliari 34, 328-31 (1964); CA 63, 5674.

999. MacDiarmid, A. G., et al., Intern. Symp. Organosilicon Chem., Sci. Commun., Prague 1965, 100-3; CA 65, 7214.

1000. Machida, K. (Tokuyama Soda Co., Ltd.), Japan. 21,793, Sept. 28, 1965; CA 64, 2192.

1001. Mackay, K. M., Sowerby, D. B., and Young, W. C., Spectrochim. Acta, Part A 24, 611-31 (1968); CA 68, 118182.

1002. Maerkl, G., and Hauptmann, H., Tetrahedron Lett. 1968, 3257-60; CA 69, 96847.

1003. --- and Lieb, F., ibid., 1967, 3489-93; CA 68, 78385.

1004. Magee, R. J., and Khattak, M. A., Microchem. J. 8, 285-94 (1964); CA 62, 22.

1005. Mague, J. T., and Mitchener, J. P., Chem. Commun. 1968, 911-12; CA 69, 92509.

1006. --- and Wilkinson, G., J. Chem. Soc. A 1966, 1736-40; CA 66, 18759.

1007. --- and Wilkinson, G., Inorg. Chem. 7, 542-6 (1968); CA 68, 74746.

1008. Maier, L. (Monsanto Co.), U.S. 3,188,294, June 8, 1965; CA 63, 13318.

1009. Mais, R. H. B., and Powell, H. M., J. Chem. Soc. 1965, 7471-81; CA 64, 5860.

1010. Malevannyi, V. A., and Shumina, V. A., Lakokrasochnye Materialy i ikh Primenenie 1964, 57; CA 62, 7999.

1011. Malinovskii, V. A., Azerb. Khim. Zh. 1963, 113-16; CA 62, 3456.

1012. Mallion, K. B., and Mann, F. G., J. Chem. Soc., Suppl. No. 1, 5716-25 (1964); CA 63, 1812.

1013. Malz, H. (Farbenfabriken Bayer A.-G.), Ger. 1,211,168, Feb. 24, 1966; CA 64, 15738.

1014. Mamiya, M., Bunseki Kagaku 11, 1314-16 (1962); CA 63, 10642.

1015. Mandzhgaladze, O. V., and Nazarenko, V. A., Soobshch. Akad. Nauk Gruz. SSR 47, 303-8 (1967); CA 68, 65335.

1016. Mann, F. G., Millar, I. T., and Baker, F. C., J. Chem. Soc. 1965, 6342-50; CA 64, 751.

1017. --- and Pragnell, M. J., ibid., 4120-7; CA 63, 6899.

1018. Manning, A. R., ibid., A 1967, 1984-7; CA 68, 29835.

1019. ---, ibid., A 1968, 651-3; CA 68, 83983.

1020. ---, ibid., 1018-21; CA 68, 118927.

1021. ---, ibid., 1135-7; CA 69, 6568.

1022. ---, ibid., 1665-7; CA 69, 47748.

1023. ---, ibid., 1670-3; CA 69, 47735.

1024. Mao, T. J. (General Motors Corp.), Brit. 1,041,949, Sept. 7, 1966; CA 65, 17078.

1025. Martin, F. J., and Price, K. R., J. Appl. Polym. Sci. 12, 143-58; CA 68, 96441.

1026. Martin, K., and Steger, E., Z. Anorg. Allgem. Chem. 345, 306-12 (1966); CA 65, 11551.

1027. Massey, A. G., Randall, E. W., and Shaw, D., Spectrochim. Acta 21, 263-73 (1965); CA 62, 7272.

1028. Mathis, R., et al., J. Mol. Struct. 1, 481-8 (1968); CA 69, 14208.

1029. Matschiner, H., and Tzschach, A., Z. Chem. 5, 144 (1965); CA 63, 7651.

1030. Matsumura, Yo., and Okawara, R., Inorg. Nucl. Chem. Lett. 4, 219-21 (1968); CA 69, 10522.

1031. ---, Shindo, M., and Okawara, R., ibid., 3, 219-22 (1967); CA 67, 73653.

1032. Matsuev, L. P., Tr. Vses. Nauchn.-Issled. Inst. Zolota i Redk. Met., Magaden 22, 421-57 (1963); CA 62, 4954.

1033. Mayer, C., Lorenz, W. J., and Fischer, H., Z. Phys. Chem. (Frankfurt) 52, 193-214 (1967); CA 66, 110955.

1034. Mazur, P. A. (Shamrock Corp.), U.S. 3,397,218, Aug. 13, 1968; CA 69, 86360.

1035. McBrearty, C. F., Jr., Irgolic, K., and Zingaro, R. A., J. Organometal. Chem. 12, 377-87 (1968); CA 69, 19272.

1036. McCall, M. A., and Blanton, R. B. (Eastman Kodak Co.), Fr. 1,401,841, June 4, 1965; CA 65, 5553.

1037. McCleverty, J. A., et al., J. Amer. Chem. Soc. 89, 6082-92 (1967); CA 68, 33719.

1038. --- and Wilkinson, G., Inorg. Synth. 8, 214-17 (1966); CA 66, 91256.

1039. McCord, R. S. (Douglas Aircraft Co., Inc.), U.S. 3,371,046, Feb. 27, 1968; CA 68, 80222.

1040. McFadden, R. T., Langner, R. R., and Wade, L. L. (Dow Chemical Co.), U.S. 3,228,830, Jan. 11, 1966; CA 65, 873.

1041. McGee, Th. W. (Dow Chemical Co.), U.S. 3,371,105, Feb. 27, 1968; CA 68, 95977.

1042. McGregor, J. K., et al., Poultry Sci. 43, 1026-30 (1964); CA 63, 7404.

1043. McInerney, E. F., Diss. Abstr. B 28, 3652 (1968).

1044. McKechnie, J. S., Miesel, S. L., and Paul, I. C., Chem. Commun. 1967, 152-3; CA 67, 15789.

1045. McKenney, R. L., and Sisler, H. H., Inorg. Chem. 6, 1178-82 (1967); CA 67, 17474.

1046. McKeown, J. J., and Wright, C. D. (Minnesota Mining and Manuf. Co.), U.S. 3,389,111, June 18, 1968; CA 69, 36706.

1047. McPhail, A. T., and Sim, G. A., Chem. Commun. 1966, 21-22; CA 64, 11980.

1048. McQueen, J. D., and Mego, J. L., J. Neurosurg. 21, 641-6 (1964); CA 62, 3303.

1049. McWhinnie, W. R., J. Chem. Soc. A 1966, 889-92; CA 65, 13022.

1050. McWhorter, C. G., Weeds 14, 191-4 (1966); CA 65, 9639.

1051. Meek, D. W., U. S. Clearinghouse Fed. Sci. Tech. Inform. AD 646013 (1966); CA 67, 70174.

1052. --- et al., Paper presented at 150th Meeting, Amer. Chem. Soc., Sept. 1965.

1053. Mego, J. L., and McQueen, J. D., J. Nucl. Med. 5, 920-8 (1964); CA 62, 15204.

1054. Mel'nikov, N. N., and Khaskin, B. A., Khim. Org. Soedin. Fosfora, Akad. Nauk SSSR, Otd. Obshch. Tekh. Khim. 1967, 274-7; CA 69, 2630.

1055. ---, Khaskin, B. A., and Elepina, L. T., USSR 180,191, Mar. 21, 1966; CA 65, 13763.

1056. --- et al., Khim. Sel. Khoz. 5, 610-14 (1967); CA 68, 21112.

1057. Merijanian, A., and Zingaro, R. A., Inorg. Chem. 5, 187-91 (1966); CA 64, 8232.

1058. Merkl, A. W., et al., J. Chem. Phys. 43, 953-7 (1965); CA 63, 5018.

1059. Merz, E., and Riedel, H. J., Chem. Effects Nucl. Transformations, Proc. Symp., Vienna, 1964, 2, 179-94; CA 63, 9314.

1060. Metzger, H., Wofsy, L., and Singer, S. J., Proc. Natl. Acad. Sci. U.S. 51, 612-18 (1964); CA 61, 3560.

1061. Meyer, J. L., U. S. At. Energy Comm. 1967, IS-T-195, 49 pp.; CA 69, 48790.

1062. Mieth, H., and Seidenath, H., Z. Tropenmed. Parasitol. 18, 53-60 (1967); CA 67, 41301.

1063. Mikulaj, V., Macasek, F., and Kopunec, R., Radiochem. Conf., Abstr. Pap. Bratislava 1966, 76-7; CA 68, 59689.

1064. Miller, F. A., and Lemmon, D. H., Spectrochim. Acta A 23, 1099-1109 (1967); CA 66, 109875.

1065. Miller, H. C., Miller, N. E., and Muetterties, E. L., J. Amer. Chem. Soc. 85, 3885-6 (1963); CA 60, 7666.

1066. Miller, N. E., Inorg. Chem. 4, 1458-63 (1965); CA 63, 16380.

1067. --- (E. I. duPont de Nemours and Co.), U.S. 3,217,023, Nov. 9, 1965; CA 64, 3103.

1068. Milyukova, M. S., and Savvin, S. B., Zh. Anal. Khim. 22, 353-8 (1967); CA 67, 29051.

1069. --- and Savvin, S. B., ibid., 751-6 (1967); CA 68, 45958.

1070. Mirviss, S. B., Dougherty, H. W., Jr., and Looney, R. W. (Esso Research and Engineering Co.), U.S. 3,310,547, March 21, 1967; CA 66, 116115.

1071. Miyazawa, E., et al., Bitamin 29, 23-5 (1964); CA 60, 16247.

1072. Mizuho, A., Niwa, T., and Soejima, A., Chikusan Shikenio Kenkyu Hokoku No. 1, 45-62 (1963); Nogyo Gijutsu Kenkyusho Hokoku, Chikusan No. 19, 1-24 (1960); CA 61, 1046.

1073. Modro, A., and Modro, T. A., Bull. Acad. Pol. Sci., Ser. Sci. Chim. 16, 63-4 (1968); CA 69, 76104.

1074. Mohai, M., and Upor, E., Magy. Kem. Foly. 74, 270-4 (1968); CA 69, 5709.

1075. Monk, R. G., and Exelby, K. A., Talanta 12, 91-100 (1965); CA 62, 3390.

1076. Moody, J. P., and Williams, R. T., Food Cosmet. Toxicol. 2, 687-93 (1964); CA 65, 11028.

1077. --- and Williams, R. T., ibid., 707-15 (1964); CA 65, 11028.

1078. --- and Williams, R. T., ibid., 695-706 (1964); CA 65, 11028.

1079. Mookherji, T., Solid State Commun. 5, 981-4 (1967); CA 68, 44457.

1080. Moore, E. P. (E. I. duPont de Nemours and Co.), U.S. 3,322,826, May 30, 1967; CA 67, 44292.

1081. Mootz, D., and Look, W., Z. Anorg. Allg. Chem. 356, 244-52 (1968); CA 68, 90717.

1082. Morait, Gh., Coada, M., and Petroniu, L., Farmacia (Bucharest) 15, 155-62 (1967); CA 67, 14911.

1083. Moreland, C. G., et al., Inorg. Chem. 7, 834-6 (1968); CA 68, 113851.

1083a. Morris, D. E., Diss. Abstr. B 29, 2340 (1969).

1084. --- and Basolo, F., J. Amer. Chem. Soc. 90, 2531-5 (1968); CA 68, 117484.

1085. Morris, D. E., and Basolo, F., J. Amer. Chem. Soc. 90, 2536-44 (1968); CA 68, 117491.

1086. Morris, M. D., Anal. Chem. 37, 977-9 (1965); CA 63, 9058.

1087. ---, McKinney, P. S., and Woodbury, E. C., J. Electroanal. Chem. 10, 85-94 (1965); CA 63, 17484.

1088. Morrison, J. L., J. Agr. Food Chem. 16, 704-5 (1968); CA 69, 42809.

1089. Mosby, W. L. (Amer. Cyanamid Co.), U.S. 3,365,478, Jan. 23, 1968; CA 68, 95978.

1090. --- and Klingsberg, E. (Amer. Cyanamid Co.), U.S. 3,397,217, Aug. 13, 1968; CA 69, 77502.

1091. Moutschen-Dahmen, J., and Degraeve, N., Experientia 21, 200-2 (1965); CA 63, 1089.

1092. --- and Degraeve, N., ibid., 706-7 (1965); CA 64, 10332.

1093. --- and Lebucq, J., Rev. Ferment. Ind. Aliment. 20, 165-8 (1965); CA 65, 9389.

1094. Moutschen, J., and Degraeve, N., Rev. Cytol. Biol. Veg. 29, 173-89 (1966); CA 66, 83179.

1095. M and T Chemicals Inc., Neth. Appl. 6,405,308, Nov. 16, 1964; CA 63, 1897.

1096. ---, Neth. Appl. 6,405,309, Nov. 16, 1964; CA 62, 16300.

1097. ---, Neth. Appl. 6,505,215, Oct. 25, 1965; CA 64, 9766.

1098. ---, Neth. Appl. 6,505,216, Oct. 25, 1965; CA 64, 9766.

1099. ---, Neth. Appl. 6,505,217, Oct. 25, 1965; CA 64, 9767.

1100. ---, Neth. Appl. 6,505,218, Oct. 25, 1965; CA 64, 9766.

1101. ---, Neth. Appl. 6,505,219, Oct. 25, 1965; CA 64, 9766.

1102. ---, Neth. Appl. 6,505,223, Oct. 25, 1965; CA 64, 14220.

1103. Mueller, A., Diemann, E., and Bollmann, F., Naturwissenschaften 55, 443-4 (1968); CA 69, 113259.

1104. Mueller, E., et al., Z. Naturforsch. 19b, 1079 (1964); CA 62, 6510.

1105. --- et al., Ann. 705, 54-65 (1967); CA 67, 64484.

1106. Mueller, R., and Dathe, C., Chem. Ber. 99, 1609-13 (1966); CA 65, 3898; see also Ger. (East) 49,607, Aug. 20, 1966; CA 66, 28892; and Ger. 1,249,865, Sept. 14, 1967; CA 68, 49772.

1107. ---, Reichel, S., and Dathe, C., Inorg. Nucl. Chem. Lett. 3, 125-33 (1967); CA 67, 21989.

1108. ---, Reichel, S., and Dathe, C., Chem. Ber. 101, 783-6 (1968); CA 68, 95909.

1109. Muk, A. A., and Radosavljevic, R. T., Bull. Boris Kidrich Inst. Nucl. Sci. 17, 9-16 (1966); CA 65, 14820.

1110. --- and Radosavljevic, R. T., ibid., 1-8 (1966); CA 65, 20251.

1111. --- and Radosavljevic, R., Croat. Chem. Acta 39, 1-9 (1967); CA 67, 26363.

1112. Mukherjee, T. K., Tetrahedron 24, 721-8 (1967); CA 68, 38911.

1113. Müller, A., and Rittner, W., Spectrochim. Acta, Part A 23, 1831-7 (1967); CA 67, 68966.

1114. Murphy, J. W., Diss. Abstr. 25, 3234 (1964).

1115. Mustafa, A., et al., Monatsh. Chem. 98, 310-14 (1967); CA 67, 2608.

1116. Myasoedov, B. F., Pal'shin, E. S., and Molochnikova, N. P., Zh. Analit. Khim. 21, 673-81 (1966); CA 65, 11424.

1117. ---, Pal'shin, E. S., and Palei, P. N., Colloq. Intern. Centre Natl. Rech. Sci. (Paris) No. 154, 293-300 (1966); CA 65, 9810.

1118. Nagarajan, G., Z. Naturforsch. A 21, 238-43 (1966); CA 65, 3018.

1119. ---, Acta Phys. Acad. Sci. Hung. 20, 323-9 (1966); CA 65, 6318.

1120. Nagasawa, M. (Ihara Agricult. Chemicals Co., Ltd.), U.S. 3,214,281, Oct. 26, 1965; CA 63, 18469.

1121. Najjar, V. A., and Griffith, M. E., Biochem. Biophys. Res. Commun. 16, 472-7 (1964); CA 61, 13751.

1122. Nakamura, K., and Fukunishi, Sh., Takamine Kenkyusho Nempo 13, 245-9 (1961); CA 63, 4880.

1123. Nanda, R. K., and Tobe, M. L., J. Chem. Soc. A 1966, 1740-4; CA 66, 18413.

1124. Nast, R., and Krueger, K. W., Z. Anorg. Allgem. Chem. 341, 189-95 (1965); CA 64, 6066.

1125. Nefedov, V. D., et al., Radiokhimiya 7, 741-4 (1965); CA 64, 13668.

1126. --- et al., ibid., 8, 376-9 (1966); CA 67, 32224.

1127. --- et al., ibid., 98-104 (1966); CA 64, 15250.

1128. ---, Vobetskii, M. E., and Borak, Ya., Radiochim. Acta 4, 104-6 (1965); CA 64, 235; see also AEC Accession No. 28364, Rept. No. UJV-1169/64; CA 64, 15279.

1129. ---, Vobecky, M., and Borak, J., AEC Accession No. 28365, Rept. No. UJV-1170/64, 5 pp. (1965); CA 64, 12719.

1130. --- et al., AEC Accession No. 28422, Rept. No. UJV-1167/64, 5 pp. (1965); CA 64, 15251.

1131. --- et al., Radiokhimiya 6, 632 (1964); CA 62, 3600.

1132. Neklesova, I. D., Egorova, N. V., and Kudrina, M. A., Dokl. Akad. Nauk SSSR 166, 1121-4 (1966); CA 64, 18334.

1133. ---, Egorova, N. V., and Kudrina, M. A., Izv. Akad. Nauk SSSR, Ser. Khim. 1967, 1998-2005; CA 68, 11911.

1134. Nemodruk, A. A., Zh. Anal. Khim. 22, 629-32 (1967); CA 67, 60730.

1135. --- and Kochetkova, N. E., ibid., 21, 427-32 (1966); CA 65, 6276.

1136. --- and Kochetkova, N. E., Radiokhimiya 10, 324-8 (1968); CA 69, 64373.

1137. ---, Palei, P. N., and Glukhova, L. P., Zh. Anal. Khim. 23, 214-20 (1968); CA 69, 5694.

1138. Nesmeyanov, A. N., et al., Izv. Akad. Nauk SSSR, Ser. Khim. 1966, 160-2; CA 64, 12720.

1139. --- and Borisov, A. E., ibid., 1968, 1922-3; CA 69, 106848.

1140. ---, Borisov, A. E., and Novikova, N. V., ibid., 1965, 763; CA 63, 2998.

1141. ---, Borisov, A. E., and Novikova, N. V., ibid., 1967, 815-18; CA 67, 90900.

1142. ---, Borisov, A. E., and Novikova, N. V., Dokl. Akad. Nauk SSSR 172, 1329-32 (1967); CA 67, 3127.

1143. --- et al., Izv. Akad. Nauk SSSR, Ser. Khim. 1966, 163-4; CA 64, 12718.

1144. ---, Mikulshina, V. V., and Reutov, O. A., J. Organometal. Chem. 13,

263-5 (1968); CA 69, 76461.

1145. Nesmeyanov, A. N., Novikov, V. M., and Reutov, O. A., Z. Organ. Khim. 2, 942-6 (1966); CA 65, 15420.

1146. ---, Pravdina, V. V., and Reutov, O. A., Zh. Organ. Khim. 3, 598-9 (1967); CA 67, 11553.

1147. --- and Reutov, O. A., ibid., 2, 1716 (1966); CA 66, 85821.

1148. Neumann, J., Czech. 118,106, Apr. 15, 1966; CA 66, 1148.

1149. Neuville, M. L., and Carroll, R. B. (Ansul Co.), U.S. 3,378,364, Apr. 16, 1968; CA 69, 9873.

1150. Nicolini, M., Pecile, C., and Turco, A., J. Amer. Chem. Soc. 87, 2379-84 (1965); CA 63, 3869.

1151. Nicpon, P. E., Diss. Abstr. B 27, 3836 (1967).

1152. --- and Meek, D. W., Chem. Commun. 1966, 398-9; CA 65, 8317; see also Abstracts of Papers 151st ACS Meeting, Pittsburgh, Pa., Mar. 28-31, 1966, H63.

1153. --- and Meek, D. W., Inorg. Chem. 6, 145-9 (1967); CA 66, 43302.

1154. Nin'o, N., Farmatsiya (Sofia) 18, 1-4 (1968); CA 69, 30135.

1155. Noelle, H. H., Mitt. Biol. Bundesanstalt Land-Forstwirtsch., Berlin-Dahlem 108, 104-8 (1963); CA 65, 4565.

1156. Norbury, A. H., and Sinha, A. I. P., Inorg. Nucl. Chem. Lett. 3, 355-7 (1967); CA 68, 45801.

1157. --- and Sinha, A. I. P., J. Chem. Soc. A 1968, 1598-603; CA 69, 48839.

1158. Norris, A. R., Breck, A., and Wan, J. K. S., Can. J. Chem. 46, 1823-4 (1968); CA 69, 23621.

1159. Novikov, Yu. V., and Abramova, L. N., Gigiena i Sanit. 31, 57-8 (1966); CA 64, 18011.

1160. Nussenzweig, V., and Benacerraf, B., Intern. Arch. Allergy Appl. Immunol. 27, 193-8 (1965); CA 64, 7204.

1161. Nyholm, R. S., Snow, M. R., and Stiddard, M. H. B., J. Chem. Soc. 1965, 6564-9; CA 64, 751.

1162. ---, Snow, M. R., and Stiddard, M. H. B., ibid., 6570-5; CA 64, 752.

1163. --- and Tobe, M. L., U. S. Clearinghouse Fed. Sci. Tech. Inform., 1967, AD 663164; CA 69, 27504.

1164. --- and Ulm, K., J. Chem. Soc. 1965, 4199-203; CA 63, 6585.

1165. --- and Vrieze, K., ibid., 5331-6; CA 63, 15833.

1166. Nyquist, R. A., et al., Appl. Spectry. 20, 90-100 (1966); CA 64, 16834.

1167. O'Brien, M. H., Doak, G. O., and Long, G. G., Inorg. Chim. Acta 1, 34-40 (1967); CA 67, 103814.

1168. Ogata, N., and Katagiri, Ya. (Toyo Rayon Co., Ltd.), Japan. 15,836, July 22, 1965; CA 63, 18372.

1168a. Okano, T., and Tsuji, K., Yakugaku Zasshi 88, 76-82 (1968); CA 69, 18416.

1169. Okawara, R., Yasuda, K., and Inoue, M., Bull. Chem. Soc. Japan 39, 1823 (1966); CA 65, 16997.

1170. Okubo, T., Nippon Kagaku Zasshi 86, 1142-5 (1965); CA 64, 11854.

1171. --- and Aoki, F., ibid., 87, 1103-4 (1966); CA 66, 41133.

1172. ---, Aoki, F., and Teraoka, T., ibid., 89, 432-3 (1968); CA 69, 22500.

1173. Olifirenko, S. P., Visn. L'vivs'k. Derzh. Univ., Ser. Khim. No. 6, 100-1 (1963); CA 63, 8401.

1174. ---, Zemlyans'kii, M. I., and Ezers'ka, L. O., ibid., 102-3 (1963); CA 63, 9982.

1175. Olson, D. C., and Bjerrum, J., Acta Chem. Scand. 20, 143-50 (1966); CA 64, 16712.

1176. Onishi, H., and Nagai, H., Anal. Chim. Acta 31, 348-51 (1964); CA 62, 1078.

1177. Onoue, K., et al., Immunochemistry 2, 401-15 (1965); CA 64, 5610.

1178. ---, Yagi, Ya., and Pressman, D., J. Immunol. 92, 173-84 (1964); CA 61, 3559.

1179. --- et al., Science 146, 404-5 (1965); CA 62, 3258.

1180. Orenberg, J. B., and Morris, M. D., Anal. Chem. 39, 1776-80 (1967); CA 68, 18435.

1181. Osborne, A. G., and Stiddard, M. H. B., J. Chem. Soc. 1964, 634-6; CA 60, 7607.

1182. Osborne, R. R., and McWhinnie, W. R., ibid., A 1968, 2153-5; CA 69, 83065.

1183. Oser, B. L., et al., Toxicol. Appl. Pharmacol. 9, 528-35 (1966); CA 66, 17620.

1184. Osipowa, M. P., Kamai, G. Kh., and Chadaeva, N. A., Zh. Obshch. Khim. 37, 1660-2 (1967); CA 68, 29811.

1185. Overby, L. R., Bocchieri, S. F., and Fredrickson, R. L., J. Assocn. Offic. Agr. Chemists 48, 17-22 (1965); CA 62, 11001.

1186. --- and Fredrickson, R. L., Toxicol. Appl. Pharmacol. 7, 855-67 (1965); CA 64, 2468.

1187. --- and Straube, L., ibid., 850-4 (1965); CA 64, 2468.

1188. Palei, P. N., Udal'tsova, N. I., and Nemodruk, A. A., Zh. Anal. Khim. 22, 1797-804 (1967); CA 68, 99233.

1189. Palenik, G. J., Acta Cryst. 20, 471-82 (1966); CA 64, 15114.

1190. Pal'shin, E. S., Myasoedov, B. F., and Novikov, Yu. P., Zh. Analit. Khim. 21, 954-60 (1966); CA 66, 4995.

1191. Panattoni, C., et al., Inorg. Chim. Acta 2, 43-8 (1968); CA 69, 22904.

1192. ---, Sindellari, L., and Volponi, L., Ric. Sci., Rend., Sez. A 8, 1149 (1965); CA 64, 13731.

1193. --- et al., Gazz. Chim. Ital. 97, 1006-11 (1967); CA 67, 104724.

1194. Parker, D. J., and Stiddard, M. H. B., J. Chem. Soc. A 1966, 695-9; CA 64, 1597.

1195. Partashnikova, M. Z., and Shafran, I. G., Zh. Analit. Khim. 20, 313-19 (1965); CA 63, 2365.

1195a. Petrovsky, V., Collection Czech. Chem. Commun. 30, 1727-30 (1965); CA 63, 4925.

1196. Paul, B., and Rao, D. V. R., Curr. Sci. 36, 291-2 (1967); CA 67, 39675.

1197. --- and Rao, D. V. R., Can. J. Chem. 46, 334-6 (1968); CA 68, 56050.

1198. --- and Rao, D. V. R., Indian J. Chem. 6, 395-7 (1968); CA 69, 102612.

1199. Paul, R. Ch., Kaushal, R., and Pahil, S. S., J. Indian. Chem. Soc. 44, 995-1000 (1967); CA 68, 74780.

1200. Pauling, P., Inorg. Chem. 5, 1498-505 (1966); CA 65, 11456.

1201. ---, Robertson, G. B., and Rodley, G. A., Nature 207, 73-4 (1965); CA 63, 9154.

1202. Pearson, R. G., Muir, M. M., and Venanzi, L. M., J. Chem. Soc. 1965, 5521-8; CA 63, 15836.

1203. ---, Sobel, H., and Songstad, J., J. Amer. Chem. Soc. 90, 319-26 (1968); CA 68, 117477.

1204. Pecile, C., Inorg. Chem. 5, 210-14 (1966); CA 64, 7535.

1205. Peloso, A., and Dolcetti, G., J. Chem. Soc. A 1967, 1944-8; CA 68, 43588.

1206. --- and Dolcetti, G., Gazz. Chim. Ital. 97, 120-34 (1967); CA 68, 63089.

1207. --- and Ettorre, R., J. Chem. Soc. A 1968, 2253-6; CA 69, 80746.

1208. ---, Ettorre, R., and Dolcetti, G., Inorg. Chim. Acta 1, 307-10 (1967); CA 68, 6792.

1209. --- and Tobe, M. L., J. Chem. Soc. 1964, 5063-74; CA 62, 4889.

1210. Penfold, B. R., and Robinson, W. T., Inorg. Chem. 5, 1758-63 (1966); CA 65, 16167.

1211. Pennsalt Chemicals Corp., Brit. 987,320, March 24, 1965; CA 63, 699.

1212. Perez-Bustamante, J. A., Radiochim. Acta 4, 61-6 (1965); CA 63, 10675.

1213. --- and Bellod, R. Parellada, An. Quim. 64, 213-18 (1968); CA 68, 100176.

1214. --- and Burriel-Marti, F., Anal. Chim. Acta 37, 49-61 (1967); CA 66, 51852.

1215. --- and Burriel-Marti, F., ibid., 62-74 (1967); CA 66, 72176.

1216. --- and Burriel-Marti, F., Inform. Quim. Anal. (Madrid) 22, 25-30 (1968); CA 69, 41029.

1217. --- and Burriel-Marti, F., ibid., 31-7 (1968); CA 69, 41043.

1218. Perry, E. (to Monsanto Co.), U.S. 3,272,786, Sept. 13, 1966; CA 66, 11294.

1219. Peterson, D. B., et al., J. Phys. Chem. 69, 2880-6 (1965); CA 63, 13008.

1220. --- et al., ibid., 71, 4506-8 (1967); CA 68, 29036.

1221. Peterson, L. K., and The, K. I., Chem. Commun. 1967, 1056; CA 68, 13122.

1222. Pettit, L. D., and Irving, H. M. N. H., J. Chem. Soc. 1964, 5336-43; CA 62, 4891.

1223. --- and Turner, D., Spectrochim. Acta, Part A 24, 999-1006 (1968); CA 69, 55916.

1224. Petukhov, G. G., and Titov, V. A., Zh. Prikl. Khim. 39, 1200-3 (1966); CA 65, 5162.

1225. Pfaffman, M. A., and Klein, R. L., Proc. Soc. Exptl. Biol. Med. 121, 539-41 (1966); CA 64, 16557.

1226. Pidcock, A., Richards, R. E., and Venanzi, L. M., J. Chem. Soc. A 1968, 1970-3; CA 69, 63385.

1227. --- and Waterhouse, C. R., Inorg. Nucl. Chem. Lett. 3, 487-9 (1967); CA 68, 45804.

1228. Phillips Petroleum Co., Neth. Appl. 6,507,150, Dec. 6, 1965; CA 64, 16010.

1229. Phillips, J. R., and Vis, J. H., Can. J. Chem. 45, 675-8 (1967); CA 66, 85835.

1230. Pietsch, R., Monatsh. Chem. 96, 138-46 (1965); CA 62, 16034.

1231. ---, Mikrochim. Acta 1967, 708-13; CA 67, 87397.

1232. --- and Ludwig, P., Mikrochim. Ichnoanal. Acta 1964, 60-6; CA 60, 13857.

1233. --- and Ludwig, P., ibid., 1082-8; CA 62, 4885.

1234. --- and Nagl, G., Z. Anal. Chem. 208, 328-32 (1965); CA 62, 11129.

1235. --- and Nagl, G., Mikrochim. Ichnoanal. Acta 1965, 1085-90; CA 64, 14936.

1236. Piovesana, O., and Furlani, C., Inorg. Nucl. Chem. Let. 3, 535-8 (1967); CA 68, 45788.

1237. Pištěk, P., Rais, J., and Kyrš, M., Z. Physik. Chem. (Leipzig) 232, 103-19 (1966); CA 65, 16138.

1238. Pligin, S. G., Zh. Anal. Khim. 22, 145-50 (1966); CA 66, 98521.

1239. ---, ibid., 21, 373-4 (1966); CA 65, 1357.

1240. ---, Farmatsiya (Moscow) 16, 55-7 (1967); CA 67, 25421.

1241. ---, Farm. Zh. (Kiev) 23, 62-8 (1968); CA 69, 109843.

1242. Polychemie-AKU G.E., N. V., Neth. Appl. 6,516,934, June 26, 1967; CA 67, 82564.

1243. Polynova, T. N., and Porai-Koshitz, M. A., Zh. Strukt. Khim. 7, 742-51 (1966); CA 66, 59746.

1244. --- and Porai-Koshitz, M. A., ibid., 8, 112-21 (1967); CA 67, 15844.

1245. Porta, P., et al., J. Chem. Soc. A 1967, 455-65; CA 66, 80255.

1246. Potts, R. A., and Allred, A. L., Inorg. Chem. 5, 1066-71 (1966); CA 65, 1746.

1247. Powell, J., and Shaw, B. L., J. Chem. Soc. 1965, 3879-81; CA 63, 10880.

1248. --- and Shaw, B. L., ibid., A 1967, 1839-51; CA 68, 2985.

1249. --- and Shaw, B. L., ibid., A 1968, 583-96; CA 68, 95924.

1250. --- and Shaw, B. L., ibid., 617-22; CA 68, 81713.

1251. --- and Shaw, B. L., ibid., 774-7; CA 68, 105339.

1252. Powell, P., and Nöth, H., Chem. Commun. 1966, 637-8; CA 65, 19667.

1253. Pugliese, A., et al., Vet. Ital. 15, 703-28 (1964); CA 63, 8964.

1254. Qaim, S. M., J. Inorg. Nucl. Chem. 28, 666-8 (1966); CA 64, 16702.

1255. Rad'ko, V. A., Yakimets, E. M., and Vladimirtsev, I. F., Zh. Analit. Khim. 20, 955-9 (1965); CA 64, 4250.

1256. Rahaman, Md. S., and Finston, H. L., Anal. Chem. 40, 1709-13 (1968); CA 69, 62038.

1257. Rahman, A. K. M., Khattak, M. A., and Magee, R. J., Proc. Conf. Appl. Phys.-Chem. Methods Chem. Anal., Budapest 1966, 3, 113-22; CA 69, 24258.

1258. Rai, R. P., Labdev (Kanpur, India) 5, 334-5 (1967); CA 68, 29805.

1259. Rais, J., Kyrs, M., and Pistek, P., Collect. Czech. Chem. Commun. 32, 2586-95 (1967); CA 67, 85467.

1260. Ramey, K. C., Lini, D. C., and Wise, W. B., J. Amer. Chem. Soc. 90, 4275-9 (1968); CA 69, 48055.

1261. Ramey, K. C., and Statton, G. L., J. Amer. Chem. Soc. 88, 4387-90 (1966); CA 65, 16999.

1262. Randall, E. W., and Shaw, D., Spectrochim. Acta, Part A 23, 1235-42 (1967); CA 67, 38141.

1263. Rapaport, L. I., and Solyanik, G. K., Farmatsevt. Zh. (Kiev) 19, 21-8 (1964); CA 66, 14071.

1264. Raver, Kh. R., USSR 202,128, Sept. 14, 1967; CA 68, 95974.

1265. ---, Bruker, A. B., and Soborovskii, L. Z., USSR 185,919, Sept. 12, 1966; CA 66, 115805.

1266. ---, Bruker, A. B., and Soborovskii, L. Z., Zh. Obshch. Khim. 35, 1162-4 (1965); CA 63, 13313.

1267. Rawson, J. W. (Hercules Inc.), U.S. 3,403,017, Sept. 24, 1968; CA 69, 105275.

1268. Ray, T. C., and Westland, A. D., Inorg. Chem. 4, 1501-4 (1965); CA 64, 5943.

1269. Razuvaev, G. A., Osanova, N. A., and Sangalov, Yu. A., Zh. Obshch. Khim. 37, 216-17 (1967); CA 66, 95139.

1270. --- et al., ibid., 35, 481-5 (1965); CA 63, 629.

1271. --- and Vyazankin, Kh. S., Intern. Symp. Organosilicon Chem., Sci. Commun., Prague 1965, 97-9; CA 65, 8944.

1272. Reichle, W. T., J. Organometal. Chem. 13, 529-32 (1968); CA 69, 106836.

1273. Remes, N. L., and Ventura, J. J. (M & T Chemicals Inc.), Brit. 1,079,658, Aug. 16, 1967; CA 68, 49770.

1274. Richardson, G. A., and Birum, G. H. (Monsanto Co.), U.S. 3,235,567, Feb. 15, 1966; CA 64, 14219.

1275. Riddle, V., and Lorenz, F. W., J. Biol. Chem. 243, 2718-24 (1968); CA 68, 111562.

1276. Rieche, A., and Dahlmann, J., Ger. (East) 44,608, Feb. 24, 1966; CA 65, 10623.

1277. Riedel, H. J., and Merz, E., Radiochim. Acta 4, 48-51 (1965); CA 63, 7828.

1278. Riepma, P., Weed Res. 5, 151-7 (1965); CA 64, 4191.

1279. Rigin, V. I., and Mel'nichenko, N. N., Zavod. Lab. 33, 3-4 (1967); CA 66, 91327.

1280. Rivest, R., Singh, S., and Abraham, C., Can. J. Chem. 45, 3138-41 (1967); CA 68, 34420.

1281. Robertson, H. H., Co., Brit. 1,030,760, May 25, 1966; CA 65, 7403.

1282. Robinson, B. H., and Fergusson, J. E., J. Chem. Soc., Suppl. No. 1, 5683-9 (1964); CA 63, 3859.

1283. Robinson, J. L., Diss. Abstr. B 27, 1762 (1966).

1284. Rodley, G. A., Goodgame, D. M. L., and Cotton, F. A., J. Chem. Soc. 1965, 1499-1505; CA 62, 8527.

1285. --- and Smith, P. W., ibid., A 1967, 1580-3; CA 67, 113296.

1286. Roebke, W., Drawe, H., and Henglein, A., Radiochim. Acta 4, 39-44 (1965); CA 63, 7831.

1287. Roesky, H. W., Chem. Ber. 100, 950-3 (1967); CA 66, 115230.

1288. ---, ibid., 1447-50 (1967); CA 67, 11557.

1289. Roesky, H. W., Z. Naturforsch. B 22, 716-18 (1967); CA 68, 59665.

1290. ---, Angew. Chem. Int. Ed. Engl. 7, 63-4 (1968); CA 68, 59044.

1291. ---, Chem. Ber. 101, 636-42 (1968); CA 68, 69087.

1292. ---, ibid., 2977-86 (1968); CA 69, 87088.

1293. --- and Bormann, D., ibid., 630-5 (1968); CA 68, 69086.

1294. --- et al., Inorg. Nucl. Chem. Lett. 3, 39-42 (1967); CA 66, 61393.

1295. Rogers, M. R., and Kaplan, A. M., Develop. Ind. Microbiol. 1967 (Pub. 1968), 9, 448-76; CA 69, 37612.

1296. Rogovin, Z. A., Gal'braikh, L. S., and Chaikina, E. A., USSR 198,635, June 28, 1967; CA 68, 60500.

1297. Roholt, O., and Pressman, D., Immunochemistry 5, 265-75 (1968); CA 69, 34275.

1298. Röhrscheid, F., Balch, A. L., and Holm, R. H., Inorg. Chem. 5, 1542-51 (1966); CA 65, 11741.

1299. Roundhill, D. M., Lawson, D. N., and Wilkinson, G., J. Chem. Soc. A 1968, 845-9; CA 68, 101360.

1300. --- and Wilkinson, G., ibid., 506-8; CA 68, 83929.

1301. Rouschias, G., and Wilkinson, G., Chem. Commun. 1967, 442; CA 67, 39686.

1302. --- and Wilkinson, G., J. Chem. Soc. A 1968, 489-96; CA 68, 83973.

1303. Roy, F., J. Gas Chromatogr. 6, 245-7 (1968); CA 69, 8287.

1304. Ruddick, J. D., and Shaw, B. L., Chem. Commun. 1967, 1135; CA 68, 29851.

1305. Rudolph, H., and Reinking, K. (Farbenfabriken Bayer A.-G.), Ger. 1,268,619, May 22, 1968; CA 69, 77507.

1306. Rüde, E., Mozes, E., and Sela, M., Biochemistry 7, 2971-5 (1968); CA 69, 58038.

1307. Ruff, J. K., Inorg. Chem. 6, 1502-4 (1967); CA 67, 77857.

1308. --- and Lustig, M., ibid., 7, 2171-3 (1968); CA 69, 102641.

1309. Russ, C. R., and MacDiarmid, A. G., Angew. Chem. Intern. Ed. Engl. 5, 418 (1966); CA 65, 2292.

1310. Ruzicka, J., and Zeman, A., Talanta 12, 997-1001 (1965); CA 63, 15546.

1311. Ryabchikov, D. I., Savvin, S. B., and Dedkov, Yu. M., Zh. Analit. Khim. 19, 1210-18 (1964); CA 62, 1058.

1312. Ryan, M. T., and Lehr, W. L., J. Organometal. Chem. 4, 455-60 (1965); CA 64, 169.

1313. Ryer, J., and Kerschner, P. M. (Cities Service Oil Co.), U.S. 3,297,421, Jan. 10, 1967; CA 66, 57641.

1314. Sabatini, A., and Bertini, I., Inorg. Chem. 4, 1665-7 (1965); CA 64, 4446.

1315. Sacconi, L., Bertini, I., and Mani, F., ibid., 7, 1417-20 (1968); CA 69, 40843.

1316. --- and Morassi, R., Inorg. Nucl. Chem. Lett. 4, 449-54 (1968); CA 69, 102620.

1317. ---, Speroni, G. P., and Morassi, R., Inorg. Chem. 7, 1521-5 (1968); CA 69, 56620.

1318. Sangal, S. P., Microchem. J. 8, 304-12 (1964); CA 62, 1078.

1319. Sangal, S. P., Microchem. J. 9, 9-15 (1965); CA 63, 6303.

1320. ---, J. Prakt. Chem. 31, 68-75 (1966); CA 64, 12177.

1321. ---, Chim. Anal. (Paris) 48, 566-70 (1966); CA 66, 34390.

1322. ---, J. Prakt. Chem. 36, 126-37 (1967); CA 67, 104828.

1323. --- and Dey, A. K., J. Indian Chem. Soc. 43, 440-2 (1966); CA 65, 14423.

1324. Sano, R., and Machihara, K. (Kansai Paint Co., Ltd.), Brit. 1,008,857, Nov. 3, 1965; CA 64, 5299.

1325. --- and Machihara, K., Shikizai Kyokaishi 38, 8-9 (1965); CA 67, 22923.

1326. Santelmann, P. W., Scifres, C. J., and Murray, J., Crop Sci. 6, 561-2 (1966); CA 66, 75175.

1327. Saraceno, A. J. (Pennsalt Chemicals Corp.), Fr. 1,395,910, Apr. 16, 1965; CA 63, 18292; U.S. 3,275,574, Sept. 27, 1966; CA 66, 29343.

1328. Sarychev, N. I., and Machinskii, A. P., Uch. Zap., Mord. Gos. Univ. No. 47, 60-6 (1965); CA 66, 73661.

1329. Sasaki, Ya., et al. (Asahi Chem. Industry Co., Ltd.), Japan. 18,021, Aug. 27, 1964; CA 62, 5298.

1330. Sastri, M. N., and Rao, T. S. R. P., J. Inorg. Nucl. Chem. 30, 1727-33 (1968); CA 69, 73505.

1331. Savvin, S. B., et al., USSR 217,683, May 7, 1968; CA 69, 73751.

1332. ---, Eliseeva, O. P., and Rozovskii, Yu. G., Dokl. Akad. Nauk SSSR 180, 374-6 (1968); CA 69, 36663.

1333. --- and Kuzin, E. L., Talanta 15, 913-21 (1968); CA 69, 90428.

1334. ---, Kuzin, E. L., and Gribov, L. A., Zh. Anal. Khim. 23, 5-12 (1968); CA 68, 100158.

1335. --- and Milyukova, M. S., ibid., 21, 1075-81 (1966); CA 66, 8078.

1336. --- and Propistsova, R. F., Zh. Anal. Khim. 23, 653-60 (1968); CA 69, 54845.

1337. ---, Rozovskii, Yu. G., and Eliseeva, O. P., Vysokomol. Soedin. Ser. B 10, 41-4 (1968); CA 68, 69638.

1338. Sazonova, V. A., Sazonova, N. S., and Plyukhina, V. N., Dokl. Akad. Nauk SSSR 177, 1352-4 (1967); CA 68, 105334.

1339. Scaife, D. E., and Wood, K. P., Inorg. Chem. 6, 358-65 (1967); CA 66, 49678.

1340. Scane, J. G., Acta Crystallogr. 23, 85-9 (1967); CA 67, 58062.

1341. --- and Stephens, R. M., Proc. Phys. Soc. 92, 833-6 (1967); CA 68, 64014.

1342. Schaefe, R., Hruska, F., and Hutton, H. M., Can. J. Chem. 45, 3143-51 (1967); CA 68, 34532.

1343. Scherer, O., Hornig, P., and Schmidt, M., J. Organometal. Chem. 6, 259-64 (1966); CA 65, 12233.

1344. Scherer, O. J., and Hornig, P., Chem. Ber. 101, 2533-47 (1968); CA 69, 52246.

1345. --- and Wokulat, J., Z. Anorg. Allg. Chem. 357, 92-102 (1968); CA 68, 78339.

1346. --- and Wokulat, J., ibid., 361, 296-305 (1968); CA 69, 96588.

1347. Schindlbauer, H., and Lass, H., Monatsh. Chem. 99, 2460-7 (1968); CA

70, 57965.

1348. Schindler, F., Schmidbauer, H., and Jonas, G., Angew. Chem. 77, 170 (1965); CA 62, 10056.

1349. Schlemper, E. O., and Britton, D., Acta Cryst. 20, 777-83 (1966); CA 65, 4763.

1350. Schmid, G., Powell, P., and Noeth, H., Chem. Ber. 101, 1205-14 (1968); CA 68, 105310.

1351. Schmidbaur, H., Armer, B., and Bergfeld, M., Z. Chem. 8, 254 (1968); CA 69, 64195.

1352. --- and Jonas, G., Angew. Chem. Int. Ed. Engl. 6, 449-50 (1967); CA 67, 43891.

1353. --- and Jonas, G., Chem. Ber. 101, 1271-85 (1968); CA 68, 105311.

1354. --- and Tronich, W., Inorg. Chem. 7, 168-9 (1968); CA 68, 56048.

1355. --- and Tronich, W., Chem. Ber. 101, 3545-55 (1968); CA 69, 106810.

1356. Schmidt, A., ibid., 3976-80 (1968); CA 70, 20182.

1357. Schmidt, U., et al., ibid., 1381-97 (1968); CA 68, 95904.

1358. --- et al., ibid., 99, 1497-501 (1966); CA 65, 2290.

1359. Schmidtke, H. H., Ber. Bunsenges. Phys. Chem. 71, 1138-45 (1967); CA 68, 34451.

1360. --- and Garthoff, D., J. Amer. Chem. Soc. 89, 1317-21 (1967); CA 66, 89818.

1361. --- and Garthoff, D., Helv. Chim. Acta 50, 1631-8 (1967); CA 67, 104688.

1362. Schmitz-DuMont, O., Rec. Chem. Progr. 29, 13-23 (1968); CA 69, 59301.

1363. --- and Ross, B., Z. Naturforsch. B 20, 72-3 (1965); CA 62, 15736.

1364. --- and Ross, B., Z. Anorg. Allg. Chem. 349, 328-36 (1967); CA 66, 76111.

1365. Schnegg, R., et al. (Farbenfabriken Bayer A.-G.), Belg. 659,510, May 28, 1965; CA 63, 18397.

1366. Schnell, H., Rudolph, H., and Reinking, K. (Farbenfabriken Bayer A.-G.), Ger. 1,270,816, June 20, 1968; CA 69, 36585.

1367. Scholven-Chemie, A.-G., Neth. Appl. 6,612,339, March 7, 1967; CA 67, 63689.

1368. Scholz, S., Explosivstoffe 11, 159-63, 181-4 (1963); CA 61, 11840.

1369. Schrauzer, G. N., and Kratel, G., Angew. Chem. 77, 130 (1965); CA 62, 11406.

1370. --- and Windgassen, R. J., Chem. Ber. 99, 602-10 (1966); CA 64, 19615.

1371. Schumann, H., Oestermann, Th., and Schmidt, M., Chem. Ber. 99, 2057-62 (1966); CA 65, 7214.

1372. ---, Oestermann, Th., and Schmidt, M., J. Organometal. Chem. 8, 105-10 (1967); CA 66, 85834.

1373. ---, Roth, A., and Stelzer, O., Angew. Chem. Int. Ed. Engl. 7, 218-19 (1968); CA 68, 95908.

1374. --- and Schmidt, M., Inorg. Nucl. Chem. Lett. 1, 1-5 (1965); CA 64, 1603.

1375. Schwartz, H., and Skaptason, J. B., Belg. 672,361, March 16, 1966; CA 65, 15282.

1376. Schweitzer, G. K., and McCarty, S. W., J. Inorg. Nucl. Chem. 27, 191-9 (1965); CA 62, 4683.

1377. Schweizer, E. E., Weeds 15, 72-6 (1967); CA 66, 54467.

1378. Schwerdle, A. (Vineland Chemical Co.), U.S. 3,269,895, Aug. 30, 1966; CA 65, 17639.

1379. ---, U.S. 3,338,780, Aug. 29, 1967; CA 67, 107660.

1379a. Scifres, C. J., and Santelmann, P. W., Proc. Southern Weed Conf. 18, 94-101 (1965); CA 63, 1167.

1380. Sckerl, M. M., Frans, R. E., and Spooner, A. E., ibid., 19, 351-7 (1966); CA 65, 6216.

1381. Scott, O. M., and Sons Co., Brit. 1,005,344, Sept. 22, 1965; CA 63, 17059.

1382. ---, Brit. 1,005,345, Sept. 22, 1965; CA 63, 18956.

1382a. Seibt, H., and Kutepow, N. v. (BASF), Belg. 635,483, Jan. 27, 1964; CA 61, 11891.

1383. Selbin, J., and Ortego, J. D., J. Inorg. Nucl. Chem. 29, 1449-56 (1967); CA 68, 56107.

1384. Sellers, C., and Suschitzky, H., J. Chem. Soc. C 1968, 2317-19; CA 69, 86488.

1385. Senning, A., Acta Chem. Scand. 19, 1755-9 (1965); CA 64, 5127.

1386. Seyferth, D., Prokai, B., and Cross, R. J., J. Organometal. Chem. 13, 169-75 (1968); CA 69, 96046.

1387. Shagidullin, R. R., et al., Izv. Akad. Nauk SSSR, Ser. Khim. 1966, 1543-6; CA 66, 89820.

1387a. --- and Grechkin, N. P., Zh. Obshch. Khim. 38, 150-6 (1968); CA 69, 66727.

1387b. ---, Lamanova, I. A., and Urazgil'deeva, P. K., Dokl. Akad. Nauk SSSR 174, 1359-62 (1967); CA 67, 77608.

1388. --- and Pavlova, T. E., Izv. Akad. Nauk SSSR, Ser. Khim. 1965, 995-8; CA 63, 10879.

1389. --- and Pavlova, T. E., ibid., 1966, 2091-6; CA 66, 89819.

1390. Shaw, B. L., and Singleton, E., J. Chem. Soc. A 1967, 1683-93; CA 68, 2984.

1391. --- and Smithies, A. C., ibid., 1047-52; CA 67, 73669.

1392. Shaw, J. T., Voorhies, J. D., and Davis, S. M. (American Cyanamid Co.), U.S. 3,260,621, July 12, 1966; CA 65, 10130.

1393. Shaw, R. A., Smith, B. C., and Thakur, C. P., Chem. Commun. 1966, 228-9; CA 65, 2291; Justus Liebig Ann. Chem. 713, 30-9 (1968); CA 69, 12970.

1394. Shell Internationale Research Maatschappij N.V., Belg. 638,289, Apr. 7, 1964; CA 62, 11937.

1395. ---, Neth. Appl. 6,601,668, Aug. 12, 1966; CA 66, 11791.

1396. Shemyakina, M. A., Vop. Prikl. Geokhim. No. 1, 128-31 (1966); CA 66, 111243.

1397. Shen, K.-W., Diss. Abstr. B 29, 1989 (1968).

1398. --- et al., J. Amer. Chem. Soc. 90, 1718-23 (1968); CA 68, 99538.

1399. Shepherd, C. G., et al., Proc. Southern Weed Conf. 19, 542-4 (1966);

CA <u>66</u>, 1793.

1400. Shepherd, T. M., Nature <u>212</u>, 745 (1966); CA <u>66</u>, 33309.

1401. ---, J. Inorg. Nucl. Chem. <u>29</u>, 2551-9 (1967); CA <u>68</u>, 7884.

1402. Shimizu, K., et al. (Toyo Rayon Co., Ltd.), Japan. 4994, Mar. 1, 1967; CA <u>67</u>, 54617.

1403. Shindo, M., Matsumura, Yo., and Okawara, R., J. Organometal. Chem. <u>11</u>, 299-305 (1968); CA <u>68</u>, 53979.

1404. --- and Okawara, R., ibid., <u>5</u>, 537-44 (1966); CA <u>65</u>, 2093.

1405. --- and Okawara, R., Inorg. Nucl. Chem. Lett. <u>3</u>, 75-7 (1967); CA <u>67</u>, 54233.

1406. Shioyama, O., et al., Noyaku Seisan Gijutsu <u>11</u>, 8-12 (1964); CA <u>62</u>, 13783.

1407. Shlyk, Yu. N., Bogolyubov, G. M., and Petrov, A. A., Zh. Obshch. Khim. <u>38</u>, 1199-1200 (1968); CA <u>69</u>, 59351.

1408. Shriver, D. F., and Johnson, M. P., Inorg. Chem. <u>6</u>, 1265-8 (1967); CA <u>67</u>, 28869.

1409. Shulman, S., Groff, J. L., and Swanborg, R. H., Int. Arch. Allergy Appl. Immunol. <u>30</u>, 417-27 (1966); CA <u>66</u>, 45054.

1410. Siemroth, J., and Hennig, I., Talanta <u>15</u>, 765-70 (1968); CA <u>69</u>, 73758.

1411. Silverthorn, W., and Feltham, R. D., Inorg. Chem. <u>6</u>, 1662-6 (1967); CA <u>67</u>, 96390.

1412. Simonnin, M. P., J. Organometal. Chem. <u>5</u>, 155-65 (1966); CA <u>64</u>, 6453.

1413. ---, Bull. Soc. Chim. France <u>1966</u>, 1774-5; CA <u>65</u>, 11576.

1414. Sindellari, L., and Centurioni, P., Ann. Chim. (Rome) <u>56</u>, 379-85 (1966); CA <u>65</u>, 2292.

1415. --- and Deganello, G., Ric. Sci. <u>35</u>, 744-7 (1965); CA <u>64</u>, 19669.

1416. Singh, J., and Burg, A. B., J. Amer. Chem. Soc. <u>88</u>, 718-22 (1966); CA <u>64</u>, 11246.

1417. Singh, P. P., and Rivest, R., Can. J. Chem. <u>46</u>, 1773-9 (1968); CA <u>69</u>, 24111.

1418. Singh, R. K. N., and Campbell, R. W., Weeds <u>13</u>, 170-1 (1965); CA <u>63</u>, 4882.

1419. Singh, S., Singh, P. P., and Rivest, R., Inorg. Chem. <u>7</u>, 1236-8 (1968); CA <u>69</u>, 23289.

1420. Sisler, H. H., Jain, S. R., Inorg. Chem. <u>7</u>, 104-7 (1968); CA <u>68</u>, 34547.

1421. --- and Stratton, C., ibid., <u>5</u>, 2003-8 (1966); CA <u>65</u>, 19977.

1422. Siuda, A., Nukleonika <u>10</u>, 393-5 (1965); CA <u>64</u>, 5735.

1423. Skroch, W. A., Proc. Southern Weed Conf. <u>19</u>, 241-2 (1966); CA <u>65</u>, 6216.

1424. Slaugh, L. H., and Mullineaux, R. D. (Shell Oil Co.), U.S. 3,239,570, March 8, 1966; CA <u>64</u>, 19420.

1425. --- and Mullineaux, R. D., J. Organometal. Chem. <u>13</u>, 469-77 (1968); CA <u>69</u>, 86254.

1426. Slavin, S. B., and Liauw, H. L., Int. Arch. Allergy Appl. Immunol. <u>31</u>, 366-79 (1967); CA <u>67</u>, 9758.

1427. Slesingr, L., Vet. Med. (Prague) <u>39</u>, 735-9 (1966); CA <u>66</u>, 62965.

1428. --- and Sevcik, V., Arch. Tierernaehr. <u>18</u>, 97-105 (1968); CA <u>68</u>, 66821.

1429. Smith, G. W., J. Mol. Spectry. <u>12</u>, 146-70 (1964); CA <u>60</u>, 7593.

1430. Smith, J. M., Morgenstern, N., and Peters, L., J. Pharmacol. Exp. Ther. 158, 451-9 (1967); CA 68, 20239.

1431. Smith, R. D., and Wilson, R. A. L. (Imperial Chemical Industries Ltd.), Brit. 970,072, Sept. 16, 1964; CA 62, 1603.

1432. Soborovskii, L. Z., Bruker, A. V., and Raver, Kh. R., USSR 150,840, Dec. 1, 1964; CA 62, 10462.

1433. Societe des Usines Chimiques Rhone-Poulenc, Fr. Addn. 88,337, Jan. 20, 1967, Addn. to Fr. 1,443,121; CA 68, 109201.

1434. ---, Fr. Addn. 89,238, May 26, 1967, Addn. to Fr. 1,443,121; CA 69, 47521.

1435. Soiebel'man, B. I., and Ruzhitskii, D. M., Farmatsiya (Moscow) 17, 78-81 (1968); CA 69, 46129.

1436. Sollott, G. P., and Peterson, W. R., Jr., J. Org. Chem. 30, 389-93 (1965); CA 62, 9172.

1437. --- et al., U. S. Dept. Com., Office Tech. Serv., A.D. 611,869 Vol. II, 441-52 (1965); CA 63, 18147.

1438. Sommer, K., and Becke-Goehring, M., Z. Anorg. Allg. Chem. 355, 182-91 (1967); CA 68, 87246.

1439. Sommer, L., and Cenek, M., Chemist-Analyst 56, 9-12 (1967); CA 66, 121713.

1440. Spiro, Th. G., Inorg. Chem. 6, 569-72 (1967); CA 66, 1440.

1441. Sprague, M. A., et al., New Jersey Agr. Expt. Sta., Bull. 803, 72 pp. (1964); CA 62, 2186.

1442. No entry.

1443. Stamm, W., J. Org. Chem. 30, 693-5 (1965); CA 62, 11842.

1444. ---, Trans. N. Y. Acad. Sci. 28, 396-401 (1966); CA 65, 18612.

1445. --- (Stauffer Chemical Co.), Ger. 1,229,529, Dec. 1, 1966; CA 66, 28896; see also Neth. Appl. 6,411,266, Apr. 5, 1965; CA 63, 11613.

1446. Starodubskaya, A. A., Shchemeleva, G. G., and Bagdasarov, K. N., Peredovye Metody Khim. Tekhnol. i Kontrolya Proizv. (Rostov-on-Don: Rostovsk. Univ.) Sb. 1964, 246-9; CA 63, 10679.

1447. Stecker, H. C., U.S. 3,189,631, June 15, 1965; CA 63, 8406.

1448. Stefcik, A., and Beiswanger, J. P. G. (GAF Corp.), U.S. 3,397,051, Aug. 13, 1968; CA 69, 75835.

1449. Steiger, N. (Hoffmann-LaRoche Inc.), U.S. 3,135,755, June 2, 1964; CA 65, 7094; see also U.S. 3,182,053, May 4, 1965; CA 63, 11441.

1450. ---, U.S. 3,184,482, May 18, 1965; CA 65, 5564.

1451. Stephenson, N. C., Acta Cryst. 17, 1517-22 (1964); CA 62, 3483.

1452. Stephenson, T. A., et al., J. Chem. Soc. 1965, 3632-40; CA 63, 4158.

1453. --- and Wilkinson, G., J. Inorg. Nucl. Chem. 28, 945-56 (1966); CA 64, 18945.

1454. --- and Wilkinson, G., ibid., 29, 2122-3 (1967); CA 67, 104717.

1455. Stevens, G. D., Proc. Southern Weed Conf. 19, 545-9 (1966); CA 65, 20769.

1456. Stevens, M., ibid., 20, 405-9 (1967); CA 67, 31847.

1457. Stevenson, D. L., and Dahl, L. F., J. Amer. Chem. Soc. 89, 3424-8 (1967); CA 67, 58056.

1458. Storey, I. D. E., Biochem. J. 95, 201-8 (1965); CA 62, 11047.

1459. Strohmeier, W., and Mueller, F. J., Z. Naturforsch. B 23, 555 (1968); CA 69, 31585.

1460. Strycker, S. J. (Dow Chemical Co.), U.S. 3,197,494, July 27, 1965; CA 63, 8407.

1461. ---, U.S. 3,197,495, July 27, 1965; CA 63, 13320.

1462. Studiengesellschaft Kohle m.b.H., Neth. Appl. 6,409,179, Feb. 11, 1965; CA 63, 5770.

1463. Sundar, D. S., Prasada Rao, T. S. R., and Sastri, M. N., Current Sci. (India) 35, 307-8 (1966); CA 65, 9724.

1464. Sundermeyer, W., and Verbeek, W., Angew. Chem. Intern. Ed., Engl. 5, 1-6 (1966); CA 64, 15918.

1465. --- and Verbeek, W. (Th. Goldschmidt A.-G.), Ger. 1,239,687, May 3, 1967; CA 68, 2989.

1466. Suzuki, T., and Fukada, Ch., Japan. J. Pharmacol. 14, 363-74 (1964); CA 62, 2144.

1467. Suzuki, Sh., Inoue, Ya., and Kishimoto, M., Sci. Rept. Res. Inst. Tohoku Univ., Ser. A 17, 293-9 (1965); CA 64, 18852.

1468. ---, Saito, M., and Inoue, Ya., ibid., 267-76 (1965); CA 64, 15296.

1469. Szabó, P., et al., J. Organometal. Chem. 12, 245-8 (1968); CA 69, 2979.

1470. Tagliavini, G., and Zanella, P., ibid., 5, 299 (1966); CA 64, 14213.

1471. --- and Zanella, P., Anal. Chim. Acta 40, 33-9 (1968); CA 68, 9172.

1472. Takahashi, Sh., Sonogashira, K., and Hagihara, N., Nippon Kagaku Zasshi 87, 610-13 (1966); CA 65, 14485.

1473. Takamura, Sh., Oyo Yakuri 1, 36-53 (1967); CA 68, 94457.

1474. Takashi, Yu., Bull. Chem. Soc. Japan 40, 612-17 (1967); CA 67, 53470.

1475. ---, ibid., 1201-4 (1967); CA 67, 54554.

1476. ---, J. Organometal. Chem. 8, 225-31 (1967); CA 67, 11560.

1477. ---, Bull. Chem. Soc. Japan 40, 1194-201 (1967); CA 68, 22025.

1478. --- and Aijima, I. (Asahi Chemical Ind. Co., Ltd.), Japan. 15,719, Aug. 29, 1967; CA 68, 22343.

1479. ---, Aijima, I., and Fujisaki, S. (Asahi Chemical Ind. Co., Ltd.), Japan. 13,603, Aug. 2, 1967; CA 68, 50302.

1480. ---, Aijima, I., and Kobayashi, Yu. (Asahi Chemical Ind. Co., Ltd.), Japan. 17,053, Aug. 18, 1964; CA 62, 2847.

1481. ---, Aijima, I., and Kobayashi, Yu. (Asahi Chemical Ind. Co., Ltd.), Japan. 19,222, Sept. 8, 1964; CA 62, 9258.

1482. ---, Aijima, I., and Kobayashi, Yu. (Asahi Chemical Ind. Co., Ltd.), Japan. 19,223, Sept. 8, 1964; CA 62, 9258.

1483. --- et al. (Asahi Chemical Ind. Co., Ltd.), Japan. 27,872, Dec. 4, 1964; CA 63, 1896.

1484. --- et al. (Asahi Chemical Ind. Co., Ltd.), Japan. 19,226, Sept. 8, 1964; CA 62, 10547.

1485. --- et al. (Asahi Chemical Ind. Co., Ltd.), Japan. 3,695, Feb. 26, 1965; CA 63, 7136.

1486. --- et al. (Asahi Chemical Ind. Co., Ltd.), Japan. 6,191, Apr. 1,

1966; CA <u>65</u>, 9051.

1487. Takashi, Yu., et al. (Asahi Chemical Ind. Co., Ltd.), Japan. 4,978, March 1, 1967; CA <u>67</u>, 44275.

1488. --- and Aishima, I., J. Organometal. Chem. <u>8</u>, 209-23 (1967); CA <u>67</u>, 11559.

1489. --- et al. (Asahi Chemical Ind. Co., Ltd.), U.S. 3,330,816, July 11, 1967; CA <u>68</u>, 30384.

1490. Takayanagi, T., Ger. 1,249,869, Sept. 14, 1967; CA <u>68</u>, 95533.

1491. Takenaka, T., and Goto, R., Proc. Intern. Symp. Mol. Struct. Spectry., Tokyo <u>1962</u> (A 211), 4 pp.; CA <u>61</u>, 1395.

1492. Takesada, M., Tamazaki, H., and Hagihara, N., Bull. Chem. Soc. Japan <u>41</u>, 270 (1968); CA <u>68</u>, 114706.

1493. Takita, K., et al., Nogyo Gijutsu Kenkyusho Hokoku, Byori Konchu No. <u>19</u>, 1-46 (1965); CA <u>63</u>, 18958.

1494. Takuma, I., Nagasaki Daigaku Fudobyo Kiyo <u>1</u>, 19-37 (1959); CA <u>62</u>, 3116.

1495. Tanaka, G., et al. (Asahi Chemical Ind. Co., Ltd.), Japan. 26,370, Dec. 16, 1963; CA <u>60</u>, 12056.

1496. Tataev, O. A., and Bagdasarov, K. N., Sb. Aspirantskikh Rabot, Dagestansk. Univ., Estestv. i Fiz.-Mat. Nauk Makhachkala <u>1964</u>, 61-8; CA <u>65</u>, 1372.

1497. Tatarenko, A. N., and Manulkin, Z. M., Zh. Obshch. Khim. <u>34</u>, 3462-5 (1964); CA <u>62</u>, 2792.

1498. --- and Manulkin, Z. M., ibid., <u>38</u>, 273-5 (1968); CA <u>69</u>, 52242.

1499. Tayim, H. A., and Bailar, J. C., Jr., J. Amer. Chem. Soc. <u>89</u>, 4330-8 (1967); CA <u>67</u>, 108103.

1500. Taylor, M. J., Odell, A. L., and Raethel, H. A., Spectrochim. Acta, Part A <u>24</u>, 1855-61 (1968); CA <u>69</u>, 111664.

1501. Taylorson, R. B., Weeds <u>14</u>, 207-10 (1966); CA <u>65</u>, 3636.

1502. Terent'ev, A. P., Volodina, M. A., and Fursova, E. G., Dokl. Akad. Nauk SSSR <u>169</u>, 851-4 (1966); CA <u>65</u>, 19304.

1503. ---, Volodina, M. A., and Fursova, E. G., USSR 188,123, Oct. 20, 1966; CA <u>66</u>, 2932.

1504. Thang, D. C., et al., J. Chem. Soc. C <u>1967</u>, 665-8; CA <u>67</u>, 11421.

1505. Theisen, C. Th., Diss. Abstr. B <u>28</u>, 3660-1 (1968).

1506. Theophanides, T., and Lafortune, Y., Can. Spectrosc. <u>13</u>, 40-4 (1968); CA <u>69</u>, 31669.

1507. Thierig, D., and Umland, F., Fresenius' Z. Anal. Chem. <u>240</u>, 19-23 (1968); CA <u>69</u>, 64379.

1508. Thompson, C. J., et al., J. Gas Chromatogr. <u>5</u>, 1-10 (1967); CA <u>66</u>, 64715.

1509. Thompson, J. T., U.S. 3,169,850, Feb. 16, 1965; CA <u>62</u>, 15362.

1510. Thompson, P. E., and Bayles, A., J. Parasitol. <u>52</u>, 674-8 (1968); CA <u>65</u>, 14302.

1511. --- et al., Amer. J. Trop. Med. Hyg. <u>16</u>, 133-45 (1967); CA <u>66</u>, 74643.

1512. Thompson, R. J., and Davis, J. C., Jr., Inorg. Chem. <u>4</u>, 1464-7 (1965); CA <u>63</u>, 14365.

1513. Thornton, D. A., J. S. African Chem. Inst. <u>17</u>, 61-70 (1964); CA <u>62</u>,

11851.

1514. Thornton, D. A., J. S. African Chem. Inst. 17, 71-8 (1964); CA 62, 11850.

1515. Thornsteinson, E. M., and Basolo, F., Inorg. Chem. 5, 1691-5 (1966); CA 65, 16111.

1516. --- and Basolo, F., J. Amer. Chem. Soc. 88, 3929-36 (1966); CA 65, 13473.

1517. Tolmachev, V. N., Kvichko, L. A., and Konkin, V. D., Zh. Anal. Khim. 22, 11-14 (1967); CA 66, 99033.

1518. --- et al., Vysokomol. Soedin., Ser. A 10, 1811-16 (1968); CA 69, 88086.

1519. Tomaschewski, G., J. Prakt. Chem. 33, 168-77 (1966); CA 66, 38013.

1520. Toyoda, S., Nogyo Gijutsu Kenkyusho Hokoku, Byori Konchu No. 18, 59-134 (1965); CA 63, 18706.

1521. Tranter, G. C., Addison, C. C., and Sowerby, D. B., J. Organometal. Chem. 12, 369-76 (1968); CA 69, 19276.

1522. Treichel, P. M., Barnett, K. W., and Shubkin, R. L., ibid., 7, 449-59 (1967); CA 66, 95158.

1523. --- and Shubkin, R. L., Inorg. Chem. 6, 1328-34 (1967); CA 67, 43905.

1524. Trivedi, J. M., J. Maharaja Sayajirao Univ. Baroda 13, 82-7 (1964); CA 64, 12719.

1525. Trotter, J., and Zobel, T., J. Chem. Soc. 1965, 4466-71; CA 63, 13038.

1526. Tsang, W. S., Meek, D. W., and Wojcicki, A., Inorg. Chem. 7, 1263-8 (1968); CA 69, 40835.

1527. Tserkovnitskaya, I. A., and Bykhovtsova, T. T., Zh. Anal. Khim. 22, 1201-9 (1967); CA 68, 55932.

1528. Tsvetkov, V. S., Byul. Eksperim. Biol. i Med. 58, 116-19 (1964); CA 62, 12310.

1529. Tuck, D. G., and Faithful, B. D., J. Chem. Soc. 1965, 5753-4; CA 64, 1604.

1530. Tzschach, A., and Deylig, W., Z. Anorg. Allgem. Chem. 336, 36-41 (1965); CA 62, 11406.

1531. --- and Deylig, W., Chem. Ber. 98, 977-82 (1965); CA 62, 14722.

1532. --- and Drohne, D., Ger. (East) 61,546, May 5, 1968; CA 70, 78150.

1533. --- and Fischer, W., Z. Chem. 7, 196-7 (1967); CA 67, 54231.

1534. --- and Haeckert, H., ibid., 6, 265-6 (1966); CA 65, 13752.

1535. --- and Pacholke, G., Z. Anorg. Allgem. Chem. 336, 270-7 (1965); CA 63, 7041.

1536. --- and Schwarzer, R., Ann. Chem. 709, 248-56 (1967); CA 68, 39750.

1537. --- and Schwarzer, R., J. Prakt. Chem. 37, 21-7 (1968); CA 68, 39749.

1538. --- and Schwarzer, R., J. Organometal. Chem. 13, 363-8 (1968); CA 69, 77375.

1539. Ueno, K., and Chang, Ch.-T., Nippon Genshiryoku Gakkaishi 4, 457-62 (1962); CA 64, 4328.

1540. Ugo, R., and Bonati, F., J. Organometal. Chem. 8, 189-92 (1967); CA 66, 115770.

1541. ---, Bonati, F., and Cenini, S., Rend. Ist. Lombardo Sci. Lett., A 98,

627-9 (1964); CA <u>66</u>, 43269.

1542. Ugo, R., et al., J. Organometal. Chem. <u>11</u>, 159-66 (1968); CA <u>68</u>, 45811.

1542a. Uhlig, E., and Maaser, M., Z. Anorg. Allg. Chem. <u>349</u>, 300-9 (1967); CA <u>66</u>, 91243.

1543. --- and Schaefer, M., ibid., <u>359</u>, 178-84 (1968); CA <u>69</u>, 48806.

1544. Ullman, E. F., and Henderson, W. A., Jr. (American Cyanamid Co.), U.S. 3,322,542, May 30, 1967; CA <u>68</u>, 8032.

1545. Unger, V. H., and Wolfrom, R. E. (Rohm and Haas Co.), U.S. 3,307,931, March 7, 1967; CA <u>66</u>, 94153; Fr. 1,465,776, Jan. 13, 1967; CA <u>67</u>, 63223.

1546. Upor, E., Szalay, J., and Klesch, K., Magy. Kem. Foly. <u>74</u>, 438-40 (1968); CA <u>69</u>, 102727.

1547. Urban, G., and Doetzer, R. (Siemens-Schuckertwerke A.-G.), Ger. 1,191,813, April 29, 1965; CA <u>63</u>, 14909.

1548. Usacheva, G. M., and Kamai, S. Kh., Izv. Akad. Nauk SSSR, Ser. Khim. <u>1968</u>, 413-14; CA <u>69</u>, 52239.

1549. --- and Kamai, G. Kh., Zh. Obshch. Khim. <u>38</u>, 365-9 (1968); CA <u>69</u>, 87135.

1550. Vadasdi, K., Magy. Kem. Lapja <u>23</u>, 344-6 (1968); CA <u>69</u>, 49002.

1551. Vallarino, L., Inorg. Chem. <u>4</u>, 161-5 (1965); CA <u>62</u>, 8646.

1552. Vannier, W. E., Bryan, W. P., and Campbell, D. H., Immunochemistry <u>2</u>, 1-12 (1965); CA <u>63</u>, 8884.

1553. Vasilenko, V. D., and Shanya, M. V., Izv. Vyssh. Ucheb. Zaved., Khim. Khim. Tekhnol. <u>11</u>, 138-41 (1968); CA <u>69</u>, 49000.

1554. Vati, G., and Shukla, S. R., Labdev (Kanpur, India) <u>5</u>, 45-7 (1967); CA <u>66</u>, 93687.

1555. Ventura, J. J. (M & T Chemicals, Inc.), U.S. 3,347,892, Oct. 17, 1967; CA <u>68</u>, 69129.

1556. Vladimirov, L. M., et al., Tr. Mosk. Khim.-Tekhnol. Inst. <u>1966</u>, No. 51, 230-3; CA <u>69</u>, 5677.

1557. Vlasov, N. A., and Morgen, E. A., Zh. Prikl. Khim. <u>38</u>, 998-1004 (1965); CA <u>63</u>, 9047.

1558. Voitenko, G. A., Farmakol. i Toksikol. <u>27</u>, 156-60 (1964); CA <u>64</u>, 11790.

1559. Volger, H. C., and Vrieze, K., J. Organometal. Chem. <u>9</u>, 527-36 (1967); CA <u>67</u>, 82257.

1560. --- and Vrieze, K., ibid., <u>13</u>, 479-93 (1968); CA <u>69</u>, 77440.

1561. ---, Vrieze, K., and Praat, A. P., J. Organometal. Chem. <u>14</u>, 429-40 (1968); CA <u>69</u>, 101518.

1562. Volpini, L., Zarli, B., and Colombini, C., Gazz. Chim. Ital. <u>98</u>, 413-20 (1968); CA <u>69</u>, 15469.

1563. Von Endt, D. W., Kearney, P. C., and Kaufman, D. D., J. Agr. Food Chem. <u>16</u>, 17-20 (1968); CA <u>68</u>, 38437.

1564. Voskuil, W., and Arens, J. F., Rec. Trav. Chim. Pays-Bas <u>83</u>, 1301-4 (1964); CA <u>62</u>, 5296.

1565. Vrchlabsky, M., and Sommer, L., Talanta <u>15</u>, 887-94 (1968); CA <u>69</u>, 90369.

1566. Vrieze, K., Proc. Int. Conf. Coord. Chem., 8th, Vienna, <u>1964</u>, 153-5;

CA <u>66</u>, 121640.

1567. Vrieze, K., et al., Rec. Trav. Chim. Pays-Bas <u>86</u>, 769-94 (1967); CA <u>67</u>, 103292.

1568. --- et al., J. Organometal. Chem. <u>6</u>, 672-6 (1966); CA <u>66</u>, 37216.

1569. --- et al., Rec. Trav. Chim. Pays-Bas <u>85</u>, 1077-98 (1966); CA <u>66</u>, 24250.

1570. ---, Praat, A. P., and Cossee, P., J. Organometal. Chem. <u>12</u>, 533-47 (1968); CA <u>69</u>, 13292.

1571. --- and Volger, H. C., ibid., <u>9</u>, 537-48 (1967); CA <u>67</u>, 82258.

1572. --- and Volger, H. C., ibid., <u>11</u>, P17-P18 (1968); CA <u>68</u>, 91656.

1573. ---, Volger, H. C., and Praat, A. P., ibid., <u>14</u>, 185-200 (1968); CA <u>69</u>, 82148.

1574. ---, Volger, H. C., and Praat, A. P., ibid., <u>15</u>, 195-208 (1968); CA <u>69</u>, 110259.

1575. Vyazankin, N. S., et al., Zh. Obshch. Khim. <u>38</u>, 205 (1968); CA <u>69</u>, 67504.

1576. --- et al., Dokl. Akad. Nauk SSSR <u>166</u>, 99-102 (1966); CA <u>64</u>, 11245.

1577. --- et al., J. Organometal. Chem. <u>6</u>, 474-83 (1966); CA <u>66</u>, 11016.

1578. Wade, L., Jr., AEC Accession No. 31569, Rept. No. TID-20812; CA <u>62</u>, 7283.

1579. Walisch, W., and Schaefer, K., Mikrochim. Acta <u>1968</u>, 765-72; CA <u>69</u>, 56836.

1580. Walton, R. A., J. Chem. Soc. A <u>1967</u>, 1485-9; CA <u>67</u>, 96389.

1581. ---, Inorg. Chem. <u>7</u>, 640-8 (1968); CA <u>68</u>, 91499.

1582. Warren, H. D., Diss. Abstr. <u>26</u>, 6342 (1966).

1583. Washburn, R. M., and Baldwin, R. A. (Amer. Potash & Chem. Corp.), U.S. 3,189,564, June 15, 1965; CA <u>63</u>, 9991.

1584. --- and Baldwin, R. A. (Amer. Potash & Chem. Corp.), U.S. 3,311,646, Mar. 28, 1967; CA <u>67</u>, 12061.

1585. --- and Baldwin, R. A. (Amer. Potash & Chem. Corp.), U.S. 3,341,477, Sept. 12, 1967; CA <u>67</u>, 100586.

1586. --- and Baldwin, R. A. (Amer. Potash & Chem. Corp.), U.S. 3,341,478, Sept. 12, 1967; CA <u>67</u>, 100585.

1587. Watkins, S. F., Chem. Commun. <u>1968</u>, 504-5; CA <u>69</u>, 22855.

1588. Wawersik, H., and Basolo, F., J. Amer. Chem. Soc. <u>89</u>, 4626-30 (1967); CA <u>67</u>, 116327.

1589. Waxdal, M. J., Diss. Abstr. <u>26</u>, 4266 (1966).

1590. Weck, A. L. de, Frey, J. R., and Geleick, H., J. Immunol. <u>100</u>, 1-6 (1968); CA <u>68</u>, 47896.

1591. Weigle, W. O., Immunology <u>7</u>, 239-47 (1964); CA <u>61</u>, 7532.

1592. ---, J. Exptl. Med. <u>121</u>, 289-308 (1965); CA <u>63</u>, 2239.

1593. ---, ibid., <u>122</u>, 1049-62 (1965); CA <u>64</u>, 7205.

1594. Weiner, M. A., and Pasternack, G., J. Org. Chem. <u>32</u>, 3707-9 (1967); CA <u>68</u>, 12252.

1595. Weingarten, H. I., and Van Wazer, J. R., J. Amer. Chem. Soc. <u>88</u>, 2700-2 (1966); CA <u>65</u>, 3329.

1596. --- and White, W. A. (Monsanto Co.), U.S. 3,366,655, Jan. 30, 1968; CA <u>68</u>, 95979.

1597. Weliky, N., et al., Immunochemistry 1, 219-29 (1964); CA 62, 5731.

1598. Wells, E. J., et al., Can. J. Chem. 46, 2733-42 (1968); CA 69, 87131.

1599. Werner, H., and Prinz, R., Chem. Ber. 99, 3582-92 (1966); CA 66, 18412.

1600. Werner, R. P. M. (Ethyl Corp.), U.S. 3,247,233, Apr. 19, 1966; CA 65, 5489.

1601. ---, U.S. 3,254,953, June 7, 1966; CA 65, 8961.

1602. Westland, A. D., J. Chem. Soc. 1965, 3060-7; CA 63, 1462.

1603. --- and Westland, L., Can. J. Chem. 43, 426-30 (1965); CA 62, 8648.

1604. Wharton, H. W., and Chapman, L. R., Anal. Chem. 36, 1679-81 (1964); CA 63, 6378.

1605. Wheeler, O. H., Trabal, J. E., and McClin, M. L., Radiochim. Acta 9, 49-51 (1968); CA 69, 15110.

1606. White, G., and Scholes, P. H., Metallurgia 70, 197-200 (1964); CA 62, 4611.

1607. Whitehouse, M. W., Biochem. Pharmacol. 16, 753-60 (1967); CA 67, 20211.

1608. Wieber, M., and Werther, H. U., Monatsh. Chem. 99, 1159-62 (1968); CA 69, 36222.

1609. Wiley, J. R., and Chang, T. S. (Whitmoyer Laboratories, Inc.), U.S. 3,264,111, Aug. 2, 1966; CA 65, 11253.

1610. Wilke, G. (Studiengesellschaft Kohle m.b.H.), Belg. 651,596, Feb. 8, 1965; CA 64, 9928; Neth. Appl. 64-08958.

1611. --- et al. (Studiengesellschaft Kohle m.b.H.), Ger. 1,191,375, Apr. 22, 1965; CA 63, 7045.

1612. Wilkinson, G., U.S. 3,092,646, June 6, 1963; CA 62, 10276.

1613. ---, Fr. 1,459,643, Nov. 18, 1966; CA 67, 53652.

1614. Willard, F. L., and Kodras, R., Appl. Microbiol. 15, 1014-19 (1967); CA 67, 98799.

1615. Williams, L. (Peter Spence & Sons Ltd.), Brit. 1,010,204, Nov. 17, 1965; CA 64, 3802.

1616. Williamson, W. I. (Imperial Chemical Industries Ltd.), Brit. 1,077,390, July 26, 1967; CA 68, 31196.

1617. Wilmshurst, E. C., and Cliffe, E. E., Absorption Distrib. Drugs, Symp. London 1963, 191-8 (Pub. 1964); CA 61, 15204.

1618. Winkhaus, G., and Singer, H., Chem. Ber. 99, 3593-601 (1966); CA 66, 18754.

1619. Winters, R. E., and Kiser, R. W., J. Organometal. Chem. 10, 7-14 (1967); CA 67, 99530.

1620. Wittig, G., and Maercker, A., ibid., 8, 491-4 (1967); CA 67, 54236.

1621. Wolf, P. A., et al., J. Paint Technol. 39, 99-103 (1967); CA 66, 116744; see also Amer. Chem. Soc., Org. Coatings Plastics Chem., Preprints 25, 23-30 (1965); CA 66, 30049.

1622. --- and Riley, W. H., Appl. Microbiol. 13, 28-33 (1965); CA 62, 4366.

1623. Woodruff, R. J., Marini, J. L., and Fackler, J. P., Jr., Inorg. Chem. 3, 687-90 (1964); CA 60, 14108.

1624. Woogerd, S. M. (Hercules Glue Co., Ltd.), U.S. 3,298,820, Jan. 17, 1967; CA 66, 54472.

1625. Workman, M. O., Dyer, G., and Meek, D. W., Inorg. Chem. 6, 1543-8

(1967); CA <u>67</u>, 96398.

1626. Wu, Ch., Biochim. Biophys. Acta <u>96</u>, 134-47 (1965); CA <u>62</u>, 8066.

1627. Wu, Ch.-L., et al., Yuan Tzu Neng <u>1</u>, 27-31 (1965); CA <u>63</u>, 14318.

1628. Yamaguchi, M., Kogyo Kagaku Zasshi <u>70</u>, 675-83 (1967); CA <u>67</u>, 99542.

1629. Yamazaki, H., et al., J. Organometal. Chem. <u>6</u>, 86-91 (1966); CA <u>65</u>, 5485.

1630. Yang, P.-Y., Hua Hsueh Tung Pao <u>1969</u>, 567-71; CA <u>60</u>, 4755.

1631. Yasukouchi, K., and Muto, H., Denki Kagaku <u>35</u>, 420-3 (1967); CA <u>68</u>, 45536.

1632. --- et al., ibid., 890-4 (1967); CA <u>69</u>, 15327.

1633. --- et al., ibid., <u>36</u>, 54-8 (1968); CA <u>69</u>, 15328.

1634. Yeager, C. C. (Scientific Chems., Inc.), U.S. 3,288,674, Nov. 29, 1966; CA <u>66</u>, 86431; Brit. 1,085,970, Oct. 4, 1967; CA <u>67</u>, 117726.

1635. ---, U.S. 3,360,431, Dec. 26, 1967; CA <u>68</u>, 40523.

1636. Yoda, K., and Kimoto, K., Bull. Chem. Soc. Japan <u>41</u>, 1687-9 (1968); CA <u>69</u>, 77829.

1637. Young, J. F., Gillard, R. D., and Wilkinson, G., J. Chem. Soc. <u>1964</u>, 5176-89; CA <u>62</u>, 3638.

1638. Yuki, H., et al., Chem. Pharm. Bull. (Tokyo) <u>15</u>, 1052-5 (1967); CA <u>68</u>, 39747.

1639. --- et al., ibid., <u>14</u>, 139-46 (1966); CA <u>64</u>, 18219.

1640. Yusupov, F. Yu., and Manulkin, Z. M., Tr. Tashkent. Farmatsevt. Inst. <u>4</u>, 531-7 (1966); CA <u>68</u>, 59676.

1641. Zaikovskii, F. V., and Sadova, G. F., USSR 167,676, Jan. 18, 1965; CA <u>62</u>, 12445.

1642. Zakharkin, L. I., Bregadze, V. I., and Okhlobystin, O. Yu., J. Organometal. Chem. <u>4</u>, 211-16 (1965); CA <u>63</u>, 11597.

1643. ---, Okhlobystin, O. Yu., and Bilevich, K. A., Tetrahedron <u>21</u>, 881-6 (1965); CA <u>62</u>, 16283.

1644. Zanella, P., and Tagliavini, G., J. Organometal. Chem. <u>12</u>, 355-62 (1968); CA <u>69</u>, 22583.

1645. Zarli, B., and Panattoni, C., Inorg. Nucl. Chem. Lett. <u>3</u>, 111-12 (1967); CA <u>67</u>, 119296.

1646. Zbiral, E., and Berner-Fenc, L., Monatsh. Chem. <u>98</u>, 666-78 (1967); CA <u>67</u>, 73648.

1647. Ziegler, K., and Lehmkuhl, H. (Karl Ziegler), Ger. 1,212,085, Mar. 10, 1966; CA <u>64</u>, 19675.

1648. --- and Lehmkuhl, H. (Karl Ziegler), U.S. 3,372,097, Mar. 5, 1968; CA <u>68</u>, 101279.

1649. Ziegler, M., and Gindl, M., Naturwissenschaften <u>54</u>, 19 (1967); CA <u>67</u>, 7729.

1650. Zimmer, H. J., Verfahrenstechnik, Fr. 1,483,585, June 2, 1967; CA <u>68</u>, 96368.

1651. Zingales, F., Graziani, M., and Belluco, U., J. Amer. Chem. Soc. <u>89</u>, 256-60 (1967); CA <u>66</u>, 45927.

1652. --- and Trovati, A., Rend. Ist. Lombardo Sci. Lett., A <u>101</u>, 527-32 (1967); CA <u>69</u>, 30540.

1653. Zingales, F., et al., Inorg. Chem. $\underline{7}$, 1653-5 (1968); CA $\underline{69}$, 61907.

1654. Zoche, G., Müller, E. W., and Korte, F. (Shell Internationale Research Maatschappij N.V.), Brit. 1,019,968, Division of Brit. 1,004,108; CA $\underline{64}$, 15742.

1655. Zolotov, Yu. A., Petrukhin, O. M., and Alimarin, I. P., Zh. Analit. Khim. $\underline{20}$, 347-50 (1965); CA $\underline{62}$, 15489.

1656. --- et al., Sovrem. Metody Analiza, Metody Issled. Khim. Sostava i Stroeniya Veshchestv, Akad. Nauk SSSR, Inst. Geokhim. i Analit. Khim. $\underline{1965}$, 238-44; CA $\underline{64}$, 2805.

1657. Zorkii, P. M., and Lazareva, S. G., Zh. Strukt. Khim. $\underline{9}$, 95-100 (1968); CA $\underline{68}$, 117818.

1658. Zorn, H., Schindlbauer, H., and Hammer, D., Monatsh. Chem. $\underline{98}$, 731-7 (1967); CA $\underline{67}$, 73656.

1659. Headley, O. St. C., Ph. D. Thesis, University of London, 1967.

1660. Abel, E. W., Honigschmidt-Grossich, R., and Illingworth, S. M., J. Chem. Soc. A $\underline{1968}$, 2623-5.

1661. ---, Crow, J. P., and Illingworth, S. M., ibid., A $\underline{1969}$, 1631-3.

Alimarin, I. P., and Savvin, S. B., "Arsenazo III and Other Azo Compounds in the Photometric Determination of Certain Elements," Pure Appl. Chem. 13, 445-56 (1966). The use of Arsenazo III and of other azo compounds is reviewed. 30 references.

Alykov, N. M., "Spectrophotometric Study of Azo Derivatives of Chromotropic Acid and Their Interaction with Metal Ions in the Scandium Subgroup," Fiz.-Khim. Issled. Prir. Sorbentov Ryada Anal. Sist. 1966 (1), 52-72. A review with 47 references, including Arsenazo I and III.

Azevedo e Siva, E. de, "Considerations on the Iatrogenic-Carcinogenic Action of Inorganic and Aliphatic Organic Arsenicals," Anais. Fac. Med. Univ. Recife 24, 111-23 (1964). A review with 29 references.

Bancari, M. B., "Chemistry and Pharmacology of Trivalent Arsenic Derivatives," Corriere Farm. 22 (3), 79-80 (1967). A review with 7 references.

Bekasova, N. I., "Organoarsenic Polymers," Usp. v Obl. Sinteza Elementoorgan. Polimerov," Akad. Nauk SSSR, Inst. Elementoorgan. Soedin. 1966, 167-72. A review with 19 references.

Booth, G., "Complexes of the Transition Metals with Phosphines, Arsines, and Stibines," Advan. Inorg. Chem. Radiochem. (H. J. Emeleus and A. G. Sharpe, editors; Academic Press) 6, 1-69 (1964). A review with 460 references.

Bozhevol'nov, E. A., "Luminescent Analysis," Zh. Anal. Khim. 22, 1692-701 (1967). A review with 107 references, including rezarson - a reagent giving fluorescent complexes.

Bregadze, V. I., and Okhlobystin, O. Yu., "Heteroorganic Derivatives of Carboranes [Carboranes(10)]," Usp. Khim. 37, 353-79 (1968). A review with 90 references, including carboranes with arsenic and antimony.

Doak, G. O., and Freedman, L. D., "Arsenic. Annual Survey Covering the Year 1967," Organometal. Chem. Rev., Sect. B 4, 405-20 (1968). A review with 76 references.

Doak, G. O., and Freedman, L. D., "Antimony. Annual Survey Covering the Year 1967," Organometal. Chem. Rev., Sect. B 4, 421-5 (1968). A review with 23 references.

Doak, G. O., and Freedman, L. D., "Bismuth. Annual Survey Covering the Year 1967," Organometal. Chem. Rev., Sect. B 4, 426 (1968). A review with 2 references.

Doak, G. O., Freedman, L. D., and Long, G. G., "Antimony Compounds," Kirk-Othmer Encycl. Chem. Technol., 2nd Ed. 2, 570-88 (1963). A review with 127 references, including stibonic acids, stibonic acids, stibines, stibonium compounds, halides, antimono compounds, and distibines.

Doak, G. O., and Long, G. G., "The Structure of Pentavalent Antimony Compounds," Trans. N. Y. Acad. Sci. 28, 402-11 (1966). A review of the stereochemistry of Sb(V).

Dolique, R., "L'Arsenic et ses Composes," Press Univ. France: Paris, 1968, 128 pp.

Edelman, G. M., and McClure, W. O., "Fluorescent Probes and the Conformation of Proteins," Accounts Chem. Res. 1 (3), 65-70 (1968). A review with 35 references; arsenic derivatives included.

Hagihara, N., Kumada, M., and Okawara, R., "Handbook of Organometallic Compounds," (translation of the Japanese edition) W. A. Benjamin, Inc., New York - Amsterdam, 1968. The coverage of organic As, Sb, and Bi derivatives is confined to 34 pages.

Hueper, W. C., "Carcinogenic Hazards from Arsenic- and Metal-Containing Drugs," Potential Carcinogenic Hazards Drugs, Proc. Symp. 1965 (Pub. 1967), 79-104, edited by R. Truhaut, Springer-Verlag, Berlin. A review of medicinal use of arsenicals and proven cancer risk.

Kamai, G., and Usacheva, G. M., "Stereochemistry of Organic Compounds of Phosphorus and Arsenic," Usp. Khim. 35, 1404-29 (1966). A review with 123 references.

Klarmann, E. G., "Antiseptics and Desinfectants," Kirk-Othmer Encycl. Chem. Technol., 2nd Ed. 2, 604-48 (1963). A review with 308 references.

Kozlovskii, M. T., Bukhman, S. P., Nosek, M. V., Dragavtseva, N. A., Nigmetova, R. Sh., and Dyuzheva, E. B., "Reduction of Arsenic, Antimony, and Germanium Compounds by Amalgams," Tr. Inst. Khim. Nauk Kaz. SSR 18, 99-109 (1967). A review with 50 references.

Marhol, M., "Ion-Exchange Resins Containing Phosphorus or Arsenic in Their Functional Groups," Chem. Listy 58, 713-33 (1964). A review with 123 references.

Meulendorff, V. J., "Arsenic Determination in Organic Arsenic Compounds According to the Netherland Pharmacopeia, "Pharm. Weekblad 100, 409-12 (1965).

Pauson, P. L., "Organometallic Chemistry," St. Martin's Press, New York, 1965. The monograph "Provides general information for the advanced undergraduate or the postgraduate student," and includes some classes of organic Sb and Bi derivatives.

Rabek, J. F., "The Newest Achievements in the Research and Synthesis of Heat Stable Polymers," Przem. Chem. <u>46</u>, 130-4 (1967). A review with 232 references including organoarsenic compounds.

Reuber, H. W., and Lux, F. A., "Veterinary Arsenicals," Iowa State Univ. Vet. <u>28</u> (1), 13-18 (1966). The As use levels in veterinary medicine are discussed; 37 references.

Savvin, S. B., "Arsenaso III. Metody Fotometricheskogo Opredeleniya Redkikh i Aktinidnykh Elementov," Atomizdat, 1966, 256 pp.

Yale, H. L., "Arsenic, Antimony, and Phosphorus Compounds of Pyridine," Pyridine Derivatives (E. Klingsberg, editor, Interscience) <u>4</u>, 439-66 (1964). 96 references.